JOURNAL OF CHROMATOGRAPHY LIBRARY - volume 22B

chromatography

fundamentals and applications of chromatographic and electrophoretic methods

part B: applications

JOURNAL OF CHROMATOGRAPHY LIBRARY

JOURNAL OF CHROMATOGRAPHY LIBRARY - volume 22B

chromatography

fundamentals and applications of chromatographic and electrophoretic methods

part B: applications

edited by
E. Heftmann

Western Regional Research Center, U.S. Department of Agriculture, Berkeley, CA 94710

ELSEVIER SCIENTIFIC PUBLISHING COMPANY
Amsterdam — Oxford — New York   1983

ELSEVIER SCIENTIFIC PUBLISHING COMPANY
Molenwerf 1
P.O. Box 211, 1000 AE Amsterdam, The Netherlands

Distributors for the United States and Canada:

ELSEVIER SCIENCE PUBLISHING COMPANY INC.
52, Vanderbilt Avenue
New York, NY 10017

ISBN 0-444-42044-4 (Vol. 22 part B)
ISBN 0-444-41616-1 (Series)
ISBN 0-444-42045-2 (Set)

Printed in The Netherlands

TO BRIGITTE

Contents

List of authors and coauthors (*)

Mr. Edward R. Adlard, Shell Research Ltd., Thornton Research Centre, P.O. Box 1, Chester CH1 3SH, Great Britain (Part B)

Dr. Robert P. Bywater, Pharmacia Fine Chemicals, Box 175, S-751 04 Uppsala, Sweden (Part A)

Dr. Nicholas Catsimpoolas, Biochemical Research Laboratories, Department of Biobehavioral Sciences, Boston University Medical Center, Boston, MA 02118, USA (Part B)

Dr. Shirley C. Churms, Department of Organic Chemistry, University of Cape Town, Rondebosch 7700, South Africa (Part B)

Dr. Graham J. Cowling, formerly Department of Biophysics, University of London, King's College, School of Biological Sciences, London WC2B 5RL, Great Britain. Present address: Searle Research & Development, G.D. Searle & Co., Ltd., P.O. Box 53, High Wycombe, Bucks HP12 4HI, Great Britain (Part B)

Dr. Carl A. Cramers, Laboratorium voor Instrumentele Analyse, Technische Hogeschool, P.O. Box 513, 5600 MB Eindhoven, The Netherlands (Part A)

Dr. Rodney Croteau, Institute of Biological Chemistry, Washington State University, Pullman, WA 99164, USA (Part B)

Dr. David H. Dolphin, Department of Chemistry, University of British Columbia, Vancouver, B.C., Canada V6T 1W5 (Part B)

Dr. Frans M. Everaerts, Laboratorium voor Instrumentele Analyse, Technische Hogeschool, Postbus 513, 5600 MB Eindhoven, The Netherlands (Part A)

Dr. Lawrence Fishbein, National Center for Toxicological Research, Food & Drug Administration, Department of Human Health Services, Jefferson, AR 72079, USA (Part B)

Dr. Jeffrey B. Harborne, Department of Botany, Plant Science Laboratories, University of Reading, Reading RG6 2AS, Great Britain (Part B)

Dr. Erich Heftmann, Western Regional Research Center, U.S. Department of Agriculture, Berkeley, CA 94710, USA. Home address: 108 Cañon Dr., P.O. Box 928, Orinda, CA 94563, USA (Parts A and B)

Dr. Csaba Horváth, Department of Chemical Engineering, Mason Laboratory, Yale University, P.O. Box 2159, New Haven, CT 06520, USA (Part A)

Dr. Jaroslav Janák, Institute of Analytical Chemistry, 662 28 Brno, Czechoslovakia (Part B)

Dr. Arnis Kuksis, Banting & Best Department of Medical Research, University of Toronto, Toronto, Ont., Canada M5G 1L6 (Part B)

Dr. Thomas Kuster, Med.-chem. Laboratorium, Kinderspital Zürich, 8032 Zürich, Switzerland (Part B)

Dr. Michael Lederer, Institut de chimie minérale et analytique, Faculté des Sciences, Collège propédeutique, Dorigny, CH-1015 Lausanne, Switzerland (Part B)

Dr. Karel Macek, Institute of Physiology, Czechoslovak Academy of Sciences, 142 20 Prague, Czechoslovakia (Part A)

* Dr. Harold M. McNair, Department of Chemistry, Virginia Polytechnic Institute and State University, Blacksburg, VA 24061, USA (Part A)

* Dr. Nigel V.B. Marsden, Institutionen för Fysiologi och Medicinsk Fysik, Uppsala Universitets Biomedicinska Centrum, Box 572, S-751 23 Uppsala, Sweden (Part A)

* Dr. Wayne R. Melander, Department of Chemical Engineering, Mason Laboratory, Yale University, P.O. Box 2159, New Haven, CT 06520, USA (Part A)

* Dr. F.E.P. Mikkers, Laboratorium voor Instrumentele Analyse, Technische Hogeschool, Postbus 513, 5600 MB Eindhoven, The Netherlands (Part A)

* Dr. Alois Niederwieser, Med.-chem. Laboratorium, Kinderspital Zürich, 8032 Zürich, Switzerland (Part B)

* Dr. Robert C. Ronald, Institute of Biological Chemistry, Washington State University, Pullman, WA 99164, USA (Part B)

Dr. Raymond P.W. Scott, Instrument Division, Perkin-Elmer Corporation, Norwalk, CT 06856, USA (Part A)

* Dr. J. Vacík, Department of Physical Chemistry, Charles University, 501 65 Hradec Králové, Czechoslovakia (Part A)

* Dr. Th.P.E.M. Verheggen, Laboratorium voor Instrumentele Analyse, Technische Hogeschool, Postbus 513, 5600 MB Eindhoven, The Netherlands (Part A)

Dr. Gerald H. Wagman, Microbiological Strain Laboratory, Schering-Plough Corporation, Kenilworth, NJ 07033, USA (Part B)

Dr. Harold F. Walton, Department of Chemistry, University of Colorado, Campus Box 215, Boulder, CO 80309, USA (Part A)

* Dr. Marvin J. Weinstein, formerly Microbiological Strain Laboratory, Schering-Plough Corporation, Kenilworth, NJ 07033, USA (Part B)

List of abbreviations

A	ampere (Amp)
Å	ångström $= 10^{-8}$ cm
AC	alternating current
AE	aminoethyl
AFID	alkali flame-ionization detector
AGR	anhydroglucose residue
ANB	2-amino-5-nitrobenzophenone
ASTM	American Society for Testing and Materials
atm	atmosphere $= 760$ torr
BD-cellulose	benzoylated DEAE-cellulose
t-BDMS	t-butyl dimethylsilyl
BHT	(2,6)-di-t-butyl-p-cresol (butylated hydroxytoluene)
b.p.	boiling point
b.r.	boiling range
BSA	N,O-bis(trimethylsilyl)acetamide
BSTFA	bis(trimethylsilyl)trifluoroacetamide
BTT	3-benzyl-1-p-tolyltriazine
C	centigrade, celsius
CA	cycloamyloses
CAM	cellulose acetate membrane (electrophoresis)
cc	cubic centimeter (cm^3)
CC	column chromatography
CCD	countercurrent distribution
CDD	chlorinated dibenzo-p-dioxins
CECD	Coulson electrolytic conductivity detector
Chap.	chapter
CI	chemical ionization
CLU	crosslinked unit structure
CM	carboxymethyl
cm	centimeter
conc.	concentrated
cpm	counts per minute
C.V.	coefficient of variation
CZE	continuous zone electrophoresis
DBM	diazobenzyloxymethyl
DC	direct current
DCC	droplet countercurrent chromatography
DEAE	diethylaminoethyl
DMSO	dimethylsulfoxide
DP	degree of polymerization
DTT	dithiothreitol

DVS   divinylsulfone
ECD   electron-capture detector
ECL   effective chain length
ECTEOLA   epichlorohydrintriethanolamine
ED   electrochemical detector
EDTA   ethylenediaminetetraacetic acid
EHV   effective hydrodynamic volume
EI   electron impact
ELS   electrophoretic light scattering
EMIT   enzyme multiplied immunoassay technique
eqn.   equation
eV   electron-volt
FD   field desorption
FFF   field-flow fractionation
fg   femtogram $= 10^{-15}$ g
FIA   fluorescent indicator analysis
FID   flame-ionization detector
fmole   femtomole $= 10^{-15}$ mole
ft   feet
FTIR   Fourier transform infrared
g   gram
GC   gas chromatography
GC^2 glass capillary gas chromatography
GLC   gas–liquid chromatography
GPC   gel-permeation chromatography
GSC   gas–solid chromatography
h   hour
HETP   height equivalent to a theoretical plate
HFB   heptafluorobutyryl
HI   hydrophobic interaction
HIC   hydrophobic interaction chromatography
HIE   hydrogen isotope exchange
HPLC   high-performance liquid chromatography (high-pressure liquid chromatography)
HPTLC   high-performance thin-layer chromatography
hR_F $R_F \times 100$
HVE   high-voltage electrophoresis
IC   Ion Chromatography
ID   inside diameter
IF   isoelectric focusing
in.   inches
IP   Institute of Petroleum
IR   infrared
ITLC   Instant Thin-Layer Chromatography
ITP   isotachophoresis

kV	kilovolt $= 10^3$ V
l	liter
LC	liquid chromatography
LCC	liquid column chromatography
LC–EC	liquid chromatography electrochemical detector
LFER	linear free energy relations
LLC	liquid–liquid chromatography
LSC	liquid–solid chromatography
M	molar
m	meter
mA	milliampere $= 10^{-3}$ A
μA	microampere $= 10^{-6}$ A
MBE	moving-boundary electrophoresis
m.d.q.	minimum detectable quantity
ME	mercaptoethanol
meq	milliequivalents $= 10^{-3}$ equivalents
μeq	microequivalents $= 10^{-6}$ equivalents
MF	mass fragmentography
mg	milligram $= 10^{-3}$ g
μg	microgram $= 10^{-6}$ g
MID	multiple-ion detection
min	minutes
ml	milliliter $= 10^{-3}$ l
μl	microliter $= 10^{-6}$ l
mM	millimolar $= 10^{-3}$ M
mm	millimeter $= 10^{-3}$ m
μm	micrometer $= 10^{-6}$ m
mmole	millimole $= 10^{-3}$ mole
μmole	micromole $= 10^{-6}$ mole
MO	methyl oxime
mol.	molecular
mp	melting point
MS	mass spectrometry
MSA	N-methyl-N-trimethylsilylacetamide
msec	milliseconds $= 10^{-3}$ sec
MSTFA	N-methyl-N-trimethylsilyltrifluoroacetamide
MTH	methylthiohydantoin
MTX	methotrexate
MU	methylene unit
mV	millivolt $= 10^{-3}$ V
MW	molecular weight
MZE	multiphasic zone electrophoresis
N	normal
ng	nanogram $= 10^{-9}$ g
nm	nanometer $= 10^{-9}$ m

nmole	nanomole $= 10^{-9}$ mole
NMR	nuclear magnetic resonance
NPD	nitrogen–phosphorus detector
OD	outside diameter
ODS	octadecylsilane or octadecylsilyl
OPA	o-phthalaldehyde
PAGE	polyacrylamide gel electrophoresis
PC	paper chromatography
PCA	polycyclic aromatic hydrocarbons
PCB	polychlorinated biphenyls
P-cellulose	phosphate cellulose
PCN	polychloronaphthalene
PCP	pentachlorophenol
PDM	programed multiple development
PE	paper electrophoresis
PEG	polyethylene glycol
PEI	polyethyleneimine
PEO	polyethylene oxides
PFB	pentafluorobenzyl
pg	picogram $= 10^{-12}$ g
PGE	pore-limit gel electrophoresis
pI	isoelectric point
PLC	preparative layer chromatography
PLOT	porous-layer open tubular (columns)
pmole	picomole $= 10^{-12}$ mole
ppb	parts per billion $= 10^{-9}$ parts
ppm	parts per million $= 10^{-6}$ parts
ppt	parts per trillion $= 10^{-12}$ parts
psi	pounds/sq.in. $= 51.77$ torr
PTFE	polytetrafluoroethylene (Teflon)
PTH	phenylthiohydantoin
RI	refractive index
R.I.	Retention Index
RIA	radioimmunoassay
RRT	relative retention time
RT	retention time
SBF	separation by flow
SCOT	support-coated open tubular (columns)
S.D.	standard deviation
SDS	sodium dodecyl sulfate
SEC	size- (or steric) exclusion chromatography
sec	seconds
SFC	supercritical-fluid chromatography
SIM	selected-ion monitoring
SM	sulfomethyl

SP	sulfopropyl
sq.	square
SSS	steady state stacking
t	tertiary
TAS	thermomicro-transfer-application-separation (technique)
TCA	trichloroacetic acid
TCD	thermal conductivity detector
TCDD	tetrachlorodibenzodioxins
TCP	tetrachlorophenol
TD	thermionic detector
TEA	thermal energy analyzer
TEAB	triethylammonium bicarbonate
TEAE	tetraethylaminoethyl
temp.	temperature
TFA	trifluoroacetyl
TFG	thermofractography
THF	tetrahydrofuran
TLC	thin-layer chromatography
TLE	thin-layer electrophoresis
TLG	thin-layer gel chromatography
TMAH	trimethylanilinium hydroxide
TMCS	trimethylchlorosilane
TMS	trimethylsilyl
TMSDE	trimethylsilyldiethylamine
Tris	tris(hydroxymethyl)aminomethane
UV	ultraviolet
V	volt
vol.	volume
v/v	vol./vol.
WCOT	wall-coated open tubular (columns)
wt.	weight
w/w	wt./wt.

Chapter 10

Amino acids and oligopeptides

THOMAS KUSTER and ALOIS NIEDERWIESER

CONTENTS

10.1. INTRODUCTION

In this chapter, emphasis will be placed on newer developments and techniques. Those which did not change very much since the last edition of this book [1] are either omitted (electrophoresis, TLC) or only updated (ion-exchange chromatography). Applications in clinical chemistry are reported in a recently published book by Bremer et al. [2].

References on p. B43

10.2. ION-EXCHANGE CHROMATOGRAPHY

10.2.1. Amino acids and oligopeptides

Analysis of amino acids by ion-exchange chromatography is based on the work of Stein and Moore [3–5], who systematically studied the problems of separation and quantitation of the common amino acids. In 1958 they presented an automatic amino acid analyzer system which allowed a complete analysis to be performed in 24 h [5]. Since that time, many modifications of this basic system have been reported, mainly with the objective of shortening the time and increasing the sensitivity of the analysis. Nowadays, picomole amounts of amino acids can be analyzed within 90 min. It should be mentioned, however, that in none of the modern systems is the resolution of all amino acids as good as in that described by Moore et al. [5]. This is due to the compression of the chromatograms, which makes it very difficult or even impossible to detect unusual amino acids.

Since the third edition of this book, methods and basic requirements of ion-exchange chromatography have not changed very much and only technical progress was recorded. In this chapter, we therefore will not review the literature concerning sample preparation, internal standards, resins, buffers, detectors, computation, and various programs. The reader interested in such questions is referred to the third edition [1]. Here, we will give only a short outline of new work in this field. Table 10.1 lists some of the recent papers concerning ion-exchange chromatography of amino acids and peptides.

10.2.2. Separation of amino acids, dipeptides, and oligopeptides into classes

Tommel et al. [38,39] reported in 1966 the separation of peptides from amino acids by ion exchange of the respective copper complexes. Following these first studies, a number of papers appeared in which the method was applied to various

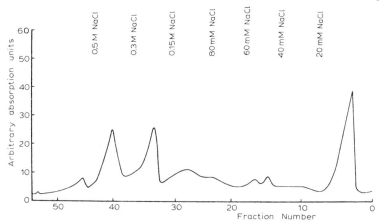

Fig. 10.1. Separation of Cu(II) complexes of oligoglycines on DEAE-Sephadex A-25. (Reproduced from *J. Chromatogr.*, 183 (1980) 9, with permission [47].)

TABLE 10.1

SURVEY OF RECENT WORK ON ION-EXCHANGE CHROMATOGRAPHY OF AMINO ACIDS AND OLIGOPEPTIDES

Abbreviations: AA = amino acid, OPA = o-phthalaldehyde derivative, TNB = trinitrobenzene.

Samples	Resins	Buffers	Detection	Remarks	Ref.
General AA	DC-4A cation	Various	Fluorimetry, OPA	—	6
General AA	DC-4A, DX 8.25	Various	Ninhydrin	Eluents with better baseline characteristics	7
General AA	Aminex A5	Pyridine acetate	TNB-sulfonic acid	—	8
General AA	DC-4A cation	Lithium buffer	Ninhydrin	Automated microbore analyzer	9
General AA	Cation	Sodium citrate	Ninhydrin	Reference values	10
General AA	Dual column	Various	Ninhydrin	Blood amino acids	11
General AA	DC-6A, XX 907 OPKU	Citrate	Ninhydrin	Short program for physiological fluids	12
γ-Abu	HCB X8	Lithium citrate	Fluorimetry, OPA	—	5, 13
β-Aib, β-Ala	Cation, dual column	Sodium citrate	Ninhydrin	Biological samples	15
β-Aib	Cation	Citrate	Ninhydrin	—	16
Pro, Hyp	DC-4A cation	Citrate	Fluorimetry, OPA	10 pmole Pro, 20 pmole Hyp	17
3-Hyp	Dowex 50 M82	Sodium citrate	Ninhydrin	Urine samples	18, 19
N-Methyl-His	Chromobeads A	Sodium citrate	Ninhydrin	Selective measurements	20
Tyr, Phe	7% crosslinked	Various	Ninhydrin	Automated system	21
Tyr, Dopa	Cation	Citrate	Ninhydrin	—	22
L-Trp, Ser, 5-HIAA, norepinephrine, dopamine	Cation	Citrate	Ninhydrin	In brain	23
HyL, glycosylated HyL	Dowex 50	Citrate	Spectrophotometry	—	24, 25
Sulfur-containing AA	Cation	Various	Ninhydrin	—	26
4-Hydroxyisoleucine isomers	Cation	Citrate	Ninhydrin	—	27
Basic AA	DC-4A	Sodium citrate	Fluorimetry, OPA	—	28
Methionine-containing peptides	Aminex A5	Pyridine acetate	Ninhydrin + liquid scintillation	—	29
Peptide fractions	Beckman W-3	Pyridine acetate	Cadmium, ninhydrin	—	30
Initiation peptides	Cation	Sodium citrate	Ninhydrin	—	31
General peptides	Cation	Aqueous buffers	Ninhydrin	—	32, 33
AA in water	Cation	Various	Ninhydrin	Trace analysis	34, 35
General AA	Cation	Aqueous buffers	Ninhydrin	Optical isomers	36
Leu-Ala-Gly-Val	Sulfonated polystyrene	Sodium citrate	Ninhydrin	Racemization test	37

References on p. B43

biological samples, e.g., wort [40], cheese [41,42,46], yeast [42], serum [43], and urine [43–45]. The Cu(II) complexes of the amino acids are eluted from an anion-exchange column at pH 8, and the peptide fraction is then recovered by acidification but, due to the remarkably lower stability of the copper complexes below pH 7, they cannot be monitored directly.

Niederwieser and coworkers [44,45] circumvented this disadvantage by using two columns at pH 8: DEAE-Sephadex A-25 separates amino acids from peptides and DEAE-Sephadex A-50 separates dipeptides from oligopeptides. They also suggested the possibility of separating tripeptides and tetrapeptides from penta- and higher peptides. Studies of the effects of varying the ionic strength, pH, copper content of Sephadex, and concentration yielded and enhanced resolution in separating peptides from large amounts of amino acids and a mixture of peptides [46]. Sampson and Barlow [47] reported the resolution of peptides into chain-length related fractions. Fig. 10.1 shows the fractionation of oligoglycines on DEAE-Sephadex A-25. The stepwise increase in ionic strength of the eluting buffer (sodium borate buffer at pH 8.5 with added sodium chloride) results in a total resolution of the five-component mixture. Hexaglycine exhibits an anomalous behavior, which prevents any resolution of this peptide. The authors tested their separation scheme on an enzymatic hydrolyzate of casein (aminosol). The larger peptides were removed by dialysis. Fig. 10.2 shows the separation of aminosol Cu(II) complexes on DEAE-Sephadex A-25. Many of the ten detected peaks are retained more strongly than any of the investigated model peptides, a fact which was ascribed to the high acidic amino acid content of the peptide fraction of aminosol.

10.3. HIGH-PERFORMANCE LIQUID CHROMATOGRAPHY OF AMINO ACIDS AND PEPTIDES

The adsorption of amino acids on nonpolar stationary phases, first investigated in the forties and used for chromatographic separation, had lost some of its significance

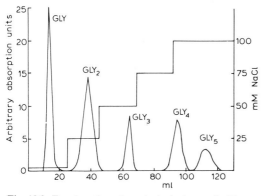

Fig. 10.2. Fractionation of casein-hydrolyzate Cu(II) complexes on DEAE-Sephadex A-25. (Reproduced from *J. Chromatogr.*, 183 (1980) 9, with permission [47].)

through the development of ion-exchange chromatography in the fifties. However, with the progress of HPLC interest in nonpolar stationary phases was reawakened. When the HPLC separation of amino acids is compared with ion-exchange and gas chromatography, the following facts should be kept in mind:

(a) automatic ion-exchange equipment is expensive and limited to the analysis of amino acids. In addition, a complete separation of the 20 naturally occurring amino acids requires about 60 min;

(b) performing separation by GC requires derivatization, which is not equally applicable to all components.

HPLC, on the other hand, offers a very versatile tool for such work. The apparatus is relatively cheap, applicable to various problems and substances, and is capable of separating 20 amino acids in less than 40 min. Various systematic studies concerning column material have revealed that chemically bonded (reversed) phases are best suited for amino acid and peptide analysis.

10.3.1. Detectors

Regarding the detection of the column effluent, it appears that no detector is now available that is as versatile and universal as one might desire and, in contrast to GC, HPLC has nothing equivalent to the TCD or FID. In the following chapter we will describe the more commonly used detectors in the field of amino acids and peptides. More general and detailed information about advantages, limitations, and characteristics of the various detection systems may be found elsewhere [48,49].

10.3.1.1. Ultraviolet detectors

UV detectors are the most commonly used detectors in HPLC today. They exhibit a high sensitivity for most solutes fulfilling the requirement that the sample must absorb in the region between 190 and 600 nm. They have good stability to flow and temperature changes, and their sensitivity lies in the nanogram range. The great dynamic range of about 10^5 allows the measurement of both bulk and trace amounts in one experiment. UV spectrometers offering a broad selection of wavelengths permit the measurement either at the absorption maximum or at a wavelength that provides maximum selectivity. A new detector type that allows measurement at 215 nm is available from Pharmacia. Some detectors permit the scanning of a peak while the flow through the detector cell is stopped. Rapid scanning devices [50–53] would yield a maximum of spectral information and could be an alternative or at least a complement to GC–MS techniques. Although such instruments are not yet commercially available, they promise to become a versatile detection system.

Somack [54] reported the separation of the amino acids from polypeptide hydrolyzates on an Ultrasphere-ODS column, monitoring the eluent at 254 nm; Fig. 10.3 shows a typical chromatogram. All 25 amino acids are separated within 32 min as phenylthiohydantoin derivatives with good resolution.

References on p. B43

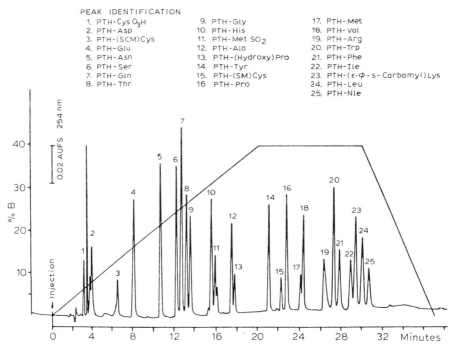

PEAK IDENTIFICATION

1. PTH-Cys O₃H	9. PTH-Gly	17. PTH-Met
2. PTH-Asp	10. PTH-His	18. PTH-Val
3. PTH-(SCM)Cys	11. PTH-Met SO₂	19. PTH-Arg
4. PTH-Glu	12. PTH-Ala	20. PTH-Trp
5. PTH-Asn	13. PTH-(Hydroxy)Pro	21. PTH-Phe
6. PTH-Ser	14. PTH-Tyr	22. PTH-Ile
7. PTH-Gln	15. PTH-(SM)Cys	23. PTH-(ε-φ-s-Carbamyl)Lys
8. PTH-Thr	16 PTH-Pro	24. PTH-Leu
		25. PTH-Nle

Fig. 10.3. Separation of phenylthiohydantoin (PTH)-amino acids by HPLC on Ultrasphere-ODS. Solvent A, 5% tetrahydrofuran in 4.2 mM acetic acid, adjusted to pH 5.16 with NaOH; Solvent B, 10% tetrahydrofuran in acetonitrile; flowrate, 1.3 ml/min; temp., 45°C; sample size, 1–2 nmole of each compound. (Reproduced from *Anal. Biochem.*, 104 (1980) 464, with permission [54].)

10.3.1.2. Fluorimetric detectors

Measuring the emission energy of solutes excited by UV radiation, fluorimeters have proved to be very sensitive and selective detectors for HPLC. In cases where the sample itself does not fluoresce, fluoroscent derivatives can be prepared by a pre- or postcolumn reaction. The sensitivity is often greater than that of UV detectors (up to $100 \times$); however, the dynamic range is lower (10^4).

A very promising form of fluorimeters uses a laser source for excitation [55]. With such devices, the low picogram range should be available for analysis. Fig. 10.4 gives an example of the HPLC separation with normal fluorimetric detection [56]. The fluorescent species are *o*-phthalaldehyde (OPA) derivatives, formed by a precolumn reaction in 1 min. The minimum detectable quantity averages 40 pg.

10.3.1.3. Electrochemical detectors

Samples that can be oxidized or reduced electrochemically may be analyzed with electrochemical detectors [ED], which measure the current between polarizable and

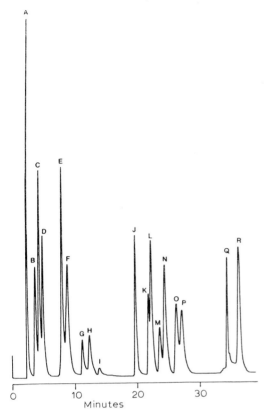

Fig. 10.4. Separation of amino acids by HPLC on a MicroSil C_8 (150×5 mm ID) column and fluorimetric detection following precolumn derivatization with o-phthalaldehyde. Solvent, 0.02 M potassium acetate (pH 5)–tetrahydrofuran–methanol; step gradient; flowrate: 2 ml/min. Peaks: A=Asp, B=Thr, C=Ser, D=Glu, E=Gly, F=Ala, G=Tyr, H=Val, I=CysH, J=His, K=Met, L=Gln, M=Ile, N=Leu, O=Phe, P=Arg, Q=Trp, R=NH$_3$. (Reproduced from J. Liquid Chromatogr., 2 (1979) 1047, with permission [56].)

reference electrodes as a function of the applied voltage. These detectors are still in the early stages of their development and, therefore, many questions and problems are not yet completely resolved. Having a sensitivity equal to – and sometimes even greater than – fluorimetric detection systems and a high dynamic range, they seem to be rather promising. Studies undertaken so far reveal that the ED are often more sensitive but less selective than fluorimetric detectors.

10.3.1.4. Mass spectrometers

A mass spectrometer as a HPLC detector fulfills in principle the main require-ments concerning versatility, sensitivity, dynamic range, and insensitivity towards temperature changes, but a liquid chromatograph is not as compatible with a mass spectrometer as a gas chromatograph. The mobile phase, flowrate, and low volatility of samples are often serious problems in coupling LC with MS. One way to circumvent these difficulties consists in off-line coupling [57,58].

Since 1974, several methods have been investigated for on-line coupling of LC with MS [59–72]. Reviews on this subject have been published by McFadden [73], Arpino and Guiochon [74], and Kenndler and Schmid [75]. So far, despite the great efforts undertaken in this field, only two interfaces are commercially available. The first, introduced by McLafferty et al. [61], leads the column effluent and the mobile phase directly into the spectrometer and uses the mobile phase as the reactant gas in chemical-ionization (CI)-MS. This device, which uses an ionization method better suited for less volatile and thermally labile compounds is relatively inexpensive and easy to construct but the low volatility of the solvents, especially in reversed-phase HPLC, limits the applicability of the method.

The second device consists of a moving belt [63], on which the solvent can be evaporated, and the sample can be analyzed by either electron-impact (EI)- or CI-MS. Takeuchi et al. [69] have constructed a jet separator for coupling the microcolumns with the spectrometer. With flowrates in the order of μl/min, the mobile gas phase again acts as the reactant gas in the CI source. However, the versatility of HPLC in the choice of the mobile phase and amount of sample is limited in this system.

A promising technique is that of Karger et al. [76]. The interface consists of a modified segmented flow extractor, which extracts the solutes in the column effluent into a volatile organic solvent subsequently evaporated via a moving belt. With such an interface, even mobile phases containing inorganic buffers can be used.

Blakley et al. [66] have used an oxyhydrogen flame for the rapid vaporization of the effluent. The detection limits for biological samples were found to be in the range of 1–10 ng for a full mass spectrum and in the subnanogram range for

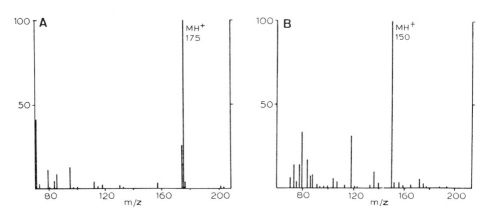

Fig. 10.5. Positive-ion mass spectra of arginine (A) and methionine (B), obtained by HPLC–MS combination after injection of 1 μg each. (Reproduced from *Clin. Chem.*, 26 (1980) 1467, with permission [66].)

TABLE 10.2

RELATIVE SENSITIVITIES OF AMINO ACIDS IN A LC–MS COUPLING [66]

Type of amino acid	Amino acid	MW	Relative response of MH^+ peak *
Aliphatic	Gly	75	100
	Ala	89	130
	Val	117	210
	Ile	131	233
	Leu	131	240
Hydroxy	Ser	105	66
	Thr	119	114
Dicarboxylic and amide	Asp	133	41
	Asn	132	65
	Glu	147	93
	Gln	146	31
Basic	Lys	146	38
	His	155	68
	Arg	174	73
Aromatic	Phe	165	136
	Tyr	181	49
	Trp	204	66
Sulfur-containing	CysH	121	13
	Met	149	15
	Pro	115	249

* Responses are expressed relative to glycine = 100.

selected-ion monitoring (SIM). The authors applied their technique to the MS analysis of amino acids. The spectra thus obtained are very simple and consist mainly of the protonated molecular ion (exceptions are Lys and His were the MH^+ are 80 and 60%, respectively). Fig. 10.5 shows the spectra of Arg and Met, obtained by injection of 20 μl of a 50-mg/l stock solution with formic acid as the mobile phase (flowrate 0.5 ml/min). The limited fragmentation of the samples allows the distinction of Glu and Lys. However, Leu and Ile cannot be distinguished. Due to the drastically reduced background, the detection limit is increased in comparison with "normal" CI.

Table 10.2 shows the relative sensitivities for 20 amino acids. These data were obtained from a series of injections by integrating the intensity of the MH^+ peak over the peak in the mass chromatogram. In scans of the entire spectra the detection limit (signal-to-noise ratio = 2) was determined to be 0.5 ng for Arg and 2.5 ng for Met. By use of the SIM technique, this limit could be reduced by a factor of ten. Desiderio et al. [77] applied the off-line combination of LC with field-desorption MS to an investigation of synthetic mixtures of hypothalamic oligopeptides containing 3 to 31 amino acids. The column effluent provides a peptide fraction which can be analyzed directly by field-desorption MS. This illustrates novel structural elucidation

References on p. B43

TABLE 10.3

COMPARISON OF SOME LC DETECTORS

Detector type	UV [*]	Fluorim-eter [*]	Electro-chem. [*]	MS [**]	
				Full scan	SIM
Application	Selective	Selective	Selective	Universal	Selective
Useful with gradients	Yes	Yes	No	Yes	Yes
Linear range	10^5	10^3	10^6	10^4	10^4
Detection limit, g	10^{-9}	10^{-12}	10^{-12}	10^{-9}	10^{-11}
Flow sensitivity	No	No	Yes	No	No
Temperature sensitivity	Low	Low	$1.5\%/°C$	No	No
Short-term stability, %	0.5	1	1	5	1

[*] Values taken from Snyder and Kirkland [48].
[**] Values taken from Blakley et al. [66].

methods for individual peptides and signals a quantitation method. Dawkins et al. [78] described the analysis of oligopeptides with LC–MS.

In conclusion, despite the recent progress in coupling LC to MS, a great many unsolved problems still severely restrict the method. A mass spectrometer would be an almost ideal detector for LC, but its high cost and the various technical difficulties do not especially encourage a potential user. Table 10.3 summarizes applications, detection limits, dynamic ranges, flow and temperature sensitivity, and stability for the described detector types.

10.3.2. Separation of amino acids

The determination of free amino acids has been reported by Schuster [79]. In 30 min he separated up to 20 amino acids in intravenous solutions on a 5-μm LiChrosorb NH_2 column, using an acetonitrile–phosphate buffer gradient. The underivatized amino acids, which were directly injected, were monitored at 200 nm. Similar to publications dealing with various derivatives, this paper states that the column temperature is an important and critical parameter; more detailed investigations of this and other parameters affecting retention times and resolution were published by Somack [54] and Kroeff and Pietrzyk [80]. The compounds were identified by means of their retention times. For a further confirmation of the peak identity, scans were performed during the chromatogram and compared with those of pure samples. Another critical parameter is the condition of the column. Schuster pointed out that Asn, Gln, and Gly may coalesce into a single peak when the column is poor or a deteriorated. While this determination of free amino acids saves time and labor and is more accurate than methods based on derivative formation or extraction, the low concentrations of amino acids in biological samples present difficulties.

This is why derivatives such as methylthiohydantoin (MTH) [81], phenylthiohy-

TABLE 10.4

VARIOUS DERIVATIVES FOR HPLC OF AMINO ACIDS

MTH = methylthiohydantoin, OPA = o-phthalaldehyde, DNP = 2,4-dinitrophenyl.

Column	Amino acids	Derivative	Detection, nm	Remarks	Pre- or postcolumn	Ref.
LiChrosorb NH₂	General	Free amino acids	UV 200			83
Amberlite XAD-2, 4, and 7	General	Free DNP and dansyl	UV 254, 208			84
μBondapak C₁₈	General	MTH	UV 254			85
LiChrosorb SI-60	General	Dansyl	Fluorimetry 340, 510	pmole/ml; effect of pH, time, and dansyl chloride concn.		87
μBondapak C₁₈, Spherisorb 50-DS	General	Dansyl	UV 250	Comparison of the two columns. 100 pmole for a single component		88
LiChrosorb RP-8	Phe	Dansyl	Fluorimetry 320, 552	250 pmole	Pre	89
Microsil C-8	General	OPA	Fluorimetry 340, 450	40 pg	Pre	90
μBondapak C₁₈, LiChrosorb RP-8	General	OPA	Fluorimetry 229, 470		Pre	91
Partisil SCX-10	Trp	OPA	Fluorimetry 370, 475	10 pg limit	Post	92
Nucleosil RP-18	General	OPA	Fluorimetry 330, 418	50 fmole detection limit	Pre	93
μBondapak C₁₈	3-Methylhis, 1-Methylhis, His	OPA	Fluorimetry 340, 440	ng level	Post	94
Micropack MCH-5	General	OPA	Fluorimetry 340, 455	pmole levels	Pre	95
NH₂-bonded Spherisorb 55-W	S-containing	DNP	UV 350–355	nmole levels	Pre	96
Zorbax ODS	General	DNP	UV 360	Clinical application	Pre	97
Knauer RP-8	General	Dimethylamino-benzene thiohydantoin	UV 436	Clinical application 8 pmole	Pre	98
μBondapak C₁₈	3-Me-His	Fluorescamine	Fluorimetry		Pre	99

References on p. B43

TABLE 10.5

HPLC OF AMINO ACID PHENYLTHIOHYDANTOIN DERIVATIVES

DETA = diethylenetriamine, TCA = trichloroacetic acid, THF = tetrahydrofuran, DMSO = dimethylsulfoxide.

Packing	Column (cm × mm)	Eluent	pH	No. of amino acids	Analysis time, min	Ref.
μBondapak C_{18}	30 × 3.9	NaOAc–MeCN	6	15	20	100
μBondapak C_{18}	30 × 3.9	NaOAc–MeOH; MeOH–H_2O	5.3	17	16	101
μBondapak	30 × 4.0	NaOAc–MeCN	4.0	7	15	102
μBondapak C_{18}	30 × 4.0	H_2O, MeOH, HOAc, Ac_2O	4.1	18	20	103
μBondapak C_{18}	30 × 0.4	NaOAc–MeOH	4.0	18	32	104
LiChrosorb RP-8, 5 μm	30 × 3.9	NaOAc–MeCN	4.6	17	60	105
LiChrosorb RP-8, 10 μm	24 × 4.6	NaOAc–MeOH	4.9	16	25	106
Hypersil ODS, 5 μm	23 × 4.6	NaOAc–H_2O		18	45	107
Spherisorb ODS, 5 μm		NaOAc–MeOH				
ODS-18, 5 μm	9 × 0.5	NaOAc–MeOH	5.3	7	25	108
		H_2O–MeCN, Me_2CHOH		6	40	
Zorbax ODS	25 × 4.6	NaOAc–MeCN	5.0	20	20	109
Tripeptide L-Val–L-Ala–L-Pro, bonded to Partisil 10	30 × 2.1	1% aqueous citric acid	2.5	9	40	110
μBondapak C_{18}	30 × 3.9	20% MeOH in 5 mM NaOAc; 38% aqueous MeOH	5.6	15	45	111
μBondapak C_{18}	30 × 4.0	8.0 mM DETA; 20 mM TCA	4.2	17	36	112
μBondapak C_{18}	30 × 4.0	90% 0.02 N NaOAc or 10% MeCN	3.7	19	26	113
μBondapak C_{18}	30 × 4.0	0.01 N NaOAc–MeCN (9:1)	4.0	20	36	114
Spherisorb-ODS, 5 μm	25 × 4.6	0.2 M LiOAc–MeCN (1:4) 0.1 M LiOAc–MeCN (1:4)	5.2 5.2	19	18	115
LiChrosorb RP-18	25 × 4.6	0.01 M NaOAc–MeCN–DMSO (5:3:2)	4.5	5	no data	116
Ultrasphere-ODS	25 × 4.6	0.0042 M HOAc brought to pH 5.16 with NaOH; 5% THF	5.16	25	32	54
LiChrosorb RP-18, 5 μm	25 × 3.0	0.01 M NaOAc. MeCN	4.3	16	32	117

dantoin (PTH) [82] dimethylaminonaphthalenesulfonyl (dans) [83], 2,4-dinitrophenyl (DNP) [84], 2,4,6-trinitrobenzenesulfonic acid [85], fluorescamine [86], and other derivatives have been used more frequently. Table 10.4 gives an overview of several papers published in the last two years, in which various derivatives were applied. The most widely used derivatives have been the PTH-amino acids. Fig. 10.3 shows an example of the separation of 25 amino acid derivatives by gradient elution from an Ultrasphere column [54]. The pH and ionic strength of the mobile phase, flowrate, and temperature of both precolumn and main column are critical factors in the resolution. Thus, it is often difficult to establish the precise chromatographic positions of basic and acidic PTH-amino acids. The retention times of PTH-Asp and PTH-Glu are very sensitive to slight pH changes, while PTH-His and PTH-Arg may be eluted together. The flowrate of 1.3 ml/min is a compromise to resolve the PTH-Met/Val and PTH-Lys/Ile pairs. An increase in pH from 3.5 to 6.0 decreases the retention times of the acidic and basic amino acids, while an increase in the acetate concentration at constant pH decreases the retention times of only the basic PTH-amino acids. The complete analysis takes only 32 min, and the coefficient of variation ranges from 0.3 (PTH-Val) to 2.9 [PTH-(S-carboxymethyl)-cysteine] on 27 repetitive injections.

Table 10.5 lists recent publications on the HPLC separation of PTH derivatives. A summary of older publications has been published by Godtfredsen and Oliver [117]. It is evident from Table 10.5 that in most of the separations hydrocarbon-bonded phases were used. We have not seen any report on the use of a microbore column, although they should be superior or at least equal to columns with ID ≥ 4 mm [118]. Very long microbore columns can be packed and, since there is a linear relation between separation efficiency and column length, the number of theoretical plates should be drastically increased. A further advantage is that the flowrates are lower and, therefore, the operating cost can be reduced by a factor of 20 by changing the column diameter from 4.6 to 1 mm ID. Scott [119] concluded that, because microbore columns provide rapid and efficient separations at low flowrates, they are probably the columns of choice. The reason why they are still seldom used is that they have only recently become commercially available and require modifications for connection to detectors and injectors. A further barrier to the application of microbore columns is the difficulty of packing them with the stationary phase. Godtfredsen and Oliver [117] have compared the separation characteristics of various columns with different ID and found that equivalent results are obtained by changing from 3 to 1.5 mm ID, provided that the operating conditions are appropriately modified.

10.3.3. Separation of enantiomers of amino acids

The separation of amino acid enantiomers is not only of theoretical interest with regard to the mechanism of interaction between chiral molecules, but it is also of practical importance in biology and in amino acid and peptide synthesis for determining the extent of racemization. Resolution of amino acid enantiomers has been accomplished by GC [120] (see Chap. 10.4.1.3), but only after derivatization,

TABLE 10.6

CAPACITY RATIOS (k') AND SELECTIVITIES $[\alpha = k'(\text{D})/k'(\text{L})]$ FOR SOME AMINO ACID ENANTIOMERS FOR THREE CONCENTRATIONS OF L-ASPARTYLCYCLOHEXYLAMIDE [140]

	$1 \cdot 10^{-3}\ M$			$2.5 \cdot 10^{-4}\ M$			$1 \cdot 10^{-4}\ M$		
	$k'(\text{L})$	$k'(\text{D})$	α	$k'(\text{L})$	$k'(\text{D})$	α	$k'(\text{L})$	$k'(\text{D})$	α
Pro	0.24	1.12	4.46	0.55	1.42	2.58	2.17	3.0	1.38
Val	0.41	0.82	2.0	1.07	1.52	1.42	1.83	2.17	1.18
Nvl	0.56	0.89	1.59	1.14	1.43	1.25	1.74	2.0	1.15
Cys	0.76	1.15	1.51	1.81	2.43	1.34	–	–	–
Met	1.41	2.06	1.46	2.52	3.29	1.30	3.7	4.4	1.19
Dopa	1.78	2.33	1.31	3.21	3.90	1.22	5.67	6.86	1.21
Ile	1.76	3.76	2.14	2.86	5.0	1.75	4.5	7.0	1.55
Leu	2.35	3.88	1.65	3.23	4.63	1.43	5.19	6.81	1.31
Nle	2.47	4.23	1.71	3.67	5.40	1.47	5.13	6.76	1.32
Tyr	3.03	4.5	1.48	5.02	6.98	1.39	7.03	8.86	1.26
Eth	4.65	6.64	1.43	7.14	9.43	1.32	10.7	12.8	1.2

and preparative separations by that technique are difficult to achieve. Therefore, various attempts have been made to resolve enantiomeric mixtures by LC [121,122]. One of the two avenues towards this goal is to convert the enantiomers to diastereomers before chromatography [123]. The second, which is now most frequently applied, consists of an in situ diastereomer formation, with an optical active reagent in either the stationary or the mobile phase.

In 1972, Snyder et al. [124] demonstrated the use of amino acids, grafted to polymeric beads, as chiral stationary phases. Metal ions, such as Zn(II) or Cu(II), were added to the mobile phase and adsorbed by the covalently bonded amino acids. The metal–amino acid complex formed exhibits a stereoselectivity for complex formation with D- or L-amino acids [125–129]. Other authors [130–133] described similar enantiomeric separation methods but without the use of metal ions.

The alternative approach, i.e. addition of the chiral reagent to the mobile phase, has been described by various authors [134–140], who used different metal complexes to achieve the enantiomeric separation. Gilon et al. [140] reported the use of an aqueous mobile phase, containing L-aspartylcyclohexylamide–Cu(II), which allows the detection of nonaromatic amino acids at 230 nm. Table 10.6 shows the capacity ratios and selectivities for some amino acid enantiomers. Lindner et al. [139], using L-2-isopropyl-4-octyldiethylenetriamine–Zn(II) as chiral chelate, could resolve all common amino acids except proline. In contrast to other workers, they separated the amino acids as their respective dansyl derivatives. The role of pH, ionic strength, and metal ions (Ni^{2+}, Cd^{2+}, Cu^{2+}, Hg^{2+}, and Zn^{2+}) upon the separation behavior was also examined.

Hare and Gil-Av [136] have separated underivatized amino acids on an ion-exchange column by using a proline–Cu(II) complex as chiral additive to the mobile

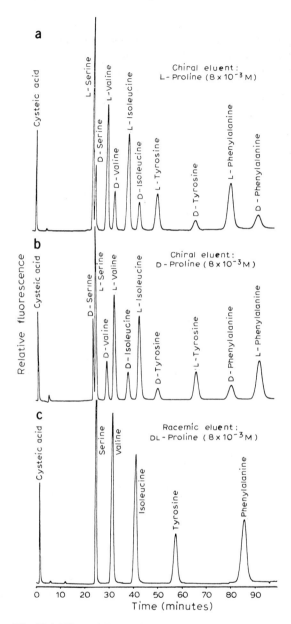

Fig. 10.6. Effect of the chirality of the eluent on the separation of D- and L-amino acid enantiomers by ligand-exchange HPLC on a 12×0.2-cm ID DC-4a cation-exchange column at 75°C. Eluent, 0.05 N sodium acetate buffer (pH 5.5), containing $4 \cdot 10^{-3}$ M CuSO$_4$ and $8 \cdot 10^{-3}$ M proline; flowrate, 10 ml/h; pressure, 200 bar; fluorimetric detection after postcolumn reaction with o-phthalaldehyde, flowrate 10 ml/h. (Reproduced from *Science*, 204 (1979) 1226, with permission [136].)

phase. Fig. 10.6 shows the chromatograms obtained with (a) L-proline, (b) D-proline, and (c) DL-proline as the eluents. It is evident from the figure that a simple change in the chirality of the eluent reverses the order of elution of the enantiomers and that the racemic eluent does not resolve the DL-pairs. For the quantitation of pmole amounts, the authors used fluorimetric determination of the postcolumn OPA derivatives. OPA has the advantage of being insensitive to proline while being highly sensitive to amino acids containing a primary amino group.

Audebert [141] has reviewed the literature concerning the direct resolution of enantiomers in CC. He points out that the technique in which the chiral reagent is contained in the mobile phase is best suited for analytical purposes. Lefebvre et al. [127] obtained a total resolution of amino acids by using porous gels based on acrylamide, grafted with L-α-amino acids and complexed with metal ions. They discussed the influence of gel structure, kinetics of liquid exchange, and nature of the complexing ion and of the chiral graft.

10.3.4. Separation of oligopeptides

In the field of peptide analysis there is great demand for chromatographic techniques capable of separating closely related analogs. Traditionally, the separation of oligopeptides has been accomplished by ion-exchange or thin-layer chromatography, but more recently, HPLC has offered some new possibilities in that it can be used to monitor reactions, check the purity of synthetic peptides, study the metabolic pathways, and separate mixtures on a preparative scale. Because of their

TABLE 10.7

EFFECT OF TETRAALKYLAMMONIUM SALTS AS ION-PAIRING AGENTS ON THE RETENTION TIMES OF DI- TO PENTAPEPTIDES

HPLC on a μBondapak alkylphenyl column (10 μm, 30 cm $\times$ 4 mm ID) in methanol–water (1:1), 2 mM reagent solution, acetate as the anion, pH 4; flowrate, 1.5 ml/min [160].

Peptide*	Retention time, min				
	R** = H	R = CH$_3$	R = CH$_2$CH$_3$	R = (CH$_2$)$_2$CH$_3$	R = (CH$_2$)$_3$CH$_3$
L–W–M–R	2.8	3.0	3.05	3.5	2.7
L–W–M–R–F	5.7	6.8	6.5	8.7	4.0
G–F	2.4	2.4	2.6	2.7	2.05
G–G–Y	2.0	2.0	2.2	1.8, 2.05 ***	1.7
M–R–F	2.6	2.7	2.8	3.0	1.7
G–L–Y	2.4	2.5	2.8	2.7	2.1
R–F–A	2.3	2.3	2.6	2.1, 2.7 ***	1.8

    * A = alanine, D = aspartic acid, F = phenylalanine, G = glycine, K = lysine, L = leucine, M = methionine, R = arginine, W = tryptophan, Y = tyrosine.
   ** Ammonium salts (R$_4$NOAc).
 *** Two peaks were observed.

TABLE 10.8

EFFECT OF METAL IONS ON THE RETENTION TIMES OF DI- TO PENTAPEPTIDES [160]

For amino acid code and conditions see Table 10.7.

Peptide	Retention time, min					
	Li^+	Na^+	K^+	Cs^+	Mg^{2+}	Ca^{2+}
L–W–M–R	2.9	2.9	3.0	3.4	2.8	3.2
L–W–M–R–F	4.7	4.7	4.8	4.8	4.5	4.9
G–F	2.2	2.2	2.35	2.5	2.2	2.5
G–G–Y	2.1	2.1	2.3	2.4	2.2	2.3
M–R–F	2.6	2.6	2.7	2.8	2.6	3.0
G–L–Y	2.3	2.3	2.5	2.7	2.3	2.5
R–F–A	2.3	2.2	2.45	2.8	2.3	2.6

polar nature, unprotected peptides are difficult to separate on silica gel phases [142–144] and, therefore, derivatized samples are most frequently used.

The possibilities for separating and isolating oligopeptides have markedly improved with the introduction of reversed-phase HPLC [145–156]. The retention of the samples to be separated depends on hydrophobic interactions between the hydrocarbon bound to the carrier and the solute. The more hydrophobic a compound, the longer will be its retention time [145]. Polar substances (e.g., alcohols) in the eluent or polar functional groups in the solutes reduce the capacity factor of a given compound. Peptides containing acidic or basic amino acids (which bear full electric charges) are poorly retarded in reversed-phase columns. Several theoretical and experimental studies have dealt with the effects on separation of manipulating secondary solution equilibria, e.g., pH and ion pairing [154,157–160]. The most useful technique for separating peptides involves the addition of hydrophobic or hydrophilic counterions to the mobile phase at low pH. Schaaper et al. [154] has studied the influence of various perfluoroalkanoic acids as lipophilic ion-pairing reagents on the retention of underivatized hexa- and heptacosapeptides by ODS-silica. The retention – and sometimes the selectivity – progressively increased from trifluoroacetic acid to perfluorodecanoic acid.

Hancock et al. [160] have described the effect of tetraalkylammonium, alkylammonium, and inorganic salts on the retention times of di- to pentapeptides, chromatographed on a μBondapak alkylphenyl column. Tables 10.7 and 10.8 show the effects for various tetraalkylammonium salts and small cations. It appears that the small, highly solvated cations in Table 10.8 give retention times similar to those obtained for the ammonium salts in Table 10.7. Going from the ammonium to the tetrapropylammonium salt, a slight increase in the retention time can be observed. However, hydrophobic cations with long or bulky carbon chains (tetrabutylammonium or dodecylammonium ions) cause substantial decreases in retention times. The authors ascribed these observations to the interplay of ion-pair partition and dynamic ion-exchange effects for the cationic reagents.

References on p. B43

Lin et al. [156] have investigated dipeptides containing acidic or basic amino acids. These, as well as the neutral samples, consisting of residues such as serine, proline, and threonine cause some problems in that they are poorly retained by reversed-phase columns. Therefore, the conditions usually employed for the separation of neutral peptides cannot be applied. The authors used aliphatic carboxylic acids as surfactants to analyze basic, acidic, and neutral dipeptides. Fig. 10.7 shows the separation of fourteen dipeptides differing in polarity. It is evident that the basic compounds are well separated from the acidic and neutral dipeptides; basic and acidic compounds would be only poorly retained if a simple aqueous solvent were used as the mobile phase. The enhancement in the capacity factors of basic dipeptides can be explained by electrostatic interactions between the carboxylic surfactant anion and the protonated amino group of the dipeptides. The electrostatic repulsion between the surfactant and the carboxylic acids, on the other hand, causes the retention times of acidic dipeptides to be low.

Analysis of peptides by HPLC would be greatly simplified if the respective retention times could be predicted by the contribution of each of the amino acids in the peptide. Meek [161] derived values (retention coefficients) representing the contribution to the retention of each of the common amino acids and end groups.

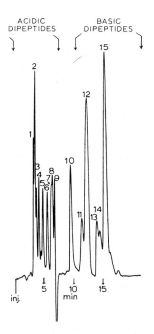

Fig. 10.7. Separation of dipeptides by HPLC on a μBondapak C_{18} column. Mobile phase, 0.038% (w/v) octanoic acid, 0.002% (w/v) pentanoic acid, and 5% (v/v) butanol in water. Peaks: 1 = Asp–Asp, 2 = Glu–Glu, 3 = Asp–Leu, 4 = Ala–Ala, 5 = Val–Ala, 6 = Glu–Leu, 7 = Ser–Ser, 8 = Val–Val, 9 = solvent disturbance, 10 = Arg–Asp, 11 = Lys–Ser, 12 = His–Ala, 13 = Arg–Val, 14 = Lys–Lys, 15 = His–Phe. (Reproduced from *J. Chromatogr.*, 197 (1980) 31, with permission [156].)

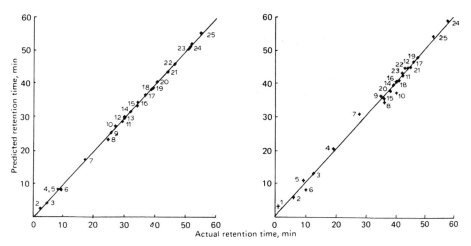

Fig. 10.8. Correlation of actual retention times versus times predicted by summing retention coefficients for the amino acids and end groups, left, pH 7.4 (correlation = 0.996); right, pH 2.1 (correlation = 0.998). For peak identification, see Table 10.9. HPLC on a Bio-Rad ODS column at room temperature with a linear gradient from $0.1\ M$ NaClO$_4$, pH 7.4 or 2.1, at 0 min to 60% acetonitrile in $0.1\ M$ NaClO$_4$ at 80 min. (Reproduced from *Proc. Nat. Acad. Sci. U.S.*, 77 (1980) 163, with permission [161].)

For this experiment, Meek used a linear gradient and found an approximately linear increase in retention time for amino acid oligomers when further equal residues are added to the dipeptide. The slope thus gives the retention added per side chain and peptide bond and extrapolation of the line to 0 residues gives a value which contains the contribution to retention of the end groups. By repeated regression analysis he calculated the retention coefficients for the 20 common amino acids. The predicted retention time for a peptide then equals the sum of the retention coefficients for the amino acids and end groups plus the time, t_0, for elution of unretained compounds. Fig. 10.8 shows a graph of the actual retention times (Table 10.9) versus the predicted time of a series of different peptides. These results show a remarkably high correlation of 0.9996. Meek tested this method with retention times taken from the literature [152]; the values are given in Table 10.10. In order to correct the predicted values for the different gradient rates and the biphasic gradient used by O'Hara and Nice [152], Meek plotted the retention times of phenylalanine oligomers in his system versus those of O'Hara and Nice. The slope of this curve (0.836) gives the factor by which the sum of the predicted coefficients has to be multiplied. Finally, the y-intercept (-12 min) was added to the above values to obtain the predicted retention times listed in Table 10.10. Again, the predicted and the actual times agree reasonably well, considering the differences in columns, mobile phases, and gradients. Thus, the data indicate that this premise is justified for peptides with up to about 20 residues.

There is a great demand for high-purity peptides in gram quantities for use in biological research. In the past, these products were purified, by ion-exchange and other liquid chromatographic techniques, methods which are slow and have low

TABLE 10.9

RETENTION TIMES OF PEPTIDES IN HPLC ON A BIO-RAD ODS COLUMN AT ROOM
TEMPERATURE WITH A LINEAR GRADIENT FROM 0.1 M NaClO$_4$, pH 7.4 OR 2.1, AT 0 MIN
TO 60% ACETONITRILE IN 0.1 M NaClO$_4$ AT 80 MIN [161]

For separations at pH 7.4, the starting buffer contained 5 mM phosphate, pH 7.4. For separations at pH
2.1, both starting and final buffers contained 0.1% phosphoric acid.

Compound	Retention time, min	
	pH 7.4	pH 2.1
1 Triglycine	2.0	3.0
2 Pentaalanine	4.6	8.1
3 Divaline	6.9	14.5
4 Dimethionine	10.5	21.0
5 TRF	11.5	11.2
6 Tuftsin	11.7	12.0
7 Trityrosine	19.5	29.7
8 [Met]-enkephalin	27.5	38.0
9 Trileucine	28.0	36.8
10 [Leu]-enkephalin	29.3	42.0
11 Ditryptophan	31.6	44.3
12 Angiotensin II	32.2	47.5
13 α-Endorphin	32.3	47.7
14 Cerulein	34.2	42.2
15 Oxytocin	36.4	37.9
16 Gastrin-(12-15)	36.5	42.4
17 Neurotensin	39.0	48.0
18 Physalaemin	41.0	43.0
19 Triphenylalanine	41.6	49.5
20 LHRF	42.8	49.5
21 α-Melanotropin	46.2	46.2
22 Bradykinin	48.0	45.0
23 Eledoisin-related peptide	53.0	44.0
24 Glucagon	53.6	60.0
25 Somatostatin	57.5	55.0

separation efficiencies. More recently, Kullmann [142] and Gabriel et al. [143] have
separated fully protected peptides in gram quantities by HPLC on silica gel columns.
Bishop et al. [153] have used a reversed-phase system of polyethylene cartridges with
octadecylsilica particles to purify underivatized peptides. Fig. 10.9 shows the elution
profile resulting from the injection of 1 g crude Leu(Gly)$_3$. In order to improve the
resolution, the leading edge of fraction 5 was recycled. Fractions 6, 7, 8, and 9 were
collected and showed the same retention time as an authentic sample of the
tetrapeptide. The sample load could be subsequently increased to 10 g of crude
product without loss of resolution. Loading a known amount onto this preparative
system gave a recovery greater than 95%. The product was found pure, as judged by
amino acid analysis after hydrolysis. Based on these results, the authors concluded
that rapid and efficient purifications of underivatized peptides in quantities up to at

TABLE 10.10

PREDICTION OF RETENTION TIMES OF A SERIES OF PEPTIDES [161]

HPLC on a Hypersil ODS column with a triphasic acetonitrile gradient containing NaH_2PO_4 [152].

Compound	No. of residues	Retention time, min		Error, min
		Predicted	Actual	
[Met]-enkephalin	5	18.4	19.0	−0.6
[Leu]-enkephalin	5	20.9	22.0	−1.1
ACTH-(5-10)	6	11.9	17.0	−5.1
ACTH-(34-39)	6	27.0	31.0	−4.0
ACTH-(4-10)	7	12.4	20.5	−8.1
ACTH-(4-11)	8	33.1	30.0	3.1
Angiotensin II	8	27.4	23.0	4.4
Substance P-(4-11)	8	33.1	30.0	3.1
Oxytocin	9	27.4	23.0	4.4
[Arg]-vasopressin	9	9.0	14.0	−5.0
[Lys]-vasopressin	9	10.2	13.0	−2.8
[Arg]-vasotocin	9	7.2	12.0	−4.8
Substance P	11	33.3	29.0	−4.3
α-MSH	13	27.3	26.0	1.3
Neurotensin	13	28.9	24.5	4.4
Somatostatin	14	25.5	32.0	−6.5
Bombesin	14	22.6	26.0	−3.4
Gastrin-1	17	26.8	28.5	−1.7
ACTH-(18-39)	22	20.2	30.5	−10.3
ACTH-(1-24)	24	34.9	21.5	13.4
Melittin	25	61.3	46.0	15.3
Glucagon	29	39.7	36.0	3.7
β-Endorphin	31	41.5	34.0	7.5

least 10 g with excellent recoveries should be possible.

Table 10.11 is a compilation of some publications dealing with HPLC of peptides that appeared between 1978 and 1980.

10.4. GAS CHROMATOGRAPHY

In the last decade, GC has gained importance as an alternative technique to ion-exchange chromatography. The advantages of GC are:

(a) detection limits low (down to 1 pmole with FID or even lower in conjunction with a mass spectrometer);

(b) equipment cheaper than for standard automated ion-exchange chromatography;

(c) analysis time short;

(d) choice of specific detection systems (e.g., FID, ECD, nitrogen or phosphorus detector, and mass spectrometer);

TABLE 10.11

PUBLICATIONS CONCERNING HPLC OF OLIGOPEPTIDES

Peptides	Stationary phase	Remarks	Ref.
Protected synthetic peptides	Silica Gel 60	Preparative separation	142
Various	Reversed-phase	High resolution, high recovery	147
Hexa- and heptacosa-peptides	Nucleosil C_{18}	Ion-pairing with perfluoroalkanoic acids	154
Somatostatin	Reversed-phase	Qualitative and quantitative determination	155
Dipeptides	μBondapak C_{18}	Ion-pairing with aliphatic acids	156
Di- to pentapeptides	μBondapak alkylphenyl	Effect of cationic reagents	160
Various	Bio-Rad ODS	Prediction of retention times	161
Di- and tripeptides	XAD-2, XAD-4, XAD-7	Investigation of retention and separation	162
Various	Reversed-phase	—	163
Thyrotropin-releasing hormone	Reversed-phase	Preparative separation	164
Oxytocin derivatives, ^{2}H-, ^{13}C-labeled	μBondapak C_{18}	Diastereomer separation	165, 166
Various	μBondapak Fatty Acid Column	Ion-pairing reagents	167
Various	Reversed-phase	Application to biochemistry	168
Dipeptides	Bonded tripeptide L-Val-L-Ala-L-Pro	pH and ionic strength dependency	169
Di- and tripeptides	C_8-bonded	Diastereomers	170
Di- to decapeptides	μBondapak C_{18}	Mobile phase containing phosphoric acid	171
Di- to pentapeptides	μBondapak alkylphenyl	Hydrophobic ion-pairing reagents	172
Insulin, glucagon, somatostatin	Corning Glycophase G/CPG-100	—	173
Peptide mixtures	μBondapak Fatty Acid Column	Peptide mapping	174
Tryptic peptides	μBondapak C_{18}	—	175
Peptide hormones	ODS-TMS silica	Ion-pairing with aliphatic acids	176
Protein hydrolyzate	Micropak AX-10	—	177
Pro-Leu-Gly-NH_2	μBondapak C_{18}	Tritiated samples	178

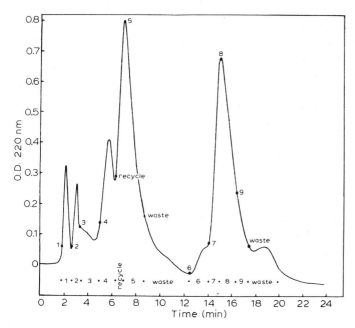

Fig. 10.9. Preparative purification of 1 g crude Leu(Gly)$_3$ by HPLC. Column, Prep Pak-500 C$_{18}$ cartridge (75 μm, 30 cm × 5.7 cm); eluent, water–methanol–trifluoroacetic acid (1900:100:1), pH 2.3; flowrate, 100 ml/min. (Reproduced from *J. Chromatogr.*, 192 (1980) 222, with permission [153].)

(e) use of a gas chromatograph in many other analytical methods.

On the other hand, the chemical heterogeneity of amino acids and peptides with such different functions as carboxyl and amino, imino, hydroxyl, sulfhydryl, imidazole, guanidine, and indole groups is a serious handicap in developing practical analysis conditions. The difficulties arise in the reproducible, quantitative, and simultaneous conversion of all these groups to volatile and stable derivatives suitable for GC. These requirements are a challenge to the analyst and are as yet not completely fulfilled. Thus, GC is not a routine method in this particular field but an extremely useful and sensitive technique for special analyses such as the monitoring of a selected number of amino acids, separation of optical isomers, and investigation of metabolic pathways, particularly in combination with MS and stable isotopes. Such applications are covered in review articles [179–181] and to some extent in refs. 2 and 182.

10.4.1. Gas chromatography of amino acids

10.4.1.1. Clean-up

Since the most widely used FID is nonspecific, a clean-up of biological samples prior to derivatization is indispensable. The method of Pocklington [183], originally intended for amino acids in sea water, is applicable to many biological preparations.

In order to obtain a large surface, the sample is lyophilized and prewashed by slurring it with dry diethyl ether into a column, 400×25 mm, filled with Dowex 50W-X12, 100–200 mesh, in the H^+ form. Extraction of the amino acids as their hydrochlorides is then performed with aqueous ethanol, acidified with HCl to pH 1.3. The extract is dried in vacuo, taken up in water and, if necessary, washed with chloroform in order to remove interfering pigments and carboxylic acids. Further purification is achieved by cation exchange. The amino acids are eluted from a sulfonated polystyrene resin column with $2\,M$ aqueous ammonia.

Blood plasma may be deproteinized with picrate [184] or better, yet, in order to avoid losses of basic amino acids, with sulfosalicylic acid [185]. The protein-free filtrate is purified by cation exchange on a 50×8-mm (1.5 g) Dowex AG 50W-X8 column. Procedures for the purification of urine samples preliminary to the separation and quantitation of amino acids and peptides are reviewed by Lou and Hamilton [182] and Bremer et al. [2].

10.4.1.2. Derivatization

The presence of both amino and carboxyl groups makes for a low volatility of the amino acids and, therefore, demands derivatization of these functional groups to make the sample suitable for GC analysis. A useful derivatization method should derivatize polar groups in the molecule without destroying sensitive structures. In 1956, Hunter et al. [186] published a GC method based on the aldehydes liberated in the reaction with ninhydrin. These aldehydes can also be produced by other reagents, and the samples can be chromatographed as amino alcohols, decarboxylated amines, nitriles, etc. [181]. The most useful derivatives, however, are those which leave the polar groups unchanged but only diminish their polarity. The literature concerning different derivatization techniques reveals an enormous number of useful derivatives, which are described in great detail in the excellent handbook of Knapp [187]. The most frequently used derivatives at present are the TMS derivatives and the acyl amino acid esters.

10.4.1.3. Trimethylsilyl derivatives

The first studies on TMS derivatives of amino acids, reported in 1960 [188], were followed by systematic investigations [189,190]. Problems are mainly due to the difficulty of silylating the amino group and the instability of the resulting TMS azanes, which are rather sensitive to traces of water. According to Gehrke and coworkers [191,192], the best procedure for the removal of water is azeotropic distillation with several portions of dichloromethane. The difficulties in silylating the nitrogen function, which cause multiple GC peaks, have led to the investigation of several silylating mixtures at various conditions. Depending on the reaction conditions, several amino acids may form double peaks [189,190,192], e.g., Gly, ω-amino acids, Arg, His, and Trp [192]. Glutamic acid may form 2-pyrrolidone-5-carboxylic acid. Although the amino acid TMS derivatives in silylating agents should be stable for at least a week in tightly closed containers [191,192], it has been reported that the

TABLE 10.12

PUBLICATIONS CONCERNING TMS DERIVATIVES OF AMINO ACIDS

Abbreviations: BSA = bis(trimethylsilyl)acetamide, BSTFA = bis(trimethylsilyl)trifluoroacetamide, TMCS = trimethylchlorosilane, TMSDEA = trimethylsilyldiethylamine, MSA = N-methyl-N-trimethylsilylacetamide, MSTFA = N-methyl-N-trimethylsilyltrifluoroacetamide.

Amino acid type	Reagent	Temperature, °C	Time	Ref.
General	BSTFA–MeCN	150	2.5 h	192
General	BSTFA–MeCN	125	15 min	194
General	BSA–TMCS–Py	60	12–14 h	195
General	TMSDEA	reflux	1 h	196
General	MSA	25 or 60–100	5 min	197
General	MSTFA–CF$_3$COOH	25	dissolve	198
General	MSTFA–TMCS–MeCN	120	1 h	199
Sulfur amino acids	BSTFA	150	5 min	200
Gly, Lys, Arg	BSTFA–MeCN	135	15 min	190, 201
Trp, tryptamines	1.CH$_2$O–NaOH	40	15 h	202
	2.BSTFA	25		
3-Methylhistidine	BSTFA–MeCN	150	2.5 h	203
Glutamic acid	BSTFA	–	–	204

concentration of the histidine derivatives drops within 2 h [194] and that no stable derivatives could be prepared from Arg, γ-Abu, Cit, Gln, His MetO, and Tau [183]. Therefore, in our experience, TMS compounds should be chromatographed almost as soon as the reaction mixture has cooled.

A few of the various trimethylsilylation methods are listed in Table 10.12. For further experimental details, the reader is referred to the handbook of Knapp [187]. The most widely applicable method of preparing TMS ethers appears to be that of Gehrke and coworkers [191,192]:

The aqueous sample solution, containing 0.5–6 mg of amino acids is evaporated in a silylating tube under a stream of nitrogen at 70°C. Traces of water are removed azeotropically with two 0.5-ml portions of dichloromethane and then 250 μl acetonitrile, containing the internal standards and 250 μ BSTFA per mg of each amino acid, are added. The tube is closed with a PTFE-lined screw cap, placed in an ultrasonic bath for 1 min, and then heated at 150°C for 2.5 h. If the procedure was properly carried out, an equal weight of standard should have a peak ratio of 15:1 for tri- to di-TMS-glycine and 10:1 for di- to tri-TMS-glutamic acid; it should contain only the tetra-TMS-lysine and have a tetra-TMS-arginine peak about half the size of the lysine peak.

GC of TMS derivatives is only performed on nonpolar liquid phases. To avoid decomposition in the injector, the sample is best injected directly onto the column. Table 10.13 shows the relative mobility of amino acid TMS derivatives, in methylene units, on two columns of different polarity: 3% OV-1 and 3% OV-17 on Chromosorb W HP. The phenyl groups in the OV-17 phase act to retain compounds with delocalized π-systems more strongly. This is illustrated by the ΔMU values in the

References on p. B43

TABLE 10.13

METHYLENE UNIT (*MU*) VALUES OF AMINO ACID TMS DERIVATIVES [201]

Column, 6 ft.×0.25-in. Pyrex tubing, packed with 3% OV-1 or 3% OV-17 on 80–100-mesh Chromosorb W HP; temperature program, 100 to 325°C at 10°C/min; initial temp., 70°C for compounds with *MU* < 12.0; carrier gas, He, flowrate 80 ml/min.

	MU_{OV-17}	MU_{OV-1}	ΔMU
N-Acetylphenylalanine	19.33	17.90	1.43
N-Acetyltryptophan *	26.00	24.18	1.82
Alanine	11.25	11.05	0.20
β-Alanine	12.30	11.90	0.40
α-Aminobutyric acid	12.00	11.77	0.23
γ-Aminobutyric acid	15.57	15.46	0.11
α-Aminoisobutyric acid	11.60	11.48	0.12
β-Aminoisobutyric acid	12.48	12.16	0.32
	14.76	14.74	0.02
Arginine **	16.40	16.32	0.08
Asparagine **	18.00	16.87	1.13
Aspartic acid	15.88	15.41	0.47
Citrulline	19.03	18.36	0.67
Cystathionine	22.84	22.35	0.49
Cysteic acid **	20.67	19.69	0.98
Cysteine	16.19	15.65	0.54
Cystine	23.95	23.12	0.83
3,4-Dihydroxyphenylalanine	21.63	21.24	0.39
1,4-Dimethylhistidine	19.59	17.35	2.24
Djenkolic acid	25.60	24.58	1.02
Glutamic acid	16.89	16.29	0.60
Glutamine **	18.87	17.73	1.14
Glycine	11.52	11.11	0.41
	13.21	13.18	0.03
Histidine	20.98	19.14	1.84
		18.00	
Homocystine	26.38	25.63	0.75
4-Hydroxylysine	18.55	18.90	− 0.35
	20.58	21.19	− 0.61
4-Hydroxyproline	15.63	15.46	0.17
5-Hydroxytryptophan	26.01	24.60	1.41
Isoleucine	13.14	13.06	0.08
Kynurenine	23.57	21.86	1.71
Lanthionine	21.57	21.33	0.42
Leucine	12.91	12.84	0.07
Lysine	19.26	19.56	− 0.30
Methionine	16.17	15.30	0.87
1-Methylhistidine	21.06	17.64	3.42
3-Methylhistine	20.40	17.24	3.16
Ornithine	18.21	18.53	− 0.32
Phenylalanine	17.24	16.25	0.99
Pipecolic acid	14.07	13.67	0.40
Proline	13.50	13.02	0.48
Sarcosine	11.73	11.43	0.30

TABLE 10.13 (continued)

	$MU_{OV\text{-}17}$	$MU_{OV\text{-}1}$	ΔMU
Serine	14.10	13.80	0.30
2-Thiohistidine	23.85	22.51	1.34
Threonine	13.97	14.04	−0.07
Tryptophan	23.77	22.15	1.62
Tyrosine	20.19	19.52	0.67
Valine	12.44	12.34	0.10

* Secondary peak on both columns.
** Secondary peak on OV-17.

table ($\Delta MU = MU_{OV\text{-}17} - MU_{OV\text{-}1}$). High ΔMU values are observed for the aromatic amino acids His, Kyn, Phe, Tyr, Trp, and for the amides and carbamides Asn, Glu, and Cit. On the other hand, low ΔMU values are found for aliphatic amino acids and those with a hydroxyl or additional amino group. A compound with a disubstituted amino group is more strongly retained than a mono-TMS derivative because of the higher molecular weight (cf., e.g., β-Ala, β-Aib, Gly, and Hyl in Table 10.13). A smaller ΔMU value is observed, however, when the amino group is more effectively shielded by two TMS groups. Similarly, Thr, having a secondary hydroxyl function, is better shielded and shows a lower ΔMU value than serine with a primary hydroxyl group. Additional hydroxyl functions increase the molecular weight and the retention index but lower the ΔMU value (Phe, Tyr, DOPA, Pro, and Hyp).

Gehrke et al. [191] used a mixed phase, containing OV-7 and OV-22 (2:1). This

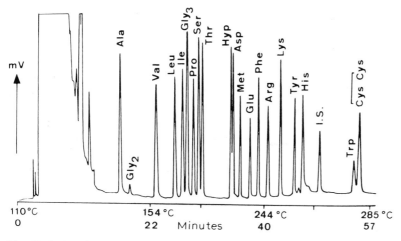

Fig. 10.10. Gas chromatogram of amino acid TMS derivatives. Column, 6 m × 2 mm, 10% OV-11 on 100–120-mesh Supelcoport. Carrier gas, 20 ml/min, nitrogen; initial temperature, 110°C/min for 22 min, then 5°C/min to 285°C; injector 275°C; detector 300°C. (Reproduced from *J. Chromatogr.*, 45 (1969) 24, with permission [191].)

was later replaced by OV-11, which contains about the same methyl-/phenylsilox-ane composition (65% methyl, 35% phenyl) [192]. Fig. 10.10 exemplifies the chromatogram of a protein hydrolyzate. Because the column used is long (6 m × 2 mm ID), difficulties in eluting His quantitatively arise. For separating Met, Glu, Phe, Arg, Lys, Tyr, His, Trp, and Cys a shorter column (2 m × 4 mm ID) has been proposed (temperature program: 190°C for 5 min, then rise to 265°C at 5°C/min).

10.4.1.4. Acyl amino acid esters

Many derivatization procedures for amino acids include esterification of the carboxyl groups. Although the resulting esters of some amino acids are volatile enough for GC analysis [205], the more generally applicable procedure is a combination of esterification of the carboxyl group and acylation of the amino function. Esters are always formed by acidic catalysis in anhydrous alcohol. In addition to HCl, which is the most widely used catalyst [206], thionyl chloride [207], dimethyl-sulfite [208], and acidic ion exchangers have been proposed. Diazomethane has the disadvantage of losses during evaporation of the excess reagent, since some methyl esters of amino acids are very volatile. Therefore, the higher esters (propyl, butyl, and amyl) have been investigated (ref. 187, p. 263). Solubility problems with the higher alcohols are circumvented by transesterification of previously formed methyl esters. Acylation of the esters is usually carried out with the acid anhydrides in dichloromethane, ethyl acetate, or acetonitrile. It should be emphasized that esterification must always be performed first, because the reverse order would lead to the deacetylation of O- and N-acyl derivatives during esterification. In a series of N-acyl esters of amino acids the effect of substitution on GC behavior was studied [209,210]. The most suitable derivatives were found to be the acetyl, propionyl, and perfluoroacyl derivatives of ethyl, propyl, and butyl esters. The fluorinated derivatives are eminently suitable for ECD and MS with negative CI. With the ECD, Zumwalt et al. [211] were able to detect cysteine and methionine derivatives at levels of 2 and 1 pg, respectively. The most frequently used derivative in recent years is the N-trifluoroacetyl-n-butyl ester, first introduced by Gehrke [212]. With n-butanol, amino acids can be derivatized directly without transesterification. Except for isoleucine, which takes 35 min [213], all amino acids are esterified within 15 min in 3 N HCl at 100°C.

A silylation study of amino acid butyl esters [214] revealed that the esterification step was not quantitative. Further examination of this reaction showed that 2.7 N HCl in butanol produces more than 2 moles of water at 150°C within 15 min and more than 0.2 mole at 100°C. Such quantities of water will shift the equilibrium significantly to the acid side.

The N-TFA amino acid butyl esters, are prepared as follows [213]. The sample solution, containing the internal standard, is evaporated to dryness in a micro reaction tube under a stream of nitrogen at 100°C. For each 100 μg of amino acids, 150 μl of 3 N HCl in n-butanol is added. This is followed by mixing in an ultrasonic mixer for at least 15 sec and heating at 100°C for 15 min. After evaporation at 100°C under nitrogen, 150 μl dichloromethane is added to remove traces of water

TABLE 10.14

VARIOUS DERIVATIVES FOR GC OF AMINO ACIDS

Derivatives	Columns	Temperature, °C	Remarks	Ref.
Heptafluorobutyryl-1-propyl esters	1% OV-1 or 3% SE-30	80–250	Amino acid composition of chymotrypsinogen A and casein	228
N-Acetyl-1-propyl esters	Mixed 0.31% Carbowax 20 M, 0.28% Silar-5CP, and 0.06% Lexan	100–250	Erythrocyte amino acids	229
N-Trifluoroacetyl-1-butyl esters	Two columns: Tabsorb and Tabsorb HAC	100–200 100–210	Amino acids in pine needles	230
N (O, S)-Heptafluorobutyryl-isobutyl esters	3% SE-30	100–250	Derivatization test	231, 232
N-Acyl esters	3% OV-17	90–230	Amino acids in rat tissues	233
N-Heptafluorobutyryl-2-propyl esters	3% SE-30	75–250	Homoserine determinations	234
N-Trifluoroacetyl-1-butyl esters	Mixed 3% QF-1–1% SE-30 (3:2, w/w)	90–250	Collagen amino acids in biopsy tissues	235
N-Acetyl-1-propyl derivatives	Carbowax 4000–NPGS–Versamide 900 (1:1:1, w/w/w)	80–250	Blood and plasma amino acids	227

References on p. B43

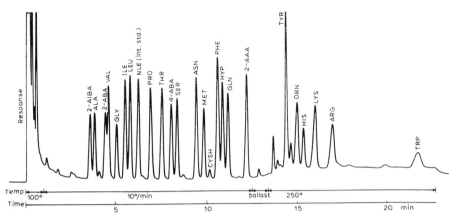

Fig. 10.11. GC separation of the N-acetyl-1-propyl esters of a mixture of 23 amino acids. Column, glass, 76 cm × 6 mm OD (2 mm ID) mixed polar packing: 0.31% Carbowax 20 M, 0.28% Silar-5CP, and 0.06% Lexan on Chromosorb W AW, 120–140 mesh; temperature program, 1 min isothermal at 100°C, followed by 10°C/min to 215°C, then ballasted to 250°C and held for 10 min; injector 250°C, detector 275°C; carrier gas, He, flowrate, 25 ml/min. Each peak represents 5 nmoles of amino acid derivative. (Reproduced from *J. Chromatogr.*, 164 (1979) 427, with permission [229].)

azeotropically. To the dry residue are added 60 μl of dichloromethane and 20 μl of trifluoroacetic anhydride per 100 μg of amino acids (minimum final volume 80 μl). Acylation proceeds for 5 min at 150°C in the tightly closed PTFE-lined screw-capped vessel. The sample is then ready for GC. Watson et al. [215] described a procedure for the simultaneous derivatization of carboxylic groups with a fluorinated alcohol and phenolic groups with a fluorinated anhydride. A one-step derivatization reaction of glutamic acid and γ-aminobutyric acid with pentafluoropropionic anhydride and hexafluoropropanol was reported by Bertilsson and Costa [216]. In our laboratory [217] we use a mixture of trifluoroacetic acid and trifluoroethanol to derivatize both amino acids and oligopeptides in 20 min at 100°C.

In 1961, Johnson et al. [218] reported the GC separation of 35 amino acids as N-acetyl *n*-amyl esters on a polyethyleneglycol phase. This method can no longer be recommended, because N-acetyl derivatives tend to show broad peaks in the chromatogram. Subsequently, extensive studies with various polar and nonpolar phases [219–225] were performed to evaluate the best separation conditions. It appears that the method of choice for the 20 protein amino acids is the dual-column system introduced by Gehrke et al. [226]. First, a 0.65% ethyleneglycol adipate phase on Chromosorb W separates all compounds, except His, Cys, and Arg. These amino acids are separated on a mixed phase, consisting of 10% (w/w) OV-210 and 2% (w/w) OV-17 on Gas Chrom Q.

Since perfluoroacyl derivatives are less stable than, e.g., acyl derivatives and since O- and S-TFA compounds may decompose on the column, only nonpolar phases can be used. Especially the diacyl derivative of His changes rapidly to the monoacyl compound. However, this problem is circumvented by simultaneous injection of the anhydride and the sample [224,227]. With the lower alcohols, the presence of heptafluorobutyric acid can cause unwanted tailing of the peaks. In this case, and for injection of amounts of around 100 μl the "solvent vent" of Zumwalt et al. [211] is advantageous. It prevents solvents and reagents from traversing the chromatographic column, enables concentration of the sample solution prior to analysis, and prevents loss of sample components during this step. A number of publications dealing with various phases and derivatives are cited in Table 10.14. Fig. 10.11 shows a recently reported example of amino acid separation on a mixed polar packing (see legend) [229].

10.4.1.5. Separation of enantiomers

Chromatography of "optical isomers" (especially of enantiomers) is of particular interest for solving two basic questions:
(a) determination of the degree of racemization in peptide synthesis,
(b) the analysis of natural peptides containing D-amino acids.
For this separation by GC, either optically active stationary phases may be used or a second asymmetric center is introduced into the molecule to form diastereomeric pairs of derivatives.

10.4.1.5.1. Separation on optically active stationary phases

The direct separation of enantiomers on a chiral phase should be theoretically possible, but in practice, the differences in retention time must be large enough to reduce the length of the column used. In 1966, Gil-Av et al. [236,237] reported the first successful separation of enantiomers. They chromatographed a series of N-trifluoroacetyl amino acid esters on a 100-m glass capillary column with S-(N-trifluoroacetyl)valyl-S-valine cyclohexyl ester as the stationary phase and adequately separated N-trifluoroacetyl alanine *t*-butyl ester enantiomers on a 2-m packed column [238]. Subsequent work was performed mainly with three types of phases:

$$CF_3\text{-}\overset{\overset{O}{\|}}{C}\text{-NH-CH-}\overset{R}{\underset{\overset{\|}{O}}{C}}\text{-O-}(CH_2)_n\text{-CH}_3 \qquad CF_3\text{-}\overset{\overset{O}{\|}}{C}\text{-NH-}\overset{R}{\underset{}{CH}}\text{-}\overset{\underset{\overset{\|}{O}}{}}{C}\text{-NH-}\overset{R}{\underset{}{CH}}\text{-}\overset{\underset{\overset{\|}{O}}{}}{C}\text{-OR} \qquad RO\text{-}\overset{\underset{\overset{\|}{O}}{}}{C}\text{-}\overset{R}{\underset{}{CH}}\text{-NH-}\overset{\overset{O}{\|}}{C}\text{-NH-}\overset{R}{\underset{}{CH}}\text{-}\overset{\underset{\overset{\|}{O}}{}}{C}\text{-OR}$$

trifluoroacetyl amino acid esters   trifluoroacetyl dipeptide esters   carbonyl bis-amino acid esters

The dipeptide phases have been particularly successful in the resolution of α-amino acid enantiomers, including the N-TFA-dipeptide cyclohexyl ester phases of L-Val–L-Val [222,238–241], L-Phe–L-Leu [221,222,242,243], L-Val–L-Leu [244], L-Leu–L-Val [245], L-Nle–L-Nle [246], L-Nval–L-Nval [246,247], L-Abu–L-Abu [246] and L-Ala–L-Ala [246].

Determinations of relative retention times, while the structure of the stationary solvent and of the enantiomeric solute is systematically varied, has led to a "three-point" model of hydrogen bonding, first postulated by Gil-Av and Feibush [238]. The complex should be a conformational stereoisomer, and a complex, resulting from a *R–S* interaction, should have physical properties different from those of the respective *S–S* complex. However, the validity of this model has been

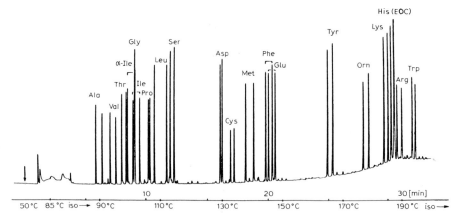

Fig. 10.12. GC separation of the enantiomers of 20 amino acids as pentafluoropropionyl amino acid isopropyl esters. Column, glass capillary (20 m × 0.28 mm), coated with Chirasil-Val; carrier gas, H_2, inlet pressure 0.38 bar; temperature program, initial 35°C, then brought to 82°C ballistically, held there for 4 min, then raised to 195°C at a rate of 4°C/min and kept there for 20 min; detector and injector at 250°C. His (EOC) = N(imidazole)-ethoxycarbonyl derivative of histidine. (Reproduced from *Clin. Chim. Acta*, 105 (1980) 201, with permission [252].)

called in question by spectroscopic work, namely, ^{1}H- and ^{13}C-NMR spectral studies. Lochmüller and coworkers [248,249] have noted that the partial double bond character of the amino bond is an important factor in determining the geometry around the association region. Furthermore, there is good evidence that the carbonyl of a peptide link most likely exhibits a *trans*-configuration to the N–H hydrogen. Therefore, many of the proposed "three point" structures are rather unlikely if not impossible. The spectroscopic results [248] indicate that there is only one significant point of attachment in forming the diastereomeric complex, i.e., only the N–H function of the amide can form a hydrogen bond with the ester carbonyl of the stationary phase [247].

Most of the work performed in this area has dealt with liquid phases having a relatively low molecular weight. Their disadvantages are: temperature problems, column bleeding, and decomposition. Frank et al. [250] employed a phase that obviates these problems, as it consists of N-propionyl-S-valine-*t*-butyl amide attached to a polysiloxane framework. With this phase, they were able to separate most protein amino acid enantiomers between 70 and 240°C. They also developed a method called "enantiomer labeling" for amino acid analysis [251,252]. Fig. 10.12 shows a chromatographic separation of the enantiomers of 20 amino acids. The respective D-antipodes of the natural L-compounds were added to the serum sample prior to clean-up and serve as internal standards.

10.4.1.5.2. Separation of diastereomers

Weygand et al. [253] were the first to separate amino acid enantiomers as diastereomers. They analyzed the N-TFA dipeptide methyl esters of L-Ala–L-Phe and L-Ala–D-Phe on a 2-m packed silicone column and then a series of diastereomeric pairs of N-TFA dipeptide methyl esters on capillary columns [254,255]. In this system, the LL- and DD-isomers are usually eluted before the LD- and DL-isomers [256,257]. Proximity of the two asymmetric centers enhances the resolution. Polar phases, like Carbowax, are superior to nonpolar ones [258,259]. The best separation was obtained with the following diasteromeric derivatives: N-TFA 3,3-dimethyl-2-butyl esters [259–261], N-TFA 2-butyl esters [257,258,261–271], N-TFA 2-octyl esters [263], and N-TFA S-prolyl amino acid methyl esters [259,272–274]. Weygand et al. [275] have tested dipeptide combinations other than proline (Ala, Val, Leu, Ile, Met and Phe). Kolb's [276] review, written in 1971, is still valid.

A systematic study by Westley et al. [259] deals with the effect of solute structure on the separation of diastereomeric esters and amides, Pollok and Kawauchi [261] have reported the resolution of racemates of Asp, Trp, hydroxy-, and sulfhydryl amino acids that are difficult to separate. The same techniques have been applied to determine amino acids in meteorites [268,269,271] and the optical purity of ^{14}C-labeled amino acids [277], as well as the configuration of the allo-Ile present in the blood plasma of patients with branched-chain ketoacidosis [270]. Employing packed columns and 3,3-dimethyl-2-butyl ester derivatives, 14 amino acids diastereomers could be separated in 93% or greater optical purity, except Asp (70%) and Pro (82%) [278].

10.4.2. Gas chromatography of peptides

Peptides exhibit the same characteristics as amino acids with regard to their high polarity and low volatility, and they likewise demand derivatization for GC. As there are only few reports in the literature dealing solely with GC of peptides, the major interest being focused on MS and GC–MS, this topic is discussed in more detail in Chap. 10.5.1.2. Here, only derivatization and some separation problems are treated.

10.4.2.1. Derivatives

Since the pioneer work of Weygand et al. [253] on the application of GC to the analysis of peptides, the N-acyl peptide derivatives have been used extensively— mainly the N-TFA methyl [254,255,272,279–285] and butyl [286] esters. The preparation of the N-pentafluoropropionyl pentafluoropropyl esters enables the use of the very sensitive ECD. The N-heptafluorobutyl peptide methyl esters [287] are more advantageous than the respective TFA derivatives in having shorter retention times. Acetyl methyl esters and permethyl peptide derivatives have been reported [286], but they are less volatile than TFA derivatives.

Prox and Sun [289] have tested a series of different N-substituted derivatives of Val-Ile-Ala methyl esters. Esters may be formed with diazomethane after acylation, by permethylation [290], by thionyl chloride in methanol, or by partial solvolysis of a peptide in methanolic HCl [282,291], in which this reaction medium causes a partial splitting of peptide bonds [292]. Trifluoroacetylation must be performed at a much lower temperature than in the case of amino acids [293,294] or, alternatively, with trifluoroacetic acid methyl ester in methanol and triethylamine as a catalyst

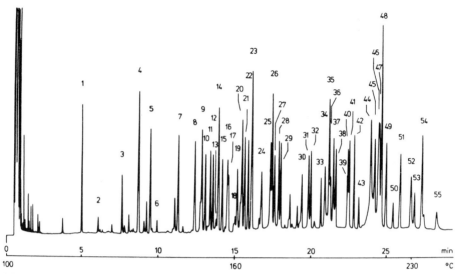

Fig. 10.13. Separation of TMS derivatives of dipeptides. Fused silica capillary column (12 m×0.2 mm), coated with SP-2100; carrier gas, He; inlet pressure, 0.75 bar; temperature program, 100 to 160°C at 4°C/min, 160 to 230°C at 6°C/min, 15 min at 230°C; injector and detector at 250°C. Each peak represents 0.1–1 ng of dipeptide. For peak identification see Table 10.15. (Reproduced from *Anal. Biochem.*, 108 (1980) 267, with permission [295].)

TABLE 10.15

PEAK IDENTIFICATION FOR FIG. 10.13 [295]

Peak No.	Dipeptide	Peak No.	Dipeptide	Peak No.	Dipeptide
1	Gly–Gly	18	Gly–L-Ser	35	L-Ser–L-Met
2	L-Val–Gly	19	L-Ala–L-Asn	36	Gly–L-Met
3	Gly–L-Leu	20	Gly–L-Thr	37	L-Phe–L-Val
	L-Leu–Gly	21	Gly–L-Phe	38	L-Val–L-Phe
4	L-Ala–L-Ala		L-Phe–Gly	39	DL-Leu–DL-Phe
5	Gly–L-Ser	22	Gly–L-Glu	40	Gly–L-Glu
6	Gly–L-Thr	23	Gly–L-Val	41	DL-Leu–DL-Phe
7	L-Ala–L-Val		L-Leu–L-Leu	42	L-Ser–L-Phe
8	L-Ala–L-Leu	24	Gly–L-Leu	43	Gly–L-Phe
9	L-Ala–L-Ile	25	L-Ala–L-Asp	44	L-His–L-Ala
	L-Leu–L-Ala	26	Gly–L-Ile	45	L-Ala–L-His
10	α-L-Asp–Gly		L-Ala–L-Asn	46	L-Met–L-Met
11	Gly–L-Met	27	L-Leu–L-Ser	47	L-Ala–L-Tyr
	L-Met–Gly	28	L-Ala–L-Met	48	L-Tyr–L-Ala
12	Gly–L-Asp	29	L-Met–L-Ala	49	L-Tyr–Gly
13	Gly–L-Ile	30	L-Ala–L-Glu	50	L-Trp–Gly
14	L-Val–L-Val	31	L-Phe–L-Ala	51	L-Met–L-Phe
	L-Ser–L-Ala	32	L-Ala–L-Phe	52	L-His–L-Ser
15	DL-Ala–DL-Ser	33	γ-L-Glu–L-Leu	53	Gly–L-His
16	DL-Ala–DL-Ser	34	Gly–L-Asp	54	L-Phe–L-Phe
17	L-Ala–L-Thr		L-His–Gly	55	L-Trp–L-Ala

[253]. A summary of some derivatives for peptides and detailed procedures can be found in Knapp's handbook [187].

10.4.2.2. Separation of peptides

In comparison with the amino acids, the GC of peptides is complicated by the higher molecular weight and the addition of polar interactions, typical for polar oligo- and polymers. Therefore, GC of peptides is at present restricted to di-, tri-, tetra-, and pentapeptides. Stationary phases for peptide separation are nonpolar (e.g., 1–5% SE-30, SE-52, OV-1, OV-17, and OV-101 [276]). Glass capillary columns are preferred, because they give better resolution. It has been found that a small amount of a highly polar phase, e.g., polyethyleneglycol, mixed with a nonpolar phase, yields a better separation of dipeptides, as a class, from the tripeptides. Retention data of numerous dipeptide derivatives are available [276].

Dizdaroglu and Simic [295] reported the separation of dipeptides on a fused silica capillary column after trimethylsilylation. Fig. 10.13 shows the chromatogram. Peak identification (Table 10.15) was obtained by comparison of retention times of the individually injected compounds. This separation is useful for peptide sequencing by means of dipeptidyl amino peptidase and GC–MS.

10.5. MASS SPECTROMETRY

In recent years, MS has made great gains in every field of analysis. It is now a powerful tool for structure elucidation, quantitation, and metabolic investigations, especially in combination with other spectroscopic methods, such as IR, UV, and NMR spectroscopy. The main features of technique are:
   (a) information on molecular weight and structure,
   (b) qualitative and quantitative analysis,
   (c) high sensitivity (down to the picogram level),
   (d) wide applicability.
With regard to the application of MS to the analysis of amino acids and peptides, an important peculiarity of the method must be borne in mind. Generally — and this is specifically true for EI and to some extent for CI — MS is predicated on the vaporization of samples, a requirement which excludes many classes of compounds. Field desorption (FD) and, very recently, fast atom bombardment (FAB), have become the methods of choice for the analysis of thermally labile and nonvolatile samples, but since, to date, there is no direct interface between chromatography and FD-MS, this and a number of other ionization methods are beyond the scope of this book. The reader is referred to the reviews by Schulten [296] on FD-MS and by Milne and Lacey [297] on various ionization techniques. In this chapter only GC–MS and LC–MS are discussed and, therefore, the ionization techniques are restricted to EI and CI.

References on p. B43

10.5.1. Gas chromatography–mass spectrometry

In connection with GC, MS may be regarded as a very sensitive and selective detector. In principle, any compound suited for GC can be analyzed by MS. General aspects of GC–MS in analytical chemistry have been reviewed by Ten Noever de Brauw [298] and the problems of interfacing GC and MS by McFadden [299].

10.5.1.1. Amino acids

In 1973, Pereira et al. [300] reported the analysis of the amino acids in soil extracts by GC–MS. Using internal standards and SIM, they were able to quantitate 12 amino acids in biological fluids [301]. Subsequently, the fragmentation of TMS amino acids was investigated to provide a spectral basis for further studies of physiological samples, and to study the possibility of quantitation of amino acids by SIM. The results indicated that TMS derivatives are suitable for these purposes [302–304]. Artigas and Gelpi [305] reported a SIM method for the simultaneous analysis of tryptophan, tryptamine, indole-3-acetic acid, serotonin, and 5-hydroxyindole-3-acetic acid in rat brain. Further studies showed the usefulness of CI for TMS derivatives. The intensive quasimolecular ion MH^+ simplified the identification of

TABLE 10.16

LABELED AMINO ACIDS USED AS INTERNAL STANDARDS IN GC–MS

Isotopes	Applications (methods)	Ref.
^2H	Phenylalanine, tyrosine (MS)	318–321
^2H	Blood analysis (CI-MS)	313
^2H	Acylglycines (MS)	322
^2H	Tryptophan (GC–MS)	323
^2H	Protein amino acids, insulin (GC–MS)	324
^2H	α-Methyldopa metabolism (CI-MS)	325
^2H	γ-Abu determination (GC–MS)	326
^2H	Glutamic acid in water (GC–SIM)	327
^{13}C	Plasma amino acids (GC–CI-MS)	328
^{13}C	Phenylalanine (GC–MS)	329
^{13}C	Glycine in brain tissues	330
^{15}N	Amino acid metabolism	331
^{15}N	Amino acids in nmole amounts	332
^{15}N	Blood amino acids	313
^{15}N	Amino acids pool sizes and turnover rates (GC–MS)	333, 334
^{15}N	Amino acids (GC–CI-MS)	335
^{15}N	Amino acids (negative-ion MS)	336
^{15}N	Glycine pool sizes and turnover rates	337
^{15}N	Plasma amino acids (CI-MS)	328
^{15}N	Glycine in brain tissues	330
^{18}O	Synthesis of amino acids for MS	338

amino acids and the determination of the number of TMS groups [306]. This and other investigations with different reagent gases (isobutane, ammonia) [307–310] revealed that the CI spectra were generally much simpler than the corresponding EI spectra and, therefore, should be better suited for identification and quantitation purposes. The methane CI spectra of some α,ω-diamino acids, ω-amino acids, cyclic and acyclic α-amino acids and their respective methyl esters were studied [311], and the prevalence of reaction was correlated to the product ion stability.

Collins and Summer [312] reported the detection of glutamine (as pyroglutamate) and glutamic acid in biological fluids. They used the N-trifluoroacetyl n-butyl esters, derivatives well known in GC analysis, which give limited fragmentation upon electron impact. Monitoring the fragments resulting from the α-cleavage easily provides identification of the respective amino acids. Following the work of Pereira and coworkers [300,301], attempts were undertaken to determine amino acids in blood samples [313,314] and to quantitate amino acids and oligopeptides in the femtomole [315] and picomole [316] range, respectively. Internal standards formerly used for quantitation were amino acid isomers, like norleucine, δ-hydroxyleucine, and 5-aminovaleric acid [317]. However, labeled amino acids are better suited for these purposes. ^{2}H, ^{13}C, ^{15}N, and ^{18}O have been used in many assays. A number of publications on the use of labeled amino acids for quantitative and metabolic analyses are listed in Table 10.16.

Bailey et al. [339] used the D-enantiomer as internal standard for the quantitation of S-methyl-L-cysteine in hemoglobin. Crossley and Ramsden [340] have determined tetraiodothyroacetate (T4A) levels in healthy euthyroid individuals by GC–SIM, and

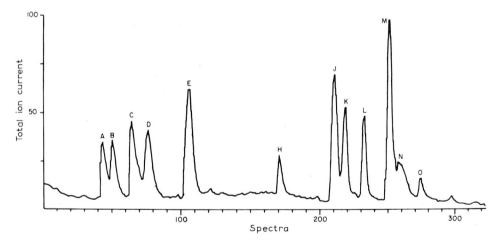

Fig. 10.14. Total ion-current chromatogram of an insulin hydrolyzate with added deuterated amino acids. Column 3.7 m × 4 mm, Pyrex, packed with 3% OV-17 on 80–100-mesh Chromosorb W HP; carrier gas, He, flowrate, 30 ml/min; temperature program, 90 to 200°C at 3°C/min; mass spectrometer, 260°C ion source, 22.5 eV, cycle time, 5 sec for m/z 0 to 410. Peak identification: A = Ala, B = Thr, C = Ser and Gly, D = Val, E = Ile and Leu, F = CysH, G = BHT, H = Pro, I = Met, J = Asp, K = Phe, L = Tyr, M = Glu, N = Lys, O = Arg. (Reproduced from *Biomed. Mass Spectrom.*, 6 (1979) 317, with permission [324].)

found values considerably lower than previously reported. Heki [341] described an analysis of r-T3 and 3,3'-T2 in biological fluids by the same method. Murphy and Lay [338] presented an elegant synthesis of ^{18}O labeled amino acids for use as internal standards in MS. The method consists of exchanging 1–4 mg amino acid in 0.1–0.2 ml acidic H_2 ^{18}O, followed by heating to 60–70°C for several days. Eighteen amino acids have thus been labeled with incorporation rates typically > 90%. These conditions did not lead to racemization and back exchange of the ^{18}O atoms, while pH and temperature dependency allowed these labeled compounds to be used in most studies.

Using deuterated analogs as internal standards, Rafter et al. [324] recently described the quantitative analysis of protein amino acids by a GC–MS–computer system. The detailed work-up procedure for protein analysis is as follows. After addition of deuterated amino acids (mixture of 16 compounds, 50–100 μg each) to 50 μg protein, the sample was hydrolyzed in 100–200 μl 6 M HCl, containing 0.5% (v/v) phenol, in vacuo at 110°C for 20 h. After drying the sample in a vacuum desiccator, it was esterified by addition of a 10% acetyl chloride solution in n-butanol (150 μl per 100 μg of total amino acids). The tube was sealed and sonicated in a water bath for 1 min and then placed in a 100 ± 2°C oil bath for 15 ± 1 min. After esterification, the tube was opened, and the contents were transferred to a fresh glass tube. The residue was dried under a stream of nitrogen at room temperature and then washed with a 200-μl portion of dry dichloromethane to remove azeotropically any remaining water. Again, the solvent was evaporated under a stream of nitrogen at room temperature. Acetylation was performed with 60 μl dichloromethane (containing BHT) and 20 μl TFAA for each 100 μg of amino acid. The sealed tube was opened immediately before GC–MS analysis; the samples were stable for a few days when kept at 0–4°C. For GC a 3.7 m × 4-mm ID coiled Pyrex column was packed with 3% OV-17 on 80–100-mesh Chromosorb W HP.

Fig. 10.14 shows the total ion-current chromatogram of an insulin hydrolyzate with added deuterated amino acids. Whole spectra were recorded and isotope ratios were determined by using one selective fragment ion for each amino acid. For quantitation, multiple internal standards were used. The method, tested on insulin, gave results which agreed well with the known composition of the protein and with a simultaneous analysis on ion exchangers.

10.5.1.2. Oligopeptides

Like the amino acids, peptides must be converted to more volatile derivatives for GC. Since complications are due to their higher molecular weight and the interactions of polar functions typical for oligomers and polymers, protective groups should not only have high thermal stability but also produce volatile derivatives without increasing the molecular weight too much. At present, GC and GC–MS are restricted to di-, tri-, and tetrapeptides. Peptides containing Arg, Asn, and Gln are still difficult to analyze.

In principle, MS should be ideally suited to peptide sequence analysis. The interesting work in this field with direct inlet systems, FD and FAB techniques, and metastable monitoring is outside the scope of this book. Readers are referred to the publications of Biemann [342], Schulten [296], and Schlunegger and coworkers [343–345].

Peptide sequencing with the aid of GC–MS can be performed in two ways: One consists in sequential cleavage of the peptide down to the individual amino acid constituents by the Edman degradation [346,347] and identification of the N-terminal residues by MS [324]. However, for MS this method seems to be of limited value. The second approach involves a partial hydrolysis, often with dipeptidyl amino peptidase. The sequence is then determined on the overlapping fragments, and these fragment sequences are again reassembled to establish the sequence of the original peptide. This method, originally considered to be an alternative to conventional

TABLE 10.17

IONS USED TO IDENTIFY DIPEPTIDE TMS DERIVATIVES

Amino acid	Mass of β-cleavage ion when residue is N-terminal	Mass added to β-cleavage ion when residue is C-terminal, to calculate $(M+1)^+$	$(ArCH_2)^+$
Gly [*]	102	175	
Ala	116	189	
Abu	130	203	
Pro	142	215	
Val	144	217	
Leu	158	231	
Ile	158	231	
Met	176	249	
Phe	192	265	91
Ser	204	277	
Thr	218	291	
Asn	231	304	
Asp	None [**]	305	
Gln	245	318	
Glu	246	319	
His	254	327	154
Cys–$(CH_2)_2NH_2$ [***]	263	336	
Cys–CH_2CONH_2	277	350	
Cys–CH_2CO_2H	278	351	
Tyr	280	353	179
Orn	303	376	
Trp	303	376	202
Lys	317	390	
Arg	142 l, 417 s [§]	490	

[*] Diketopiperazines formed with Gly–Abu, –Ala, –Arg, –Gln, –Glu, –Gly, –Met, and –Lys.
[**] N-terminal Asp cyclizes to five-membered imide; add 158 to C-terminal value to calculate $(M+1)^+$.
[***] Recent data show that over 90% is monosubstituted on ε-NH_2 group of aminoethyl Cys.
[§] l = large, s = small.

References on p. B43

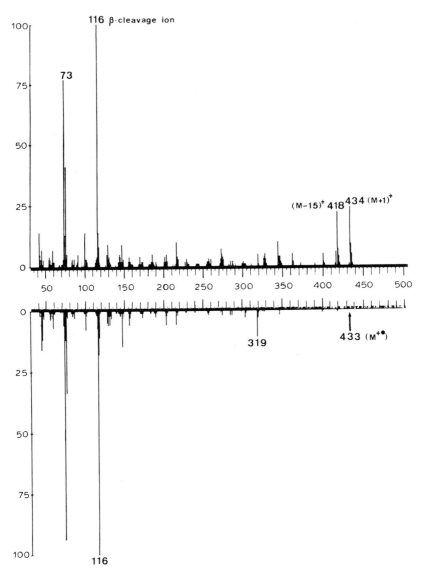

Fig. 10.15. CI (top) and EI (bottom) mass spectra of TMS Ala–Gln. CI with $5 \cdot 10^{-5}$ Torr isobutane as reagent gas; box current, 500 μA; box voltage, 150 eV; accelerating voltage, 3.5 kV. EI with an ionizing current of 50 μA; ionizing voltage, 70 eV; accelerating voltage, 3.5 kV. Samples were introduced either by GC or by direct probe. (Reproduced from *Anal. Biochem.*, 92 (1979) 525, with permission [352].)

sequencing techniques, can be a powerful complement to them. Thus, particular sequences that are difficult to analyze with the Edman method may be well determined by MS and vice versa. This has been examplified by Hudson and Biemann [348] in a study on monellin and by Rees et al. [349].

Krutzsch and Pisano [350] have investigated the fragmentation patterns of the TMS derivatives of about 200 dipeptides in order to identify all possible 400 dipeptides. They [351,352] also compared CI and EI mass spectra of over 40 TMS dipeptides. Due to the higher amount of transferred energy in the ionization process, EI yields more fragment ions than the soft CI technique. Therefore, the $(M^{+\cdot}-15)$ ion in EI ionization is very weak or absent (cf. Fig. 10.15) and the determination of the molecular weight of the sample may be very difficult, if the amounts are in the 0.1-nmole range. In contrast, CI yields three ions of high abundance, and this allows an unambiguous identification, as shown below:

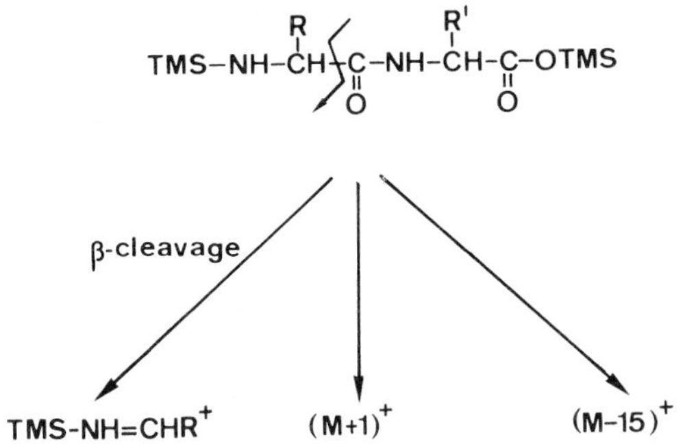

$$\text{TMS-NH-CH}\overset{R}{\diagdown}\text{C-NH-CH-C-OTMS}$$

The molecular weight and the sequence of the respective dipeptides are determined from three ions:

(a) the $(M + 1)^+$ ion typical for CI with isobutane as reagent gas,
(b) the $(M - 15)^+$ ion arising from loss of methyl from a TMS group,
(c) the sequence-determining ion from the β-cleavage of a C–C = O bond.

Fig. 10.15 shows the EI and CI spectra of Ala-Gln as TMS derivatives, and Table 10.17 lists the sequence and MW relevant ions for the common amino acids. Since β-cleavage ion is also present in the EI mode, CI extends the lower limit of dipeptide identification.

Seifert et al. [353] have studied the GC–MS behavior of 120 pentafluoropropionyl methyl esters of dipeptides with CI and EI. Dipeptides in mixture could be clearly identified, and in suitable cases mass spectral analysis of peptides containing up to seven amino acids was possible. However, the peptides still have an undesirably low volatility, because the N–H bond remains free and can form hydrogen bridges. Permethylation yields more volatile derivatives [187,354]. In another derivatization method, introduced by Biemann and coworkers [355,356], the N-acetyl peptides are reduced to the respective polyamino alcohols, and the hydroxyl groups are trimethyl-silylated. These derivatives are even more volatile than the permethylated ones. In a modification of this procedure, the N-heptafluorobutyryl derivatives have been prepared prior to reduction [357–362], yielding the most volatile peptide derivatives so far produced.

Frank and Desiderio [363] described a new reduction method for oligopeptides to amino alcohols with borane, allowing as little as 10–100 nmole of these peptide derivatives to be sequenced [364]. A modified procedure according to Nau et al. [358,362] has been applied to the oligopeptide determination in urine by Steiner and

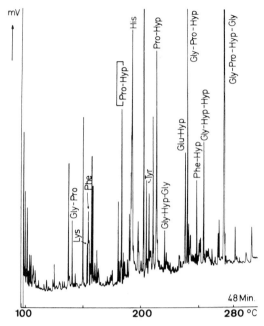

Fig. 10.16. Total ion-current chromatogram of a urine sample. Glass capillary column (25 m×0.28 mm), coated with SE-30; carrier gas, He; inlet pressure, 1.2 bar; temperature program, 3 min at 100°C, then raised to 280°C at 4°C/min; injector, 300°C; transfer line, 300°C; ion source, 215°C; ionizing current, 50 μA; ionizing voltage, 20 eV, scan time, 1 sec/decade. (Reproduced from *Clin. Chim. Acta*, 92 (1979) 431, with permission [365].)

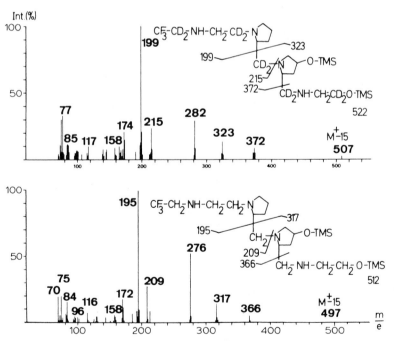

Fig. 10.17. EI mass spectra of deuterated (top) and hydrogenated (bottom) Gly–Pro–4 Hyp–Gly derivatives. For GC and MS conditions see Fig. 10.16. (Reproduced from *Clin. Chim. Acta*, 92 (1979) 431, with permission [365].)

Niederwieser [365]. Since the esterification with HCl–methanol may cause cleavage of peptide bonds, Steiner and Niederwieser [365] have used $SOCl_2$–methanol. Reductions were performed with $LiAlH_4$ as well as $LiAl^2H_4$ in order to increase the information content for the identification of the peptides. Fig. 10.16 shows the total ion-current chromatogram of a urine sample, and Fig. 10.17 represents the EI spectra of the hydrated and the deuterated tetrapeptide Gly-Pro-4 Hyp-Gly. The molecular weights of the compounds were verified by CI (isobutane as reagent gas).

As far as we are informed, all GC–MS experiments on oligopeptides have been performed by monitoring positive ions. However, the use of negative ions should be favorable, especially in cases where the derivatives are highly halogenated compounds. The method of pulsed positive-ion negative-ion chemical ionization (PPINICI), introduced by Hunt et al. [366] should not only enhance the sensitivity of mass spectrometric analysis, but also combine the advantages of both ionization techniques.

REFERENCES

1 A. Niederwieser, in E. Heftmann (Editor), *Chromatography, a Laboratory Handbook of Chromatographic and Electrophoretic Methods,* Van Nostrand-Reinhold, New York, 1975, p. 393.

2 H.J. Bremer, M. Duran, J.P. Kamerling, H. Przyrembel and S.K. Wadman, *Clinical Chemistry and Diagnosis,* Urban and Schwarzenberg, Baltimore, MD, 1981.

3 W.H. Stein and S. Moore, *Cold Spring Harbor Symp. Quant. Biol.,* 14 (1949) 179.

4 S. Moore and W.H. Stein, *J. Biol. Chem.,* 192 (1951) 663.

5 S. Moore, D.H. Spackman and W.H. Stein, *Anal. Chem.,* 30 (1958) 1185 and 1190.

6 J.R. Cronin, S. Pizzarello and W.E. Gandy, *Anal. Biochem.,* 93 (1979) 174.

7 Y. Kobayashi, *Anal. Biochem.,* 105 (1980) 48.

8 A.C.C. Spadaro, W. Draghetta, S.N. Del Lama, A.C.M. Camargo and L.J. Greene, *Anal. Biochem.,* 96 (1979) 317.

9 G.R. Beecher, *Adv. Exp. Med. Biol.,* 105 (1978) 827.

10 P. Parvy, Y. Huang and P. Kamoun, *J. Clin. Chem. Clin. Biochem.,* 17 (1979) 205.

11 J.C. Castets, P. Parvy, D. Allard and Y. Huang, *Ann. Biol. Clin.,* 36 (1978) 143.

12 K. Olek, S. Uhlhaas and P. Wardenbach, *J. Clin. Chem. Clin. Biochem.,* 16 (1978) 283.

13 T.A. Hare and N.V. Bala Manyam, *Anal. Biochem.,* 101 (1980) 349.

14 P. Böhlen, P.J. Schechter, W. van Damme, G. Coquillat, J.C. Dosch and J. Koch-Weser, *Clin. Chem.,* 24 (1978) 256.

15 K.C. Kuo, T.F. Cole, C.W. Gehrke, T.P. Waalkes and E. Borek, *Clin. Chem.,* 24 (1978) 1373.

16 D.F. Evered and J.F. Barley, *Clin. Chim. Acta,* 84 (1978) 339.

17 P. Böhlen and M. Mellet, *Anal. Biochem.,* 94 (1979) 313.

18 A. Szymanovicz, G. Poulin, A. Randoux and J.P. Borel, *Clin. Chim. Acta,* 91 (1979) 141.

19 A. Szymanovicz, A. Malgras, A. Randoux and J.P. Borel, *Biochim. Biophys. Acta,* 576 (1979) 253.

20 L.C. Ward, *Anal. Biochem.,* 88 (1978) 598.

21 I.F. Tarbit, J.P. Richardson and G. Dale, *J. Chromatogr.,* 181 (1980) 337.

22 Y. Menezo, C. Voulot, J.P. Ortonne and C. Khatchadourian, *Arch. Dermatol. Res.,* 263 (1978) 267.

23 J.L. Marini, S.P. Williams and M.H. Sheard, *Pharmacol. Biochem. Behav.,* 11 (1979) 183.

24 W.C. Bisbee and P.C. Kelleher, *Clin. Chim. Acta,* 90 (1978) 29.

25 W.C. Bisbee and P.C. Kelleher, *J. Chromatogr.,* 145 (1978) 473.

26 M. Friedman, A.T. Noma and J.R. Wagner, *Anal. Biochem.,* 98 (1979) 293.

27 R. Hardman and I.M. Abu-Al-Futuh, *Planta Med.,* 36 (1979) 79.

28 V.R. Villanueva and R.C. Adlakha, *Anal. Biochem.,* 91 (1978) 264.

29 J. Degen and J. Kyte, *Anal. Biochem.,* 89 (1978) 529.

30 P. Johnson, *J. Chromatogr. Sci.*, 17 (1979) 406.

31 L. Bachner, J.-P. Boissel and H. Wajcman, *J. Chromatogr.*, 193 (1980) 491.

32 S. Takeuchi, K. Fuzita, F. Nakaziwa and Y. Arikawa, *Nippon Kagaku Kaishi*, (1978) 64.

33 J.C. Wright and R.F. Evilia, *J. Liquid Chromatogr.*, 2 (1979) 719.

34 W.S. Gardner, *Marine Chem.*, 6 (1978) 15.

35 R. Dawson and R.G. Pritchard, *Marine Chem.*, 6 (1978) 27.

36 R.A.A. Muzarellli, F. Tanfani, M.G. Muzarelli, G. Scarpini and R. Rocchetti, *Separ. Sci. Technol.*, 13 (1978) 869.

37 S.B.H. Kent, A.R. Mitchell, G. Barany and R.B. Merrifield, *Anal. Chem.*, 50 (1978) 155.

38 D.K.J. Tommel, J.F.G. Vliegenthart, T.J. Penders and J.F. Arens, *Biochem. J.*, 99 (1966) 48.

39 D.K.J. Tommel, J.F.G. Vliegenthart, T.J. Penders and J.F. Arens, *Biochem. J.*, 107 (1967) 335.

40 F.J. Clapperton, *J. Inst. Brew.*, 77 (1971) 177.

41 K.P. Polzhofter and K.H. Ney, *Tetrahedron*, 28 (1972) 1721.

42 E. Rothenbühler, R. Waibel and J. Solms, *Anal. Biochem.*, 97 (1979) 367.

43 W. Lutz, K. Markewicz and K. Klyszejko-Stefanowicz, *Clin. Chim. Acta*, 39 (1972) 319, 425.

44 A. Niederwieser and H.Ch. Curtius, *J. Chromatogr.*, 51 (1970) 491.

45 P. Giliberti and A. Niederwieser, *J. Chromatogr.*, 66 (1972) 261.

46 H.E. Gallo-Torres, O.N. Miller and J. Ludorf, *Anal. Biochem.*, 64 (1975) 260.

47 B. Sampson and G.B. Barlow, *J. Chromatogr.*, 183 (1980) 9.

48 L.R. Snyder and J.J. Kirkland, *Introduction to Modern Liquid Chromatography*, 2nd Edn., Wiley, New York, 1979, p. 125.

49 R.P.W. Scott, *Liquid Chromatography Detectors*, Elsevier, New York, 1977.

50 M.J. Milano, S. Lam, M. Savonis, D.B. Pautler, J.W. Pav and E. Grushka, *J. Chromatogr.*, 149 (1978) 599.

51 R.J. Hamilton and P.A. Sewell, *Introduction to High-Performance Liquid Chromatography*, Chapman & Hall, London, 1977, Chap. 3.

52 M.S. Denton, T.P. De Angelis, A.M. Yacynych, W.R. Heineman and T.W. Gilbert, *Anal. Chem.*, 48 (1976) 20.

53 R.E. Dessey, W.G. Nunn, C.A. Titus and W.R. Renolds, *J. Chromatogr. Sci.*, 14 (1976) 195.

54 R. Somack, *Anal. Biochem.*, 104 (1980) 464.

55 G.J. Diebold and R.N. Zare, *Science*, 196 (1977) 1439.

56 J.C. Hodgin, *J. Liquid Chromatogr.*, 2 (1979) 1047.

57 J.F.K. Huber, A.M. van Urk-Schoen and G.B. Sieswerda, *Z. Anal. Chem.*, 264 (1973) 257.

58 E.R. Schmid, I. Fogy, F. Heresch, E. Kenndler and J.F.K. Huber, *Mikrochim. Acta*, (1979) 207.

59 D.I. Caroll, I. Dzidic, R.N. Stillwell, K.D. Haegele and E.C. Horning, *Anal. Chem.*, 47 (1975) 2369.

60 P.R. Jones and S.K. Yang, *Anal. Chem.*, 47 (1975) 1000.

61 F.W. McLafferty, R. Knutti, R. Venkataakhavan, P.J. Arpino and B.G. Dawkins, *Anal. Chem.*, 47 (1975) 1503.

62 R.P.W. Scott, C.G. Scott, M. Munroe and J. Hess, Jr., *J. Chromatogr.*, 122 (1976) 389.

63 W.H. McFadden, H.L. Schwartz and S. Evans, *J. Chromatogr.*, 122 (1976) 389.

64 W.L. Erdahl and O.S. Privett, *Lipids*, 12 (1977) 797.

65 C.R. Blakley and M.L. Vestal, *25th Ann. Conf. Mass Spectrom.*, Washington, DC, 1977, p. 376.

66 C.R. Blakley, J.C. Carnody and M.L. Vestal, *Clin. Chem.*, 26 (1980) 1467.

67 J.W. Serum and A. Melera, *26th Ann. Conf. Mass Spectrom.*, St. Louis, MO, 1978, p. 655.

68 P.J. Arpino and P. Krien, *26th Ann. Conf. Mass Spectrom.*, St. Louis, MO, 1978, p. 426.

69 T. Takeuchi, Y. Hiraka and Y. Okumura, *Anal. Chem.*, 50 (1978) 659.

70 H.R. Udseth, R.G. Orth and J.H. Futrell, *26th Ann. Conf. Mass Spectrom.*, St. Louis, MO, 1978, p. 659.

71 W.H. McFadden, D.C. Bradford, D.E. Games and J.L. Gauer, *Amer. Lab.*, 9 (1977) 55.

72 P.J. Arpino, H. Colin and G. Guiochon, *25th Ann. Conf. Mass Spectrom.*, Washington, DC, 1977, p. 189.

73 W.H. McFadden, *J. Chromatogr. Sci.*, 18 (1980) 97.

74 P.J. Arpino and G. Guiochon, *Anal. Chem.*, 51 (1979) 683 A.

75 E. Kenndler and E.R. Schmid, in J.F.K. Huber (Editor), *Instrumentation in High-Performance Liquid Chromatography*, Elsevier, Amsterdam, 1978, p.163.

76 B.L. Karger, D.P. Kirby, P. Vonros, R.L. Foltz ad B. Hidy, *Anal. Chem.*, 51 (1979) 2324.

77 D.M. Desiderio, J.L. Stein, M.D. Cunningham and J.Z. Sabbatini, *J. Chromatogr.*, 195 (1980) 369.

78 B.G. Dawkins, P.J. Arpino and F.W. McLafferty, *Biomed. Mass Spectrom.*, 5 (1978) 1.

79 R. Schuster, *Anal. Chem.*, 52 (1980) 617.

80 E.P. Kroeff and D.J. Pietrzyk, *Anal. Chem.*, 50 (1978) 502.

81 M.J. Horn, P.A. Hargrave and J.K. Wang, *J. Chromatogr.*, 180 (1979) 111.

82 C.C. Zimmermann, E. Apella and J.J. Pisano, *Anal. Biochem.*, 77 (1977) 569.

83 T. Yamabe, N. Takai and H. Nakamura, *J. Chromatogr.*, 104 (1975) 359.

84 H. Beyer and U. Schenk, *J. Chromatogr.*, 39 (1969) 482.

85 J. Harmeyer, H.-P. Sallmann and L.J. Ayoub, *J. Chromatogr.*, 32 (1968) 258.

86 W. Voelter and K. Zech, *J. Chromatogr.*, 112 (1975) 643.

87 E. Bayer, E. Grom, B. Kaltenegger and R. Uhmann, *Anal. Chem.*, 48 (1976) 1106.

88 J.M. Wilkinson, *J. Chromatogr. Sci.*, 16 (1978) 547.

89 G.J. Schmidt, D.C. Olson and W. Slavin, *Perkin-Elmer Chromatogr. News Lett.*, 7 (1979) 10.

90 J.C. Hodgin, *J. Liquid Chromatogr.*, 2 (1979) 1024.

91 D.W. Hill, F.H. Walters, T.D. Wilson and J.D. Stuart, *Anal. Chem.*, 51 (1979) 1338.

92 J.P. Garnier, B. Bousquet and C. Dieux, *Analusis*, 7 (1979) 355.

93 P. Lindroth and K. Mopper, *Anal. Chem.*, 51 (1979) 1667.

94 Z. Friedman, H.W. Smith and W.S. Hancock, *J. Chromatogr.*, 182 (1980) 414.

95 W.S. Gardner and W.H. Miller, *Anal. Biochem.*, 101 (1980) 61.

96 D.J. Reed, J.R. Babson, P.W. Beathy, A.E. Brodie, W.W. Ellis and D.W. Potter, *Anal. Biochem.*, 106 (1980) 55.

97 E.W. Bachmann and J. Frei, *Chromatographia*, 12 (1979) 345.

98 J.Y. Chang, A. Lehmann and B. Wittman-Liebold, *Anal. Biochem.*, 102 (1980) 380.

99 S.J. Wassner, J.L. Schlitzer and J.B. Li, *Anal. Biochem.*, 104 (1980) 284.

100 G. Klapper, C.E. Wilde and J.D. Capra, *Anal. Biochem.*, 85 (1978) 126.

101 R. Zeeuws and A.D. Strosberg, *FEBS Lett.*, 85 (1978) 68.

102 M.N. Margolies and A. Brauer, *J. Chromatogr.*, 148 (1978) 429.

103 A.S. Bhown, J.E. Mole, A. Weissinger and J.C. Bennett, *J. Chromatogr.*, 148 (1978) 532.

104 J. Elion, M. Downing and K. Mann, *J. Chromatogr.*, 155 (1978) 436.

105 M. Abrahamsson, K. Gröningsson and S. Castensson, *J. Chromatogr.*, 154 (1978) 313.

106 M. Strickland, W.N. Strickland, W.F. Brandt, C. van Holt, B. Wittmann-Liebold and A. Lehmann, *Eur. J. Biochem.*, 89 (1978) 443.

107 G. Oestvold, E. Jensen and T. Greibrokk, *Medd. Norsk. Farm. Selsk.*, 40 (1978) 173.

108 J. van Beenmann, J. van Damme and J. Deley, *FEBS Lett.*, 85 (1978) 68.

109 M.W. Hahnkapiller and L.E. Hood, *Biochemistry*, 17 (1978) 2124.

110 G.W. Fong and E. Grushka, *Anal. Chem.*, 50 (1978) 1154.

111 W.F. Moo-Penn, M.H. Johnson, K.C. Bechtel and D.L. Jue, *J. Chromatogr.*, 172 (1979) 476.

112 W.D. Annan, *J. Chromatogr.*, 173 (1979) 194.

113 J.U. Harris, D. Robinson and A.J. Johnson, *Anal. Biochem.*, 105 (1980) 239.

114 M.J. Horn, P.A. Hargrave and J.K. Wang, *J. Chromatogr.*, 180 (1979) 111.

115 P.W. Moser and E.E. Rickli, *J. Chromatogr.*, 176 (1979) 451.

116 R. Spatz and E. Roggendorf, *Z. Anal. Chem.*, 299 (1979) 267.

117 S.E. Godtfredsen and R.W.A. Oliver, *Carlsberg Res. Commun.*, 45 (1980) 35.

118 R.P.W. Scott and P. Kucera, *J. Chromatogr.*, 169 (1979) 51.

119 R.P.W. Scott, *J. Chromatogr. Sci.*, 18 (1980) 49.

120 S. Weinstein, B. Feibush and E. Gil-Av, *J. Chromatogr.*, 126 (1976) 97.

121 V.A. Davankov, S.V. Rogozhin and A.V. Semechkin, *J. Chromatogr.*, 91 (1974) 493.

122 I.S. Krull, *Advan. Chromatogr.*, 16 (1977) 175.

123 T. Nambara, S. Ikegawa, M. Hasegawa and J. Goto, *Anal. Chim. Acta*, 101 (1978) 111.

124 R.V. Snyder, R.J. Angelici and R.B. Meck, *J. Amer. Chem. Soc.*, 94 (1972) 2660.

125 V.A. Davankow and Y.A. Zolotarev, *J. Chromatogr.*, 155 (1978) 303.

126 G. Gubitz, W. Jellenez, G. Lofler and W. Santi, *J. High Resolut. Chromatogr. Chromatogr. Commun.*, 2 (1979) 145.

127 B. Lefebvre, R. Audebert and C. Quivron, *J. Liquid Chromatogr.*, 1 (1978) 761.

128 E. Tsuchida, H. Nishikawa and E. Terada, *Eur. Polym. J.*, 12 (1976) 611.

129 F.K. Chow and E. Grushka, *Anal. Chem.*, 50 (1978) 1346.

130 R.J. Baczuk, G.K. Landram, R.J. Dubois and H.C. Dehm, *J. Chromatogr.*, 60 (1971) 351.

131 L.R. Sousa, G.D.Y. Sogah, D.H. Hoffman and D.J. Cram, *J. Amer. Chem. Soc.*, 100 (1978) 4569.

132 W.H. Pirkle and D.W.J. House, *J. Org. Chem.*, 44 (1979) 1957.

133 S. Hara and A.J. Dobashi, *J. Liquid Chromatogr.*, 2 (1979) 883.

134 H. Yoneda and T. Yoshizawa, *Chem. Lett.*, (1976) 707.

135 F.K. Chow and E. Grushka, *J. Chromatogr.*, 185 (1979) 361.

136 P.E. Hare and E. Gil-Av, *Science*, 204 (1979) 1226.

137 C. Gilon, R. Leshem, Y. Tapuhi and E. Grushka, *J. Amer. Chem. Soc.*, 101 (1979) 7612.

138 J.N. LePage, J.W. Lindner, G. Davies, D.E. Seitz and B.L. Karger, *Anal. Chem.*, 51 (1979) 433.

139 W. Lindner, J.N. LePage, G. Davies, D.E. Seitz and B.L. Karger, *J. Chromatogr.*, 185 (1979) 323.

140 C. Gilon, R. Leshem and E. Grushka, *Anal. Chem.*, 52 (1980) 1206.

141 R. Audebert, *J. Liquid Chromatogr.*, 2 (1979) 1063.

142 W. Kullmann, *J. Liquid Chromatogr.*, 2 (1979) 1017.

143 T.F. Gabriel, J. Michalewsky and J. Meienhofer, *J. Chromatogr.*, 129 (1976) 287.

144 E. Gil-Av, in Y. Wolman (Editor), *Peptides, 1974*, Wiley, New York, 1975, p. 247.

145 I. Molnár and Cs. Horváth, *J. Chromatogr.*, 142 (1977) 623.

146 W.S. Hancock, C.A. Bishop, R.L. Prestidge, D.R. Harding and M.T.W. Hearn, *Science*, 200 (1978) 1168.

147 J. Rivier, *J. Liquid Chromatogr.*, 1 (1978) 343.

148 E. Lundanes and T. Greibrokk, *J. Chromatogr.*, 149 (1978) 241.

149 E.C. Nice and M.J. O'Hare, *J. Chromatogr.*, 162 (1979) 401.

150 M.T.W. Hearn, C.A. Bishop, W.S. Hancock, D.R.K. Harding and G.D. Reynolds, *J. Liquid Chromatogr.*, 2 (1979) 1.

151 S. Terabe, R. Konaka and K. Inouye, *J. Chromatogr.*, 172 (1979) 163.

152 M.J. O'Hare and E.C. Nice, *J. Chromatogr.*, 171 (1979) 209.

153 C.A. Bishop, D.R.K. Harding, L.J. Meyer, W.S. Hancock and M.T.W. Hearn, *J. Chromatogr.*, 192 (1980) 222.

154 W.M.M. Schaaper, D. Voskamp and C. Olieman, *J. Chromatogr.*, 195 (1980) 181.

155 M. Abrahamsson and K. Gröningsson, *J. Liquid Chromatogr.*, 3 (1980) 495.

156 S.-N. Lin, L.A. Smith and R.M. Caprioli, *J. Chromatogr.*, 197 (1980) 31.

157 M.T.W. Hearn, B. Grego and W.S. Hancock, *J. Chromatogr.*, 185 (1979) 429.

158 F. Nachtmann, *J. Chromatogr.*, 176 (1979) 391.

159 D. Voskamp, C. Olieman and H.C. Beyermann, *Rec. Trav. Chim. Pays-Bas*, 99 (1980) 105.

160 W.S. Hancock, C.A. Bishop, J.E. Battersby, D.R.K. Harding and M.T.W. Hearn, *J. Chromatogr.*, 168 (1979) 377.

161 J.L. Meek, *Proc. Nat. Acad. Sci. U.S.*, 77 (1980) 1632.

162 E.P. Kroeff and D.J. Pietrzyk, *Anal. Chem.*, 50 (1978) 502.

163 E. Lundanes and T. Greibrokk, *J. Chromatogr.*, 149 (1978) 241.

164 W. Voelter, H. Bauer, S. Fuchs and E. Pietrzik, *J. Chromatogr.*, 153 (1978) 433.

165 B. Larsen, V. Viswanatha, S.Y. Chang and V.J. Hruby, *J. Chromatogr. Sci.*, 16 (1978) 207.

166 B. Larsen, B.L. Fox and V.J. Hruby, *Int. J. Pept. Protein Res.*, 13 (1979) 12.

167 W.S. Hancock, C.A. Bishop, R.L. Prestidge and D.R.K. Harding, *Science*, 200 (1978) 1168.

168 B.L. Karger and R.W. Giese, *Anal. Chem.*, 50 (1978) 1048A.

169 G.W.K. Fong and E. Grushka, *Anal. Chem.*, 50 (1978) 1154.

170 E.P. Kroeff and D.J. Pietrzyk, *Anal. Chem.*, 50 (1978) 1353.

171 W.S. Hancock, C.A. Bishop, R.L. Prestidge, D.R.K. Harding and M.T.W. Hearn, *J. Chromatogr.*, 153 (1978) 391.

172 W.S. Hancock, C.A. Bishop, L.J. Meyer, D.R.K. Harding and M.T.W. Hearn, *J. Chromatogr.*, 161 (1978) 291.

173 L.J. Fischer, R.L. Thies and D. Charkowski, *Anal. Chem.,* 50 (1978) 2143.
174 W.S. Hancock, C.A. Bishop, R.L. Prestidge and M.T.W. Hearn, *Anal. Biochem.,* 89 (1978) 203.
175 H. Kratzin, C. Yang, J.U. Krusche and N. Hilschmann, *Hoppe-Seyler's Z. Physiol. Chem.,* 361 (1980) 1591.
176 W.A. Schroeder, J.B. Shelton, J.R. Shelton and D. Powars, *J. Chromatogr.,* 174 (1979) 385.
177 M. Dizdaroglu and M.G. Simic, *J. Chromatogr.,* 195 (1980) 119.
178 A. Witter, H. Scholtens and J. Verhoeff, *Neuroendocrinology,* 30 (1980) 377.
179 J.R. Coulter and C.S. Hann, in A. Niederwieser and G. Pataki (Editors), *New Techniques in Amino Acid, Peptide and Protein Analysis,* Ann Arbor Sci. Publ., Ann Arbor, M1, 1971, p. 75.
180 K. Blau, in H.A. Szymanski (Editor), *Biomedical Application of Gas Chromatography,* Plenum, New York, 1968, p. 1.
181 P. Hušek and K. Macek, *J. Chromatogr.,* 113 (1975) 139.
182 M.F. Lou and P.B. Hamilton, in D. Glick (Editor), *Methods of Biochemical Analysis,* Vol. 25, Wiley-Interscience, New York, 1977, p. 203.
183 R. Pocklington, *Anal. Biochem.,* 45 (1972) 403.
184 R.W. Zumwalt, D. Roach and C.W. Gehrke, *J. Chromatogr.,* 53 (1970) 171.
185 A.M. Lewis, C. Waterhouse and L.S. Jacobs, *Clin. Chem.,* 26 (1980) 271.
186 I.R. Hunter, K.P. Dimick and J.W. Corse, *Chem. Ind.,* 16 (1956) 294.
187 D.R. Knapp, *Handbook of Analytical Derivatization Reactions,* Wiley, New York, 1979.
188 R. Birkhofer and A. Ritter, *Chem. Ber.,* 93 (1960) 424.
189 E.D. Smith and K.L. Shewbart, *J. Chromatogr. Sci.,* 7 (1969) 704.
190 C.W. Gehrke and K. Leimer, *J. Chromatogr.,* 53 (1970) 201.
191 C.W. Gehrke, H. Nakamoto and R.W. Zumwalt, *J. Chromatogr.,* 45 (1969) 24.
192 C.W. Gehrke and K. Leimer, *J. Chromatogr.,* 57 (1971) 219.
193 K. Bergstrom, J. Gurtler and R. Blomstrand, *Anal. Biochem.,* 34 (1970) 74.
194 J.L. Laseter, J.D. Weete, A. Albert and C.H. Walkinshaw, *Anal. Lett.,* 4 (1971) 671.
195 F.P. Abramson, *Anal. Biochem.,* 57 (1974) 482.
196 E.D. Smith and K.L. Shewbart, *J. Chromatogr. Sci.,* 7 (1969) 704.
197 L. Birkofer and M. Donike, *J. Chromatogr.,* 26 (1967) 270.
198 M. Donike, *J. Chromatogr.,* 85 (1973) 1.
199 M. Donike, H. Suberg and L. Jaenicke, *J. Chromatogr.,* 85 (1973) 9.
200 F. Shahrokhi and C.W. Gehrke, *J. Chromatogr.,* 36 (1968) 31.
201 W.C. Butts, *Anal. Biochem.,* 46 (1972) 187.
202 B.S. Middleditch, *J. Chromatogr.,* 126 (1976) 581.
203 H. Vielma and J. Mendez, *J. Chromatogr.,* 196 (1980) 166.
204 H.B. Conacher, J.R. Iyenegar, W.F. Miles and H.G. Botting, *J. Ass. Offic. Anal. Chem.,* 62 (1979) 604.
205 K. Biemann, J. Seibl and F. Gapp, *J. Amer. Chem. Soc.,* 83 (1961) 3795.
206 E. Bayer, *Z. Naturforsch.,* 22B (1967) 924.
207 P.B. Hagen and W. Black, *Can. J. Biochem.,* 43 (1965) 309.
208 P.A. Cruickshant and J.C. Sheehan, *Anal. Chem.,* 36 (1964) 1191.
209 S.-C.J. Fu and D.S.H. Mak, *J. Chromatogr.,* 54 (1971) 205.
210 S.-C.J. Fu and D.S.H. Mak, *J. Chromatogr.,* 78 (1973) 211.
211 R.W. Zumwalt, K. Kuo and C.W. Gehrke, *J. Chromatogr.,* 57 (1971) 193.
212 C.W. Gehrke, *Quantitative Gas – Liquid Chromatography of Amino Acids in Proteins and Biological Substances,* Anal. Biochem. Labs., Columbia, MO, 1968.
213 D. Roach and C.W. Gehrke, *J. Chromatogr.,* 44 (1969) 269.
214 J.P. Hardy and S.L. Kerrin, *Anal. Chem.,* 44 (1972) 1497.
215 E. Watson, S. Wilk and J. Roboz, *Anal. Biochem.,* 59 (1974) 441.
216 L. Bertilsson and E. Costa, *J. Chromatogr.,* 118 (1976) 395.
217 N. Blau, personal communication.
218 D.E. Johnson, S.J. Scott and A. Meister, *Anal. Chem.,* 33 (1961) 669.
219 C.W. Gehrke and H. Takeda, *J. Chromatogr.,* 76 (1973) 63.

220 G.E. Pollock, *Anal. Chem.*, 39 (1967) 1194.
221 W. Parr, C. Yang, J. Pleterski and E. Bayer, *J. Chromatogr.*, 50 (1970) 510.
222 J.P. Zanetta and G. Vincendon, *J. Chromatogr.*, 76 (1973) 91.
223 C.W. Moss, M.A. Lambert and F.J. Diaz, *J. Chromatogr.*, 60 (1971) 134.
224 J. Jonnson, J. Eyem and J. Sjöquist, *Anal. Biochem.*, 51 (1973) 204.
225 D. Roach and C.W. Gehrke, *J. Chromatogr.*, 43 (1969) 303.
226 C.W. Gehrke, K. Kuo and R.W. Zumwalt, *J. Chromatogr.*, 57 (1971) 209.
227 D. Roach, C.W. Gehrke and R.W. Zumwalt, *J. Chromatogr.*, 43 (1969) 311.
228 T. Yoneda, *Anal. Biochem.*, 104 (1980) 247.
229 W.P. Leighton, S. Rosenblatt and J.D. Chanley, *J. Chromatogr.*, 164 (1979) 427.
230 S.K. Sarkar and S.S. Malhotra, *J. Chromatogr.*, 170 (1979) 371.
231 S.L. MacKenzie and D. Tenaschuk, *J. Chromatogr.*, 171 (1979) 195.
232 S.L. MacKenzie and D. Tenaschuk, *J. Chromatogr.*, 173 (1979) 53.
233 J. Gabryś and J. Konecki, *J. Chromatogr.*, 182 (1980) 147.
234 M.A. Kirkman, M.M. Burell, P.J. Lea and W.R. Mills, *Anal. Biochem.*, 101 (1980) 364.
235 C. Perier, M.C. Ronzière, A. Rattner and J. Frey, *J. Chromatogr.*, 182 (1980) 155.
236 E. Gil-Av, B. Feibush and R. Charles-Sigler, in A.B. Littlewood (Editor), *Gas Chromatography, Sixth Intern. Symp.*, Bartholomew, Dorking, 1967, p. 227.
237 E. Gil-Av, B. Feibush and R. Charles-Sigler, *Tetrahedron Lett.*, (1966) 1009.
238 E. Gil-Av and B. Feibush, *Tetrahedron Lett.*, (1967) 3345.
239 B. Feibush and E. Gil-Av, *Tetrahedron*, 26 (1970) 1361.
240 S. Nakaparksin, P. Birell, E. Gil-Av and J. Oro, *J. Chromatogr. Sci.*, 8 (1970) 177.
241 E. Bayer, E. Gil-Av, W.A. Koenig, S. Nakaparksin, J. Oro and W. Parr, *J. Amer. Chem. Soc.*, 92 (1970) 1738.
242 W. Koenig, W. Parr, H. Lichtenstein, E. Bayer and J. Oro, *J. Chromatogr. Sci.*, 8 (1970) 183.
243 W. Parr, C. Yang, E. Bayer and E. Gil-Av, *J. Chromatogr. Sci.*, 8 (1970) 591.
244 W. Parr and P. Howard, *Chromatographia*, 4 (1971) 162.
245 J.A. Corbin, J.E. Rhoad and L.B. Rogers, *Anal. Chem.*, 43 (1971) 327.
246 W. Parr and P. Howard, *J. Chromatogr.*, 66 (1972) 141.
247 W. Parr and P.Y. HowarD, *J. Chromatogr.*, 67 (1972) 227.
248 C.H. Lochmüller, J.M. Harris and R.W. Souter, *J. Chromatogr.*, 71 (1972) 405.
249 C.H. Lochmüller and P.M. Gross, *Separ. Purif. Methods*, 8 (1979) 21.
250 H. Frank, G.J. Nicholson and E. Bayer, *J. Chromatogr. Sci.*, 15 (1977) 174.
251 H. Frank, G.J. Nicholson and E. Bayer, *J. Chromatogr.*, 167 (1978) 187.
252 H. Frank, A. Rettenmeier, H. Weicker, G.J. Nicholson and E. Bayer, *Clin. Chim. Acta*, 105 (1980) 201.
253 F. Weygand, B. Kolb, A. Prox, M.A. Tilak and I. Tomida, *Hoppe Seyler's Z. Physiol. Chem.*, 322 (1960) 38.
254 F. Weygand, A. Prox, L. Schmidhammer and W. Koenig, *Angew. Chem.*, 75 (1963) 282.
255 F. Weygand, A. Prox, L. Schmidhammer and W. Koenig, in H.C. Beyerman, A. van de Linde and W. Maasen van den Brink (Editors), *Peptides*, Pergamon, Oxford, 1963, p. 97.
256 E. Gil-Av, R. Charles-Sigler, G. Fischer and D. Nurok, *J. Gas Chromatogr.*, 4 (1966) 51.
257 G.E. Pollok and V.I. Oyama, *J. Gas Chromatogr.*, 4 (1966) 126.
258 G.E. Pollok, V.I. Oyama and R.D. Johnson, *J. Gas Chromatogr.*, 3 (1965) 174.
259 J.W. Westley, B. Halpern and B.L. Karger, *Anal. Chem.*, 40 (1968) 2046.
260 G.S. Ayers, R.E. Monroe and J.H. Mossholder, *J. Chromatogr.*, 63 (1971) 259.
261 G.E. Pollok and A.H. Kawauchi, *Anal. Chem.*, 40 (1968) 1356.
262 R. Charles-Sigler, G. Fisher and E. Gil-Av, *Isr. J. Chem.*, 1 (1963) 234.
263 E. Gil-Av, R. Charles and G. Fischer, *J. Chromatogr.*, 17 (1965) 408.
264 K. Kvenholden, E. Peterson and F. Brown, *Science*, 169 (1970) 1079.
265 K. Kvenholden, J. Lawless, K. Pering, E. Peterson, J. Flores, C. Ponnamperuma, I.R. Kaplan and C. Moore, *Nature (London)*, 228 (1970) 923.
266 J. Lawless, K. Kvenholden, E. Peterson, C. Ponnamperuma and C. Moore, *Science*, 173 (1971) 626.

267 G.E. Pollok and L.H. Frommhagen, *Anal. Biochem.*, 24 (1968) 18.

268 G.E. Pollok, *Anal. Chem.*, 44 (1972) 2368.

269 J.G. Lawless, K. Kvenholden, E. Peterson, C. Ponnamperuma and E. Jarosewich, *Nature (London)*, 236 (1972) 66.

270 B. Halpern and G.E. Pollok, *Biochem. Med.*, 4 (1970) 352.

271 C. Ponnamperuma, *Ann. N.Y. Acad. Sci.*, (1972) 56.

272 B. Halpern and J.W. Westley, *Tetrahedron Lett.*, 21 (1966) 2283.

273 J.C. Dabrowiak and D.W. Cooke, *Anal. Chem.*, 43 (1971) 791.

274 H. Iwase and A. Murai, *Chem. Pharm. Bull.*, 22 (1974) 8.

275 F. Weygand, B. Kolb and P. Kirchner, *Z. Anal. Chem.*, 181 (1961) 396.

276 B. Kolb, in A. Niederwieser and G. Pataki (Editors), *New Techniques in Amino Acid, Peptide and Protein Analysis*, Ann Arbor Sci. Publ., Ann. Arbor, MI, 1971, p. 129.

277 A.V. Barooshian, M.J. Lautenschleger and W.G. Harris, *Anal. Biochem.*, 44 (1971) 543.

278 F. Raulin and B.N. Khare, *J. Chromatogr.*, 75 (1973) 13.

279 F. Weygand, A. Prox, E.C. Jorgensen, R. Axen and P. Kirchner, *Z. Naturforsch.*, 18b (1963) 93.

280 F. Weygand, *Z. Anal. Chem.*, 243 (1968) 2.

281 I. Tomida, T. Tokuda, J. Ohashi and M. Nakajima, *J. Agr. Chem. Soc. Japan*, 39 (1965) 391.

282 A. Prox and F. Weygand, in H.C. Beyerman, A. van den Linde and W. Maassen van den Brink (Editors), *Peptides*, North Holland, Amsterdam, 1967, p. 158.

283 E. Bayer and W.A. Koenig, *J. Chromatogr. Sci.*, 7 (1969) 95.

284 F. Weygand, D. Hoffmann and A. Prox, *Z. Naturforsch.*, 23b (1968) 279.

285 A. Prox, J. Schmid and H. Ottenheym, *Liebigs Ann. Chem.*, 722 (1969) 179.

286 J.M.L. Mee, *J. Chromatogr.*, 87 (1973) 258.

287 B.A. Andersson, *Acta Chem. Scand.*, 21 (1967) 2906.

288 D.H. Calam, *J. Chromatogr.*, 70 (1972) 146.

289 A. Prox and K.K. Sun, *Z. Naturforsch.*, 21b (1966) 1028.

290 P.A. Leclerq and D.M. Desiderio, *Anal. Lett.*, 4 (1971) 305.

291 T. Wieland, G. Lueben, H. Ottenheym, J. Faesel, J.X. de Fries, W. Konz, A. Prox and J. Schmid, *Angew. Chem.*, 80 (1968) 209.

292 R.T. Aplin, I. Eland and J.H. Jones, *Org. Mass Spectrom.*, 2 (1969) 795.

293 P.A. Cruickshank and J.C. Sheehan, *Anal. Chem.*, 36 (1964) 1191.

294 F. Weygand and R. Geiger, *Chem. Ber.*, 89 (1956) 647.

295 M. Dizdaroglu and M.G. Simic, *Anal. Biochem.*, 108 (1980) 269.

296 H.-R. Schulten, *Int. J. Mass Spectrom. Ion Phys.*, 32 (1979) 97.

297 G.W.A. Milne and M.J. Lacey, *Critical Reviews in Analytical Chemistry*, CRC, Cleveland, OH, 1974, p. 45.

298 M.C. ten Noever de Brauw, *J. Chromatogr.*, 165 (1979) 207.

299 W.H. McFadden, *J. Chromatogr. Sci.*, 17 (1979) 2.

300 W.E. Pereira, Y. Hoyano, W.E. Reynolds, R.E. Summons and A.M. Duffield, *Anal. Biochem.*, 55 (1973) 236.

301 R.E. Summons, W.E. Pereira, W.E. Reynolds, T.C. Rindfleisch and A.M. Duffield, *Anal. Chem.*, 46 (1974) 582.

302 H. Iwase, Y. Takeuchi and A. Murai, *Chem. Pharm. Bull.*, 27 (1979) 1307.

303 K.R. Leimer, R.H. Rice and C.W. Gehrke, *J. Chromatogr.*, 141 (1977) 355.

304 C. Wiecek, B. Halpern, A.M. Sargeson and A.M. Duffield, *Org. Mass Spectrom.*, 14 (1979) 281.

305 F. Artigas and E. Gelpi, *Anal. Biochem.*, 92 (1979) 233.

306 M. Takimoto, T. Takeda, S. Takahashi and T. Murata, *Shimazu Hyoron*, 34 (1977) 159; *C.A.*, 89 (1978) 163926h.

307 H. Budzikiewicz and G. Meissner, *Org. Mass Spectrom.*, 13 (1978) 608.

308 K. Okada and A. Sakuno, *Org. Mass Spectrom.*, 13 (1978) 535.

309 J.J. Shieh, K. Leung and D.M. Desiderio, *Anal. Lett.*, 10 (1977) 575.

310 D.-H. Jo, J. Desgrès and P. Padieu, *J. Chromatogr.*, 146 (1978) 413.

311 R.J. Weinkam, *J. Org. Chem.*, 43 (1978) 2581.

312 F.S. Collins and G.K. Summer, *J. Chromatogr.*, 145 (1978) 456.

313 J.M.L. Mee, J. Korth, B. Halpern and L.B. James, *Biomed. Mass Spectrom.*, 4 (1977) 178.

314 M.F. Schulman and F.P. Abramson, *Biomed. Mass Spectrom.*, 2 (1975) 9.

315 F.P. Abramson, M.W. McCaman and R.E. McCaman, *Anal. Chem.*, 51 (1974) 482.

316 W. Frick, D. Chang, K. Folkers and G.D. Daves, Jr., *Anal. Chem.*, 49 (1977) 1241.

317 P.S. Callery, M. Stogniew and L.A. Geelhaar, *Biomed. Mass Spectrom.*, 6 (1979) 23.

318 F.K. Trefz, D.J. Byrd, M.E. Blaskovics, W. Kochen and P. Lutz, *Clin. Chim. Acta*, 73 (1976) 431.

319 M.-J. Zagalak, H.-Ch. Curtius, W. Leimbacher and U. Redweik, *J. Chromatogr.*, 142 (1977) 523.

320 V. Fell, J.A. Hoskins and R.J. Pollit, *Clin. Chim. Acta*, 83 (1978) 259.

321 J.A. Hoskins and R.J. Pollit, in T.A. Baillie (Editor), *Stable Isotopes: Applications in Pharmacology, Toxicology and Clinical Research*, Macmillan, London, 1978, p. 253.

322 H.S. Ramsdell, B.H. Baretz and K. Tanaka, *Biomed. Mass Spectrom.*, 4 (1977) 220.

323 H.-Ch. Curtius, H. Wegmann, U. Redweik and W. Leimbacher, in E.R. Klein and P.D. Klein (Editors), *Stable Isotopes: Proc. 3rd Intern. Conf.*, Academic Press, New York, 1979, p. 573.

324 J.J. Rafter, M. Ingelmann-Sundberg and J.A. Gustafson, *Biomed. Mass Spectrom.*, 6 (1979) 317.

325 N. Castagnoli, K.L. Melmon, C.R. Freed, M.M. Ames, A. Kalir and R. Weinkam, in T.A. Baillie (Editor), *Stable Isotopes*, Univ. Park Press, Baltimore, MD, 1978, p. 261.

326 B.N. Colby and M.W. McCaman, *Biomed. Mass Spectrom.*, 5 (1978) 215.

327 R.T. Coutts, G.R. Jones and S.-F. Lin, *J. Chromatogr. Sci.*, 17 (1979) 551.

328 D.E. Mathews, E. Ben-Galim and D.M. Bier, *Anal. Chem.*, 51 (1979) 80.

329 D. Halliday, M. Madigan, S. Ell, P. Richards, J. Bergström and P. Fürst, in E.R. Klein and P.D. Klein (Editors), *Stable Isotopes, Proc. 3rd Intern. Conf.* Academic Press, New York, 1979, p. 583.

330 A. Lapin and M. Karobath, *J. Chromatogr.*, 193 (1980) 95.

331 S. Hirano, *Taisha*, 14 (1977) 1495.

332 J.H. McReynold and M. Anbar, *Anal. Chem.*, 49 (1977) 1832.

333 C.S. Irving, I. Nissim and A. Lapidot, *Biomed. Mass Spectrom.*, 5 (1978) 117.

334 C.S. Irving, I. Nissim and A. Lapidot, *Monogr. Hum. Genet.*, 9 (1978) 50.

335 J.R. Robinson, A.N. Starrat and E.E. Schlahetka, *Biomed. Mass Spectrom.*, 5 (1978) 648.

336 D. Voight and J. Schmidt, *Biomed. Mass Spectrom.*, 5 (1978) 44.

337 A. Lapidot, I. Nissim, C.S. Irving, U.A. Liberman and R. Samuel, in E.R. Klein and P.D. Klein (Editors), *Stable Isotopes: Proc. 3rd Intern. Conf.*, Academic Press, New York, 1979, p. 599.

338 R.C. Murphy and K.L. Lay, *Biomed. Mass Spectrom.*, 6 (1979) 309.

339 E. Bailey, P.B. Farmer and J.H. Lamb, *J. Chromatogr.*, 200 (1980) 145.

340 D.N. Crossley and D.B. Ramsden, *Clin. Chim. Acta*, 94 (1979) 267.

341 N. Heki, *Nippon Naibumpi Gakkai Zasshi*, 54 (1978) 1157.

342 K. Biemann, in G.R. Waller and O.C. Dermer (Editors), *Biomedical Applications of Mass Spectrometry, 1st Suppl.*, J. Wiley, New York, 1980, p. 469.

343 U.P. Schlunegger, *Angew. Chem.* 87 (1975) 731.

344 U.P. Schlunegger and P. Hirter, *Isr. J. Chem.*, 17 (1978) 168.

345 R. Steinauer, H.J. Walther and U.P. Schlunegger, *Helv. Chim. Acta*, 63 (1980) 610.

346 J. Rosmus and Z. Deyl, *J. Chromatogr.*, 70 (1972) 221.

347 J. Rosmus and Z. Deyl, *Chromatogr. Rev.*, 13 (1971) 163.

348 G. Hudson and K. Biemann, *Biochem. Biophys. Res. Commun.*, 71 (1976) 212.

349 M.W. Rees, R. Casey, M.N. Short, J.J. Sexton, J.F. March, J. Eagles, K.R. Parsley and R. Self, *Biomed. Mass Spectrom.*, 7 (1980) 132.

350 H.C. Krutzsch and J.J. Pisano, *Methods Enzymol.*, 47 (1977) 391.

351 H.C. Krutzsch and J.J. Pisano, *Biochemistry*, 17 (1978) 2791.

352 H.C. Krutzsch and T.J. Kindt, *Anal. Biochem.*, 92 (1979) 525.

353 W.E. Seifert, R.L. McKee, C.F. Beckner and R.M. Caprioli, *Anal. Biochem.*, 88 (1978) 149.

354 B.C. Das and E. Lederer, in A. Niederwieser and G. Pataki (Editors), *New Techniques in Amino Acid, Peptide and Protein Analysis*, Ann Arbor Sci. Publ., Ann Arbor, MI, 1971.

355 K. Biemann, F. Gapp and J. Seibl, *J. Amer. Chem. Soc.*, 81 (1959) 2274.

356 K. Biemann and W. Vetter, *Biochem. Biophys. Res. Commun.*, 3 (1960) 578.

357 F. Weygand and R. Geiger, *Chem. Ber.,* 89 (1956) 647.
358 H. Nau, *Biochem. Biophys. Res. Commun.,* 59 (1974) 1088.
359 H. Nau, *Angew. Chem. Int. Ed. Eng.,* 15 (1976) 75.
360 H. Nau and K. Bieman, *Anal. Biochem.,* 73 (1976) 139, 154, 175.
361 J.A. Kelley, H. Nau, H.-J. Foerster and K. Biemann, *Biomed. Mass Spectrom.,* 2 (1975) 313.
362 H. Nau, H.-J. Foerster, J.A. Kelley and K. Biemann, *Biomed. Mass Spectrom.,* 2 (1975) 326.
363 H. Frank and D.M. Desiderio, *Anal. Biochem.,* 90 (1978) 413.
364 V.K. Mahajan and D.M. Desiderio, *Biochem. Biophys. Res. Commun.,* 82 (1978) 1104.
365 W. Steiner and A. Niederwieser, *Clin. Chim. Acta,* 92 (1979) 431.
366 D.F. Hunt, G.C. Stafford Jr., F.M. Crow and J.W. Russel, *Anal. Chem.,* 48 (1976) 2098.

Chapter 11

Proteins

NICHOLAS CATSIMPOOLAS

CONTENTS

11.1. INTRODUCTION

The use of chromatographic and electrophoretic techniques for the separation of proteins has had an immense impact on the progress of modern biochemistry over the past twenty years. Since proteins exhibit diverse biological functions (e.g., enzymatic, hormonal, regulatory, and immunological functions) there is no field in the biological and biomedical sciences where chromatography and electrophoresis have not been utilized advantageously, either for research or clinical purposes.

The separation of proteins requires caution in regard to the pH, ionic strength, temperature, and nature of the electrolytes and supporting material to be employed. The physico-chemical and biological properties of each particular protein depend on a combination of all of these variables. The three levels of structure (viz., secondary, tertiary, and quaternary) as well as molecular interactions depend on ionic, hydrophobic, and hydrogen bonding. The same forces are also involved in the separation processes. Thus, the environmental physical conditions have to be chosen in such a manner that the separated product maintains certain desirable properties, usually associated with the preservation of the native state and biological activity. However,

References on p. B72

for the determination of the physical properties of protein subunits it is often necessary to denature the protein, and this may require the combination of harsh treatment (e.g., urea, guanidine hydrochloride) and chemical modification (e.g., cleavage of disulfide bonds and blockage of sulfhydryl groups). Thus, the separation problem at hand will determine the different experimental methods required to achieve the final objective. Another point worth mentioning is that for proteins in their "native" state the chemical groups on the surface of the particles participate in the separation process. However, when the proteins are denatured, either partially or completely, new groups are exposed that were previously hidden in the interior of the macromolecule. These newly exposed groups may alter the magnitude and probably also the nature of the separation force effect. Such complications are nonexistent in the separation of small molecules, but have to be considered seriously in the separation of proteins.

A prerequisite for subjecting a protein to a separation process is that it be soluble under the conditions of the experient. This is a much neglected caveat. In many instances, only the "soluble" part of the sample is analyzed, the rest remaining unresolved in chromatographic columns or electrophoresis gels. Since the solubility characteristics of many proteins depend on both pH and ionic strength, this problem is more prevalent in gradient (pH or ionic strength) chromatography. Also, it is more obvious in polyacrylamide gel electrophoresis, where the aggregated material cannot enter the sieving gel and stays at the origin.

Another special consideration in comparing the chromatographic behavior of small molecules with that of proteins is the use of reference standards. These are well defined in the case of small molecules. Although certain purified proteins are used as "molecular weight" standards in gel filtration and polyacrylamide gel electrophoresis, these cannot be considered as "universal", because in their native form they are assumed to have a spherical shape and in their denatured form a completely unfolded coil. This may not be the case for a large number of proteins, especially the ones associated with carbohydrates. Thus, the available "standards" do not apply to all cases.

The use of organic solvents in the chromatography and electrophoresis of proteins is ruled out by the denaturing effects of solvents, which renders the macromolecules insoluble. Other organic compounds may have similar effects. Therefore, most protein separations are performed in carefully selected aqueous buffer systems. However, the need to solubilize proteins—mainly those associated with biological membranes—has led to the introduction of detergents, both ionic and non-ionic, and hydrogen–hydrophobic bond-breaking reagents, such as urea and guanidine HCl. Although denatured, most proteins remain soluble in solutions of these compounds, especially after cleavage of the disulfide bonds with dithiothreitol (DTT) or mercaptoethanol (ME).

11.2. CHROMATOGRAPHY OF PROTEINS

11.2.1. Gel chromatography

The theory and methodology of gel chromatography have been discussed extensively in several reviews and books [1–9] (cf. Chap. 8). Basically, this technique separates proteins according to their molecular size, the larger molecules being eluted first from the column, while smaller molecules are retarded. Proteins that are capable of diffusing into the liquid present in the pores of the matrix are retarded in their migration through the column. The pore size of the matrix controls the range of macromolecular sizes that can effectively move in and out of the gel particles. Polyacrylamide, agarose, and dextran gel particles of various exclusion limits are all suitable for the fractionation of proteins.

By far the most common application of gel chromatography to protein fractionation is for preparative purposes. Usually, the technique is employed as the first stage of a separation strategy. Because its revolving power is low, chromatographic peaks may consist of several components, and further fractionation by other techniques is necessary. For an unknown mixture, one starts with a gel matrix that provides a wide size range of separation. The peak that contains the protein of interest is then collected and subjected to gel chromatography with a narrower size range of gel. Selected fractions may then be fractionated by other techniques, such as ion-exchange chromatography and electrophoresis.

Another important application of gel chromatography in biochemistry is the removal of low-molecular-weight contaminants from proteins. The contaminants may be substances (e.g., amino acids, sugars, and steroids) that interfere with an analytical assay, or they may be reagents and products used for chemical modification of a protein. Products of protein radiolabeling and fluorescent labeling are most often removed by gel chromatography. Desalting or exchange of buffer, required in certain fractionation schemes, is also carried out faster and more effectively by chromatography than by dialysis, and this is also true of the removal of cofactors and inhibitors necessary for the study of enzyme kinetics. Furthermore, binding of low-molecular-weight substances, such as drugs, metal ions, and dyes to proteins can be easily studied by this technique [10].

The partition coefficient, K_d, of "standard" proteins of known molecular weight can be employed for the determination of the size of similarly shaped proteins. Porath [11] derived a calibration curve, which predicts a linear relationship between the cube root of K_d and the square root of the molecular width. The equation is based on the assumption that the gel pores are of different shape (spheres, cones, or crevices), and that the effective radius of the macromolecule is the radius of gyration. This model has provided a good basis for the interpretation of results of gel chromatography for several years. Others [12,13] have used the relationship

$$erfc^{-1} \cdot K_{av} = a + b \log R_s \qquad (11.1)$$

where K_{av} is the partition coefficient, a and b are regression coefficients, R_s is the Stokes' radius (or $MW^{1/3}$) and $erfc^{-1}$ is the inverse error function complement. The

References on p. B72

above model assumes a Gaussian distribution for the effective radii of the gel pores and an effective radius of the protein that is identical to its hydrodynamic Stokes' radius. If one assumes a "logistic" distribution of the pore sizes, the following relationship is applicable:

$$K_{av} = 1/[1 + (MW/c)^b] \qquad (11.2)$$

where MW is the molecular weight and c and b are empirical constants. This can be linearized [14] by use of a plot of logit K_{av} vs. log MW

$$\text{logit } K_{av} = a + b \log MW \qquad (11.3)$$

where

$$\text{logit } Y = \log_e [Y/(1 - Y)] \qquad (11.4)$$

The logit transformations and inverse error functions introduce severe nonuniformity of variance so that a weighted regression must be utilized [15]. In the case of random coils (unfolded proteins), the effective radius of gyration (R_g) is taken as being proportional to the square root of molecular weight, $R_g \propto MW^{0.5}$ [11].

Molecular weight estimations can be performed only with purified proteins, except in the case of enzymes (or other proteins that can be determined by bioassays) where their partition coefficient can be determined from their distribution in specifically assayed fractions. Rapid analytical gel chromatography of proteins and peptides and molecular-weight distribution analysis has been described in detail by Catsimpoolas and Kenney [16–18] and by Sosa [19].

11.2.2. Ion-exchange chromatography

Ion-exchange chromatography of proteins [20] is performed with packing materials that have a hydrophilic matrix, such as cellulose and dextran. Anion exchangers, especially DEAE-cellulose and DEAE-Sephadex, are more frequently used than cation exchangers, such as CM-cellulose. The cellulosic ion exchangers provide an open network structure with ionized sites readily accessible to proteins. The number of electrostatic bonds formed between the exchanger and the protein depends not only on the materials used, but also to a large extent on the pH and ionic strength of the buffer. These ionic bridges are continuously dissociating and reforming, as ions of the milieu compete for the exchange sites. The exchangers suitable for the fractionation of proteins provide a low charge density, so that the number of ionic linkages is low enought to allow the elution of the macromolecule. Although the ionic linkages constantly dissociate and reform, at the start of the chromatographic experiment the protein remains relatively immobile in the column, because some of the bonds are always present. However, when the concentration of the competing small ions increases to the level that all the bonds dissociate at the same time, the protein is eluted and moves down the column. When elution with an ionic strength gradient is used, the protein at the starting zone migrates more slowly than the liquid, and it is overtaken by the gradient. Thus, the medium surrounding the protein progressively increases in eluting power and, therefore, the macromolecule moves faster and faster, catching up with the velocity of the liquid at the salt concentration which prevents effectively its interaction with the exchanger. The shape of the gradient for each particular separation is important, because its slope should be rather flat in certain areas of the chromatogram to increase resolution, but also steep enough in other areas to avoid band broadening.

The size of proteins is also a factor in band broadening during elution. A large protein–often composed of multisubunits—offers various regions of attachment to the exchanger. These regions—each containing several ionic sites—may exhibit variable strengths of interaction. As the protein moves down the column, the exchanger binds to different regions at different times and, thus, contributes to spreading.

The properties and use of several ion exchangers have been discussed extensively in previous reviews [20–22]. Carboxymethyl (CM)-cellulose, Sephadex phosphate (P)-cellulose, Sulfomethyl (SM)-cellulose, and Sulfopropyl (SP)-Sephadex are typical cation exchangers, suitable for the fractionation of basic proteins. Anion exchangers include aminoethyl (AE)-cellulose, diethylaminoethyl (DEAE)-cellulose and -Sephadex, ECTEOLA-cellulose, triethylaminoethyl (TEAE)-cellulose, and 2-hydroxypropylamine (QAE)-Sephadex [20]. The acid–base capacity of these materials ranges from 0.1 to 4.5 meq/g, except for P-cellulose which may range up to 7.4 meq/g. Both strong (e.g., QAE, SP) and weak (e.g., DEAE, CM) exchangers are available.

Suitable buffers (0.001 to 0.01M) are, for cation exchangers: acetate, citrate, glycine, ammonium acetate (volatile), and phosphate; and for anion exchangers: Tris hydrochloride, Tris phosphate, phosphate, and ammonium carbonate (volatile). Other nonbuffering salts (e.g., NaCl, KCl) can be added to the buffer to form an ionic strength gradient. If a pH gradient is employed, this should decrease in pH with anion exchangers and increase with cationic ones. In both cases, the ionic strength is increasing.

Excellent practical hints for performing ion-exchange chromatography experiments with proteins are given by Peterson [20].

11.2.3. Affinity chromatography

The technique of affinity chromatography [23–27] has played an important role in the isolation of biologically active proteins during the past decade. A ligand having biological affinity for the protein to be isolated is covalently coupled to an insoluble matrix and packed into a column. When a solution of various substances passes through this matrix, only the protein of interest is sorbed and the impurities are washed out. Subsequently, the specified protein is desorbed, either by perturbing the interaction of the macromolecule with the immobilized ligand or by using a competitive ligand in the elution buffer. Examples of specific ligand–protein interactions include: enzymes and their inhibitors or cofactors, antigens and antibodies, hormones and receptors, lectins and glycoproteins, small molecules and their transport or binding proteins.

Although many methods exist for covalently coupling the ligand to the polymeric matrix, the cyanogen bromide method is the one most commonly used [28]. The reaction produces an "activated" product of the matrix (e.g., agarose, polyacrylamide) which reacts with the amino groups of the ligand. Often "spacer" groups (aliphatic chains up to ten carbons long) are coupled between the matrix and the ligand in order to facilitate complexing with large proteins or small ligands [29,30]. In selecting the appropriate ligand for a specific isolation problem, consideration must be given to the affinity of the ligand for the protein, the mode of attachment, the concentrations of the ligand and protein, the nature of the sorption isotherm, and temperature effects.

Once the protein is bound to the affinity column and the impurities are removed, elution of the sorbed macromolecule can be achieved either by nonspecific or specific procedures. Nonspecific methods of elution require changing either the ionic strength, pH, dielectric constant of the buffer or using protein denaturants, such as urea and guanidine HCl. All of these procedures aim to decrease the affinity of the proteins for the ligand by means of some conformational change. It is anticipated that the protein is nor irreversibly denatured by the elution process. Specific methods of elution usually involve competition of a new ligand or of a similar protein for the binding site.

A special kind of affinity chromatography involves the utilization of hydrophobic interactions between accessible hydrophobic binding sites on the protein and hydrophobic ligands as the matrix [31–34]. Desorption takes place in the presence of decreasing ionic strength and increasing pH, or by elution with ethylene glycol solutions and detergents.

11.2.4. Adsorption chromatography

The most useful form of this technique for the separation of proteins involves adsorption on calcium phosphate, especially hydroxyapatite [35–38]. A common problem with this type of chromatography is excessive "tailing", which is due to the nature of the adsorption isotherm. The line obtained by plotting the amount of protein adsorbed against its concentration in the mobile phase curves strongly toward the saturation concentration. As the protein moves down the column, the "tail", which contains protein at lower concentrations, binds more strongly than the peak. Thus, the use of low starting concentrations may improve the separation, provided the adsorption is not very strong.

11.2.5. High-performance liquid chromatography

Rapid chromatographic separations of proteins were originally performed with Sephadex G-25 and G-50 (crosslinked dextran materials) [16–18]. Since then, additional sorbents have been produced which exhibit mechanical stability at high buffer flowrates and high load capacity [39]. The new sorbents for high-performance gel chromatography consist of controlled-pore silica gels, glass, and bonded-phase sorbents on either glass or silica [40–44]. Typical analytical protein separations are accomplished in less than 20 min at flowrates of 0.5–1.0 ml/min.

Ion exchangers include covalently and noncovalently bonded stationary phases [45] and macroporous, hydrophilic polymers [46]. Affinity HPLC has been accomplished by use of a ligand–glycosil (silica) support [47]. Undoubtedly, this technique of protein fractionation will find increasing applications in clinical and analytical laboratories.

11.3. ELECTROPHORESIS OF PROTEINS

11.3.1. General classification of techniques

The physical process involved in the separation of electrically charged particles subjected to a DC electric field is known as electrophoresis (cf. Chap. 9). The net charge on the surface of proteins originates primarily from the ionization of groups such as carboxyl, amino, imidazolium, phenoxy, and sulfhydryl, as well as secondarily from ion binding. Numerous electrophoretic techniques have been applied to the separation of proteins, and their classification is based on the particular electrolyte systems used, the choice of supporting media, dimensionality of separation, and even on apparatus design and mode of detection.

With respect to the type of electrolyte systems, we may distinguish:

(a) continuous (uniform) pH zone electrophoresis (CZE) or simply electrophoresis [48–57],

(b) isoelectric focusing (IF) in pH gradients [58–64],

(c) use of isotachophoresis (ITP) with leading and terminating ions [58,61,62,65],

(d) multiphasic zone electrophoresis (MZE) in discontinuous electrolyte phases [66,67].

Stabilizing media in electrophoretic procedures either exhibit negligible "macromolecular sieving" effects, e.g., media employed in density gradient [68], cellulose acetate [69], and agarose electrophoresis [51], or they impart considerable retardation in molecular migration, e.g., the media used in starch gel [54,70] and polyacrylamide gel [66,67,70,71] electrophoresis (PAGE). Additionally, zone stabilization in electrophoretic separations may be achieved either by capillary forces, or by mechanical rotation in specially designed apparatus. These techniques are known as free-flow [72], free-zone [73], endless-fluid belt [74], and continuous-flow [75] electrophoresis. It is also useful in distinguishing various electrophoretic techniques according to dimensionality. Thus, names such as column, block, slab, and thin-layer electrophoresis refer to the shape or dimensions of the stabilizing medium used [49,54,55]. These names may be used in conjunction with two-dimensional gel electrophoresis [76]. Immunoelectrophoresis [53,56,57] and immunoelectrofocusing [77] also belong in this category, because they involve immunodiffusion in a second dimension.

Other, primarily instrumental techniques are concerned with the mode of detection of separated species and the meaurement of their electrophoretic mobility. Thus, the electrophoretic light scattering [78] method utilizes the Doppler shift in the frequency of scattered light from the moving particles to determine their velocity.

References on p. B72

TRANS electrophoresis [79] involves repetitive optical scanning of the separated species to produce kinetic data from which physico-chemical parameters of the macromolecules can be estimated.

11.3.2. Analytical electrophoresis instrumentation

The principle of the moving boundary technique [80,81] for measuring the electrophoretic mobility of proteins is discussed in Chap. 9. This method, which has been replaced by other analytical procedures, is now only of historical interest.

The laser electrophoretic light scattering (ELS) method was introduced in 1971 by Ware and Flygare [82]. Doppler velocimetry and electrophoresis in free solution (Fig. 11.1) were combined to obtain mobility measurements of relatively pure proteins in a few seconds. Hjertén [83] designed an apparatus which eliminates protein zone convection in free solution by rotating a horizontal electrophoresis tube along its longitudinal axis (Fig. 11.2). The separated zones are monitored by optical scanning of the transparent quartz column. Catsimpoolas [79] used density gradients in vertical quartz columns, followed by repetitive scanning, to follow the kinetics of electrophoretic phenomena (Fig. 11.3). Stabilization by means of a capillary-thin gap

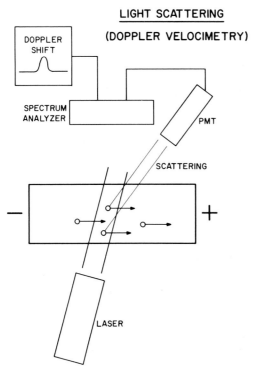

Fig. 11.1. Schematic diagram of the electrophoretic light scattering (ELS) technique. PMT = photomultiplier.

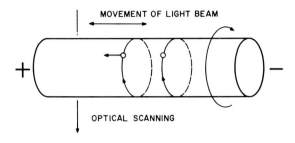

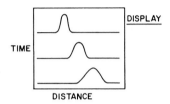

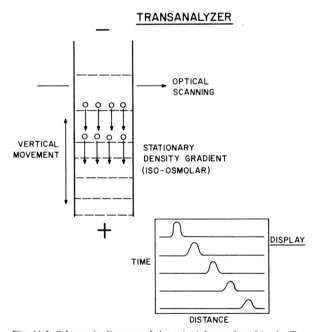

Fig. 11.2. Schematic diagram of the free-zone electrophoresis method.

Fig. 11.3. Schematic diagram of the principle employed in the Transanalyzer instrument.

References on p. B72

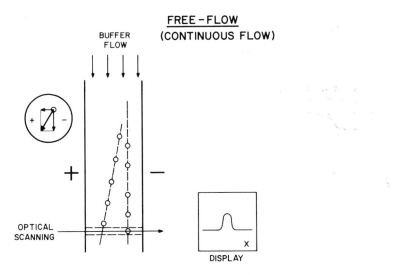

Fig. 11.4. Schematic diagram of the free-flow electrophoresis principle.

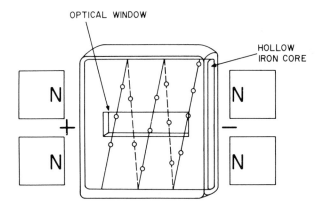

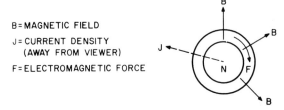

Fig. 11.5. Schematic diagram of the endless-fluid belt electrophoresis.

between two plates, which is subjected to a high-voltage electric field perpendicular to the laminar buffer flow, is the basis of the analytical free-flow electrophoresis instrument designed by Hannig et al. [84] (Fig. 11.4). Kolin [85] stabilized zones in an endless fluid belt by ingenious application of electromagnetic forces which rotate the liquid in an annular cell while the charged particles migrate in the electric field along helical paths (Fig. 11.5).

11.3.3. Analytical apparatus for supporting media

The types of apparatus needed for electrophoresis in supporting media such as paper, cellulose acetate membranes, agarose, starch, and polyacrylamide gels are very simple (Fig. 11.6). Paper and cellulose acetate membranes rest on two support rods, agarose and starch gels on glass plates, and polyacrylamide gels are cast in

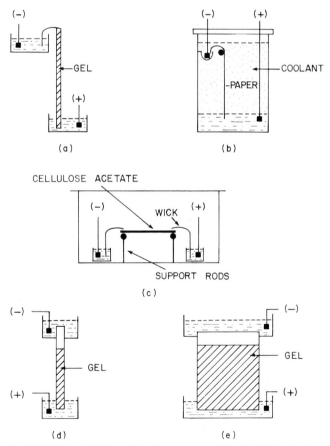

fig. 11.6. Schematic illustration of some types of analytical apparatus used for electrophoresis in various supporting media. (a) Starch gel, (b) high-voltage electrophoresis on paper, (c) paper, cellulose acetate, or agarose, (d) polyacrylamide gel in tubes (rods), (e) polyacrylamide gels in slabs.

CONTINUOUS FLOW
VERTICAL ROTATION
(CONCENTRIC CYLINDERS)

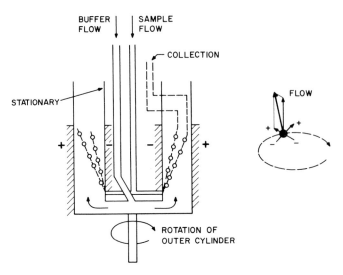

Fig. 11.7. Schematic diagram of the principle involved in a large-scale preparative continuous-flow electrophoresis apparatus.

glass tubes or plates (slabs). The sample is applied to the supporting medium which is in contact with the buffer, contained in the positive and negative electrode vessels. After electrophoresis, the separated components may be visualized by specific staining methods, immunodiffusion, autoradiography, etc. Because of their simplicity, these methods have found widespread use in biological and clinical laboratories.

11.3.4. Preparative apparatus for liquid media

In preparative techniques, the same basic design is utilized on a larger scale with provision for efficient cooling and fraction collection. Preparative-scale electrophoresis in liquid media can be performed either as a continuous-flow process or in batchwise operation.

Free-flow electrophoresis [72] and STAFLO electrophoresis [86] are examples of continuous-flow systems, as described above. Large-scale electrophoresis of proteins can also be performed continuously within two vertical, concentric cyliders (Fig. 11.7) [75]. The outer cylinder is rotated to provide stabilization of the upward-flowing charged species. The electrode vessels being perpendicular to the direction of flow, the electric field deflects the compounds, which are then collected through openings at the top of the apparatus.

Stabilization of the separated components in batch operation is achieved by a

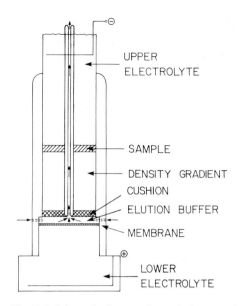

UPPER
ELECTROLYTE

SAMPLE

DENSITY GRADIENT

CUSHION

ELUTION BUFFER

MEMBRANE

LOWER
ELECTROLYTE

Fig. 11.8. Schematic diagram of a typical preparative density gradient electrophoresis apparatus.

density gradient (of sucrose, Ficoll, etc.), which is in contact with the electrolyte in the electrode vessels (Fig. 11.8). The density gradient is formed on top of a "cushion" of concentrated solution of the material used for the gradient. Cooling is provided by both an external water jacket and an internal "cold finger" of circulating water at 4°C. Thus, the cross-section of the separation compartment has the form of an annulus cooled on both sides. Fractions are collected after electrophoresis either by draining the density gradient from the bottom of the column or by pumping the gradient through a capillary tube in the center of the cold finger [87].

11.3.5. Electrolyte systems

The four previously specified electrolyte systems (Chap 11.3.1) are used not only for the electrophoretic separation of proteins but also for their physical characterization, i.e. the determination of the isoelectric point, molecular weight, and valence at a specific pH and ionic strength.

The characteristic pH vs. mobility curve of each protein determines its behavior in a particular type of electrophoresis. When at a given pH value the sum of all positive charges is equal to the sum of all negative charges—the net charge therefore being zero— the protein is at its isoelectric point (pI). At any other pH value, the net charge of the protein causes movement toward the negative or positive electrode. By using suitable buffer systems, proteins may be made to migrate with different velocities at a certain pH (CZE), can be stacked next to each other in sequence according to their constituent mobilities (ITP), or focused at their isoelectric point (IF). The combination of the above techniques to form two-dimensional maps can

References on p. B72

dramatically improve the resolution. In addition, the "sieving" effect of porous gels may be utilized to exploit the size differences that exist among various proteins. Furthermore, the interaction of proteins with an ionic detergent, such as sodium dodecyl sulfate (SDS), can be used to equalize the charges on the molecules, and thus, to produce separations according to size only.

11.3.6. Continuous-pH zone electrophoresis

When the pH of the buffer is the same throughout the system, i.e., in the positive and negative electrolyte reservoirs, the separation path, and the sample zone, the term continuous-pH zone electrophoresis (CZE) can be used to describe the method (Fig. 11.9). The pH is not supposed to change during the course of the electrophoresis experiment. Proteins exhibiting widely different net charge and, therefore, electrophoretic mobility at a specified pH are best separated in this system. Since most proteins are maximally charged at high pH values, alkaline buffers are usually favored for CZE separations. A number of buffer systems for CZE have been described with exactly determined pH, ionic strength, and conductivity [67]. These buffers cover the pH range 2.5–11 in 0.5 pH-unit intervals.

Cellulose acetate [26] and agarose gels [4] are the supporting media most commonly used in CZE. The resolution in these system is very low, due to the excessive spreading caused by diffusion during migration. Starch and polyacrylamide gels used as supporting media [70,71] considerably improve the resolution in CZE, primarily due to their sieving effects. Immunoelectrophoresis [56] is also a form of CZE, where after separation the proteins are allowed to diffuse against specific antisera to form immunoprecipitin arcs. A semi-quantitative estimation of each antigenic component can be performed by the technique of electroimmunodiffusion [57]. After separation by CZE, the antigenic proteins migrate electrophoretically in a

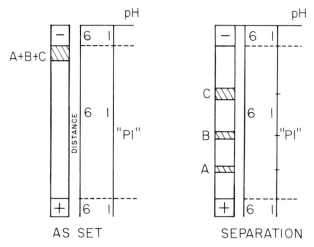

Fig. 11.9. Stages of continuous-pH zone electrophoresis.

direction perpendicular to their original path into a gel containing a uniformly distributed antiserum. The height of the immuno-loops formed gives a measure of the relative concentration of the proteins in the mixture.

11.3.7. Isotachophoresis

ITP is the process whereby proteins are separated into a series of consecutive zones, arranged in the order of their constituent mobilities [58,61,62,65]. The protein zones under constant current are in immediate contact with each other (i.e. "stacked") and exhibit differences in pH. The protein stack is formed between the leading and the terminating ions in the presence of a common counterion having the opposite sign of charge (cf. Chap. 9). The concentration of the leading ion governs the velocity at which the protein stack migrates and controls the ionic concentrations between the boundaries. This occurs because the velocity at which the terminating ion migrates is controlled by the migration velocity of the leading ion, i.e. the ion with the lowest constituent mobility is accelerated to the constituent mobility of the ion with the highest constituent mobility due to the increasing voltage gradient—front to back. Proteins within the stack are concentrated into thin zones, since the concentration of ions on either side of a moving boundary is fixed, as described by the Kohlrausch principle [65]. "Spacers" with mobilities intermediate between those of the species to be separated can be used to achieve separation of the protein zones within the stack. Carrier ampholytes intended for use in IF have been successfully employed as spacers [58].

As shown schematically in Fig. 11.10, the protein sample is placed between the upper and lower buffer systems. The upper buffer (Phase alpha) contains the terminating ion (Constituent 1) and the counterion (Constituent 6). The lower buffer (Phase beta) contains the leading ion (Constituent 2) and the counterion (Constituent 6). Constituent 1 can be a monovalent weak electrolyte. Any monovalent,

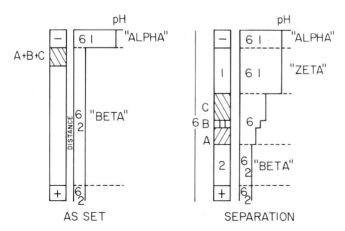

Fig. 11.10. Stages of isotachophoresis. 1 = Terminating ion, 2 = leading ion, 6 = counterion.

References on p. B72

divalent, or polyvalent weak electrolyte can be used as Constituent 2. Constituent 6, common to all phases, can be either an ion or a weak electrolyte. The sample is dissolved in the lower buffer along with "spacers", if they are to be used. Poly-acrylamide gel of low concentration is usually used as the supporting medium for these separations to allow unrestrictive migration of proteins. At equilibrium, after application of the electric field, all the components are stacked between the beta and the zeta phases. The stack migrates as a moving pH gradient of constant width. Phases alpha and zeta are identical in terms of constituents, but they are operation-ally distinguishable, because Phase zeta is formed inside the gel matrix, where mobilities of Constituents 1 and 6 may be slightly different from those in free solution. In its most popular form, ITP is used in disc electrophoresis as the stacking phase [66,67].

11.3.8. Multiphasic zone electrophoresis

MZE consists of three stages: (a) stacking, (b) unstacking, and (c) resolution [67]. The overall process involves a programed sequence of developing phases which are determined by the selected boundary conditions (see Fig. 11.11). The stacking phase (SSS) is the same as in ITP without spacers. The stack migrates out of the stacking gel and into the resolving gel during the unstacking phase. In the resolving gel, the components of the stack are subjected to sieving and, therefore, mobility retardation because of its higher concentration of acrylamide. The steady-state conditions which maintain the ITP stack are abruptly removed by this discontinuity. The resolving gel is made in the resolving buffer (gamma phase) which contains Constituent 3 and Counterion 6 and has a pH different (higher for anodic migration) from that of the beta phase. As the beta phase migrates into the resolving gel, a new phase, lamda, is generated. When Constituent 1 of the zeta phase enters the resolving gel, as the moving boundary (lamda/gamma) migrates further, its mobility changes so that it overtakes the protein components in the stack and migrates ahead of them. Thus, a new phase (pi) and a new moving boundary (pi/lamda) is generated through which

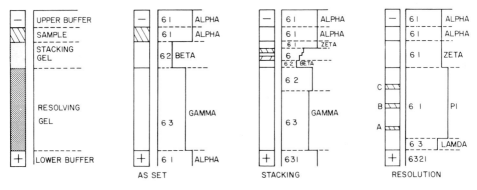

Fig. 11.11. Stages of multiphasic zone electrophoresis.

all the protein zones migrate. They are separated according to their individual velocities, directed by the pH of the pi phase, conductance, and applied current. Several factors, therefore, contribute to a high-resolution pattern of separated protein bands:

(a) The starting zone is thin, due to steady stacking;

(b) spreading due to diffusion is decreased, because the pore size of the resolving gel is smaller;

(c) in addition to charge separation, separation is according to molecular size, due to gel sieving effects. In fact, the size of globular proteins can be determined by utilizing this effect under controlled conditions of gel polymerization [88].

11.3.9. Electrophoretic retardation in polyacrylamide gels

The position of the front (pi/lamda boundary) in MZE can be visualized by using a tracking dye. The relative mobility value, R_F, is defined as the ratio of the distance traveled by a protein zone to that of the tracking dye. The retardation coefficient, K_R, of several proteins can be determined simultaneously by preparing gels with varying acrylamide concentrations (%T) but a constant percent ratio of crosslinking

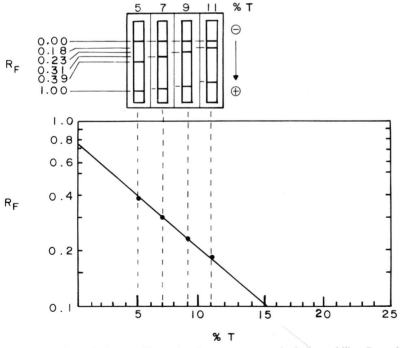

Fig. 11.12. Schematic diagram illustrating the measurement of relative mobility, R_F, values of a protein relative to the tracking dye ($R_F = 1.0$) at different percent acrylamide concentrations (%T). The plot of log R_F vs. %T gives the retardation coefficient, K_R, of a protein, which corresponds to the numerical value of the slope.

References on p. B72

(%C). This is accomplished by plotting the $\log R_F$ vs. %T for each species (Fig. 11.12). The numerical value of the slope of the straight line obtained from this plot corresponds to the retardation coefficient, K_R. The y-intercept gives the relative free mobility, Y_0. These two parameters give a measure of molecular size and molecular net charge, respectively. By utilizing a series of globular protein standards of known molecular radius, $\bar{R}$, a linear relation between the square root of K_R and $\bar{R}$ can be obtained. The molecular radius and corresponding molecular weight (MW) of an unknown globular protein can thus be determined by measuring its K_R and by referring to a plot of $\bar{R}$ vs. $\sqrt{K_R}$ [88]. Similarly a linear relationship can be obtained between K_R and MW for random-coil proteins (e.g., in SDS) [88,89]. The MW of dissociated subunits, rather than that of the native, undissociated protein, is determined in the case of polyacrylamide gel electrophoresis in the presence of SDS (SDS-PAGE). Thus, the number of subunits in a protein molecule can be estimated by electrophoretic retardation techniques with a combination of PAGE and SDS-PAGE data.

11.4. ISOELECTRIC FOCUSING

The isoelectric point (pI) of a protein is that pH at which it has no net charge. IF involves the electrophoretic migration of a protein in a pH gradient until it reaches the pH corresponding to its pI [90–92]. The protein will be concentrated in that region, because its net charge is zero. Any movement of the protein by diffusion away from its pI produces a restoration of electric charge and, consequently, electrophoretic movement toward the pI. Thus, in IF, the electric mass transport and the zonal diffusion combine to produce a steady state process. Proteins are focused at different regions of the pH gradient, and this results in their separation, if they have different pI values (Fig. 11.13). In addition, the pI of a particular protein

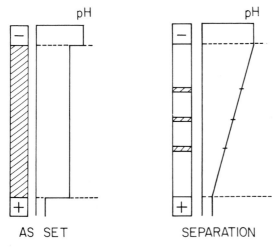

Fig. 11.13. Stages of isoelectric focusing.

species can be determined directly by measuring the pH at the focusing location. Density gradients and polyacrylamide or granular gels can be used as supporting media for IF.

Application of a constant-voltage electric field to a mixture of carrier ampholytes (low-molecular-weight amphoteric compounds of closely spaced pI values) causes the pH gradient to form. The carrier ampholytes themselves are focused and exhibit a continuous distribution of pI values, thus forming a pH gradient with sufficient buffering capacity and conductance. Since the net charge of a protein is governed by the pH of its surroundings, a protein placed at any position in the pH gradient will acquire a positive or negative charge and will migrate electrophoretically toward the electrode of opposite charge until it reaches the pI. Thus, for IF it is not necessary to apply the sample as a zone of infinite thinness.

At the steady state, the width of the zone of a focused protein depends on the diffusion coefficient, D, the magnitude of the electric field, E, and the electrofocusing parameter, p, which in turn depends on the steepness of the pH gradient [d(pH)/dx] and the slope of the pH–mobility curve of the protein at the pI [dM/d(pH)], where M is the electrophoretic mobility and x is the distance [93]. The standard deviation, σ, of the Gaussian zone is given by:

$$\sigma = (D/pE)^{1/2} \tag{11.12}$$

where

$$p = -[dM/d(pH)][d(pH)/dx] \tag{11.13}$$

Eqn. 11.12 shows that low values of D and high values of dM/d(pH), normally encountered in proteins, will produce small values of σ and, therefore, sharply focused zones. The zone can be further sharpened by increasing the steepness of the pH gradient [i.e., large values of d(pH)/dx] and increasing the field strength, E. However, Joule heating effects, which cause convective disturbances of the migration path in addition to possible denaturation of the protein, limit the magnitude of the field strength.

The highest resolution ever achieved in charge separation of proteins is offered by IF [58–64,92]. Separation of protein molecules differing by as little as 0.01 pH unit in their pI can be achieved by this method. Often, a difference of not more than one charged amino acid between two structures is enough for this separation. However, focusing may also detect other charge differences, which are not strictly related to the gross homogeneity of the proteins. Some of these factors are: postsynthetic modification of primary structures (e.g., deamidation), ligand binding, chemical modification, variation in nonprotein components (e.g., lipids, carbohydrates, and other prosthetic groups), association and dissociation phenomena, and changes of redox states of metalloenzymes. If these factors are kept in mind, interpretation of IF patterns may be rewarding, as they reveal the microheterogeneity of protein structures. At the present time, IF is being used in combination with other electrophoretic techniques, such as SDS-PAGE, to produce two-dimensional maps of separated components. Separation by charge (pI) is provided by the first dimension, while the second dimension involves separation by size differences relative to MW.

References on p. B72

Mixtures containing thousands of protein components can be characterized by the use of such techniques [94]. A few years ago, such high-resolution separations were unthinkable, but now they are increasingly applied to the investigation of a variety of biochemical problems with overwhelming success.

REFERENCES

1 G.K. Ackers, *Biochemistry*, 3 (1964) 723.
2 M. Joustra, in T. Gerritsen (Editor), *Modern Separation Methods of Macromolecules and Particles*, Wiley-Interscience, New York, 1969, p. 183.
3 H. Determann, *Gel Chromatography*, Springer, New York, 2nd Edn., 1969.
4 L. Fisher, *An Introduction to Gel Chromatography*, Elsevier, New York, 1969.
5 S. Hjertén, in A. Niederwieser and G. Pataki (Editors), *New Techniques in Amino Acid, Peptide, and Protein Analysis*, Ann Arbor Sci. Publ., Ann Arbor, MI, 1971.
6 W.W. Fish, J.A. Reynolds and C. Tanford, *J. Biol. Chem.*, 245 (1970) 5166.
7 T.C. Laurent and J. Killander, *J. Chromatogr.*, 14 (1964) 317.
8 G.K. Ackers, in H. Neurath and R.L. Hill (Editors), *The Proteins*, Vol. 1, Academic Press, New York, 3rd Edn., 1975, p. 2.
9 G.K. Ackers, in N. Catsimpoolas (Editor), *Method of Protein Separation*, Vol. 2, Plenum, New York, 1976, p. 1.
10 G.K. Ackers, *Methods Enzymol.*, 27 (1973) 441.
11 J. Porath, *Pure Appl. Chem.*, 6 (1963) 233.
12 G.K. Ackers, *J. Biol. Chem.*, 242 (1967) 3237.
13 W. Ostrowski and Z. Wasyl, *Biochim. Biophys. Acta*, 181 (1969) 479.
14 D. Rodbard, in N. Catsimpoolas (Editor), *Methods of Protein Separation*, Vol. 2, Plenum, New York, 1976, p. 145.
15 D. Rodbard, in N. Catsimpoolas (Editor), *Methods of Protein Separation*, Vol. 2, Plenum, New York, 1976, p. 181.
16 N. Catsimpoolas and J. Kenney, *J. Chromatogr.*, 64 (1972) 77.
17 N. Catsimpoolas and J. Kenney, *J. Chromatogr.*, 71 (1972) 573.
18 N. Catsimpoolas, *Anal. Biochem.*, 61 (1974) 101.
19 J.M. Sosa, *Anal. Chem.*, 52 (1980) 910.
20 E.A. Peterson, *Cellulosic Ion Exchangers*, North-Holland, Amsterdam, 1970.
21 E.A. Peterson and H.A. Sober, *Methods Enzymol.*, 5 (1962) 3.
22 E.A. Peterson and H.A. Sober, in A. Meister (Editor), *Biochemical Preparations*, Wiley, New York, 1961, pp. 8, 39, 43, 45, 47.
23 P. Cuatrecasas, in G.R. Stark (Editor), *Biochemical Aspects of Reactions on Solid Supports*, Academic Press, New York, 1971, p. 79.
24 P. Cuatrecasas and C.B. Anfinsen, *Ann. Rev. Biochem.*, 40 (1971) 259.
25 J. Porath and T. Kristiansen, in H. Neurath, R.L. Hill and C.-L. Boeder (Editors), *The Proteins*, Vol. 1, Academic Press, New York, 3rd Edn., 1975, p. 95.
26 C.R. Lowe and P.D.G. Dean, *Affinity Chromatography*, Wiley-Interscience, New York, 1974.
27 I. Parikh and P. Cuatrecasas, in N. Catsimpoolas (Editor), *Methods of Protein Separation*, Vol. 1, Plenum, New York, 1975, p. 255.
28 R. Axén, J. Porath and S. Ernback, *Nature (London)*, 214 (1967) 1302.
29 P. Cuatrecasas, *J. Biol. Chem.*, 245 (1970) 3059.
30 C.R. Lowe, M.J. Harvey, D.B. Craven and P.D.G. Dean, *Biochem. J.*, 133 (1973) 499.
31 B.H.J. Hofstee, *Anal. Biochem.*, 52 (1973) 430.
32 Z. Er-El, Y. Zaidenzaig and S. Shaltiel, *Biochem. Biophys. Res. Commun.*, 49 (1972) 383.
33 S. Hjertén, *J. Chromatogr.*, 87 (1973) 325.
34 S. Hjertén, in N. Catsimpoolas (Editor), *Methods of Protein Separation*, Vol. 2, Plenum, New York, 1976, p. 233.

35 A. Tiselius, S. Hjertén and O. Levin, *Arch. Biochem. Biophys.*, 65 (1956) 326.
36 E. Hayek and W. Stadelmann, *Angew. Chem.*, 67 (1956) 326.
37 W. Anacker and V. Stoy, *Biochem. Z.*, 330 (1958) 141.
38 G. Bernardi, M.-G. Giro and C. Gaillard, *Biochim. Biophys. Acta*, 278 (1972) 409.
39 F.E. Regnier and K.M. Gooding, *Anal. Biochem.*, 103 (1980) 1.
40 J.J. Kirkland, *J. Chromatogr.*, 125 (1976) 231.
41 J.J. Kirkland and P.E. Antle, *J. Chromatogr. Sci.*, 15 (1977) 137.
42 K. Unger, J. Schick-Kalb and K.-F. Krebs, *J. Chromatogr.*, 83 (1973) 5.
43 F.E. Regnier and R. Noel, *J. Chromatogr. Sci.*, 14 (1976) 316.
44 Y. Kato, K. Komiya, Y. Sawada, H. Sasaki and T. Hashimoto, *J. Chromatogr.*, 190 (1980) 305.
45 S.H. Chang, K.M. Gooding and F.E. Regnier, *J. Chromatogr.*, 125 (1976) 103.
46 O. Mikeš, P. Štrop, J. Zbrožek and J. Čoupek, *J. Chromatogr.*, 119 (1976) 339.
47 S. Ohlson, L. Hansson, P.-O. Larsson and K. Mosbach, *FEBS Lett.*, 93 (1978) 5.
48 M. Bier (Editor), *Electrophoresis*, Vol. 1, Academic Press, New York, 1959.
49 H. Bloemendal, *Zone Electrophoresis in Blocks and Columns*, Elsevier, Amsterdam, 1963.
50 I. Smith (Editor), *Chromatographic and Electrophoretic Techniques*, Vol. 2, Heinemann, London, 1968.
51 R.J. Wieme, *Agar Gel Electrophoresis*, Elsevier, Amsterdam, 1965.
52 D.J. Shaw, *Electrophoresis*, Academic Press, New York, 1969.
53 L.P. Cawley, *Electrophoresis and Immunoelectrophoresis*, Little, Brown, Boston, MA, 1969.
54 J.R. Whitaker, *Electrophoresis in Stabilizing Media*, Academic Press, New York, 1967.
55 J.R. Sargent, *Methods in Zone Electrophoresis*, British Drug Houses, Poole, 1965.
56 P. Grabar and P. Burtin (Editors), *Immunoelectrophoretic Analysis*, Elsevier, Amsterdam, 1964.
57 C.-B. Laurell (Editor), *Electrophoretic and Electro-Immunochemical Analysis of Proteins, Scand. J. Lab. Invest.*, Vol. 29, Suppl. 124, 1972.
58 N. Catsimpoolas (Editor), *Isoelectric Focusing and Isotachophoresis, Ann. N.Y. Acad. Sci.*, Vol. 209, 1973.
59 N. Catsimpoolas (Editor), *Isoelectric Focusing*, Academic Press, New York, 1976.
60 J.P. Arbuthnott and J.A. Beeley (Editor), *Isoelectric Focusing*, Butterworths, London, 1975.
61 P.G. Righetti (Editor), *Progress in Isoelectric Focusing and Isotachophoresis*, North-Holland, Amsterdam, 1975.
62 B.J. Radola and D. Graesslin (Editors), *Electrofocusing and Isotachophoresis*, De Gruyter, Berlin, 1977.
63 R.C. Allen and H.R. Maurer (Editors), *Electrophoresis and Isoelectric Focusing in Polyacrylamide Gel*, De Gruyter, Berlin, 1974.
64 N. Catsimpoolas and J.W. Drysdale (Editors), *Biological and Biomedical Applications of Isoelectric Focusing*, Plenum, New York, 1977.
65 F.M. Everaerts, J.L. Beckers and T.P.E.M. Verheggen, *Isotachophoresis: Theory, Instrumentation, and Applications*, Elsevier, Amsterdam, 1976.
66 L. Ornstein, in J.F. Frederick (Editor), *Gel Electrophoresis, Ann. N.Y. Acad. Sci.*, Vol. 121/2, 1964, p. 321.
67 A. Chrambach, T.M. Jovin, P.J. Svendsen and D. Rodbard, in N. Catsimpoolas (Editor), *Methods of Protein Separation*, Vol. 2, Plenum, New York, 1976, p. 27.
68 N. Catsimpoolas, in B.J. Radola (Editor), *Electrophoresis '79*, De Gruyter, Berlin, 1980, p. 503.
69 H.P. Chin, *Cellulose Acetate Electrophoresis: Techniques and Applications*, Ann Arbor Sci. Publ., Ann Arbor, MI, 1970.
70 A.H. Gordon, *Electrophoresis of Proteins in Polyacrylamide and Starch Gels*, Elsevier, Amsterdam, 1969.
71 H.R. Maurer, *Disc Electrophoresis*, De Gruyter, New York, 1971.
72 K. Hannig, in M. Bier (Editor), *Electrophoresis*, Vol. 2, Academic Press, New York, 1967, p. 423.
73 S. Hjertén, *Free Zone Electrophoresis*, Almqvist & Wiksells, Uppsala, 1967.
74 A. Kolin, in N. Catsimpoolas (Editor), *Methods of Cell Separation*, Vol. 2, Plenum, New York, 1979, p. 93.
75 A.R. Thomson, P. Mattock and G.A. Aitchison, in H.E. Sandberg (Editor), *Proc. Intern. Workshop Technol. Protein Separ. Improvement Blood Plasma Fract.*, DHEW Publication No. (NIH) 78-1422, Bethesda, MD, 1978.

76 C.W. Wrigley, in N. Catsimpoolas (Editor), *Isoelectric Focusing*, Academic Press, New York, 1976, p. 93.
77 N. Catsimpoolas, *Ann. N.Y. Acad. Sci.*, 209 (1973) 144.
78 B.A. Smith and B.R. Ware, in D.M. Hercules, G.M.H. Hieftje, J.R. Snyder and M.A. Evenson (Editors), *Contemporary Topics in Analytical and Clinical Chemistry*, Vol. 2, Plenum, New York, 1978, p. 29.
79 N. Catsimpoolas, in N. Catsimpoolas (Editor), *Methods of Protein Separation*, Vol. 1, Plenum, New York, 1975, p. 27.
80 A. Tiselius, *Trans. Faraday Soc.*, 33 (1937) 524.
81 L.G. Longsworth, *Chem. Rev.*, 30 (1942) 323.
82 B.R. Ware and W.H. Flygare, *Chem. Phys. Lett.*, 12 (1971) 81.
83 S. Hje tén, *Ark. Kemi*, 13 (1958) 151.
84 K. Hannig, H. Wirth, R.K. Schindler and K. Spiegel, *Hoppe-Seyler's Z. Physiol. Chem.*, 358 (1977) 753.
85 A. Kolin, *Proc. Nat. Acad. Sci. U.S.*, 46 (1960) 509.
86 R.D. Tippetts, H.C. Mel and A.V. Nichols, in D. Hershey (Editor), *Chemical Engineering in Biology and Medicine*, Plenum, New York, 1967, p. 505.
87 N. Catsimpoolas and A.L. Griffith, in N. Catsimpoolas (Editor), *Methods of Cell Separation*, Vol. 1, Plenum, New York, 1977, p. 1.
88 D. Rodbard, in N. Catsimpoolas (Editor), *Methods of Protein Separation*, Vol. 2, Plenum, New York, 1976, p. 145.
89 K. Weber and M. Osborn, in H. Neurath and R.L. Hills (Editors), *The Proteins*, Vol. 1, Academic Press, New York, 1975, p. 180.
90 A. Kolin, *J. Chem. Phys.*, 23 (1955) 407.
91 H. Svensson, *Acta Chem. Scand.*, 16 (1962) 456.
92 O. Vesterberg and H. Svensson, *Acta Chem. Scand.*, 20 (1966) 820.
93 H. Rilbe, in N. Catsimpoolas (Editor), *Isoelectric Focusing*, Academic Press, New York, 1976, p. 14.
94 P.H. O'Farrell, *J. Biol. Chem.*, 250 (1975) 4007.

Chapter 12

Lipids

ARNIS KUKSIS

CONTENTS

12.1. INTRODUCTION

Over the last decade, interest in lipids has passed from a curiosity about their composition revealed by advances made in chromatography to a genuine concern based on knowledge of their importance in biological structure and function. As a result, many laboratories have undertaken lipid analyses on a wide variety of tissue, cellular, and subcellular samples. With this interest has arisen the necessity of analyzing quantitatively minute amounts of lipids, and this frequently taxes the capabilities of most analytical systems. Recent developments in chromatography of lipids have therefore been directed towards increasing the sensitivity of the analytical methods and increasing the accuracy of data analysis. Both of these goals have been achieved to a large extent by microprocessor-controlled GLC and HPLC systems, which have allowed the realization of the full resolving power of these chromatographic methods.

TLC and GLC have provided the most successful routines, while HPLC has also made a significant contribution wherever it has been possible to combine the resolution with a sensitive detector. The literature on the application of these methods of chromatography to lipids is now so vast that it is impossible to give a comprehensive review of the subject in the space allowed. Therefore, I have limited the coverage of the topic to describing the more successful routines currently employed in quantitative analyses of glycero- and sphingolipids including molecular species, and to free fatty acids and prostaglandins from natural sources. Within the areas identified, I have cited either the references concerned with appreciable advances in technique or papers which are typical of the application of these techniques. In fairness to original authors I have included a number of references to work done prior to the 1975–1980 period covered here.

For a detailed coverage of various other aspects of the chromatography of lipids reference may be made to reviews of single chromatographic techniques, e.g., TLC [1–4], GLC [5–7], and HPLC [8,9], and to books [10–13] as well as general review articles [14–18]. A number of more specific reviews are cited in the text.

12.2. PREPARATION OF LIPID EXTRACTS

12.2.1. Extraction of lipids from tissues

There are two basic routines which yield essentially quantitative extraction of all lipid classes, when applied to homogenates of whole tissue or tissue subfractions.

The most popular extraction method is that described by Folch et al. [19], which employs a chloroform–methanol (2:1) mixture in a solvent–tissue ratio of 20:1. This method gives excellent recoveries for neutral lipids, the diacylglyc-erophospholipids and sphingolipids. Lysophospholipids are only partially recovered, and the more polar acidic phospholipids may be lost during backwashing with salt solutions and water [20]. Repeated extractions and limited backwashing, however, can yield also quantitative extractions of the lysolipids [21,22].

The second most popular extraction method is that of Bligh and Dyer [23], which effects a single-phase solubilization of the lipids with chloroform–methanol (1:1) in a solvent/tissue ratio 2:1. Subsequent partitioning of the extracts between chloroform and water results in losses of the more polar acidic phospholipids and lysophospholipids, as already noted for the procedure of Folch et al. The advantages and disadvantages of these methods of lipid extraction have been discussed in great detail by Nelson [20] and by Zahler and Niggli [24], who have proposed improvements relating to the purification of the initial extracts. Scott et al. [25] have improved the extraction of the more polar lipids by inclusion of 2% glacial acetic acid in the methanol phase of the chloroform–methanol system, while Arthur and Sheltawy [26] have substituted partitioning with chloroform–methanol–1 M HCl (4:2:3) for the same purpose. The plasmalogens would be expected to be lost during the latter procedure, as they are known to be destroyed during extraction of polyphosphoinositides, which requires the addition of dilute hydrochloric acid to the chloroform–methanol system [27] for complete recovery. Hanson and Lester [28] have shown that slightly alkaline mixtures of both ethanol–water and ethanol–diethyl ether–water at elevated temperatures effectively extract inositol-containing phospholipids from intact yeast cells.

Prostaglandins and other oxygenated fatty acids are effectively extracted with diethyl ether following acidification with dilute citric or formic acid [29,30]. Use of the organic acids obviates the need for extensive washing of extracts with water, and thus the loss of polar prostaglandins. Min et al. [31] have recommended the recovery of prostaglandins from 1 M phosphate buffers (pH 3.0) with benzene–dichloro-methane (9:1). Pace-Asciak et al. [32] have used chloroform–methanol as the initial extractant of tissue homogenates.

After filtration of the initial lipid extract, some 25–75% of the total mass of the extract may represent nonlipid contaminants. The chloroform–methanol extract can be freed of essentially all nonlipid material by chromatography on a dextran gel column [33]. This technique has been critically reviewed [34]. There is a general lack of simple and reliable methods for the isolation of gangliosides from tissue extracts. Most of them are based upon that of Folch et al., where total lipids are extracted with a chloroform–methanol mixture and the gangliosides are partitioned into the aqueous phase, which is then dialyzed. Randell and Pennock [35] have recently reviewed the numerous modifications of these routines and have proposed an improved, simple method for the extraction of gangliosides, based on the earlier techniques and including either a dialysis against Carbowax or the use of Millipore filter cones.

Phillips and Privett [36] have recently described a new method for the quantitative extraction of lipids from brain tissue with chloroform–methanol that eliminates secondary purification of the lipid extract by dextran gel chromatography or aqueous washing of the organic extract. Nonlipid substances that generally contaminate chloroform–methanol lipid extracts are separated by pre-extraction of the tissue with dilute (0.25%) aqueous acetic acid. When the residual tissue is extracted twice with 40 volumes of chloroform–methanol (1:1), ca. 97% of the lipid is recovered. A third extraction, which yields about 1% more lipid, is performed if the process is terminated at this stage in a shortened version of the method. The remainder of the lipid is recovered after treatment of the tissue with 1 N HCl by two additional extractions, the first with 40 volumes of chloroform–ethanol (1:2) and the second with 40 volumes of methanol. The method, applied to pig brain, gave a complete extraction of the lipids, including gangliosides, free of nonlipid material.

Hydrolytic enzymes, particularly phospholipase D, must be destroyed by heat treatment to avoid the formation of artifacts during the extraction of lipids from plant tissues. Transphosphatidylation of phosphatidylcholine with water yields phosphatidic acid [37,38]; and phosphatidylethanol [38] and phosphatidylmethanol [39] are produced upon extraction of fresh tissue with the corresponding alcohol.Phillips and Privett [40] have extracted lipids from immature soybeans without artifact formation. Their method, which involves pretreatment and extraction of the beans with hot dilute (0.25%) acetic acid, followed by chloroform–methanol extraction, provides complete recovery of the lipid, including highly polar glycolipids, free of nonlipid substances.

Post-mortem lipolytic activity during lipid extraction [41,42] and lipolysis due to cell disruption by freezing [43] is reflected in increased amounts of free fatty acids and lysophospholipids. Freezing and pulverization at dry-ice temperatures [41,44] appears to minimize lipolysis and generation of free fatty acids, but pretreatment of animal tissues with dilute acetic acid, as suggested by Phillips and Privett [40] for plant tissues, may be helpful.

12.2.2. Extraction of lipids from adsorbents

During preparative isolation, each lipid class is generally recovered from the adsorbent with the solvent system used for developing the chromatogram. Since the recovery usually also involves evaporation of the solvent, not all solvent systems are equally suitable. All neutral lipids may be recovered from silica gel with chloroform containing some methanol. Chloroform–methanol (2:1) has been used most frequently [16,42], but pure chloroform [45], diethyl ether [46] diethyl ether–methanol (4:1) [16] and chloroform–methanol (9:1) [45,47] have also been employed, depending on the nature of the neutral lipid to be recovered. Digalactosyldiacylglycerols may be recovered with chloroform–methanol–water–28% ammonia (100:50:6:1) [48]; gangliosides require chloroform–methanol (1:2) for recovery [49]. Phospholipids are usually recovered with chloroform–methanol–actic acid–water (50:39:1:10) [50], (25:15:4:2) [51], or (49:49:2:1) [28], or with chloroform–methanol–formic acid (1:1:2) [52].

Impregnation of silica gel with sodium carbonate, magnesium acetate, oxalic acid, or boric acid does not present any additional difficulties in recovering neutral or polar lipids. However, silica gel impregnated with silver nitrate requires treatment of the extraction solvents with a few drops of dilute (1.5 N) hydrochloric acid to precipitate the silver salt [53].

12.2.3. Sample protection

Since most common lipids contain fatty acids with one or more double bonds, care must be taken to avoid autoxidation of the sample during manipulating and storage. Autoxidation can be minimized by working with oxygen-free solvents and by performing all manipulations under nitrogen [42]. Purified lipid extracts may be stored in tightly closed vials at low temperatures ($-20°$C or lower) in the presence of inert solvents and gases for short periods [48,54]. Antioxidants such as 2,6-di-*t*-butyl-*p*-cresol (BHT), added to the extracting solvents at concentrations of $<0.005\%$ effectively prevent oxidative degradation of unsaturated lipids [55]. This antioxidant is easily removed by chromatography [34] but it should be remembered that BHT is eluted from a number of GLC liquid phases with the retention time of methyl myristate [48].

Tissue samples must be protected against the action of degradative enzymes; enzymatic hydrolysis may be significant during prolonged storage, even at $-20°$C [43]. Finally, the sample must be protected from contamination with lipids or other impurities in solvents, reagents, and on equipment.

12.3. SEPARATION OF LIPID CLASSES

All animal and plant tissues contain lipid mixtures of such complexity that no single analytical method is adequate for the isolation and identification of all components, and a combination of several techniques is required. A preliminary separation into two or three well-defined groups of lipid classes has been found to facilitate greatly the ultimate resolution of the mixtures into pure components and is now routinely applied whenever time and sample size permit. There are several excellent systems for this purpose, involving both column and thin-layer separations.

12.3.1. Separation of nonpolar and polar lipids

This separation is accomplished by means of adsorption or partition chromatography. It generally results in the separation of neutral lipids plus free fatty acids from phospholipids. Glycolipids, when present, may be recovered as a fraction of intermediate polarity.

12.3.1.1. Column chromatography

By far the most popular procedure for preliminary separation of total lipid extracts is chromatography on silicic acid columns according to Borgström [56] and

References on p. B130

Rouser and coworkers [57,58]. According to this routine, neutral lipids are isolated by elution with chloroform, glycolipids with acetone, and phospholipids with methanol. This method has been successfully applied to the separation of total lipid extracts from plants [59–61], microorganisms [62–64], and animal tissues [65–71]. Minor modifications have been found to be advantageous for special applications, e.g., the use of acetone–methanol (9:1) instead of acetone for the elution of the glycolipids [72,73], the initial elution of triacylglycerols with hexane [74,75], and the elution of phospholipids with chloroform–methanol (1:9) instead of methanol [76,77]. In many instances, a separate acetone fraction (glycolipids) was collected [78–83]. Total lipid extracts containing plasmalogens have been isolated by a silicic acid microcolumn method [84,85]. Detailed procedures for the isolation of glycolipids by column chromatography on Florisil have been described by various authors [12].

An alternate procedure to column adsorption chromatography for the separation of nonpolar and polar lipids is available in anion-exchange chromatography on either DEAE- or TEAE-cellulose. Rouser and coworkers [57,58] and Nelson [20] have detailed the practical aspects of column preparation and solvent selection for this purpose. The neutral lipids are eluted with chloroform, and the choline-containing phospholipids, cerebrosides, and glycosyldiacylglycerols with chloroform–methanol (9:1). Chloroform–methanol (2:1) containing 1% acetic acid will elute the ethanolamine phosphatides, but acetic acid is required to remove the phosphatidylserine. The acidic phospholipids are eluted with chloroform–methanol (4:1) containing 0.1 M acetate and ammonia. The ceramide polyhexosides may be removed as a separate fraction with chloroform–methanol (2:1), following the removal of the choline phosphatides and sphingomyelin. There have been numerous early [12] and recent [86,87] applications of the method.

Recently, HPLC has provided effective separation of polar and nonpolar lipids in small amounts. Both adsorption [88] and reversed-phase partition [89] chromatography have been utilized. A preparative application has also been described [90].

12.3.1.2. Thin-layer chromatography

It is convenient to separate small samples into nonpolar and polar lipid classes by TLC [14–17,91]. A phospholipid solvent system carries the neutral lipids near the solvent front, while a neutral lipid system leaves the phospholipids and other polar lipids at the origin. The neutral and the polar lipid fractions may be recovered from the TLC plate by elution with appropriate solvent mixtures. In principle, any one of the neutral lipid solvent systems is suitable for the isolation of the phospholipids and other polar lipids, as is any one of the polar lipid solvent systems for the isolation of the neutral lipids and other nonpolar lipids as a group. In practice, those solvent systems are preferred which are most easily evaporated. For the isolation of nonpolar lipids the most commonly used solvent systems are chloroform–methanol–water (65:25:4) [92,93] and chloroform–methanol–28% ammonia (13:7:1) [34,58], and for the recovery of the polar lipids light petroleum–diethyl ether (7:3 to 9:1) [58,94], light petroleum–diethyl ether–acetic acid (80:20:1)

[51,95], and heptane–isopropyl ether–acetic acid (60:40:3) [96]. This type of separation is not very well suited for the isolation of the glycolipids as a subclass from a total lipid extract. In the neutral lipid solvents the glycolipids remain at the origin along with the polar lipids, while in polar solvent systems the glycolipids overlap with one or more classes of the phospholipids when the latter are also present. However, it has been possible to isolate by TLC various brain galactolipids as separate subfractions of the total lipid extract [97]. Fischer and coworkers [98,99] have described a large number of other TLC systems for the separation of glycolipids and glycerophospholipids.

12.3.2. Isolation of individual nonpolar lipid classes

The isolation of individual classes of neutral lipids and free fatty acids may be achieved by refining the chromatographic systems employed in the initial resolution of nonpolar and polar lipids. This is usually accomplished by means of TLC, but in many instances adsorption and partition column chromatography may also be employed, especially when large amounts of material are involved.

12.3.2.1. Column chromatography

The separation of the neutral lipids into individual lipid classes by adsorption column chromatography is usually accomplished by using either silicic acid or Florisil as adsorbent. Carroll [100] has described in great detail the use of both silicic acid and Florisil columns for neutral lipid separation and has called attention to various subtle differences between the two adsorbents which may be exploited for practical separations. The order of elution of different neutral lipid classes is about the same for both adsorbents, but Florisil retains the monoacylglycerols and free fatty acids much more strongly than does silicic acid. A linear gradient of diethyl ether in hexane resolves the neutral lipids into hydrocarbons (100% hexane), cholesteryl and wax esters (1% ether in hexane), triacylglycerols (5%), free fatty acids (8%), free cholestrol (15%), diacylglycerols (15%), and monoacylglycerols (100% diethyl ether). On silicic acid, the cholesteryl esters overlap the wax esters [101], but they can be separated on magnesium oxide [102,103]. Free cholesterol may overlap the X-1,2-diacylglycerols, but they can be resolved on silicic acid columns impregnated with boric acid [104]. Glyceryl ethers are not sufficiently different in polarity from the corresponding glyceryl esters to be resolved by adsorption columns [105]. Alternatively, cholesteryl esters may be eluted with 30% benzene in hexane and triacylglycerols with 100% benzene [106]. Wax esters and triacylglycerols can also be resolved by a gradient of diethyl ether in light petroleum on a silicic acid column previously deactivated with 10% water [107]. An initial elution with hexane, followed by a linear gradient of 0–25% diethyl ether in hexane has been used for the isolation of dolichol esters from silicic acid columns [108].

Although adsorption chromatography tends to separate lipids into classes having the same number and type of polar groups, the separations are affected to some extent by the chain length and degree of unsaturation of the fatty acid substituents.

Thus, synthetic triacylglycerols of short and long chain length may be resolved by means of silicic acid column chromatography [109] as may be milk fat triacylglycerols containing short- and long-chain fatty acids [110,111]. Improved separation of short-, medium- and long-chain triacylglycerols of milk fat is obtained by a 0–9% gradient of diethyl ether in hexane [112,113]. The presence of unsaturation has a tendency to retard the elution of lipids from silicic acid, and this effect is most pronounced during elution with the less polar solvents. Thus, squalene is eluted after squalane [114], and so are unsaturated esters of cholesterol after saturated esters of cholesterol [115].

Monoacylglycerols and diacylglycerols are isomerized on adsorption columns, but this can be prevented by incorporating 10% boric acid (w/w) into the adsorbents [104,116].

The presence of additional oxygen functions on fatty acids greatly increases their affinity for the adsorbent and results in an effective resolution between oxygenated and ordinary fatty acids. Thus, while the methyl esters of stearate and palmitate are eluted with 5% diethyl ether in benzene, the hydroxy fatty acids require diethyl ether and methanol or 40% ethyl acetate in benzene for elution [117,118]. In other instances, hydroxylated and ordinary fatty acids have been resolved with a gradient of diethyl ether in hexane [119,120]. Methyl dihydroxypalmitate has been eluted with diethyl ether–hexane–methanol (10:5:1) [121].

Silicic acid column chromatography has been extensively utilized in the separation of prostaglandins according to their oxygen content. For chromatography of prostaglandins as free fatty acids, the use of acid-washed silica gel is recommended [122] and for work with Florisil a hot acid wash is absolutely necessary [123]. Several different solvent systems may be used for the resolution of the free acids, including increasing concentrations of ethyl acetate in hexane [124], in cyclohexane [125], or in benzene [126,127] as well as of methanol in ethyl acetate–benzene [128] and methanol in chloroform [129, 130]. Depending on the source of the prostaglandins, a stepwise elution [131] may yield fractions containing PGA_1, PGA_2, PGB_1, and PGB_2 [ethyl acetate–benzene (3:7)]; PGE_1, PGE_2, and PGE_3 [ethyl acetate–benzene (3:2)]; 19-hydroxy derivatives of PGA_1, PGA_2, PGB_2, as well as $PGF_{1\alpha}$ and $PGF_{2\alpha}$ [ethyl acetate–benzene (4:1)]; and more polar compounds [ethyl acetate–methanol (1:1)].

A comparable group separation is obtained [129,130] with chloroform–methanol (49:1) (PGA), chloroform–methanol (96:5) (PGE), and chloroform–methanol (9:1) (PGF, 19-hydroxy PG). The ethyl acetate–benzene and ethyl acetate–methanol systems with minor modifications are also applicable to the resolution of the prostaglandin methyl esters [132]. Silicic acid column chromatography has been used in characterizing metabolites of $PGF_{1\alpha}$ in the guinea pig [133], $PGF_{2\alpha}$ in the rabbit [134], and PGF_2 in the rat [131]. A rapid procedure for the isolation of the PGA, PGE, and PGF series of prostaglandins on a microcolumn of acid-washed Florisil has been described by Bansbach and Love [123]. The column is eluted with benzene–ethyl acetate (3:2) (neutral lipids); benzene–ethyl acetate–methanol (30:20:1) (PGA and PGB); benzene–ethyl acetate–methanol (15:10:1) (PGE) and benzene–ethyl acetate–methanol (3:2:1) (PGF). PG, PG analogs and thromboxane

B_2 have been eluted from silicic acid columns with diethyl ether–benzene (7:3), ethyl acetate–methanol (9:1), and methanol, respectively [135].

Triacylglycerols containing epoxy [136], keto [137], hydroxy [138], and acyloxy [139] fatty acids may be separated from the corresponding normal-chain glyceryl esters and autoxidized standards of glyceryl esters may be purified by silicic acid adsorption chromatography [137,138]. Glyceryl ethers containing oxygenated fatty acids are retained longer than the ethers of normal fatty acids on silicic acid or Florisil columns [140].

Alumina is rarely used as an adsorbent for column chromatography of neutral lipids because it may hydrolyze glyceryl esters. However, alumina has been used as an adsorbent for column chromatography of mercuric acetate adducts [13]. Thus, the hydrocarbon fraction is readily isolated by eluting such a column with hexane [141].

A major drawback to the use of adsorption columns for efficient separation and isolation of pure lipid classes is the difficulty of detecting the substances in the effluent. Usually, the eluted fractions are either evaporated to dryness and weighed, or the composition of the eluent is monitored by TLC. Continuous monitoring of the column effluent would be preferred, but it has been achieved in only a few instances [142,143].

12.3.2.2. Thin-layer chromatography

TLC on a variety of silica gels gels provides neutral lipid class separations that are essentially identical to those obtained on silicic acid and Florisil columns. Preparative TLC [14–17] is frequently employed in place of column chromatography, because it offers greatly increased speed, increased resolution, effective detection of minor constituents, and convenient access to all components. Most separations of complex mixtures of neutral lipids and free fatty acids are still carried out by means of the mixtures of petroleum ether (b.r. 60–70°C)–diethyl ether–acetic acid (90:10:1 to 35:15:1) originally proposed by Mangold and Malins [144]. In many instances, subsequent workers have substituted *n*-pentane, *n*-hexane, or *n*-heptane for the petroleum ether and formic acid for acetic acid without materially affecting the nature of the resolution. These neutral lipid systems retain the monoacylglycerols and phospholipids at the origin while carrying the least polar hydrocarbon fraction near the solvent front and resolving more or less completely all neutral lipid fractions of intermediate polarity. The hydrocarbons are followed in descending order by cholesteryl esters, fatty acid methyl ethers, triacylglycerols, free fatty acids, diacylglycerols and free cholesterol [145]. Sharaf et al. [146] have obtained increased resolution of a complex mixture of neutral lipids by successive developments with hexane (to 19 cm), benzene (to 19 cm), and hexane–diethyl ether–acetic acid (70:30:1) (to 9 cm). Under these conditions, separate bands are obtained for squalene, cholesteryl esters, wax esters, diol esters and methyl esters, glyceryl ether acetonides, free fatty acids, free cholesterol, and diacylglycerols. Van der Vusse et al. [147] have suggested developing the plate with chloroform–methanol–water–acetic acid (10:10:1:1) until the lipid front has reached a level of 1 cm above the site of sample application prior to resolving the neutral lipids with hexane–diethyl ether–acetic acid (240:50:3).

References on p. B130

Effective separations of 1,3- and 1,2-diacylglycerols and free cholesterol were achieved with diethyl ether–isooctane (4:1) [148], with chloroform–acetone–methanol (96:4:1) [66,149] and with chloroform–methanol–water (65:25:4) followed by chloroform–hexane (3:1) [150]. Mancha et al. [151] and Oo and Stumpf [152] have resolved mono-, di-, and tricylglycerols and free fatty acids with diethyl ether–benzene–ethanol–acetic acid (400:500:20:1).

Impregnation of silica gel plates with borate prevents the isomerization of both monoacylglycerols and diacylglycerols and effects the resolution of their isomers [153]. Thus, isomeric monoacylglycerols are readily resolved on 5–10% borate plates with chloroform–acetone (24:1) as the developing solvent. A more complicated solvent system for the purification of 2-monoacylglycerols consists of chloroform–acetone–methanol–acetic acid (170:25:5:1) [154]. Renkonen [155] has separated neutral lipid classes on 0.1 M borate plates with chloroform–methanol–3.5 M ammonia (65:35:8) and Lamb et al. [156] have accomplished similar separations with hexane–diethyl ether–acetic acid (75:25:2). Isomerization of acylglycerols on TLC plates may be also prevented by incorporating trimethylborate in the developing solvent [157].

Isomerization of the partial acylglycerols on the silica gel is best avoided by derivatizing them prior to chromotography. The mono- and diacylglycerols may be converted into acetates or the t-butyldimethylsilyl ethers. Both of them are chromatograpically stable and may be separated from other neutral lipids in a variety of neutral lipid solvent systems [16]. The acetates of the mono- and diaglycerols behave like short-chain triacylglycerols. The triacylglycerols and diacylglycerol acetates of short- and long-chain fatty acids have been resolved with petroleum ether–diethyl ether (4:1) [158–160]. Improved resolution of triacylglycerol fractions containing short-chain and long-chain acids may be obtained by two developments of 18 cm in petroleum ether–diethyl ether–acetic acid (80:20:1), followed by one development of 2 cm in diisopropyl ether–acetic acid (24:1) [161]. The carbonates of 1- and 3-monoacylglycerols have been purified by TLC using 1,2-dichloroethane as the solvent [162].

In most of the above solvent systems the alkyl and alkenyl derivatives of acylglycerols overlap the pure acylglycerols of corresponding chain length. Myher [16] has recently surveyed the solvent systems applicable to the resolution of neutral acylglycerols, alkylglycerols, and alkylacylglycerols. Triacylglycerols and their alkyl ether analogs may be completely resolved with hexane–diethyl ether (19:1) [163]. Effective resolution of 1-alk-1-enyl-2-acylglycerol acetates, 1-alkyl-2-acylglycerol acetates and 1,2-diacylglycerol acetates has been obtained with hexane–diethyl ether (1:1) followed by a second development in toluene [164,165], with benzene–hexane–diethyl ether–acetic acid (45:50:5:1), or with benzene [166]. Besides the acetates, other derivatives of diradylglycerols (e.g., t-butyldimethylsilyl ethers) may be used to separate the alkenylacyl-, alkylacyl-, and diacylglycerols in comparable solvent systems [167]. Free cholesterol may be separated from diacylglycerols and alkylacylglycerols with chloroform–methanol–acetic acid (98:2:1) [163] or with heptane–isopropyl ether–acetic acid (15:15:1) following acetylation of the mixture [16]. A separation of 1,2- and 1,3-isomers of diacylglycerols and their ether analogs can be prevented by the use of hexane–diethyl ether–28% ammonia (40:60:1) [163]. The isomeric monoalkylglycerols are readily resolved on 10% borate plates with chloroform–acetone (24:1) [153]. However, in this system, the alkyl ethers run faster than the corresponding mono- and diacylglycerols [168]. The borate system also yields an excellent separation of 1,3- and 1,2-alkylacylglycerols. The elution order of the 1-alkyl- and 2-alkylglyceryl ethers is reversed on arsenate-impregnated silica gel, when chloroform–ethanol (9:1) is used as a solvent [163]. A separation of 1- and 2-monoalkylglycerols should be possible on plain silica gel with the 12-component solvent system of Kuntz [169]. The acidic solvents, which isomerize the acyl esters, do not affect the alkyl ethers.

The presence of additional functional groups, such as hydroxyl groups, retard the migration of the fatty acids and of the acylglycerols containing them in relation to the corresponding ordinary fatty acids and their esters. The epoxy, keto, hydroxy, and dihydroxy derivatives of methyl stearate can be readily separated from the ordinary fatty acid esters with hexane–diethyl ether (17:3) as the solvent system [170,171]. Petroleum ether–diethyl ether–acetic acid mixtures (80:20:1 to 90:10:1) have given satisfactory resolution of ordinary and α-hydroxy fatty acids [172]. Position isomers of mono- and dihydroxy fatty acids have also been resolved by TLC with diethyl ether–hexane (4:1) as the solvent [173]. Matsuda et al. [174] have separated 9- and 13-hydroxystearates by preparative TLC with diethyl ether–petroleum ether (1:1) or hexane–diethyl ether (3:2). The *threo*- and *erythro*-isomers of vicinal dihydroxyesters can be resolved on silica gel impregnated with boric acid [175]. Other examples are discussed in Chap. 12.4.1.

The fatty acids are also subject to resolution based on chain length. Christiansen et al. [176] purified erucic acid by TLC with petroleum ether–diethyl ether–acetic acid (113:20:1), and Litchfield et al. [177] isolated a new C_{30} fatty acid from a marine sponge with hexane–diethyl ether (19:1) as the solvent. The mycolic acids, which are high-molecular-weight 3-hydroxy acids with a long alkyl branch at C-2, are readily separated from the common fatty acids and from mycolic acids containing carbonyl, carboxyl, and methoxy groups by a two-step development: development to 8 cm with petroleum ether–acetone (7:3) is followed by development to 15 cm with petroleum ether–diethyl ether (17:3) [178]. Complex mixtures of the mycolic acids were resolved by means of two-dimensional TLC. Best results were obtained by triple development with petroleum ether–acetone (19:1) in the first dimension, followed in the second dimension by either single development with a 97:3 mixture of toluene–acetone or a triple development with a 99:1 mixture. With various methanolyzates of bacteria containing mycolic acid, separate spots were obtained for methyl mycolate, methoxymycolate, ketomycolate, ω-carboxymycolate, nonhydroxylated fatty acid methyl esters, as well as for numerous unknown lipids [178].

The separation of oxygenated fatty acids from ordinary fatty acids is of particular interest in the isolation of prostaglandins. The systematic studies of Green and Samuelsson [179] and of Anderson [180] have provided a large variety of solvent systems. Daniels [122] has reviewed these systems along with a few others, developed more recently. The most frequently used solvent system for the group separation of

PGE, PGF$_\alpha$, and PGF$_\beta$ is benzene–dioxane–acetic acid (20:20:1). The compounds with three hydroxyl groups (PGF) are more strongly retained than the compounds with two hydroxyl groups and one keto group (PGE), which in turn are more strongly retained than those with one hydroxyl group and one keto group (PGA and PGB). The number of double bonds has no significant effect on the elution order under these conditions. Excellent resolution of most prostaglandins and their methyl esters is also obtained with chloroform–methanol–acetic acid (18:1:1 to 8:1:1) [181,182]. These solvent systems allow the isolation of prostaglandins from relatively crude lipid extracts, as they either retain the prostaglandins close to the origin or allow them to migrate close to the solvent front. Group separation of prostaglandin A and prostaglandin B compounds is not possible on plain silica gel, but by impregnating the plates with ferric chloride and developing them with a mixture of ethyl acetate–acetic acid–hexane (30:1:19) PGA$_1$ and PGB$_1$ are easily separated [183]. Prostaglandins of various chain lengths have been separated by developing the plates twice with chloroform–methanol–acetic acid–water (450:40:5:4) [184]. This solvent system has been especially widely employed in recent work [29,186–191].

Separations of prostaglandins in the form of free acids and methyl esters have also been obtained with systems made up of ethyl acetate–benzene–formic acid (25:25:1) (PGA$_2$) and ethyl acetate–benzene–formic acid (80:20:1) (PGE$_2$) [192]. Daniels [122] has tabulated the R_F values for a large number of metabolites of natural and synthetic prostaglandins in the above systems. Lapetina and Cuatrecasas [193] have reported very good separations of arachidonic acid, 12L-hydroxy-5,8,10,14-eicosatetraynoic acid, 12L-hydroxy-5,8,10-heptadecatrienoic acid, thromboxane B$_2$, and phosphatidic acid with a system of ethyl acetate–trimethylpentane–acetic acid–water (9:5:2:10). All other phospholipids remained at the origin. Vick et al. [194] have reported the TLC isolation of an oxygenated cyclic fatty acid from a flax-seed extract by chloroform–acetic acid (50:1). Hamberg et al. [195] and Graff et al. [191] have used TLC to analyze the cyclooxygenase reaction products before and after purification on silicic acid columns. The separations were made by developing the plates with ethyl acetate–trimethylpentane–petroleum ether–acetic acid (500:500:200:3) and ethyl acetate–trimethylpentane–acetic acid (500:500:3) at 4°C. Hamilton and Tobias [196] and Graff et al. [191] have used isopropyl ether–2-butanone–acetic acid (50:40:1), while Nugteren and Hazelhof [197] and Graff et al. have used ethyl acetate–acetic acid (100:1) for the same purpose. Matsuda et al. [174] have separated 9- and 13-hydroperoxides with isooctane–diethyl ether–acetic acid (50:50:1) as the developing solvent. The methyl esters of fatty acids and dimethylacetals of corresponding chain length may be separated with 1,2-dichloroethane, followed by diethyl ether–water (200:1) [198].

The presence of oxygenated fatty acids in the acylglycerol molecules retards the migration of these esters in relation to the corresponding nonoxygenated acyl esters [199]. Mikolajczak and Smith [200] and Smith et al. [201] have used petroleum ether–diethyl ether (2:3) to resolve triacylglycerols with up to three hydroxyl groups per molecule, as well as their acetoxy derivatives. Yazicioglu et al. [202] have recently resolved the triacylglycerols of ordinary, epoxy, and hydroxy fatty acids by TLC, using petroleum ether–diethyl ether–acetic acid (35:15:1) as the developing solvent.

A number of oxygenated acylglycerol ethers have also been resolved by TLC [203,204].

The multiacylglycerols or estolides containing four or more ester groups are not easily separated from common triacylglycerols by TLC, but separations of estolide triacylglycerols by multidevelopment TLC have been reported [205]. Payne-Wahl et al. [206] have reexamined the TLC migration characteristics of synthetic multi-acylglycerols on silica gel by multiple development with benzene. Of the estolides examined, the tetraacylglycerol migrated furthest, followed by pentaacyl and then by hexaacylglycerols. The migration order on silica gel layers was similar to the order of elution from the μPorasil HPLC columns: increasing the number of acyl groups retarded migration, while increasing the chain length accelerated migration. Although the trivernolin content of the epoxy oils can be determined by TLC, the procedure is time-consuming [207].

Oette and Tschung [208] have used TLC to resolve the natural mono-, di-, and triacylglycerols and their 2-N-acylglycerol analogs on 45-cm long plates. The plates were first developed for 25 cm with chloroform–acetone–acetic acid (47:2:1) and then for 40 cm with petroleum ether–diethyl ether (9:1). In all instances, the 2-N-acylglycerol analogs moved much more slowly than the corresponding natural acylglycerols.

Fears et al. [209] have separated triacylglycerols containing natural and unnatural synthetic fatty acids (e.g., 4-benzyloxybenzoate and other aromatic acids) with petroleum ether–diethyl ether–acetic acid (70:30:1). Hancock et al. [210] have employed TLC with neutral lipid solvents for the separation of synthetic pseudotri-acylglycerols (tris-homoacyl derivatives of cyclopentane 1,2,3-triols).

12.3.3. Isolation of individual phospholipid classes

The phospholipid fractions recovered from the initial lipid group separation may be resolved into individual phospholipid classes by further column or thin-layer chromatography. Not all of the chromatographic systems are equally well suited for this purpose and a choice must be frequently made between representative recovery of the desired components and the yield of the product, as well as the efficiency of the separation.

12.3.3.1. Column chromatography

Adsorption chromatography on columns does not allow a resolution of the entire spectrum of phospholipid classes, but it can be effectively employed for isolating the major phospholipids in fairly pure condition, as well as for purifying phospholipid classes prepared by synthetic methods. An effective separation of the neutral phospholipids may be obtained by silicic acid column chromatography [57]. Chloroform–methanol (4:1) plus 1% water elutes phosphatidylcholine, while chloroform–methanol (4:1) plus 1.5% water elutes sphingomyelin. Methanol plus 2% water elutes lysophosphatidylcholine and any oxidation products present. The silicic acid column is also suitable for resolving phosphatidylethanolamine [eluted with chloroform–

methanol (4:1)] and phosphatidylserine (eluted with methanol), provided the adsorbent has been previously treated with ammonia [211]. A full account of the experimental details of both techniques may be found in a review [58]. Similar solvent systems have yielded separate fractions for cardiolipin, phosphatidylethanolamine plus phosphatidylserine, phosphatidylinositol, phosphatidylcholine, sphingomyelin, and lysophosphatidylcholine [105,212].

Crawford and Wells [213] employed silicic acid chromatography to separate phosphatidylcholine and sphingomyelin by elution with chloroform–methanol (6:1, 3:1, and 1:1). Phosphatidylcholine was eluted as a pure compound with chloroform–methanol (3:1), and phosphatidylethanolamine was recovered with chloroform–methanol (8:1). Effective separations of the neutral and acidic phospholipids on silicic acid may be obtained by elution with a discontinuous gradient of methanol in chloroform [214–216]. Cardiolipin and phosphatidylglycerol are eluted with 25–50% methanol in chloroform [216], phosphatidylethanolamine with 80% methanol in chloroform, and phosphatidylcholine plus sphingomyelin with methanol [214]. Silicic acid columns have also been utilized for a large-scale isolation of lysophospholipids. Elution with chloroform–methanol–water (25:75:4) gave 1-O-alkyl- and 1-O-acylglycerophosphorylethanolamine, while chloroform–methanol–2 N ammonia (25:75:3) eluted lysophosphatidylcholine and sphingomyelin [217]. Chakrabarti and Khorana [218] have employed silicic acid for the purification of synthetic phospholipids containing photosensitive groups (e.g., 4-azido-2-nitrophenyl), while Stoffel and Michaelis [219] and Stuhne-Sekalec et al. [220] have used it to purify phospholipids containing fatty acids, labeled with the fluorescent anthracene groups and the free radical nitroxyl groups, respectively.

Alumina is rarely used as an adsorbent for column chromatography of complex lipids because of its basic properties. Nonacidic phospholipids may be recovered with mixtures of 1:1 [67] or 2:1 [221] chloroform–methanol, and monophosphoinositides by use of ethanol–chloroform–water (5:2:2) [222], but for the recovery of other acidic phospholipids it is necessary to add ammonium salts to the solvent [223].

Some of the phospholipid classes may be obtained in a relatively pure state by elution of DEAE- or TEAE-cellulose columns with chloroform–methanol mixtures and glacial acetic acid [57,58]. This technique has been extensively utilized for the preliminary isolation of different phospholipid classes [20,76,224]. According to the original procedure, phosphatidylethanolamine is eluted with chloroform–methanol (7:3) and phosphatidylserine with glacial acetic acid. Rooney et al. [76] have used DEAE-cellulose in the acetate form to isolate phosphatidylcholine and sphingomyelin by elution with chloroform–methanol (4:1) as well as the acidic phospholipids, phosphatidylinositol and phosphatidylglycerol, by elution with chloroform–methanol–28% ammonia–0.05 M ammonium acetate (180:720:9:3.46). Likewise, Okano and coworkers [224–226] eluted phosphatidylcholine and phosphatidylethanolamine with methanol, phosphatidylserine with acetic acid, and phosphatidylinositol plus phosphatidylglycerol with chloroform–methanol (2:1), saturated with ammonia containing 50 mM ammonium acetate. Crawford and Wells [213] employed a DEAE column to separate choline- and ethanolamine-containing phospholipids by using chloroform–methanol (12:1) to elute phosphatidylcholine and sphingomyelin and chloroform–methanol (9:1 and 4:3) to elute phosphatidyl-ethanolamine. Other phospholipid classes, phosphatidylcholine and lysophosphatidylcholine, have been resolved and purified by chromatography on Sephadex LH-20 by stepwise or gradient elution with chloroform–methanol [227]. Schacht [228] has purified polyphosphoinositides by chromatography on a column of neomycin, immobilized on glass beads.

Recent developments in HPLC have shown that silicic acid columns have great possibilities for separating phospholipids into individual classes, but the quantities that can be effectively resolved are small, and truly preparative separations may not be practical. Jungalwala et al. [229] have resolved phosphatidylethanolamine, phosphatidylserine, lysophosphatidylethanolamine, and ethanolamine plasmalogens as the biphenylamides by HPLC on a MicroPak SI-100 column with dichloromethane–methanol–15 M ammonia (92:8:1 and 80:15:3). The derivatives were detected at 280 nm. Subsequently, Jungalwala et al. [230] were able to obtain good separations of sphingomyelin and phosphatidylcholine on a silica gel column with acetonitrile–methanol–water (65:21:14) as the eluent and detection at 210 nm. Similar HPLC methods but with FID have been described by Kiuchi et al. [231] and Rainey and Purdy [232]. The latter reported a successful separation of all the common phospholipid classes on a 180-cm Corasil II column with a solvent mixture of chloroform–methanol–ammonia (500:359:70). The best solvent ratio for the separation of phosphatidylcholine and sphingomyelin was derived by a two-dimensional simplex procedure, capable of expansion and contraction. The simplex method, frequently used in analytical chemistry [233], is a statistical design for finding the optimum of a response surface. A more effective separation of the common phospholipids has been obtained by HPLC on a LiChrosorb Si-60 column in combination with a gradient of *n*-hexane–2-propanol–water (6:8:0.75 to 6:8:1.4) and detection at 200 nm [234,235]. However, phosphatidylcholine and long-chain sphingomyelins were not resolved. The lipids in extracts of human erythrocyte membranes were eluted in the following order: cholesterol, phosphatidic acid, phosphatidylethanolamine, lysophosphatidylethanolamine, phosphatidylinositol, phosphatidylserine, phosphatidylcholine and long-chain sphingomyelins, lysophosphatidylcholine, and short-chain sphingomyelins.

Blom et al. [236] have also used a LiChrosorb Si-60 column for separating some 10 to 12 different phospholipid classes with chloroform–propanol–acetic acid–water (20:22:1:2, 40:44:2:7, and 10:11:1:2). The detection of the components was facilitated by the use of radioactive phosphorus as a marker, the radioactivity being monitored continuously by means of a flow-through cell. Effective separations were obtained (in order of increasing retention time) for: phosphatidic acid, diphosphatidylglycerol, phosphatidylethanolamine, phosphatidylglycerol, lysophosphatidylethanolamine, phosphatidylserine, phosphatidylinositol, phosphatidylcholine, sphingomyelin, diphosphatidylinositol. The results indicate that HPLC yields better phospholipid class separations than conventional silicic acid column chromatography. A critical factor is the flowrate, which should be high enough to establish an effective gradient but not so high that resolution is impaired. The choice of ^{32}P as a

References on p. B130

marker allowed the detection of minor compounds that are not detectable in total phospholipid extracts with conventional detection systems. Semipreparative HPLC allows unidentified phospholipids to be separated and collected for further study. Gross and Sobel [237] have described a rapid isocratic HPLC procedure with UV detection separating the common glycerophospholipids and lysophospholipids with acetonitrile–methanol–water (200:50:17) as the solvent system. However, phosphatidylserine and phosphatidylethanolamine could not be resolved when this solvent system was used in combination with a Whatman PXS 10/25 SCX cation-exchange column, and there was considerable tailing of peaks. When a Porasil (Waters) column is eluted with acetonitrile–methanol–water (65:21:4), phosphatidylethanolamine has an extremely short retention time [238].

12.3.3.2. Thin-layer chromatography

The most useful solvents for TLC of the major phospholipid classes in animal and plant tissues are mixtures of chloroform, methanol, and water, with or without added ammonia or acetic acid, as originally described by Skipski and Barclay [94] and by Rouser et al. [58]. A very satisfactory one-dimensional system of this type consists of chloroform–methanol–acetic acid–water (25:15:4:2) and silica gel, impregnated with 0.001 M Na$_2$CO$_3$ [239]. Complete separations are achieved for phosphatidylethanolamine, phosphatidylserine, phosphatidylinositol, phosphatidylcholine, sphingomyelin, and lysophosphatidylcholine. This system has been extensively employed for the separation of the common phospholipids in both qualitative and quantitative analyses [66,240–246]. Neutral lipids, acylglycerols, sterols, steryl esters, waxes, and hydrocarbons migrate with the solvent front. Minor acidic phospholipids, cardiolipin, phosphatidic acid, bis-phosphatidic acid, and their partial deacylation products are not satisfactorily resolved in this system, which also carries them near the solvent front. The latter group is resolved more satisfactorily with chloroform–methanol–acetic acid–water (50:35:4:2) [247–250]. This system gives separate peaks for phosphatidylethanolamine, phosphatidylglycerol, lyso-bis-phosphatidic acid, sphingomyelin, phosphatidylcholine, and phosphatidylserine plus phosphatidylinositol. A 17:25:15:6 mixture of chloroform–methanol–acetic acid–water has been used to purify synthetic phosphatidylcholine containing a pyrine-decanoyl group in the sn-2-position of the molecule [251], while a 53:25:2:3 mixture has been specifically employed to resolve the lysophospholipids and free fatty acids [252].

Acidic solvent systems that have been frequently used for one-dimensional TLC include a 10:4:2:2:1 mixture of chloroform–acetone–methanol–acetic acid–water [51,156,253–255] and a 40:15:13:12:8 mixture [256,257], which has been employed to resolve the mono-, di-, and triphosphoinositides, as well as others [258,259]. Lin et al. [166] have isolated the choline and ethanolamine phosphatides, including plasmalogens, by preparative TLC with chloroform–acetone–acetic acid–water (25:15:4:1) and have tested their purity with chloroform–acetone–methanol–acetic acid–water (8:6:2:2:1). A new acidic solvent system for one-dimensional TLC is isooctane–ethyl acetate–acetic acid (16:3:1) [260] and mention must also be made of two systems used in succession in the same direction. Nishihara and Kito [261] have separated phosphatidylcholine, phosphatidylethanolamine, phosphatidylinositol, and phosphatidylmonomethylethanolamine by development with chloroform–acetone–methanol–water–acetic acid (50:50:25:5:2), followed by chloroform–methanol–acetic acid–water (18:15:3:1) in the same direction.

The minor acidic phospholipids may also be analyzed effectively with less polar solvent mixtures, containing smaller amounts of methanol and water, e.g., chloroform–methanol–water (40:10:1) [262,263]. The solvent system chloroform–methanol–water (65:25:4) and similar formulations proposed earlier [212,253] are also suitable for the resolution of all common phospholipid classes in the presence of neutral lipids. This solvent system is most effective when used along with silica gel plates prepared with magnesium silicate instead of the calcium sulfate as the binder. This neutral solvent system has been extensively employed for quantitative one-dimensional TLC [262–264] and for preparative TLC [265,266]. It is well suited for the isolation of the plasmalogenic phospholipids [267,268], which may decompose during solvent extraction and evaporation when acidic solvents are used. Separate bands are obtained for lysophosphatidylcholine, sphingomyelin, phosphatidylcholine plus phosphatidylinositol, phosphatidylethanolamine plus phosphatidylserine, phosphatidylglycerol plus cardiolipin plus phosphatidic acid, and neutral lipids. In this system, the thioester analogs of phosphatidylcholine and phosphatidylethanolamine migrate slightly ahead of the corresponding normal choline and ethanolamine phosphatides [269]. Preparative separations of phosphatidylcholine and other phospholipids have been made on 2-mm thick layers with chloroform–methanol–petroleum ether–water (8:8:6:1) as the developing solvent [54,270].

An excellent resolution of the phospholipid classes may be obtained by one-dimensional TLC with chloroform–methanol–28% ammonia (13:7:1 or 13:5:1) [253]. This type of system has been extensively utilized in the qualitative and quantitative separation of the common phospholipids [271–276]. The methanol content of the basic solvent system has a marked effect on the relative rates of migration of the acidic phospholipids [277]. Thus, in a system with relatively low methanol content [chloroform–methanol–28% ammonia (13:7:1)] phosphatidylserine and phosphatidylinositol run behind phosphatidylcholine [253], while in a solvent with a high methanol content [chloroform–methanol–28% ammonia (13:17:1)] the phosphatidylserine and phosphatidylinositol run much faster than the phosphatidylcholine [253]. These solvent systems have been utilized in one-dimensional TLC largely for the separation of the choline- and ethanolamine-containing phospholipids from other phospholipids, but such mixtures of acidic lipids as cardiolipin, cytidine diphosphate diacylglycerol, phosphatidylglycerol, phosphatidylinositol and phosphatidylserine plus phosphatidic acid may also be separated. For two-dimensional TLC, chloroform–methanol–14 N ammonia (13:7:1) or chloroform–methanol–7 N ammonia–water (26:14:1:1) have been most extensively utilized, as indicated below, but other ratios, e.g.. chloroform–methanol–14 N ammonia (75:25:4) [278,279] and chloroform–methanol–7 N ammonia (65:20:4) [1,280,281] have also been used. The basic solvent systems have the advantage of not decomposing the alkenylacyl derivatives of phosphatidylcholine and phosphatidylethanolamine (plasmalogens), a problem encountered when thin-layer chromatograms developed with acidic solvents are dried.

Vitiello and Zaneta [97] have devised a one-dimensional HPTLC system that allows a complete separation of the major galactolipids, neutral lipids, and phospholipids found in the brain. The separations are performed on precoated plates with methyl acetate–1-propanol–chloroform–methanol–0.5% aqueous KCl (25:25:25:10:9) as the developing solvent. Separate bands are obtained (in ascending order) for: sphingomyelin, phosphatidylcholine, phosphatidylserine, phosphatidylinositol, phosphatidic acid plus diphosphatidylglycerol, phosphatidylethanolamine, sulfatides, ceramides, and neutral lipids. The sulfatides and ceramides give two bands each, corresponding to the derivatives of hydroxy and ordinary fatty acids. A comparative HPTLC system has been used by Sherma and Touchstone [282]. Touchstone et al. [283] have reported the mobilities of phospholipids in various solvent systems on prepared HPTLC plates. Excellent separations of the choline, ethanolamine, serine, and inositol phosphatides and of lysophosphatidylcholine and sphingomyelin were obtained with chloroform–methanol–2-propanol–0.25% aqueous KCl–triethylamine (30:9:25:6:18). Triethylamine has the same solvent strength as ethyl acetate. HPTLC has the advantages that the development is faster, the resolution more effective, and the samples required are smaller, but the disadvantages are that this highly efficient system cannot be scaled up for preparative purposes and that the plates are expensive.

The differences in the relative rates of migration of the different phospholipid classes are best exploited by two-dimensional TLC with two complementary solvent systems, as demonstrated by Rouser et al. [253] and Skipski and Barclay [94]. The most effective combinations have been those involving basic and acidic or basic and neutral solvent systems, but neutral and acidic or two acidic systems have also been reported. An excellent distribution of different phospholipid classes over the area of the thin-layer plate is obtained by combining the basic solvent system chloroform–methanol–28% ammonia (13:7:1) in the first development with the acidic system chloroform–acetone–methanol–acetic acid–water (10:4:2:2:1) in the second development [253]. This system has been very widely utilized for both qualitative and quantitative analysis of phospholipids of animal tissues [254,259,283–288]. For a separation of total lipid extracts, Nishihara and Kito [261] have varied this routine by using chloroform–acetone–methanol–acetic acid–water (40:20:30:3:1) for the second development, while Renkonen et al. [1] and other investigators [280,289,290] have substituted chloroform–methanol–ammonia (65:20:4) for the first development. A comparable separation is obtained without the acetone in the second solvent [291–295].

A slightly better distribution of the phospholipid classes in two-dimensional development is obtained by combining a basic solvent system made up of chloroform–methanol–water–28% ammonia (70:30:3:2) in the first development with a neutral solvent system made up of chloroform–methanol–water (13:7:1) in the second development [296]. Because phosphatidylglycerol and phosphatidylethanolamine frequently overlap in both of the above systems, their quantitative estimation is difficult or impossible. Poorthuis et al. [297] have proposed the use of

0.4 M boric acid in the adsorbent and development with chloroform–methanol–water–28% ammonia (70:30:3:2) in the first dimension and with chloroform–methanol–water (13:7:1) in the second dimension. The relative positions of the phospholipids in these chromatograms are essentially similar to those obtained in the widely used two-dimensional system of Rouser et al. [296] except that the positions of phosphatidylinositol and phosphatidylserine are reversed, and the phosphatidyl-glycerol is more distinctly separated from phosphatidylethanolamine. Both of these differences are due to the presence in the silica gel of boric acid, which forms complexes with compounds having vicinal hydroxyl groups and thus retards their mobility [278,298]. This improved system is currently gaining favor for separating hydroxy from nonhydroxy glycerophospholipids [81,299,300].

A two-dimensional system employing Na_2CO_3-impregnated plates in combination with chloroform–methanol–28% ammonia–water (26:14:2:1) for the first development and chloroform–methanol–butanol–acetic acid–water (18:12:8:4:3) for the second development has been proposed by Cho et al. [301] for the resolution of phospholipid mixtures containing lysophospholipids. This system also has found general application, although butanol is difficult to remove. A two-dimensional TLC system based on magnesium acetate-impregnated plates and chloroform–methanol–28% ammonia–water (120:60:4:3) for the first and chloroform–acetone–methanol–acetic acid–water (6:8:2:2:1) for the second development has been proposed by Hostetler et al. [302,303].

Two-dimensional TLC analysis of phospholipids frequently results in poor resolution of many sample components due to the variability in atmospheric humidity. Yavin and Zutra [304] have proposed a simple and rapid method, which they claim obviates these problems. Their innovation is based primarily on shortening the running distance of the developing solvents (10 × 10-cm plates) and the selection of more efficient solvent systems. For the first dimension they propose chloroform–methanol–40% methylamine (26:12:3), and for the second development an intermediate run with diethyl ether–acetic acid (19:1), followed by chloroform–acetone–methanol–acetic acid–water (10:4:2:3:1). Excellent separations of phosphatidylinositol, phosphatidylserine, phosphatidic acid, sphingomyelin, lysophosphatidylcholine, phosphatidylcholine, lysophosphatidylethanolamine, phosphatidylethanolamine, and cardiolipin were obtained. The new technique was tested in other laboratories and found suitable [258]. A chloroform–methanol–40% methylamine–water (65:31:5:5) system had been used earlier by Getz et al. [305] and Eichberg et al. [306] as a second solvent after development with chloroform–methanol–acetic acid–water (52:20:7:3). Between the two developments, the plate was washed with either diethyl ether or acetone as an intermediate development. Excellent separations were obtained for cardiolipin, phosphatidylglycerol, phosphatidic acid, phosphatidylethanolamine, dimethylethanolamine phosphatide, phosphatidylcholine, phosphatidylserine, phosphatidylinositol, lysophosphatidylethanolamine, lysophosphatidylcholine, and sphingomyelin.

Other combinations of basic and acidic solvents have also been employed: chloroform–methanol–7 M ammonia (90:54:11), followed by chloroform–methanol–acetic acid–water (90:40:12:1) along with silica gel, impregnated with 2%

magnesium hydroxide carbonate [212]; chloroform–methanol–28% ammonia–water (50:35:3:3), followed by chloroform–methanol–acetic acid–water (50:25:8:4) [307]; chloroform–methanol–28% ammonia (65:25:2), followed by chloroform–methanol–acetic acid–water (85:15:10:3) [39,308]; chloroform–methanol–28% ammonia (13:5:1), followed by chloroform–methanol–acetic acid–water (81:10:45:5) in the second direction [309]. In many instances, the basic system was used for the second development. Thus, chloroform–methanol–acetic acid (50:25:8), followed by chloroform–methanol–28% ammonia (13:7:1), gave a complete separation of N-isopropylphosphatidylethanolamine from all other glycerophospholipids [310–313].

A neutral system, chloroform–methanol–water (65:25:4) in the first dimension, was used in combination with tetrahydrofuran–dimethoxymethane–methanol–2 N ammonia (10:5:5:1) in the second dimension [314,315]. A three-solvent system for two-dimensional TLC has been described by Chapman and Robertson [316] and Chapman [317]. In this method the different phospholipid classes are resolved by first developing the plate with chloroform, then in the same direction with chloroform–methanol–7 N ammonia (65:30:4), and finally in the second direction with chloroform–methanol–acetic acid–water (170:25:25:4). Effective separations were obtained for choline and ethanolamine phosphatides, and phosphatidic acid. Two-dimensional TLC with two acidic solvent systems has also been employed for the separation of phospholipid classes [318]: chloroform–acetone–methanol–acetic acid–water (25:10:5:5:2) and chloroform–methanol–acetic acid–water (25:15:4:2). It produces a good separation of sphingomyelin, phosphatidylcholine, phosphatidylinositol, phosphatidylserine, phosphatidylethanolamine, and cardiolipin. Vaskovsky and Terekhova [319] have employed two-dimensional HPTLC for the resolution of phospholipid mixtures containing phosphatidylglycerol. Satisfactory separation of all components was obtained with chloroform–methanol–benzene–28% ammonia (65:30:10:6) in the first dimension and chloroform–methanol–benzene–acetone–acetic acid–water (70:30:10:5:4:1) in the second dimension. The addition of benzene to the first system and benzene plus acetone to the second one improved the shape of the spots and the separation of components compared to the conventional chloroform–methanol–ammonia or chloroform–methanol–acetic acid–water systems.

The alkylacyl- or alkenylacyl- and diacylglycerophospholipids are not resolved in any of the one- or two-dimensional TLC systems, but a differentiation can be made between them if the plate is treated with acid prior to the development with the second solvent. The plate is developed in the first direction with chloroform–methanol–28% ammonia (65:25:4), then exposed to HCl fumes, and finally developed in the second direction with chloroform–methanol–28% ammonia (50:25:6) [320]. This technique has found extensive application in the qualitative and quantitative evaluation of brain lipid extracts [49,321]. Subsequently, Horrocks and Sun [322] have proposed the use of chloroform–methanol–15 N ammonia (26:13:2) for the first development, followed by exposure to HCl fumes to cleave the alkenyl ether bonds of the plasmalogens, and development in the second direction with chloroform–methanol–acetone–acetic acid–0.1 M ammonium acetate (26:10:11:7:2).

This method has been employed extensively in the field [323,324]. The alkenyl derivatives can also be converted to the corresponding lysophospholipids by treating the plate with mercuric chloride [291] following the first development. Separate spots are obtained for the unaffected diacyl- and alkylacylglycerophospholipids, and for the lysophospholipids arising from the plasmalogens. A resolution of the molecular species of the normal and the plasmalogenic phospholipids may be obtained following dephosphorylation (see Chap. 12.3.2.2). Phosphatidylcholine and 2-aminoethylphosphonolipids may be resolved with chloroform–acetic acid–methanol–water (375:125:25:11) [325].

Conventional TLC systems usually also fail to separate the phosphoryl and phosphono analogs of the phospholipids. These types of phospholipid classes can be separated with solvents containing a high proportion (85–92%) of chloroform and acetic acid [325,326], in which the phosphono analogs migrate faster than the corresponding phosphatidylethanolamines and phosphatidylcholines. Oette and Tschung [208] have used chloroform–methanol–water (65:25:4) as a solvent to effect a partial resolution of normal phosphatidylcholine and phosphatidyl-ethanolamine and the corresponding serinol 1-N-acyl and 2-N-acyl analogs. In both instances, the normal phosphatide was found to migrate between the faster-moving 1-N-palmitoyl derivative and the slower-moving 2-N-linoleoyl derivative. Mason et al. [327] and Rooney et al. [76] have used triple development with chloroform–methanol–8 M ammonia (150:70:7) to separate saturated and unsaturated species of phosphatidylcholine and phosphatidylglycerol, following reaction of the unsaturated species with mercuric acetate. Recently, TLC on Chromarods has attracted much interest for quantitative analysis of lipid classes [2–4]. Reference mixtures of oilseed phospholipids have been resolved by Ackman and Woyewoda [328] with chloroform–methanol–acetic acid–water (60:30:9:3) following removal of neutral lipids with acetone. Phosphatidylcholine and phosphatidylethanolamine have been resolved by Herslof [329], and Taguchi et al. [330] have resolved purified preparations of common phospholipids on Chromarods with chloroform–methanol–water (80:35:3) or chloroform–methanol–acetic acid–water (80:15:10:4).

12.3.4. Isolation of individual glycolipid classes

The glycolipid fractions recovered from the initial lipid group separation may be resolved by either adsorption or partition chromatography. TLC and HPLC have provided the best routines for the separation of this highly variable group of compounds, but certain separations and purifications may also be advantageously performed by conventional adsorption and partition column chromatography.

12.3.4.1. Column chromatography

Rouser et al. [331] have shown that ceramides may be eluted from Florisil columns with 19:1:1 mixtures of chloroform–methanol–dimethoxypropane and cerebrosides and cerebroside sulfates with 14:6:1 mixtures. Vorbeck and Marinetti [332] have demonstrated that silicic acid columns are useful for separating plant lipid extracts to yield monogalactosyldiacylglycerols by elution with chloroform-acetone (1:1) and digalactosyldiacylglycerols, sulfolipid, and sulfatides by elution with acetone. Silicic acid column chromatography has also been employed by Saito and Hakomori [333] to isolate acetylated glycolipids by elution with 1,2-dichloroethane–acetone (1:1). These methods have been used by various investigators as means of concentrating particular glycolipids rather than for obtaining pure fractions [334–336]. Gellerman et al. [59] have applied silicic acid columns to the isolation of monogalactosyldiacylglycerols from mosses with acetone–chloroform–water (15:30:1) as the eluent, and Peters and coworkers [337,338] have used silicic acid columns to isolate and purify glucocerebrosides from Gaucher spleen by elution with chloroform–methanol (19:1). Likewise, Wilson et al. [60] have isolated plant glycolipids by eluting silicic acid columns with chloroform–methanol (9:1). Extensive fractionations of glycolipids have been performed by Hirabayashi et al. [339] with CC on a mixture of silicic acid and Hyflo Supercel in a 2:1 ratio. The column was eluted with increasing concentrations of methanol in chloroform from 9:1 to 2:3 to yield five subfractions, ranging in polarity from glucosylceramide to GM_3. Silicic acid columns have also been used for the separation of gangliosides into three fractions [340,341]. An elution with chloroform–methanol–water (65:25:4) yields GM_3, followed by GD_3 and GM_2. The remainder is eluted with chloroform–methanol–water (60:35:8). Narasimhan and Murray [342] have used the silicic acid column method of Vance and Sweeley [72] as modified by Yogeeswaran et al. [343] to separate neutral glycosphingolipids and gangliosides of human lung.

A simple method for the isolation of hematoside NeuNGl–Lac–Cer from horse erythrocytes has been described by Maget-Dana and Michalski [344]. An aliquot of the crude ganglioside fraction was labeled by tritiated sodium borohydride after mild periodate oxidation. The compounds obtained were used as radioactive tracers in column chromatography. Gangliosides were applied to a silicic acid column and eluted stepwise by solvents of increasing polarity. The major ganglioside, NeuNGl–Lac–Cer, was eluted in high yield by the solvent mixture chloroform–methanol–water (60:35:8). This procedure is apparently suitable for the separation of low-polarity gangliosides from neutral lipids without repeated elutions.

Ando et al. [345] have obtained good preparative separations of neutral glycolipids of human erythrocytes by adsorption chromatography on totally porous silica spheres, Iatrobeads (Iatron Lab., Tokyo, Japan), by applying a linear gradient of chloroform–methanol–water from 166:32:1 to 110:84:6. Two subfractions were obtained for each ceramide di-, tri-, and tetrahexoside as well as for the globoside and paragloboside fractions. The subfractionation was due to chain-length differences in the component fatty acids and nitrogenous bases. Ueno et al. [346] have used the Iatrobead columns for the resolution of gangliosides recovered from a DEAE-Sephadex column. Prior to sample application the Iatrobead column was washed with chloroform–methanol–2.5 N ammonia (3:6:1) and then with chloroform–methanol–water (3:6:1) in order to remove contaminants. Elution was started with chloroform–methanol (17:3), which removed sulfatides; pure gangliosides were then eluted with chloroform–methanol (1:2). Suzuki et al. [347] have employed Iatrobeads to separate acetylated globotriaosylceramide, eluted with dichloro-ethane–acetone (4:1), from acetylated lactotriaosylceramide, eluted with dichloro-ethane–acetone (7:3).

Ion-exchange chromatography on DEAE-cellulose is an effective alternative to adsorption chromatography for the separation of complex lipids [20,58]. Simple lipids, with the exception of free fatty acids, are eluted with chloroform. Chloro-form–methanol (9:1) elutes the choline-containing phospholipids, while chloro-form–methanol (1:1) yields ethanolamine phosphatides and ceramide di- and poly-hexosides. The most popular routines are those described by Rouser et al. [58] and Nichols and James [348]. There are numerous variations of both of these techniques, some of which materially alter the separations obtained [349–351]. Extensive appli-cations of this technique have been made in the analysis of tissue glycolipids [86,349–354]. The neutral and acidic glycosphingolipids have also been separated by DEAE-Sephadex [355–358]. Bremer et al. [350] have recommended a method for large-scale isolation of gangliosides based on DEAE-Sephadex A-25 chromatogra-phy. After sample application, the column is first eluted with chloroform–methanol–water (15:30:4). The gangliosides are then recovered with chloroform–methanol–0.8 M sodium acetate (15:30:4). For small-scale isolation of gangliosides from plasma they recommend solvent partition with chloroform–methanol (1:1 to 2:1) and water, and reversed-phase HPLC with a C_{18} Sep-Pak Cartridge (Waters Assoc.). Gangliosides were eluted with methanol. Iwamori and Nagai [359] and Fredman et al. [351] have compared various DEAE-Sephadex anion exchangers for their ability to separate gangliosides. DEAE-Spherosil was best, giving good separa-tion of mono-, di-, tri-, tetra, and pentasialogangliosides.

Most of the HPLC separations of glycolipids thus far described have been performed on cerebrosides and ceramide polyhexosides following perbenzoylation, which greatly facilitates their detection in the column effluent by means of UV spectrometry. McCluer and Evans [360] have prepared benzoylated cerebrosides and separated the derivatives containing hydroxy and ordinary fatty acids by HPLC with 0.13% methanol in pentane as the eluent and a Zipax (pellicular silica gel, Du Pont) column. Pure perbenzoyl galactosylceramide was isolated from crude chloroform–methanol extracts of adult brain with 7% ethyl acetate in hexane. Sulfatides did not interfere with the resolution. Benzoylated glucosyl- and galatosylceramides were resolved on a MicroPak NH_2 column with 1.5% 2-pentanol in cyclopentane. Jungalwala et al. [361] have subsequently described conditions for the quantitative analysis of perbenzoylated ceramides containing ordinary and hydroxy fatty acids in the 0.5- to 10-nmole range. The ceramides were separated on a Zipax column with 2.8 to 5.5% dioxane in hexane or 2 to 7% ethyl acetate in hexane. In another study, Evans and McCluer [362] and, subsequently, Ullman and McCluer [363] reported the separation of the perbenzoylated derivatives of mono-, di-, tri-, and tetraglyco-sylceramides of plasma on Zipax columns. The most effective separations were obtained by using a linear gradient of 2 to 17% ethyl acetate in hexane. Subse-quently, the same authors [364] described a quantitative analysis of the perbenzo-ylated derivatives of glycosphingolipids in the picomole range by HPLC on Zipax with a linear gradient of 1 to 20% dioxane in hexane and detection at 230 nm. Ceramide polyhexosides with up to four sugar residues were resolved. Suzuki et al. [347] have obtained improved separation of O-acetyl-N-p-nitrobenzoyl derivatives of neutral glycosphingolipids by HPLC with linear and nonlinear gradients of 1 to 5% isopropanol in hexane–dichloroethane (2:1), while Watanabe and Arao [365] have used a 2-propanol–hexane–water gradient without sample derivatization.

Bremer et al. [350] have described a quantitative HPLC method for the analysis of monosialogangliosides as their perbenzoyl derivatives. The HPLC analysis was performed with a LiChrosphere SI 4000 column and a linear gradient of 7 to 23% dioxane in hexane in 18 min. Detection at 230 nm allowed as little as 50 pmole of injected material to be analyzed. Separate peaks were obtained for GM_1, GM_2, GM_3, and GM_4, which were eluted in this order. Complete analyses were obtained on as little as 1 ml of plasma. HPLC of polysialogangliosides showed that the perbenzoyl derivatives of GD_3 migrated more slowly than GM_1 and did not interfere with GM_1 quantitation. Likewise, GD_{1a} and GD_{1b} migrated more slowly than GD_3 and did not interfere with the analysis of the monosialogangliosides. Under the benzoylation conditions used, GD_{1a} and GD_{1b} did not yield a single peak but multiple or asymmetrical broad peaks.

12.3.4.2. Thin-layer chromatography

The solvent systems best suited for TLC resolution of the individual classes of the glycolipids are based on acetone, pyridine, and tetrahydrofuran [314,366]. These solvents act as electron donors and form hydrogen bonds with the numerous hydroxyl groups of the glycolipids. Thus, an effective separation of the mono- and digalactosyldiacylglycerols from each other and from many other lipids is obtainable with the solvent system acetone–benzene–water (91:30:8) [367,368]. It has been used for the preparative isolation of these glycolipids, improved resolution being obtained by combination with ammonium sulfate-impregnated silica gel [369]. Another effective solvent system for the separation of plant glycolipids is chloroform–acetone–methanol–acetic acid (146:50:3:1) [370]. It resolves 6-O-acyl derivatives of monogalactosyldiacylglycerols and steryl glucosides, which are well separated from 2-monoacylglycerol and free fatty acids. Effective separation is also realized of monogalactosyldiacylglycerols from N-acylphosphatidylethanolamine and other more polar glycolipids, which remain at the origin. Increase of the acetone content to chloroform–acetone–water (15:30:1) leads to an increase in the mobility of the polar glycolipids. A separation is now obtained for monogalactosylceramide, monogalactosylmonoacylglycerol, digalactosyldiacylglycerol, and digalacto-sylmonoacylglycerol. Excellent resolution of the glycolipids in plants is also realized with other phospholipid solvent systems, such as chloroform–methanol–28% ammonia–water 24:14:2:1 [370] or 160:20:7:4 and 800:100:15:7 [371]. In other instances, chloroform–methanol–ammonia–2-propylamine (130:70:10:1) has been used for the isolation of monogalactosyldiacylglycerol [371–373]. Comparable separations of the glycolipids are also obtained with somewhat simpler solvent systems [72,374,375]. Glucosylceramides and galactosylceramides, which migrate together on silica gel plates, can be separated on borate-impregnated silica gel [298,376].

Repeated development with systematically selected solvent systems greatly increases the separation of glycolipids. A most useful procedure for the separation of glycolipids from animal tissues consists of an initial development with acetone–pyridine–chloroform–water (40:60:5:4), followed, after drying of the plate, by diethyl ether–pyridine–ethanol–2 M ammonia (65:30:8:2), and a final development in the same direction with diethyl ether–acetic acid (100:3) [347]. The initial development moves only the neutral lipids and the glycolipids along the plate, while any phospholipids remain stationary. The second development washes the neutral lipids away, the glycolipids remaining stationary. The third development washes away the free fatty acids. In this system, neutral lipids migrate further than a ceramide monohexoside, and all phospholipids move more slowly than a ceramide tetrahexoside. This system yields an excellent separation of ceramides, ceramide monohexosides, sulfatides, ceramide dihexosides, psychosine, ceramide trihexosides, ceramide N-acetylhexosamine trihexoside, and cardiolipin. The glycolipids are separated largely according to the number of monohexoside units per molecule. The sphingoglycolipids are also resolved to some extent according to differences in chain length and number of hydroxyl groups in the hydrocarbon chains [377]. The cerebrosides can be separated from the sulfatides by developing twice with chloroform–methanol–water (720:125:14) [378].

The glyceroglycolipids of plant [379] and microbial [380] origin are conveniently separated with chloroform–methanol–water and chloroform–water–acetone mixtures [370]. Even the more complex bacterial glycolipids, like the glycerophosphoryldiglucosyldiacylglycerols of streptococci, are easily separated by TLC with chloroform–methanol–water (65:35:8) [381]. Tadano and Ishizuka [382] have used chloroform–methanol–water (65:25:4), chloroform–methanol–28% ammonia–water (60:35:1:7), and chloroform–acetone–methanol–acetic acid–water (10:4:2:2:1) to separate galactosylceramide, lactosylceramide, $GbOSe_3$ ceramides, $GbOSe_4$ ceramides, and GM_3. Sung and Sweeley [383] have used chloroform–methanol–water (50:21:3) to separate glycosylsphingolipids GL_2, GL_3, GL_4, and a trihexosylceramide. A phosphoglycolipid was isolated from *Acholeplasma* with chloroform–methanol (9:1) [64], and the glycolipids from bovine epididymal spermatozoa were resolved with chloroform–methanol–water (40:10:1) [71]. The more complex neutral ceramide oligosaccharides containing five to eight monosaccharide residues are conveniently separated after conversion to the fully acetylated form [384].

Sialic acid-containing sphingolipids or gangliosides are commonly separated by use in TLC of propanol–water or chloroform–water–methanol mixtures with or without added ammonia [385–387]. An effective solvent system for the separation of six gangliosides (made up of ceramide tetrasaccharides llinked to one or more sialic acid molecules) is propanol–water (7:3) [385]. The ganglioside fraction from bovine mammary tissue was resolved by TLC using 1-propanol–28% ammonia–water (14:61:1). Separate bands were obtained for the GM_3, GM_2, and GM_1 components, in the order of decreasing mobility [388,389]. Preparative separations were effected with a solvent system of chloroform–methanol–28% ammonia–water

(60:35:7:3) [389]. The latter system was also used for the preparative isolation of gangliosides from rat mammary carcinoma cell lines [390]. Smith [391] and Svennerholm et al. [376] have resolved glycolipids with chloroform–methanol–water (65:25:4). Additional solvents systems for the separation of carbohydrate-containing lipids have been described elsewhere [11,99].

Seyfried et al. [392] and Ando et al. [393] have shown that gangliosides can be effectively separated by TLC or HPLTC with chloroform–methanol–0.02% aqueous $CaCl_2$ (60:35:8) or chloroform–methanol–2.5 N ammonia (60:35:8). Iwamori and Nagai [359] have resolved the subclasses of gangliosides from human and animal brains by TLC with chloroform–methanol–2.5 M ammonia (60:40:9), chloroform–methanol–water (65:35:8), and chloroform–methanol–0.25% KCl (60:35:8) as the developing solvents. By combining TLC with preliminary DEAE-Sephadex column chromatography they were able to separate at least 25 unidentified gangliosides in addition to the well-known compounds.

Yates et al. [394] have used chloroform–methanol–0.02% aqueous $CaCl_2$ (5:4:1) as the developing solvent for separating gangliosides from human neural tumors and the cells of two gliomas cultured in vitro. Excellent separations were obtained for GQ, GT_{1b}, GD_{1b}, GD_{1a}, GD_3, GM_1, GM_2, and GM_3 classes of gangliosides on precoated plates. Aliquots of purified gangliosides were prepared on 0.5-mm thick silica gel layers with chloroform–methanol–water–28% ammonia (60:35:7:1) as the developing solvent. Schwarzmann [395] separated gangliosides by TLC on precoated plates with chloroform–methanol–water (60:35:8) containing 20 mg $CaCl_2$ per 100 ml solvent, and chloroform–methanol–water (65:25:4) for separating the neutral glycosphingolipids. The location of the various components on the thin-layer plates was determined by radioautography, tritium having been specifically introduced into the ceramide portion of gangliosides, neutral glycosphingolipids, and sphingomyelins by means of [^{3}H]borohydride in the presence of Pd as catalyst.

The separation of glycolipids from each other and from potential contaminants is enhanced by two-dimensional TLC. The solvent systems chloroform–methanol–water (65:25:4), followed by n-butanol–acetic acid–water (3:1:1); and chloroform–methanol–28% ammonia (13:7:1), followed by chloroform–acetone–methanol–acetic acid–water (10:4:2:2:1), produced good resolution of cerebrosides with ordinary fatty acids, cerebrosides with hydroxy fatty acids, ceramide dihexosides, sulfatides with ordinary fatty acids, sulfatides with hydroxy fatty acids, gangliosides, as well as monogalactosyldiacylglycerols, digalactosyldiacylglycerols, and sulfolipids [58]. Gray [314] as well as Karli and Lewis [86] have used two-dimensional TLC to separate the glycosphingolipids of the thyroid gland with chloroform–methanol–water (65:24:4) in the first dimension and tetrahydrofuran–methylal–methanol–4 M ammonia (10:5:4:1) in the second dimension. Separate spots were obtained for monohexosylceramide, dihexosylceramide, trihexosylceramide, an aminoglycolipid, and a sulfatide. Ohashi [396] resolved gangliosides by developing twice in the first direction, with chloroform–methanol–28% ammonia–water (60:40:3:6), and with either 1-propanol–28% ammonia–water (15:1:5) or chloroform–methanol–water (60:40:9) in the second direction. This complementary pair of separations proved very useful for a preliminary identification of gangliosides on a microscale (4–5 μg of sialic acid per spot).

12.4. RESOLUTION OF MOLECULAR SPECIES

For the purpose of the present discussion any separation within a lipid class will be considered to be resolution of molecular species, although single molecular species may not necessarily be obtained in any one chromatographic system, or even in a combination of systems. The most effective methods presently available for the resolution of fatty acids, and of the fatty acid esters and amides within a class of lipids are TLC with silver nitrate, GLC on polar liquid phases, and HPLC on reversed-phase columns. In most instances, the efficiency of the separation is greatly increased by a conversion to derivatives that are less polar and thermally more stable.

12.4.1. Fatty acids

Fatty acids are commonly separated into molecular species by means of GLC on polar liquid phases, following their conversion to methyl esters. However, for the identification of unknown fatty acids it is useful to effect an initial fractionation of the acids into groups of uniform degree of unsaturation and geometric configuration by argentation chromatography.

12.4.1.1. Argentation chromatography

The argentation principle has been utilized in the resolution of fatty acids in combination with CC [397], HPLC [398], and, especially, TLC [399]. It is based on the formation of a reversible charge-transfer complex involving the silver ion and an olefinic double bond [400,401]. Barrett et al. [399] and Dudley and Anderson [402] have shown that $AgNO_3$-TLC is effective in fractionating fatty acid ester mixtures according to the number of double bonds. However, reproducible results are obtained only within the relative humidity limits of 42–44%. The recovery of highly unsaturated esters (5 or 6 double bonds) was about 80%, presumably because of oxidative losses. The *cis-* and *trans-*isomers and certain positional isomers are also resolved [399,402,403].

Minnikin et al. [404] have used argentation TLC for further fractionation of mycolic acids that were homogeneous by adsorption TLC, and Rao et al. [405] have isolated long-chain fatty acids in preparative amounts by argentation TLC with hexane–diethyl ether (4:1) as the developing solvent. Ilinov [406] has recommended the use of silver sulfamate-impregnated silica gel for argentation TLC of fatty acid methyl esters. The sulfamate causes a more intense darkening of the zones upon charring than does silica gel alone. Development is accomplished with *n*-hexane–petroleum ether–diethyl ether–acetic acid (35:12:2:1) at 12–15°C. The sulfamic acid does not interfere with the separation of the lipid classes.

Because of decreased peroxidation, greater recovery of polyunsaturated fatty acids is obtained when adsorption [397] or partition [398] columns are used for argentation chromatography. Originally, De Vries [397] reported the separation of saturated and unsaturated fatty acids by argentation chromatography on columns of silicic acid. A disadvantage of columns with silver nitrate-impregnated silica is their limited life time [407,408]. Nevertheless, they have found practical application in numerous instances. Neville et al. [409] have used AgNO$_3$-Unisil columns for separating methyl cis-vaccenate from traces of other methyl esters. Fractions were eluted with 15% benzene in petroleum ether, then 50% benzene in petroleum ether, and finally diethyl ether. Most of the cis-vaccenate was eluted by 50% benzene in petroleum ether. Because silver nitrate has a limited solubility in the solvents used for HPLC, the AgNO$_3$-containing columns are stable for as long as two months under daily use [398]. Ozcimder and Hammers [398] have shown that argentation HPLC is a rapid semipreparative prefractionation method for highly unsaturated fatty acids with 3 to 6 double bonds. The separations are performed on Partisil 10 containing 5% (w/w) silver nitrate with 0.4% acetonitrile in n-hexane as the eluent. The acetonitrile has a large influence on the elution of the unsaturated esters due to competitive 1:1 complex formation with Ag$^+$ sites [410]. Lam and Grushka [411] had earlier shown that the p-bromophenacyl esters of fatty acids give excellent resolution in AgNO$_3$-HPLC; even the cis- and trans-isomers are resolved. The trans-isomers are eluted earlier than the corresponding cis-isomers.

Adlof et al. [412] have investigated partial argentation resin chromatography for the separation of octadecadienoate ester isomers. In comparison to saturated silver resin chromatography, the time necessary to elute methyl cis,cis-octadecadienoates was dramatically shortened when columns containing sulfonic acid ion-exchange resin, silvered in the range of 60–90% of theoretical, were used. Mixtures of cis,trans-; trans,trans-; trans,cis-; and cis,cis-methyl-12,15-octadecadienoates were separated in 20-g batches on a 91% silvered column. Adlof and Emken [413] used partial argentation resin chromatography on XN-1010 (Rohm & Haas) to separate a mixture of saturated and monoenoic, dienoic, trienoic, and tetraenoic fatty acid esters. However, application of these columns to the separation of mixtures of fatty acids and mixtures of triacylglycerols was not feasible. The separation of monoenoic and dienoic fatty ester isomers on silver-saturated macroreticular cation-exchange resin has been reported [414]. Although the separation of 10- to 20-g samples of monoenoic ester isomers was easily accomplished, the cis,cis-dienoic and cis,cis,cis-trienoic esters required long elution times and large volumes of methanol, and gave poor peak shapes. Addition of 1-hexene to the eluent greatly shortened the retention time, due to competition for the silver ion binding [415].

Schofield [416] has combined the technique of argentation chromatography on a macroreticular ion-exchange resin with HPLC to separate fatty acid methyl esters and their isomers. Elution of methyl linoleate from the column and hence rapid separation of dienes was facilitated by programing column temperatures from 25 to 70°C and eluting with methanol. Mordret et al. [417] have used Chromarods, impregnated with AgNO$_3$, for the separation of the fatty acid methyl esters of sunflower seed oil. The separation of the geometric isomers of unsaturated fatty acid methyl esters by countercurrent distribution, argentation TLC, and silver resin chromatography has been reviewed by Schofield [418]. Glass et al. [419] found that the furanoid fatty acids are not retarded appreciably during argentation TLC. Their methyl esters migrate with those of the saturated and monoene esters, when the plates are developed with hexane–diethyl ether (3:1). It should be noted that hydroxy-*cis*-enynoic acids are readily converted to furan-type acids by treatment with alkali or by argentation chromatography [420].

12.4.1.2. Gas–liquid chromatography

GLC constitutes the most powerful single method of separating complex mixtures of natural fatty acids, and useful correlations exist between the retention time and structure of an acid [421,422]. However, the large number of isomers potentially present in natural fatty acid mixtures prevents their positive identification except for the simplest mixtures. Preliminary argentation [423] or preparative [424] GLC substantially increases the certainty of identification for most fatty acids and must be included in any systematic separation and identification scheme. Effective separations of long- and short-chain fatty acids, and of ordinary and oxygenated fatty acids may be obtained by adsorption chromatography, which usually precedes argentation and partition chromatography. More complete identification is obtained by GC–MS [425]. Kuksis [15] and Lie Ken Jie [17] have recently reviewed the GLC methods commonly employed in systematic separation and identification schemes for natural and synthetic fatty acids and, along with others [426], they have made recommendations for the preparation of representative samples of fatty acid methyl esters for analysis. In addition, Emken [427] has provided a short description of methods and techniques normally used for analyzing isotope-labeled fatty acids, including radio-GLC.

The short-chain free fatty acids may be separated by GLC on a variety of liquid phases, fortified with nonvolatile organic or inorganic acids [5,7] to block adsorption sites on the support and column material. Ackman [422] has demonstrated that the addition of formic acid to the carrier gas is necessary to achieve adequate separation and recovery of free fatty acids from GLC columns. Esterification of the short-chain fatty acids greatly improves their GLC properties but does not affect the relative order of elution, which remains that of the molecular weights, the lower homologs emerging first. Either conventional [428] or capillary [429] GLC columns may be used for separating and identifying the simple esters of short-chain acids in complex mixtures. Ashes and Haken [430] have examined in great detail the structure–retention time relationships of saturated and unsaturated short-chain fatty acids in the form of various saturated and unsaturated simple alkyl and isoalkyl esters on a variety of polar and nonpolar liquid phases.

The saturated normal and branched medium- and long-chain fatty acid esters, isolated as a group by argentation TLC along with any cyclic-chain derivatives, are separated by GLC in the ester form. The free long-chain fatty acids are usually unsatisfactory for resolving any but the simplest of mixtures [7,431]. Kuksis [15], Lie Ken Jie [17], and Ashes et al. [432] have discussed in detail the separation of the normal, monomethyl branched, multimethylbranched and cyclic medium- and long-chain fatty acids on both polar and nonpolar columns. The resolution of the normal and monomethyl branched fatty esters of the same carbon number depends on the location of the branching on the hydrocarbon chain, and a distinction may be made

References on p. B130

among iso-, anteiso- and neo-isomers [422]. The GLC behavior of fatty esters with two or more methyl branches in the hydrocarbon chain abides by the concept of additivity of fractional chain-length values [432]. The major group of cyclic esters consists of cyclopropanoic fatty acids. All natural forms are *cis*; they possess slightly longer retention times than the *trans*-forms, which have been prepared synthetically [433]. Other long-chain fatty acids with cyclic structures are the alicyclic fatty acids [434]. Oshima and Ariga [435] have used GLC to identify the 11-cyclo-hexylunde-canoate and 13-cyclo-hexyltridecanoate in acidophilic–thermophilic bacteria, and Glass and coworkers [419,436] have identified furanoid fatty acids in fish lipids.

The ethylenic components of the common unsaturated fatty acids, recovered according to the total number of double bonds by argentation TLC, are examined separately as the mono-, di-, tri-, tetra,- penta-, and hexa-unsaturated fatty acid subclasses. However, cross-contamination may occur either because of incomplete resolution or because of the presence of positional and chain-length isomers. Although *cis*- and *trans*-isomers can be resolved by $AgNO_3$-TLC [437,438], there are also liquid stationary phases enabling this separation by GLC. Thus, methyl oleate and methyl elaidate can be separated, along with other saturated and unsaturated C_{18} fatty acid methyl esters on 15% OV-275 provided sufficiently long columns are used (20 ft.) [439]. The ethylenic fatty acids are commonly resolved on conventional GLC columns, containing polar liquid phases (10% EGSS-X, 10% DEGS, 10% EGS) or certain nonpolar liquid phases (10% Apiezon L). These separations are performed isothermally at about 200°C, although in certain instances higher temperatures (220°C) have been used [15,17]. Cyanoalkylsiloxane (Silar-10C) is a liquid phase with somewhat higher temperature stability and yields much better resolution of the *cis*- and *trans*-isomers than do the polyester liquid phases [440,441]. The natural *cis*- and *trans*-monoethylenic [442,443] and polyethylenic [444,445] fatty acids have been effectively resolved by glass capillary GLC.

The polyethylenic fatty acid esters are retained longer on polar liquid phases than the corresponding mono- and diethylenic analogs of the same carbon number. The natural dimethylene-interrupted methyl *cis,cis*-octadecadienoates, chromatographed on several polar and nonpolar columns, showed unique differences in the effective chain-length (ECL) values, depending on the location of double bond [446]. In addition to the normal 1,4-methylene-interrupted (nonconjugated) polyethylenic acids, there occur in nature and in industrial preparations the 1,3-methylene-inter-rupted (conjugated) polyethylenic fatty acids. The presence of conjugated ethylenic bonds increases the retention times considerably compared to methylene-interrupted isomers on both polar and nonpolar columns. The ECL values for a series of methyl *cis,cis*- and *trans,trans*-octadecadienoates have been published [446,447]. The all-*cis*-isomers have the shortest and the all-*trans*-isomers have the longest retention times on both nonpolar and polar columns, with the following exceptions: the conjugated isomers on DEGS, the conjugated isomers, and those with the methylene groups between the two double bonds. On Apiezon L the *cis,cis*-compounds are eluted before the corresponding *trans,trans*-isomers, but on XE-60 the reverse order of elution occurs.

The methyl esters of simple long-chain acetylenic acids are recovered from argentation TLC in mixture with the olefinic fatty acids of comparable degree of overall unsaturation. GLC on both polar and nonpolar liquid phases in conventional columns is effective for tentative identification, provided proper reference compounds are available. Long-chain acetylenic compounds are retained more strongly than the corresponding olefinic esters on both polar and nonpolar columns, but positional isomers introduce uncertainty, and rigid orders of elution cannot be given. The ELC values of acetylenic and *cis*-ethylenic undecanoic acids on Apiezon L, DEGS, and Silar-10C have been reported [448]. In subsequent studies, Lie Ken Jie [449] has examined the GLC behavior of the isomeric methyl undecynoates and *cis*-undecenoates on other polar and nonpolar liquid phases. FFAP, Carbowax 20 M, and XE-60 were found to be more efficient than DEGS in these separations, while OV-101 and SE-30 were comparable to Apiezon L. The ECL values for the complete series of dimethylene-interrupted methyl octadecadiynoic acids on polar and nonpolar liquid phases have also been reported [450]. Silar-10C was superior to all other phases examined in separating the isomers. In another study, Lie Ken Jie [451] has examined the GLC behavior of the trimethylene-interrupted methyl octadecadiynoates and the corresponding *cis,cis*-octadecadienoates. This again showed the superiority of Silar-10 C in the resolution of the positional isomers and of the ethylenic and acetylenic esters. It has been pointed out elsewhere [452,453] that the contribution of the acetylenic group to the relative retention time is similar to that of three ethylenic groups. The GLC properties of fatty acids with ethylenic and acetylenic unsaturation sites at various positions of the fatty chain have been reviewed [17].

Mycolic acid esters that are inseparable by adsorption TLC have been isolated and further fractionated by GLC on nonpolar columns [454,455]. Takagi et al. [456] have separated cholesteryl esters of fatty acids according to molecular weight and degree of unsaturation on Silar-10C columns by temperature programing from 240 to 270°C. ECL values were recorded for cholesteryl esters with 14 to 22 acyl carbons and 0 to 6 double bonds. König and Benecke [457] have reported the GLC resolution of enantiomeric O-TFA and O-TMS derivatives of 2-hydroxy and of branched carboxylic acids as their diastereoisomeric esters of (+)-3-methyl-2-butanol on SE-30 capillary columns. Separations were obtained for the enantiomeric 2-hydroxypalmitates, 2-hydroxymyristates, and 2-hydroxylaurates, as well as of many shorter-chain hydroxy acids. TFA and TMS derivatives gave improved peak shape and component recovery.

12.4.1.3. High-performance liquid chromatography

In the last few years, numerous methods have been described for the separation of fatty acids by means of HPLC. Because the UV absorption by the small amounts of acids analyzed is limited, the preparation of the UV adsorbing derivatives has yielded the best results [458–461], but RI [8,9] and FID [398] have also been satisfactory. The phenacyl [411,460], nitrobenzyl [460], and 2-naphthacyl esters [462] have been used most often. The separations are usually accomplished on reversed-phase columns [458–462], but adsorption columns have also been effective [463].

Excellent separations of the fatty acid methyl esters in cod liver oil have been obtained on LiChrosorb 10 RP-18 with acetonitrile as the eluent [398]. The separations are based on the partition numbers of the fatty acids and result in an overlap of esters of corresponding partition numbers. The peaks were collected and analyzed by GLC. This sytem also separates *cis*- and *trans*-isomers of fatty acids; previously, *cis*- and *trans*-fatty acids have been resolved by HPLC with 90–80% aqueous methanol [461,464,465]. Owing to the influence of steric effects on the eluent–double bond interactions, the *cis*-isomers are eluted first. Although the $18:3\,\omega 3$ and $18:3\,\omega 6$ *p*-bromophenacyl esters have been separated by HPLC [464], this chromatographic system is not very well suited for resolving positional isomers. The short-chain fatty acid naphthacyl esters have been resolved on reversed-phase C_{18} LiChrosorb columns with a convex gradient of 38% to 75% aqueous acetonitrile [462].

HPLC on a μBondapak C_{18} Porasil column with methanol–water and RI detection has been used by Warthen [465] to separate methyl oleate and methyl elaidate. Schofield [466] used a C_{18} Corasil column with aqueous acetonitrile to separate the esters of fatty acids according to chain length and degree of unsaturation. He also used a Bondapak C_{18} Porasil (37–75 μm) column with aqueous acetonitrile to separate the esters of fatty acids in 200-mg batches. Schofield [467] was able to separate *cis*- and *trans*-isomers within an hour. Lam and Grushka [411] have separated certain *cis*- and *trans*-fatty acids as their phenacyl ester derivatives, which were detected by UV absorption.

Bailie et al. [468] have reported a 50-min HPLC analysis with RI detection of the free fatty acids extracted from margarine. The distribution of the *cis*- and *trans*-isomers of the octadecenoates was studied and eight of the fatty acids were quantitated. The column used was the reversed-phase Fatty Acid Column (10 μm, Waters Assoc.) and the eluent was tetrahydrofuran–acetonitrile–water, (5:7:9), containing 0.1% acetic acid to improve peak symmetry. This column did not separate palmitic acid from linoelaidic acid ($18:2\,\Delta 9t$, 12t). A LiChrosorb Hibar-II RP-8, 10-μm, reversed-phase column, eluted with tetrahydrofuran–acetonitrile–water (3:67:30) containing 0.1% acetic acid, provided the best resolution between adjacent species. In later work, a Porasil A precolumn was used to saturate the mobile phase with silica in order to lengthen the column life. The use of pure tetrahydrofuran to dissolve the fatty acids was found to destroy the resolution of the peaks. The RP-8 column did not separate the $18:2\,\Delta 9c$, 12t and $18:2\,\Delta 9t$, 12c acids.

12.4.2. Oxygenated fatty acids and prostaglandins

The subclasses of oxygenated fatty acids and prostaglandins of uniform polarity, recovered by adsorption chromatography, may be subfractionated or resolved into individual molecular species by means of argentation TLC, GLC, and HPLC. For this purpose the acids are usually derivatized but in certain instances analysis in the free form has offered special advantages.

12.4.2.1. Argentation chromatography

Unsaturated epoxy acids may be separated by argentation TLC with benzene–chloroform–diethyl ether (25:25:1) into saturated, monoenoic, and dienoic esters [469]. Hydroxy fatty esters of different degrees of unsaturation are resolved by argentation TLC with benzene–chloroform–diethyl ether (10:10:3) [469] or benzene–diethyl ether (1:1) [470]. Dihydroxy fatty esters have been separated on the basis of both ethylenic unsaturation and the *threo-* or *erythro*-configuration of the vicinal glycol groups on silica gel impregnated with both silver nitrate and boric acid [471].

Argentation TLC permits separation of individual members of each series of prostaglandins, based on the number of olefinic double bonds available for complexing with silver ions. Ramwell and Daniels [472] and Daniels [122] have tabulated the relative migration rates for a large variety of prostaglandins, obtained by argentation TLC with a variety of solvent systems. Complete separations were obtained with the upper phase of ethyl acetate–2,2,4-trimethylpentane–acetic acid–methanol–water (22:2:6:7:20) (PGE_1, PGE_2, PGE_3, $PGF_{1\alpha}$, $PGF_{1\beta}$, $PGF_{2\alpha}$ + $PGF_{2\beta}$, $PGF_{3\alpha}$ + $PGF_{3\beta}$, PGA_1 + PGA_2 + PGB_1, PGB_2); upper phase of ethyl acetate–2,2,4-trimethylpentane–acetic acid–water (9:5:2:10) (PGE_1, PGE_2, $PGF_{1\alpha}$ + $PGF_{1\beta}$, $PGF_{2\alpha}$ + $PGF_{2\beta}$, PGA_1 + PGB_1, PGA_2, PGB_2, $MePGE_1$, $MePGE_2$, $MePGE_{1\alpha}$, $MePGE_{2\beta}$, $MePGA_1$ + $MePGA_2$); chloroform–methanol–acetic acid (18:1:1) (PGE_1, PGE_2, $PGF_{1\alpha}$ + $PGF_{1\beta}$, $PGF_{2\alpha}$ + $PGF_{2\beta}$, PGA_1 + PGB_1, PGA_2 + PGB_2, $MePGE_1$, $MePGE_2$, $MePGE_{1\alpha}$, $MePGE_{2\alpha}$, $MePGA_1$ + $MePGA_2$); chloroform–methanol–acetic acid (8:1:1) (PGE_1, PGE_2, $PGF_{1\alpha}$ + $PGF_{1\beta}$, $PGF_{2\alpha}$ + $PGF_{2\beta}$, PGA_1 + PGB_1, PGA_2 + PGB_2, $MePGE_1$, $MePGE_3$, $MePGE_{1\alpha}$, $MePGE_{2\alpha}$, $MePGA_1$ + $MePGA_2$); upper phase of ethyl acetate–methanol–water (4:1:1) ($MePGE_1$, $MePGE_2$, $MePGE_3$, $MePGE_{1\alpha}$, $MePGE_{2\alpha}$, $MePGE_{3\alpha}$).

Argentation TLC has also been employed for the resolution of certain prostaglandin metabolites [122]. Thus, the upper phase of ethyl acetate–2,2,4-trimethylpentane–acetic acid–water (10:3:1:10) resolved PGE_1 + 11α,15-dihydroxy-9-oxoprost-5-enoic acid, PGE_2, 11α,15-dihydroxy-9-oxoprostanoic acid, 11α-hydroxy-9,15-dioxoprostanoic acid, and 11α-hydroxy-9,15-dioxoprost-5-enoic acid. Furthermore, the upper phase of ethyl acetate–2,2,4-trimethylpentane–acetic acid–water (11:3:2:10) resolved 5α,7α,11,15-tetrahydroxytetranorprostanoic acid + 5α,7α,11,16-tetrahydroxytetranorprostanoic acid and 5α,7α,11,15-tetrahydroxynorprost-9-enoic acid + 5α,7α,11,16-tetrahydroxytetranorprost-9-enoic acid. For this purpose, the silica gel is impregnated either by incorporating silver nitrate in the slurry used for preparing the plates or by dipping precoated plates in solutions of 5 to 20% silver nitrate. Recovery of prostaglandins from the silica gel, especially in subnanogram amounts, remains a major problem; it is not uncommon to obtain less than 50% recovery. Because methyl esters can generally be recovered in higher yields than the free fatty acids, the esters are preferable for quantitiative work. Argentation TLC completely separates HHT and 12-HETE, while PGE_2 migrates with TXB_2 in this TLC procedure.

The methyl esters of prostaglandins have also been chromatographed on Amberlyst-15 (Rohm & Haas) ion-exchange resin in the silver salt cycle to yield PGE_1 (eluted with 95% ethanol) and PGE_2 (eluted with 5% cyclohexene in 95% ethanol) [129]. A similar procedure has been used to separate PGA_2 and 5-*trans*-PGA_2 [473]. Merritt and Bronson [474] have used silver ion-loaded microparticulate cation-exchange resin columns for the separation of the *p*-nitrophenacyl esters of selected prostaglandins by HPLC.

12.4.2.2. Gas–liquid chromatography

Tulloch [475] and Tulloch and Mazurek [476] have devised a scheme for the identification of most of the isomeric hydroxy stearates by GLC with three liquid phases: EGS, QF-1, and SE-30. The 12-hydroxy-, acetoxy-, and oxostearates derived from hydrogenated castor oil were used as standards. O'Brien and Rouser [477] have separated all the isomeric methyl hydroxypalmitates on EGS, DEGS, and Apiezon L, and the acetoxypalmitates on EGS. Karlsson and Pascher [478] have described the resolution and the chromatographic analysis of configuration of 2-hydroxy fatty acids by use of various derivatives and 3% OV-1. The ECL values for a number of keto fatty acids and related compounds obtained during chemical transformations of certain natural fatty acids have been recorded [479]. Wood et al. [480] have reported the GLC separation of polyhydroxy methyl ester of TMS ether derivatives of fatty acids on capillary GLC with Apiezon L as the stationary liquid phase. Partial separations of methyl *threo*- and *erythro*-9,10-dihydroxystearate were obtained, as were partial separations of the diastereoisomeric methyl 9,10,12-trihydroxystearates and the methyl 9,10,12,13-tetrahydroxystearates. Wood [481] has shown that packed columns can be used for the separation of the isopropylidene derivatives of the dihydroxy compounds derived from oleic and elaidic acids. The isopropylidene derivatives of the eight diastereoisomers from linoleic acid hydroxylation were resolved into five peaks on an EGSS-X column. The isopropylidene–TFA derivatives of the four diastereoisomeric trihydroxy acids from alkaline permanganate oxidation of ricinoleic and ricinoelaidic acid gave four peaks in GLC.

Systematic studies of the GLC behavior of the primary prostaglandins have been reported by several groups of investigators, and these studies have been summarized in several reviews [122,482,483]. With FID, the PGF compounds are usually chromatographed as either the TMS ether methyl esters or the acetoxy methyl esters. As yet, no entirely suitable derivative for PGE compounds has been reported. To avoid the problem of dehydration and of double peak formation during oximization, the PGE compounds have been converted into either PGA and PGB series, which are then derivatized and chromatographed [484]. Improved sensitivity has been observed with the ECD. Thus, Jouvenaz and coworkers [485,486] have analyzed PGE_1 and PGE_2 with ECD by converting them to bromosilylated derivatives of the PGB_1 and PGB_2, respectively, which have subnanogram detection limits. The TMS ether pentafluorobenzyl ester of $PGF_{2\alpha}$ has low-picogram detection limits [487,488] and the oximated pentafluorobenzoyl esters of D and E prostaglandins are also highly sensitive to ECD [489,490].

Skrinska and Butkus [491] have described the use of the pentafluorobenzoyl esters of PGE_1, PGE_2, $PGF_{1\alpha}$, and $PGF_{2\alpha}$. During derivatization of the mixture, the PGE_1 and PGE_2 are converted to PGB_1 and PGB, $PGF_{1\alpha}$ and $PGF_{2\alpha}$ TMS ethers. A similar conversion allowed the analysis of the A prostaglandins as the B prostaglandin derivatives. The separations were performed on a conventional column, containing a 3% QF-1 packing. The detector was a ^{63}Ni ECD and the carrier gas was argon–methane (19:1). The TFA [122,486] and heptafluorobutyryl [492,493] derivatives have given conflicting results, apparently due to instability. Most of the GLC separations of the prostaglandins have been performed on 1% SE-30 columns at temperatures between 190 and 230°C, but the less polar polyester liquid phases (DMCS) have also been occasionally used. Daniels [122] has compiled an extensive collection of chromatographic conditions for the determination of primary prostaglandins in the form of acetates, acetate methoximes, TMS ethers and TMS ether methoximes. Nugteren [494] has reported the use of $NaBH_4$ and Zn/HCl to reduce prostaglandins to the ultimate hydrocarbon skeleton. The degradation procedure has the advantage that dehydration products of PGE compounds or lactones of different prostaglandins are all normalized to a single chemical species. Pace-Asciak and Wolfe [495] used n-butyl boronate, which reacts specifically with vicinal hydroxyl groups to separate E from F prostaglandin by GLC. The persilylated derivative of $PGF_{2\alpha}$ has been employed for sensitive detection [496], and the t-BDMS ethers have advantages for mass spectrometry [497].

The application of capillary columns to prostaglandin research promises renewed interest in the GLC technique. Rigaud et al. [498] have obtained excellent separation of 11-epi-$PGF_{2\alpha}$, 13,14-dihydro-$PGF_{2\alpha}$, 11-epi-$PGF_{1\alpha}$, and 15-epi-$PGF_{1\alpha}$ in six min on a 20-m polysiloxane column at 235°C. Maclouf et al. [499], using polysiloxane capillary columns and conventional columns, have resolved some complex prostaglandin mixtures, including the methyl ester TMS ethers of $PGF_{2\beta}$, $PGF_{1\beta}$, 13,14-dihydro-$PGF_{1\beta}$, 15-epi-$PGF_{2\alpha}$, 15-epi-$PGF_{1\alpha}$, $PGF_{3\alpha}$, 5-trans-$PGF_{2\alpha}$, 15-keto-$PGF_{2\alpha}$, and $PGF_{1\alpha}$. Prostaglandin and thromboxane separations by capillary GLC have been described by Fitzpatrick [500].

The E prostaglandins have poor GLC properties and, consequently, they are analyzed as oxime derivatives or they are first converted to the respective B prostaglandins. Nicosia and Galli [484] have developed a procedure for simultaneously silylating and methylating PGE_1 and PGE_2 and for converting them to PGB_1 and PGB_2, respectively, with a mixture of TMS imidazole and piperidine. The most effective identification and quantitation of molecular species of various prostaglandins is obtained by a combination of GLC with selected-ion detection mass spectrometry, including appropriate stable isotope-labeled internal standards [501–503].

12.4.2.3. High-performance liquid chromatography

Liquid–liquid chromatography is also well suited for the resolution of molecular species of the oxygenated fatty acids and prostaglandins, but because most of these compounds lack significant absorption above 220 nm, they cannot be detected in the nanogram range without conversion to UV-absorbing derivatives.

Hydroxy fatty acid methyl esters, derived from free-radical oxidation of polyunsaturated phosphatidylcholine, were separated by HPLC on a Porasil (10 μm, Waters Assoc.) column with 0.5% ethanol in hexane as the eluent and detected at 235 nm [504]. Free hydroxy fatty acids were analyzed with acetic acid–2-propanol–hexane (1:16:983) as the mobile phase. The air-oxidation products of arachidonic acid have been similarly purified by HPLC [505,506]. Fitzpatrick [507] has determined $PGF_{2\alpha}$, PGE_2, and PGD_2 prostaglandins as the p-nitrophenacyl esters by HPLC with UV detection. Merritt and Bronson [474,508] have chromatographed the p-nitrophenacyl esters on a silver-loaded cation-exchange resin, while Oesterling et al. [509] chose the p-nitrobenzyl esters for the detection of PGE_2, $PGF_{2\alpha}$ and PGA_2, after separation on a Zipax column by elution with 31% aqueous 2-propanol or 30% aqueous methanol. This procedure is not suitable for PGI_2, as it spontaneously decomposes to 6-oxo-$PGF_{1\alpha}$ during resolution on silica gel or on reversed-phase columns [510]. Turk et al. [511] have used the fluorescent 4-bromomethyl-7-methoxy-coumarin derivatives of prostaglandins and thromboxane for HPLC. In a single experiment, complete separations were obtained of PGD_2, PGE_2, $PGF_{2\alpha}$, 6-keto-$PGF_{1\alpha}$, and thromboxane B_2 by programed gradient elution with two solvent systems: chloroform–isooctane–methanol (35:65:1) and chloroform–methanol (80:1).

Where insufficient material is available or where derivative preparation is not possible, UV detection below 220 nm has proved satisfactory. Whorton et al. [512] have described an HPLC method for the separation of PGE_1 from PGE_2, 13,14-dihydro-PGE_2, and other related prostaglandins. They used a Fatty Acid Column with a solvent system of water–acetonitrile–benzene–acetic acid (767:230:2:1). The method of Whorton et al. [513] required the inconvenient use of liquid scintillation spectrometry, and retention times were exceedingly long (1 h). Hill [514] claimed that he can separate prostacyclin ($PGFI_2$) from its hydrolysis product, 6-oxo-prostaglandin $PGF_{1\alpha}$ and from other prostaglandin impurities with a laboratory-prepared reversed-phase packing and water–methanol (3:2), containing 2.5 g/l H_3BO_3 and 3.8 g/l $Na_2B_4O_7$, as eluent. Wynalda et al. [515] reinvestigated this method and concluded that the presence of methanol in the eluent leads to methyl ketal formation. These products appear as distinct peaks in the elution profile and complicate the assay. They recommend instead the use of 20% aqueous acetonitrile, buffered at pH 9.3 with 0.009 M H_3BO_3 and 0.004 M $Na_2B_4O_7$ and a μBondapak C_{18} column. Inayama et al. [516] separated 6-keto-$PGF_{1\alpha}$, $PGF_{2\alpha}$, PGE_2, PGE_1, $PGA_2 + PGB_2$, and $PGA_1 + PGB_1$ by HPLC and UV detection in 10 min, using water–acetonitrile–tetrahydrofuran (35:15:1) as the eluent. The method was also suitable for the determination of thromboxane B_2 (equivalent to thromboxane A_2) and the leukotrienes under neutral and mild conditions in place of an alkaline medium [514,515].

Van Rollins et al. [517] have described a HPLC procedure, which provides baseline resolution for all major cyclooxygenase and lipoxygenase products from arachidonate. Complete separations were obtained of the underivatized 6-keto-$PGF_{1\alpha}$, TXB_2, $PGF_{2\alpha}$, PGE_2, PGD_2, HHF, and 12-HETE. The chromatographic system consisted of a reversed-phase Ultrasphere ODS (5 μm, Altex, Berkeley, CA, USA) column and a gradient of 30.5 to 95% acetonitrile in aqueous H_3PO_4 (pH 2.0). The effluent was monitored at 192 nm. Underivatized prostaglandins have also been partially resolved by reversed-phase HPLC by Russell and Deykin [518] and by Nagayo and Mizuno [519]. Earlier, Hansen and Bukhave [520] had obtained excellent separations of PGE and PGF_α prostaglandins as the free acids on Sephadex LH-20 with 1,2-dichloroethane–heptane–methanol (25:25:2). Complete separations were realized for PGA prostaglandins, 15-keto-dihydro-PGE, PGE_1, 15-keto-PGE_1, 15-keto-dihydro-$PGF_{1\alpha}$, 15-keto-$PGF_{1\alpha}$, and $PGF_{1\alpha}$. This system does not resolve dihydro-PGE_1 and PGE_2, which can be separated, however, with 1,2-dichloroethane–methanol (50:1). It is anticipated that combinations of HPLC and MS will soon find general use in the identification and quantitation of molecular species of prostaglandins, just as the GC–MS combination has in the past [521].

12.4.3. Neutral acylglycerols

Neutral acylglycerols are effectively separated into groups of uniform degree of unsaturation by argentation TLC and into groups of uniform carbon number or molecular weight by GLC, while reversed-phase HPLC leads to a resolution that has aspects of both of the above. For this purpose the partial acylglycerols are usually converted to appropriate chemical derivatives.

12.4.3.1. Argentation chromatography

Argentation chromatography, especially argentation TLC is one of the most important techniques for the resolution of acylglycerols within a single class of compounds. Triacylglycerols previously isolated by conventional adsorption chromatography can be further separated by argentation TLC into fractions having the same total number of double bonds [53,522]. On silica gel impregnated with 10 to 20% silver nitrate, mixtures of natural triacylglycerols form up to 20 subfractions in a single dimension when diethyl ether or benzene–diethyl ether (9:1) [522], hexane–diethyl ether (7:3) or chloroform–methanol (47:3) [53] are used. Oligoenoic triacylglycerols may be readily resolved with 1 to 2% methanol in chloroform [53,523]. Polyunsaturated triacylglycerols that remain crowded near the origin with 1% methanol in chloroform are separated by redeveloping the plate with 5% methanol in chloroform. By this means the triacylglycerols, containing saturated (0), monounsaturated (1), diunsaturated (2), and triunsaturated (3) acids are separated into molecular species as follows: 000, 001, 011, 002, 111, 012, 112, 022, 003, 122, 013, 222, 113, 023, 123, 223, 033, 233, 033, 133, 233, 333 [522].

According to Takagi and Itabashi [524], improved resolution of triacylglycerols is obtained in argentation TLC by double development with benzene–chloroform (9:1) or chloroform–methanol (97:3), or by triple development with chloroform–methanol (19:1). Repeated development with the same or different mixtures have been employed by other workers on a more limited scale [525,526]. Similar resolutions may be obtained with triacylglycerols containing fatty acids other than those listed above. Thus, triacylglycerols containing hydroxy [522], epoxy [527], and keto

[528] acids can be resolved into molecular species by argentation chromatography, but require more polar solvent systems than the ordinary esters [529]. Cyclopropene fatty acids react with silver nitrate to form a mixture of products [530]. Triacylglycerols containing epoxy fatty acids have been resolved with benzene–diethyl ether (3:1) [527], and triacylglycerols containing hydroxy acids with diethyl ether [528]. Ether analogs, where one or more of the acyl groups has been replaced by an ether linkage, may also be resolved by argentation TLC [163,531] with essentially the same solvent systems as those used for the triacylglycerols.

Argentation chromatography can also be used to resolve either the sn-1,2 (2,3)- or the sn-1,3-diacylglycerols. A convenient system for the resolution of free sn-1,2-diacylglycerols consists of chloroform–ethanol (93:7), which yields separate fractions for the molecular species 00, 01, 11, 02, 12, 03, and 04, where 0 to 4 represent individual fatty acids with 0 to 4 double bonds per molecule [532,533]. Since the fractionation depends on the total number of double bonds and the distribution of double bonds in the 1- and 2-positions within the component fatty acids, more than one fraction may be obtained for any one degree of unsaturation. Likewise, diacylglycerols containing cis- and trans-monoenoic fatty acids are resolved in most solvent systems [534]. Since free diacylglycerols may isomerize during argentation chromatography, it is advisable to derivatize them prior to chromatography. Acetate [535,536], t-BDMS [537] and sometimes the unstable TMS [471,538] ether derivatives have been used for argentation TLC of diacylglycerols.

The fractionation of diacylglycerol acetates by argentation TLC allows separations of molecular species containing 0 to 12 ethylenic double bonds. Many positional isomers are separable as well [539]. The separations obtained are similar to those described above for the free diacylglycerols and triacylglycerols. The diacylglycerol acetates with 0 to 4 double bonds are effectively resolved with chloroform–methanol (99:1) [540,541], but more unsaturated mixtures of diacylglycerols may require more polar solvents, e.g., chloroform–methanol (19:1) [539]. The 1,3-diacylglycerols and their derivatives may be resolved into molecular species by similar solvent systems. The sn-1,2-diacylglycerol acetates derived from lung phosphatidylcholine have been resolved by argentation TLC and a stepwise development with chloroform–methanol (19:1) and chloroform as the solvents [542,543]. Alk-1-enylacylglycerol acetates or alkylacylglycerol acetates can also be separated according to the degree of unsaturation by argentation TLC with benzene–chloroform (9:1) [164,544] or chloroform–methanol (49:1) [545]. The saturated alk-1-enylacylglycerol acetates migrate more slowly than the saturated alkylacylglycerol acetates, because of the retarding effect of the double bond in the alkyl-1-enyl moiety. Diacylglycerol acetates of the same degree of fatty acid unsaturation migrate midway between the two types of ether lipids on AgNO₃-TLC plates [537].

Monoalkylglycerols and monoacylglycerols with 0, 1, and 2 double bonds may be separated by argentation TLC with chloroform–methanol (9:1) [166,546,547]. Saturated and monounsaturated 2-methoxyalkylglycerols have been separated by argentation TLC with trimethylpentane–ethyl acetate–methanol (50:40:7) [548]. The advantage of submitting the monoacylglycerols, rather than the corresponding fatty acids, to argentation TLC is that the samples are less likely to be contaminated with foreign monoacylglycerols than with foreign free fatty acids.

Recently, Chromarods impregnated with silver nitrate and developed with benzene–diethyl ether (49:1) have been demonstrated to be suitable for the resolution and quantitation of synthetic tripalmitoyl-, trioleoyl-, and trilinoleoylglycerols [549]. Smith et al. [550] have used benzene as a mobile phase for a preparative separation of triacylglycerol mixtures, including the positional isomers 2-unsaturated-1,3-disaturated (SUS) and 1-unsaturated-2,3-disaturated (SSU) by argentation HPLC. The optimum loading of silver nitrate was 5% on the Partisil 5 packing, and RI detection was employed.

12.4.3.2. Gas–liquid chromatography

GLC is the most powerful tool for resolving molecular species of neutral lipids, yielding separations based on molecular weight, degree of unsaturation, and in some instances, resolution of geometric isomers. Its effectiveness is further increased in combination with argentation TLC.

Long-chain triacylglycerols previously isolated by adsorption chromatography may be separated according to carbon number by temperature programing GLC (200 to 300°C, 2 to 4°C/min) on short (30 to 50 cm × 0.24 cm OD) columns, prepared with packings containing 1–3% nonpolar liquid phase (methyl or phenyl siloxane) [91,199,551]. This constitutes a practical routine method for high-temperature GLC of natural fats and oils [6,16,199,552]. Although longer columns may be used to obtain increased resolution, they generally reduce the recoveries of the longer-chain components relative to the shorter-chain components [6,199]. Short-chain triacylglycerols are readily resolved by temperature-programed GLC on conventional columns (180 cm × 0.25 cm OD), packed with the above liquid phases [160]. Triacylglycerols may also be resolved by means of capillary GLC [553–557]. However, on 20-m columns the relative recoveries of the higher-molecular-weight species have been low and have required the use of hydrogen as a carrier gas [553,555]. In contrast, 4- to 6-m capillary columns have given excellent resolution and recovery of both short- and long-chain triacylglycerols in the temperature range 280 to 350°C with He [554] or H_2 [556] as the carrier gas. The short capillary columns yield complete resolution of triacylglycerols differing by one methylene unit, but otherwise the resolutions do not differ significantly from those obtained on the short packed columns in use earlier [551].

Monseigny et al. [554] tested short capillary columns under various conditions for the chromatography of neutral acylglycerols. Grob et al. [555] reported separations of a variety of triacylglycerol mixtures on 14-m OV-1 capillary columns with hydrogen as a carrier gas in the temperature range of 240 to 340°C. On-column injection at 60°C was used for the samples in hexane. Although separations of all isomers within a group of triacylglycerols with identical carbon numbers were not feasible, capillary columns give more information than can be obtained with packed columns. Persilylated columns coated with nonpolar silicone gum phases are still useable for about a year after hundreds of injections. With both packed and capillary columns, natural triacylglycerols show peak broadening due to a relatively earlier elution of the unsaturated species within a given carbon number, but true resolutions of saturated and unsaturated species are not achieved. Peak broadening may be eliminated by hydrogenation of the sample [199]. Hydrogenation may also be used to prevent thermal degradation of the more highly unsaturated species [199,558].

Prior argentation TLC is required for a resolution according to carbon numbers within each fraction of triacylglycerols of uniform degree of unsaturation [558]. The presence of branched-chain acids also leads to decreases in the relative retention time of the triacylglycerols, and their presence causes peak broadening or shouldering [559]. Triacylglycerols containing fatty acids with cyclopentane rings are eluted sufficiently behind their straight-chain analogs to allow their resolution [559]. Triacylglycerols containing acetoxy fatty acids are readily separated by GLC [560]. Epoxy acid-containing triacylglycerols may be resolved by GLC as such, or as their 1,2-dioxolane derivatives [561]. The 1,3-dioxolane derivatives formed by condensing the epoxytriacylglycerols with cyclopentane in the presence of boron trifluoride are thermally stable and allow a simultaneous separation of both native and derivatized triacylglycerols. The estolides [199,560] are not eluted during GLC under normal conditions.

GLC may also be used to resolve alkyldiacylglycerols according to carbon number [562]. Because these compounds possess one carbonyl oxygen less than the triacylglycerols, they are eluted approximately one methylene unit earlier than the corresponding triacylglycerols. The resolution of species having different carbon numbers is comparable to that obtained for long-chain triacylglycerols. In general, triradyl (alkylacyl) glycerol species having the same carbon number are eluted approximately one methylene unit earlier for each O-alkyl group present. A mixture of C_{48}-glyceryl ethers and esters is eluted in the order: tripalmityl-, 1,2-dipalmityl-3-palmitoyl-, 1-palmityl-2,3-dipalmitoyl-, and tripalmitoyl-glycerol [6]. Like triacylglycerols, alkylacylglycerols containing branched-chain or unsaturated fatty acids yield partial splitting or shouldering of carbon number peaks, when present in sufficiently high concentration [563].

Short-chain triacylglycerols may be resolved by GLC at moderate temperatures according to both carbon number and degree of unsaturation when thin films of polar liquid phases are used. Triacylglycerols of a molecular distillate of bovine milk fat have been resolved into a complex mixture of molecular species on 3% EGSS-X [160] and Silar-5CP [564]. The elution pattern included numerous partially overlapping peaks. The superposition resulted from the presence of a high proportion of butyric, caproic, and caprylic acids, which interact significantly with the polar liquid phase, due to their higher polarity.

Takagi and Itabashi [524] have reported an attempt to separate triacylglycerols according to carbon number and degree of unsaturation on a Silar-10C column. Although the liquid phase had to be used at its temperature limit, considerable fractionations were obtained for triacylglycerols containing one to four double bonds per molecule within the 36 to 54 acyl carbon range. The data obtained suggest that work with capillary columns, prepared with comparable liquid phases, may accomplish complete resolution of triacylglycerols based on carbon number and degree of unsaturation of the acyl groups.

The sn-1,2 (2,3)- and 1,3-diacylglycerol isomers as well as their alkylacyl, alk-1-enylacyl, and dialkylglycerol analogs, recovered from adsorption chromatography, may be separately resolved into molecular species by GLC. The TMS [565,566], *t*-BDMS [537,567], and acetate [536,568] derivatives, as well as the free diradylglycerols [566] may be resolved on the basis of carbon number on short columns containing nonpolar liquid phases of the type employed for triacylglycerol resolution. The positional isomers of acetoglycerols [569], butyroglycerols [570,571], or diacylglycerol acetates [16] are not readily resolved by the short GLC columns. Because their molecular weight is lower, these separations can usually be achieved below 300°C. Lohninger and Nikiforov [572] have reported the capillary GLC analysis of the TMS ethers of 1,2- and 1,3-diacylglycerols. Resolution based on carbon number was obtained on a 12-m column containing SE-30 in the temperature range 240 to 300°C with hydrogen as the carrier gas. Like triacylglycerols, the unsaturated diacylglycerols were emerged slightly ahead of the corresponding saturated species and this led to partial splitting and shouldering of the peaks. Effective separations of saturated and unsaturated diacylglycerols were not obtained on the nonpolar columns. Therefore, a complete resolution of molecular species by this method requires a prior resolution of the diacylglycerols according to degree of unsaturation by means of argentation TLC [536]. The combination of $AgNO_3$-TLC and GLC of the diacylglycerol acetates has been widely employed in the fractionation of molecular species of diacylglycerols derived from diacylglycerol phospholipids [536,542,543,573–577]. The short nonpolar GLC columns are also suitable for a partial resolution of mixtures of alkylacylglycerols. The TMS ethers of the diacylglycerols overlap the dialkylglycerols having two methylene units less, but the alkylacylglycerols are well separated from the diacylglycerols [545,578]. The acetate derivatives are superior for this separation, because combinations of dialkyl-, alkylacyl-, and diacylglycerols or diacyl-, alkylacyl-, and alk-1-enylacylglycerols may be resolved. Like triacylglycerols, the diacylglycerols yield more compact peaks and improved resolution according to carbon number after hydrogenation [16].

Effective resolutions of saturated and unsaturated diacylglycerols and of the analogous alkylacylglycerols may be obtained on conventional GLC columns containing thin films (3% packings) of polar liquid phases, such as ethyleneglycol succinate [579], cyclohexanedimethanol succinate [580], and Silar-5CP [581], when they are chromatographed in the form of TMS ethers. These columns have given essentially complete resolution of molecular species of diacylglycerols composed of saturated, monounsaturated, diunsaturated, and triunsaturated C_{16} and C_{18} fatty acids. Only the TMS ethers of oleoyllinoleoyl- and dilinoleoylglycerols remained unresolved, whereas the TMS ethers of stearoyllinoleoyl- and dioleoylglycerols were partially resolved. Furthermore, 1-stearoyl-2-linoleoyl-sn-glycerol is only partially separated from 1-palmitoyl-2-arachidonoyl-sn-glycerol [6]. Reverse isomers, such as the TMS ethers of 1-oleoyl-2-linoleoyl- and 1-linoleoyl-2-oleoyl-sn-glycerols and enantiomeric pairs are also unresolved.

Dyer and Klopfenstein [582] have obtained improved separations of saturated and unsaturated diacylglycerols as the TMS ethers on 10% Silar-10C by isothermal GLC at 265°C. Diacylglycerols ranging from 32:0 to 38:4 gave essentially baseline resolution on a 2.5 m × 3-mm ID column. However, the anticipated 40:6 species was not recovered. Itabashi and Takagi [541] reported excellent separations of diacylglycerols, varying from dipalmitates to dilinolenoates, as the TMS ethers by use of a 1.5-m column, containing 3% Silar-10C and temperature programing in the range of 180 to 270°C. They compared the behavior of the diacylglycerol acetates and TMS ethers on 3–5% Silar-5CP, Silar-7GP, Silar-10C, and OV-275 liquid phases. The best separations for the diacylglycerols, derived from the common plant oils and animal fats, have been obtained with the TMS ethers on SP-2330 at 250–270°C isothermally [583]. In some instances the TMS ethers gave resolution of positional isomers, which were not obtainable with the acetate derivatives.

Monoacylglycerols are readily resolved by carbon number on conventional non-polar columns in the form of acetates [555,584], isopropylidenes [585], TMS ethers [552,586], or *t*-BDMS ethers [567]. Good separations of monoacyl- and monoal-kylglycerols have been obtained with the TMS ethers [584], acetates [584], trifluoro-acetates [546], and isopropylidenes [585]. The monoacylglycerols give excellent separations of molecular species on polar columns, such as EGSS-X when they are in the form of acetates [584], TMS ethers [540], or isopropylidenes [585,587,588], and on Silar-5CP columns when they are chromatographed as acetates [584] or TMS ethers [584]. Silar-5CP has a higher thermal stability than the other polar liquid phases and is therefore preferable for operation at elevated temperatures. Even higher thermal stability is exhibited by Silar-10C, which has provided excellent separations for the acetate and TMS derivatives of monoacylglycerols according to carbon number and degree of unsaturation [541]. Since there is only one fatty acid group per molecule, the GLC separations of monoacylglycerols and their derivatives parallel those of fatty acid methyl esters. The TMS ethers and acetates of monoalkyl- and monoalkenylglycerols are also readily separated on the basis of chain length [166] and unsaturation [584,589], as are the methaneboronates of monoalkylglycerols [528,590]. The resolutions of the alkyl-, alkenyl-, and acylglycerols on polar columns have been discussed at length by Myher [16].

Glyceryl ethers having substituents on the alkyl chain have also been separated by GLC [203,204,591]. Standard 1-(9/10-hydroxy) octadecylglycerol and the 1-(11/12-hydroxy) alkylglycerols were separated on a 183-cm 10% EGGS-X column at 200°C as the TMS ethers, TMS/isopropylidene and acetate/isopropylidene derivatives [204,591]. A number of derivatives of 1-O-(2-hydroxylalkyl)- and 1-O-(2-ketoalkyl)glycerols have also been separated by GLC as the acetate/isopropylidene derivatives, when chromatographed on a 10% SP-1000 column [203]. The alkyl- and 2-methoxyalkylglycerols have been resolved on a 1% Apiezon L plus 0.1% polyethyl-ene glycol column at 218°C and on a 1% EGS column at 194°C [548]. The unsaturated species are eluted earlier than the saturated species on the nonpolar Apiezon L column and vice versa on the polar EGS column. GLC separations have also been effected of the four isomers formed by acid-catalyzed cyclization of 1-O-*cis*-alk-1-enyl-sn-glycerol: *cis*-2-alkyl-5-hydroxy-1,3-dioxane, *trans*-2-alkyl-5-hy-droxy-1,3-dioxane, *cis*-2-alkyl-4-hydroxymethyl-1,3-dioxane, and *trans*-2-alkyl-4-hydroxymethyl-1,3-dioxane, on a 190-cm 18% HI-EFF-2BP (EGS) column [592]. It is also possible to separate the 1-S-alkylglycerol ether from the corresponding 1-O-alkylglycerol ether in the form of the isopropylidene derivatives on SE-30, EGSS-X, and EGS columns [593]. Oette and Tschung [208] have reported an isothermal (210°C) GLC resolution of the 2-N-acylserinol mixtures as the TMS ethers on 1% OV-1. Molecular species of wax esters have also been resolved by GLC on 5% Silar-10C at 200–260°C [594].

12.4.3.3. High-performance liquid chromatography

An effective resolution of triacylglycerols according to chain length was obtained by reversed-phase liquid partition chromatography on columns containing silanized Celite [595]. Subsequent work [596–599] dealt with reversed-phase separation of triacylglycerols on columns of hydroxyalkoxypropyl Sephadex. An improved method of triacylglycerol separation was described by Lindqvist et al. [600]. Using a linear gradient of two solvents, consisting of isopropanol–chloroform–heptane–water (115:15:2:35) and heptane–acetone–water (5:15:1), they were able to obtain an extensive separation of saturated C_9- to C_{54}-triacylglycerols. Unsaturated triacylglycerols were eluted about 1.42 methylene units earlier for each double bond in the molecule and overlapped the corresponding saturated triacylglycerols. Curstedt and Sjövall [601,602] have separated 1,2-diacyl-3-trimethylsilyl-sn-glycerols on similar columns with acetone–water–heptane (87:13:10) as the eluent. In order to prevent acid hydrolysis of the ethers, the solvent also contained 1% pyridine. From a mixture of the sn-1,2-diacylglycerols, derived from rat liver phosphatidylcholines, they obtained two minor peaks, corresponding to 38:6 and to 32:2 + 34:3 + 34:4, and two major peaks, corresponding to 32:1 + 34:2 + 36:3 and to 32:0 + 34:1 + 36:2. The chief disadvantage of these techniques is that they are too tedious and time-consuming (e.g., 20 h for the resolution of C_9- to C_{54}-triacylglycerols).

Rapid separation of triacylglycerols by chain length and degree of unsaturation has been achieved by HPLC on a μBondapak C_{18} column with acetonitrile–acetone (2:1) [603]. For saturated triacylglycerols, a linear relationship was observed between the carbon number and the logarithm of the retention volume, as previously noted for conventional partition columns [596]. Each double bond increment in the triacylglycerol decreased the retention volume to approximately that of a saturated triacylglycerol with two methylene units less. The fractions were collected and analyzed by GLC. The triacylglycerols represented carbon numbers 32 to 52. Methanol-containing solvents gave partial separation of triacylglycerols within one carbon number, including a partial resolution of tripalmitoyl- and trioleoylglycerols. There appeared to be a solubility problem in chromatographing tristearoylglycerol or higher fully saturated triacylglycerols. Occasionally, they precipitated during sample injection. As an example of the separation of natural oils, soybean oil was resolved into 7 to 8 peaks, representing C_{54}-triacylglycerols with from eight to two total double bonds, C_{52}-triacylglycerols with six to one total double bonds, and C_{50}-triacylglycerols with two to zero total double bonds.

Earlier, Pei et al. [604] had demonstrated the potential usefulness of reversed-phase partition chromatography for the resolution of triacylglycerols with 60% aqueous methanol as eluent. Bezard and Ouedraogo [605] have described an effective isocratic HPLC resolution of natural triacylglycerols with acetonitrile–acetone (43:58). A total of eight fractions were obtained for peanut oil triacylglycerols. They were collected and identified as triacylglycerols with integral partition numbers, calculated by assuming that a double bond is exactly equivalent to a reduction of two carbon atoms (two methylene units), except for the last fraction, which gave a number of 55:02 instead of 56. The presence of hydroxy or epoxy functional groups decreases the retention volume dramatically [603,606]. Thus, trivernoylglycerol is eluted far ahead of normal triacylglycerols [606]. Triacylglycerols containing hydroxy fatty acids, such as ricinoleic acid, are eluted slightly before the normal triacylglycerols [603]. When triple bonds and cyclopropenyl groups are present in an oil, they have about the same effect as two double bonds [603].

The multiacylglycerols or estolides, containing one, two, or three monohydroxy fatty acid moieties have been separated by conventional and reversed-phase HPLC [206]. In both systems, triacylglycerols were eluted first, followed in sequence by tetra-, penta-, and hexa-acylglycerols. Within each acylglycerol class, further separation occurred due to variations in chain length and degree of unsaturation among the component fatty acids. Separations were obtained on 30 cm $\times$ 7.8 mm μBondapak C_{18} columns, eluted with acetonitrile–acetone (2:1) at 1 ml/min. Reversed-phase HPLC was also performed on the Triacylglycerol column (Waters Assoc.), eluted with acetonitrile–tetrahydrofuran (3:1) at 2 ml/min. Adsorption HPLC was carried out on μPorasil with isooctane–diethyl ether–acetic acid (98:2:1) at 1 ml/min as the developing solvent. Separations between and within classes on the Triacylglycerol column were similar to those obtained on the μBondapak C_{18} column. On the μPorasil column, separations between multiacylglycerol classes were again similar, the compounds having longer acyl groups being retained longer. Within each class, however, separations on the μPorasil column were the reverse of those found on the μBondapak C_{18} and the Triacylglycerol column. When components containing the same number of acyl groups were compared, those with longer chain length and less unsaturation were eluted first. The component eluted earliest from the Triacylglycerol column was the component eluted last from the μPorasil column. The effects of chain length and degree of unsaturation in HPLC of triacylglycerols have been examined in detail and are discussed in a separate paper [607]. HPLC techniques still need to be designed for more complex estolides, containing free hydroxyl functions.

Fallon and Shimizu [608] have used HPLC to fractionate the sorbic acid-containing triacylglycerols of aphids. These triacylglycerols, differing by two methylene groups, were easily separated by adsorption chromatography on a Zorbax-Sil column. The sorbic acid esters absorb in the UV and are easily monitored in the column effluent.

12.4.4. Ceramides

The free ceramides occurring naturally and the ceramide moieties derived from natural sphingolipids are resolved into groups of uniform degree of unsaturation by argentation TLC and into groups of uniform molecular weight by GLC, while reversed-phase HPLC leads to a resolution that has features of both of the above techniques. For HPLC the ceramides are converted to appropriate derivatives, but prior to this the ceramides must be resolved according to their number of hydroxyl groups by adsorption chromatography.

12.4.4.1. Adsorption chromatography

Karlsson and Pascher [609] have published extensive studies on TLC of ceramides. There may be from two to four hydroxyl groups in the molecule (two or three in the base and zero or one in the fatty acid), and TLC on silica gel or on silica gel containing diol-complexing agents, such as sodium tetraborate or sodium arsenite, can be used to effect separations that depend on the number and configuration of the hydroxyl groups present. In addition, ceramides having a *trans*-double bond in position 4 of the long-chain base are also separable on silica gel, impregnated with sodium borate, where saturated compounds migrate ahead of unsaturated compounds, although the reason for this effect is not understood [610]. From a comprehensive study of a wide variety of synthetic ceramides, Karlsson and Pascher [609] have concluded that four groups of ceramides can be separated on the arsenite plates with chloroform–methanol (19:1): dihydroxy base-normal fatty acid and dihydroxy base-hydroxy acid, and trihydroxy base-normal fatty acid and trihydroxy base-hydroxy acid. Derivatives of dihydroxy bases isolated in this manner can then be separated on borate-impregnated plates with the same solvent system into those containing long-chain bases with *trans*-double bonds in position 4 and those which do not. A similar separation is achieved with the 2-acetoxy isomers, with the exception of the trihydroxy base-containing ceramides. No separation is obtained for the corresponding fully acetylated derivatives. Ordinarily ceramides with trihydroxy base migrate more slowly than ceramides with dihydroxy base; however, a reversal of migration is obtained for fully acetylated dihydroxy base and trihydroxy base. Ceramides with short and long chains may also be separated on Silica Gel G with chloroform–methanol (19:1), where the compounds with longer chains travel further. A clear-cut separation of the C_{18}- from C_{24}-fatty acid ceramides is obtained. Acetylation does not improve the separation. Similar separations of short- and long-chain ceramides have been effected on silicic acid columns [611]. Bouhours and Guignard [612] have used chloroform–methanol (9:1) as the mobile phase for ceramide subfractionation on borate-impregnated plates, while Ballio et al. [613] have accomplished ceramide purification on ordinary silica gel with benzene–acetone (1:1). Vunnam and Radin [614] have recently reported R_F values for homologous amides of DL-2-hydroxy acids and DL-sphingoid bases. Effective separations were obtained of the various chain-lengths with chloroform–methanol–water (155:25:28) as the developing solvent.

12.4.4.2. Argentation chromatography

Ceramides of uniform number of hydroxyl groups can be further separated into groups according to the number of double bonds of the component fatty acids, together with the number of *cis*-double bonds in the long-chain bases on silver nitrate-impregnated plates [615–617]. The *trans*-double bonds in the nitrogenous base have little or no effect. Best results are obtained following acetylation of the hydroxyl groups with acetic anhydride–pyridine. Chloroform–benzene–acetone (8:2:1) is a suitable solvent system [615,616]. With increasing number of double bonds , the ceramide acetates move more slowly. Free ceramides may be resolved by developing the thin-layer plates three times with chloroform–methanol (19:1). Excellent separations of ceramides according to the degree of unsaturation are also obtained with the *t*-BDMS ethers [556]. The latter have the advantage of combining chemical stability with ideal MS fragmentation patterns. TLC on 20% $AgNO_3$-Silica Gel G with chloroform, followed by chloroform–methanol (200:3) has been used as a preparative separation of *t*-BDMS ethers of the ceramides of natural sphingomyelins. The separations are based on the unsaturation of both the nitrogenous base and the fatty acid.

12.4.4.3. Gas–liquid chromatography

Ceramides prefractionated by TLC and argentation TLC can be completely resolved in the free form [617] or in the form of the acetate [566], TMS ether [566,615,618–620], and permethylether [621] derivatives, according to the chain lengths of the aliphatic components by high-temperature GLC under conditions similar to those used in the separation of the diacylglycerols. Thus, the TMS ethers have been resolved on 1.2 m × 3-mm ID glass U-tubes, containing 1% OV-1, at 270°C [615]. Comparable separations of the acetates and the TMS ethers have been achieved on the short (50 cm × 2 mm ID) nonpolar columns employed for triacylglycerol resolution [566]. The use of temperature programing improves the recovery of the higher-molecular-weight components without significant loss of resolution. These columns have also given effective separations of the free ceramides [617]. Casparrini et al. [619] have successfully recovered both TMS and heptafluorobutyryl derivatives of 2-hydroxy fatty acid ceramides from 2.7- and 3.6-m long columns, containing a 1% SE-30 packing. For this purpose, temperature programing from 250°C (heptafluorobutyrates) or 270°C (TMS ethers) at 2°C/min was used. Excellent separations of ceramides were also obtained with the *t*-BDMS ethers, which have desirable MS properties [622]. Greatly improved resolution of the ceramide TMS and *t*-BDMS ethers is obtained on 5- to 15-m capillary columns, coated with a nonpolar liquid phase [623]. Under these conditions, a nearly complete separation is also obtained for the saturated and monounsaturated C_{24}-acid amides.

12.4.4.4. High-performance liquid chromatography

Sugita et al. [624] determined the conditions for the benzoylation of ceramides containing ordinary and hydroxylated fatty acids and demonstrated that these derivatives are suitable for a sensitive and efficient resolution by HPLC on a Zipax (Du Pont) column with 0.05% methanol in *n*-pentane or 2.5% ethyl acetate in hexane as the eluents. A similar HPLC system was used by Iwamori and Moser [625] for the resolution of urinary ceramides in Farber's disease. Subsequetly, Iwamori et al. [626] resolved the benzoyl derivatives of ceramides containing hydroxylated and ordinary fatty acids with mixtures of hexane and ethyl acetate (94:6, 95:5, and 97:3) as eluents. The separation of molecular species of ceramides in the form of benzoyl and *p*-nitrobenzoyl derivatives by HPLC on silica and reversed phase columns has been reported [627]. Underivatized ceramides from sphingomyelins were separated into ten components on a reversed-phase column. The separations were based primarily on the carbon number and degree of unsaturation, ceramides with shorter chains and more unsaturation being eluted earlier. Ceramides containing monounsaturated fatty acids were eluted together with ceramides containing saturated fatty acids having three carbons less. Such pairs are resolved by combined argentation and reversed-phase HPLC [628].

12.4.5. Glycerophospholipids and sphingomyelins

The molecular species of glycerophospholipids and sphingomyelins are best determined following the removal of the polar head groups, either by enzymatic or chemical methods. For instance, the phosphorylcholine moieties of the choline-containing phospholipids may be readily removed by phospholipase C [544,629]. This enzyme can also be used to dephosphorylate phosphatidylethanolamine and phosphatidylserine [630], although at a slower rate than phosphatidylcholine. The enzymatic hydrolysis is also applicable to the alkylacyl- and alk-1-enylacylglycerophospholipids [544,631]. The analyses of the molecular species of the neutral lipid moieties of the various phospholipid classes are performed as described above (see Chap. 12.4.3). However, extensive resolution of molecular species may also be obtained with native phospholipids or with phospholipids having chemically modified polar head groups.

12.4.5.1. Argentation chromatography

Intact phosphatidylcholine, phosphatidylethanolamine, and phosphatidylinositol may be separated into molecular species according to the number of total double bonds per molecule by argentation TLC. Thus, phosphatidylcholine and phosphatidylethanolamine may be fractionated into saturated, monoenoic, dienoic, trienoic, and tetraenoic species, while phosphatidylethanolamine usually also gives an additional band for hexaenes [632,633]. For this fractionation, the thin-layer plates were developed with chloroform–methanol–water (60:35:4). In a similar solvent system, chloroform–methanol–water (65:35:5), phosphatidylinositol yields bands for saturates, monoenes, dienes, trienes, and tetraenes [634]. However, many of the finer fractionations, observed for the corresponding diacylglycerols, are not obtained. Neither is there a separation of the alkylacyl-, alkenylacyl-, and diacyl-glycerolphospholipids [1]. Intact phosphatidylcholine and phosphatidylglycerol may be resolved into molecular species by direct argentation TLC, using chloroform–methanol–water (65:25:4) as the solvent system [635,636].

The dimethyl esters of phosphatidic acids are readily resolved by argentation TLC according to the number and position of double bonds in their molecules [155]. Molecules containing up to 12 double bonds can be separated with chloroform–methanol–water (90:10:1) [539]. For the determination of molecular species, the dimethyl esters are prepared with diazomethane. Other glycerophospholipids may be converted into phosphatidic acids by hydrolysis with phospholipase D [637]. Diazomethanolysis produces phosphatidic acid dimethyl esters from phosphatidylserine [638]. Diazomethylation is also suitable for masking the phosphate groups in the alkylacyl- or alkenylacylglycerophosphates. Equally effective masking of the polar groups is accomplished by N-dinitrophenylation of phosphatidylethanolamine [639].

Argentation TLC of these derivatives separates individual molecular species according to number, geometry, and position of the ethylenic double bonds. For this separation, chloroform–methanol (49:1) is used as solvent. Dinitrophenylation is also applicable to the countercurrent separation of the molecular species of phosphatidylethanolamine and phosphatidylserine [640]. For this purpose, the carboxyl group may also be methylated by brief exposure to diazomethane. The dinitrophenylphosphatidylethanolamines [641] and dinitrophenylphosphatidylserines [640] are resolved by distribution between petroleum and 85% aqueous ethanol. The phosphatidylinositols are resolved into molecular species by argentation TLC after the inositol moiety has been acetylated to reduce the polarity of the molecule [642]. The modified phosphatidylinositols may be resolved into saturates + monoenes, dienes, trienes, tetraenes, and polyenes by means of chloroform–acetone (3:1 and 1:1). Since the saturated and monoenoic species are not separated from each other by this method, they are subjected to permanganate–periodate oxidation, followed by TLC, to isolate the saturated species [643]. Shaw and Bottino [644] and Bottino [645] have fractionated polyunsaturated phosphatidylcholines and phosphatidylethanolamines with chloroform–methanol–1-propanol–0.5% acetic acid (55:30:5:7). The distinctive feature of these separations was the development of the plates at $-10°C$ for 14 to 16 h (phosphatidylcholine) and at $4°C$ for 11 to 12 h (phosphatidylethanolamine).

Chromarods, impregnated with silver nitrate, have also been applied to the separation of molecular species of the choline and ethanolamine phosphatides [646]. For this application, the rods were developed with chloroform–methanol–water (65:25:3).

12.4.5.2. High-performance liquid chromatography

One of the first successful separations of molecular species of phosphatidylcholine by liquid chromatography was reported by King and Clements [647], who separated the mercuric acetate addition products of unsaturated phosphatidylcholines from each other and from saturated analogs on a Sephadex LH-20 column by elution with chloroform–methanol (1:1), containing increasing amounts (from 0.01 to 0.1%) of glacial acetic acid. Four species of phosphatidylcholines were resolved with an overall recovery of 65 to 90% and reproducibility of 10%. Separations of unmodified molecular species of egg yolk phosphatidylcholines have been obtained by Arvidson [648] on an alkylated derivative of Sephadex with methanol–water as the eluent.

Much more rapid and more efficient are the separations by HPLC. Certain molecular species of phosphatidylcholine and of sphingomyelin may be resolved on HPLC adsorption columns. Thus, 1,2-dipalmitoyl- and 1,2-didocosa-13'-cis-enoyl-glycerol-3-phosphorylcholines and long- and short-chain sphingomyelins can be separated [234]. Much more effective and more extensive separations have been described by Porter et al. [649], who used reversed-phase HPLC on μBondapak C_{18} columns and Fatty Acid Columns. The Bondapak column was eluted with chloroform–water–methanol (1:1:10), the Fatty Acid Column was eluted with chloroform–water–methanol (10:19:70), and an RI dector was used. Synthetic phosphatidylcholines were fractionated into Effective Carbon Numbers of 28 to 36, containing compounds with 1 to 4 double bonds per molecule. The Effective Carbon Number is defined as the total carbon number minus the number of double bonds in the phosphatidylcholine molecule. Egg yolk phosphatidylcholine gave 16:0 18:1, 16:0 18:2, 18:0 18:2, and 18:0 18:1 as the major species. Similarly, Crawford et al. [650] have separated soy phosphatidylcholines into the component molecular species by reversed-phase HPLC on a μBondapak C_{18} column eluted with an aqueous methanol gradient (91 to 95%) and UV detection at 206 nm. Separate peaks were obtained for molecular species corresponding to di-18:3, 18:2 18:3, di-18:2, 16:2 18:2, 18:1 18:2, and 18:0 18:2. This technique was also used to separate and purify dilinoleoylphosphatidylcholine from each species of soy phosphatidylcholine. Oxidized species of di-18:2 phosphatidylcholines were retained longer by the column and were detected by UV measurements at 234 nm. Subsequently, Porter et al. [504] have used reversed-phase HPLC for the separation of the oxidation products of 1-palmitoyl-2-linoleoylphosphatidylcholine with methanol–water–chloroform (10:1:1) as the solvent system. The 1-stearoyl-2-arachidonoylphosphatidylcholine oxidation products were resolved with methanol–water–chloroform (50:6:5) on the same reversed-phase C_{18} column.

Patton et al. [88] have reported the separation of all major glycerophospholipids into 30 to 35 molecular species by HPLC on reversed-phase Ultrasphere ODS (Altex) columns with 20 mM choline chloride in methanol–water–acetonitrile (181:14:5). Comparable separations of phosphatidylcholine have been obtained by Smith and Jungalwala [89] on a Nucleosil-5-C_{18} column with methanol–1 mM phosphate buffer (pH 7.4) (19:1). In both instances the eluate was monitored at 205 nm.

12.4.6. Glycolipids

The molecular species of the glycolipids, including those of the ceramide polyhexosides, can be resolved by argentation chromatography and GLC. Partial resolution of the long- and short-chain components may also be obtained by liquid–liquid partition and adsorption chromatography.

References on p. B130

12.4.6.1. Argentation chromatography

No enzyme or chemical reaction known selectively hydrolyzes the glycosidic linkage in glycosyldiacylglycerols or ceramides, and molecular species of these compounds cannot be determined from the corresponding diacylglycerol or ceramide moieties. However, many classes of the intact glycolipids may be separated into smaller groups or individual molecular species by argentation TLC. Thus, monogalactosyldiacylglycerols have been separated into distinct fractions, containing one to six double bonds per molecule, by argentation TLC with chloroform–methanol–water (60:21:4) as the solvent system [651,652]. That system was used by Siebertz and coworkers [653,654] for preparative silver nitrate TLC to obtain molecular species of monogalactosyl- and digalactosyldiacylglycerols. The resolution is improved when the free hydroxyl groups are acetylated. The acetates are then resolved with chloroform–methanol (197:3) [655].

Siebertz et al. [635] have subsequently separated the molecular species of the glycolipids in chloroplast envelopes and thylakoids by argentation TLC with the following solvent systems: chloroform–methanol–water (60:21:4) for monogalactosyldiacylglycerols, chloroform–methanol–water (65:35:4) for digalactosyldiacylglycerols, chloroform–methanol–acetic acid–water (50:25:4:4) for trigalactosyldiacylglycerols, chloroform–methanol–acetic acid–water (25:15:4:2) for tetragalactosyldiacylglycerols, and chloroform–methanol–water (65:25:4) for sulfoquinovosyldiacylglycerol. For this procedure precoated Kieselgel G plates were impregnated with silver nitrate by spraying with 10% AgNO$_3$ in acetonitrile (10 ml per 20 $\times$ 20-cm plate), activated for at least 3 h at 130°C, and developed at 4°C. Excellent separations were obtained for molecules containing zero to six double bonds.

12.4.6.2. Gas–liquid chromatography

High-temperature GLC of the TMS derivatives of the mono- and digalactosyldiacylglycerols (after hydrogenation) has been used to separate species differing in molecular weight under conditions similar to those used for triacylglycerols [656]. Comparable resolutions of the mono- and digalactosyldiacylglycerols have been obtained without hydrogenation by high-temperature GLC of the TMS ethers by Kuksis [617] and Williams et al. [368].

Intact monoglycosylceramides (cerebrosides) have been subjected to high-temperature GLC after conversion of the free hydroxyl groups of the hexose and aliphatic residues to the TMS derivatives. Compounds are separated according to the combined chain lengths and number of hydroxyl groups in the long-chain base and fatty acid components, but the GLC recorder tracings obtained are complex, and components are not readily identified unless the column effluent is processed by MS [657,658]. Likewise, Kuksis [617] has demonstrated that ceramide mono- and diglycosides can be readily resolved on the basis of carbon number by means of GLC of the TMS ethers. In a test of the capability of the chromatographic system, apparently correct retention times were obtained for ceramide glycosides, containing up to six hexoside units, as well as for the corresponding degradation (pyrolysis) products.

12.4.6.3. High-performance liquid chromatography

The HPLC conditions designed to detect the perbenzoyl derivatives of various glycolipids at 230 nm as the primary goal provide baseline resolution only for the lipid classes. However, certain molecular species, differing in the chain length of the component fatty acids, are also partially resolved. Nonaka and Kushimoto [659] have described a HPLC procedure for the separation of gluco- and galactocerebroside classes in the form of benzoates without preliminary purification. The Spherisorb (5–10 μm) column was eluted with a gradient of 0.5 to 10% 2-propanol in hexane, and the effluent was monitored at 230 nm. The method allowed the separation of cerebrosides and sulfatides, containing hydroxylated and ordinary fatty acids, straight lines being obtained when retention times of individual homologs were plotted on a logarithmic scale against fatty acid carbon numbers. This HPLC system has been applied to the resolution of the cerebroside III sulfatides with comparable efficiency [660].

Evidence for further resolution of the gangliosides by HPLC is based on shoulders observed for peaks corresponding to GH_3 and GM_4 [382]. Although these shoulders could be the result of incomplete benzoylation or degradation of the products, a quantitative evaluation of the reaction mixture is not consistent with such a possibility. A more likely explanation is that the shoulders represent molecular heterogeneity. Differences in the fatty acid composition of GM_3 have been shown to alter TLC behavior. GM_3 from neural tumors on TLC splits into two bands, which corresponds to differences between long- (C_{22} and C_{24}) and short-chain (C_{16} and C_{18}) fatty acids [661]. Human liver GM_3 has also been reported to separate on TLC according to fatty acid composition [374]. In these chromatograms the upper band contains mainly ordinary long-chain fatty acids, while the lower band contains a mixture of hydroxy fatty acids and ordinary short-chain fatty acids. Similar findings have been recorded for GD_3 gangliosides from bovine optic nerve [662].

12.5. DETERMINATION OF TOTAL LIPID PROFILES

Chromatography of lipids is aimed at the resolution of different lipid classes and molecular species for the ultimate purpose of complete identification and quantitation. In the course of chromatographic studies, it has become apparent that many organisms, tissues, cells, and subcellular components possess characteristic lipid compositions that can be recognized without complete resolution and determination of individual molecular species. Quantitative estimation of a partially resolved lipid profile is frequently sufficient for establishing the origin of the total lipid extract and for assessing its relationship to normal or abnormal metabolic states. Essentially all chromatographic techniques are suitable for this purpose, but those which combine rapid resolution with effective quantitation are preferable. The various applications of this approach range from a recording of total lipid in plasma to the characterization of bacterial species based on their fatty acid methyl ester patterns or on the profile of their pyrolysis products.

12.5.1. Thin-layer chromatography

The most useful technique for obtaining the qualitative profile of a total lipid extract is TLC. Despite much ingenuity, effective quantitation of the lipid fractions by ordinary TLC has not been obtained. Transmethylation of the various fatty esters, followed by quantitative GLC analysis of the fatty acid methyl esters, is satisfactory [45,663–668], but time-consuming. This technique has been extensively utilized for the quantitation of the proportions of glycerolipids in the various TLC fractions as well as for further characterization of the composition of these fractions. Numerous applications have been reported to the lipids in man [668,669] and experimental animals [670–672], to the characterization of tissue and organ lipids [673,674], and to lipids of subcellular fractions [656,675–677].

An excellent TLC procedure for the determination of the total lipid profiles of plasma or serum involves developing the plate first with petroleum–diethyl ether (9:1) and then with petroleum–diethyl ether–acetic acid (400:100:1) in the same direction, but to a second front 5 cm below the first. The polar lipids remaining at the origin consist almost entirely of phospholipids [678,679]. The content of the cholesteryl esters, triacylglycerols, free fatty acids, free cholesterol, and phospholipids is determined by charring and densitometry [680,681].

Pollet et al. [682] have described a microanalysis of brain lipids based on a multiple two-dimensional TLC method. Total lipid extracts are first chromatographed with chloroform–methanol–water (35:15:2), which separates most lipid classes, except the choline, serine, and inositol phosphatides, forming one group, and the gangliosides and proteolipids, forming another group. Subsequent development at right angle with chloroform–methanol (1:4), followed by chloroform–methanol (2:1) and a final development with the latter in the first direction resolves the remaining components. The zones in the chromatogram are located by iodine staining and then quantitated by chemical analyses. These determinations can be performed without conversion of the phosphate into inorganic phosphorus [683,684]. Charring of lipids, followed by densitometry, is widely practiced [681,685,686], but both steps require standardization [687,688].

Bitman et al. [689] have developed a routine TLC analysis of lipid classes in blood, milk, tissue, and egg yolk, giving rapid and reproducible separations suitable for in situ quantitation by densitometry. A two-step development, starting with chloroform–methanol–acetic acid (98:2:1) to 16 cm, is followed by a second development with hexane–diethyl ether–acetic acid (470:30:1) to 20 cm. Prior to charring, the plate is dipped into a solution of 3% cupric acetate in 8% phosphoric acid for 3 sec.

Segura and Gotto [690] have described a procedure for the detection and quantitation of organic compounds on TLC plates after inducing their transformation into fluorescent derivatives by thermal treatment of the chromatoplate in the presence of NH_4HCO_3. The method has been applied to the determination of lipids, requiring only microliter quantities of plasma [691,692]. More recently, Segura and Navarro [693] have observed that heating the chromatoplates in the presence of $SiCl_4$ induces the fluorescent derivative formation of all classes of compounds much more reproducibly and gives fluorophores that are stable for months.

The possibility of classifying microorganisms on the basis of their lipid composition [694] has been significantly advanced through sensitive analytical techniques, such as TLC and GLC. Analyses of methanolyzates of mycolic acid-containing bacteria by TLC have given good indication of the general mycolic acid composition of these organisms. Minnikin et al. [178] have reexamined the use of TLC for the typing of bacteria containing mycolic acids. One-dimensional TLC with petroleum–diethyl ether (17:3) was inadequate, but distinct patterns of mycobacterial mycolates were produced by two-dimensional TLC. After multiple development with petroleum–acetone (19:1) in the first direction, the plates were developed with toluene–acetone (99:1 to 97:3) in the second direction.

The most successful method of quantitating the lipid components resolved by TLC takes advantage of the sintered coating on quartz rods, developed by Okumura and coworkers [3,695,696]. This method, now commercialized under the name Iatroscan TLC/FID system, combines the technique of TLC with automated quantitative detection, based on the flame-ionization principle employed in GLC. Most separations currently performed by conventional TLC can be duplicated on the patented Chomarods, and the FID is applicable to all lipids. Chromarods can be impregnated with silver nitrate or boric acid to enhance certain types of separations. The resolution of neutral and glycerophospholipids on the basis of their degree of unsaturation further expands the total lipid profile. Sample loads range from 2 to 20 μg in 1 to 2 μl of solution. The neutral lipids may be resolved with light petroleum–diethyl ether–formic acid (97:3:1) [697,698], 1,2-dichloroethane–chloroform–acetic acid (920:80:1) [699], light petroleum–diethyl ether (17:3) [700,701], light petroleum–diethyl ether–formic acid (850:150:1) [699], and by multiple developments with hexane–acetic acid (100:1), followed by hexane–acetic acid–2-propanol (200:2:1) [702], or with light petroleum–diethyl ether–acetic acid (175:25:2) [703], while the phospholipids remain at the origin. In other applications, the neutral lipids were separated from the phospholipids (at the origin) in 30 species of yeast with light petroleum–diethyl ether–acetic acid (90:10:1) [704,705], while the neutral lipids, including glycolipids, in microorganisms have been resolved with chloroform–methanol–water (40:10:1) [706].

Determination of the phosphatidylcholine (lecithin) to sphingomyelin ratio (L/S) provides a means for assessing the development of the fetus in pregnancy and yields reliable information about pulmonary maturity. Glueck et al. [707] separated the phospholipids in chloroform–methanol–water (65:25:4) and charred the spots with sulfuric acid. Many modifications of the original method have been suggested since, including the quantitation of the L/S ratio [708,709]. Chromarod analyses have also been applied to this determination [710]. Ackman [4] has recently reviewed the potential of the Iatroscan system for routine and research analyses of lipids.

References on p. B130

12.5.2. Gas–liquid chromatography

A procedure that is well suited for a qualitative and quantitative assessment of total lipid extracts is high-temperature GLC. It so happens that the common natural free fatty acids and their mono-, di-, and tri-esters with glycerol differ greatly in molecular weight from each other, from free cholesterol, and from the fatty acid esters of cholesterol. With few exceptions, therefore, effective resolution of the neutral lipids may be obtained on the basis of molecular weight even after trimethyl-silylation of the free hydroxyl and carboxyl groups [566,711]. Most total lipid extracts of natural products are rich in ionic lipids such as the glycerophosphatides, sphingolipids, and glycolipids, which cannot be passed through the GLC column without prior removal of the phosphoric and sialic acid moieties. This may be accomplished by enzymic [566] or chemical [711,712] methods, although not without a loss of identity of individual lipid classes.

Although the direct GLC techniques are suitable for the handling of most natural lipid mixtures, the greatest progress has been made in the analysis of total lipid extracts of plasma [713–718] and of individual lipoprotein classes [714,719]. This is due to the advantageous distribution of the molecular weights of the component classes of lipids and to their relative proportions in the mixture. The most successful GLC separations of total lipid extracts have been achieved on the short, nonpolar siloxane columns originally described for work with natural triacylglycerols [551]. Stainless-steel or glass tubes (30 to 50 cm × 0.2 to 0.3 mm ID) have generally proved satisfactory when packed with 1–3% methyl siloxane or equivalent high-temperature silicone polymers coated on an inert support. These columns are used after preconditioning for 2–3 h at 350°C and a satisfactory recovery test. The column temperature is programed to rise linearly at the rate of 4 to 8°C/min in the range 175 to 350°C during analysis, depending on the composition of the lipid mixture. Extensive applications to neutral lipid mixtures of plasma have also been made [721–723].

Recently, capillary columns have been adopted for this purpose [554,556]. The best separations and nearly quantitative recoveries have been obtained with capillary columns, 5 m in length, and operated with hydrogen as the carrier gas. Under these conditions, effective baseline separation is achieved for molecular species of glycerolipids differing by one methylene unit and between certain saturated and unsaturated homologs. In 15-m columns, a significant degradation of long-chain triacylglycerides [724] and cholesteryl esters [725] has been observed.

12.5.3. High-performance liquid chromatography

A reversed-phase system for the liquid–liquid chromatographic profiling of total lipid extracts was described by Hirsch in 1963 [726]. Subsequently, chromatography on Sephadex columns held out considerable promise [599,648]. Despite much progress in some laboratories [601,602], these methods were plagued by experimental difficulties, excessive slowness, and generally poor reproducibility. The HPLC systems for both adsorption and reversed-phase chromatography of lipid extracts now promise to provide ideal means of profiling most lipid extracts. Thus far, however, there have been few efforts in this direction. Duncan et al. [727] have described the HPLC resolution of free cholesterol and various cholesteryl esters on a reversed-phase column with 2-propanol–acetonitrile (1:1 or 5:2) as the mobile phase. Using the same methodology, Smith et al. [728] have obtained characteristic neutral lipid profiles of plasma for normal and diabetic rats. From the total lipid extract, separate estimates were obtained for 13 chromatographic fractions, including separate estimates for various subfractions of triacylglycerols and cholesteryl esters. The plasma phospholipids were apparently eluted so early that they were not separately quantitated. For this analysis, a 25 cm × 4.5-mm OD Zorbax ODS column (5 μm, Du Pont) and a UV detector at 215 nm were used. With 2-propanol–acetonitrile (5:2) at a flowrate of 1 ml/min, the relative standard deviation varied from 4 to 8%, depending on the solute and its peak size.

Coupek and Mareš [729] have compared the resolution of the same sample of neutral lipids by HPLC and GLC. For the HPLC separation of the C_{14} to C_{20} esters of cholesterol and glycerol a reversed-phase column (Separon SI VSIL, produced in Czechoslovakia) was used. Since lipid esters possess only a limited absorption in the UV, few solvent systems are compatible with the UV detector. The use of RI detectors is also limited in its applicability to the HPLC of lipids. Therefore, much effort has been expended in the development of other detectors. Erdahl et al. [730] have proposed a combination of HPLC separation of total lipid extracts with MS identification and quantitation of peaks, followed by computerized evaluation of the data. The interface system for coupling HPLC to the mass spectrometer is based on the endless-chain principle in which a belt of special construction is used as the transport device (cf. Chap. 4.10.2). After removal of the solvent, the sample is transported by the belt to a reactor, where it is evaporated if it is volatile, or converted to hydrocarbons if it is not volatile, and introduced into the source of the mass spectrometer by a carrier gas. This system was tested with model compounds of common lipid classes and found to be practical. Compton and Purdy [731] have proposed a combination of the HPLC separation with a universal glycerolipid detector, which is based on the colorimetric estimation of formaldehyde, for which they have developed a sensitive reagent.

12.6. SUMMARY AND CONCLUSIONS

A review of the chromatography of lipids as practiced over the last few years reveals much progress despite relatively limited advances in methodology. There has been a remarkable increase in the variety of samples analyzed by the various chromatographic methods without much change in techniques. The major development in GLC has been the introduction of the glass capillary column, especially the flexible quartz tubing, which has rendered the system much more durable for routine work. The greatest advance has been the development and application of HPLC to the lipid field, but a sensitive universal detector is still needed. The areas of lipid chromatography which have been benefited most form this development have been

References on p. B130

those that deal with polar, high-molecular-weight components of low volatility and low thermal stability. This advance must be considered complementary to, rather than competitive with, GLC. In those instances where both techniques are applicable, GLC appears to have the advantage of being somewhat faster than HPLC. This is due to differences in the properties of the mobile phases and, thus, there is no reason to believe that this advantage of GLC will ever be matched by HPLC. Furthermore, GLC has the advantage of a highly sensitive detector for lipids, the hydrogen FID. This leaves a continuing need for better volatile derivatives of the polar and thermally unstable or high-boiling lipids for GLC.

In TLC, the major advance has been the development of the Chromarod system. Chromatography on porous quartz surfaces allows them to be used repeatedly, and the hydrogen FID makes the quantitation of the resolved components convenient. Thus, a routine quantitative tool for lipid analysis has become available, and in many laboratories it has effectively replaced the less accurate charring and densitometry technique previously used for the evaluation of chromatoplates. The impact of HPTLC has been less significant, although several improved resolutions of notably difficult mixtures have been reported. Despite these advances, no single technique of chromatography has completely displaced any other chromatographic technique for the analytical or preparative separation of lipids. It is obvious that combinations of complementary analytical techniques will continue to be required for the foreseeable future. This includes the combinations of GLC or HPLC with MS, which are unable to cope individually with the identification and quantitation of complex mixtures of lipids. The introduction of automated systems of chromatographic analysis, including computerized data processing, has resulted in a considerable time saving and has increased the accuracy of quantitative analyses. As a result, the methods of sample preparation and preservation now appear to be the major remaining uncontrolled variables in interlaboratory comparisons of analytical data on lipids.

12.7. ACKNOWLEDGEMENTS

The studies by the author and his collaborators referred to in this review were supported by funds from the Medical Research Council of Canada, Ottawa, Canada; The Ontario Heart Foundation, Toronto, Canada; and the Heart and Lung Institute, NIH-NHLI-72-917, Bethesda, MD, USA.

REFERENCES

1 O. Renkonen and A. Luukkonen, in G.V. Marinetti (Editor), *Lipid Chromatographic Analysis*, Vol. 1, Marcel Dekker, New York, 2nd Edn., 1976, p. 1.
2 A. Zlatkis and R.E. Kaiser (Editors), *HPTLC — High Performance Thin-Layer Chromatography*, J. Chromatogr. Library Series, Vol. 9, Elsevier, Amsterdam, 1977.
3 T. Okumura, *J. Chromatogr.*, 184 (1980) 37.
4 R.G. Ackman, *Methods Enzymol.*, 72 (1981) 205.
5 G.C. Cochrane, *J. Chromatogr. Sci.*, 13 (1975) 440.

6 A. Kuksis, in G.V. Marinetti (Editor), *Lipid Chromatographic Analysis,* Vol. 1, Marcel Dekker, New York, 2nd Edn., 1976, p. 215.

7 A. Kuksis, *Separ. Purif. Methods,* 6 (1977) 353.

8 W.C. Hubbard, J.T. Watson and B.J. Sweetman, in G. Hawk (Editor), *Biological Applications of Liquid Chromatography,* Vol. 10, Marcel Dekker, New York, 1979, p. 31.

9 K. Aitzetmüller, *Lipids,* 17 (1982) in press.

10 A. Pryde and M.T. Gilbert, *Applications of High Performance Liquid Chromatography,* Chapman & Hall, London, 1979.

11 J.C. Touchstone and J. Sherma (Editors), *Densitometry in Thin-Layer Chromatography, Practice and Applications,* Wiley, Somerset, NJ, 1979.

12 L.A. Whitting (Editor), *Glycolipid Methodology,* Amer. Oil Chem. Soc., Champaign, IL, 1976.

13 L.D. Bergelson (Editor), *Lipid Biochemical Preparations,* Elsevier, Amsterdam, 1980.

14 A. Kuksis, *J. Chromatogr.* 143 (1977) 3.

15 A. Kuksis, in A. Kuksis (Editor), *Fatty Acids and Glycerides,* Vol. 1, Plenum, New York, 1978, p. 1.

16 J.J. Myher, in A. Kuksis (Editor), *Fatty Acids and Glycerides,* Vol. 1, Plenum, New York, 1978, p. 123.

17 M.S.F. Lie Ken Jie, *Advan. Chromatogr.,* 18 (1980) 1.

18 M.S.J. Dallas, L.J. Morris and B.W. Nichols, in E. Heftmann (Editor), *Chromatography,* Van Nostrand-Reinhold, New York, 3rd Edn., 1975, p. 527.

19 J. Folch, M. Lees and G.H. Sloane-Stanley, *J. Biol. Chem.,* 226 (1957) 497.

20 G.J. Nelson, in E.G. Perkins (Editor), *Analysis of Lipids and Lipoproteins,* Amer Oil Chem Soc, Champaign, IL, 1975, p. 1.

21 N.H. Shaikh and E. Downar, *Circ. Res.,* 49 (1981) 316.

22 S. Mogelson, G.E. Wilson, Jr. and B.E. Sobel, *Biochim. Biophys. Acta,* 619 (1980) 680.

23 E.G. Bligh and W.J. Dyer, *Can. J. Biochem.,* 37 (1959) 911.

24 P. Zahler and V. Niggli, in E.D. Korn (Editor), *Methods in Membrane Biology,* Vol. 8, Plenum, New York, 1977, p. 1.

25 C.C. Scott, C.A. Ackerman and F. Snyder, *Biochim. Biophys. Acta,* 575 (1979) 215.

26 G. Arthur and A. Sheltawy, *Biochem. J.,* 191 (1980) 523.

27 S. Spanner, in G.B. Ansell, J.N. Hawthorne and R.M.C. Dawson (Editors), *Form and Function of Phospholipids,* Biochim. Biophys. Acta Library, Vol. 3, Elsevier, Amsterdam, 1973, p. 43.

28 B.A. Hanson and R.L. Lester, *J. Lipid Res.,* 21 (1980) 309.

29 L. Taylor, P. Polgar, J.A. McAteer and W.H.J. Douglas, *Biochim. Biophys. Acta,* 572 (1979) 502.

30 C.B. Struijk, R.K. Berthuis, H.J. Pabon and D.A. van Dorp, *Rec. Trav. Chim. Pays-Bas,* 85 (1966) 2.

31 B.H. Min, J. Pao, W.A. Garland, J.A.F. de Silva and M. Parsonnet, *J. Chromatogr.,* 183 (1980) 411.

32 C. Pace-Asciak, K. Morawska and L.S. Wolfe, *Biochim. Biophys. Acta,* 218 (1970) 288.

33 M.A. Wells and J.C. Dittmer, *Biochemistry,* 2 (1963) 1259.

34 G.J. Nelson, in G.J. Nelson (Editor), *Blood Lipids and Lipoproteins,* Wiley-Interscience, New York, 1972, p. 25.

35 J.A.J. Randell and C.A. Pennock, *J. Chromatogr.,* 195 (1980) 257.

36 F. Phillips and O.S. Privett, *Lipids,* 14 (1979) 590.

37 M. Kates and F.M. Eberhardt, *Can. J. Bot.,* 35 (1957) 895.

38 S.F. Yang, S. Faure and A.A. Benson, *J. Biol. Chem.,* 242 (1967) 477.

39 P.G. Roughan, C.R. Slack and R. Holland, *Lipids,* 13 (1978) 497.

40 F.C. Phillips and O.S. Privett, *Lipids,* 14 (1979) 949.

41 D. Fairbairn, *J. Biol. Chem.,* 157 (1945) 645.

42 M. Kates, in T.S, Work and E. Work (Editors), *Laboratory Techniques in Biochemistry and Molecular Biology,* Vol. 3, *Techniques of Lipidology,* North-Holland, Amsterdam, 1972, p. 347.

43 G. Rouser, G.J. Nelson, S. Fleischer and G. Simon, in D. Chapman (Editor), *Biological Membranes,* Academic Press, New York, 1968, p. 5.

44 J.K.G. Kramer and H.W. Hulan, *J. Lipid Res.,* 19 (1978) 103.

45 W.W. Christie, *Lipid Analysis,* Pergamon, Oxford, 1973.

46 R.E. Pitas, G.J. Nelson, E.M. Jaffe and R.W. Mahley, *Lipids,* 14 (1979) 469.

47 R.E. Pitas, M.M. Hagerty and R.G. Jensen, *Lipids*, 13 (1978) 844.
48 J.P. Williams, *Biochim. Biophys. Acta*, 618 (1980) 461.
49 J. Robert, P. Mandel and G. Rebel, *Lipids*, 14 (1979) 852.
50 G.A.E. Arvidson, *J. Lipid Res.*, 8 (1967) 155.
51 M.W. Smith, Y. Callan, M.W. Kahng and B.F. Tramp *Biochim. Biophys. Acta*, 618 (1980) 192.
52 K. Koizumi, K. Kano-Tanaka, S. Shimizu, K. Nishida, N. Yamanaka and K. Ota, *Biochim. Biophys. Acta*, 619 (1980) 344.
53 J.N. Roehm and O.S. Privett, *Lipids*, 5 (1970) 353.
54 E.A. Emken, W.K. Rohwedder, H.J. Dutton, W.J. Dejarlais, R.O. Adlof, J.F. McKin, R.M. Daugherty and J.M. Iacono, *Lipids*, 14 (1979) 547.
55 J.T. Dodge and G.B. Phillips, *J. Lipid. Res.*, 7 (1966) 387.
56 B. Borgström, *Acta Physiol. Scand.*, 25 (1952) 101.
57 G. Rouser, G. Kritchevsky, D. Heller and E. Lieber, *J. Amer. Oil Chem. Soc.*, 40 (1963) 425.
58 G. Rouser, G. Kritchevsky and A. Yamamoto, in G.V. Marinetti (Editor), *Lipid Chromatographic Analysis*, Vol. 3, Marcel Dekker, New York, 2nd Edn., 1976, p. 713.
59 J.L. Gellerman, W.H. Anderson, D.G. Richardson and H. Schlenk, *Biochim. Biophys. Acta*, 388 (1975) 277.
60 A.C. Wilson, M. Kates and A.I. de la Roche, *Lipids*, 13 (1978) 504.
61 Y. Fujino and T. Miyazawa, *Biochim. Biophys. Acta*, 572 (1979) 442.
62 H. Meyer, L. Provasoli and I. Meyer, *Biochim. Biophys. Acta*, 573 (1979) 464.
63 L.L. Yang and A. Haug, *Biochim. Biophys. Acta*, 573 (1979) 308.
64 P.F. Smith, K.R. Patel and A.J.N. Al-Shammaki, *Biochim. Biophys. Acta*, 617 (1980) 419.
65 S.K. Das, M.E. Steen, M.S. McCullough and D.K. Bhattacharya, *Lipids*, 13 (1978) 679.
66 J. Marion and L.S. Wolfe, *Biochim. Biophys. Acta*, 574 (1979) 25.
67 H. Ogino, T. Matsumara, K. Satouchi and K. Saito, *Biochim. Biophys. Acta*, 574 (1979) 57.
68 B.E. Dwyer and J. Bersohn, *Biochim. Biophys. Acta*, 575 (1979) 309.
69 R.I. Karney and G.A. Dhopeshwarkar, *Lipids*, 14 (1979) 257.
70 R. Wood, F. Chandler, M. Matocha and A. Zoeller, *Lipids*, 14 (1979) 789.
71 D.P. Selivonchick, P.C. Schmid, V. Natarojan and H.H.O. Schmid, *Biochim. Biophys. Acta*, 618 (1980) 242.
72 D.E. Vance and C.C. Sweeley, *J. Lipid Res.*, 8 (1967) 621.
73 Y. Jigami, O. Suzuki and S. Nakasato, *Lipids*, 14 (1979) 937.
74 C.P. Burns, D.G. Luttenegger, S.-P.L. Wei and A.A. Spitzer, *Lipids*, 12 (1977) 747.
75 W.B. Im, J.T. Deutchler and A.A. Spector, *Lipids*, 14 (1979) 1003.
76 S.A. Rooney, P.M. Canavan and E.K. Motoyama, *Biochim. Biophys. Acta*, 360 (1974) 56.
77 L.L. Nardone and S.B. Andrews, *Biochim. Biophys. Acta*, 573 (1979) 276.
78 J.I. Rabinowitz, S.E. Askins and F.E. Luddy, *Lipids*, 13 (1978) 317.
79 R.G. Johnson, M.A. Lugg and T.E. Nicholas, *Lipids*, 14 (1979) 555.
80 M.A. Haas and W.J. Longmore, *Biochim. Biophys. Acta*, 573 (1979) 166.
81 G. Okano, H. Matsuzaka and T. Shimojo, *Biochim. Biophys. Acta*, 619 (1980) 167.
82 K.F. Buchler and R.A. Rhodes, *Biochim. Biophys. Acta*, 619 (1980) 186.
83 J.F. Oram, E. Shafrir and E.L. Bierman, *Biochim. Biophys. Acta*, 619 (1980) 214.
84 R. Wood and F. Snyder, *Lipids*, 3 (1968) 129.
85 T.-C. Lee, V.F.N. Stephens and F. Snyder, *Biochim. Biophys. Acta*, 619 (1980) 420.
86 J.N. Karli and G.M. Lewis, *Lipids*, 9 (1974) 819.
87 J.M. McKibbin, *J. Lipid Res.*, 19 (1978) 131.
88 G.M. Patton, J.M. Fasulo and S.J. Robins, *J. Lipid Res.*, 23 (1982) 190.
89 M. Smith and F.B. Jungalwala, *J. Lipid Res.*, 22 (1981) 697.
90 W.S.M. Geurts van Kessel, *J. Amer. Oil Chem. Soc.*, 57 (1980) No. 2, Abstr. No. 84.
91 A. Kuksis, in R.T. Holman (Editor), *Progress in the Chemistry of Fats and Other Lipids*, Vol. 12, Pergamon, Oxford, 1972, p. 1.
92 H. Wagner, L. Hörhammer and P. Wolff, *Biochem. Z.*, 334 (1961) 175.
93 A. Kuksis, W.C. Breckenridge, L. Marai and O. Stachnyk, *J. Amer Oil Chem. Soc.*, 45 (1968) 537.

94 V.P. Skipski and M. Barclay, *Methods Enzymol.*, 14 (1969) 530.

95 J.K. Beckman and J.G. Coniglio, *Lipids*, 14 (1979) 262.

96 W.C. Breckenridge and A. Kuksis, *J. Lipid Res.*, 8 (1967) 473.

97 F. Vitiello and J-.P. Zanetta, *J. Chromatogr.*, 166 (1978) 637.

98 W. Fisher, R.A. Laine and M. Nankano, *Biochim. Biophys. Acta*, 528 (1978) 298.

99 W. Fischer, D. Schuster and R.A. Laine, *Biochim. Biophys. Acta*, 575 (1979) 389.

100 K.K. Carroll, in G.V. Marinetti (Editor), *Lipid Chromatographic Analysis*, Vol. 1, Marcel Dekker, New York, 2nd Edn., 1976, p. 173.

101 N.W. Withers and J.C. Nevenzel, *Lipids*, 12 (1977) 989.

102 N. Nicolaides, *J. Chromatogr. Sci.*, 8 (1970) 717.

103 N. Nicolaides, H.C. Fu, M.N.A. Ansari and G. Rice, *Lipids*, 7 (1972) 506.

104 B. Serdarevich, *J. Amer. Oil Chem. Soc.*, 44 (1967) 381.

105 S. Ruggieri and A. Fallani, *Lipids*, 14 (1979) 323.

106 G. Friedman, O. Stein and Y. Stein, *Biochim. Biophys. Acta*, 573 (1979) 521.

107 C.M. Lok and B. Folkersma, *Lipids*, 14 (1979) 872.

108 G. van Dessel, A. Lagrou, H.J. Hilderson, R. Dommisse, E. Esmans and W. Dierick, *Biochim. Biophys. Acta*, 573 (1979) 296.

109 J. Hirsch and E.H. Ahrens, Jr., *J. Biol. Chem.*, 233 (1958) 311.

110 M.L. Blank and O.S. Privett, *J. Dairy Sci.*, 47 (1964) 481.

111 A.A.Y. Shehata, J.M. de Man and J.C. Alexander, *Can. Inst. Food Technol. J.*, 4 (1971) 61.

112 M.W. Taylor and J.C. Hawke, *N.Z. J. Dairy Sci. Technol.*, 10 (1975) 40.

113 M. Morrison and J.C. Hawke, *Lipids*, 12 (1977) 994.

114 C. Lintas, A.M. Balduzzi, M.P. Bernardini and A. Di Muccio, *Lipids*, 14 (1979) 298.

115 P.D. Klein and E.T. Janssen, *J. Biol. Chem.*, 234 (1959) 1417.

116 B. Serdarevich and K.K. Carroll, *J. Lipid Res.*, 7 (1966) 277.

117 E.N. Frankel and W.E. Neff, *Lipids*, 14 (1979) 39.

118 M. Rigaud, J. Durand and J.C. Breton, *Biochim. Biophys. Acta*, 573 (1979) 408.

119 D.J. Sessa, H.W. Gardner, R. Kleiman and D. Weisleder, *Lipids*, 12 (1977) 613.

120 W. Grosch and G. Laskawy, *Biochim. Biophys. Acta*, 575 (1979) 439.

121 K.E. Espelie and P.E. Kolattukudy, *Lipids*, 13 (1978) 832.

122 E.G. Daniels, in G.V. Marinetti (Editor), *Lipid Chromatographic Analysis*, Vol. 2, Marcel Dekker, New York, 2nd Edn., 1976, p. 611.

123 M.W. Bansbach and P.K. Love, *Prostaglandins*, 17 (1977) 193.

124 S. Bergstrom and B. Samuelsson, *Acta Chem. Scand.*, 17 (1963) 5282.

125 J.E. Pike, F.H. Lincoln and W.P. Schneider, *J. Org. Chem.*, 34 (1969) 3553.

126 B. Samuelsson, *J. Biol. Chem.*, 238 (1963) 3229.

127 E. Anggard, *Biochem. Pharmacol.*, 14 (1965) 1507.

128 R. Chanderbhan, V.A. Hodges, C.R. Treadwell and G.V. Vahouny, *J. Lipid. Res.*, 20 (1979) 116.

129 E.G. Daniels and J.E. Pike, in P.W. Ramwell and J.E. Shaw (Editors), *Prostaglandin Symp. Worcester Found. Exptl. Biol.*, Wiley-Interscience, New York, 1968, p. 379.

130 D.P. Wallach and E.G. Daniels, *Biochim. Biophys. Acta*, 231 (1971) 445.

131 M. Hamberg and B. Samuelsson, *J. Biol. Chem.*, 241 (1966) 257.

132 K. Green, *Biochim. Biophys. Acta*, 231 (1971) 419.

133 H. Kindahl and E. Granstrom, *Biochim. Biophys. Acta*, 280 (1972) 466.

134 K. Svanborg and M. Bygdeman, *Eur. J. Biochem.*, 28 (1972) 127.

135 W.S. Powell, *Biochim. Biophys. Acta*, 575 (1979) 335.

136 J.F. Fioriti, N. Buide and R.J. Sims, *J. Amer. Oil Chem. Soc.*, 46 (1969) 108.

137 C.D. Evans, D.G. McConnell, R.L. Hoffman and H. Peters, *J. Amer. Chem. Soc.*, 44 (1967) 281.

138 J. Pokorny, J. Hlakid and I. Zeman, *Pharmazie*, 23 (1968) 332.

139 R. Kleiman, G.F. Spencer, F.R. Earle, H.J. Nieschlag and A.S. Barclay, *Lipids*, 7 (1972) 660.

140 K. Kasama, N. Uezumi and K. Itoh, *Biochim. Biophys. Acta*, 202 (1970) 56.

141 A. Ferreti and V.P. Flanagan, *Lipids*, 12 (1977) 198.

142 E. Haahti and T. Nikkari, *Acta Chem. Scand.*, 17 (1963) 2565.

143 W.L. Erdahl and O.S. Privett, *Lipids*, 12 (1977) 797.

144 H.K. Mangold and D.C. Malins, *J. Amer. Oil Chem. Soc.*, 37 (1960) 383.

145 W.C. Vogel, W.M. Doizaki and L. Zieve, *J. Lipid Res.*, 3 (1962) 138.

146 D.M. Sharaf, S.J. Clark and D.T. Downing, *Lipids*, 12 (1977) 786.

147 G.J. van der Vusse, T.H.M. Roemen and R.S. Reneman, *Biochim. Biophys. Acta*, 617 (1980) 347.

148 O.S. Privett, M.L. Blank, D.W. Codding and E.C. Nickell, *J. Amer. Oil Chem. Soc.*, 42 (1965) 381.

149 G. MacDonald, R.R. Baker and W. Thompson, *J. Neurochem.*, 24 (1975) 655.

150 J.-F. Pernes, Y. Nurit and M. De Heaulme, *J. Chromatogr.*, 181 (1980) 254.

151 M. Mancha, G.B. Stokes and P.K. Stumpf, *Anal. Biochem.*, 68 (1976) 600.

152 K.C. Oo and P.K. Stumpf, *Lipids*, 14 (1979) 132.

153 A.E. Thomas, J.E. Sharoun and H. Ralston, *J. Amer Oil Chem. Soc.*, 42 (1965) 789.

154 H.V. Ammon, P.J. Thomas and S.F. Phillipis, *Lipids*, 14 (1979) 395.

155 O. Renkonen, *Biochim. Biophys. Acta*, 152 (1968) 114.

156 R.G. Lamb, T.G. Gardner and H.J. Fallon, *Biochim. Biophys. Acta*, 619 (1980) 385.

157 J.D. Pollack, D.S. Clark and N.L. Somerson, *J. Lipid Res.*, 12 (1971) 563.

158 O.S. Privett, L.J. Nutter and R.A. Gross, in M.F. Brink and D. Kritchevsky (Editors), *Symposium: Dairy Lipids and Lipid Metabolism*, Avi, Westport, CT, 1968, p. 99.

159 W.C. Breckenridge and A. Kuksis, *J. Lipid Res.*, 9 (1968) 388.

160 A. Kuksis, L. Marai and J.J. Myher, *J. Amer. Oil Chem. Soc.*, 50 (1973) 193.

161 M.O. Marshall and J. Knudsen, *Eur. J. Biochem.*, 81 (1977) 259.

162 J. Oehlenschlager and G. Gercken, *Lipids*, 13 (1978) 557.

163 F. Snyder, *J. Chromatogr.*, 82 (1973) 7.

164 O. Renkonen, in A. Niederwieser and G. Pataki (Editors), *Progress in Thin-Layer Chromatography and Related Methods*, Vol. 2, Ann Arbor Sci. Publ., Ann Arbor, MI, 1971, p. 143.

165 T. Sugiura, Y. Masuzawa and K. Waku, *Lipids*, 15 (1980) 475.

166 H.J. Lin, M.S.F. Lie Ken Jie, C.L.H. Lie and D.H.S. Lee, *Lipids*, 12 (1977) 620.

167 J.J. Myher and A. Kuksis, *J. Amer. Oil Chem. Soc.*, 79 (1982) in press.

168 R. Wood and F. Snyder, *Lipids*, 2 (1967) 89.

169 F. Kuntz, *Biochim. Biophys. Acta*, 296 (1973) 331.

170 E. Vioque and R.T. Holman, *J. Amer. Oil Chem. Soc.*, 39 (1962) 63.

171 L.J. Morris, D.M. Wharry and E.W. Hammond, *J. Chromatogr.*, 33 (1968) 471.

172 B.R. Jordan and J.L. Harwood, *Biochim. Biophys. Acta*, 573 (1979) 218.

173 L.J. Morris and D.M. Wharry, *J. Chromatogr.*, 20 (1965) 27.

174 Y. Matsuda, T. Beppu and K. Arima, *Biochim. Biophys. Acta*, 530 (1978) 439.

175 L.J. Morris, *J. Chromatogr.*, 12 (1963) 321.

176 E.N. Christiansen, M.S. Thomassen, R.Z. Christiansen, H. Osmundsen and K.R. Norum, *Lipids*, 14 (1979) 829.

177 C. Litchfield, J. Tyszkiewicz, E.E. Marcantonio and G. Noto, *Lipids*, 14 (1979) 619.

178 D.E. Minnikin, I.G. Hutchinson, A.B. Caldicott and M. Goodfellow, *J. Chromatogr.*, 188 (1980) 221.

179 K. Green and B. Samuelsson, *J. Lipid Res.*, 5 (1964) 117.

180 N.H. Anderson, *J. Lipid Res.*, 10 (1969) 316.

181 K. Crawshaw, *Prostaglandins*, 3 (1973) 607.

182 R. Chanderbhen, V.A. Hodges, C.R. Treadwell and G.V. Vahouny, *J. Lipid Res.*, 20 (1979) 116.

183 J.A.F. Wickramasinghe and S.R. Shaw, *Prostaglandins*, 4 (1973) 903.

184 R.K. Berthuis, D.H. Nugteren, H.J.J. Pabon and D.A. van Dorp, *Rec. Trav. Chim. Pays-Bas*, 87 (1968) 461.

185 S.I. Murota, Y. Mitsui and M. Kawamura, *Biochim. Biophys. Acta*, 574 (1979) 351.

186 G. Graff, J.H. Stephenson, R.R. Winget and N.D. Goldberg, *Lipids*, 14 (1979) 212.

187 G. Graff, J.H. Stephenson, D.B. Glass, M.K. Haddox and N.D. Goldberg, *J. Biol. Chem.*, 253 (1978) 7662.

188 M.W. Anderson, D.J. Crutchley, A. Chaudhari, A.G.E. Wilson and T.E. Eling, *Biochim. Biophys. Acta*, 573 (1979) 40.

189 C.R. Pace-Asciak and M.C. Carrara, *Biochim. Biophys. Acta*, 574 (1979) 177.

190 C.R. Pace-Asciak, A. Rosenthal and Z. Domazet, *Biochim. Biophys. Acta*, 574 (1979) 182.

191 G. Graff, E.W. Dunham, T.P. Krick and N.D. Goldberg, *Lipids*, 14 (1979) 334.

192 R.J. Light and B. Samuelsson, *Eur. J. Biochem.*, 28 (1972) 232.
193 E.G. Lapetina and P. Cuatrecasas, *Biochim. Biophys. Acta*, 573 (1979) 394.
194 B.A. Vick, D.C. Zimmerman and D. Weisleder, *Lipids*, 14 (1979) 734.
195 M. Hamberg, J. Svensson, T. Waka Bayashi and B. Samuelsson, *Proc. Nat. Acad. Sci. U.S.*, 71 (1974) 345.
196 J.G. Hamilton and L.D. Tobias, *Prostaglandins*, 13 (1977) 1019.
197 D.H. Nugteren and E. Hazelhof, *Biochim. Biophys. Acta*, 326 (1973) 448.
198 H. Herrmann and G. Gereken, *Lipids*, 15 (1980) 179.
199 C. Litchfield, *Analysis of Triglycerides*, Academic Press, New York, 1972.
200 K.L. Mikolajczak and C.R. Smith, Jr., *Lipids*, 2 (1967) 261.
201 C.R. Smith, Jr., R.V. Madrigal and R.D. Plattner, *Biochim. Biophys. Acta*, 572 (1979) 314.
202 T. Yazicioglu, A. Karaali and J. Gokcen, *J. Amer. Oil Chem. Soc.*, 55 (1978) 412.
203 T. Muramatsu and H.H.O. Schmid, *Chem. Phys. Lipids*, 9 (1972) 123.
204 K. Kasama, W.T. Rainey and F. Snyder, *Arch. Biochem. Biophys.*, 154 (1973) 648.
205 L.J. Morris and S.W. Hall, *Lipids*, 1 (1966) 188.
206 K. Payne-Wahl, R.D. Plattner, G.F. Spencer and R. Kleiman, *Lipids*, 14 (1979) 601.
207 R. Kleiman, C.R. Smith and S.G. Yates, *J. Amer. Oil Chem. Soc.*, 42 (1965) 169.
208 K. Oette and T.S. Tschung, *Z. Physiol. Chem.*, 361 (1980) 1179.
209 R. Fears, K.H. Baggaley, R. Alexander, B. Morgan and R.M. Hindley , *J. Lipid Res.*, 19 (1978) 3.
210 A.J. Hancock, S.M. Greenwald and H.Z. Sable, *J. Lipid Res.*, 16 (1975) 300.
211 G. Rouser, J. O'Brien and D. Heller, *J. Amer. Oil Chem. Soc.*, 38 (1961) 14.
212 C.A. Demopoulos, R.N. Pinckard and D.J. Hanahan, *J. Biol. Chem.*, 254 (1979) 935.
213 C.G. Crawford and M.A. Wells, *Lipids*, 14 (1979) 757.
214 A. Horigane, M. Horiguchi and T. Matsumoto, *Biochim. Biophys. Acta*, 572 (1979) 385.
215 G.J. Merkel and J.J. Perry, *Biochim. Biophys. Acta*, 619 (1980) 68.
216 J.A. Kalin and C.M. Allen, *Biochim. Biophys. Acta*, 619 (1980) 76.
217 U.H. Do and S. Ramachandran, *J. Lipid Res.*, 21 (1980) 888.
218 P. Chakrabarti and H.G. Khorana, *Biochemistry*, 14 (1975) 5021.
219 W. Stoffel and G. Michaelis, *Z. Physiol. Chem.*, 357 (1976) 21.
220 L. Stuhne-Sekalec, N.Z. Stanacev, L. Marai and A. Kuksis, *Can. J. Biochem.*, 57 (1979) 408.
221 E.L. Pugh, M. Kates and D.J. Hanahan, *J. Lipid Res.*, 18 (1977) 710.
222 B.J. Holub, A. Kuksis and W. Thompson, *J. Lipid Res.*, 11 (1970) 558.
223 M.G. Luthra and A. Sheltawy, *Biochem. J.*, 126 (1972) 251.
224 G. Okano, T. Kawamoto and T. Akino, *Biochim. Biophys. Acta*, 528 (1978) 385.
225 G. Okano and T. Akino, *Lipids*, 14 (1979) 541.
226 T. Kawamoto, G. Okano and T. Akino, *Biochim. Biophys. Acta*, 619 (1980) 20.
227 C.M. Gupta, R. Radhakrishnan and H.G. Khorana, *Proc. Nat. Acad. Sci. U.S.*, 74 (1977) 4315.
228 J. Schacht, *J. Lipid Res.*, 19 (1978) 1063.
229 F.B. Jungalwala, R.J. Turel, J.E. Evans and R.H. McCluer, *Biochem. J.*, 145 (1975) 517.
230 F.B. Jungalwala, J.E. Evans and R.H. McCluer, *Biochem. J.*, 155 (1976) 55.
231 K. Kiuchi, T. Ohta and H. Ebine, *J. Chromatogr.*, 133 (1977) 226.
232 M.L. Rainey and W.C. Purdy, *Anal. Chim. Acta*, 93 (1977) 211.
233 S. Deming and S.L. Morgan, *Anal. Chem.*, 45 (1973) 278A.
234 W.S.M. Geurts van Kessel, W.M.A. Hax, R.A. Demel and J. de Gier, *Biochim. Biophys. Acta*, 486 (1977) 524.
235 L.G. Abood, N. Salem, M. MacNeil and M. Butler, *Biochim. Biophys. Acta*, 530 (1978) 35.
236 C.P. Blom, F.A. Deierkauf and J.C. Riemersma, *J. Chromatogr.*, 171 (1979) 331.
237 R.W. Gross and B.E. Sobel, *J. Chromatogr.*, 197 (1980) 79.
238 W.J. Hurset and R.A. Martin, Jr., *J. Amer. Oil Chem. Soc.*, 57 (1980) 307.
239 V.P. Skipski, R.F. Peterson and M. Barclay, *Biochem. J.*, 90 (1964) 374.
240 P.C. Sen and T.K. Ray, *Biochim. Biophys. Acta*, 618 (1980) 300.
241 R.Z. Christiansen, J. Norseth and E.N. Christiansen, *Lipids*, 14 (1979) 614.
242 B.J. Holub, J.A. McNaughton and J. Pikarski, *Biochim. Biophys. Acta*, 572 (1979) 413.

243 J.C. O'Kelly and S.C. Mills, *Lipids,* 14 (1979) 983.

244 S.K. Fischer and C.R. Rowe, *Biochim. Biophys. Acta,* 618 (1980) 231.

245 C.L. Parker, M. Waite and L. King, *Biochim. Biophys. Acta,* 620 (1980) 142.

246 L. Taylor, P. Polgar, J.A. McAteer and W.H.J. Douglas, *Biochim. Biophys. Acta,* 572 (1979) 502.

247 S.R. Rooney, T.S. Wai-Lee, L. Gobran and E.K. Motoyama, *Biochim. Biophys. Acta,* 431 (1976) 447.

248 S.R. Rooney, L.I. Gobran, P.A. Marino, W.M. Maniscalco and I. Gross, *Biochim. Biophys. Acta,* 572 (1979) 64.

249 I. Gross, C.M. Wilson, L.D. Ingleson, A. Brehier and S.R. Rooney, *Biochim. Biophys. Acta,* 575 (1979) 375.

250 S.A. Rooney, L.L. Nardone, D.L. Shapiro, E.K. Motoyama, L. Gobran and N. Zaehringer, *Lipids,* 12 (1977) 438.

251 H.J. Galla and W. Hartmann, *Chem. Phys. Lipids,* 27 (1980) 199.

252 C.P. Burns, S.P.L. Wei, D.G. Luttenegger and A.A. Spence, *Lipids,* 14 (1979) 144.

253 G. Rouser, C. Kritchevsky, A. Yamamoto, G. Simon, C. Galli and A.J. Bauman, *Methods Enzymol.,* 14 (1969) 272.

254 S.C. Jamdar, *Lipids,* 14 (1979) 463.

255 B.J. Holub, *Biochim. Biophys. Acta,* 618 (1980) 255.

256 N.H. Shaikh and F.B.St.C. Palmer, *J. Neurochem.,* 28 (1977) 395.

257 C.J. Daniels, and F.B.St.C. Palmer, *Biochim. Biophys. Acta,* 618 (1980) 263.

258 J.E. Bleasdale, P. Wallis, P.C. MacDonald and J.M. Johnston, *Biochim. Biophys. Acta,* 575 (1979) 135.

259 T. Kawamoto, T. Akino, M. Nakamura and M. Mori, *Biochim. Biophys. Acta,* 619 (1980) 35.

260 R.J. Chenery and A.E.M. McLean, *Biochim. Biophys. Acta,* 572 (1979) 9.

261 M. Nishihara and M. Kito, *Biochim. Biophys. Acta,* 531 (1978) 25.

262 D.R. Body and G.M. Gray, *Chem. Phys. Lipids,* 1 (1967) 254.

263 D.E. Epps, V. Natarajan, P.C. Schmid and H.H.O. Schmid, *Biochim. Biophys. Acta,* 618 (1980) 420.

264 A.I. Leikin, A.M. Nervi and R.R. Brenner, *Lipids,* 14 (1979) 102.

265 J. Lecerf, *Biochim. Biophys. Acta,* 617 (1980) 398.

266 K. Suyama, S. Adachi, H. Sugawara and H. Honjon, *Lipids,* 14 (1979) 707.

267 W.D. Blaker and E.A. Moscatelli, *Lipids,* 14 (1979) 1027.

268 M. Brunetti, A. Gaiti and G. Porcellati, *Lipids,* 14 (1979) 925.

269 J.W. Cox, W.R. Snyder and L.A. Horrocks, *Chem. Phys. Lipids,* 25 (1979) 369.

270 H.W. Peter and H.U. Wolf, *J. Chromatogr.,* 82 (1973) 15.

271 T.A. Eisele, R.S. Parker, T.K. Yoss, J.E. Nixon, N.E. Powlovski and R.O. Sinnhuber, *Lipids,* 14 (1979) 523.

272 R. Anderson, M. Kates and B.E. Volcani, *Biochim. Biophys. Acta,* 573 (1979) 557.

273 K. Kobayashi and H. Kanoh, *Biochim. Biophys. Acta,* 575 (1979) 350.

274 K. Morimoto and H. Kanoh, *Biochim. Biophys. Acta,* 531 (1978) 16.

275 J. Baranska, *Biochim. Biophys. Acta,* 619 (1980) 258.

276 W. Stoffel and I. Melzner, *Z. Physiol. Chem.,* 361 (1980) 755.

277 W.D. Skidmore and C. Entenmann, *J. Lipid Res.,* 3 (1962) 471.

278 L.A. Horrocks, *J. Amer. Oil Chem. Soc.,* 40 (1963) 235.

279 C. Chang, R.L. Pike and C.O. Clagett, *Lipids,* 13 (1978) 167.

280 A. Joutti, *Biochim. Biophys. Acta,* 575 (1979) 10.

281 A. Joutti and O. Renkonen, *J. Lipid Res.,* 20 (1979) 230.

282 J. Sherma and J.C. Touchstone, *J. High Resolut. Chromatogr. Chromatogr. Commun.,* 2 (1979) 199.

283 J.C. Touchstone, J.C. Chen and K.M. Bearer, *Lipids,* 15 (1980) 61.

284 E. Wodtke, *Biochim. Biophys. Acta,* 529 (1978) 280.

285 R.S. Finkel and J.J. Volpe, *Biochim. Biophys. Acta,* 572 (1979) 461.

286 R.G. Perkins and R.E. Scott, *Lipids,* 13 (1978) 653.

287 H. Hasegawa-Sasaki and K. Ohno, *Biochim. Biophys. Acta,* 617 (1980) 205.

288 G. Daum, G. Gameri and F. Paltauf, *Biochim. Biophys. Acta,* 573 (1979) 413.

289 M. Hallman and P. Kankare, *Lipids,* 14 (1979) 435.

290 R. Bligny and R. Douce, *Biochim. Biophys. Acta,* 617 (1980) 254.

291 R.M. Broeckhuyse, *Biochim. Biophys. Acta,* 152 (1968) 307.
292 A. Montfort and W.A.M. Boere, *Lipids,* 13 (1978) 580.
293 R.F.A. Zwaal, R. Fluckinger, S. Moser and P. Zahler, *Biochim. Biophys. Acta,* 373 (1974) 416.
294 J. Jimeno-Abendano and P. Zahler, *Biochim. Biophys. Acta,* 573 (1979) 266.
295 J.S. Tou, *Biochim. Biophys. Acta,* 572 (1979) 307.
296 G. Rouser, S. Fleischer and A. Yamamoto, *Lipids,* 5 (1970) 494.
297 B.J.H.M. Poorthuis, P.J. Yazaki and K.Y. Hostetler, *J. Lipid Res.,* 17 (1976) 433.
298 E.L. Kean, *J. Lipid Res.,* 7 (1966) 449.
299 B.J.H.M. Poorthuis and K.Y. Hostetler, *J. Lipid Res.,* 19 (1978) 309.
300 M. Post, J.J. Batenburg and L.L.M. van Golde, *Biochim. Biophys. Acta,* 618 (1980) 308.
301 B.H.S. Cho, D.M. Irvine and J.B.M. Rattray, *Lipids,* 12 (1977) 983.
302 K.Y. Hostetler, B. Zenner and H.P. Morris, *Biochim. Biophys. Acta,* 441 (1976) 231.
303 K.Y. Hostetler, B. Zenner and H.P. Morris, *J. Lipid Res.,* 20 (1979) 607.
304 E. Yavin and A. Zutra, *Anal. Biochem.,* 80 (1977) 430.
305 G.S. Getz, S. Jakovcic, J. Heywood, J. Frank and M. Rabinowitz, *Biochim. Biophys. Acta,* 218 (1970) 441.
306 J. Eichberg and J. Gates, *Biochim. Biophys. Acta,* 573 (1979) 90.
307 C.P. Burns, S.P.L. Wei and A.A. Spector, *Lipids,* 13 (1978) 666.
308 P.R. Roughan and C.R. Slack, *Biochim. Biophys. Acta,* 431 (1976) 86.
309 J.S. Tou, *Biochim. Biophys. Acta,* 531 (1978) 167.
310 T.C. Lee, M.L. Blank, C. Piantodosi, K.S. Ishay and F. Snyder, *Biochim. Biophys. Acta,* 409 (1975) 218.
311 M.L. Blank, T.C. Lee, C. Piantodosi, K.S. Ishay and F. Snyder, *Arch. Biochem. Biophys.,* 177 (1976) 317.
312 C. Moore, T.C. Lee, N. Stephens and F. Snyder, *Biochim. Biophys. Acta,* 531 (1978) 125.
313 M.L. Blank, F. Snyder, L.W. Byers, B. Brooks and E.E. Muirhead, *Biochem. Biophys. Res. Commun.,* 90 (1979) 1194.
314 G.M. Gray, *Biochim. Biophys. Acta,* 144 (1967) 511.
315 O. Colard, D. Bard, G. Bereziat and J. Polonovski, *Biochim. Biophys. Acta,* 618 (1980) 88.
316 G.W. Chapman, Jr. and J.A. Robertson, *J. Amer. Oil Chem. Soc.,* 54 (1977) 195.
317 G.W. Chapman, Jr., *J. Amer. Oil Chem. Soc.,* 57 (1980) 299.
318 R. Hoffmann, H.J. Ristow, H. Pachowski and W. Frank, *Eur. J. Biochem.,* 49 (1974) 317.
319 V.E. Vaskovsky and T.A. Terekhova, *J. High Resolut. Chromatogr. Chromatogr. Commun.,* 2 (1979) 671.
320 L.A. Horrocks, *J. Lipid. Res.,* 9 (1968) 469.
321 D.P. Selivonchick and B.I. Roots, *Lipids,* 14 (1979) 66.
322 L.S. Horrocks and G.Y. Sun, in N. Marks and R. Rodnight (Editors), *Research Methods in Neurochemistry,* Vol. 1, Plenum, New York, 1972, p. 223.
323 G.Y. Sun, *Lipids,* 14 (1979) 918.
324 A. Radominska-Pyrek, Z. Dabrowiecki and L.A. Horrocks, *Biochim. Biophys. Acta,* 574 (1979) 248.
325 Y. Nozawa and G.A. Thompson, *J. Cell Biol.,* 49 (1971) 712.
326 V.M. Kapoulas, *Biochim. Biophys. Acta,* 176 (1969) 324.
327 R. Mason, G. Huber and M. Vaughan, *J. Clin. Invest.,* 51 (1972) 51.
328 R.G. Ackman and A.D. Woyewoda, *J. Chromatogr. Sci.,* 17 (1979) 514.
329 B. Herslof, in L. Appelqvist and C. Liljienberg (Editors), *Advances in Biochemistry and Physiology of Plant Lipids,* Elsevier, Amsterdam, 1979, p. 301.
330 R. Taguchi, Y. Asahi and H. Ikezawa, *Biochim. Biophys. Acta,* 619 (1980) 48.
331 G. Rouser, A.J. Bauman, G. Kritchevsky, D. Heller and J. O'Brien, *J. Amer. Oil Chem. Soc.,* 38 (1961) 554.
332 M.L. Vorbeck and G.V. Marinetti, *J. Lipid Res.,* 6 (1965) 3.
333 T. Saito and S.I. Hakomori, *J. Lipid Res.,* 12 (1971) 257.
334 R. Anderson, B.P. Livermore, M. Kates and B.E. Volcani, *Biochim. Biophys. Acta,* 528 (1978) 77.
335 R. Anderson, M. Kates and B.E. Volcani, *Biochim. Biophys. Acta,* 528 (1978) 89.

336 J.S. Ericson and N.S. Radin, *J. Lipid Res.*, 14 (1973) 133.
337 S.P. Peters, L. Aquino, W.F. Naccarato, J.R. Gilbertson, W.F. Diven and R.H. Glew, *Biochim. Biophys. Acta*, 575 (1979) 27.
338 S.P. Peters, R.H. Glew and R.E. Lee, in R.H. Glew and S.P. Peters (Editors), *Practical Enzymology of Sphingolipidoses*, Liss, New York, 1977, p. 71.
339 Y. Hirabayashi, T. Taki, M. Matsumoto and K. Kojima, *Biochim. Biophys. Acta*, 529 (1978) 96.
340 L. Svennerholm, A. Bruce, J.E. Mansson, B.M. Rynmark and M.T. Vanier, *Biochim. Biophys. Acta*, 280 (1972) 626.
341 G.M. Levis, J.N. Karli and S.D. Moulopoulos, *Lipids*, 14 (1979) 9.
342 R. Narasimhan and R.K. Murray, *Biochem. J.*, 179 (1979) 199.
343 G. Yogeeswaran, R. Sheinin, J.R. Wherrett and R.K. Murray, *J. Biol. Chem.*, 247 (1972) 5146.
344 R. Maget-Dana and J.C. Michalski, *Lipids*, 15 (1980) 682.
345 S. Ando, M. Isobe and Y. Nagai, *Biochim. Biophys. Acta*, 424 (1976) 98.
346 K. Ueno, S. Ando and R.K. Yu, *J. Lipid Res.*, 19 (1978) 863.
347 A. Suzuki, S.K. Kundu and D.M. Marcus, *J. Lipid Res.*, 21 (1980) 473.
348 B.W. Nichols and A.T. James, *Fette, Seifen, Anstrichm.*, 66 (1964) 1003.
349 C.C. Winterbourn, *J. Neurochem.*, 18 (1971) 1153.
350 E.G. Bremer, S.K. Gross and R.H. McCluer, *J. Lipid Res.*, 20 (1979) 1028.
351 P. Fredman, O. Nilsson, J.L. Tayot and L. Svennerholm, *Biochim. Biophys. Acta*, 618 (1980) 42.
352 P.H. Fishman, R.M. Bradley, J. Moss and V.C. Manganiello, *J. Lipid Res.*, 19 (1978) 77.
353 T. Ariga, S. Ando, A. Takahashi and T. Miyatake, *Biochim. Biophys. Acta*, 618 (1980) 480.
354 K. Kojima, A. Slomiany, V.L.N. Murty, N.I. Galicki and B.L. Slomiany, *Biochim. Biophys. Acta*, 619 (1980) 403.
355 R.W. Ledeen, R.K. Yu and L.F. Eng, *J. Neurochem.*, 21 (1973) 829.
356 G.M. Levis, J.N. Karli and N.J. Crumpton, *Biochem. Biophys. Res. Commun.*, 68 (1976) 336.
357 A. Slomiany, C. Annese and B.L. Slomiany, *Biochim. Biophys. Acta*, 441 (1976) 316.
358 J. Koscielak, W. Maslinski, J. Zielenski, E. Zdebska, T. Brudzynski, H. Miller-Podraza and B. Cedergren, *Biochim. Biophys. Acta*, 530 (1978) 385.
359 M. Iwamori and Y. Nagai, *Biochim. Biophys. Acta*, 528 (1978) 257.
360 R.H. McCluer and J.E. Evans, *J. Lipid Res.*, 14 (1973) 611.
361 F.B. Jungalwala, L. Hayes and R.H. McCluer, *J. Lipid Res.*, 18 (1977) 285.
362 J.E. Evans and R.H. McCluer, *Biochim. Biophys. Acta*, 270 (1972) 565.
363 M.D. Ullman and R.H. McCluer, *J. Lipid Res.*, 18 (1977) 371.
364 M.D. Ullman and R.H. McCluer, *J. Lipid Res.*, 19 (1978) 910.
365 K. Watanabe and Y. Arao, *J. Lipid Res.*, 22 (1981) 1020.
366 V.P. Skipski, A.F. Smolove and M. Barclay, *J. Lipid Res.*, 8 (1967) 295.
367 P. Pohl, H. Glasl and H. Wagner, *J. Chromatogr.*, 49 (1970) 488.
368 J.P. Williams, G.R. Watson, M. Kahn, S. Leung, A. Kuksis, O. Stachnyk and J.J. Myher, *Anal. Biochem.*, 66 (1975) 110.
369 M.-U. Khan and J.P. Williams, *J. Chromatogr.*, 140 (1977) 179.
370 T.A. Clayton, T.A. MacMurray and W.R. Morrison, *J. Chromatogr.*, 47 (1970) 277.
371 N.S. De Silva and H.W. Fowler, *Phytochemistry*, 15 (1976) 1735.
372 N. Sato, N. Murata, Y. Miura and N. Ueta, *Biochim. Biophys. Acta*, 572 (1979) 19.
373 N. Sato and N. Murata, *Biochim. Biophys. Acta*, 619 (1980) 353.
374 E. Svennerholm and L. Svennerholm, *Biochim. Biophys. Acta*, 70 (1963) 432.
375 M.T. Coleman and A.J. Yates, *J. Chromatogr.*, 166 (1978) 611.
376 L. Svennerholm, M.-T. Vanier and J.-E. Manson, *J. Lipid Res.*, 21 (1980) 53.
377 G.J.M. Mooghwinkel, P. Borri and J.G. Riemersma, *Rec. Trav. Chim. Pays-Bas*, 83 (1964) 576.
378 I.K. Grundt, E. Stensland and T.L.M. Syversen, *J. Lipid Res.*, 21 (1980) 162.
379 H.E. Carter, R.H. McCluer and E.D. Slifter, *J. Amer. Chem. Soc.*, 78 (1956) 3735.
380 N. Shaw and J.Baddiley, *Nature, (London)* 217 (1968) 142.
381 W. Fischer, I. Ishizuka, H.R. Landgraf and J. Herrmann, *Biochim. Biophys. Acta*, 296 (1973) 527.
382 K. Tadano and I. Ishizuka, *Biochim. Biophys. Acta*, 575 (1979) 421.

383 S.S.J. Sung and C.C. Sweeley, *Biochim. Biophys. Acta*, 575 (1979) 295.
384 S.I. Hakomori and H.D. Andrews, *Biochim. Biophys. Acta*, 202 (1979) 225.
385 R. Kuhn and H. Wiegandt, *Chem. Ber.*, 96 (1963) 866.
386 J.R. Wherrett and J.N. Cummings, *Biochem. J.*, 86 (1963) 378.
387 K. Puro, P. Maury and J.K. Huttunen, *Biochim. Biophys. Acta*, 187 (1969) 230.
388 T.W. Keenan, *Biochim. Biophys. Acta*, 337 (1974) 255.
389 A.A. Bushway and T.W. Keenan, *Lipids*, 13 (1978) 59.
390 T.W. Keenan, E. Schmid and W.W. Franke, *Lipids*, 13 (1978) 451.
391 P.F. Smith, *Biochim. Biophys. Acta*, 619 (1980) 367.
392 T.N. Seyfried, E.J. Weber and R.K. Yu, *Lipids*, 12 (1977) 979.
393 S. Ando, N.C. Chang and R.K. Yu, *Anal. Biochem.*, 89 (1978) 437.
394 A.J. Yates, D.K. Thompson, C.P. Boesel, C. Albrightson and R.W. Hart, *J. Lipid Res.*, 20 (1979) 428.
395 G. Schwarzmann, *Biochim. Biophys. Acta*, 529 (1978) 106.
396 M. Ohashi, *Lipids*, 14 (1979) 52.
397 B. de Vries, *J. Amer. Oil Chem. Soc.*, 40 (1963) 184.
398 M. Özcimder and W.E. Hammers, *J. Chromatogr.*, 187 (1980) 307.
399 C.B. Barrett, M.S.J. Dallas and F.B. Padley, *Chem. Ind. (London)*, (1962) 1050.
400 F.R. Hartley, *Chem. Rev.*, 73 (1973) 163.
401 O.K. Guha and J. Janák, *J. Chromatogr.*, 68 (1972) 325.
402 P.A. Dudley and R.A. Anderson, *Lipids*, 10 (1975) 113.
403 M.S.F. Lie Ken Jie and C.H. Lam, *J. Chromatogr.*, 124 (1976) 147.
404 D.E. Minnikin, P.V. Patel and M. Goodfellow, *FEBS Lett.*, 39 (1974) 322.
405 G.A. Rao, R.L. Kilpatrick, S.C. Goheen and E.C. Larkin, *Lipids*, 15 (1980) 686.
406 P.P. Ilinov, *Lipids*, 14 (1979) 598.
407 R.R. Heath, J.H. Tomlinson, R.E. Doo Little and A.T. Proveaux, *J. Chromatogr. Sci.*, 13 (1975) 380.
408 J.D. Warthen, Jr., *J. Chromatogr. Sci.*, 14 (1976) 513.
409 M.E. Neville, Jr., T.C. Miin and K.A. Ferguson, *Biochim. Biophys. Acta*, 573 (1979) 201.
410 W.E. Hammers, M.C. Spanjer and C.L. de Ligny, *J. Chromatogr.*, 174 (1979) 291.
411 S. Lam and E. Grushka, *J. Chromatogr. Sci.*, 15 (1977) 234.
412 R.O. Adlof, H. Rakoff and E.A. Emken, *J. Amer. Oil Chem. Soc.*, 57 (1980) 273.
413 R.O. Adlof and E.A. Emken, *J. Amer. Oil Chem. Soc.*, 57 (1980) 276.
414 C.R. Schofield and T.L. Mounts, *J. Amer. Oil Chem. Soc.*, 54 (1977) 319.
415 E.A. Emken, J.C. Hartman and C.R. Turner, *J. Amer. Oil Chem. Soc.*, 55 (1978) 561.
416 C.R. Schofield, *J. Amer. Oil Chem. Soc.*, 57 (1980) 331.
417 F. Mordret, A. Prevot, N. Le Barbachon and C. Barbati, *Rev. Trans. Corps Gras*, 24 (1977) 467.
418 C.R. Schofield, in E.A. Emken and H.J. Dutton (Editors), *Geometrical and Positional Fatty Acid Isomers*, Amer. Oil Chem. Society, Champaign, IL, 1979, p. 17.
419 R.L. Glass, T.P. Krick, D.L. Olson and R.L. Thorson, *Lipids*, 12 (1977) 828.
420 E. Crundwell and A.L. Cripps, *Chem. Phys. Lipids*, 16 (1976) 161.
421 A.T. James, in D. Glick (Editor), *Methods of Biochemical Analysis*, Vol. 8, Interscience, New York, 1960, p. 1.
422 R.G. Ackman, *Progr. Chem. Fats Other Lipids*, 7 (1972) 41.
423 A. Kuksis, *Fette, Seifen, Anstrichm.*, 71 (1971) 130.
424 N. Nicolaides, J.M.B. Apon and D.H. Wong, *Lipids*, 11 (1976) 781.
425 J.J. Myher, L. Marai and A. Kuksis, *Anal. Biochem.*, 62 (1974) 188.
426 A.J. Sheppard, J.L. Iverson and J.L. Weihrauch, in A. Kuksis (Editor), *Fatty Acids and Glycerides*, Vol. 1, Plenum, New York, 1978, p. 341.
427 E.A. Emken, in A. Kuksis (Editor), *Fatty Acids and Glycerides*, Vol. 1, Plenum, New York, 1978, p. 77.
428 K. Tanaka and G.M. Yu, *Clin. Chim. Acta*, 43 (1973) 151.
429 G.R. Allen and M.J. Saxby, *J. Chromatogr.*, 37 (1968) 312.
430 J.R. Ashes and J.K. Haken, *J. Chromatogr.*, 111 (1975) 171.
431 D.M. Ottenstein and W.R. Supina, *J. Chromatogr.*, 91 (1974) 119.

432 J.R. Ashes, J.K. Haken and S.C. Mills, *J. Chromatogr.*, 187 (1980) 297.

433 W.W. Christie and R.T. Holman, *Lipids*, 1 (1966) 176.

434 A.K. Sen Gupta and H. Peeters, *Chem. Phys. Lipids*, 3 (1969) 371.

435 M. Oshima and T. Ariga, *J. Biol. Chem.*, 250 (1975) 6963.

436 R.L. Glass, T.P. Krick, D.M. Sand, C.H. Rahn and H. Schlenk, *Lipids*, 10 (1975) 695.

437 E.A. Emken, *Lipids*, 7 (1972) 459.

438 E.A. Emken and H.J. Dutton, *Lipids*, 9 (1974) 272.

439 H. Rakoff and E.A. Emken, *Lipids*, 12 (1977) 760.

440 D.M. Ottenstein, D.A. Bartley and W.R. Supina, *J. Chromatogr.*, 119 (1976) 401.

441 R.V. Golovnya, V.P. Uralets and T.E. Kuzmenko, *J. Chromatogr.*, 121 (1976) 118.

442 H. Jaeger, H.U. Kloer and H. Ditschuneit, *J. Lipid Res.*, 17 (1976) 185.

443 T. Kobayashi, *J. Chromatogr.*, 194 (1980) 404.

444 H.T. Slover and E. Lanza, *J. Amer. Oil Chem. Soc.*, 56 (1979) 933.

445 G.M. Patton, S. Cann, H. Brunengraber and J.M. Lowenstein, *Methods Enzymol.*, 72 (1981) 8.

446 C.H. Lam and M.S.F. Lie Ken Jie, *J. Chromatogr.*, 117 (1976) 365.

447 C.H. Lam and M.S.F. Lie Ken Jie, *J. Chromatogr.*, 121 (1976) 303.

448 M.S.F. Lie Ken Jie and C.H. Lam, *J. Chromatogr.*, 97 (1974) 165.

449 M.S.F. Lie Ken Jie, *J. Chromatogr.*, 111 (1975) 189.

450 C.H. Lam and M.S.F. Lie Ken Jie, *J. Chromatogr.*, 115 (1975) 559.

451 M.S.F. Lie Ken Jie, *J. Chromatogr.*, 109 (1975) 81.

452 G.R. Jamieson, A.L. McMinn and E.H. Reid, *J. Chromatogr.*, 178 (1979) 555.

453 I. Zeman, *J. Gas Chromatogr.*, 3 (1965) 18.

454 I. Yano, K. Kageyama, Y. Ohno, M. Masui, E. Kusonose, M. Kusonose and N. Akimori, *Biomed. Mass Spectrom.*, 5 (1978) 14.

455 S. Toriyama, I. Yano, M. Kusunose, and E. Kusunose, *FEBS Lett.*, 15 (1978) 111.

456 T. Takagi, A. Sakai, Y. Itabashi and K. Hayashi, *Lipids*, 12 (1977) 228.

457 W.A. König and I. Benecke, *J. Chromatogr.*, 195 (1980) 292.

458 H. Engelhard and H. Elgass, *J. Chromatogr.*, 158 (1978) 249.

459 M.J. Cooper and M.W. Anders, *Anal. Chem.*, 46 (1974) 1849.

460 E. Grushka, H.D. Durst and E.J. Kikta, Jr., *J. Chromatogr.*, 112 (1975) 673.

461 H.C. Jordi, *J. Liquid Chromatogr.*, 1 (1978) 215.

462 W. Distler, *J. Chromatogr.*, 192 (1980) 240.

463 J. Weatherston, L.M. MacDonald, T. Blake, M.H. Benn and Y.Y. Huang, *J. Chromatogr.*, 161 (1978) 347.

464 P.T.S. Pei, W.C. Kossa, S. Ramachandran and R.S. Henly, *Lipids*, 11 (1976)) 814.

465 J.D. Warthen, Jr., *J. Amer. Oil Chem. Soc.*, 52 (1975) 151.

466 C.R. Schofield, *J. Amer. Oil Chem. Soc.*, 52 (1975) 36.

467 C.R. Schofield, *Anal. Chem.*, 47 (1975) 1417.

468 A.G. Bailie, Jr., T.D. Wilson, J.M. Beebe and J.D. Stuart, *J. Chromatogr. Sci.*, (1982) in press.

469 R. Kleiman, C.F. Spencer, L.W. Tjarks and F.R. Earle, *Lipids*, 6 (1971) 617.

470 M.B. Bohannon and R. Kleiman, *Lipids*, 10 (1975) 703.

471 L.J. Morris, *J. Lipid Res.*, 7 (1969) 717.

472 P.W. Ramwell and E.G. Daniels, in G.V. Marinetti (Editor), *Lipid Chromatographic Analysis*, Vol. 2, Marcel Dekker, New York, 1968, p. 313.

473 G.L. Bundy, E.G. Daniels, F.H. Lincoln and J.E. Pike, *J. Amer. Chem. Soc.*, 94 (1972) 2124.

474 M.V. Merritt and G.E. Bronson, *Anal. Biochem.*, 80 (1977) 392.

475 A.P. Tulloch, *J. Amer. Oil Chem. Soc.*, 41 (1964) 833.

476 A.P. Tulloch and M. Mazurek, *Lipids*, 11 (1976) 228.

477 J.S. O'Brien and G. Rouser, *Anal. Biochem.*, 7 (1964) 288.

478 K.A. Karlsson and J. Pascher, *Chem. Phys. Lipids*, 12 (1974) 65.

479 H.B.S. Conacher and F.D. Gunstone, *Chem. Phys. Lipids*, 3 (1969) 203.

480 R. Wood, E.L. Beaver and F. Snyder, *Lipids*, 1 (1966) 399.

481 R. Wood, *Lipids*, 2 (1967) 199.

482 F.A. Fitzpatrick, in J.C. Frohlich (Editor), *Advances in Prostaglandin and Thromboxane Research,* Vol. 5, Raven, New York, 1978, p. 95.

483 F.A. Fitzpatrick, D.A. Stringfellow, J. Maclouf and M. Rigaud, *J. Chromatogr.,* 177 (1979) 51.

484 S. Nicosia and G. Galli, *Anal. Biochem.,* 61 (1974) 192.

485 G.H. Jouvenaz, D.H. Nugteren, R.K. Berthuis and D.A. van Dorp, *Biochim. Biophys. Acta,* 202 (1970) 231.

486 G.H. Jouvenaz, D.H. Nugteren and D.A. van Dorp, *Prostaglandins,* 3 (1973) 175.

487 J.A.F. Wickramasinghe, W. Morozowich, W.E. Hamlin and S.R. Shaw, *J. Pharm. Sci.,* 62 (1973) 1428.

488 J.A.F. Wickramasinghe and S.R. Shaw, *Biochem. J.,* 141 (1974) 179.

489 E.A.M. DeDechere, D.H. Nugteren and F. ten Hoor, *Nature (London),* 268 (1977) 160.

490 F.A. Fitzpatrick, M.A. Wynalda and D.G. Kaiser, *Anal. Chem.,* 49 (1977) 1032.

491 V.A. Skrinska and A. Butkus, *Prostaglandins,* 16 (1978) 571.

492 M.J. Levitt, J.B. Josimovich and K.D. Broskin, *Prostaglandins,* 1 (1972) 121.

493 B.S. Middleditch and D.M. Desiderio, *Prostaglandins,* 2 (1972) 195.

494 D. Nugteren, *J. Biol. Chem.,* 250 (1975) 2808.

495 C. Pace-Asciak and L.S. Wolfe, *J. Chromatogr. Sci.,* 56 (1971) 129.

496 F. Szederkenyi and G. Kovacs, *Prostaglandins,* 8 (1974) 285.

497 R.W. Kelly and P.L. Taylor, *Anal. Chem.,* 48 (1976) 465.

498 M. Rigaud, P. Chebroux, J. Durand, J. Maclouf and C. Mandani, *Tetrahedron Lett.,* 44 (1976) 3935.

499 J. Maclouf, M. Rigaud, J. Durand and P. Chebroux, *Prostaglandins,* 11 (1976) 999.

500 F.A. Fitzpatrick, *Anal. Chem.,* 50 (1978) 47.

501 K. Green, M. Hamberg and B. Samuelsson, in B. Samuelsson and R. Paoletti (Editors), *Advances in Prostaglandin and Thromboxane Research,* Vol. 1, Raven, New York, 1976, p. 47.

502 K. Green, M. Hamberg, B. Samuelsson, M. Smigel and J.C. Frohlich, in J.C. Frolich (Editor), *Advances in Prostaglandin and Thromboxane Research,* Vol. 5, Raven, New York, 1978, p. 39.

503 A. Ferretti and V.P. Flanagan, *Lipids,* 14 (1979) 483.

504 N.A. Porter, R.A. Wolf and H. Weenen, *Lipids,* 15 (1980) 163.

505 N.A. Porter, R.A. Wolf, E.M. Yarbro and H. Weenen, *Biochem. Biophys. Res. Commun.,* 89 (1979) 1058.

506 N.A. Porter, J. Logan and V. Kontoyiannidou, *J. Org. Chem.,* 44 (1979) 3177.

507 F.A. Fitzpatrick, *Anal. Chem.,* 48 (1976) 499.

508 M.V. Merritt and G.E. Bronson, *Anal. Chem.,* 48 (1976) 1851.

509 T.O. Oesterling, W. Morozowitch and T.J. Roseman, *J. Pharm. Sci.,* 61 (1972) 1861.

510 M.J. Cho and M.A. Allen, *Prostaglandins,* 15 (1978) 943.

511 J. Turk, S.J. Weiss, J.E. Davis and P. Needleman, *Prostaglandins,* 16 (1978) 291.

512 A.R. Whorton, K. Carr, M. Smigel, L. Walker, K.Ellis and J.A. Oates, *J. Chromatogr.,* 163 (1979) 64.

513 A.R. Whorton, B.J. Sweetman and A.J. Oates, *Anal. Biochem.,* 98 (1979) 455.

514 G.T. Hill, *J. Chromatogr.,* 176 (1979) 407.

515 M.A. Wynalda, F.H. Lincoln and F.A. Fitzpatrick, *J. Chromatogr.,* 176 (1979) 413.

516 S. Inayama, H. Hori, T. Shibata, Y. Ozawa, K. Yamagami, M. Imazu and H. Hayashida, *J. Chromatogr.,* 194 (1980) 85.

517 M. Van Rollins, S.H.K. Ho, J.E. Greenwald, M. Alexander, N.J. Dorman, L.K. Wong and L.A. Horrocks, *Prostaglandins,* 20 (1980) 571.

518 F.A. Russell and D. Deykin, *Prostaglandins,* 18 (1979) 11.

519 K. Nagayo and N. Mizuno, *J. Chromatogr.,* 178 (1979) 347.

520 H.S. Hansen and K. Bukhave, *Prostaglandins,* 16 (1978) 311.

521 E. Granstrom, *Ann. Clin. Biochem.,* 16 (1979) 354.

522 F.D. Gunstone and F.B. Padley, *J. Amer. Oil Chem. Soc.,* 42 (1965) 957.

523 S.H. Fatemi and E.G. Hammond, *Lipids,* 12 (1977) 1037.

524 T. Takagi and Y. Itabashi, *Lipids,* 12 (1977) 1062.

525 J.J. Myher, L. Marai, A. Kuksis and D. Kritchevsky, *Lipids,* 12 (1977) 775.

526 H. Wessels and N.S. Rajagopalan, *Fette, Seifen, Anstrichm.,* 71 (1969) 543.

527 H.B.S. Conacher, F.D. Gunstone, G.M. Hornby and F.B. Padley, *Lipids*, 5 (1970) 434.

528 F.D. Gunstone and M.I. Qureshi, *J. Sci. Food Agr.*, 19 (1968) 356.

529 R.J. van der Wal, *J. Amer. Oil Chem. Soc.*, 42 (1965) 1155.

530 A.R. Johnson, K.E. Murray, A.C. Fogerty, B.H. Kennett, J.A. Pearson and F.S. Schenstone, *Lipids*, 2 (1967) 316.

531 F. Snyder, E.A. Cress and N. Stephens, *Lipids*, 1 (1966) 381.

532 L.M.G. van Golde and L.L.M. van Deenen, *Biochim. Biophys. Acta*, 125 (1966) 496.

533 L.M.G. van Golde, W.A. Pietersen and L.L.M. van Deenen, *Biochim. Biophys. Acta*, 152 (1968) 84.

534 A. Kuksis, *Fette, Seifen, Anstrichm.*, 73 (1971) 332.

535 O. Renkonen, *Biochim. Biophys. Acta*, 125 (1966) 288.

536 A. Kuksis and L. Marai, *Lipids*, 2 (1967) 217.

537 J.J. Myher, A. Kuksis, L. Marai and S.K.F. Yeung, *Anal. Chem.*, 50 (1978) 430.

538 L.J. Morris, *Biochem. Biophys. Res. Commun.*, 20 (1965) 340.

539 O. Renkonen, *Lipids*, 3 (1968) 191.

540 A. Kuksis, L. Marai, W.C. Breckenridge, D.A. Gornall and O. Stachnyk, *Can. J. Physiol. Pharmacol.*, 46 (1968) 511.

541 Y. Itabashi and T. Takagi, *Lipids*, 15 (1980) 205.

542 M. Nakamura, T. Kawamoto and T. Akino, *Biochim. Biophys. Acta*, 620 (1980) 24.

543 G. Ikano, T. Kawamoto and T. Akino, *Biochim. Biophys. Acta*, 528 (1978) 285.

544 O. Renkonen, *J. Amer. Oil Chem. Soc.*, 42 (1965) 298.

545 S.K.F. Yeung and A. Kuksis, *Can. J. Biochem.*, 52 (1974) 830.

546 R. Wood and F. Snyder, *Lipids*, 1 (1966) 62.

547 H.J. Lin, M.S.F. Lie Ken Jie and F.C.S. Ho, *J. Lipid Res.*, 17 (1976) 53.

548 B. Hallgren and G. Stallberg, *Acta Chem. Scand.*, 21 (1967) 1519.

549 T. Itoh, M. Tanaka and H. Kaneko, *J. Amer. Oil Chem. Soc.*, 56 (1979) 191A.

550 E.C. Smith, A.D. Jones and E.W. Hammond, *J. Chromatogr.*, 188 (1980) 205.

551 A. Kuksis and M.J. McCarthy, *Can. J. Biochem.*, 40 (1962) 679.

552 W. Eckert, *Fette, Seifen, Anstrichm.*, 79 (1977) 360.

553 P.P. Schmid, M.D. Muller and W. Simon, *J. High Resolut. Chromatogr. Chromatogr. Commun.*, 2 (1979) 675.

554 A. Monseigny, P.V. Vigneron, M. Levacq and I. Zwoboda, *Rev. Corps Gras*, 26 (1979) 107.

555 K. Grob, Jr., P. Neukom and B. Battaglia, *J. Amer Oil Chem. Soc.*, 75 (1980) 282.

556 A. Kuksis, J.J. Myher, K. Geher, W.C. Breckenridge, G.J.L. Jones and J.A. Little, *J. Chromatogr.*, 224 (1981) 1.

557 R.P. D'Alonzo, W.J. Kozarek and H.W. Wharton, *J. Amer. Oil Chem. Soc.*, 58 (1981) 215.

558 A. Kuksis, in G.V. Marinetti (Editor), *Lipid Chromatographic Analysis*, Vol. 1, Marcel Dekker, New York, 1967, p. 239.

559 C. Litchfield, R.D. Harlow and R. Reiser, *Lipids*, 2 (1969) 363.

560 R.G. Powell, R. Kleiman and C.R. Smith, Jr., *Lipids*, 4 (1969) 450.

561 J.A. Fioriti, M.J. Kanuk and R.J. Sims, *J. Chromatogr. Sci.*, 7 (1969) 448.

562 R. Wood and F. Snyder, *J. Lipid Res.*, 8 (1967) 494.

563 M.L. Blank, K. Kasama and F. Snyder, *J. Lipid Res.*, 13 (1972) 390.

564 A. Kuksis, in E.G. Perkins (Editor), *Analysis of Lipids and Lipoproteins*, Amer. Oil Chem. Society, Champaign, IL, 1975, p. 36.

565 M.G. Horning, G. Casparrini and E.C. Horning, *J. Chromatogr. Sci.*, 7 (1969) 267.

566 A. Kuksis, O. Stachnyk and B.J. Holub, *J. Lipid Res.*, 10 (1969) 660.

567 K. Saito, H. Ogino and K. Satouchi, in M. Kates and A. Kuksis (Editors), *Membrane Fluidity*, Humana, Clifton, NJ, 1980, p. 33.

568 W.C. Breckenridge and A. Kuksis, *Can. J. Biochem.*, 53 (1975) 1184.

569 C.R. Smith, Jr., R.V. Madrigal, D. Weisleder and R.D. Plattner, *Lipids*, 12 (1977) 736.

570 A. Kuksis and W.C. Breckenridge, *J. Amer. Oil Chem. Soc.*, 42 (1965) 978.

571 R. Watts and R. Dils, *J. Lipid Res.*, 9 (1968) 40.

572 A. Lohninger and A. Nikiforov, *J. Chromatogr.*, 192 (1980) 185.

573 A. Kuksis, L. Marai, W.C. Breckenridge, D.A. Gornall and O. Stacknyk, *Can. J. Physiol. Pharmacol.*, 46 (1968) 511.

574 B.J. Holub, W.C. Breckenridge and A. Kuksis, *Lipids*, 6 (1971) 307.

575 B.J. Holub and A. Kuksis, *Can. J. Biochem.*, 49 (1971) 1347.

576 M. Nakamura, T. Onodera and T. Akino, *Lipids*, 15 (1980) 616.

577 K. Ishidate and P.A. Weinhold, *Biochim. Biophys. Acta*, 664 (1981) 133.

578 O. Renkonen, *Biochim. Biophys. Acta*, 137 (1967) 575.

579 A. Kuksis, *Can. J. Biochem.*, 49 (1971) 1245.

580 A. Kuksis, *J. Chromatogr. Sci.*, 10 (1972) 53.

581 J.J. Myher and A. Kuksis, *J. Chromatogr. Sci.*, 13 (1975) 138.

582 A. Dyer and E. Klopfenstein, *Lipids*, 12 (1977) 889.

583 J.J. Myher and A. Kuksis, *Can. J. Biochem.*, 60 (1982) in press.

584 J.J. Myher and A. Kuksis, *Lipids*, 9 (1974) 382.

585 F. Snyder, in G.V. Marinetti (Editor), *Lipid Chromatographic Analysis*, Vol. 1, Marcel Dekker, New York, 1976, p. 111.

586 R. Wood, P.K. Raju and R. Reiser, *J. Amer. Oil Chem. Soc.*, 42 (1965) 161.

587 S.J. Friedberg and M. Halpert, *J. Lipid Res.*, 19 (1978) 57.

588 S.J. Gaskell, C.G. Edmonds and C.J.W. Brooks, *Anal. Lett.*, 9 (1976) 325.

589 J.J. Myher, L. Marai and A. Kuksis, *J. Lipid Res.*, 15 (1974) 586.

590 C.F. Poole and A. Zlatkis, *J. Chromatogr.*, 184 (1980) 99.

591 C.O. Rock and F. Snyder, *Arch. Biochem. Biophys.*, 171 (1975) 631.

592 W.J. Bauman, T.H. Madson and B.J, Weseman, *J. Lipid Res.*, 13 (1972) 640.

593 R. Wood, C. Piantadosi and F. Snyder, *J. Lipid Res.*, 10 (1969) 370.

594 T. Takagi, Y. Itabashi, K. Ota and K. Hayashi, *Lipids*, 11 (1976) 354.

595 E.C. Nickel and O.S. Privett, *Separ. Sci.*, 2 (1967) 307.

596 E. Nyström and J. Sjövall, *Anal. Chem.*, 6 (1973) 155.

597 J.E. Ellingboe, J.F. Nyström and J. Sjövall, *Biochim. Biophys. Acta*, 152 (1968) 803.

598 J.E. Ellingboe, J.F. Nyström and J. Sjövall, *J. Lipid Res.*, 11 (1970) 266.

599 E. Nyström and J. Sjövall, *Methods Enzymol.*, 35 (1975) 378.

600 B. Lindqvist, I. Sjögren and R. Nordin, *J. Lipid Res.*, 15 (1974) 65.

601 T. Curstedt and J. Sjövall, *Biochim. Biophys. Acta*, 360 (1974) 24.

602 T. Curstedt and J. Sjövall, *Biochim. Biophys. Acta*, 369 (1974) 173.

603 R.D. Plattner, G.F. Spencer and R. Kleiman, *J. Amer. Oil Chem. Soc.*, 54 (1977) 511.

604 P.T.S. Pei, R.S. Henly and S. Ramachandran, *Lipids*, 10 (1975) 152.

605 J.A. Bezard and M.A. Ouedraogo, *J. Chromatogr.*, 196 (1980) 279.

606 R.D. Plattner, K. Wade and R. Kleiman, *J. Amer. Oil Chem. Soc.*, 55 (1978) 381.

607 R.D. Plattner, *J. Amer. Oil Chem. Soc.*, 58 (1981) 638.

608 W.E. Fallon and Y. Shimizu, *Lipids*, 12 (1977) 765.

609 K.A. Karlsson and I. Pascher, *J. Lipid Res.*, 12 (1971) 466.

610 W.R. Morrison, *Biochim. Biophys. Acta*, 176 (1969) 537.

611 K.A. Karlsson, *Biochem. J.*, 92 (1969) 39P.

612 J.F. Bouhours and H. Guignard, *J. Lipid Res.*, 20 (1979) 897.

613 A. Ballio, C.G. Casinovi, M. Framondino, G. Marino, G. Nota and B. Santurbani, *Biochim. Biophys. Acta*, 573 (1979) 51.

614 R.R. Vunnam and N.S. Radin, *Biochim. Biophys. Acta*, 573 (1979) 73.

615 B. Samuelsson and K. Samuelsson, *J. Lipid Res.*, 10 (1969) 47.

616 K. Samuelsson, *Scand. J. Clin. Invest.*, 27 (1971) 371.

617 A. Kuksis, *Fette, Seifen, Anstrichm.*, 75 (1973) 317.

618 B. Samuelsson and K. Samuelsson, *Biochim. Biophys. Acta*, 164 (1968) 421.

619 G. Casparrini, E.C. Horning and M.G. Horning, *Chem. Phys. Lipids*, 3 (1969) 1.

620 S. Hammarstrom, *J. Lipid Res.*, 11 (1970) 175.

621 R.T.C. Huang, *Z. Physiol. Chem.*, 352 (1971) 1306.

622 A. Kuksis, J.J. Myher, W.C. Breckenridge and J.A. Little, in K. Lippel (Editor), *Report on the High*

Density Lipoprotein Methodology Workshop, NIH Publ. No. 79-1661, U.S. Dept. of Health, Education and Welfare, Bethesda, MD, 1979, p. 142.

623 J.J. Myher, A. Kuksis, W.C. Breckenridge and J.A. Little, *Can. J. Biochem.*, 59 (1981) 626.

624 M. Sugita, N. Iwamori, J.E. Evans, R.H. McCluer, J.T. Dulaney and H.W. Moser, *J. Lipid Res.*, 15 (1979) 223.

625 M. Iwamori and H.W. Moser, *Clin. Chem.*, 21 (1975) 725.

626 M. Iwamori, C. Costello and H.W. Moser, *J. Lipid Res.*, 20 (1979) 86.

627 U.H. Do, P.T. Pei and R.D. Minard, *Lipids*, 16 (1981) 855.

628 M. Smith, P. Monchamp and F.B. Jungalwala, *J. Lipid Res.*, 22 (1981) 714.

629 K. Waku and Y. Nakazawa, *J. Biochem. (Tokyo)*, 72 (1972) 149.

630 R. Wood and R.D. Harlow, *Arch. Biochem. Biophys.*, 135 (1969) 272.

631 L. Marai and A. Kuksis, *Can. J. Biochem.*, 51 (1973) 1365.

632 G.A.E. Arvidson, *J. Lipid Res.*, 6 (1965) 574.

633 G.A.E. Arvidson, *Eur. J. Biochem.*, 4 (1968) 478.

634 B.J. Holub and A. Kuksis, *Lipids*, 4 (1969) 466.

635 H.P. Siebertz, E. Heinz, J. Joyard and R. Douce, *Eur. J. Biochem.*, 108 (1980) 177.

636 E. Heinz and J.L. Harwood, *Z. Physiol. Chem.*, 358 (1977) 897.

637 C.F. Wurster and J.H. Copenhaver, *Lipids*, 1 (1966) 422.

638 E. Baer and J. Maurukas, *J. Biol. Chem.*, 212 (1955) 29.

639 O. Renkonen, *J. Lipid Res.*, 9 (1968) 34.

640 F.D. Collins, in A.T. James and L.J. Morris (Editors), *New Biochemical Separations*, Van Nostrand, London, 1965, p. 380.

641 F.A. Shamgar and F.D. Collins, *Biochim. Biophys. Acta*, 409 (1975) 104.

642 M.G. Luthra and A. Sheltawy, *Biochem. J.*, 126 (1972) 1231.

643 T. Shimojo, M. Abe and M. Ohta, *J. Lipid Res.*, 15 (1975) 525.

644 J.M. Shaw and N.R. Bottino, *J. Lipid Res.*, 15 (1974) 317.

645 N.R. Bottino, *Lipids*, 13 (1978) 18.

646 M. Tanaka, T. Itoh and H. Kaneko, *Yakugaku*, 28 (1979) 96.

647 R.J. King and J.A. Clements, *J. Lipid Res.*, 11 (1970) 381.

648 G.A.E. Arvidson, *J. Chromatogr.*, 103 (1975) 201.

649 N.A. Porter, R.A. Wolf and J.R. Nixon, *Lipids*, 14 (1979) 20.

650 C.G. Crawford, R.D. Plattner, D.J. Sessa and J.J. Rackis, *Lipids*, 15 (1980) 91.

651 B.W. Nichols and R. Moorhouse, *Lipids*, 4 (1969) 311.

652 T.R. Eccleshall and J.C. Hawke, *Phytochemistry*, 10 (1971) 3035.

653 M. Siebertz and E. Heinz, *Z. Physiol. Chem.*, 358 (1977) 27.

654 H.P. Siebertz, E. Heinz, M. Linsheid, J. Joyard and R. Douce, *Eur. J. Biochem.*, 101 (1979) 429.

655 R.D. Araunga and W.R. Morrison, *Lipids*, 6 (1971) 768.

656 G. Auling, E. Heinz and A.P. Tulloch, *Z. Physiol. Chem.*, 352 (1971) 905.

657 M. Oshima, T. Ariga and T. Murata, *Chem. Phys. Lipids*, 19 (1977) 289.

658 S. Hammarstrom and B. Samuelson, *J. Biol. Chem.*, 247 (1972) 1001.

659 G. Nonaka and Y. Kishimoto, *Biochim. Biophys. Acta*, 572 (1979) 423.

660 S. Yahara, I. Singh and Y. Kishimoto, *Biochim. Biophys. Acta*, 619 (1980) 177.

661 K.H. Chou, L.S.A. Ambers and F.B. Jungalwala, *J. Neurochem.*, 23 (1979) 863.

662 S.K. Das and M.S. McCullough, *Lipids*, 15 (1980) 932.

663 D.E. Bowyer, W.M.F. Leat, A.N. Howard and G.A. Gresham, *Biochim. Biophys. Acta*, 70 (1963) 423.

664 H. Ko and M.E. Royer, *J. Chromatogr.*, 88 (1974) 253.

665 A. Kuksis, N. Kovacevic, D. Lau and M. Vranic, *Fed. Proc., Fed. Amer. Soc. Exp. Biol.*, 34 (1975) 2238.

666 W.C. Breckenridge and A. Kuksis, *Can. J. Biochem.*, 53 (1975) 1170.

667 A. Kuksis, *Chromatogr. Rev.*, 8 (1966) 172.

668 J.H. Shand and R.C. Noble, *Anal. Biochem.*, 101 (1980) 427.

669 V. Rogiers, *J. Chromatogr.*, 182 (1980) 27.

670 V. Rogiers, R. Crokaert and H.L. Vis, *Clin. Chim. Acta*, 105 (1980) 105.

671 D.A. Gornall and A. Kuksis, *J. Lipid Res.*, 14 (1973) 197.
672 S. Mookerjea, C.E. Park and A. Kuksis, *Lipids*, 10 (1975) 374.
673 L.W. Daniel, L.S. Kucera and M. Waite, *J. Biol. Chem.*, 255 (1980) 5697.
674 J.K.G. Kramer and H.W. Hulan, *Lipids*, 12 (1977) 159.
675 W. Stremmel and H. Debuch, *Biochim. Biophys. Acta*, 573 (1979) 301.
676 H. Debuch, *J. Neurol.*, 215 (1977) 261.
677 B. Akesson, S. Gronowitz, B. Herslof and R. Ohlson, *Lipids*, 13 (1978) 338.
678 T. Takatori, F.C. Phillips, H. Shimasaki and O.S. Privett, *Lipids*, 11 (1976) 272.
679 M. Sano and O.S. Privett, *Lipids*, 15 (1980) 337.
680 O.S. Privett, K.A. Doughtery and J.D. Castell, *Amer. J. Clin. Nutr.*, 24 (1971) 1265.
681 O.S. Privett, K.A. Doughtery and W.L. Erdahl, in J.C. Touchstone (Editor), *Quantitative Thin-Layer Chromatography*, Wiley, New York, 1973, p. 57.
682 S. Pollet, S. Ermidou, F. Le Saux, M. Monge and V. Baumann, *J. Lipid Res.*, 19 (1978) 916.
683 R.K. Raheja, C. Kaur, A. Singh and I.S. Bhatia, *J. Lipid Res.*, 14 (1973) 695.
684 A.B. Awad, *Lipids*, 13 (1978) 850.
685 J.L. Hojnacki, R.J. Nicolosi and K.C. Hayes, *J. Chromatogr.*, 128 (1976) 133.
686 J. Sherma and J.C. Touchstone, *J. High Resolut. Chromatogr. Chromatogr. Commun.*, 1 (1979) 199.
687 D.T. Downing and A.M. Stranieri, *J. Chromatogr.*, 192 (1980) 208.
688 J.J. Kabara and J.S. Chen, *Anal. Chem.*, 48 (1976) 814.
689 J. Bitman, D.L. Wood and J.M. Ruth, *J. Amer. Oil Chem. Soc.*, 57 (1980) Abstr. No. 384.
690 R. Segura and A.M. Gotto, Jr., *J. Chromatogr.*, 99 (1974) 643.
691 R. Segura and A.M. Gotto, *Clin. Chem.*, 21 (1975) 991.
692 I.R. Kupke and S. Zeugner, *J. Chromatogr.*, 146 (1978) 261.
693 R. Segura and X. Navarro, *J. Amer. Oil Chem. Soc.*, 57 (1980) Abstr. No. 387.
694 W.E.C. Moore, *Int. J. Syst. Bacteriol.*, 20 (1970) 535.
695 T. Okumura and T. Kadono, *Bunseki Kagaku*, 22 (1973) 980.
696 T. Okumura, T. Kadono and A. Iso'o, *J. Chromatogr.*, 108 (1975) 329.
697 J.C. Sipos and R.G. Ackman, *J. Chromatogr. Sci.*, 16 (1978) 443.
698 N.N. Ranayake and R.G. Ackman, *Lipids*, 14 (1979) 795.
699 W.W. Christie and M.L. Hunter, *J. Chromatogr.*, 171 (1979) 517.
700 D. Vandamme, V. Blaton and H. Peeters, *J. Chromatogr.*, 145 (1978) 151.
701 D. Vandamme, G. Vankerckhoven, R. Vercaemst, F. Soetewey, V. Blaton, H. Peeters and M. Rosseneau, *Clin. Chim. Acta*, 89 (1978) 231.
702 D.M. Bradley, C.R. Richards and N.S.T. Thomas, *Clin. Chim. Acta*, 92 (1979) 293.
703 N. Shishido, T. Isobe, I. Horii, N. Shishido and K. Udaka, *J. Toxicol. Sci.*, 1 (1976) 95.
704 T. Itoh, H. Waki and H. Kaneko, *Agr. Biol. Chem.*, 39 (1975) 2365.
705 H. Kaneko, M. Hosohara, M. Tanaka and T. Itoh, *Lipids*, 11 (1976) 837.
706 M. Tanaka, T. Itoh and H. Kaneko, *Yakugaku*, 26 (1977) 454.
707 L. Glueck, M.V. Kulovich, R.C. Borer, P.H. Brenner, G.G. Anderson and V.N. Spellacy, *Amer. J. Obst. Gynecol.*, 109 (1971) 440.
708 E.B. Olson and S.N. Graven, *Clin. Chem.*, 20 (1974) 1408.
709 V.R. Mallikarjuneswara, *Clin. Chem.*, 21 (1975) 260.
710 A. Martin-Pouthier, N. Porchet, J.C. Fruchart, G. Sezille, P. Dewailly, X. Codaccioni and M. Delecour, *Clin. Chem.*, 25 (1979) 31.
711 A. Kuksis, L. Marai and D.A. Gornall, *J. Lipid Res.*, 8 (1967) 352.
712 N.J.H. Mercer and B.J. Holub, *Lipids*, 14 (1979) 1009.
713 A. Kuksis, *Fette, Seifen, Anstrichm.*, 75 (1973) 517.
714 A. Kuksis, J.J. Myher, L. Marai and K. Geher, *J. Chromatogr. Sci.*, 13 (1975) 423.
715 R.B. Watts, T. Carter and S. Taylor, *Clin. Chem.*, 22 (1976) 1692.
716 A. Kuksis, J.J. Myher, K. Geher, A.G.D. Hoffman, W.C. Breckenridge, G.J.L. Jones and J.A. Little, *J. Chromatogr.*, 146 (1978) 393.
717 A. Kuksis, J.J. Myher, K. Geher, N.A. Shaikh, W.C. Breckenridge, G.J.L. Jones and J.A. Little, *J. Chromatogr.*, 182 (1980) 1.

718 R. Hedlin, A. Kuksis and K. Geher, *Obst. Gynecol.*, 52 (1978) 430.

719 E. Griffin, W.C. Breckenridge, A. Kuksis, M.H. Bryan and A. Angel, *J. Clin. Invest.*, 64 (1979) 1703.

720 A. Kuksis, J.J. Myher, K. Geher, G.J. Jones, J. Shepherd, C.J. Packard, J.D. Morisett, O.D. Taunton and A.M. Gotto, *Atherosclerosis*, 41 (1982) 221.

721 P. Mareš, E.Tvrzická and V. Tamchyna, *J. Chromatogr.*, 146 (1978) 241.

722 J. Skořepa, P. Mareš, J. Rubličová and S. Vinogradov, *J. Chromatogr.*, 162 (1979) 177.

723 P. Mareš, E. Tvrzická and J. Skořepa, *J. Chromatogr.*, 164 (1979) 331.

724 K. Grob, Jr., *J. Chromatogr.*, 205 (1981) 289.

725 J.J. Myher and A. Kuksis, (1981) unpublished results.

726 J. Hirsch, *J. Lipid Res.*, 4 (1963) 1.

727 I.W. Duncan, P.H. Culbreth and C.A. Burtis, *J. Chromatogr.*, 162 (1979) 281.

728 S.L. Smith, M. Novotný, S.A. Moore and D.L. Felten, *J. Chromatogr.*, 221 (1980) 19.

729 J. Coupek and P. Mareš, *J. Amer. Oil Chem. Soc.*, 57 (1980) Abstr. No. 89.

730 O.S. Privett and W.L. Erdahl, *Methods Enzymol.*, 72 (1981) 56.

731 B.J. Compton and W.C. Purdy, *Anal. Chim. Acta*, (1981) in press.

Chapter 13

Terpenoids

RODNEY CROTEAU and ROBERT C. RONALD

CONTENTS

13.1. INTRODUCTION

The terpenes constitute one of the largest and most structurally varied groups of natural products, and they exhibit a tremendous diversity in chromatographic properties, both within and between the various classes. While it may have been preferable in some respects to organize this chapter on the basis of chromatographic properties or major separation techniques, we have instead followed the precedent set in earlier editions and have subdivided the material according to the number of isoprenoid units per molecule, because this is the approach most useful to readers seeking information on the chromatography of specific classes of terpenoids. In this connection, it is important to note, while the vapor pressures differ considerably between the terpenoid classes, certain chromatographic properties (and separation problems) of related structural types (e.g., olefins of the mono-, sesqui-, and diterpene series) are quite similar. As we have attempted to minimize repetition, the reader may thus find it profitable to review the discussion on the chromatography of relevant higher and lower isoprenalogs. The terpenoid vitamins (A, E, and K) have each been dealt with separately; the vitamins D are covered in Chap. 14.3. Separate sections are devoted to phosphorylated terpenes and glycosylated terpenes, as these derivatives exhibit profound chromatographic differences relative to their parent terpenyl moieties. The meroterpenes, such as terpenoid alkaloids, are not covered here.

The chapter contents are based principally on studies published since the 1975 edition of this book. Because of space limitations, it has not been possible to include much of the material appearing in the terpenoid chapter of earlier editions. Relevant earlier studies, especially those forming the basis of now generally accepted procedures, have been cited, as have earlier publications containing useful tabular data. The body of recent literature relevant to the chromatography of terpenoids is far too large to attempt comprehensive coverage here. We have tried, instead, to cite primarily descriptions of general methods and representative applications. Because the application of HPLC to the analysis of terpenoids and other natural products has increased dramatically over the last few years [1–4], emphasis has been placed on this increasingly important analytical and preparative technique. As most terpenoid separations are made in a biological context (e.g., analysis of biological fluids or tissues, or purification to radiochemical homogeneity in biosynthetic studies), these applications remain the major focus of this chapter, as in earlier editions. General information on the chromatography of terpenoids can be found in earlier editions of this text and in a recent CRC Handbook [5].

13.2. HEMITERPENES, MEVALONIC ACID, AND RELATED COMPOUNDS

Chromatographic isolation of mevalonic acid (or mevalonolactone) is often required in the preparation of this isotopically labeled substrate for studies on isoprenoid biosynthesis [6], and in the assay of 3-hydroxy-3-methylglutaryl coenzyme A reductase (HMG CoA reductase), a key regulatory enzyme of the isoprenoid pathway [7]. The enzyme assay generally involves the separation of the product, mevalonic acid (as the derived lactone), from the substrate, HMG CoA (often determined as HMG). Several simple TLC methods have been devised for this purpose, including chromatography on silica gel with acetone–benzene (1:1) and other developing solvents [7–11], or on Super-Cel–CaSO$_4$ (3:1) [12] or cellulose [8]. Because of the high levels of interfering salts present in the assay mixtures, some procedures employ preliminary ether extraction of the lactone and HMG before TLC analysis [8,13]. However, such procedures require the use of internal standards to correct for variations in the extraction [14,15]. In some instances, the ether extraction may be by-passed, and the deproteinized assay mixture may be applied directly to the adsorbent layer [9,16]. A simple one-step isolation procedure, based on the selective extraction of mevalonolactone into benzene, has been developed [7], while another rapid method involves the separation of mevalonolactone from HMG by absorbing the mixture on filter paper, followed by ascending development with toluene [17].

GLC offers another rapid method for assaying the formation of mevalonic acid, both with and without isotope labeling [18]. Thus, mevalonic acid can be separated on various polyester phases either directly as the lactone [19] or after conversion to the corresponding methyl ester [20] or TMS derivative. However, methyl mevalonate may undergo hydrolysis under GLC conditions [22].

Column chromatography on Celite–H$_2$SO$_4$ with chloroform as the developer has been used to purify isotopically labeled mevalonate [6,10], and ion-exchange chromatography on Dowex-1 has been shown to be equally effective [23–25]. Such ion-exchange procedures appear to be widely applicable to the purification of mevalonic acid, as in many instances the need is the separation of mevalonic acid from neutral substances, or of the neutral lactone from charged species. Mevalonic acid is readily lactonized by warming (37°C) in 0.1 N HCl for 30 min [26], yet dehydration may also occur, giving rise to the by-product Δ^2-3-CH$_3$-mevalonolactone [27]. Paper chromatography, e.g., on Whatman No. 3MM paper with 1-propanol–conc. NH$_4$OH–water (6:3:1), is often used to monitor the purity of mevalonic acid [5,17].

The hemiterpene alcohols, isopentenol, dimethylallyl alcohol, and dimethylvinylcarbinol, are readily separated by argentation TLC (Silica Gel G containing 25% AgNO$_3$ with ethyl acetate as developer) [28], and by various GLC procedures [29,30]. Separations on both analytical and preparative scale appear to be feasible by low-pressure CC on C$_{18}$ phase-bonded silica [31]. Reversed-phase TLC, on silicone oil- or mineral oil-impregnated silica gel, offers another useful means of separating hemiterpene alcohols from higher-molecular-weight terpenols [32,33].

Isoprene (2-methyl-1,3-butadiene) has been identified in the volatile complex emitted by a variety of higher plants [34,35]. This gaseous natural product, which is also a pyrolysis product of natural rubber [36], is readily separated from other C$_5$H$_8$ isomers by GLC on Tenax-GC [37].

References on p. B177

13.3. MONOTERPENES

13.3.1. Gas–liquid chromatography

The volatile nature of the monoterpenes makes this class well-suited for analysis by GLC, and GLC separation of monoterpene mixtures, particularly complex essential oils, is now fairly routine [38–44] and used as an industrial process operation [45]. The most powerful tool for the resolution and identification of monoterpene mixtures is the GLC–MS combination, which provides retention data and spectra in one operation at a sensitivity unrivaled by virtually any other technique [46–48]. The addition of a computer for treatment of both GLC and MS data, as well as library searching, has greatly facilitated the identification of individual monoterpenes in the complex mixtures often encountered in chemosystematic studies and flavor analyses [49,50]. A large number of monoterpene-containing oils have been analyzed over the years, and the composition of many can be found in various published compendia [46,51,52].

The monoterpenes are rather prone to degradation and consequent artifact formation during both isolation procedures and subsequently GLC analysis [38,53,54]. While steam distillation is probably the most common method for removal of volatile oil from biological tissues (see ref. 38 for especially suitable techniques), solvent extraction is generally preferred when degradative modifications must be minimized [54]. However, this procedure, unless combined with fractional distillation [54,55], generally produces a total lipid extract that must be subjected to preliminary clean-up methods (CC, TLC, etc.) before GLC analysis. A procedure which eliminates preliminary isolation of the sample is the direct injection technique. In this method, the tissue (generally plant material) is placed directly into the heated injector of the gas chromatograph to release the volatile monoterpenes for analysis. A variety of devices for accomplishing this have been described [38,56–58]. The technique is particularly useful when a large number of samples must be analyzed. However, because the sample is small, it is not always representative.

Problems that may be encountered during GLC analysis of monoterpenes include various isomerizations, dehydrations, and polymerizations occurring in the hot injection port [38,39], on the solid support [39,59] or in the liquid phase itself. For instance, sabinene is reportedly degraded on silicone oil [60], and isomenthone may undergo enolization on alcohol–amine phases [61]. Decomposition and artifact formation can be greatly minimized by maintaining the lowest possible injector temperature needed for complete volatilization of the sample, and by the use of all-glass systems, on-column injection, and silanized solid supports. These precautions are considered to be almost standard in the field [38,39]. While decomposition of the sample after leaving the chromatographic column seldom causes trouble when either a FID or a TCD is used in analytical work, decomposition in the thermal conductivity cell [62] may be a problem in preparative applications.

A variety of polar and nonpolar stationary phases have been utilized to separate monoterpenes, yet the Carbowaxes (polyethylene glycols) are probably the most versatile [38,39,42] (Table 13.1). Tentative identification of monoterpenes can often

TABLE 13.1

RETENTION VOLUMES OF MONOTERPENES RELATIVE TO THAT OF CAMPHOR

SE = 0.125 in.×10 ft., 10% SE-30 on 80–100-mesh Gas-Chrom Q, 130°C, 35 ml/min. CW = 0.125 in.×10 ft., 10% Carbowax 20 M on 80–100-mesh Gas-Chrom Q, 165°C, 35 ml/min.

Compound	SE	CW	Compound	SE	CW
α-Thujene	375	109	Umbellulone	1102	1318
Tricyclene	392	104	Neomenthol	1122	1043
α-Pinene	404	111	Isopinocamphone	1126	1073
Camphene	434	148	Borneol	1126	1490
α-Fenchene	443	147	Menthol	1143	1173
Sabinene	460 *	197	Isopinocampheol	1182	1561
β-Pinene	465	188	Terpinen-4-ol	1184	1111
Myrcene	490	242	Myrtenal	1224	1331
α-Phellandrene	530	251	Neoisomenthol	1224	1161
Δ-3-Carene	551	233	trans-Dihydrocarvone	1241	1223
α-Terpinene	559	272	cis-Dihydrocarvone	1286	1341
p-Cymene	563	448	Neodihydrocarveol	–	1527
cis-Ocimene	567	350	Dihydrocarveol	1249	1663
Limonene	596	303	Isodihydrocarveol	–	1837
trans-Ocimene	602	391	Neoisodihydrocarveol	–	1999
1,8-Cineole	605	402	Isomenthol	1241	1279
γ-Terpinene	673	422	α-Terpineol	1241	1415
trans-Sabinene hydrate	694	695	Myrtenol	1278	1945
Fenchone	763	691	trans-Verbenol	1310	1701
Linalool	767	807	trans-Piperitol	1318	1626
Terpinolene	775	484	trans-Carveol	1392	2148
cis-Sabinene hydrate	808	894	Citronellol	1428	1614
3-Isothujone	833	720	Nerol	1441	1868
3-Thujone	869	757	cis-Carveol	1453	2384
Fenchol	882	1007	Neral	1486	1409
3-Thujanol	939	1229	Pulegone	1526	1392
Citronellal	967	747	Carvone	1531	1775
Camphor	1000	1000	Geraniol	1608	2138
3-Neothujanol	1004	1160	Piperitone	1628	1788
3-Isothujanol	1004	1291	Geranial	1690	1630
Menthone	1030	819	Thymol	1886	6673
Isomenthone	1085	888	Carvacrol	1967	7114
Isoborneol	1089	1359	p-Menth-1-en-9-ol	2016	3103
3-Neoisothujanol	1093	1390	p-Cymene-7-ol	2265	5214
Menthofuran	1102	844	Piperitenone	2437	3267

* Sabinene does not decompose under the conditions of analysis.

be made by comparing relative retention characteristics on two or more columns of differing polarity, and so analyses on Carbowax are often carried out in conjuction with analyses on a nonpolar phase such as Apiezon (a paraffin grease) or silicone oil [38,39,60,63–69] (Table 13.1). An alternative approach to multiple column analysis for compounds of a single class is the chromatographic characterization of a series of derivatives (e.g., TMS ethers and acetates of monoterpene alcohols [70]). Depending

References on p. B177

on the particular needs of the analysis, other stationary phases can be quite useful [39], including nitriles [60], aromatic polyethers [71,72], and polyesters [73–76]. Over sixty liquid phases suitable for monoterpene analysis have been described [77]. Von Rudloff [39] has utilized mixed liquid phases to achieve particularly difficult resolutions of monoterpene pairs. Stationary phases that have been recently applied to monoterpene analysis include Tenax-GC [78] and graphitized carbon black [79]. Useful adjuncts to the classical GLC analysis of monoterpenes include hydrogenolysis of the sample in the injection port [80] and pressure programing [81].

For preparative GLC of monoterpenes, all-glass systems are preferred. Because sample decomposition may occur in the TCD of the preparative chromatograph, we use a glass effluent splitter (with a FID, receiving 2–5% of the gas stream) and find it to be far superior to the TCD for many purposes. Virtually any stationary phase and solid support used for analytical separations is suitable for preparative work on columns of larger diameter, although a somewhat larger support size (60–80 mesh) is commonly substituted for the finer supports (80–100 mesh) used in analytical work. For strict reproducibility in automated operation, isothermal conditions are generally employed. It is therefore often desirable to prefractionate the sample according to boiling range or chemical class (e.g., to separate monoterpene olefins from sesquiterpene olefins or olefins from alcohols). Such prefractionation, which is also a desirable adjunct to analytical studies, may involve column chromatography, fractional distillation, or the preparation of readily separated derivatives followed by regeneration (e.g., 2,4-dinitrobenzoates of terpene alcohols which are later hydrolyzed) [82–88].

The direct coupling of the chromatograph to a radioactivity monitor has proved to be a very powerful technique for biochemical tracer studies, particularly in the terpene area [89–92]. Coincidence of radioactivity with a particular terpene peak suggests identity, but while the approach is often useful in preliminary studies (or in routine assays), such identifications should be independently verified, preferably by the preparation of derivatives and their recrystallization to constant specific activity [93,94]. GLC-based enzyme assays with unlabeled substrates have also been employed, MS serving for product verification [95].

GLC has proven to be an especially useful technique for the separation of monoterpene isomers [61,96–100], and is, thus, a valuable tool for stereochemical analysis [100–102]. The six positional isomers of p-menthene were completely separated on a column of 30% AgNO$_3$–glycol [100], and the four diastereomeric menthols were separated on Carbowax 400 [61]. The menthol isomers can also be resolved as their TMS ethers on SE-30 [103]. Racemates have been separated by GC after their conversion to diastereomers. Thus, (±)-menthone can be resolved as its (+)-tartaric acid derivative [104], (±)-menthol and (±)-borneol as the acetyl-D-glucoside [105], and (±)-camphor as the ketal of D-(−)-2,3-butanediol [106,107]. The alternative approach to the preparation of diastereomers is the use of optically active stationary phases [108]. Resolution of racemates has also been achieved on standard stationary phases by simultaneous injection of the racemate and a volatile resolving agent [109].

Since the major problem in the chromatography of monoterpenes is that of resolving very complex mixtures into individual components, the more efficient WCOT or SCOT capillary columns are rapidly replacing conventional packed columns for most types of analyses [66,69,110–115]. WCOT columns are easily overloaded—much easier than SCOT columns—and require an injection splitter, which may discriminate between high- and low-boiling terpenes. However, modern inlet splitters have largely overcome this disadvantage, while permitting multiple injection modes. With the commercial availability of fused silica capillary columns having various lengths and coatings [116,117], capillary GC is likely to become the preferred analytical method. Capillary columns of up to 300 m have been employed for fractionating monoterpene mixtures [118], but shorter columns provide resolution adequate for many applications [119].

13.3.2. Thin-layer chromatography

TLC of monoterpenes is commonly employed as a semipreparative method preliminary to GLC [120–122], as a rapid screening method for the analysis of essential oils [44,123–125], or as a monitor of reactions or LC separations [126,127], and as an assay technique in radiotracer experiments [128–130]. In the later application, internal standards must be added to minimize evaporative losses of the labeled monoterpenes. A useful discussion of the combination of TLC with other chromatographic techniques has been contributed by Kubeckza [131].

A variety of adsorbents have been used for TLC of monoterpenes, but silica gel is the most common. Various combinations of hexane–ethyl acetate provide a flexible, general-purpose developing system for oxygenated monoterpenes on silica gel [132], but even low proportions of ethyl acetate afford monoterpene hydrocarbons as a single unresolved band. One drawback of silica gel chromatography of monoterpenes is the possibility of artifact formation [133,134], such as the oxidation of α-terpinene to p-cymene or the isomerization of sabinene to α-thujene and p-menthadienes. Such problems can be minimized by preparing the gel with dilute base instead of water, and by predeveloping the plates to remove impurities and to deactivate catalytic sites. Polyethylene glycol can also be used to deactivate silica gel for some applications [135]. Improved resolution can often be obtained by chromatographing monoterpene alcohols as the TMS ether, acetate, or dinitrobenzoate derivatives [136–138]. TLC is a useful tool in the conformational analysis of monoterpene alcohols and ketones [102,139]. It was recently employed to resolve (±)-fenchone oxime via the diastereomeric (−)-menthyloxymethyl ethers [140].

A powerful technique for the separation of monoterpenes, especially olefins [141], is argentation TLC on AgNO$_3$-impregnated silica gel. Thus, sabinene can be separated from β-pinene, and α-thujene from α-pinene by chromatography on silica gel impregnated with 6–8% AgNO$_3$ and developed with hexane or benzene [136,142]. With 1% diethyl ether in cyclohexane as the developing solvent, γ-terpinene can be separated from α-terpinene, and α-terpinene from limonene [142]. In a similar fashion, cis/trans-isomeric monoterpene alcohols can be well separated by argentation TLC by using more polar solvents [28,137]. The light-sensitivity and short storage life of AgNO$_3$-impregnated plates have led to attempts to substitute other heavy metals that also form π-complexes with double bonds, but these techniques [143,144] offer little advantage. Reversed-phase TLC, a technique generally used for resolving terpenols of differing chain length [145], has been applied to the separation of monoterpene alcohols [132].

Monoterpenes and higher terpenes may be detected on TLC plates with a variety of reagents, including iodine, conc. H_2SO_4, vanillin in H_2SO_4, phosphomolybdic and phosphotungstic acid, and the chlorides of tin, arsenic and antimony [146–148]. Chemical reactions carried out on the thin-layer plate before development [126], as well as derivatization on the plate (e.g., formation of dinitrophenylhydrazones) [123], are also useful techniques in terpene analysis. A variety of nondestructive methods for locating terpenes on TLC plates are also available [149,150], the most common being the use of fluorescent dyes, such as dichlorofluorescein and rhodamine. Aromatic monoterpenes and certain oxygenated monoterpenes quench the UV light-induced fluorescence, thus providing a means of distinguishing chromophoric substances. Most monoterpenes can be recovered from silica gel by elution with dry diethyl ether, a procedure which leaves the fluorescent dye on the gel.

13.3.3. Column chromatography

Although CC finds considerable use in the separation of terpenoid substances [151], the volatile monoterpenes do not readily lend themselves to this technique, because losses may occur when the eluates are concentrated. In spite of this, preparative column separations are commonly used for fractionating essential oil samples prior to GLC analysis [82–88]. Hydrocarbons can be isolated by chromatography on silica or alumina with pentane or hexane as eluents, and oxygenated monoterpenes can be separated into classes (ethers, ketones, alcohols, etc.) by sequential elution with various combinations of dichloromethane in pentane, or diethyl ether in pentane. CC on $AgNO_3$–alumina has been used to advantage in fractionating unsaturated monoterpenes [81,152].

HPLC has also been used as a prefractionation technique in conjunction with the GLC analysis of essential oils [153], and in several other preparative or semi-preparative applications [154,155]. Thus far, analytical applications have largely been restricted to monoterpenes containing a chromophore, which permit the use of UV detectors [156–160]. The most promising use for HPLC would appear to be in the separation of thermally labile terpenes not readily analyzed by GLC. Monoterpene mixtures have been separated on columns of silica [156,159], phase-bonded C_{18}-silica [157,159] and several gel permeation matrices [157,158]. Thallium nitrate-coated silica has been used to separate monoterpene olefins [161]. Because $TlNO_3$ is less soluble than $AgNO_3$, it may offer an advantage in HPLC by permitting the use of more polar solvents. A number of monoterpenoid acids have been resolved as their (+)-1-(1-naphthyl)ethylamides by HPLC on microparticulate silica [162]. The technique has potential for such applications, since many diastereomeric derivatives of monoterpenes are of relatively high molecular weight and may be thermally unstable on GLC analysis.

13.4. SESQUITERPENES

Many of the chromatographic techniques described for the separation of mono-terpenes apply equally well to the sesquiterpenes, because the chromatographic properties of the two classes, aside from the obvious boiling range difference, are quite similar [39,40]. Monoterpenes and sesquiterpenes generally occur together in essential oils, and they are necessarily analyzed together. Like the monoterpenes, the sesquiterpenes are prone to thermal and acid-catalyzed rearrangements [163–167], and this complicates their isolation [168,169] and analysis, particurlarly GLC analysis. There are far more naturally occurring sesquiterpenes than monoterpenes, and the existence of many groups of related geometric isomers and stereoisomers greatly increases the separation and identification difficulties.

13.4.1. Gas–liquid chromatography

The physical properties of sesquiterpene hydrocarbons make them ideally suited for separation and isolation by GLC. Analyses of complex terpene mixtures are not

TABLE 13.2

MODIFIED KOVÁTS INDICES OF SESQUITERPENES [175]

AL = Apiezon L, CW = Carbowax 20M. For further details on operating conditions and indices on other liquid phases see ref. 175. (Reprinted with permission.)

Compound	AL (155°C)	CW (165°C)	Compound	AL (155°C)	CW (165°C)
Cubebene	1368	–	Selina-4(14),7-diene	1491.9	1694
α-Ylangene	1401.5	1538.5	δ-Selinene	1504.9	1728.5
β-Elemene	1410	–	γ-Muurolene	1505.7	1725
α-Bourbonene	1410	–	γ-Amorphene	1506.4	1724
α-Copaene	1410.2	1551.3	α-Himachalene	1508	1704.5
Cyclosativene	1411.9	1549	α-Amorphene	1509.5	1724.5
Longicyclene	1417.1	1554	Zizaene	1511.6	1706.3
Cyclocopacamphene	1417.8	1555.4	β-Bisabolene	1512.9	1745.5
β-Bourbonene	1418.3	1586.5	β-Curcumene	1513.6	1756
β-Farnesene	1429.2	1668	α-Zingaberene	–	1738
Sativene	1434.7	1594.5	Valencene	1525.6	1760
Cyperene	1446.6	1606	β-Himachalene	1529.7	1752.5
Caryophyllene	1451.7	1655.5	β-Selinene	1530.2	1766.5
Longifolene	1464	1643	γ-Bisabolene	1531.3	1765.5
Isosativene	1464.4	1639	α-Muurolene	1531.3	1725.5
Calarene	1466	1655.5	α-Pyrovetivene	1533.9	1817
α-Cedrene	1473.4	1640	α-Selinene	1534.5	–
Thujopsene	1476.1	1684.2	ε-Bulgarene	1538	–
γ-Curcumene	1481.9	–	δ-Cadinene	1546.4	1784
β-Cedrene	1482.4	1670	γ-Cadinene	1554.9	1792.3
α-Curcumene	1483	1787.5	Selina-4(14),7(11)-diene	1572	1816.3
ε-Muurolene	1484.8	1713.8	Selina-3,7(11)-diene	1580	–
Humulene	1487.2	1719	β-Vetivenene	1583	1885

References on p. B177

without difficulty, however, because of the possibilities of decomposition referred to above, and because the retention times of sesquiterpene hydrocarbons overlap the range for oxygenated monoterpenes on most columns [38,39,82]. Acyclic, mono-cyclic, bicyclic, and tricyclic sesquiterpene hydrocarbons have been successfully separated on a large variety of polar and nonpolar liquid phases [170–174]. An extensive table of retention data has been compiled by Andersen and Falcone [175] (Table 13.2). Relatively less attention has been paid to the separation of oxygenated sesquiterpenes by GLC [39], although the technique has proved to be very suitable where utilized, e.g., for the elemol–eudesmol group [39], for the tricothecane mycotoxins and their TMS derivatives [176], for furanosesquiterpenes [177–180], and for several other structural types [181,182]. The various geometric isomers of farnesol [183] and juvenile hormone [184] have been separated on polyester and silicone oil columns. GLC (with FID or ECD) forms the basis of several assays for abscisic acid [185–188].

A very useful adjunct to the GLC separation of sesquiterpenoids is the prepara-tion and analysis of the corresponding dehydrogenation [189,190] and hydrogena-tion (or hydrogenolysis) [175,191] products, procedures which can be performed in the injection port of the gas chromatograph. Coupled GLC–MS has greatly ex-panded the capability of the chromatographic technique for sesquiterpene analysis [192–195], and this method is now used routinely for the determination of abscisic acid and its metabolites [196–201]. The coupled GLC–radioactivity monitor has been employed in studies on the biosynthesis of isomeric farnesols [202].

13.4.2. Thin-layer and column chromatography

Silica gel is the most widely used adsorbent for analytical and preparative TLC and CC of sesquiterpenes. A major use of the latter technique is the separation of sesquiterpene hydrocarbons from oxygenated substances (e.g., oxygenated mono-terpenes) that interfere in subsequent GLC analysis [192,194,203,204]. Chromatogra-phy on silica gel has also been utilized in the separation of oxygenated sesquiter-penes, including alcohols [205–208], ketones and aldehydes [209,210], various lac-tones [211–215], furanoid types [179,193,216], and abscisic acid [187,188]. Mixtures of sesquiterpene hydrocarbons have been resolved on $AgNO_3$–silica gel layers [172,217] and columns [172,218], and this argentation technique is also applicable to isomeric sesquiterpene alcohols [28]. Adsorption chromatography on alumina col-umns (both with and without $AgNO_3$) has been satisfactory for several types of sesquiterpenes [178,219,220], while for polar derivatives, such as polyoxygenated lactones, partition CC (e.g., on cellulose) is effective [221,222].

HPLC on silica gel has recently been employed in the separation of sesquiterpene lactones [223,224], furanosesquiterpenes [225], and unsaturated sesquiterpene al-cohol isomers [226–228]. Silver-loaded ion-exchange columns [229] and $AgNO_3$-coated silica columns [230] show considerable promise for resolving geometrical isomers. Acyclic, bicyclic, and tricyclic sesquiterpene hydrocarbons and alcohols are easily separated by gel permeation chromatography on the basis of their molecular size [158], while sesquiterpene alcohols are well separated from other terpenol classes by low-pressure reversed-phase CC on C_{18} phase-bonded silica [31]. HPLC (normal- and reversed-phase partition, as well as ion exchange) forms the basis of several rapid and sensitive assays for abscisic acid [185,231–235].

13.5. DITERPENES

13.5.1. Gas–liquid chromatography

Eglinton et al. [236] were among the first to demonstrate the feasibility of GLC analysis for diterpenoid substances, and the technique has since found very wide use. Acyclic diterpene hydrocarbons [237,238], and cyclic olefins of the labdane [239], pimarane [172,240], abiatane [241–243], and kaurane [244] type are readily separated by GLC, as are acyclic diterpene acids (phytanic acid and its derivatives) [245,246] and the various resin acids (generally analyzed as the methyl or TMS esters) [247–251]. The retention data for many resin acids (as methyl and TMS esters) have been compiled [250–253]. The feasibility of using other resin acid ester derivatives for GLC analysis has also been examined [254]. The difficult separation of levo-pimaric from palustric acid was accomplished by chromatographing them as either the methyl or t-butyl esters on a cyanosilicone column. Diterpene alcohols have been analyzed directly [238,240,255], but preliminary conversion to the acetate or TMS ether is often preferred [238,251]. Packed columns with relatively low loading of either nonpolar (Apiezon L, SE-30) or polar (polyester) liquid phases are generally employed for the separation of diterpenoids, in order to reduce the operating temperature and thereby minimize the possibility of isomerization [38]. For this application, the solid support must be thoroughly deactivated (e.g., by silanization), or else isomerization will occur [250,256]. Diterpenes are also prone to degradation during distillation procedures [257]. Recently, capillary GLC (with both polar and nonpolar phases) and coupled GLC–MS techniques have been very successfully applied to the analysis of diterpenoid hydrocarbons and acids [238,245,246,257–261].

As the number of naturally occurring gibberellins continues to grow (approximately 50 are now known [262]), increasingly powerful analytical tools are required for the separation, identification and quantitation of these diterpenoid plant growth substances [263]. An excellent review of the application of the GC and GC–MS techniques to the analysis of gibberellins has been written by Gaskin and MacMillan [264]. Very sensitive electron-capture GLC techniques (at nanogram levels) have been described for the analysis of gibberellins as their trifluoroacetate esters [265]. Gas–liquid radiochromatography finds considerable use in studies on gibberellin metabolism [266–268].

13.5.2. Thin-layer chromatography

TLC techniques (with silica gel, alumina, or kieselguhr as adsorbents) are extensively employed in the preliminary fractionation of oxygenated diterpenes for structural studies, and in biosynthetic assays. Procedures for the separation of resin acids [269,270], geranylgeranyl esters [271], diterpene alcohols (272–275] and highly oxygenated diterpenoids (e.g., phorbol esters) [276–279] have been described. The gibberellin precursors, kaurene, kaurenol, kaurenal, and kaurenoic acid are well separated by TLC on silica gel [280].

Diterpene hydrocarbons are particularly well suited for separation by argentation TLC on $AgNO_3$–silica gel. A procedure described by Robinson and West [281] employs a combination silica gel plate, partially impregnated with $AgNO_3$, and developed with hexane–benzene (7:3). Polar compounds remain at the origin of the plain portion of the plate, the macrocyclic diterpene casbene is immobilized where the $AgNO_3$ starts, and other polycyclic diterpene hydrocarbons (sandaracopimaradiene, kaurene, beyerene, trachylobane) migrate to various zones in the $AgNO_3$-containing region. Acyclic diterpene olefins have also been separated by argentation TLC [238], as have unsaturated diterpene alcohols [33] and acids [282]. Reversed-phase TLC systems allow the separation of diterpenoids from other isoprenalogs on the basis of chain length [32,33].

TLC is an important analytical method for studies on gibberellins, being both rapid and sensitive (0.1 μg). Both adsorption [283,284] and partition [283,285,286] have been utilized, often in conjunction with radiotracer experiments [287]. To separate gibberellin pairs differing only by the presence or absence of a double bond, argentation TLC of the corresponding p-nitrobenzyl esters has been employed [288]. The technique is very sensitive (10 ng), and accomplishes the difficult resolutions of pairs, such as GA_4–GA_7, GA_1–GA_3 and GA_5–GA_{20}.

13.5.3. Column chromatography

CC on silica or alumina is a routine purification procedure for virtually all types of diterpenoid substances [274,275,279,281,289–294]. $AgNO_3$-impregnated silicic acid columns and gradient elution techniques are effective for fractionating oxygenated diterpenes [295,296] and diterpene hydrocarbons [281]. Ion-exchange chromatography of resin acids has been described [297]. It is a useful technique for separating resin acids from their neutral congeners which may interfere in the GLC analysis of resin acid methyl esters. Recently, reversed-phase HPLC techniques have been utilized to separate highly oxygenated diterpenoids [298–300].

For the preliminary purification of gibberellins, a variety of CC procedures may be employed, including those based on charcoal–Celite [301], polyvinylpyrrolidone [302], gel permeation [303], ion exchange [304], and adsorption [305] and partition [306] on silicic acid. HPLC is a powerful technique for separating plant hormones [307–309] and is particularly well suited to the analysis of gibberellins, either directly or after derivatization. Argentation HPLC of gibberellin p-nitrobenzyl esters on $AgNO_3$-impregnated silica has provided a very useful means of separating congeners differing in the degree of unsaturation [310]. This derivatization procedure has the added advantage of facilitating detection by enhanced UV absorption. A similar advantage is provided by the conversion of gibberellins to the bromophenacyl esters, which may be separated by reversed-phase HPLC [311], or by conversion to the benzyl esters which have been separated on silica [312]. Underivatized gibberellins have been separated directly on silica [312] and on reversed-phase columns [313]. An HPLC system for the analysis of gibberellins with an on-stream radioactivity monitor has been described [314], and a thorough review of both analytical and preparative HPLC procedures for gibberellins has been published [312].

13.6. SESTERTERPENES

The sesterterpenes are a relatively recent addition to the terpenoid family. This C_{25} group is rarely encountered, although the few known representatives occur across a wide biological spectrum (i.e., higher and lower plants, and insects). There are at least six distinct structural types of sesterterpenes, and a variety of derivatives, some of which are highly oxygenated. The diversity in the chromatographic properties of this class has, in part, limited any systematic attempt to determine how widespread in nature the sesterterpenes actually are. What is known about the chromatographic properties of the sesterterpenes has been described by Cordell [315,316] in a general review of this terpenoid class. For the most part, the sesterterpenes have been subjected to the same adsorption and partition techniques employed for diterpenes and triterpenes of comparable polarity.

13.7. TRITERPENES

13.7.1. Gas–liquid chromatography

The triterpenes are now quite routinely analyzed by GLC procedures similar to those employed for steroidal substances (Chap. 14) which the triterpenes closely resemble. Ikekawa [317] has tabulated retention data for numerous triterpenes, including some sapogenins, on SE-30, and has evaluated the influence of the various substituents on retention properties. In order to improve the separation of polar triterpenes and minimize thermal degradation during GLC analysis, derivatization procedures are generally employed. As in the case of related diterpenoid substances, acidic triterpenes are analyzed as the methyl esters, and the triterpenols (particularly polyols) as TMS ethers or short-chain esters (acetates, butyrates, etc.) [317–321]. Esters of triterpenols with higher fatty acids have also been separated on silicone oil-based liquid phases [321]. Triterpene monohydroxy compounds have been resolved in underivatized form on both nonpolar silicone oil and polar polyester phases [317,318,322,323]. GLC has been extensively employed in the separation of triterpanes (and steranes) derived from petroleum and other geological materials [324,325]. The ability of highly selective silicone phases to distinguish subtle structural differences in triterpenoid substances [323,326] suggests that GLC is a useful, if relatively unexplored, technique for stereochemical analysis of this group.

References on p. B177

13.7.2. Thin-layer chromatography

Lisboa [327] has determined the chromatographic mobilities of many tetra- and pentacyclic triterpenes and triterpenyl glycosides on several silica gel-based TLC systems, while Ikan et al. [328] have tabulated the mobilities of some 30 triterpenes on alumina layers. Separations according to functional groups are readily achieved by TLC, yet the technique is also suitable for the resolution of epimeric triterpenols [328]. Squalene is readily separated from the corresponding 2,3-oxide and 2,3-diol, and from presqualene alcohol, by TLC [329,330]. Polar triterpenes have been separated directly, with polar solvent systems, as well as after their conversion to less polar derivatives [331,332].

Argentation TLC on $AgNO_3$-impregnated silica gel layers has been utilized extensively for the resolution of unsaturated triterpenes. Mixtures of triterpene alcohols have been separated directly by such means [333,334], as well as after their conversion to the corresponding acetates [329,333], benzoates [335], or TMS ethers [336]. The technique is also useful for triterpene olefins and triterpene acid esters.

Reactions carried out on the chromatoplate, either before or after development [327,337], and the use of specific spray reagents [327,328], find considerable use in triterpene analysis. Polar triterpenes, such as soyasapogenols and cucurbitacins, have been separated on cellulose [338,339].

13.7.3. Column chomatography

CC on silica gel [333,340,341] or alumina [342] is a routine step in the isolation of triterpenes from other lipids, and in the fractionation of various triterpene classes. This technique, including argentation CC [343], has been used successfully for virtually all triterpene types, including very polar substances [344,345].

Recently, analytical and preparative-scale HPLC has been enlisted for the separation of a variety of triterpenes, including ketones and acids as well as monools and polyols [346–352]. Most separations, thus far, have utilized adsorption chromatography, although both polar and nonpolar mixtures are well resolved by reversed-phase partition techniques [350,351]. HPLC shows considerable promise for the separation of triterpenoid substances, particularly polar derivatives such as the limonoids [352] and soyasapogenols [349,351], which are not readily analyzed by GLC techniques.

13.8. TETRATERPENES

The highly unsaturated tetraterpenes usually occur in higher plants, fungi, photosynthetic bacteria, and algae in complex mixtures associated with photosynthetic pigments. These pigments, the so-called "chloroplast pigments", are the objects of the oldest applications of chromatography, principally because these highly colored substances are readily visible. A variety of chromatographic sorbents have been employed to fractionate this terpenoid group, the most common being silica, alumina, $ZnCO_3$, $CaCO_3$, MgO, $Ca(OH)_2$, cellulose, and sucrose. Chromatograms are usually developed with hydrocarbons containing a more polar solvent such as ethyl ether, acetone, methanol, or propanol.

13.8.1. High-performance liquid chromatography

Major advances have occurred in the applications of HPLC to the separation of chloroplast pigments and other plant pigments (see also Chap. 19). Owing to its speed and reproducibility, HPLC has nearly supplanted the older methods. Chromatograms on adsorption columns of microparticulate silica [353] are developed with mixtures consisting of hexane and either 1-propanol [354] or 2-propanol [355]. A μPartisil 10 column, developed with a light petroleum–acetone–DMSO–diethylamine (300:93:6:1) solvent system, separates β-carotene, pheophytin, and chlorophylls a and b; and with light petroleum–acetone–methanol–DMSO (30:40:27:3), it resolves chlorophyll c and pheophorbide a [356]. Silica has also been used in low-pressure high-speed CC of the chloroplast pigments from tobacco mutants with heptane, ether, and acetone as eluents [357]. This is reported to be an excellent preparative method.

Reversed-phase HPLC has also been used effectively in the separation of plant pigments. For these lipophilic compounds, silica treated with octadecylsilane (ODS), eluted with aqueous methanol, appears to be most effective [358,359]. A column of 1% ODS Permaphase on Zipax (Du Pont), eluted with 95% aqueous methanol, separates phytoene from β-carotene and retinol [360]. The separation of various carotenes with acetonitrile as the mobile phase has also been reported. A Partisil-5 ODS (C_{18}) (Whatman) column, eluted with 8% chloroform in acetonitrile has been employed to isolate lycopene and the α- and β-carotenes from tomatoes [361]. A Spherisorb ODS column has similarly been utilized for the separation of the α-, β-, and γ-cartotenes, lycopene, and various diterpenes [362]. An excellent summary of the separation of carotenoid pigments by HPLC methods, including a comparison of HPLC performance with TLC and PC, has been published (Table 13.3) [363].

An interesting variant of the HPLC technique is centrifugal chromatography in which columns of microparticulate silica or MgO are spun at high speed to drive the eluent through the stationary phase. With light petroleum and methanol as a solvent system, mixtures containing up to 0.5 mg of each component can be chromatographed in 2–6 min [364].

Gel chromatography on Sephadex LH-20 with a chloroform–methanol–hexane solvent system has proved to be particularly effective in the separation of carotenoids from steroids [365,366]. The separation of carotenes from plant proteins and other pigments can be achieved on DEAE-cellulose with 0.5 M NaCl [367]. A modification of the official method for the analysis of carotenes and xanthophylls has been recently reported [368]. When a mixture of Silica Gel 60 and Hyflo Super-Cel was substituted separations were completed within 30 min. The columns could be reused four times.

References on p. B177

TABLE 13.3

SEPARATIONS OF CAROTENOIDS [363]

	HPLC [a]			Silica TLC	Circular PC
	R_t (min)	Eluent [b]	Gradient rate (%/min)	R_F	R_F [c]
β, ε-Carotene	4.2	0	0	0.71 [e]	0.70 [d]
β, β-Carotene	4.2			0.71 [e]	0.70 [d]
β, ψ-Carotene	4.6			0.64 [e]	0.55 [d]
Lycopene	5.0			0.57 [e]	0.35 [d]
2,2'-Diol	8.0	0–30	10	0.67 [j]	0.85 [d]
Lutein	9.5			0.61 [j]	0.77 [d]
Zeaxanthin	9.7			0.61 [j]	0.63 [d]
Lutein		0–40	3		
Neo B	12.4				
Neo A	13.1				
All-*trans*	14.0			0.70 [k]	0.70 [g]
Neo U	14.8			0.62 [k]	0.55 [g]
Neo V	15.4			0.52 [k]	0.40 [g]
Bacterioruberin		20–60	3		
Neo A	14.1			0.61 [k]	0.55 [i]
All-*trans*	14.6			0.41 [k]	0.50 [i]
Neo U	15.1				
Neo V	16.0				} 0.40 [i]
Neo W	16.3				
Lutein 3'-ether		0–30	1		
Epimer 1	17.5			} 0.54 [j]	} 0.57 [f]
Epimer 2	17.9				
Auroxanthin		0–40	1		
Epimer 1	25.5			} 0.20 [j]	} 0.70 [h]
Epimer 2	25.9				
Neochrome		0–40	3		
Epimer 1	17.5			} 0.63 [k]	} 0.80 [h]
Epimer 2	17.8				
Mono-*cis*	18.4			0.60 [k]	0.70 [h]

[a] Column, Spherisorb (5 μm) 250×4.6 mm; pressure, 300 psi; flowrate, 1.4 ml/min.
[b] Percentage of acetone in hexane containing 0.1% of methanol.
[c] Schleicher and Schüll No. 287 (silica-filled).
[d] Schleicher and Schüll No. 288 (alumina-filled); mobile phase, hexane.
[e] Mobile phase: 2% acetone in hexane.
[f] Mobile phase: 5% acetone in hexane.
[g] Mobile phase: 10% acetone in hexane.
[h] Mobile phase: 20% acetone in hexane.
[i] Mobile phase: 30% acetone in hexane.
[j] Mobile phase: 40% acetone in hexane.
[k] Mobile phase: 60% acetone in hexane.

13.8.2. Thin-layer chromatography

Hyflo Super-Cel has also been applied in TLC. A mixture of Hyflo Super-Cel, MgO, and $CaSO_4$ produces durable layers that can be rapidly developed and have superior resolving power for nonpolar carotenes [369]. A single development separates all carotenoids in carrot. A method for resolving the chloroplast pigments in *Capsicum* uses cellulose layers to separate the chlorophylls and xanthophylls, followed by TLC on a mixture of MgO and Hyflo Super-Cel to resolve the carotenoid hydrocarbons [370]. Layers made from MgO can be developed with light petroleum and, when spectroscopic detection is used, they form the basis for the quantitative analysis of α- and β-carotenes in biomass analyses [371]. An assay suitable for low concentrations of β-carotene is based on a layer formed from a mixture of $CaCO_3$, MgO, and $Ca(OH)_2$ and development with acetone–light petroleum–chloroform (5:5:4) [372]. Layers prepared from cornstarch, cellulose, and microcrystalline cellulose have been compared for their ability to separate primary pigments. With a heptane–ethyl acetate–propanol solvent system complete separations were obtained on starch layers [373].

Silica gel layers have also been employed for the separation of plant pigments [374], but this stationary phase appears to be more suitable for the isolation of chloroplast pigments as a group than for their separation into individual components. Thus, silica plates, prepared with 10% $(NH_4)_2SO_4$, separated the carotenes and pheophytins from the chlorophylls, but the carotenes were only poorly resolved from each other [375]. A high-speed video-densitometric technique, similar to that used for the TLC of amino acids, was used for detection, and gave results similar to conventional spectrometric measurements. Rapid and facile separation of chlorophylls and their derivatives from yellow carotenoid pigments was achieved with either *t*-butanol–benzene (1:9) or *t*-butanol–pentane–acetone (1:18:1) [376]. A solvent system of dichloromethane–ethyl acetate–diethyl ether (8:2:1) gave excellent resolution of the zeaxanthin, lutein, and lutein diester [377]. Individual components were eluted from the silica gel with ethanol and determined photometrically. Similarily, astaxanthin and canthaxanthin from flamingo feathers, resolved on Silufol UV-254 foils with dichloromethane–diethyl ether (9:1), gave R_F values of 0.6 and 0.7, respectively [378].

An interesting modification of the thin-layer method is reversed-phase TLC on a chemically bonded C_{18} (ODS) stationary phase [379]. The use of a methanol–acetone–water (20:4:3) solvent system enabled the separation of neoxanthin, chlorophylls *a* and *b*, carotene, violaxanthin, and luteins in spinach chloroplasts. Citrus carotenoids have also been separated by TLC [380].

13.8.3. Gas–liquid chromatography

GLC is of limited value as an analytical technique, due to the lability of these highly unsaturated terpenoids. Only nonconjugated types are sufficiently stable to survive the thermal conditions required to chromatograph such relatively large molecules. Conjugated systems must be hydrogenated before injection, and fre-

TABLE 13.4

GLC RETENTION TIMES OF HYDROGENATED CAROTENOIDS AND RELATED
TERPENOIDS RELATIVE TO SQUALENE OR PERHYDRO-β-CAROTENE [381]

Under isothermal conditions (275°C), retention times are relative to perhydro-β-carotene. Under programed conditions (3 min at 225°C followed by 3°C/min to 300°C), retention times are relative to squalene.

	Relative retention times			
	Isothermal		Programed	
	SE-52	OV-17	SE-52	OV-17
Geranyllinalool	−	−	0.15	0.13
Phytol	−	−	0.18	0.10
Geranylgeraniol	−	−	0.21	0.15
Squalene	0.14	0.15	1.00	1.00
Cholestane	0.16	0.18	1.00	1.00
Cholesterol	0.28	0.49	1.69	1.67
Ergosterol	0.35	0.59	1.98	1.80
Stigmasterol	0.38	0.61	2.14	1.95
Lanosterol	0.44	0.72	2.31	2.11
Lycopersene	1.03	1.38	3.95	2.76
Hydrogenation products of:				
Retinol	−	−	0.11	0.07
Retinaldehyde	−	−	0.10	0.06
Crocetin	−	0.02	0.13	0.13
Dimethylcrocetin	0.03	0.05	0.34	0.40
Diethylcrocetin	0.06	0.07	0.45	0.50
Bixin	0.15	0.20	1.14	1.20
Methylbixin	0.14	0.19	1.08	1.18
β-Apo-10'-carotenal	0.06	0.05	0.42	0.41
Azafrin	0.27	0.45	2.08	1.80
Methylazafrin	0.21	0.37	1.70	1.73
Squalene	0.09	0.07	0.66	0.54
4,4'-Diapophytoene	0.09	0.07	0.67	0.53
4,4'-Diapophytofluene	0.09	0.06	0.67	0.54
4,4'-Diapo-ζ-carotene	0.09	0.07	0.67	0.53
4,4'-Diaponeurosporene	0.09	0.07	0.67	0.53
β-Apo-8'-carotenal	0.11	0.10	0.82	0.70
3,4-Dehydro-β-apo-8'-carotenal	0.11	0.10	0.82	0.69
β-Apo-8'-carotenoic acid	0.23	0.24	1.57	1.40
β-Apo-8'-carotenoic acid methyl ester	0.24	0.26	1.61	1.46
β-Apo-8'-carotenoic acid ethyl ester	0.26	0.28	1.77	1.50
β-Apo-4'-carotenal	0.30	0.26	2.10	1.44
Lycopersene	0.59	0.46	3.15	1.76
Phytoene	0.60	0.45	3.14	1.77
Phytofluene	0.60	0.46	3.15	1.75
ζ-Carotene	0.61	0.46	3.13	1.78
Neurosporene	0.60	0.46	3.14	1.77
Lycopene	0.60	0.46	3.15	1.75
γ-Carotene	0.78	0.67	3.66	2.24

TABLE 13.4 (continued)

	Relative retention times			
	Isothermal		Programed	
	SE-52	OV-17	SE-52	OV-17
β-Zeacarotene	0.79	0.67	3.66	2.24
β-Carotene	1.00	1.00	3.81	2.43
α-Carotene	1.00	1.00	3.81	2.43
Dehydro-β-carotene	1.00	1.00	3.81	2.43
Carotinin	1.00	1.00	3.81	2.43
Echininenone	1.71	2.25	4.65	3.19
Canthaxanthin	2.85	3.01	5.90	3.66
β-Carotenone	2.69	2.55	5.33	3.09
Torularhodin	1.07	1.26	4.23	2.58
Capxanthin	1.87	1.49	5.13	2.81
Astacene	1.70	1.46	5.79	3.00
Cryptoxanthin	1.27	1.43	4.14	2.55
Cryptoxanthin, Ac	1.23	1.76	4.10	2.53
Cryptoxanthin, TMS	1.47	1.82	4.12	2.54
Isocryptoxanthin	1.20	1.40	3.92	2.53
Isocryptoxanthin, Ac	1.21	1.74	3.78	2.44
Isocryptoxanthin, TMS	1.43	1.80	4.07	2.51
Zeaxanthin	1.68	2.21	4.68	3.30
Zeaxanthin, diAc	1.63	1.93	5.04	2.98
Zeaxanthin, diTMS	1.66	1.91	5.07	3.05
Isozeaxanthin	1.48	1.90	4.16	2.59
Isozeaxanthin, diAc	1.54	1.64	4.31	2.47
Isozeaxanthin, diTMS	1.58	1.58	4.35	2.53
Dimethoxyzeaxanthin	2.02	2.68	5.15	3.44
Dimethoxyisozeaxanthin	1.23	1.46	4.02	2.37
Fucoxanthin	2.24	2.56	6.24	3.42
Decapreno-β-carotene	6.35	6.87	7.54	5.32
Retention time (min) of:				
Squalene	0.70	2.48	5.50	11.5
Perhydro-β-carotene	5.13	16.65	20.95	27.09

quently require acetylation or silylation of hydroxyl groups. However, hydrogenation irretrievably destroys the *cis/trans*-isomer difference. An extensive study of the GLC retention behavior of hydrogenated carotenoids has been reported (Table 13.4) [381,382].

13.9. POLYPRENOLS

Interest in the long-chain polyprenols, such as undecaprenol and the dolichols (prenols containing 16 to 22 isoprene units), stems mainly from the role of the

References on p. B177

corresponding prenyl phosphates in the synthesis of complex polysaccharides and glycoproteins [383–385]. Polyprenols also occur in relatively high amounts in several plants as mixtures of groups with shorter chains (e.g., betulaprenols, containing 6 to 9 isoprene units; and ficaprenols, containing 10 to 13 isoprene units) [383,386]. The various polyprenol groups differ not only in the chain length, but also in the relative proportion of *cis-* and *trans-*double bonds. Some polyprenols (dolichols) contain saturated α-isoprene units, while in others this grouping is allylic. The polyprenols of a given species generally occur as isoprenalogs rather than as geometrical isomers. Polyprenols may occur either in free form, in esterified form, or as the more polar phosphoryl or sugar-phosphoryl derivatives (see Chap. 13.10 and 13.11).

13.9.1. Gas–liquid chromatography

Polyprenyl acetates, perhydropolyprenyl acetates, and the derived hydrocarbons have been separated on columns of 1% SE-30 on Celite [387,388]. For a series of prenyl acetates from a single species, the relationship between the number of isoprene residues and the logarithm of the retention time is linear, suggesting that the same geometrical arrangement of double bonds is present in each isoprenalog [388].

13.9.2. Thin-layer chromatography

TLC provides one of the most useful techniques for the separation of polyprenol homologs. Although TLC on silica gel is generally inadequate, except for very simple mixtures [389,390], the technique is useful for separating the polyprenols from lower prenols (e.g., geranylgeraniol, presqualene alcohol) and from the corresponding aldehydes, acids, phosphates and anhydro derivatives, which appear as by-products in biosynthetic experiments [391–393]. Complete resolution of homologous polyprenols is best achieved by reversed-phase TLC on paraffin-impregnated kieselguhr [394,395] or cellulose [393], with aqueous acetone as the mobile phase. The technique has been successfully applied to a wide range of polyprenol types [393,396–401]. A straight-line relationship is obtained when the R_M values are plotted against the chain lengths of the polyprenols of a given species. The slopes vary for the different series of homologs [394,395]. Some resolution of geometric isomers has been obtained by reversed-phase TLC [397,399], and the technique is also useful for resolving ubiquinone isoprenalogs [402]. A number of detecting reagents have been suggested for polyprenols, anisaldehyde–sulfuric acid and iodine being among the more useful ones [393,394].

13.9.3. Column chromatography

The isolation of polyprenols or polyprenol esters frequently involves their preliminary separation from other lipids by CC on alumina with alkane–diethyl ether mixtures [394,403,404]. Other column materials used for prefractionation include silicic acid [387,400] and Sephadex LH-20 [389].

For the isolation of individual, pure polyprenols from polyprenol mixtures, reversed-phase CC on a paraffin-impregnated support has been employed [405]. However, for this purpose hydroxy (C_{15})alkoxypropyl-Sephadex (Lipidex-5000) is far superior to the impregnated support. Using Lipidex-5000 columns and various acetone–water gradients, Chojnacki et al. [406] succeeded in resolving betulaprenol, ficaprenol, and dolichol mixtures into individual polyprenols with quantitative recovery.

Recently, HPLC on porous polymer columns [393] and on μBondapak C_{18} [404,407] has been applied to the separation of polyprenols. The reversed-phase technique provides excellent resolution of homologous juniperoprenols, containing from 14 to 21 isoprene residues [404], and of dolichols [407]. In cases where geometrical isomers must be separated (as in some biosynthetic experiments), HPLC of polyprenol acetates on silver-loaded columns [229] may offer a useful approach.

13.10. PHOSPHORYLATED TERPENES

Addition of a phosphoryl or pyrophosphoryl moiety converts terpenoids to ionic compounds with profoundly different chromatographic properties. The extent of the influence of the phosphoryl group depends on the molecular weight and polarity of the original terpene. Thus, whereas the effect of the phosphoryl function is greatest in the case of hemiterpenes, it is much less significant in the case of polyprenols. Many naturally occurring phosphorylated terpenes are pyrophosphate esters of allylic alcohols (e.g., geraniol, farnesol) or cyclopropyl carbinols (presqualene and prephytoene alcohols), which are extremely sensitive to acid and heat and, therefore, require certain precautions. However, not all phosphorylated terpenoids are sensitive to acid, e.g., saturated compounds, like phosphomevalonate and bornyl pyrophosphate, isopentenyl pyrophosphate (homoallylic), and dolichol phosphates (saturated α-isoprene residue). The chromatography of phosphorylated terpenoids is useful mainly in the separation of terpenyl pyrophosphates from the corresponding monophosphates (a problem encountered in substrate preparation) and in the resolution of terpenyl pyrophosphates of differing chain lengths (commonly needed in biosynthetic experiments).

13.10.1. Paper chromatography

The polar nature of phosphorylated terpenes makes these compounds very suitable candidates for separation by PC. Good separations of phosphorylated hemiterpenes (phosphate and pyrophosphate esters of mevalonic acid, isopentenol, and dimethylallyl alcohol) and phosphoesters of higher-molecular-weight terpenols have been achieved on Whatmann No. 1 or 3MM paper by ascending or descending development with mixtures of an aqueous alcohol with an acid or base (for acid-labile derivates) [408–413]. The R_F values of a number of phosphorylated terpenoids with several solvent systems have been tabulated [412–414]. A major limitation of PC is its low capacity, which essentially limits the technique to analytical separations.

13.10.2. Thin-layer chromatography

The more stable phosphorylated terpenoids can be chromatographyed on Silica Gel G layers with chloroform–methanol–water mixtures [409–411,415–417]. To minimize degradation of unstable allylic pyrophosphates, the adsorbent layer (Silica Gel H) can be buffered by slurrying the gel with 0.1 M ammonium phosphate (e.g., pH 6.5–7.0) instead of water [205,418]. Using buffered silica gel and chloroform–methanol–water (6:4:1) as the developing solvent, the following approximate R_F values were obtained for terpenyl pyrophosphates of differing chain length: C_5, 0.07; C_{10}, 0.15; C_{15}, 0.29; C_{20}, 0.41; C_{25}, 0.52; C_{30}, 0.60; and C_{40}, 0.68. A similar solvent system has been used to separate retinoyl phosphate from retinyl phosphate [419]. Alternative solvent systems for TLC of labile pyrophosphates on silica gel are 1-propanol–ammonium hydroxide–water [393,420] and chloroform–methanol–ammonium hydroxide–water [421], the particular mixture depending on the relative polarity of the class of compounds separated. With the solvent 1-propanol–ammonium hydroxide–water (6:3:1), the R_F values for the triphosphate, pyrophosphate, and monophosphate of the C_{20} prenol, geranylgeraniol, were 0.17, 0.35, and 0.52, respectively [296]. Another suitable sorbent for phosphorylated terpenoids is cellulose. Both cellulose and Silica Gel H are generally superior to Silica Gel G for labile substances [422–424]. Stable polyprenyl pyrophosphates have been separated on silica gel with diisobutyl ketone–acetic acid–water (8:5:1) [425].

13.10.3. Column chromatography

Preliminary fractionation of phosphorylated prenols has been carried out on silicic acid columns [392,415,422,423]. The separation of phosphate from pyrophosphate esters (of both short- and long-chain prenols) is readily achieved on silica by elution with 1-propanol–ammonium hydroxide–water [296,426,427].

Ion-exchange chromatography on DEAE-cellulose [408,410,411,417,419,427,428] or DEAE-Sephadex [30,400] is often employed to separate phosphate from pyrophosphate esters and, less commonly, to separate prenyl pyrophosphates of differing size [412,414]. Depending on the solubility of the phosphorylated prenol, aqueous or methanolic solutions are utilized as eluents with gradients of ammonium carbonate, ammonium bicarbonate, ammonium formate or ammonium acetate. Gradients of KCl [429] and K_2HPO_4 [412] have found occasional use, although these salts cannot be subsequently removed by lyophilization, an important consideration in preparing labeled phosphorylated substrates. However, triethylammonium bicarbonate (TEAB) is very readily removed by lyophilization, and Laskovics and Poulter [430] have used TEAB gradients on DEAE-Sephadex A-25 columns to purify a variety of pyrophosphorylated substrates for biosynthetic studies.

Ion-exchange chromatography on Dowex-1 provides another useful means of separating terpenyl phosphates and pyrophosphates. Elution with ammonium formate–formic acid gradients is useful for separating phosphorylated derivatives of mevalonic acid and isopentenol [431–433], but because of the low pH of the eluent, this system is not suitable for acid-labile allylic derivatives [408,434]. However, alcoholic ammonium hydroxide [433] and gradients of ammonium formate in methanol [205,435] and aqueous LiCl at pH 8 [436] are suitable for the resolution of allylic phosphate ester mixtures. The positional isomers, farnesyl and nerolidyl pyrophosphate, have been resolved on Dowex [205]. The separation of phosphorylated prenols of various chain lengths has also been achieved by gel chromatography on Sephadex G-25 [425] and Sephadex LH-20 [437]. Chromatography on polystyrene resin [438] would appear to offer a useful alternative for the fractionation of terpenyl phosphates. A reversed-phase HPLC technique has recently been applied to the separation of retinyl phosphate [439]. An important consideration in the preparation of pyrophosphorylated substrates is the removal of inorganic phosphates which may be generated in the phosphorylation reaction and which may also interfere in purification [436,440]. Preliminary treatment of the reaction mixture with inorganic pyrophosphatase may thus be useful in eliminating the influence of inorganic pyrophosphate [441]. Inorganic phosphates can also be removed from crude preparations before CC [427,436].

13.11. GLYCOSYLATED TERPENES

The polar terpenyl glycosides need not be chromatographed directly if hydrolysis affords stable aglycones that are readily separated and if hydrolysis does not lead to loss of structural information, such as the position of glycosidic attachment. In the case of one of the largest families of terpenyl glycosides, the iridoids and secoiridoids, hydrolysis is generally not applicable, because the aglycones are labile dialdehydes [442,443]. Similar instability of the aglycone (sapogenin) is sometimes encountered in triterpenoid saponins [444]. Such compounds must, necessarily, be chromatographed as such or after conversion (e.g., by acetylation or silylation) to less polar derivatives. For complex mixtures of these and other terpenyl glycosides, successive chromatography of the free and derivatized forms is often most effective in the isolation of individual components [443–450]. The isolation [443] and chromatographic properties [442] of iridoid glucosides have been reviewed. Recent work on the isolation of triterpenoid saponins has also been described [444].

13.11.1. Gas–liquid chromatography

GLC techniques have been applied to essentially all of the glycosides that can be converted to derivatives of sufficient volatility for such analyses [451]. The GLC separation and MS analysis of the TMS ethers of iridoid and secoiridoid glucosides have been described [452]. Columns with low loading of silicone oil phases (e.g., SE-30) are generally employed for this work [452–454]. The applicability of GLC to the separation of other types of monoterpenyl glycosides has been demonstrated by Sakata and Koshimizu [105], who successfully resolved the peracetylated derivatives of (±)-menthyl- and (±)-bornyl-β-D-glucosides on silicone oil phases. Gibberellin-O-glucosides have also been separated by GLC of the TMS ether-TMS ester or TMS ether-methyl ester derivatives [264,455]. This demonstrates the utility of the technique for terpenyl glycosides of even this molecular weight range.

13.11.2. Thin-layer chromatography

Analytical and preparative TLC (generally on silica gel) is widely employed in the separation of iridoid and secoiridoid glucosides in both derivatized and underivatized form [442,443,447,448]. Inouye et al. [456] have utilized layers of 17% $AgNO_3$-impregnated silica gel to resolve a mixture of iridoid glucoside tetraacetates differing only in the number of double bonds. The technique has also been used to separate other types of monoterpenyl glycosides [457], and is sufficient to resolve diastereomers, such as the β-D-glucosides of menthol and neomenthol [458]. Gibberellin glycosides (and their acetates) can be separated on silica gel [254,459], as can triterpenyl glycosides and glycoside esters [327,460]. Tetraterpenyl (carotenoid) glucosides and their acetate derivatives have been resolved on kieselguhr [461]. TLC on silica gel and kieselguhr is commonly used in the isolation of many different types of polyprenyl phosphate sugars [399,416,462–464].

13.11.3. Column chromatography

CC is a well-established technique for the preparative isolation of iridoid and secoiridoid glucosides [443] and other monoterpenyl glycosides [446,465]. Sorbents utilized for CC of iridoid and secoiridoid glucosides include kieselguhr, polyamide, cellulose, magnesium silicate, Sephadex LH-20, various anion exchangers (for acidic types), and silica gel [443]. The latter also finds use in the fractionation of derivatized monoterpenyl glucosides [443,448,449]. Iridoid and secoiridoid glucosides have been separated by HPLC on μBondapak C_{18} [466].

A variety of CC techniques are used in the preliminary fractionation of diterpene glycosides [467,468] and gibberellin glucosides [264,469], ion-exchange procedures being particularily effective for this latter group of conjugates [304]. HPLC is an effective approach to the separation of gibberellin conjugates [469]. It is finding increasing use in the fractionation of triterpenoid glycosides [470] and related saponins [471,472], which are also purified by CC on silica gel [473,474], alumina [475], charcoal–Celite [476], polyamide [477], and DEAE-Sephadex [478]. Carotenoid glucosides have been separated on cellulose columns, and the corresponding acetates on alumina columns [461]. Polyprenyl phosphate sugar derivatives are generally separated by ion-exchange chromatography on DEAE-cellulose [392,399,479–482], or by gel permeation chromatography on Sephadex LH-20 [415,479,482,483]. Recently, reversed-phase HPLC was applied to the separation of mannosylretinyl phosphate [439].

13.11.4. Paper chromatography and other methods

Although the use of PC for the analysis of terpenyl glycosides has declined as the use of TLC increased, PC is still employed in the separation and characterization of iridoid glucosides [442,484] and polyprenyl phosphate sugars [485–487]. Two-dimensional PC provides a general screening procedure for the detection of iridoid glucosides [488].

An effective means of separating complex irridoid and secoiridoid glucoside mixtures is countercurrent distribution [489–491]. An extension of this classical technique, droplet countercurrent chromatography, is finding widespread application for resolving glycosides [492,493], especially the iridoids [494] and saponins [495–499].

13.12. TERPENOID VITAMINS

13.12.1. Vitamin A

Vitamin A and its derivatives are of great interest, not only in relation to vision, but also because vitamin A analogs may be effective in the prevention and treatment of cancer and of skin disorders. All vitamin A derivatives are characterized by their extreme lability to air, heat, and light. Since the various *cis*-isomers are important in vision research, much effort has been expended in the separation and purification of these isomers [500–503].

The extreme sensitivity of these compounds has made the more traditional analytical methods less useful for any but routine analyses. With special precautions, vitamin A compounds can be chromatographed on silica gel thin-layer plates by using mixtures of ethers and hydrocarbons as developing solvents [504,505]. Generally, plates should be activated for 30–60 min at 110–120°C, and chromatograms should be developed in the dark [506] below room temperature. The addition of antioxidants is beneficial, and plates may be sprayed with a solution of BHT prior to development [507]. Detection at 366 nm avoids interference from BHT, which absorbs below 300 nm. With acetone–light petroleum (9:41) at 8°C, the following R_F values were observed: retinyl palmitate, 0.71; retinyl acetate, 0.54; retinal, 0.40; retinol, 0.26; retinoic acid, 0.21. Recoveries of 75–94% were reported [507].

GC of vitamin A compounds is complicated by the thermal lability and low volatility of these substances [508]. Clearly, all-glass systems are mandated for these compounds. The best separations of retinoate esters are obtained on columns with low loadings of SE-30. This liquid phase is preferred over XE-52 or QF-1. A 3% coating of SE-30 on 80–100-mesh Chromosorb W HP (Supelco) at 230°C is reported to give a good separation of all-*trans* methyl retinoate from 13-*cis* methyl retinoate. Interfacing with MS enables detection of as little as 1 ng/ml in human blood [509].

CC is by far the most useful means of separating vitamin A compounds with minimum decomposition and maximum resolution. Among the many types of sorbents that have employed, silica gel ranks first. Basic alumina [504] has occasionally been used, particularly when the analysis is to include carotenes and xanthins as well [510]. When Celite 345 is eluted with hexane, saturated with PEG 400 and containing antioxidants, separation of the all-*trans*- from the 9,13-di-*cis*-vitamin A isomers is achieved, but the 9-*cis*- and 13-*cis*-isomers are not separated in this system [511].

Rapid advances in HPLC had an enormous impact on vitamin A analysis, and by now HPLC has superseded all other separation methods in this field. Its rapidity, sensitivity, and resolving power has made HPLC the technique of choice. Adsorption HPLC is suitable for the elimination of carotenoid interference [512], for the isolation of all the geometric isomers of vitamin A and its esters [513–516] from biological fluids [515] and food samples [517], and for preparative work [513,516]. Reversed-phase HPLC is equally suitable for biological fluids [518–520], and food products [521–525] and for the separation of vitamin A derivatives [526,527]. Vitamin A esters of unsaturated acids are readily resolved in reversed-phase systems with methanolic $AgNO_3$ in the mobile phase [527].

Vitamin A is readily separated from vitamins D_2 and E on a Micro-Pak CH-10 column [528]. Adsorption chromatography on Micro-Pak (5 μm) with a hexane–dioxane solvent system separates all of the isomeric *cis/trans–syn/anti*-retinal oximes obtained by treatment of eye pigments with hydroxylamine [505,529]. Partisil columns have been especially useful for the separation of retinoic acids and esters. Isomeric retinoates can be separated on Partisil 10 ODS-2 columns by elution with mixtures of acetonitrile–1% aqueous ammonium acetate (3:2) [530,531], or with dichloromethane–acetic acid (199:1) [532]. Chromatography on Partisil-ODS columns is reported to be slower than on Spherisorb ODS (5 μm) or on Partisil (10 μm) adsorption columns [532]. With mixtures of heptane, acetonitrile and DMSO, all of the *cis/trans*-isomers of methyl retinoate can be resolved on Partisil ODS-2 [533].

Chromatography of vitamin A compounds from margarines can be achieved on LiChrosorb Si 60 with a heptane–isopropyl ether (19:1) solvent mixture. Crude preparations can be injected directly, as the column retains glycerides, tocopherols, lecithins, and other polar compounds, and 20 to 50 analyses can be performed before the column requires regeneration [534]. Similarly, LiChrosorb Si-60 has been applied to the analysis of retinol in serum [535] and to the separation of retinoic acid analogs and their metabolites with either hexane–tetrahydrofuran–acetic acid (980:15:6) or hexane–methyl benzoate–propionic acid (1750:250:7) as the solvent system [536]. A LiChrosorb (10 μm) reversed-phase column, eluted with 90% aqueous methanol, has been employed for the analysis of milk and infant foods [537]. Carotenes are nicely separated with 99% aqueous methanol.

Retinal and retinol *cis/trans*-isomers are readily resolved on a Zorbax Sil column with diethyl ether–hexane, ethyl acetate–hexane, or ethyl acetate–dichloromethane–hexane mixtures. This column is reported to be especially advantageous for retinal photoisomers when eluted with 12% diethyl ether in hexane, and it is capable of resolving the four isomeric retinols in the presence of the corresponding retinals [538,539]. Retinal isomers and 3-dehydroretinal isomers have been chromatographed on μPorasil. Isocratic elution with 2% diethyl ether in light petroleum effects the separation of the 9-*cis*-, 11-*cis*-, 13-*cis*-, and all-*trans*-isomers [540]; and 6–8% diethyl ether separates the 11-*cis*-retinal from the 11-*cis*-3-dehydroretinal [541].

Coupled reversed-phase columns have been employed for the simultaneous determination of vitamin A acetate (or other vitamin A compounds), vitamin D_2, and vitamin E acetate. A methanol–water gradient on coupled μBondapak Phenyl–μBondapak C_{18} columns achieves excellent results within 50 min [542]. This technique can also be used for retinoic acid *cis/trans*-isomers, and has a sensitivity of about 1 ng [543].

Gel chromatography for vitamin A compounds has also been investigated. It is particularly useful for the group separation of retinoids from large amounts of other lipophilic materials in products such as margarines and other prepared foods [544–546]. In low-pressure chromatography, Sephadex LH-20, eluted with chloroform–light petroleum mixtures, affords good separations of retinol, retinyl esters, retinal, and retinoic acids and esters, although this gel matrix does not appear capable of resolving *cis/trans*-isomers [546,547]. HPLC gel permeation applications have also been reported. Hitachi Gel 3010 has been used for the determination of vitamin A and β-carotene in fats and oil [548]. μStyragel (100 Å) has been used in conjunction with reversed-phase HPLC on μBondapak C_{18} to separate retinol from β-carotene [544].

13.12.2. Vitamin E

Vitamin E actually represents a class of eight naturally occurring phenols, known as the tocopherols and tocotrienols. The four tocopherols possess a saturated isoprenoid side chain and differ from each other in the number and position of aromatic methyl groups. Because these compounds possess varying degrees of biological activity, it is important to differentiate among them. Earlier colorimetric assays for vitamin E were nonspecific and relied heavily on extensive clean-up procedures to remove reducing interferences [549–551]. The determination of vitamin E has been reviewed extensively [551–559].

The chromatographic separation of this vitamin group is rather straightforward, since these compounds are relatively stable, and only minimal precautions are needed. They are easily extracted with common organic solvents (hexane, diethyl ether, dichloromethane), and saponification of crude extracts affords a fraction of nonsaponifiable lipids that can be readily chromatographed by a variety of techniques. The only serious interferences are encountered in foods fortified with vitamin A esters. Such materials definitely require saponification, as it is easy to separate retinol itself from tocopherols chromatographically.

TLC has seen extensive application in vitamin E determinations in many different materials such as food products, animal tissues, and pharmaceutical preparations [551,560–562]. Most tocopherols separate readily on silica gel plates, developed with either hexane–ethyl acetate (37:3) [563], cyclohexane–diethyl ether (4:1) [564], or dichloromethane [565]. TLC methods do not resolve the positional isomers β- and γ-tocopherol [563].

Tocopherols are readily detected on TLC plates by oxidation to the tocopherolquinones [565]. The oxidation is facile and gives zones suitable for quantitation by densitometric techniques [562]. Oxidation is accomplished by spraying the plates with Ce^{4+}–TCA, followed by neutralization with ammonia vapor [562], or with 2,6-dichloro-*p*-benzoquino-4-chlorimine (Gibb's reagent) [566], or by merely heating the developed plates in air at 110°C for 18 h [567]. The oxidation products themselves can be chromatographed on silica gel with benzene or light petroleum–diethyl ether–acetone (94:5:1) solvent systems [568].

References on p. B177

Reversed-phase TLC on starch, cellulose, or talc layers, impregnated with mineral oil, with acetone–acetic acid (3:2) as the mobile phase has been used to separate α-tocopherol from α-tocopheryl acetate [569]. A column modification of this technique has been employed for the separation of vitamins A and D from E [570]. Sephadex G-50 and LH-20, eluted with chloroform, can be used similarly [571].

Thin-layer techniques are frequently combined with GLC analysis because the latter is more sensitive and easier to quantitate [572]. On plates sprayed with (or containing) Rhodamine 6G or 2,7-dichlorofluorescein, the vitamin E fractions are readily visible in UV light, and can be scraped off, extracted and analyzed by GLC [573,574]. GLC appears to be a good general method for the analysis of tocopherols and tocopheryl esters in a variety of materials [575–578]. Most analysts have used nonpolar phases at low loadings, such as 0.3% Apiezon L [578,579], 2–5% SE-30 on Gas-Chrom Q or Chromosorb HP [572,574,576], or OV-1 on Gas-Chrom Q [580]. A 1% OV-1 column at 240°C was used in GLC–MS applications [580]. SE-30 is reported to be superior to OV-17 in that the retention times are shorter [572,573]. A limitation of GLC with packed columns is the difficulty in separating β- from γ-tocopherol. A WCOT capillary column, deactivated with Silanox and coated with a new polar phase (polyphenyl ether sulfone), is reported to afford excellent resolution of tocopherols as their TMS ethers [581].

By far the most significant advances in the analysis of vitamin E compounds have been in the area of HPLC. Although for some applications pretreatments of extracts such as saponification or precipitation are indicated [582–585], in most cases the crude extracts can be injected directly on the column. For example, crude lipid extracts of cereal products can be directly analyzed on μPorasil, eluted with hexane–chloroform (17:3), which yields a complete separation of retinyl palmitate from α-tocopheryl acetate [586]. Generally, other lipids do not cause problems, but in one report β-carotene was said to interfere with the quantitation of ubiquinone [587]. A further aid to the elimination of interfering substances is fluorescence detection. Excitation at 298 nm and emission at 325–335 nm allows detection of as little as 0.03 μg tocopherol [588–594]. When the excitation wavelength is decreased to 205 nm, a 20-fold increase in sensitivity results [595].

Reversed-phase columns appear to be advantageous for routine determinations of vitamin E, but adsorption columns are required to separate all of the tocopherols and their corresponding acetates [596]. The separation of β- from γ-tocopherol is the most difficult [580,597]. For most general applications, μBondapak C_{18}, eluted with methanol–water (19:1), appears satisfactory [598–600], but for the separation of all eight vitamin E compounds, adsorption columns are required. From most adsorbents tocopherols are eluted, typically with low concentrations of ethers in hydrocarbons as the mobile phase, in the order: α-, β-, γ-, δ-. Eluents containing 0.2–4% isopropyl ether in hexane have been utilized with LiChrosorb Si-60 (5 μm) [592], JASCO-PACK WC-03-500 [593], and Corasil I [601], while 0.5% THF in hexane with Corasil II is effective for preparative applications [602]. A Spherisorb (5 μm) column, eluted with hexane–isopropanol (399:1) at 0.8 ml/min, separated all four tocopherols with the following relative retention ratios: α-, 1.00; β-, 1.44; γ-, 1.59; δ-, 2.46. Under these conditions, α-tocopherol had a retention time of 4.5 min.

13.12.3. Vitamin K

Vitamin K occurs in two forms which differ in the degree of saturation of the isoprenoid side chain attached to the benzoquinoid nucleus. Phylloquinone has only a single unsaturation, which occurs in the isoprenoid chain, while menaquinone has double bonds in each of the isoprene units of the side chain. The number of isoprene residues in the side chains of some of the menaquinones also differs, and thus there are only subtle differences in the total interaction with the chromatographic matrix [603]. Vitamin K is very sensitive to photooxidation, which yields a variety of products, presumably arising from the interactions with singlet oxygen [604]. These photooxidation products have been analyzed by GLC on 3% OV-1 on Chromosorb W at 260°C [605] and by GLC–MS [604].

TLC methods have been applied to the vitamin K group [606–609], but no single TLC system appears to be wholly satisfactory for all compounds. The layers used include: alumina and alumina impregnated with β,β'-oxydipropionitrile, silica and silica impregnated with PEG 200 [607], kieselguhr impregnated with paraffin [608], AgNO₃-impregnated [608,610], and C_{18}-bonded [609] plates. Reversed-phase partition appears to offer greater selectivity in TLC [Table 13.5].

Reversed-phase HPLC also appears preferable to adsorption chromatography, although LiChrosorb (5 μm), eluted with 0.03% isopropanol in hexane [611], and silica columns, eluted with acetonitrile–hexane or dichloromethane–hexane [612,613] have been advocated. Most vitamin K analyses have been performed with the "reversed-phase" techniques on μBondapak C_{18} with aqueous acetonitrile [608], Permaphase ODS with 63% aqueous dioxane or 89% aqueous methanol [613,614], Nucleosil C_{18} with methanol–ethanol (7:3) [615], Zorbax ODS with methanol–ethyl

TABLE 13.5

TLC OF VITAMIN K AND SOME DERIVATIVES [608]

Adsorption = Silica Gel G/6% diethyl ether in light petroleum. Reversed-phase = paraffin-impregnated kieselguhr/90% aqueous acetone. Argentation = 10% (w/w) AgNO₃-Silica Gel G/60% diisopropyl ether in light petroleum.

	TLC system		
	Adsorption (R_F)	Reversed-phase (R_F)	Argentation (R_F)
Menaquinone-4	0.29	0.66	0.48
2,3-Epoxymenaquinone-4	0.27	0.81	0.46
Demethylmenaquinone-4	0.27	0.72	0.45
Menaquinone-3	0.28	0.79	0.67
2,3-Epoxymenaquinone-3	0.27	0.89	0.59
Phylloquinone	0.35	0.38	0.97
2,3-Epoxyphylloquinone	0.33	0.59	0.91
Menadione	0.12	0.96	0.62

References on p. B177

TABLE 13.6

RETENTION TIMES FOR VITAMIN K AND DERIVATIVES ON HPLC [608]

System A = column, 300×3 mm ID μBondapak C_{18}; eluent, 85% aqueous acetonitrile. System B = column, 200×2 mm ID μBondapak C_{18}; eluent, methanol. Retention time in min.

	HPLC system	
	A	B
Menaquinone-4	27.5	–
2,3-Epoxymenaquinone-4	20.0	–
Menaquinone-3	17.0	–
2,3-Epoxymenaquinone-3	12.0	–
Phylloquinone	–	9.6
2,3-Epoxyphylloquinone	–	7.2
Menadione	5.0	–

TABLE 13.7

CAPACITY RATIOS OF VITAMIN K COMPOUNDS IN ADSORPTION, REVERSED-PHASE, AND CYANO-BONDED PHASE CHROMATOGRAPHY [612]

A = Silica column/0.2% acetonitrile in hexane; B = Silica column/10% dichloromethane in hexane; C = C_{18} HL column/methanol; D = C_{18} HL column/50% methanol in acetonitrile; E = C_{18} HL column/acetonitrile; F = C_8 column/acetonitrile; G = Cyano-bonded column/0.05% acetonitrile in hexane.

	A	B	C	D	E	F	G
cis-Vitamin K_1	4.4	5.8	7.6	8.7	13.0	5.6	4.2
trans-Vitamin K_1	5.2	6.9	7.6	8.7	13.0	5.6	4.6
Vitamin K_1 epoxide	6.1	6.5	4.7	5.2	7.3	4.2	5.5
Menaquinone-2	8.7	–	1.8	1.7	1.9	1.7	–
Menaquinone-4	8.8	8.8	4.3	4.2	5.1	3.0	7.5
Menaquinone-9	9.5	10.3	–	–	–	27.7	9.0

acetate (43:7) [616], and methanol–dichloromethane (4:1) or acetonitrile–dichloromethane (7:3) [617]. Table 13.6 illustrates the separation of vitamin K and derivatives on μBondapak C_{18} columns. Analysis of biological fluids is facilitated by preliminary separation of the vitamin K fraction on Partisil 5 with hexane–diethyl ether (97:3), followed by analysis on a reversed-phase column [617]. In milk analysis, a Partisil 10 PAC column was found superior for preliminary separations [617]. A comparison of various HPLC systems is presented in Table 13.7.

REFERENCES

1 W.J. Baas and G.J. Niemann, *J. High Resolut. Chromatogr. Chromatogr. Commun.*, 1 (1978) 18.
2 M.A. Adams and K. Nakanishi, *J. Liquid Chromatogr.*, 2 (1979) 1097.
3 A. Pryde, M.A.G. Dielsdorf and M.T. Gilbert, *Applications of High Performance Liquid Chromatography*, Wiley, New York, 1979.
4 D.G.I. Kingston, *J. Nat. Prod.*, 42 (1979) 237.
5 G. Zweig and J. Sherma (Editors), *CRC Handbook Series in Chromatography, Terpenoids*, CRC Press, Boca Raton, FL, 1982.
6 J.W. Bardenheier and G. Popják, *Biochem. Biophys. Res. Commun.*, 74 (1977) 1023.
7 C.D. Goodwin and S. Margolis, *J. Lipid Res.*, 17 (1976) 297.
8 D.J. Shapiro, R.L. Imblum and V.W. Rodwell, *Anal. Biochem.*, 31 (1969) 383.
9 D.J. Shapiro, J.L. Nordstrom, J.J. Mitschelin, V.W. Rodwell and R.T. Schimke, *Biochim. Biophys. Acta*, 370 (1974) 369.
10 S. Shefer, S. Hauser, V. Lapar and E.H. Mosbach, *J. Lipid Res.*, 13 (1972) 402.
11 H.R. Murthy, S. Moorjani and P.-J. Lupien, *Anal. Biochem.*, 81 (1977) 65.
12 T.C. Linn, *J. Biol. Chem.*, 242 (1967) 984.
13 A.M. Fogelman, J. Edmond, J. Seager and G. Popják, *J. Biol. Chem.*, 250 (1975) 2045.
14 S. Goldfarb and H.C. Pitot, *J. Lipid Res.*, 12 (1971) 512.
15 M.S. Brown, S.E. Dana, J.M. Dietschy and M.D. Siperstein, *J. Biol. Chem.*, 248 (1973) 4731.
16 J.D. Brooker and D.W. Russell, *Arch. Biochem. Biophys.*, 167 (1975) 723.
17 H.R. Murthy, P.-J. Lupien and S. Moorjani, *Anal. Biochem.*, 89 (1978) 14.
18 T.R. Green and D.J. Baisted, *Anal. Biochem.*, 38 (1970) 130.
19 R.G. Guchait and J.W. Porter, *Anal. Biochem.*, 15 (1966) 509.
20 M.D. Siperstein, V.M. Fagan and J.M. Dietschy, *J. Biol. Chem.*, 241 (1966) 597.
21 B. Hamprecht and F. Lynen, *Eur. J. Biochem.*, 14 (1970) 323.
22 J.W. Cornforth, R.H. Cornforth, G. Popják and L. Yengoyan, *J. Biol. Chem.*, 241 (1965) 3970.
23 L.W. White and H. Rudney, *Biochemistry*, 9 (1970) 2713.
24 J. Huber, S. Latzin and B. Hamprecht, *Hoppe-Seyler's Z. Physiol. Chem.*, 354 (1973) 1645.
25 J. Avigan, S.J. Bhathena and M.E. Schreiner, *J. Lipid Res.*, 16 (1975) 151.
26 P.A. Edwards, G. Popják, A.M. Fogelman and J. Edmond, *J. Biol. Chem.*, 252 (1977) 1057.
27 A. Sanghvi and B. Parikh, *Biochim. Biophys. Acta*, 444 (1976) 727.
28 E. Cardemil, J.R. Vicuña, A.M. Jabalquinto and O. Cori, *Anal. Biochem.*, 59 (1974) 636.
29 E. Jedlicki, G. Jacob, F. Faini and O. Cori, *Arch. Biochem. Biophys.*, 152 (1972) 590.
30 E. Cardemil and O. Cori, *J. Labelled Compd.*, 9 (1973) 15.
31 T.G. McCloud and P. Heinstein, *J. Chromatogr.*, 174 (1979) 461.
32 G.P. McSweeney, *J. Chromatogr.*, 17 (1965) 183.
33 M.O. Oster and C.A. West, *Arch. Biochem. Biophys.*, 127 (1968) 112.
34 G.A. Sanadze, *Fiziol. Rast.*, 8 (1971) 555.
35 R.A. Rasmussen, *J. Air Pollut. Control. Assoc.*, 22 (1972) 537.
36 J. Naveau, *J. Chromatogr.*, 174 (1979) 109.
37 E.G. DeMaster and H.T. Hagasawa, *Life Sci.*, 22 (1978) 91.
38 E. Von Rudloff, *Recent Advan. Phytochem.*, 2 (1969) 127.
39 E. Von Rudloff, *Advan. Chromatogr.*, 10 (1974) 173.
40 S. Hayashi, *Method. Chim.*, 11 (1978) 44.
41 J.A. Rogers, *Perfum. Flavor.*, 3 (1979) 41.
42 Anal. Meth. Comm., *Analyst (London)*, 96 (1971) 887; 105 (1980) 262.
43 E. Ziegler, in K.-H. Kubeczka (Editor), *Vorkommen und Analytik ätherischer Öle*, Thieme, Stuttgart, 1979, p. 124.
44 J.S. Prithi, in C.O. Chichester, E.M. Mrak and G.F. Stewart (Editors), *Spices and Condiments: Chemistry, Microbiology, Technology*, Academic Press, New York, 1980, p. 69.
45 R. Bonmati and G. Guiochon, *Perfum. Flavor.*, 3 (1978) 17.
46 Y. Masada, *Analysis of Essential Oils by Gas Chromatography and Mass Spectrometry*, Wiley, New York, 1976.

47 M. von Schantz, in K.-H. Kubeczka (Editor), *Vorkommen und Analytik ätherischer Öle*, Thieme, Stuttgart, 1979, p. 88.

48 H. Hendriks and A.P. Bruins, *J. Chromatogr.*, 190 (1980) 321.

49 S.R. Srinivas and D.C. Borecki, *Proc. 6th Int. Congr. Essent. Oils, San Francisco*, Allured Publ., Oak Park, 1974, p. 34.

50 R.P. Adams, M. Granat, L.R. Hogge and E. von Rudloff, *J. Chromatogr. Sci.*, 17 (1979) 75.

51 G. Gilbertson and R.T. Koenig, *Anal. Chem.*, 51 (1979) 183R.

52 B.M. Lawrence, *Perfum. Flavor.*, 5 (1980) 27.

53 J. Cermak, M. Penka and K. Tesarik, *Acta Univ. Agric. Ser. C (Brno)*, 47 (1978) 3.

54 A. Koedam, J.J.C. Scheffer and A. Baerheim Svendsen, *Chem. Mikrobiol. Technol. Lebensm.*, 6 (1979) 1.

55 H. Maarse and R.E. Kepner, *J. Agr. Food Chem.*, 18 (1970) 1095.

56 A. Baerheim Svendsen and J. Karlsen, *Planta Med.*, 14 (1966) 376.

57 W. Henderson, J.W. Hart, P. How and J. Judge, *Phytochemistry*, 9 (1970) 1219.

58 U.M. Senanayake, R.A. Edwards and T.H. Lee, *J. Chromatogr.*, 116 (1976) 468.

59 J.P. Minyard, J.H. Tumlinson, P.A. Hedin and A.C. Thompson, *J. Agr. Food Chem.*, 13 (1965) 599.

60 M.H. Klouwen and R. ter Heide, *J. Chromatogr.*, 7 (1962) 297.

61 D.G. Gillen and J.T. Scanlon, *J. Chromatogr. Sci.*, 10 (1972) 729.

62 E. von Rudloff, *Can. J. Chem.*, 38 (1960) 631.

63 S. Geyer and R. Mayer, *Z. Chem.*, 5 (1965) 308.

64 R. ter Heide, *Z. Anal. Chem.*, 236 (1968) 215.

65 R. ter Heide, *J. Chromatogr.*, 129 (1976) 143.

66 T. Saeed, G. Redant and P. Sandra, *J. High Resolut. Chromatogr. Chromatogr. Commun.*, 2 (1979) 75.

67 W. Jennings and T. Shibamoto, *Qualitative Analysis of Flavor and Fragrance Volatiles by Glass Capillary Gas Chromatography*, Academic Press, New York, 1980.

68 A.A. Swigar and R.M. Silverstein, *Monoterpenes*, Aldrich Chemical Co., Milwaukee, WI, 1981.

69 F. Morishita, T. Okano and T. Kojima, *Bunseki Kagaku*, 29 (1980) 48.

70 R.B. Watts and R.G.O. Kekwick, *J. Chromatogr.*, 88 (1974) 15.

71 B.J. Tyson, *J. Chromatogr.*, 111 (1975) 419.

72 A.R. Vinutha and E. von Rudloff, *Can. J. Chem.*, 46 (1968) 3743.

73 B.V. Bapat, B.B. Ghatge and S.C. Bhattacharyya, *J. Chromatogr.*, 18 (1965) 308.

74 B.V. Bapat, B.B. Ghatge and S.C. Bhattacharyya, *J. Chromatogr.*, 23 (1966) 227.

75 B.V. Bapat, B.B. Ghatge and S.C. Bhattacharyya, *J. Chromatogr.*, 23 (1966) 363.

76 B.V. Bapat, T.K. Sankpal, B.B. Ghatge and S.C. Bhattacharyya, *J. Chromatogr.*, 29 (1967) 333.

77 I.I. Bardyshev, V.I. Kulikov, L.A. Naumovich and T.V. Tumysheva, *Zh. Anal. Khim.*, 23 (1968) 1893.

78 G. Marcelin, *J. Chromatogr.*, 174 (1979) 208.

79 A. Di Corcia, A. Liberti, C. Sambucini and R. Samperi, *J. Chromatogr.*, 152 (1978) 63.

80 R.E. Kepner and H. Maarse, *J. Chromatogr.*, 66 (1972) 229.

81 B.M. Lawrence, J.W. Hogg and S.J. Terhune, *J. Chromatogr.*, 50 (1970) 59.

82 J.J.C. Scheffer and A. Baerheim Svendsen, *J. Chromatogr.*, 115 (1975) 607.

83 J.J.C. Scheffer, A. Koedam and A. Baerheim Svendsen, *Sci. Pharm.*, 44 (1976) 119.

84 J.J.C. Scheffer, A. Koedam, M.T.I.W. Schuesler and A. Baerheim Svendsen, *Chromatographia*, 10 (1977) 669.

85 C. Jequier, G. Nicollier, R. Tabacchi and J. Garnero, *Phytochemistry*, 19 (1980) 461.

86 R. Tressl, L. Friese, F. Fendesack and H. Köppler, *J. Agr. Food Chem.*, 26 (1978) 1422 and 1426.

87 B.C. Clark, C.C. Powell and T. Radford, *Tetrahedron*, 33 (1977) 2187.

88 M.S.F. Ross, *J. Chromatogr.*, 106 (1975) 392.

89 C. George-Nascimento and O. Cori, *Phytochemistry*, 10 (1971) 1803.

90 L. Chayet, C. Rojas, E. Cardemil, A.M. Jabalquinto, R. Vicuña and O. Cori, *Arch. Biochem. Biophys.*, 180 (1977) 318.

91 R. Croteau, *Plant Physiol.*, 59 (1979) 519.

92 A.J. Poulose and R. Croteau, *Arch. Biochem. Biophys.*, 187 (1978) 307.

93 A.R. Battersby, D.G. Laing and R. Ramage, *J. Chem. Soc. (Perkin I)*, (1972) 2743.

94 D.V. Banthorpe, B.V. Charlwood and M.J.O. Francis, *Chem. Rev.*, 72 (1972) 115.

95 R. Croteau, A.J. Burbott and W.D. Loomis, *Biochem. Biophys. Res. Commun.*, 50 (1973) 1006.
96 V.V. Bazylchik, B.G. Udarov and N.P. Polyakova, *Zh. Anal. Khim.*,31 (1976) 604.
97 T. Norin, *Acta Chem. Scand.*, 16 (1962) 640.
98 B.M. Lawrence and J.W. Hogg, *Perfum. Essent. Oil. Rec.*, 59 (1968) 515.
99 K.L. McDonald and D.M. Cartlidge, *J. Chromatogr. Sci.*, 9 (1971) 440.
100 J. Herling, J. Shabtai and E. Gil-Av, *J. Chromatogr.*, 8 (1962) 349.
101 C. Paris and P. Alexandre, *J. Chromatogr. Sci.*, 10 (1972) 402.
102 H. Rothbächer and F. Suteu, *J. Chromatogr.*, 100 (1974) 236.
103 P. Bournot, B.F. Maume and C. Baron, *J. Chromatogr.*, 57 (1971) 55.
104 I. Abe and S. Musha, *Bunseki Kagaku*, 23 (1974) 755.
105 I. Sakata and K. Koshimizu, *Agr. Biol. Chem.*, 43 (1979) 411.
106 J. Casanova and E.J. Corey, *Chem. Ind. (London)*, (1961) 1664.
107 R. Croteau and F. Karp, *Arch. Biochem. Biophys.*, 184 (1977) 77.
108 M. Nakajima and T. Ueno, *Chemistry*, 24 (1969) 469.
109 P.D. Maestas and C.J. Morrow, *Tetrahedron Lett.*, (1976) 1047.
110 E. Kugler, R. Langlais, W. Halang and M. Hufschmidt, *Chromatographia*, 8 (1975) 468.
111 W. Averill and E.W. March, *Chromatogr. Newsl.*, 4 (1976) 20.
112 C.W. Wilson and P.E. Shaw, *J. Agr. Food Chem.*, 28 (1980) 919.
113 E. Kugler, in K.-H. Kubeczka (Editor), *Vorkommen und Analytik ätherischer Öle*, Thieme, Stuttgart, 1979, p. 114.
114 W. Jennings, *J. Chromatogr. Sci.*, 17 (1979) 636.
115 R. Hiltunen and S. Räisänen, *Planta Med.*, 41 (1981) 174.
116 R. Dandeneau, P. Bente, T. Rooney and R. Hiskes, *Amer. Lab.*, 11 (1979) 61.
117 Anon., *Gas Chromatography Applications Notes*, GCA-5, Perkin-Elmer, Norwalk, CT.
118 L.A. Smedman, K. Snajberk, E. Zavarin and T.R. Mon, *Phytochemistry*, 8 (1969) 1471.
119 T.A. Rooney, L.H. Altmayer, R.R. Freeman and E.H. Zerenner, *Amer. Lab.*, 11 (1979) 81.
120 M.S. Karawya, S.I. Balbaa and M.S. Hifnawy, *J. Pharm. Sci.*, 60 (1971) 381.
121 M.Y. Haggag, A. Shalaby and G. Verzar-Petri, *Planta Med.*, 27 (1975) 361.
122 V.K. Kaul, S.S. Nigam and A.K. Banerjee, *Ind. Perfum.*, 23 (1979) 1.
123 A.R. Sen, P.K. Sardar and P. Sengupta, *J. Ass. Offic. Anal. Chem.*, 60 (1977) 235.
124 P. Baslas and R.K. Baslas, *Ind. Perfum.*, 22 (1977) 125.
125 M.S. Siddiqui, F. Mohammad, S.K. Srivastava and R.K. Gupta, *Perfum. Flavor.*, 4 (1979) 19.
126 M.E. Younes and A. Mokhtar, *Bull. Chem. Soc. Jap.*, 47 (1974) 2588.
127 A.C. Thompson, P.A. Hedin, R.C. Gueldner and F.M. Davis, *Phytochemistry*, 13 (1974) 2029.
128 D.V. Banthorpe, G.A. Bucknall, H.J. Doonan, S. Doonan and M.G. Rowan, *Phytochemistry*, 15 (1976) 91.
129 R. Croteau and F. Karp, *Arch. Biochem. Biophys.*, 176 (1976) 734.
130 R. Croteau, C.L. Hooper and M. Felton, *Arch. Biochem. Biophys.*, 188 (1978) 182.
131 K.-H. Kubeczka, *Planta Med.*, 35 (1979) 291.
132 J. Battaile, R.L. Dunning and W.D. Loomis, *Biochim. Biophys. Acta*, 51 (1961) 538.
133 R.E. Wrolstad and W.G. Jennnings, *J. Chromatogr.*, 18 (1965) 318.
134 J.J.C. Scheffer, A. Koedam and A. Baerheim Svendsen, *Chromatographia*, 9 (1976) 425.
135 E. von Rudloff and F.M. Couchman, *Can. J. Chem.*, 42 (1964) 1890.
136 B.M. Lawrence, *J. Chromatogr.*, 38 (1968) 535.
137 T.D. Meehan and C.J. Coscia, *Biochem. Biophys. Res. Commun.*, 53 (1973) 1043.
138 J.W. Adcock and T.J. Betts, *J. Chromatogr.*, 34 (1968) 411.
139 D.V. Banthorpe and K.W. Turnbull, *J. Chromatogr.*, 37 (1968) 366.
140 R. Croteau, M. Felton and R.C. Ronald, *Arch. Biochem. Biophys.*, 200 (1980) 524.
141 R.S. Prasad, A.S. Gupta and S. Dev, *J. Chromatogr.*, 92 (1974) 450.
142 R. Croteau and M. Johnson, unpublished results.
143 D.A. Baines and R.A. Jones, *J. Chromatogr.*, 47 (1970) 130.
144 A.M. Bambagiotti, F.F. Vincieri and G. Cosi, *Phytochemistry*, 11 (1972) 1455.
145 T. Shinka, K. Ogura and S. Seto, *J. Biochem.*, 78 (1975) 1177.
146 B.C. Madsen and H.W. Latz, *J. Chromatogr.*, 50 (1970) 288.

147 J.C. Kohli, *J. Chromatogr.*, 105 (1975) 193.
148 J.C. Kohli, *J. Chromatogr.*, 121 (1976) 116.
149 A. Martinek, *J. Chromatogr.*, 56 (1971) 338.
150 G.C. Barrett, *Advan. Chromatogr.*, 11 (1974) 145.
151 O. Motl, in Z. Deyl, K. Macek and J. Janák (Editors), *Liquid Chromatography. A Survey of Modern Techniques and Applications*, J. Chromatogr. Library Series, Vol. 3, Elsevier, Amsterdam, 1975, p. 623.
152 B.M. Lawrence, J.W. Hogg and S.J. Terhune, *Perfum. Essent. Oil Record*, 60 (1969) 88.
153 B.Y. Meklati and A.Y. Ahmed, *Riv. Ital.*, 61 (1979) 302.
154 B.B. Jones, B.C. Clark, Jr. and G.A. Iacobucci, *J. Chromatogr.*, 178 (1979) 575.
155 N.A. Bergman and B. Hall, *Acta Chem. Scand.*, B33 (1979) 148.
156 Anon., *Application Highlights*, No. AH 313 (Essential Oils), Waters Assoc., Framingham, MA, 1974.
157 H. Komae and N. Hayashi, *J. Chromatogr.*, 114 (1975) 258.
158 N. Hayashi, Y. Yamamura and H. Komae, *Chem. Ind. (London)*, (1977) 34.
159 M.S.F. Ross, *J. Chromatogr.*, 160 (1978) 199.
160 J. Schwanbeck, in K.-H. Kubeczka (Editor), *Vorkommen und Analytik ätherischer Öle*, Thieme, Stuttgart, 1979, p. 72.
161 A.M. Siouffi, J.C. Traynard and G. Guiochon, *J. Chromatogr. Sci.*, 15 (1977) 469.
162 B.J. Bergot, R.J. Anderson, D.A. Schooley and C.A. Henrick, *J. Chromatogr.*, 155 (1978) 97.
163 G. Ohloff, G. Uhde and K.H. Schulte-Elte, *Helv. Chim. Acta*, 50 (1967) 561.
164 W.G. Dauben, *J. Agr. Food Chem.*, 22 (1974) 156.
165 W.G. Dauben, J.P. Hubbell and N.D. Vietmeyer, *J. Org. Chem.*, 40 (1975) 479.
166 I.A. Southwell, *Phytochemistry*, 9 (1970) 2243.
167 A.J. Weinheimer, W.W. Youngblood, P.H. Washecheck, T.K.B. Karns and L.S. Ciereszko, *Tetrahedron Lett.*, (1970) 497.
168 J.A. Pickett, J. Coates and F.R. Sharpe, *Chem. Ind. (London)*, (1975) 571.
169 J.A. Pickett, F.R. Sharpe and T.L. Peppard, *Chem. Ind. (London)*, (1977) 30.
170 V. Lukes and R. Komers, *Collect. Czech. Chem. Commun.*, 29 (1964) 1598.
171 I.C. Nigam and L. Levi, *J. Chromatogr.*, 23 (1966) 217.
172 L. Westfelt, *Acta Chem. Scand.*, 20 (1966) 2829.
173 J.A. Wenninger and R.L. Yates, *J. Ass. Offic. Anal. Chem.*, 52 (1969) 1155.
174 N.H. Andersen and D.D. Syrdal, *Phytochemistry*, 9 (1970) 1325.
175 N.H. Andersen and M.S. Falcone, *J. Chromatogr.*, 44 (1969) 52.
176 C.O. Ikediobi, I.C. Hsu, J.R. Bamburg and F.M. Strong, *Anal. Biochem.*, 43 (1971) 327.
177 Y. Mori and M. Nakamura, *Chem. Pharm. Bull.*, 19 (1971) 499.
178 W.D. Hamilton, R.J. Park, G.J. Perry and M.D. Sutherland, *Aust. J. Chem.*, 26 (1973) 375.
179 G. Wood and A. Huang, *J. Agr. Food Chem.*, 23 (1975) 239.
180 H. Pyysalo, E.-L. Seppä and K.-G. Widén, *J. Chromatogr.*, 190 (1980) 466.
181 T. Furuya and H. Kojima, *J. Chromatogr.*, 29 (1967) 341.
182 E.G. Heisler, J. Siciliano and D.D. Bills, *J. Chromatogr.*, 154 (1978) 297.
183 C. George-Nascimento, R. Pont-Lezica and O. Cori, *Biochem. Biophys. Res. Commun.*, 45 (1971) 119.
184 A. Pfiffner in T.W. Goodwin (Editor), *Aspects of Terpenoid Chemistry and Biochemistry*, Academic Press, New York, 1971, p. 95.
185 J. Velasco, G.R. Chandra and N. Mandava, *J. Agr. Food Chem.*, 26 (1978) 1061.
186 P.F. Saunders, in J.R. Hillman (Editor), *Isolation of Plant Growth Substances*, Cambridge Univ. Press, Cambridge, 1978, p. 115.
187 S.A. Quarrie, *Anal. Biochem.*, 87 (1978) 148.
188 K.T. Hubick and D.M. Reid, *Plant Physiol.*, 65 (1980) 523.
189 I.C. Nigam and L. Levi, *J. Chromatogr.*, 17 (1965) 466.
190 N.H. Andersen, M.S. Falcone and D.S. Syrdal, *Phytochemistry*, 9 (1970) 1341.
191 H. Maarse, *J. Chromatogr.*, 106 (1975) 369.
192 A. Matsuo, S. Uto, M. Nakayama and S. Hayashi, *Z. Naturforsch.*, 31 (1976) 401.
193 T. Takemoto, M. Uchida and G. Kusano, *Chem. Pharm. Bull.*, 24 (1976) 531.
194 C.W. Wilson and P.E. Shaw, *Phytochemistry*, 17 (1978) 1435.

195 R. Fluckiger, Y. Kato and S. Hishida, in A. Frigerio (Editor), *Mass Spectrometry in Biochemistry and Medicine,* Raven, New York, 1974, p. 277.
196 I.D. Railton, D.M. Reid, P. Gaskin and J. MacMillan, *Planta,* 117 (1974) 179.
197 L. Rivier, H. Milon and P.E. Pilet, *Planta,* 134 (1977) 23.
198 C.H.A. Little, J.K. Heald and G. Browning, *Planta,* 139 (1978) 133.
199 B. Andersson, N. Häggström and K. Andersson, *J. Chromatogr.,* 157 (1978) 303.
200 D. Tietz, K. Doerffling, D. Woehrle, I. Erxleben and F. Liemann, *Planta,* 147 (1979) 168.
201 A.A. Adesomoju, J.I. Okogun, D.E. Ekong and P. Gaskin, *Phytochemistry,* 19 (1980) 223.
202 G. Jacob, E. Cardemil, L. Chayet, R. Tellez, R. Pont-Lezica and O. Cori, *Phytochemistry,* 11 (1972) 1683.
203 A. Salveson and A. Baerheim Svendsen, *Planta Med.,* 30 (1976) 93.
204 L. Piovetti, G. Combaut and A. Diara, *Phytochemistry,* 19 (1980) 2117.
205 S.S. Sofer and H.C. Rilling, *J. Lipid Res.,* 10 (1969) 183.
206 S. Hayashi and A. Matsuo, *Experientia,* 26 (1970) 347.
207 K. Weinges and W. Baehi, *Justus Liebigs Ann. Chem.,* 759 (1972) 158.
208 F. Bohlmann, M. Grenz, R.K. Gupta, A.K. Dhar, M. Ahmed, R.M. King and H. Robinson, *Phytochemistry,* 19 (1980) 2391.
209 I.C. Nigam, *J. Pharm. Sci.,* 54 (1965) 1823.
210 E.W.B. Ward, C.H. Unwin, G.L. Rock and A. Stoessl, *Can. J. Bot.,* 54 (1976) 25.
211 R.W. Doskotch and F.S. El-Feraly, *J. Pharm. Sci.,* 58 (1969) 877.
212 H. Yoshioka, T.J. Mabry and B.N. Timmerman, *Sesquiterpene Lactones,* Univ. Tokyo Press, Tokyo, 1973.
213 F.S. El-Feraly and D.A. Benigni, *J. Nat. Prod.,* 43 (1980) 527.
214 Y. Asakawa, M. Toyota and T. Takemoto, *Phytochemistry,* 19 (1980) 2141.
215 A.K. Picman, G.H.N. Towers and P.V. Subba Rao, *Phytochemistry,* 19 (1980) 2206.
216 E.E. Waldner, C. Schlatter and H. Schmid, *Helv. Chim. Acta,* 52 (1969) 15.
217 A.S. Gupta and S. Dev, *J. Chromatogr.,* 12 (1963) 189.
218 W. Fenical, J.J. Sims, R.M. Wing and P.C. Radlick, *Phytochemistry,* 11 (1972) 1161.
219 S.J. Terhune, J.W. Hogg, A.C. Bromstein and B.M. Lawrence, *Can. J. Chem.,* 53 (1975) 3285.
220 J. Pliva, M. Horak, V. Herout and F. Sorm, *Die Terpene, Teil 1, Sesquiterpene,* Akademie Verlag, Berlin, 1960.
221 R.W. Doskotch and C.D. Hufford, *J. Pharm. Sci.,* 58 (1969) 186.
222 L.A.P. Anderson, W.T. de Kock, W. Nel and K.G.R. Pachler, *Tetrahedron,* 24 (1968) 1687.
223 M.S.F. Ross, *J. Chromatogr.,* 118 (1976) 273.
224 W.M. Daniewski, M. Kocor and J. Krol, *Roczn. Chem.,* 50 (1976) 2095.
225 M.R. Boyd, L.T. Burka, T.M. Harris and B.J. Wilson, *Biochim. Biophys. Acta,* 337 (1974) 184.
226 M. Koreeda, G. Weiss and K. Nakanishi, *J. Amer. Chem. Soc.,* 95 (1973) 239.
227 S. Hara, A. Ohsawa, J. Endo, Y. Sashida and H. Itokawa, *Anal. Chem.,* 52 (1980) 428.
228 J.D. Warthen, Jr., *J. Liquid Chromatogr.,* 3 (1980) 279.
229 N.W.H. Houx and S. Voerman, *J. Chromatogr.,* 129 (1976) 456.
230 R.R. Heath, J.H. Tumlinson, R.E. Doolittle and A.T. Proveaux, *J. Chromatogr. Sci.,* 13 (1975) 380.
231 P.B. Sweetser and A. Vatvars, *Anal. Biochem.,* 71 (1976) 68.
232 A. Rapp, A. Ziegler, O. Bachmann and H. Duering, *Chromatographia,* 9 (1976) 44.
233 D.R. Reeve and A. Crozier, *Phytochemistry,* 15 (1976) 68.
234 A.J. Ciha, M.L. Brenner and W.A. Brun, *Plant. Physiol.,* 59 (1977) 821.
235 N.L. Cargile, R. Borchert and J.D. McChesney, *Anal. Biochem.,* 97 (1979) 331.
236 R. Eglinton, R.J. Hamilton, R. Hodges and R.A. Raphael, *Chem. Ind. (London),* (1959) 955.
237 A.F. Shlyakhov, R.I. Koreshkova and M.S. Telkova, *J. Chromatogr.,* 104 (1975) 337.
238 D.R. Body, *Lipids,* 12 (1977) 204.
239 C.L. Bower and J.W. Rowe, *Phytochemistry,* 6 (1967) 151.
240 T. Norin and L. Westfelt, *Acta Chem. Scand.,* 17 (1963) 1828.
241 R.M. Carman, *Aust. J. Chem.,* 19 (1966) 1535.
242 R.M. Carman and N. Dennis, *Aust. J. Chem.,* 20 (1967) 157.
243 R.M. Carman and N. Dennis, *Aust. J. Chem.,* 20 (1967) 163.

244 R.T. Aplin, R.C. Cambie and P.S. Rutledge, *Phytochemistry*, 2 (1963) 205.

245 M. Kates, A.J. Hancock and R.G. Ackman, *J. Chromatogr. Sci.*, 15 (1977) 177.

246 B. Pileire, P. Beaune, M.H. Laudat and P. Cartier, *J. Chromatogr.*, 182 (1980) 269.

247 J.A. Hudy, *Anal. Chem.*, 31 (1959) 1754.

248 C.R. Enzell and B.R. Thomas, *Acta Chem. Scand.*, 19 (1965) 913.

249 B.R. Thomas, *Acta Chem. Scand.*, 20 (1966) 1074.

250 F.H.M. Nestler and D.F. Zinkel, *Anal. Chem.*, 39 (1967) 1118.

251 D.F. Zinkel, M.B. Lathrop and L.C. Zank, *J. Gas Chromatogr.*, 6 (1968) 158.

252 D.F. Zinkel, L.C. Zank and M.F. Wesolowksi, *Diterpene Resin Acids—A Compilation of Infrared, Mass, Nuclear Magnetic Resonance, Ultraviolet Spectra and Gas Chromatographic Retention Data,* USDA Forest Service, Madison, WI, 1971.

253 I.I. Bardyshev and A.N. Bulgakov, *Kromatogr. Anal. Khim. Drev.*, (1975) 101.

254 D.F. Zinkel and C.C. Engler, *J. Chromatogr.*, 136 (1977) 245.

255 K. Mori, M. Matsui, N. Ikekawa and Y. Sumiki, *Tetrahedron Lett.*, (1966) 3395.

256 G. Valkanas and N. Iconomou, *Pharm. Acta Helv.*, 41 (1966) 209.

257 B. Holmbom, E. Avela and S. Pekkala, *J. Amer Oil Chem. Soc.*, 51 (1974) 397.

258 B. Holmbom, *J. Amer Oil Chem. Soc.*, 54 (1977) 289.

259 J.C. Hasson and M.V. Kulkarni, *Anal. Chem.*, 44 (1972) 1586.

260 K. Jacob, W. Vogt, A. Clauss, G. Schwertfeger and M. Knedel, in A.P. De Leenheer, R.R. Roncucci and C. Van Peteghem (Editors), *Quantitative Mass Spectrometry in Life Sciences II*, Elsevier, Amsterdam, 1978, p. 201.

261 J. Borthwick, C.J.W. Brooks, W.W. Reid and R.A. Russell, *Ann. Tab.*, 12 (1975) 19.

262 N. Takahashi, *Seikagaku*, 50 (1978) 184.

263 P. Hedden, *Proc. Plant Growth Regul. Work. Group*, 5 (1978) 33.

264 P. Gaskin and J. MacMillan, in J.R. Hillman (Editor), *Isolation of Plant Growth Substances*, Cambridge Univ. Press, Cambridge, 1978, p. 79.

265 G. Kuellertz, H. Eckert and G. Schilling, *Biochem. Physiol. Pflanz.*, 173 (1978) 186.

266 L. Rappaport and D. Adams, *Phil. Trans. Roy. Soc. London*, B284 (1978) 521.

267 D.R. Reeve, A. Crozier, R.C. Durley, D.M. Reid and R.P. Pharis, *Plant Physiol.*, 55 (1975) 42.

268 R.L. Wample, R.C. Durley and R.P. Pharis, *Physiol. Plant*, 35 (1975) 273.

269 S.B. Challen and M. Kučera, *J. Chromatogr.*, 32 (1968) 53.

270 F.W. Blechinger, *Fette, Seifen, Anstrichm.*, 72 (1974) 275.

271 R. Ekman, *Phytochemistry*, 19 (1980) 321.

272 H.J. Nicholas, *Biochim. Biophys. Acta*, 84 (1964) 80.

273 M. Ruddat, E. Heftmann and A. Lang, *Arch. Biochem. Biophys.*, 110 (1965) 496.

274 A. Sato, A. Ogiso and H. Kuwano, *Phytochemistry*, 19 (1980) 2267.

275 R.K. Bansal, M.C. Garcia-Alvarez, K.C. Joshi, B. Rodriguez and R. Patni, *Phytochemistry*, 19 (1980) 1979.

276 F.J. Evans, R.J. Schmidt and A.D. Kinghorn, *Biomed. Mass Spectrom.*, 2 (1975) 126.

277 M. Hergenhahn, W. Adolf and E. Hecker, *Tetrahedron Lett.*, (1975) 1595.

278 T. Fujita, I. Masuda, S. Takao and E. Fujita, *J. Chem. Soc. (Perkin I)*, (1976) 2098.

279 M.C. Garcia-Alvarez and B. Rodriguez, *Phytochemistry*, 19 (1980) 2405.

280 D.T. Dennis and C.A. West, *J. Biol. Chem.*, 242 (1967) 3293.

281 D.R. Robinson and C.A. West, *Biochemistry*, 9 (1970) 70.

282 J.H. Baxter and G.W. Milne, *Biochim. Biophys. Acta*, 176 (1969) 265.

283 J. MacMillan and P. Suter, *Nature (London)*, 197 (1962) 790.

284 L.G. Paleg, *Ann. Rev. Plant Physiol.*, 16 (1965) 291.

285 K.C. Jones, *J. Chromatogr.*, 52 (1970) 512.

286 M.-T. LePage-Degivry, G. Garello, A. Isaia, P. Barthe and C. Bulard, *Physiol. Veg.*, 15 (1977) 343.

287 L.J. Davies and L. Rappaport, *Plant Physiol.*, 55 (1975) 620.

288 E. Heftmann and G.A. Saunders, *J. Liquid Chromatogr.*, 1 (1978) 333.

289 J.E. Graebe, D.T. Dennis, C.D. Upper and C.A. West, *J. Biol. Chem.*, 240 (1965) 1847.

290 V. Benesova, I. Benes, H.M. Chau and V. Herout, *Collect. Czech. Chem. Commun.*, 40 (1975) 658.

291 P. Ruedi and C.H. Eugster, *Helv. Chim. Acta*, 58 (1975) 1899.

292 W. Herz and R.P. Sharma, *J. Org. Chem.*, 40 (1975) 192.

293 Y. Asakawa, M. Toyota and T. Takemoto, *Phytochemistry*, 19 (1980) 1799.

294 T. Okuda, T. Yoshida, S. Koike and N. Toh, *Phytochemistry*, 14 (1975) 509.

295 A.J. Aasen, J.R. Hlubucek and C.R. Enzell, *Acta Chem. Scand.*, B29 (1975) 589.

296 Y. Gafni and I. Shechter, *Anal. Biochem.*, 92 (1979) 248.

297 D.F. Zinkel and J.W. Rowe, *Anal. Chem.*, 36 (1964) 1160.

298 G.H.R. Rao, A.A. Jachimowicz and J.G. White, *J. Chromatogr.*, 96 (1974) 151.

299 G.K. Trivedi, I. Kubo and T. Kubota, *J. Chromatogr.*, 179 (1979) 219.

300 I. Ganjian, I. Kubo and T. Kubota, *J. Chromatogr.*, 200 (1980) 250.

301 D.H. Bowen, A. Crozier, J. MacMillan and D.M. Reid, *Phytochemistry*, 12 (1973) 2935.

302 J.L. Glenn, C.C. Kuo, R.C. Durley and R.P. Pharis, *Phytochemistry*, 11 (1972) 345.

303 D.R. Reeve and A. Crozier, *Phytochemistry*, 15 (1976) 791.

304 G. Gräbner, G. Schneider and G. Sembdner, *J. Chromatogr.*, 121 (1976) 110.

305 G. Schneider, G. Sembdner, I. Focke and K. Schreiber, *Phytochemistry*, 10 (1971) 3009.

306 L.E. Powell and K.J. Tautvydas, *Nature (London)*, 213 (1967) 292.

307 M.G. Carnes and M.L. Brenner, in C.R. Andersen (Editor), *Proc. 8th Int. Conf. Plant Growth Substances*, Hirokawa, Tokyo, 1974, p. 99.

308 A. Crozier and D.R. Reeve, in P.E. Pilet (Editor), *Proc. 9th Int. Conf. Plant Growth Substances*, Springer, Berlin, 1977, p. 67.

309 P.B. Sweetser, *Proc. Plant Growth Regul. Work. Group*, 5 (1978) 1.

310 E. Heftmann, G.A. Saunders and W.F. Haddon, *J. Chromatogr.*, 156 (1978) 71.

311 R.O. Morris and J.B. Zaerr, *Anal. Lett.*, 11 (1978) 73.

312 D.R. Reeve and A. Crozier, in J.R. Hillman (Editor), *Isolation of Plant Growth Substances*, Cambridge Univ. Press, Cambridge, 1978, p. 41.

313 M.G. Jones, J.D. Metzger and J.A.D. Zeevaart, *Plant Physiol.*, 65 (1980) 218.

314 D.R. Reeve, T. Yokota, L.J. Nash and A. Crozier, *J. Exp. Bot.*, 27 (1976) 1243.

315 G.A. Cordell, *Phytochemistry*, 13 (1974) 2343.

316 G.A. Cordell, *Progr. Phytochem.*, 4 (1977) 209.

317 N. Ikekawa, *Methods Enzymol.*, 15 (1969) 200.

318 T. Ohmoto, M. Ikuse and S. Natori, *Phytochemistry*, 9 (1970) 2137.

319 H.E. Nordby and S. Nagy, *J. Chromatogr.*, 75 (1973) 187.

320 G.A. Fokina and N.V. Belova, *Khim. Prir. Soedin.*, 11 (1975) 735.

321 B. Wilkomirski and Z. Kasprzyk, *J. Chromatogr.*, 103 (1975) 376.

322 R. Ikan and R. Gottlieb, *Isr. J. Chem.*, 8 (1970) 685.

323 T. Itoh, T. Uetsuki, T. Tamura and T. Matsumoto, *Lipids*, 15 (1980) 407.

324 J.G. Pym, J.E. Ray, G.W. Smith and E.V. Whitehead, *Anal. Chem.*, 47 (1975) 1617.

325 E.J. Gallegos, *Anal. Chem.*, 47 (1975) 1524.

326 Y. Tsuda, K. Isobe, S. Fukushima, H. Ageta and K. Iwata, *Tetrahedron Lett.*, (1967) 23.

327 B.P. Lisboa, *Methods Enzymol.*, 15 (1969) 3.

328 R. Ikan, J. Kashman and E.D. Bergmann, *J. Chromatogr.*, 14 (1964) 275.

329 C. Anding, R.D. Brandt and G. Ourisson, *Eur. J. Biochem.*, 24 (1971) 259.

330 A.A. Qureshi, E. Beytia ad J.W. Porter, *J. Biol. Chem.*, 248 (1973) 1848.

331 R. Tschesche, F. Lampert and G. Snatzke, *J. Chromatogr.*, 5 (1961) 217.

332 J. Zieliński and J. Konopa, *J. Chromatogr.*, 36 (1968) 540.

333 Z. Kasprzyk and J. Pyrek, *Phytochemistry*, 7 (1968) 1631.

334 R. Ikan, *J. Chromatogr.*, 17 (1965) 591.

335 F. Fischer and R. Hertel, *J. Chromatogr.*, 38 (1968) 274.

336 B.O. Lindgren and C.M. Svahn, *Acta Chem. Scand.*, 20 (1966) 1763.

337 A.M. Osman, M.E.-G. Younes and F.M. Ata, *Bull. Chem. Soc. (Japan)*, 47 (1974) 2056.

338 B. Gestetner, *J. Chromatogr.*, 13 (1964) 259.

339 J.C. Schabort and D.J.J. Potgieter, *J. Chromatogr.*, 31 (1969) 235.

340 R.F. Severson, J.J. Ellington, R.F. Arrendale and M.E. Snook, *J. Chromatogr.*, 160 (1978) 155.

341 C. Betancor, R. Freire, A.G. Gonzalez, J.A. Salazar, C. Pascard and T. Prange, *Phytochemistry*, 19 (1980) 1989.

342 W. Lawrie, J. McLean and M.E.-G. Younes, *Chem. Ind. (London)*, (1966) 1720.
343 S.H. Rizvi, A. Shoeb, R.S. Kapil and S.P. Popli, *Phytochemistry*, 19 (1980) 2409.
344 S.S. Kang and W.S. Woo, *J. Nat. Prod.*, 43 (1980) 510.
345 D.A. Mulholland and D.A.H. Taylor, *Phytochemistry*, 19 (1980) 2421.
346 K. Seifert, H. Budzikiewicz and K. Schreiber, *Pharmazie*, 31 (1976) 81.
347 J.F. Fisher, *J. Agr. Food Chem.*, 23 (1975) 1199.
348 S. Amagaya and Y. Ogihara, *Shoyaku Zasshi*, 33 (1979) 166.
349 E. Heftmann, R.E. Lundin, W.F. Haddon, I. Peri, U. Mor and A. Bondi, *J. Nat. Prod.*, 42 (1979) 410.
350 S.L. Smith, J.W. Jorgenson and M. Novotný, *J. Chromatogr.*, 187 (1980) 111.
351 J.-T. Lin, W.D. Nes and E. Heftmann, *J. Chromatogr.*, 207 (1981) 457.
352 J.F. Fisher, *J. Agr. Food Chem.*, 26 (1978) 497.
353 S.K. Hajbrahim, P.J.C. Tibbets, C.D. Watts, J.R. Maxwell, G. Eglinton, H. Colin and G. Guiochon, *Anal. Chem.*, 50 (1978) 549.
354 J.E. Paanakker and G.M. Hallegraeff, *Comp. Biochem. Physiol.*, 60B (1978) 51.
355 K. Iriyama, M. Yoshiura and M. Shiraki, *J. Chromatogr.*, 154 (1978) 302.
356 J.K. Abaychi and J.P. Riley, *Anal. Chim. Acta*, 107 (1979) 1.
357 D.W. DeJong and W.G. Woodlief, *J. Agr. Food Chem.*, 26 (1978) 1281.
358 G. Calabro, G. Micali and P. Curro, *Essenze Deriv. Agrum.*, 48 (1978) 359.
359 T. Braumann and L.H. Grimme, *J. Chromatogr.*, 170 (1979) 264.
260 C. Puglisi and J.A.F. de Silva, *J. Chromatogr.*, 120 (1976) 457.
361 M. Zakaria, K. Simpson, P.R. Brown and A. Krstulovic, *J. Chromatogr.*, 176 (1979) 109.
362 H. Pfander, H. Schurtenberger and V.R. Meyer, *Chimia*, 34 (1980) 179.
363 A. Fiksdahl, J.T. Mortensen and S. Liaaen-Jensen, *J. Chromatogr.*, 157 (1978) 111.
364 H. Pfander, F. Haller, F.J. Levenberger and H. Thommen, *Chromatographia*, 9 (1976) 630.
365 T. Suzuki and K. Hasegawa, *Agr. Biol. Chem.*, 38 (1974) 871.
366 K. Hasegawa, *Methods Enzymol.*, 67 (1980) 261.
367 S. Wieckowski and M. Droba, *Plant Sci. Lett.*, 13 (1978) 397.
368 A. Gimeno, *J. Ass. Offic. Anal. Chem.*, 63 (1980) 653.
369 A.K. Baloch, K.A. Buckle and R.A. Edwards, *J. Chromatogr.*, 139 (1977) 149.
370 K.A. Buckle and F.M.M. Rahman, *J. Chromatogr.*, 171 (1979) 385.
371 M. Piasecki, *Rocz. Akad. Roln. Poznaniu*, 89 (1976) 125.
372 E. Siklosi-Rajki and L. Farago, *Acta Aliment. Acad. Sci. Hung.*, 3 (1974) 389.
373 S.M. Petrović, L.A. Kolarov and N.U. Perišić-Janjić, *J. Chromatogr.*, 171 (1979) 522.
374 H.T. McKone, *J. Chem. Educ.*, 56 (1979) 676.
375 I. Csorba, Z. Buzás, B. Polyák and L. Boross, *J. Chromatogr.*, 172 (1979) 287.
376 M. Shirak, M. Yoshiura and K. Iriyama, *Chem. Lett.*, (1978) 103.
377 J. Washuettl, *Qual. Plant. Plant Foods Hum. Nutr.*, 23 (1974) 369.
378 H. Thielemann, *Z. Anal. Chem.*, 271 (1974) 285.
379 J. Sherma and M. Latta, *J. Chromatogr.*, 154 (1978) 73.
380 J. Cross, *Chromatographia*, 13 (1980) 572.
381 R.F. Taylor and B.H. Davies, *J. Chromatogr.*, 103 (1975) 327.
382 R.F. Taylor, *Methods Enzymol.*, 67 (1980) 233.
383 F.W. Hemming, in T.W. Goodwin (Editor), *Biochemistry of Lipids*, MTP Intern. Rev. Sci., Vol. 4, Butterworths, London, 1974, p. 39.
384 C.J. Waechter and W.J. Lennarz, *Ann. Rev. Biochem.*, 45 (1976) 95.
385 F.W. Hemming, *Phil. Trans. Roy. Soc. London*, B 284 (1978) 559.
386 F.W. Hemming, in T.W. Goodwin (Editor), *Natural Substances Formed Biologically from Mevalonic Acid*, Academic Press, New York, 1970, p. 105.
387 B.O. Lindgren, *Acta Chem. Scand.*, 19 (1965) 1317.
388 A.R. Wellburn and F.W. Hemming, *J. Chromatogr.*, 23 (1968) 51.
389 I.F. Durr and M.Z. Habbal, *Biochem. J.*, 127 (1972) 345.
390 K.J. Stone and F.W. Hemming, *Biochem. J.*, 104 (1967) 43.
391 L. DeLuca, G. Rosso and G. Wolf, *Biochem. Biophys. Res. Commun.*, 41 (1970) 615.
392 J.P. Frot-Coutaz, C.S. Silverman-Jones and L.M. DeLuca, *J. Lipid Res.*, 17 (1976) 200.

393 H. Sagami, K. Ogura and S. Seto, *Biochemistry*, 16 (1977) 4616.
394 P.J. Dunphy, J.D. Kerr, J.F. Pennock, K.J. Whittle and J. Feeney, *Biochim. Biophys. Acta*, 136 (1967) 136.
395 A.R. Wellburn, J. Stevenson, F.W. Hemming and R.A. Morton, *Biochem. J.*, 102 (1967) 313.
396 T. Suga, T. Shishibori and K. Nakaya, *Phytochemistry*, 19 (1980) 2327.
297 D.P. Gough, A.L. Kirby, J.B. Richards and F.W. Hemming, *Biochem, J.*, 118 (1970) 167.
398 C.M. Allen, M.V. Keenan and J. Sack, *Arch. Biochem. Biophys.*, 175 (1976) 236.
399 W. Sasak and T. Chojnacki, *Arch. Biochem. Biophys.*, 181 (1977) 402.
400 T. Baba and C.M. Allen, *Biochemistry*, 17 (1978) 5598.
401 K. Hannus and G. Pensar, *Phytochemistry*, 13 (1974) 2563.
402 A.S. Beedle, M.J. Walton and T.W. Goodwin, *Insect Biochem.*, 5 (1975) 465.
403 K.J. Stone, A.R. Wellburn, F.W. Hemming and J.F. Pennock, *Biochem. J.*, 102 (1967) 325.
404 W.Sasak, T. Mankowksi, T. Chojnacki and W.M. Daniewski, *FEBS Lett.*, 64 (1976) 55.
405 W. Jankowksi, T. Mankowksi and T. Chojnacki, *Acta Biochim. Polon.*, 22 (1975) 67.
406 T. Chojnacki, W. Jankowski, T. Mankowksi and W. Sasak, *Anal. Biochem.*, 69 (1975) 114.
407 D.J. Freeman, C.A. Rupar and K.K. Carroll, *Lipids*, 15 (1980) 191.
408 D.N. Skilleter and R.G. Kekwick, *Anal. Biochem.*, 20 (1967) 171.
409 R. Goldman and J.L. Strominger, *J. Biol. Chem.*, 247 (1972) 5116.
410 A. Parodi, N.H. Behrens, L.F. Leloir and M. Dankert, *Biochim. Biophys. Acta*, 270 (1972) 529.
411 R. Pont-Lezica, C.T. Brett, P.R. Martinez and M.A. Dankert, *Biochem. Biophys. Res. Commun.*, 66 (1975) 980.
412 R.E. Dugan, E. Rasson and J.W. Porter, *Anal. Biochem.*, 22 (1968) 249.
413 E. Garcia-Peregrin, M.D. Suarez, M.C. Aragon anf F. Mayor, *Phytochemistry*, 11 (1972) 2495.
414 B.H. Davies, A.F. Rees and R.F. Taylor, *Phytochemistry*, 14 (1975) 717.
415 M. Scher, W.J. Lennarz and C.C. Sweeley, *Proc. Nat. Acad. Sci. U.S.*, 59 (1968) 1313.
416 J.S. Tkacz, A. Herscovic, C.D. Warren and R.W. Jeanloz, *J. Biol. Chem.*, 249 (1974) 6372.
417 L.M. DeLuca, J.P. Frot-Coutaz, C.S. Silverman-Jones and P.R. Roller, *J. Biol.Chem.*, 252 (1977) 2575.
418 S.C. Kushwaha, M. Kates, R.L. Renaud and R.E. Subden, *Lipids*, 13 (1978) 352.
419 J.P. Frot-Coutaz and L.M. DeLuca, *Biochem. J.*, 159 (1976) 799.
420 H. Sagami, K. Ogura and S. Seto, *Biochem. Biophys. Res. Commun.*, 85 (1978) 572.
421 N.H. Behrens and L.F. Leloir, *Proc. Nat. Acad. Sci. U.S.*, 66 (1970) 153.
422 K. Stone and J.L. Strominger, *J. Biol. Chem.*, 247 (1972) 5107.
423 R.M. Barr and F.W. Hemming, *Biochem. J.*, 126 (1972) 1203.
424 D.P. Gough and F.W. Hemming, *Biochem. J.*, 117 (1970) 309.
425 M.V. Keenan and C.M. Allen, *Arch. Biochem. Biophys.*, 161 (1974) 375.
426 H. Pleininger and H. Immel, *Chem. Ber.*, 98 (1965) 414.
427 R.H. Cornforth and G. Popják, *Methods Enzymol.*, 15 (1969) 359.
428 M. Lahov, T.H. Chin and W.J. Lennarz, *J. Biol. Chem.*, 244 (1969) 5890.
429 C.D. Upper and C.A. West, *J. Biol. Chem.*, 242 (1967) 3285.
430 F.M. Laskovics, and C.D. Poulter, *Biochemistry*, 20 (1981) 1893.
431 K. Bloch, S. Chaykin, A.H. Phillips and A. DeWaard, *J. Biol. Chem.*, 234 (1959) 2595.
432 G. Suzue, K. Orihara, H. Morishita and S. Tanaka, *Radioisotope*, 13 (1964) 300.
433 P. Valenzuela, E. Beytia, O. Cori and A. Yudelevich, *Arch. Biochem. Biophys.*, 113 (1966) 536.
434 D.M. Logan, *J. Lipid Res.*, 13 (1972) 137.
435 W.W. Epstein and H.C. Rilling, *J. Biol. Chem.*, 245 (1970) 4597.
436 B.K. Tidd, *J. Chem. Soc. B,*, (1971) 1168.
437 M.A. Dankert, A. Wright, W.S. Kelley and P.W. Robbins, *Arch. Biochem. Biophys.*, 116 (1966) 425.
438 T. Uematsu and R.J. Suhadolnik, *J. Chromatogr.*, 123 (1976) 347.
439 P.V. Bhat, L.M. DeLuca and M.L. Wind, *Anal. Biochem.*, 102 (1980) 243.
440 T. Fukuda, T. Nakamura and S. Ohashi, *J. Chromatogr.*, 128 (1976) 212.
441 G. Del Campo, J. Puente, M.A. Valenzuela, A. Traverso-Cori and O. Cori, *Biochem. J.*, 167 (1977) 525.

442 J.M. Bobbitt and K.-P. Segebarth, in W.I. Taylor and A.R. Battersby (Editors), *Cyclopentanoid Terpene Derivatives,* Dekker, New York, 1969, p. 1.
443 O. Sticher and V. Junod-Busch, *Pharm. Acta Helv.,* 50 (1975) 127.
444 R.S. Chandel and R.P. Rastogi, *Phytochemistry,* 19 (1980) 1889.
445 C.J. Coscia, L. Botta and R. Guarnaccia, *Arch. Biochem. Biophys.,* 136 (1970) 498.
446 I. Sakata and T. Mitsui, *Agr. Biol. Chem.,* 39 (1975) 1329.
447 O. Sticher and A. Weisflog, *Pharm. Acta Helv.,* 50 (1975) 394.
448 A. Bianco, M. Guiso, C. Iavarone and C. Trogolo, *Gazz. Chim. Ital.,* 105 (1975) 175, 185, and 195.
449 Y. Takeda, H. Nishimura and H. Inouye, *Chem. Pharm. Bull.,* 24 (1976) 1216.
450 G. Kinast and L.-F. Tietze, *Chem. Ber.,* 109 (1976) 3640.
451 E.M. Martinelli, *Eur. J. Mass Spectrom. Biochem. Med. Environ. Res.,* 1 (1980) 33.
452 H. Inouye, K. Uobe, M. Hirai, Y. Masada and K. Hashimoto, *J. Chromatogr.,* 118 (1976) 201.
453 T. Furoya, *J. Chromatogr.,* 18 (1965) 152.
454 C.J. Coscia, R. Guarnaccia and L. Botta, *Biochemistry,* 8 (1969) 5036.
455 G. Schneider, S. Jänicke and G. Sembdner, *J. Chromatogr.,* 109 (1975) 409.
456 H. Inouye, S. Ueda, Y. Aoki and Y. Takeda, *Chem. Pharm. Bull.,* 20 (1972) 1287.
457 D.V. Banthorpe and J. Mann, *Phytochemistry,* 11 (1971) 2589.
458 R. Croteau and C. Martinkus, *Plant Physiol.,* 64 (1979) 169.
459 K. Schreiber, J. Weiland and G. Sembdner, *Tetrahedron,* 25 (1969) 5541.
460 Z.A. Wojciechowski, *Phytochemistry,* 14 (1975) 1749.
461 G.W. Francis, S. Hertzberg, K. Andersen and S. Liaaen-Jensen, *Phytochemistry,* 9 (1968) 629.
462 T. Mankowski, W. Sasak, E. Janczura and T. Chojnacki, *Arch. Biochem. Biophys.,* 181 (1977) 393.
463 A.K.A. Kerr and F.W. Hemming, *Eur. J. Biochem.,* 83 (1978) 581.
464 S. Yamamori, N. Murazumi, Y. Araki and E. Ito, *J. Biol. Chem.,* 253 (1978) 6516.
465 R. Tschesche, F. Ciper and E. Breitmaier, *Chem. Ber.,* 110 (1977) 3111.
466 B. Meier and O. Sticher, *J. Chromatogr.,* 138 (1977) 453.
467 S. Hasegawa and Y. Hirose, *Phytochemistry,* 19 (1980) 2479.
468 J. Sakakibara and N. Shirai, *Phytochemistry,* 19 (1980) 2159.
469 I. Yamaguchi, T. Yokota, S. Yoshida and N. Takahashi, *Phytochemistry,* 18 (1979) 1699.
470 S.E. Chen and E.J. Staba, *Lloydia,* 41 (1978) 361.
471 F. Erni and R.W. Frei, *J. Chromatogr.,* 149 (1978) 561.
472 F. Erni and R.W. Frei, *J. Chromatogr.,* 130 (1977) 169.
473 B. Abegaz and B. Tecle, *Phytochemistry,* 19 (1980) 1553.
474 R.D. Tripathi and K.P. Tiwari, *Phytochemistry,* 19 (1980) 2163.
475 K.P. Tiwari, S.D. Srivastava and S.K. Srivastava, *Phytochemistry,* 19 (1980) 980.
476 I. Kitagawa, K.S. Im and I. Yosioka, *Chem. Pharm. Bull.,* 24 (1976) 1260.
477 S. Yahara, O. Tanaka and T. Komori, *Chem. Pharm. Bull.,* 24 (1976) 2204.
478 S. Shany, B. Gestetner, Y. Birk, A. Bondi and I. Kirson, *Isr. J. Chem.,* 10 (1972) 881.
479 W. Jankowski, T. Mankowski and T. Chojnacki, *Biochim. Biophys. Acta,* 337 (1974) 153.
480 P. Zatta, *Experientia,* 32 (1976) 693.
481 W.T. Forsee and A.D. Elbein, *Proc. Nat. Acad. Sci. U.S.,* 73 (1976) 2574.
482 E.L. Kean, *J. Biol. Chem.,* 252 (1977) 5622.
483 R.W. Keeman, M. Kruczek and L. Fusinato, *Arch. Biochem. Biophys.,* 167 (1975) 697.
484 J.H. Wieffering, *Phytochemistry,* 5 (1966) 1053.
485 C.J. Waechter, J.L. Kennedy and J.B. Harford, *Arch. Biochem. Biophys.,* 174 (1976) 726.
486 P.A. Peterson, L. Rask, T. Helting, L. Ostberg and Y. Fernstedt, *J. Biol. Chem.,* 251 (1976) 4986.
487 G.R. Daleo and R. Pont-Lezica, *FEBS Lett.,* 74 (1977) 247.
488 T.J. Danielson, E.M. Hawes and C.A. Bliss, *J. Chromatogr.,* 103 (1975) 216.
489 A.R. Battersby, A.R. Burnett and P.G. Parsons, *J. Chem. Soc. C,* (1969) 1187.
490 H. Inouye, S. Ueda and Y. Nakamura, *Chem. Pharm. Bull.,* 18 (1970) 1856.
491 K. Weinges, P. Kloss and W. Henkels, *Justus Liebigs Ann. Chem.,* 759 (1972) 173.
492 K. Hostettmann, M. Hostettmann-Kaldas and O. Sticher, *J. Chromatogr.,* 186 (1979) 529.
493 K. Hostettmann, M. Hostettmann-Kaldas and K. Nakanishi, *J. Chromatogr.,* 170 (1979) 355.
494 K. Hostettmann, M. Hostettmann and O. Sticher, *Helv. Chim. Acta,* 62 (1979) 2079.

495 K. Kawai, T. Akiyama, Y. Ogihara and S. Shibata, *Phytochemistry*, 13 (1974) 2829.
496 S. Yahara, R. Kasai and O. Tanaka, *Chem. Pharm. Bull.*, 25 (1977) 2041.
497 K. Kawai and S. Shibata, *Phytochemistry*, 17 (1978) 287.
498 H. Ishii, K. Tori, T. Tozyo and Y. Yoshimura, *Chem. Pharm. Bull.*, 26 (1978) 674.
499 Y. Ogihara, O. Inoue, H. Otsuka, K.-I. Kawai, T. Tanimura and S. Shibata, *J. Chromatogr.*, 128 (1978) 218.
500 G.W.T. Groenendijk, P. Jansen, S. Bonting and F. Daemen, *Methods Enzymol.*, 67 (1980) 203.
501 A. McCormick, J. Napoli and H. DeLuca, *Methods Enzymol.*, 67 (1980) 220.
502 M. Arnaboldi, M. Motto, K. Tsujimoto, V. Balogh-Nair and K. Nakanishi, *J. Amer. Chem. Soc.*, 101 (1979) 7082.
503 M.A. Adams and K. Nakanishi, *J. Liquid Chromatogr.*, 2 (1979) 1097.
504 Y. Takashima, T. Nakajima, S. Tanaka, M. Washitake, T. Anmo and H. Matsumaru, *Chem. Pharm. Bull.*, 27 (1979) 1553.
505 G.W.T. Groenendijk, W. de Grip and F. Daemen, *Anal. Biochem.*, 99 (1979) 304.
506 H. Pařízková and J. Blattná, *J. Chromatogr.*, 191 (1980) 301.
507 Y.K. Fung, G.R. Rahwan and R. A. Sams, *J. Chromatogr.*, 147 (1978) 528.
508 P. Dunagin and J. Olson, *Methods Enzymol.*, 15 (1969) 289.
509 T.-C. Chiang, *J. Chromatogr.*, 182 (1980) 335.
510 S. Reeder and G.L. Park, *J. Ass. Offic. Anal. Chem.*, 58 (1975) 595.
511 A. Mariani and R. Guanitolini, *Bull. Chim. Farm.*, 114 (1975) 626.
512 D.B. Dennison and J.R. Kirk, *J. Food Sci.*, 42 (1972) 1376.
513 J.E. Paanakker and G.W.T. Groenendijk, *J. Chromatogr.*, 168 (1979) 125.
514 D. Egberk, J. Heroff and R. Potter, *J. Agr. Food Chem.*, 25 (1977) 1127.
515 M. De Ruyter and A. De Leenheer, *Clin. Chem.*, 22 (1976) 1593.
516 K. Tsukida, A. Kodama and M. Ito, *J. Nutr. Sci. Vitaminol.*, 24 (1978) 593.
517 I. Macleod and R. Wiggins, *Proc. Anal. Div. Chem. Soc.*, 15 (1978) 329.
518 A. Roberts, M. Nichols, C. Frolik, D. Newton and M. Sporn, *Cancer Res.*, 38 (1978) 3327.
519 A.P. De Leenheer, V.O.R.C. De Bevere, M.G.M. De Ruyter and A.E. Claeys, *J. Chromatogr.*, 162 (1979) 408.
520 M. De Ruyter and A. De Leenheer, *Clin. Chem.*, 24 (1978) 1920.
521 R. Kwok, T. Hudson and S. Subramanian, *Appl. HPLC Meth. Det. Fat. Sol. Vitam. A, D, E, K Food Pharm.*, *Symp. Proc.*, Assoc. Vitam. Chem., Chicago, IL, 1978, p. 34.
522 W.P. Rose, *Appl. HPLC Meth. Det. Fat. Sol. Vitam. A, D, E, K Foods Pharm.*, *Symp. Proc.*, Assoc. Vitam. Chem., Chicago, IL, 1978, p. 109.
523 M.-H. Bui-Nguyen and B. Blanc, *Experientia*, 36 (1980) 374.
524 P. Soderhjeld and B. Andersson, *J. Sci. Food Agr.*, 29 (1978) 697.
525 S. Henderson and A. Wickroski, *Appl. HPLC Meth. Det. Fat. Sol. Vitam. A, D, E, K Foods Pharm.*, *Symp. Proc.*, Assoc. Vitam. Chem., Chicago, IL, 1978, p. 48.
526 R.M. McKenzie, D.M. Hellwege, M.L. McGregor, N.L. Rockley, P.J. Riquetti and E.C. Nelson, *J. Chromatogr.*, 155 (1978) 379.
527 M. De Ruyter and A. De Leenheer, *Anal. Chem.*, 51 (1979) 43.
528 M. Eriksson, T. Eriksson and B. Sorensen, *Acta Pharm. Suecica*, 15 (1978) 274.
529 G. Groenendijk, W. de Grip and F. Daemen, *Biochim. Biophys. Acta*, 617 (1980) 430.
530 C. Frolik, T. Tavela and M. Sporn, *J. Lipid Res.*, 19 (1978) 32.
531 C. Frolik, T. Tavela, G. Peck and M. Sporn, *Anal. Biochem.*, 86 (1978) 743.
532 C. Puglisi and J.A.F. de Silva, *J. Chromatogr.*, 152 (1978) 421.
533 B.A. Halley and E.C. Nelson, *J. Chromatogr.*, 175 (1979) 113.
534 K. Aitzetmueller, J., Pilz and R. Tasche, *Fette, Seifen, Anstrichm.*, 81 (1979) 40.
535 R. Fankel, *Mikrochim. Acta*, 1 (1978) 359.
536 R. Hänni, D. Hervouet and A. Busslinger, *J. Chromatogr.*, 162 (1979) 615.
537 J. Thompson and W. Maxwell, *J. Ass. Offic. Anal. Chem.*, 60 (1977) 766.
538 K. Tsukida, A. Kodama and M. Ito, *J. Chromatogr.*, 134 (1977) 331.
539 K. Tsukida, A. Kodama, M. Ito and M. Kawamoto, *J. Nutr. Sci. Vitaminol.*, 23 (1977) 263.
540 J. Rotmans and A. Kropf, *Vision Res.*, 15 (1975) 1301.

541 K. Tsukida, R. Masahara and M. Ito, *J. Chromatogr.*, 192 (1980) 395.

542 S. Barnett and L. Frick, *Anal. Chem.*, 51 (1979) 641.

543 A. McCormick, J. Napoli and H. DeLuca, *Anal. Biochem.*, 86 (1978) 25.

544 W. Landen and R. Eitenmiller, *J. Ass. Offic. Anal. Chem.*, 62 (1979) 283.

545 W. Landen, *J. Ass. Offic. Anal. Chem.*, 63 (1980) 131.

546 M. Holasová and J. Blattná, *J. Chromatogr.*, 123 (1976) 225.

547 Y. Ito, M. Zile, H. Ahrens and H. DeLuca, *J. Lipid Res.*, 15 (1974) 517.

548 T. Maruyama, T. Ushigusa, H. Kanematsu, I. Niiya and M. Imamura, *Shokuhin Eiseigaku Zasshi,* 18 (1977) 487.

549 J.C. Bieri, R.K.H. Poukka and E.L. Prival, *J. Lipid Res.*, 11 (1970) 118.

550 S.R. Ames, *J. Ass. Offic. Anal. Chem.*, 54 (1971) 1.

551 R.H. Bunnel, *Lipids,* 6 (1971) 245.

552 A.J. Sheppard, A.R. Prosser and W.D. Hubbard, *J. Amer. Oil. Chem. Soc.*, 49 (1972) 619.

553 R. Stillman and T.S. Ma, *Mikrochim. Acta,* (1974) 641.

554 A.A. Christie, *Chem. Ind. (London),* (1975) 492.

555 J. Davídek, in Z. Deyl, K. Macek and J. Janák (Editors), *Liquid Column Chromatography. A Survey of Modern Techniques and Applications,* J. Chromatogr. Library Series, Vol. 3, Elsevier, Amsterdam, 1975, p. 953.

556 J.N. Thompson, *Appl. HPLC Meth. Det. Fat. Sol. Vitam. A, D, E, K Foods Pharm., Symp. Proc.,* Assoc. Vitam. Chem., Chicago, IL, 1978, p. 62.

557 R.A. Wiggins, in T.G. Taylor (Editor), *Proc. 4th Kellogg Nutr. Symp.,* MTP Int. Rev. Sci., Butterworths, Lancaster, 1979, p. 73.

558 T. Aratani, Y. Asakawa and J.C. Touchstone, in J. Sherma (Editor), *Tocopherols,* Wiley, New York, 1979, p. 695.

559 G. Cheralambous (Editor), *Liquid Chromatographic Analysis of Foods and Beverages,* Academic Press, New York, 1979.

560 M. Gogolewski, M. Nogala and A. Luczynski, *Rocz. Akad. Roln. Poznaniu,* 89 (1976) 59.

561 M. Toulova, *Acta Vet.,* 44 (1975) 163.

562 M. Koleva, M. Dzhoneidi and O. Budevski, *Pharmazie,* 30 (1975) 168.

563 W. Mueller-Mulot, *J. Amer. Oil Chem. Soc.*, 53 (1976) 732.

564 H. Thielemann, *Sci. Pharm.,* 42 (1974) 221.

565 W.V. Dompert and H. Beringer, *Fette, Seifen, Anstrichm.,* 78 (1976) 108.

566 G.F.M. Ball and P.W. Ratcliff, *J. Food Technol.,* 13 (1978) 433.

567 J. Hess, M. Pallansch, K. Harich and G.E. Bunce, *Anal. Biochem.,* 83 (1977) 401.

568 N.-T. Luan, J. Pokorný, J. Čoupek and S. Pokorný, *J. Chromatogr.,* 130 (1977) 378.

569 N. Perisic-Janjic, S. Petrovic and P. Hadzic, *Chromatographia,* 9 (1976) 130.

570 J.W. Strong, *J. Pharm. Sci.,* 65 (1976) 968.

571 F. Ueda, T. Higashi and Y. Ayukawa, *Vitamins,* 48 (1974) 439.

572 F. Mordret and A.M. Laurent, *Rev. Fr. Corps Gras,* 25 (1978) 245.

573 M. Chiarotti and G.V. Giusti, *J. Chromatogr.,* 147 (1978) 481.

574 P.W. Meijboom and G.A. Jongenotter, *J. Amer. Oil. Chem. Soc.*, 56 (1979) 33.

575 D.K. Feeter, *J. Amer. Oil Chem. Soc.*, 51 (1974) 184.

576 P.E. Sheppard and M.J. Stutsmen, *J. Soc. Cosmet. Chem.*, 28 (1977) 115.

577 A.J. Sheppard and W.D. Hubbard, *J. Pharm. Sci.,* 68 (1979) 98.

578 J. Lehman, *Lipids,* 13 (1978) 616.

579 J. Lehman, *Lipids,* 15 (1980) 135.

580 A.P. De Leenheer, V.O. De Bevere, A.A. Cruyl and A.E. Claeys, *Clin. Chem.*, 24 (1978) 585.

581 S.-N. Lin and E.C. Horning, *J. Chromatogr.,* 112 (1975) 465.

582 J.W. DeVries, D.C. Egberg, J.C. Heroff, in G. Charalambous (Editor), *Liquid Chromatographic Analysis of Foods and Beverages,* Vol. 2. Academic Press, New York, 1979, p. 477.

583 O.H. Rueckemann and K. Ranfft, *Z. Lebensm.-Unters. Forsch.,* 166 (1978) 151.

584 L. Pickston, *N.Z.J. Sci.,* 21 (1978) 383.

585 K. Abe and G. Katsui, *Vitamins,* 49 (1975) 259.

586 W. Widicus and J.R. Kirk, *J. Ass. Offic. Anal. Chem.*, 62 (1979) 637.

587 K. Abe, M. Ohmae, K. Kawabe and G. Katsui, *Vitamins*, 53 (1979) 385.

588 A.P. De Leenheer, V.O. De Bevere and A.E. Claeys, *Clin. Chem.*, 25 (1979) 425.

589 G.T. Vatassery and D.F. Hagen, *Anal. Biochem.*, 79 (1977) 129.

590 C.C. Tangney, J.A. Driskell, J.A. Apffel and H.M. McNair, *VIA, Varian Instr. Appl.*, 13 (1979) 14.

591 C.C. Tangney, J.A. Driskell and H.M. McNair, *J. Chromatogr.*, 172 (1979) 513.

592 J.N. Thompson, G. Hatina and W.B. Maxwell, *Appl. HPLC Meth. Det. Fat. Sol. Vitam. A, D, E, K Foods Pharm*, Assoc. Vit. Chem., Chicago, IL, 1978, p. 84.

593 K. Abe, Y. Yuguchi and G. Katsui, *J. Nutr. Sci. Vitaminol.*, 21 (1975) 183.

594 L. Jansson, B. Nilsson and R. Lindgren, *J. Chromatogr.*, 181 (1980) 242.

595 L. Hatam and H.J. Kayden, *J. Lipid Res.*, 20 (1979) 639.

596 T. Eriksson and B. Sorensen, *Acta Pharm. Suecica*, 14 (1977) 475.

597 G.T. Vatassery, V.R. Maynard and D.F. Hagen, *J. Chromatogr.*, 161 (1978) 299.

598 H. Cohen and M. Lapointe, *J. Agr. Food Chem.*, 26 (1978) 1210.

599 J.G. Bieri, T. Tolliver and G.L. Catignani, *Amer. J. Clin. Nutr.*, 32 (1979) 2143.

600 C.H. McMurray and W.J. Blanchflower, *J. Chromatogr.*, 178 (1979) 525.

601 B. Nilsson, B. Johansson, L. Jansson and L. Holmberg, *J. Chromatogr.*, 145 (1978) 169.

602 J.F. Cavins and G.E. Inglett, *Cereal Chem.*, 51 (1974) 605.

603 J.W. Suttlie, in H.H. Draper (Editor), *Handbook of Lipid Research*, Vol. 1, Plenum, New York, 1978, p. 211.

604 J.M.L. Mee, C.C. Brooks and K.H. Yanagihara, *J. Chromatogr.*, 110 (1975) 178.

605 Y. Nakata, Y. Mita and S. Khono, *Yakugaku Zasshi*, 96 (1976) 53.

606 C.L. Silvi and T. Ioneda, *Rev. Microbiol.*, 8 (1977) 39.

607 B. Rittich, M. Krska, M. Simek and J. Coupek, *Rocz. Nauk. Zootech.*, 5 (1978) 33.

608 P.L. Donnahey, V.T. Burt, H.H. Rees and J.F. Pennock, *J. Chromatogr.*, 170 (1979) 272.

609 M.D. Collins, H.N. Shah and D.E. Minnikin, *J. Appl. Bacteriol.*, 48 (1980) 277.

610 D.O. Mack, *J. Liquid Chromatogr.*, 3 (1980) 1005.

611 J.N. Thompson, G. Hatina and W.B. Maxwell, *Trace Org. Anal.: New Front Anal. Chem.*, NBS Spec. Publ., No. 519 (1979) 279.

612 M.F. Lefevere, A.P. De Leenheer and A.E. Claeys, *J. Chromatogr.*, 186 (1979) 749.

613 Y. Yamano, S. Ikenoya, T. Tsuda and M. Ohmae, *Yakugaku Zasshi*, 97 (1977) 486.

614 Y. Yamano, S. Ikenoya, M. Anze, M. Ohmae and K. Kawaba, *Yakugaku Zasshi*, 98 (1978) 774.

615 O. Hiroshima, K. Abe, S. Ikenoya, M. Ohmae and K. Kawabe, *Yakugaku Zasshi*, 99 (1979) 1007.

616 S.A. Barnett, L.W. Frick and H.M. Baine, *Anal. Chem.*, 52 (1980) 610.

617 M.J. Sheaver, V. Allan, Y. Haroon and P. Barkhan, in J.W. Suttie (Editor), *Proc. 8th Steenbock Symp.*, Univ. Park Press, Baltimore, MD, 1980, p. 317.

Chapter 14

Steroids

ERICH HEFTMANN

CONTENTS

14.1. INTRODUCTION

Having published a book on steroid chromatography [1] in 1976, I believe that it would not be profitable to recapitulate the literature of the 1964–1975 period. Instead, I should like to concentrate on the most recent addition to our armamentarium, HPLC, and treat other chromatographic methods more cursorily. Other references may be found in my book [1] and in other current reviews of steroid chromatography [2–4]. A discussion of the behavior of stereoisomeric steroids in LC is also published [5].

With the exception of a few isolated applications to very polar conjugated steroids, the use of PC in the steroid field has now ended. For rapid, inexpensive comparisons with reference material, TLC continues to be used [6], but research activity has practically come to a standstill. Its usefulness in quantitative and radiometric analysis is limited [7]. Ion exchange and electrophoresis are not useful for the nonionic steroids.

This leaves only GC and LCC. CC of steroids has been reviewed by Procházka [8]; Sjövall and Axelson [9,10] have more recently reviewed the analysis of steroids by gel chromatography and by GC–MS. Nyström and Sjövall [11] have summarized their research on the use of lipophilic Sephadex, and Vestergaard [12] has published a summary of his work on multiple columns. In addition to a book [13], we now have reviews on HPLC of steroids [14] and on natural products, including steroids

References on p. B215

TABLE 14.1

RELATIVE RETENTION TIMES OF STEROLS [26] AND THEIR TMS ETHERS [27]

All columns glass. All supports 80–100 mesh.

Sterols	Free sterols *				TMS ethers ***							
	UCCW-982 **, 265°C	SE-30, 265°C	OV-17, 265°C	OV-25, 265°C	QF-1, 225°C	Dexsil 300 §, 285°C	SE-30, 253°C	OV-17, 265°C	OV-25, 260°C	HI-EFF-8BP, 238°C	Silar 5CP, 240°C	SP-100, 240°C
Cholesterol	1.00	1.00	1.00	1.00	1.00	1.00	1.00	1.00	1.00	1.00	1.00	1.00
5α-Cholestanol	1.01	1.01	1.00	1.00	1.07	1.05	1.00	1.00	1.00	1.00	1.00	0.95
4-Cholestenol	1.00	1.00	1.00	1.00	1.00	1.01	1.00	1.00	1.00	1.00	1.00	1.00
6-Cholestenol	0.99	0.99	1.01	1.01	0.99	1.01	0.99	1.01	1.01	1.01	1.01	1.01
7-Cholestenol	1.11	1.12	1.17	1.18	1.12	1.13	1.12	1.17	1.20	1.22	1.23	1.20
14-Cholestenol	0.99	0.99	0.99	0.99	0.99	1.00	0.99	0.99	0.99	0.99	0.99	0.99
5,7-Cholestadienol	–	–	–	–	1.10	1.09	1.08	1.14	1.17	1.16	1.27	1.22
Desmosterol	1.07	1.08	1.19	1.24	1.08	1.08	1.08	1.28	1.29	1.30	1.31	1.33
Campesterol	1.26	1.29	1.30	1.28	1.33	1.28	1.29	1.30	1.30	1.31	1.30	1.29
5α-Campestanol	1.27	1.30	1.30	1.28	1.40	1.33	1.30	1.30	1.30	1.31	1.31	1.23
7-Campestenol	1.40	1.43	1.52	1.50	1.48	1.48	1.45	1.52	1.55	1.58	1.60	1.56
7-Ergostenol	1.40	1.43	1.52	1.50	1.48	1.48	1.45	1.52	1.55	1.58	1.60	1.56
8-Ergostenol	1.25	1.27	1.30	1.31	1.33	1.35	1.34	1.35	1.37	1.33	1.37	1.34
8(14)-Ergostenol	–	–	–	–	1.31	1.24	1.27	1.31	1.34	1.25	1.35	1.26
14-Ergostenol	1.25	1.29	1.29	1.27	1.32	1.28	1.28	1.29	1.29	1.30	1.29	1.28
22-Ergostenol	1.12	1.15	1.14	1.12	–	–	–	–	–	–	–	–
24-Methylenecholesterol	1.24	1.26	1.34	1.35	1.31	1.26	1.26	1.39	1.41	1.43	1.45	1.44
Brassicasterol	1.10	1.12	1.12	1.12	1.11	1.09	1.12	1.12	1.12	1.10	1.10	1.11
5,7-Ergostadienol	–	–	–	–	1.47	1.41	1.40	1.51	1.55	1.65	1.68	1.56
7,9-Ergostadienol	–	–	–	–	1.40	1.39	1.39	1.49	1.54	1.55	1.60	1.56
7,14-Ergostadienol	–	–	–	–	1.34	1.33	1.31	1.44	1.44	1.46	1.51	1.51
8,14-Ergostadienol	–	–	–	–	1.31	1.32	1.31	1.39	1.44	1.46	1.47	1.49
7,22-Ergostadienol	1.23	1.24	1.33	1.32	1.25	1.27	1.26	1.33	1.36	1.36	1.38	1.36

Compound												
8(14),22-Ergostadienol	1.08	1.15	1.04	1.16	1.13	1.10	1.05	1.09	–	–	–	–
14,22-Ergostadienol	1.10	1.11	1.10	1.11	1.11	1.12	1.10	1.07	–	–	–	–
Ergosterol	1.47	1.49	1.43	1.36	1.32	1.21	1.20	1.24	1.18	1.21	1.32	1.35
7,9(11),22-Ergostatrienol	1.36	1.40	1.33	1.35	1.31	1.20	1.19	1.17	1.17	1.18	1.31	1.33
Sitosterol	1.56	1.57	1.59	1.61	1.60	1.60	1.56	1.63	1.52	1.60	1.60	1.54
5α-Stigmastanol	1.48	1.57	1.59	1.61	1.60	1.60	1.62	1.75	1.54	1.63	1.60	1.54
7-Stigmastenol	1.90	1.93	1.91	1.91	1.87	1.79	1.80	1.81	1.68	1.79	1.86	1.81
8(14)-Stigmastenol	1.51	1.62	1.51	1.67	1.61	1.57	1.50	1.59	–	–	–	–
14-Stigmastenol	1.55	1.56	1.58	1.60	1.59	1.59	1.56	1.62	–	–	–	–
Stigmasterol	1.36	1.36	1.36	1.41	1.40	1.40	1.34	1.37	1.35	1.40	1.40	1.38
5,7-Stigmastadienol	1.90	2.02	1.98	1.90	1.85	1.75	1.73	1.78	–	–	–	–
7,9(11)-Stigmastadienol	1.89	1.91	1.85	1.86	1.81	1.71	1.71	1.73	1.61	1.71	1.82	1.79
7,14-Stigmastadienol	1.82	1.83	1.77	1.78	1.71	1.63	1.62	1.68	1.57	1.67	1.73	1.73
8,14-Stigmastadienol	1.76	1.79	1.77	1.78	1.71	1.63	1.61	1.64	–	–	–	–
14,22-Stigmastadienol	1.35	1.36	1.36	1.41	1.40	1.40	1.38	1.38	–	–	–	–
7,22-Stigmastadienol	1.73	1.72	1.68	1.69	1.67	1.60	1.58	1.55	1.49	1.57	1.65	1.62
5,25-Stigmastadienol	1.68	1.70	1.67	1.67	1.64	1.54	1.48	1.55	–	–	–	–
7,25-Stigmastadienol	2.04	2.06	2.02	1.96	1.90	1.73	1.72	1.72	–	–	–	–
7,24-Stigmastadienol	2.24	2.27	2.22	2.19	2.16	1.93	1.96	1.98	1.80	1.93	2.19	2.25
5,24(28)E-Stigmastadienol	1.76	1.74	1.74	1.76	1.70	1.61	1.58	1.65	1.53	1.61	1.68	1.68
5,24(28)Z-Stigmastadienol	1.86	1.83	1.83	1.84	1.75	1.65	1.61	1.68	1.55	1.65	1.77	1.76
7,24(28)Z-Stigmastadienol	2.20	2.19	2.17	2.13	2.04	1.85	1.85	1.86	1.72	1.84	2.04	2.06
5,7,22-Stigmastatrienol	–	–	–	–	–	–	–	–	1.44	1.50	1.63	1.63
5,22,25-Stigmastatrienol	1.55	1.56	1.55	1.53	1.48	1.35	1.28	1.40	–	–	–	–
7,22,25-Stigmastatrienol	1.92	1.92	1.87	1.85	1.80	1.56	1.52	1.58	–	–	–	–
Lanosterol	1.52	1.62	1.54	1.66	1.57	1.55	1.53	1.60	1.53	1.62	1.68	1.71
24-Dihydrolanosterol	1.14	1.24	1.19	1.31	1.31	1.40	1.39	1.49	1.44	1.48	1.40	1.37

* Columns, 3 m × 3 m ID.

** 3.8% on Diatoport, all others 3% on Supelcoport.

*** Columns, 3.5 m × 4 mm ID.

§ 2% on Chromosorb W AW, all others 3% on Chromosorb W DMCS.

[15], available. Reviews of more restricted topics of steroid chromatography will be listed in the subsequent sections.

14.2. STEROLS

This section deals with genuine sterols having one oxygen function. The seco-sterols are treated in Chap. 14.3 and the more polar insect-molting hormones in Chap. 14.11. The genuine sterols and, especially, their esters are difficult to separate, because this class of steroids includes homologs (C_{26} to C_{30}) and stereoisomers (C-3, C-5, and C-24 isomers). It also includes double-bond isomers, and the 24, 28-double bond may exhibit geometric isomerism, but the 22, 23-double bond usually has the *trans*-configuration.

Chromatographic methods for sterols have been reviewed by Nerlo and Kosior [16], and their densitometric analysis was reviewed by Teng and Smith [17]. Review articles on GC of cholesterol and its precursors [18] and on GC of plant sterols [19] have been published. Articles on LC of lipids also contain summaries of research on sterols [20,21].

Isomeric sterols are poorly resolved by TLC on most silica gel plates, but Knights [22] has found that Anasil B plates have uncommon resolving power, affording a separation of, e.g., the acetates of campesterol and 5-ergostenol by continuous development. For the fractionation of unsaturated sterols, Sliwowski and Caspi [23] described a generally useful improvement in the preparation of silver nitrate-impregnated thin-layer plates. Nelson and Natelson [24] improved the staining characteristics of cholesterol and its esters by halogenation. HPTLC was reported to allow the estimation of nanogram amounts of cholesterol in blood [25].

For the separation of free sterols by GC, the phenylmethylsilicones OV-17 and OV-25 are more suitable than the less polar stationary phases UCCW-982 and SE-30 (cf. Table 14.1) [26]. The relative retention times (RRT) of homologous sterols are in the order $C_{27} < C_{28} < C_{29}$. The effect of double bonds depends on the location in the molecule. Whereas conversion of sterols to their acetyl derivatives increases the RRT on all four of the stationary phases listed in Table 14.1, conversion to the TMS ethers increases RRT in UCCW-982 and SE-30 but decreases them in OV-17 and OV-25 columns.

The TMS ethers and other derivatives are generally preferred to the free sterols in GC because the derivatives exhibit less tailing and tend to exaggerate stereochemical differences. Table 14.1 [27] shows the retention times of various TMS ethers relative to cholesterol TMS ether. Generally, homologous sterols are well resolved, the higher homologs showing greater RRT. As a rule, double bonds, especially at C-7 and C-24, tend to increase the RRT, but the effect again depends on their location in the molecule and on the nature of the stationary phase. The separation of saturated from 5-unsaturated sterols is difficult. A double bond at C-22 always shortens the RRT.

The behavior of products of the autoxidation and metabolism of cholesterol under various GLC conditions was also investigated [28]. Cholesteryl esters of fatty acids may be analyzed directly, without hydrolysis, by GC on cyanosiloxane Silar 10C [29] or, with temperature programing, on Poly-S-179 [30] columns.

A useful derivatization method for hydroxylated steroids is the oxidation to ketones by cholesterol oxidase [31]. It makes many sterols separable by GC [32]. 4-Methyl- and 4,4-dimethylsterols may be analyzed by GC on OV-17 or QF-1 columns either in the free form [33] or in the form of their acetates [34]. The ΔR_{Ac} values, i.e. the ratio of the RRT for the steryl acetates to those for the corresponding free sterols on 1.5% OV-17, decrease in the order: 4-desmethylsterols > 4-monomethylsterols > 4,4-dimethylsterols. The ΔR_{oxo} values, i.e. the ratio of the RRT for the 3-oxo derivatives to those for the corresponding free sterols on 1.5% QF-1, are also generally in that order, the values for 4-desmethyl-Δ^5-sterols being distinctly larger than those of other 4-desmethylsterols [35].

The GC–MS combination is extremely valuable in the analysis of complex sterol mixtures [36]. Thermostabile and selective stationary phases, such as PZ-176 [37] and a C_{87} hydrocarbon [38], are useful for such work. Glass capillary columns provide high resolution [38,39], but computer analysis of repetitively scanned spectra should make the analysis of unresolved peaks even more accurate [40]. Mass fragmentography yields highly specific data on sterols [41] and, in combination with isotope-labeled internal standards, it is capable of remarkable accuracy [42].

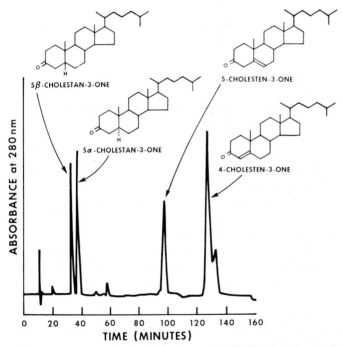

Fig. 14.1. Separation of isomeric 3-oxo C_{27} sterols by HPLC. Column, LiChrosorb Si60-10, 2 ft. × 1/4 in. OD; eluent, dichloromethane–n-hexane–ethyl acetate (94:5:1); flowrate, 9.5 ml/min; pressure, 550 p.s.i.; detector at 280 nm, range 0.16; recorder, speed 6 cm/h, span 20 mV. (Reproduced from *Lipids*, 14 (1979) 689, with permission [46].)

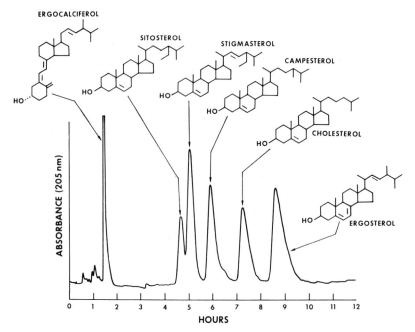

Fig. 14.2. Separation of C_{27}, C_{28}, and C_{29} sterols by HPLC. Column, Bondapak C_{18}–Porasil B (37–75 μm), 16 ft.×1/8 in. ID; eluent, 0.5% 2-propanol in *n*-hexane; flowrate, 0.4 ml/min; pressure 190 p.s.i.; detector at 205 nm. (Reproduced from *J. Chromatogr.*, 153 (1978) 60 [55].)

Among the methods of LCC, two must be mentioned: the isolation of sterols from water by Amberlite XAD-2 resin [43], which is important in pollution control, and from the unsaponifiable matter of animals by lipophilic gels [44,45], which is an important preparatory method.

In HPLC, several approaches have been used to accomplish the resolution of sterol mixtures. Adsorption chromatography on silica gel, in conjunction with either isocratic [46–48], gradient [49], or recycling [50] elution, is capable of great resolving power. Isomeric 3-oxosterols [46], epoxides [49], acetates [47,48], and epimeric 26-hydroxycholesterols [50] have been separated in this way. Fig. 14.1 [46] is an example of the simplest approach, i.e. isocratic elution. The A/B-*cis*-sterol is eluted before the A/B-*trans*-isomer. A double bond increases the retention time, especially if it is conjugated. Silver nitrate-impregnated silica [51] or alumina [52] have also been effective in the separation of unsaturated steryl acetates or benzoates.

Another approach is reversed-phase partition chromatography of either free [48,53] or esterified [54] sterols, usually on μBondapak C_{18} columns. However, better results are obtained by eluting such columns with anhydrous organic solvents [55,56]. Fig. 14.2 [55] shows a sample chromatogram in which homologous sterols are separated in the free form. The higher homologs are eluted before the lower homologs, and unsaturation increases retention time.

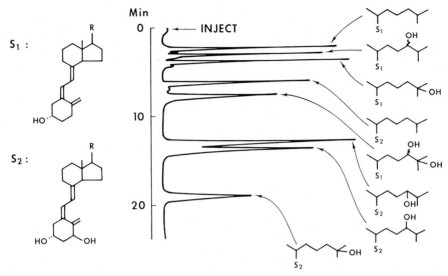

Fig. 14.3. Separation of vitamin D_3 metabolites and their analogs by HPLC. Column, Zorbax SIL, 250×2.1 mm ID; eluent, 2% methanol in dichloromethane; flowrate, 0.4–0.42 ml/min; pressure, 90 kg/cm^2; detector at 254 nm. Structures are shown in two parts, e.g., the first peak is vitamin D_3 and the last one is $1\alpha,25$-dihydroxyvitamin D_3. (Reproduced from *J. Chromatogr.*, 119 (1976) 230, with permission [63].)

14.3. VITAMINS D

Because the vitamins D (D) are sensitive to light, air, heat, and strong adsorbents, their chromatographic resolution has been even more difficult than that of their sterol precursors. The most satisfactory methods for the compounds in this group are based on CC, where detection is facilitated by their strong UV absorption. Some use has been made of hydrophobic gel chromatography [57,58], but the best separations are obtained by HPLC. A comparison of HPLC with the combination of TLC and GLC showed that quantitative analysis of D_2 by HPLC is not only faster and easier but also more accurate [59]. Results obtained by HPLC agree with the current chemical and biological assays of the Association of Official Analytical Chemists, and HPLC is suggested to replace them as the official method for D [60,61].

Silica gel columns effectively separate D metabolites and photoisomers. Using a Zorbax SIL column and 2.5% 2-propanol in Skellysolve B, Jones and DeLuca [62] observed the following elution sequence: D_2 plus D_3, 24-hydroxy-D_2, 24-hydroxy-D_3, 25-hydroxy-D_2, and 25-hydroxy-D_3. Thus, D_2 and D_3 did not separate but their hydroxylated analogs did. Elution with 10% instead of 2.5% 2-propanol separated the mono- and dihydroxy metabolites in the following order: D_3, 25-hydroxy-D_3, 24,25-dihydroxy-D_3, 1α-hydroxy-D_3, 25,26-dihydroxy-D_2, 25,26-dihydroxy-D_3, $1\alpha,25$-dihydroxy-D_2, and $1\alpha,25$-dihydroxy-D_3.

Fig. 14.3 shows a chromatogram of D_3 metabolites, similarly prepared by Ikekawa and Koizumi [63] with 2% methanol in dichloromethane as the eluent. Evidently, neither the epimeric 24-hydroxy derivatives of D_3 nor its epimeric 24,25-dihydroxy derivatives were resolved, but separation of the $1\alpha,24\alpha$- and $1\alpha,24\beta$-epimers was accomplished. The effect of hydroxyl groups on the polarity of D increases in the order: $24 < 25 < 1\alpha < 24,25 < 1,24 < 1,25 < 1,24,25$.

The separation of photoisomers of D [64–66] and of D_2 from D_3 [67,68] are examples of the resolving power of HPLC. Practical applications are exemplified by the analysis of D and its metabolites in blood [58,69–71], milk [72,73], egg yolk [74], animal tissues [75], pharmaceutical formulations [76,77], and animal feed [78,79].

14.4. SAPOGENINS AND ALKALOIDS

Some of the steroidal sapogenins are structural analogs of alkaloids, found in the same plant, and behave similarly in chromatography. Compounds differing in A/B-ring junction or in the nature or number of oxygen functions are relatively easy to separate. However, the 5α- and Δ^5-analogs are notoriously difficult to resolve, and isomerism at C-22 and C-25 complicates the isolation of individual aglycones from plant extracts. For analytical work, TLC generally gives satisfactorily results [5], but HPLC is ideally suited to the isolation of these plant steroids on a larger scale.

Fig. 14.4 [80] illustrates the separation of two pairs of sapogenin acetates, epimeric at C-5 and at C-25, by adsorption chromatography. The 25α-sapogenins are eluted ahead of the 25β-epimers, and the Δ^5-compounds emerge ahead of their

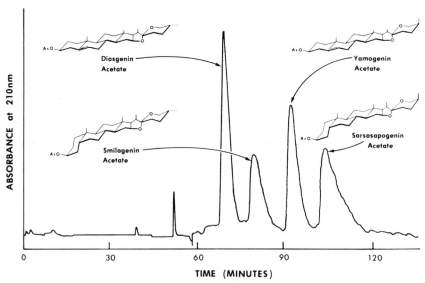

Fig. 14.4. Separation of sapogenin acetates by HPLC. Column, Zorbax SIL, 100×4.6 mm ID; eluent, 1% acetone in *n*-hexane; flowrate, 2.0 ml/min; pressure, 3200 p.s.i.; detector at 210 nm. (Reproduced from *J. Nat. Prod.*, 44 (1981) 245, with permission [80].)

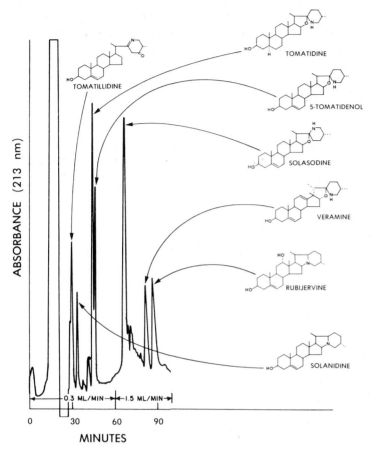

ABSORBANCE (213 nm)

Fig. 14.5. Separation of steroidal alkaloids by HPLC. Column, Zorbax SIL, 500×4.6 mm ID; eluent, *n*-hexane–methanol–acetone (18:1:1); flowrate, 0.3 ml/min for 1 h, then 1.5 ml/min; pressure, 800 psi; detector at 213 nm, range 0.2; recorder, speed 6 cm/h, span, 10 mV. (Reproduced from *J. Chromatogr.*, 198 (1980) 363 [85].)

25β-analogs. Not shown are the 5α-sapogenins tigogenin and neotigenin, which have conformations similar to diosgenin and yamogenin, respectively, and are not separable from them under these conditions. To accomplish such a separation, hydrophobic adsorption chromatography on Zorbax ODS with an eluent mixture of acetonitrile–*n*-hexane–tetrahydrofuran (17:2:1) was used. In each instance, the Δ^5-sapogenin acetate was eluted ahead of the 5α-analog.

TLC continues to serve for the analysis of glycoalkaloids [81–83], but HPLC is beginning to replace it [84]. A recent modification of our HPLC method for steroidal alkaloids (Fig. 14.5) [85] shows that finer adsorbents increase resolving power and speed. Not only the 22-epimers, 5-tomatidenol and solasodine, were separated but also the Δ^5-steroid, 5-tomatidenol, was separated from its saturated 5α-analog, tomatidine. The flowrate was increased after 1 h. A mixture of the more polar

References on p. B215

Veratrum alkaloids was resolved on the same columns by a further increase in flowrate. Quantitative assays based on HPLC [86] and centrifugal chromatography, followed by HPLC [87], have also been developed.

14.5. ESTRANE DERIVATIVES

The estrogens are characterized by the phenolic A ring, which provides a convenient handle for their detection by UV absorption or derivatization. Thus, coupling with diazonium salts results in a specific and sensitive test for estrogens separated by TLC [88–90]. Their phenolic character also facilitates their sorption on columns of Amberlite XAD-2 [91], DEAE-Sephadex [92,93], Sephadex LH-20 [94] and in liquid–liquid ion-pair chromatographic systems [95,96].

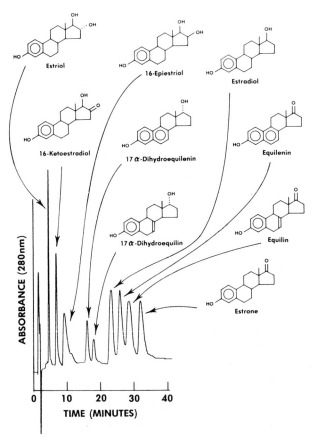

Fig. 14.6. Separation of estrogens by HPLC. Column, Zorbax BP-ODS, 250 × 4 mm ID; eluent, 35% aqueous acetonitrile; flowrate, 2 ml/min; pressure, 700 p.s.i.; detector at 280 nm. (Reproduced from *J. Chromatogr.*, 212 (1981) 239 [103].)

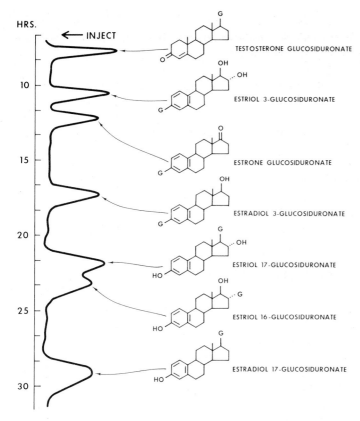

Fig. 14.7. Separation of glucosiduronates by HPLC. Column, Cellex E (9 μm), 250×3 mm ID; eluent, 0.5 *M* Cl⁻ +0.05 *M* OAc⁻ at pH 4.5; 40°C; detector at 223 nm. (Reproduced from *J. Chromatogr.*, 102 (1974) 373, with permission [107].)

For HPLC, a variety of column materials have been used: ETH-Permaphase (a chemically bonded ether) [97] and silica [98,99] for adsorption; and octadecylsilane-bonded phases, such as Partisil-10 ODS [98], Zorbax ODS [100], μBondapak C_{18} [99,101], LiChrosorb RP-8-10A [102], and Zorbax BP-ODS [103] for reversed-phase partition chromatography of free estrogens. An example of the results obtained by reversed-phase partition chromatography is shown in Fig. 14.6 [103]. As expected, the 17-ketones are less polar than the 17-hydroxy steroids, and the triols are more polar than the diols. This system also separates the estrogens according to the degree of unsaturation, but epimers are better resolved by adsorption chromatography [103]. In every case, the 17β-hydroxy compounds are more polar than the 17α-epimers.

The sensitivity of HPLC is greatly increased by the use of an electrochemical detector [104]. The detection limit for, e.g., estriol is 1 ng. Conversion of the estrogens to their dansyl (5-dimethylaminonaphthalene-1-sulfonyl) derivatives permits the detection of as little as 0.05 ng of estradiol by its fluorescence [105,106].

References on p. B215

The separation of estrogen conjugates by HPLC on cellulose ion exchangers has been studied extensively by Van der Wal and Huber [107,108]. Fig. 14.7 shows a chromatogram of various glucosiduronates on the anion exchanger Cellex E (ECTEOLA cellulose) [107]. Similar results were reported by Keravis et al. [109]. Cellulose and polystyrene anion exchangers were found to have the greatest selectivity for the site of conjugation, but hydrophobic adsorption chromatography on LiChrosorb RP-18 or RPZ, coated with a liquid anion exchanger, are also suitable for this fractionation [110]. A strong anion exchanger, μPartisil 10-SAX, also gave satisfactory results [111].

Poor results in conventional GLC of estrogens were traced to improper silanization [112]. When open-tubular columns were used instead of packed columns, the specificity of estrogen assays in normal urine was said to be greatly increased [113]. The most specific assay methods are based on mass fragmentography [114]; e.g. urinary estriol [115] and serum estradiol [116,117] have been analyzed that way.

14.6. ANDROSTANE DERIVATIVES

Lafosse et al. [118] have studied the separation of 17-ketosteroids as well as their sulfates and glucosiduronates by reversed-phase partition chromatography on Micropak CH with methanol–water mixtures and an RI detector. The order of elution for the free and conjugated steroids was: dehydroepiandrosterone, epiandrosterone, etiocholanolone, and androsterone. Thus, the equatorial hydroxyl group makes the first three steroids more polar than the axial hydroxyl group of androsterone. Among the steroids with equatorial hydroxyl groups, the most polar one is dehydroepiandrosterone, with a double bond at C-5, followed by the A/B-*trans* compound epiandrosterone and then the A/B-*cis* compound etiocholanolone.

Adsorption chromatography on a silica column gave quite a different picture (Fig. 14.8) [119]. The A/B-*trans* steroid androstanedione was eluted before the A/B-*cis* steroid etiocholanedione. The saturated epiandrosterone was more "polar" than the unsaturated dehydroepiandrosterone, and both of them, having the equatorial hydroxyl group, were eluted before the axial epimer, androsterone. Also contrary to expectation (cf. Chap. 14.5), the 17β-hydroxy steroid, testosterone, was less "polar" than its 17α-epimer.

HPLC and TLC procedures for separating testosterone metabolites were recently published by Shaikh et al. [120]. HPLC has also been used for the specific determination of nanogram quantities of testosterone in blood [121], of the anabolic drug methandione in athletes' urine and cattle feed [122], testosterone and trenbolone residues in meat [123], and of the antifertility drug 17α-ethynyl-17β-acetoxy-19-norandrost-4-en-3-one oxime in pharmaceuticals [124].

Some of the results of adsorption chromatography on silica layers agree with those shown in Fig. 14.8. Thus, Tinschert and Träger [125], who used chloroform–ethyl acetate–light petroleum (10:9:1) as the mobile phase, and Lisboa and Hoffmann [126], who used continuous development with hexane–ethyl acetate or cyclohexane–ethyl acetate mixtures, also found that etiocholanediol is more strongly

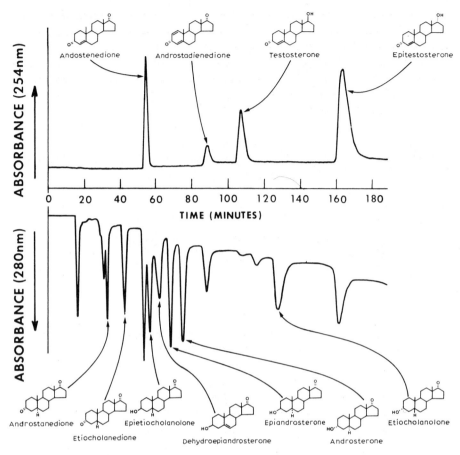

Fig. 14.8. Separation of androstane derivatives by HPLC. Column, Partisil 5, 600×4.6 mm ID; eluent, dichloromethane–acetonitrile–2-propanol (179:20:1); flowrate, 0.4 ml/min; pressure, 800 p.s.i. Two UV detectors in series: No. 1 set at 280 nm, range 0.1; No. 2 set at 254 nm, range 1.28. Dual-pen recorder: top at 20 mV for 254 nm, bottom at 5 mV for 280 nm. Chart speed, 10 min/cm. (Reproduced from *J. Chromatogr.*, 176 (1979) 487 [119].)

adsorbed than androstanedione. Sunde et al. [127], who used multiple development with dichloromethane–ethyl acetate (9:1), also observed that androsterone is more polar than epiandrosterone. However, in the two systems cited above [125,126] the sequence of androsterone and epiandrosterone was reversed.

TLC forms the basis of a fluorimetric assay of the antiarrhythmic steroid 2β-hydroxy-3α-amino-5α-androstan-17-one [128] and of anabolic steroids, such as tenbolone, in meat [129–131]. It also serves as a purification step in the radioimmunoassay of testosterone conjugates in urine and serum [132] and of 5α-androst-16-en-3α-ol in plasma [133].

In many cases, epimeric pairs of C_{19} steroids which are unresolved by TLC can be separated by GLC, either in the free form or as TMS ethers [126]. Struckmeyer [134] made a study of the derivatization of androsterone with tritiated silylating agents. Enzymic oxidation of 3β-hydroxyandrostenes also leads to useful derivatives [135]. For the determination of dehydroepiandrosterone in plasma with the ECD, it may be converted to the O-pentafluorobenzyloxime TMS ether [136], iodomethyl-dimethylsilyl ether [137], or heptafluorobutyrate [138]. Etiocholanolone was assayed in plasma with the alkali-flame ionization detector after esterification with dimethylthiophosphinic acid [139]. Metabolites of [4-^{14}C]androstene-3,17-dione in tissues were estimated with the aid of a gas-flow proportional counter [140].

The combination of SCOT columns with a mass spectrometer has greatly enhanced the performance of gas chromatography of androgens [141]. Various $C_{19}O_3$ steroids [142] and synthetic anabolics [143] have been analyzed that way. Increased specificity, obtained by selected-ion monitoring, has been exploited in the analysis of androgens in human prostatic [144] and mammary [145] tissues as well as pork products [146]. Finally, mass fragmentography has supplied an extremely sensitive and specific assay for testosterone in blood [147,148].

14.7. PREGNANE DERIVATIVES

This section deals with all pregnane derivatives except the adrenocortical hormones, which are treated in Chap. 14.8. The reduction of progesterone gives rise to a multitude of C_{21} steroids, isomeric at C-3, C-5, and C-20. Their resolution challenges the ingenuity of the analyst. Egert [149] has accomplished difficult separations on thin-layer plates by repeated development.

Lin et al. [150] found that most of the isomeric reduction products of progesterone can be separated by HPLC with a combination of adsorption and reversed-phase partition systems. Fig. 14.9 [150 illustrates the efficiency of a 60-cm column of 5-μm silica. The order of polarity for the diones is: $5\alpha < 5\beta < \Delta^4$, and for the 3-hydroxy-20-ketones it is: $3\beta,5\beta < 3\alpha,5\alpha < 3\beta,\Delta^5 < 3\beta,5\alpha < 3\alpha,5\beta$. While adsorption chromatography is superior to reversed-phase partition chromatography in the resolution of the epimeric diketones, the reverse is true of the resolution of the 20-epimers (cf. Fig. 14.10) [150].

From a 50-cm octadecylsilane column, 60% aqueous acetonitrile eluted the 20α-pregnane derivatives ahead of their 20β-epimers. An RI detector was needed to monitor the saturated alcohols. In the A/B-*trans* series, the 3-epimers were better resolved by reversed-phase partition than by adsorption, whereas in the A/B-*cis* series, adsorption chromatography proved to be superior. For the separation of Δ^5-steroids from their 5α-analogs, the reversed-phase system gave better results than the adsorption system [150]. Purdy et al. [151] have made similar observations on the relative merits of adsorption and reversed-phase partition HPLC in the analysis of pregnane derivatives.

Vandenheuvel [152,153] has further refined his theoretical treatment of the relationships between the structure of pregnane derivatives and their behavior in GLC. An excellent derivative for the detection of progesterone by electron capture is

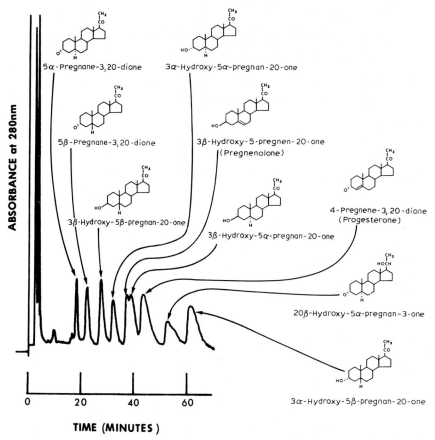

Fig. 14.9. Separation of three diones and six monohydroxymonoketones of the pregnane series by adsorption HPLC. Column, Partisil 5, 600×2 mm ID; eluent, 0.25% ethanol in dichloromethane; flowrate, 1 ml/min; pressure, 3200 p.s.i.; UV detector at 280 nm, range 0.05; recorder, speed 6 cm/h, span 10 mV. (Reproduced from *J. Chromatogr.*, 190 (1980) 171 [150].)

the dipentafluorobenzoyloxime [154], but analogs of 17-acetoxyprogesterone, such as melengestrol acetate, do not require derivatization [155]. Mass fragmentography is hardly a routine method for the determination of progesterone in serum, but it provides a highly specific and sensitive method for validating other assay procedures [156]. Baillie et al. [157] have used specific deuterium labeling and computerized GC–MS in studies of the metabolism of pregnane derivatives.

14.8. CORTICOSTEROIDS

The chromatography of adrenocorticosteroids has recently been reviewed [158]. Minor improvements in their analysis continue to appear in the literature, e.g., the

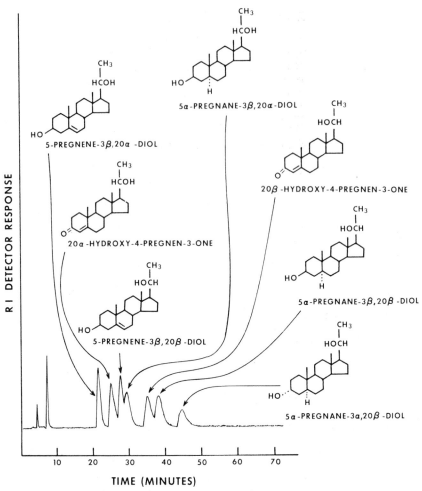

Fig. 14.10. Separation of two monohydroxymonoketones and five diols of the pregnane series by reversed-phase HPLC. Column, Zorbax BP-ODS, 500 × 4 mm ID; eluent, 60% aqueous acetonitrile; flowrate, 1 ml/min; pressure, 1390 p.s.i.; RI detector, sensitivity 8 ×; recorder, speed 12 cm/h, span 10 mV. (Reproduced from *J. Chromatogr.*, 190 (1980) 173 [150].)

separation of various pregnane glucosiduronic acid methyl esters with liquid ion exchangers on filter paper [159], TLC on polyamide sheets [160], and the fluorimetric scanning of thin-layer chromatograms [161,162].

Among the methods used to purify corticosteroid extracts, gel chromatography constitutes a significant advance [163–165]. Sephadex LH-20 is used extensively for aldosterone [166–168] and other cortical hormones [169,170].

Adrenocortical hormones for which new methods of analysis by GLC have been devised include 18-hydroxycorticosterone [171,172], 18-hydroxy-11-deoxycorticosterone [173,174], aldosterone [175], aldadiene [176], and prednisone and prednisolone [177]. Mass fragmentography [178] and capillary column GC–MS [179] have been applied to corticosteroid analysis (cf. Chap. 14.9).

The most active progress in research on the chromatography of corticosteroids is taking place in the field of HPLC. This is because they are easily detected with fixed-wavelength detectors at 254 nm and require no derivatization. Only a few of the ca. 50 publications dealing specifically with corticosteroids can be cited here. The work of Cavina et al. [180] will serve as an example of silicic acid chromatography. With a 5-μm LiChrosorb Si 100 column (250 × 4.6 mm ID) and a linear gradient of methanol in chloroform from 0.5 to 5% in 20 min, followed by isocratic elution, these workers obtained accurate analyses of a mixture of the following natural adrenocortical hormones (in order of elution): deoxycorticosterone, dehydrocorticosterone, deoxycortisol, corticosterone, cortisone, aldosterone, cortisol. Earlier procedures utilized isocratic elution with chloroform (or dichloromethane)–methanol (or ethanol) mixtures [181–184] (cf. also Fig. 14.11). For the specific determination of nanogram quantities of corticosteroids, a fluorimetric assay of the dansyl hydrazones may be used [185].

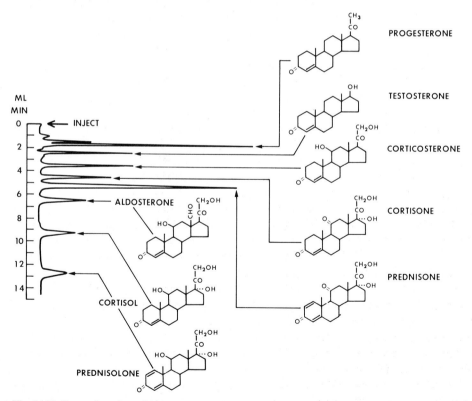

Fig. 14.11. Separation of steroid hormones by HPLC. Column, Spherosil XOA (4–8 μm), 300 × 3 mm ID; eluent, dichloromethane–ethanol–water (936 : 47 : 17); flowrate, 1.14 ml/min; detector at 240 nm. (Reproduced from *Chromatographia*, 6 (1973) 347, with permission [231].)

References on p. B215

The method of Scott and Dixon [186] is a recent example of the use of reversed-phase partition HPLC. When a 120×4.5-mm ID Hypersil octadecylsilane column was developed with 45% aqueous methanol, the natural and synthetic corticosteroids were eluted in the order of decreasing polarity. Octadecyl-bonded silica columns are the most popular [187–192], but other nonpolar packing materials, having, e.g., octyl [193], cyanopropyl [194], nitro [195], or phenyl [196] groups bonded to silica, have also been used. Cortoic acids, in the form of their *p*-bromophenacyl esters, were separated by reversed-phase HPLC [197].

HPLC is now being used extensively for the analysis of such corticoid analogs as prednisolone [198,199], methylprednisolone [200,201], triamcinolone acetonide [202,203], budesonide [204], dexamethasone [205], betamethasone [206], and canrenone [207]. Some drug analysis procedures have been automated [208].

14.9. MISCELLANEOUS HORMONES

This section records progress in the chromatography of biological extracts, drug preparations, and other mixtures containing more than one class of steroid hormones. Douglas [209] has surveyed GLC and HPLC and Fitzpatrick [210] has reviewed the HPLC of steroid hormones. Gregorová et al. [211] made a comparative study of PC, TLC, and GC in the analysis of neutral urinary steroids. As advantages of the planar chromatography methods, they mentioned savings of time and money, the specificity of detection methods, and the applicability to preliminary purification and group separations. When GC is combined with MS, its main advantage is the detailed quantitative information about urinary metabolites.

Chemically bonded alkylsilyl silica gel has been used for the reversed-phase TLC of steroid hormones [212]. The effectiveness of the programed multiple development technique was demonstrated with steroid hormones [213]. The resolving power of various solvent systems in the TLC of numerous steroids has been the subject of extensive studies by Hara and Mibe [214] and by Mattox et al. [215]. Their conclusions are equally applicable to HPLC. Novel detection methods for TLC of steroid hormones include the spray reagents: molybdovanadophosphoric acids [216], chloramine T–sulfuric acid [217], stannous chloride–monochloroacetic acid [218], and arsenic formate [219], and the incubation of thin-layer plates with a reagent containing 3β-hydroxy steroid oxidase [220].

The nonionic resin Amberlite XAD-2 is used extensively for the isolation of steroid hormones and their conjugates [221]. Lipophilic gel chromatography effectively fractionates the conjugates of neutral steroids prior to GC–MS [9,222,223]. Relationships between steroid structures and behavior in gel chromatography were studied by Vose et al. [224], and applications of Sephadex LH-20 [225,226] and Lipidex 5000 [227,228] to the analysis of steroid hormones have been published.

Hara and Hayashi [229] have studied the retention behavior of steroid hormones in adsorption and reversed-phase HPLC, and more recently, Hara et al. [230] have devised two-component solvent systems for HPLC on silica gel. Earlier work by Hesse and Hövermann [231] on the design of three-component two-phase solvent

systems for HPLC of steroid hormones on silica gel (Fig. 14.11) was extended by
Van den Berg et al. [232] to an investigation of the role of the sorption mechanism in
effecting optimal separation. An extensive study by O'Hare et al. [233] has dealt with
the use of gradient elution in reversed-phase HPLC of numerous steroid hormones.
Later work by Nice and O'Hare [234] probed the selectivity of different reversed-
phase packing materials in this application.

The hormones from ovarian tissues have been purified by HPLC [235]. Comparison of various HPLC systems for the purification of adrenal and gonadal steroids
showed that DIOL columns, which contain bonded hydroxyl groups, eluted with an
n-hexane–dioxane gradient, give the best results [236]. Another procedure [237]
required two steps: Celite column chromatography and reversed-phase HPLC.

Nanogram amounts of steroid hormones are easily detected by their UV absorption or with a UV detector in tandem with an RI detector [238]. However,
conversion to the isonicotinylhydrazones affords a highly specific assay in the
picomole range with the aid of a fluorescence detector [239].

For GLC and GC–MS, steroid hormones are converted to various thermostabile
and volatile derivatives [240], such as methyloximes [241,242], O-ω-haloalkyloximes
[243], and dimethylthiophosphinic esters [244]. The most popular derivatives are still
the various silyl ethers [245], including some sterically crowded trialkylsilyl [246], the
allyldimethylsilyl [247], and the t-butyldimethylsilyl [248–251] ethers.

Notable recent contributions to the GLC of steroid hormones include the work of
Edwards on the prediction of the relative molar FID response [252] and on the
temperature dependence of methylene-unit values [253]. Mention should also be
made of pioneering work on the use of mixed stationary phases [254] and of nematic
liquid crystals for the resolution of epimeric steroids [255,256].

The most significant progress in resolution and sensitivity of GLC has come
about with the introduction of SCOT capillary columns [257]. Some steroids may be
chromatographed in unmodified form [258], but usually derivatives are applied with
special injectors [259] to the coated [260] capillary columns. The effect of column
temperature has been studied by Krupčik et al. [261]. The complex chromatogram,
commonly referred to as steroid profile, is now routinely [262] used by endocrinolo-
gists in the examination of blood [263,264] and urine [265–271] specimens. The
second most significant development in GLC of steroid hormones, which, however,
is not yet a routine method, is mass fragmentography. Examples of its applicability
to this area are studies of the hormones circulating in the blood [271,273] and
secreted by the ovaries [274,275]. Other techniques used in hormone studies include
radio GC [276] and pyrolysis GLC [277].

14.10. BILE ACIDS

Sjövall [278] has reviewed analytical methods for bile acids in 1977, and the TLC
of bile acids and alcohols was reviewed by Eneroth [279] in 1976. A two-dimensional
TLC procedure was published by Ikawa and Goto [280], and improved solvent
systems for free [281] and methylated [281,282] bile acids as well as their keto

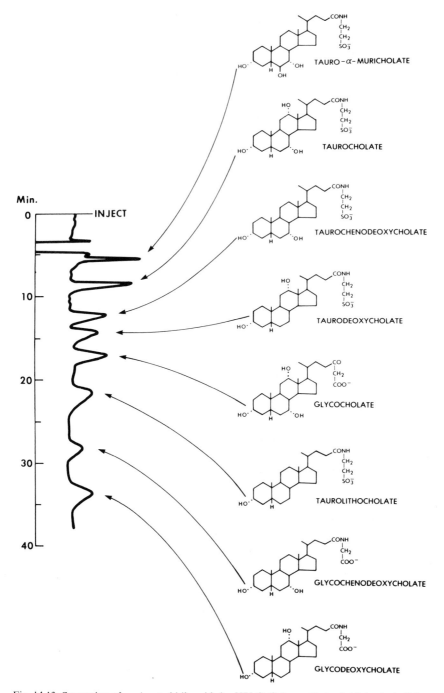

Fig. 14.12. Separation of conjugated bile acids by HPLC. Column, Fatty Acid Analysis Column (Waters Assoc.), 300×4 mm ID; eluent, 8.8 mM phosphate buffer (pH 2.5)–2-propanol (17:8); flowrate 1 ml/min; RI detector, 8×. (Reproduced from *Lipids*, 13 (1978) 972, with permission [305].)

derivatives [283,284] have been devised. Systems are now available for separating free bile acids, glyco-, and tauroconjugates from each other [285–287]. In addition, most conjugated bile acids [288–291] and bile salt sulfates [292,293] can be differentiated by TLC or reversed-phase TLC. Specific color reagents [294,295] provide additional evidence of identity. For quantitative estimation, direct fluorimetry may be used after treating the chromatogram with sulfuric acid [296,297]. Alternatively, the zones may be collected separately and assayed spectrophotometrically [298] or radiometrically [299] after the sulfuric acid reaction. Chromarods are also applicable to bile acid analysis [300].

For HPLC of bile acids, the RI detector may be used when adequate amounts are chromatographed, but spectrometry at 210 nm is more sensitive and specific [301]. Although the reversed-phase systems are generally more satisfactory [301–304], adsorption chromatography on silica gel must be used for the resolution of some pairs, e.g. the 5-epimers [305]. A representative reversed-phase partition chromatogram of conjugated bile acids is shown in Fig. 14.12 [305]. The taurine conjugates are more polar than the corresponding glycine conjugates. Reversed-phase partition also separates the sulfated bile acids [306]. Parris [307] has determined the free and conjugated bile acids as UV-absorbing ion pairs by reversed-phase HPLC.

Prior to the application of HPLC to biological specimens, a group fractionation into free, glycine-, and taurine-conjugated bile acids may be accomplished by TLC [308] or anion-exchange chromatography on diethylaminopropyl [309] or piperidinohydroxypropyl [310] Sephadex LH-20 or TSK Gel IEX 540 DEAE [311]. Sulfated bile acids are detected in HPLC by their absorption at 210 nm [312]. Conversion of bile acids to their *p*-nitrobenzyl [313,314] or phenacyl [315] esters permits their detection at 254 nm.

The GC of bile acids has been reviewed by Kuksis [316]. For GLC, bile acids must be converted to derivatives. The silyl ethers are the most desirable derivatives for GC–MS [317,318] and capillary GLC [319]. Partial TMS ether derivatives are prepared by use of N,O-bis(trimethylsilyl)trifluoroacetamide as silylating agent [320]. Methyl ester acetates [321] and ethyl ester trifluoroacetates [322] are most useful for GC–MS. Stationary phases suitable for such work are Poly S-179 [323] and cyanopropylphenylsiloxane [324]. Polymetaphenoxylene has been recommended for both acetate and TMS derivatives [325]. The hexafluoroisopropyl ester-trifluoroacetyl derivatives have been advocated for GC–MS [326] and the heptafluorobutyrate derivatives for a simplified assay of both free and conjugated bile acids with the ECD [327].

Among numerous applications of GLC to bile acids, mention should be made of a combination with TLC, which permits the resolution of saturated and unsaturated cholanic acids [328], the determination of sulfated and nonsulfated bile acids in serum [329], and new GLC methods for secondary bile acids [330] and for bile alcohols [331].

References on p. B215

14.11. ECDYSTEROIDS

Ecdysteroids differ chromatographically from other sterols in their polar nature, which is due to their highly functionalized structures, and in their UV absorption around 254 nm, which is due to their conjugated carbonyl group. In TLC, they are easily detected in short-wave UV light [332]. The UV absorption of ecdysteroids eluted from preparative thin-layer chromatograms [333] or the quenching effect on the fluorescence of thin-layer plates [334] can be used for quantitative analysis. The fluorescence of their phenanthreneboronates provides a selective detection method [335].

The HPLC of ecdysteroids has been reviewed by Hashimoto [336] and compared with GLC by Wilson et al. [337]. Because they are so easily detected by their UV absorption, the first application of a forerunner of HPLC to steroids was Hori's classical work on insect-molting hormones [338]. Fig. 14.13 is taken from his paper. A mixture of ecdysteroids was slowly eluted from a column of powdered macroreticular styrene–divinylbenzene copolymer with a linear gradient of ethanol in water, the hormones emerging in the order of decreasing polarity. With modern packing materials, the analysis does not take as long, but the resolution is not necessarily

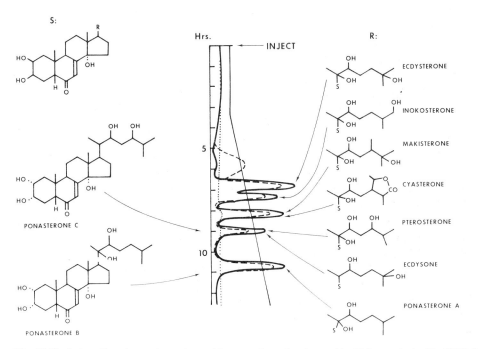

Fig. 14.13. Automatic column chromatographic separation of ecdysteroids. Column, Amberlite XAD-2 (200–400-mesh), 1500×9 mm ID; eluent, linear gradient of ethanol in water from 20 to 70%; temperature, 20°C; flowrate 1 ml/min; detector response at — — — 230 nm, ———— 250 nm, ······ 300 nm. (Reproduced from *Steroids*, 14 (1969) 38, with permission [338].)

better. Corasil II [339] and Zorbax SIL [340] have been used as column packing materials for partition chromatography; Bondapak phenyl [339], Bondapak C_{18} [341], Zorbax ODS [340], and Zorbax C-8 [340] for reversed-phase partition; and Permaphase ETH [342] for hydrophobic adsorption chromatography.

In contrast to most steroids, ecdysteroids give TMS ethers which are detectable at the picogram level by electron capture, but the preparation of these derivatives requires care [343]. They are suitable for quantitative analysis by mass fragmentography [344].

14.12. LACTONE DERIVATIVES

This section deals with polar steroids containing a lactone ring, and includes the withanolides, cardenolides, and bufadienolines.

The withanolides have both an α,β-unsaturated δ-lactone group and an α,β-unsaturated carbonyl group in the molecule and are therefore readily detected in HPLC by their UV absorption. Following a report on HPLC of withaferin A metabolites with RI detection [345], Hunter et al. [346] studied the separation of 12

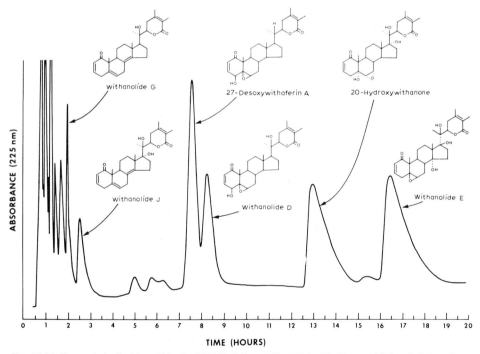

Fig. 14.14. Separation of withanolides by HPLC. Column, Porasil A (37–75 μm), 12 ft. $\times 1/8$ in.; eluent, n-hexane–2-propanol (9:1); flowrate, 0.2 ml/min; pressure, 1100 p.s.i.; detector at 225 nm; range, 0.2; recorder speed, 2 cm/h; span, 10 mV. (Reproduced from *J. Chromatogr.*, 170 (1979) 441 [346].)

References on p. B215

withanolides with the use of a UV detector. Fig. 14.14 gives an example of the results obtained with a 12-ft. coiled column of Porasil A. The polarity of the withanolides depends not only on the number and position of the oxygen functions in the molecule, but also on their nature and increases in the order: tertiary < secondary < primary.

There are essentially only two satisfactory chromatographic methods for cardiac

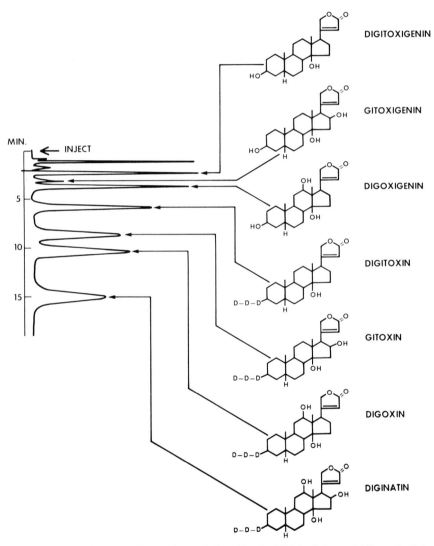

Fig. 14.15. Separation of cardiac genins and glycosides by HPLC. Column, LiChrosorb Si60 (10 μm), 250×3 mm ID; eluent, n-pentanol–acetonitrile–isooctane–water (35 : 12 : 124 : 2); flowrate, 1.3 ml/min; detector at 220 nm; recorder speed, 1 cm/min. (Reproduced from *J. Chromatogr.*, 117 (1976) 85, with permission [354].)

glycosides: TLC and HPLC. Silica gel plates previously exposed to boric acid vapors have the ability to retard selectively the mobility of cardiac glycosides containing *cis*-1,2-diol groups while increasing that of 1,3-diols [347]. Increased resolution has been achieved by repeated [348] or continuous [349] solvent development. HPTLC has been applied to the *Digitalis* glycosides [350]. When programed multiple development is performed with HPTLC plates, superior resolution is obtained [351], and application of fluorodensitometry permits the quantitative analysis of digitoxin in serum [352]. Another method for digitoxin determination in blood and urine is based on double development followed by elution of the glycoside and ^{86}Rb assay [353].

The separation of cardiac genins and glycosides by partition HPLC on a silica column is illustrated by Fig. 14.15 [354]. The genins precede the glycosides, and in each of the two groups, the polarity depends on the number and position of the hydroxyl groups. Cobb [355] observed the same order of elution in adsorption HPLC on silica. Comparison of normal and reversed-phase partition indicates some superiority of the latter [356]. Castle [357], using a μBondapak C$_{18}$ column in conjunction with an elution gradient from 25 to 40% aqueous acetonitrile, observed, as expected, that digoxigenin glycosides are eluted before the digitoxigenin glycosides, but within each group the genin was eluted first, followed by the mono-, then the bis-, and finally the tris-glycoside. A μBondapak C$_{18}$ column, developed with aqueous methanol or tetrahydrofuran, has also been employed in the separation of bufadienolide half-esters [358]. A similar method, based on reversed-phase partition on a LiChrosorb 5 RP-8 column and detection of the bufadienolides at 280 nm, is also published [359]. The cardenolides are detected by UV absorption at 220 nm. For the detection of nanogram quantities at 254 nm, they are converted to the 4-nitrobenzoate derivatives [360]. As little as 0.5 ng may be detected by post-column derivatization with hydrochloric acid, which produces a fluorescence [361].

REFERENCES

1 E. Heftmann, *Chromatography of Steroids*, J. Chromatogr. Library Series, Vol. 8, Elsevier, Amsterdam, 1976.
2 E. Heftmann, in E. Heftmann (Editor), *Chromatography*, Van Nostrand Reinhold, New York, 3rd Edn., 1974, p. 610.
3 B.P. Lisboa, in R.M. Burton and F.C. Guerra (Editors), *Fundamentals of Lipid Chemistry*, Bi-Science Publications Div., Webster Groves, MO, 1974, p. 439.
4 S. Görög and Gy. Szász, *Analysis of Steroid Hormone Drugs*, Akademiai Kiadó, Budapest, 1978.
5 E. Heftmann, *J. Liquid Chromatogr.*, 2 (1979) 1137.
6 B.P. Lisboa, in G.V. Marinetti (Editor), *Lipid Chromatographic Analysis*, Vol. 2, Dekker, New York, 2nd Edn., 1976, p. 339.
7 J.C. Touchstone and M.F. Dobbins, in J.C. Touchstone and J. Sherma (Editors), *Densitometry in Thin Layer Chromatography*, Wiley, New York, 1979, p. 633.
8 Z. Procházka, in Z. Deyl, K. Macek and J. Janák (Editors), *Liquid Column Chromatography*, J. Chromatogr. Library Series, Vol. 3, Elsevier, Amsterdam, 1975, p. 593.
9 J. Sjövall and M. Axelson, *J. Steroid Biochem.*, 11 (1979) 129.
10 J. Sjövall, *J. Steroid Biochem.*, 6 (1975) 227.
11 E. Nyström and J. Sjövall, *Methods Enzymol.*, 35B (1975) 378.
12 P. Vestergaard, *Acta Endocrinol. (Copenhagen)*, 88 (1978) Suppl. 217.

13 M.P. Kautsky (Editor), *Recent Applications of HPLC to Steroid Analysis*, Chromatogr. Sci. Series, Vol. 16, Dekker, New York, 1980.

14 E. Heftmann and I.R. Hunter, *J. Chromatogr.*, 165 (1979) 283.

15 D.G.I. Kingston, *J. Nat. Prod.*, 42 (1979) 237.

16 H. Nerlo and A. Kosior, *Herba Pol.*, 21 (1975) 324.

17 J.I. Teng and L.L. Smith, in J.C. Touchstone and J. Sherma (Editors), *Densitometry in Thin-Layer Chromatography*, Wiley, New York, 1979, p. 661.

18 R. Fumagalli, in G.V. Marinetti (Editor), *Lipid Chromatographic Analysis*, Vol. 3, Dekker, New York, 2nd Edn., 1976, p. 791.

19 E. Homberg, *Fette, Seifen, Anstrichm.*, 79 (1977) 234.

20 K. Aitzetmüller, *J. Chromatogr.*, 113 (1975) 231.

21 O.S. Privett and W.L. Erdahl, *Chem. Phys. Lipids*, 21 (1978) 361.

22 B.A. Knights, *Anal. Biochem.*, 80 (1977) 324.

23 J.K. Sliwowski and E. Caspi, *J. Steroid Biochem.*, 8 (1977) 47.

24 W.R. Nelson and S. Natelson, *Clin. Chem.*, 23 (1977) 835.

25 R. Wintersteiger and G. Gubitz, *Sci. Pharm.*, 46 (1978) 269.

26 A. Seher and H. Vogel, *Fette, Seifen, Anstrichm.*, 78 (1976) 106.

27 E. Homberg, *J. Chromatogr.*, 139 (1977) 77.

28 E. Homberg and B. Bielefeld, *J. Chromatogr.*, 180 (1979) 83.

29 T. Takagi, A. Sakai, K. Hayashi and Y. Itabashi, *J. Chromatogr. Sci.*, 17 (1979) 212.

30 R.D. Schwartz, R.G. Mathews, S. Ramachandran, R.S. Henly and J.E. Doyle, *J. Chromatogr.*, 112 (1975) 111.

31 C.J.W. Brooks and A.G. Smith, *J. Chromatogr.*, 112 (1975) 499.

32 C.G. Edmonds, A.G. Smith and C.J.W. Brooks, *J. Chromatogr.*, 133 (1977) 372.

33 T.M. Jeong, T. Itoh, T. Tamura and T. Matsumoto, *Lipids*, 10 (1975) 634.

34 T. Itoh, T. Tamura, T. Iida and T. Matsumoto, *Steroids*, 26 (1975) 93.

35 T. Itoh, T. Tamura, S. Ogawa and T. Matsumoto, *Steroids*, 25 (1975) 729.

36 D.R. Idler, M.W. Khalil, J.D. Gilbert and C.J.W. Brooks, *Steroids*, 27 (1976) 155.

37 J.A. Ballantine and K. Williams, *J. Chromatogr.*, 148 (1978) 504.

38 J.A. Ballantine, K. Williams and R.J. Morris, *J. Chromatogr.*, 166 (1978) 491.

39 M. Novotny, M.L. Lee, C.E. Low and M.P. Maskarinec, *Steroids*, 27 (1976) 665.

40 J.M. Halket, *J. Chromatogr.*, 175 (1979) 229.

41 M. Tohma, Y. Nakata and T. Kurosawa, *J. Chromatogr.*, 171 (1979) 469.

42 P. Gambert, C. Lallemant, A. Archambault, B.F. Maume and P. Padieu, *J. Chromatogr.*, 162 (1979) 1.

43 C.K. Wun, R.W. Waler and W. Litsky, *Water Res.*, 10 (1976) 955.

44 H.U. Melchert, *Deut. Lebensm.-Rundsch.*, 71 (1975) 400.

45 G.W. Patterson, M.W. Khalil and D.R. Idler, *J. Chromatogr.*, 115 (1975) 153.

46 I.R. Hunter, M.K. Walden, G.F. Bailey and E. Heftmann, *Lipids*, 14 (1979) 687.

47 J.R. Thowsen and G.J. Schroepfer, Jr., *J. Lipid Res.*, 20 (1979) 681.

48 E. Hansbury and T.J. Scallen, *J. Lipid Res.*, 21 (1980) 921.

49 G.A.S. Ansari and L.L. Smith, *J. Chromatogr.*, 175 (1979) 307.

50 J. Redel, *J. Chromatogr.*, 168 (1979) 273.

51 W.B. Smith and L. Hogle, *Rev. Latinoam. Quim.*, 7 (1976) 20.

52 R.A. Pascal, Jr., C.L. Farris and G.J. Schroepfer, Jr., *Anal. Biochem.*, 101 (1980) 15.

53 I.W. Duncan, P.H. Culbreth and C.A. Burtis, *J. Chromatogr.*, 162 (1979) 281.

54 H.H. Rees, P.L. Donnahey and T.W. Goodwin, *J. Chromatogr.*, 116 (1976) 281.

55 I.R. Hunter, M.K. Walden and E. Heftmann, *J. Chromatogr.*, 153 (1978) 57.

56 H. Colin, G. Guiochon and A. Siouffi, *Anal. Chem.*, 51 (1979) 1661.

57 T. Kobayashi, K. Mizuno, M. Yasumura and T. Okano, *Vitamins*, 52 (1978) 217.

58 P.W. Lambert, E.A. Lindmark and T.C. Spelsberg, *Chromatogr. Sci.*, 10 (1979) 637.

59 E. Albertario and A. Gnocchi, *Chron. Chim.*, 55 (1978) 12.

60 H. Hofsass, N.J. Alicino, A.L. Hirsch, L. Ameika and L.D. Smith, *J. Ass. Offic. Anal. Chem.*, 61 (1978) 735.

61 F.J. Mulder, E.J. de Vries and B. Borsje, *J. Ass. Offic. Anal. Chem.*, 62 (1979) 1031.
62 G. Jones and H.F. DeLuca, *J. Lipid Res.*, 16 (1975) 448.
63 N. Ikekawa and N. Koizumi, *J. Chromatogr.*, 119 (1976) 227.
64 K.A. Tartivita, J.P. Sciarello and B.C. Rudy, *J. Pharm. Sci.*, 65 (1976) 1024.
65 K. Tsukida, A. Kodama and K. Saiki, *J. Nutr. Sci. Vitaminol.*, 22 (1976) 15.
66 R. Vanhaelen-Fastré and M. Vanhaelen, *J. Chromatogr.*, 153 (1978) 219.
67 R.J. Tscherne and G. Capitano, *J. Chromatogr.*, 136 (1977) 337.
68 M. Osadca and M. Araujo, *J. Ass. Offic. Anal. Chem.*, 60 (1977) 993.
69 G. Jones, *Clin. Chem.*, 24 (1978) 287.
70 K.T. Koshy and A.L. VanDerSlik, *Anal. Biochem.*, 85 (1978) 283.
71 B. Pelc and A.L. Holmes, *J. Chromatogr.*, 173 (1979) 403.
72 J.N. Thompson, W.B. Maxwell and M. L'Abbé, *J. Ass. Offic. Anal. Chem.*, 60 (1977) 998.
73 S.K. Henderson and A.F. Wickroski, *J. Ass. Offic. Anal. Chem.*, 61 (1978) 1130.
74 K.T. Koshy and A.L. VanDerSlik, *J. Agr. Food Chem.*, 27 (1979) 180.
75 K.T. Koshy and A.L. VanDerSlik, *J. Agr. Food Chem.*, 25 (1977) 1246.
76 E.J. de Vries, J. Zeeman, R.J. Esser, B. Borsje and F.J. Mulder, *J. Ass. Offic. Anal. Chem.*, 62 (1979) 129.
77 R. Vanhaelen-Fastré and M. Vanhaelen, *J. Chromatogr.*, 179 (1979) 131.
78 A.C. Ray, J.N. Dwyer and J.C. Reagor, *J. Ass. Offic. Anal. Chem.*, 60 (1977) 1296.
79 H. Cohen and M. Lapointe, *J. Chromatogr. Sci.*, 17 (1979) 510.
80 I.R. Hunter, M.K. Walden, G.F. Bailey and E. Heftmann, *J. Nat. Prod.*, 44 (1981) 245.
81 V.N. Borisov, L.A. Pikova, G.L. Zachepilova and A.I. Ban'kovskii, *Pharm. Chem. J.*, 10 (1977) 1694.
82 L.S. Cadle, D.A. Stelzig, K.L. Harper and R.J. Young, *J. Agr. Food Chem.*, 26 (1978) 1453.
83 R. Jellema, E.T. Elema and T.M. Malingré, *J. Chromatogr.*, 189 (1980) 406.
84 R.J. Bushway, E.S. Barden, A.W. Bushway and A.A. Bushway, *J. Chromatogr.*, 178 (1979) 533.
85 I.R. Hunter, M.K. Walden and E. Heftmann, *J. Chromatogr.*, 198 (1980) 363.
86 P.G. Crabbe and C. Fryer, *J. Chromatogr.*, 187 (1980) 87.
87 W.D. Nes, E. Heftmann, I.R. Hunter and M.K. Walden, *J. Liquid Chromatogr.*, 3 (1980) 1687.
88 P.A. Weiss, R. Winter, F. Scherr and H. Bayer, *Geburtshilfe Frauenheilkd.*, 36 (1976) 256.
89 G. Carignan, M. Lanouette and B.A. Lodge, *J. Chromatogr.*, 135 (1977) 523.
90 B. Wortberg, R. Woller and T. Chulamorakot, *J. Chromatogr.*, 156 (1978) 205.
91 E.A. Yapo, V. Barthelemy-Clavey, A. Racadot and J. Mizon, *J. Chromatogr.*, 145 (1978) 478.
92 P.I. Musey, D.C. Collins and J.R.K. Preedy, *Steroids*, 29 (1977) 657.
93 P. Jarvenpaa, T. Fotsis and H. Adlercreutz, *J. Steroid Biochem.*, 11 (1979) 1583.
94 F.E. Newsome and W.D. Kitts, *Steroids*, 26 (1975) 215.
95 B. Fransson, K.-G. Wahlund, I.M. Johansson and G. Schill, *J. Chromatogr.*, 125 (1976) 327.
96 J. Hermansson, *J. Chromatogr.*, 152 (1978) 437.
97 R.W. Roos, *J. Chromatogr. Sci.*, 14 (1976) 505.
98 R.J. Dolphin and P.J. Pergande, *J. Chromatogr.*, 143 (1977) 267.
99 R.W. Roos, *J. Ass. Offic. Anal. Chem.*, 63 (1980) 80.
100 H. Fukuchi, S. Tsukiai and M. Inoue, *Yakuzaigaku,*, 38 (1978) 102.
101 G. Capitano and R. Tscherne, *J. Pharm. Sci.*, 68 (1979) 311.
102 L.F. Krzeminski, B.L. Cox and G.H. Dunn, III, *J. Agr. Food Chem.*, 26 (1978) 891.
103 J.-T. Lin and E. Heftmann, *J. Chromatogr.*, 212 (1981) 239.
104 K. Shimada, T. Tanaka and T. Nambara, *J. Chromatogr.*, 178 (1979) 350.
105 G.J. Schmidt, F.L. Vandemark and W. Slavin, *Anal. Biochem.*, 91 (1978) 636.
106 R.W. Roos, *J. Pharm. Sci.*, 67 (1978) 1735.
107 S. van der Wal and J.F.K. Huber, *J. Chromatogr.*, 102 (1974) 353.
108 S. van der Wal and J.F.K. Huber, *J. Chromatogr.*, 135 (1977) 305.
109 G. Keravis, M. Lafosse and M.H. Durand, *Chromatographia*, 10 (1977) 678.
110 S. van der Wal and J.F.K. Huber, *J. Chromatogr.*, 149 (1978) 431.
111 P.I. Musey, D.C. Collins and J.R. Preedy, *Steroids*, 31 (1978) 583.
112 R. Roman, C.H. Yates and J.F. Millar, *J. Chromatogr. Sci.*, 15 (1977) 555.

113 J.P. Fels, L. Dehennin and R. Scholler, *J. Steroid Biochem.*, 6 (1975) 1201.

114 H. Breuer and L. Siekmann, *J. Steroid Biochem.*, 6 (1975) 685.

115 I. Björkhem, R. Blomstrand, L. Svensson, F. Tietz and K. Carlström, *Clin. Chim. Acta*, 62 (1975) 385.

116 D.W. Wilson, B.M. John, G.V. Groom, C.G. Pierrepoint and K. Griffiths, *J. Endocrinol.*, 74 (1977) 503.

117 J. Zamecnik, D.T. Armstrong and K. Green, *Clin. Chem.*, 24 (1978) 627.

118 M. Lafosse, G. Keravis and M.H. Durand, *J. Chromatogr.*, 118 (1976) 283.

119 I.R. Hunter, M.K. Walden and E. Heftmann, *J. Chromatogr.*, 176 (1979) 485.

120 B. Shaikh, M.R. Hallmark and H.J. Issaq, *J. Liquid Chromatogr.*, 2 (1979) 943.

121 R.C. Cochran, K.J. Darney, Jr. and L.L. Ewing, *J. Chromatogr.*, 173 (1979) 349.

122 C.G.B. Frischkorn and H.E. Frischkorn, *J. Chromatogr.*, 151 (1978) 331.

123 H.-J. Stan and F.W. Hohls, *Z. Lebensm.-Unters.-Forsch.*, 169 (1979) 266.

124 R.E. Huettemann and A.P. Schroff, *J. Chromatogr. Sci.*, 13 (1975) 357.

125 W. Tinschert and L. Träger, *J. Chromatogr.*, 152 (1978) 447.

126 B.L. Lisboa and U. Hoffmann, *J. Chromatogr.*, 115 (1975) 177.

127 A. Sunde, P. Stenstad and K.B. Eik-Nes, *J. Chromatogr.*, 175 (1979) 219.

128 I. Søndergaard and E. Steiness, *J. Chromatogr.*, 162 (1979) 422.

129 K.-L. Oehrle, K. Vogt and B. Hoffmann, *J. Chromatogr.*, 114 (1975) 244.

130 J.M. Wal, J.C. Peleran and G. Bories, *J. Chromatogr.*, 136 (1977) 165.

131 R. Verbecke, *J. Chromatogr.*, 177 (1979) 69.

132 C.M. Puah, J.M. Kjeld and G.F. Joplin, *J. Chromatogr.*, 145 (1978) 247.

133 D.C. Bicknell and D.B. Gower, *J. Steroid Biochem.*, 7 (1976) 451.

134 H.F. Struckmeyer, *Z. Anal. Chem.*, 286 (1977) 244.

135 A.G. Smith, G.E. Joannou, M. Mák, T. Uwajima, O. Terada and C.J.W. Brooks, *J. Chromatogr.*, 152 (1978) 467.

136 T. Nambara, T. Kigasawa, T. Iwata and M. Ibuki, *J. Chromatogr.*, 114 (1975) 81.

137 E.K. Symes and B.S. Thomas, *J. Chromatogr.*, 116 (1976) 163.

138 T. Fehér, L. Bodrogi, K.G. Fehér and E. Poteczin, *Chromatographia*, 10 (1977) 86.

139 W. Vogt, K. Jacob, I. Fischer and M. Knedel, *Z. Anal. Chem.*, 279 (1976) 158.

140 H.T. Schneider, B.P. Lisboa and H. Breuer, *Z. Anal. Chem.*, 279 (1976) 161.

141 C.G. Edmonds and C.J.W. Brooks, *J. Chromatogr.*, 116 (1976) 173.

142 A. Kerebel, R.F. Morfin, F.L. Berthou, D. Picart, L.G. Bardou and H.H. Floch, *J. Chromatogr.*, 140 (1977) 229.

143 H.W. Dürbeck, I. Buker, B. Scheulen and B. Telin, *J. Chromatogr.*, 167 (1978) 117.

144 D.S. Millington, M.E. Buoy, G. Brooks, M.E. Harper and K. Griffiths, *Biomed. Mass Spectrom.*, 2 (1975) 219.

145 P.V. Maynard, A.W. Pike, A. Weston and K. Griffiths, *Eur. J. Cancer*, 13 (1977) 971.

146 R.H. Thompson and A.M. Pearson, *J. Agr. Food Chem.*, 25 (1977) 1241.

147 I. Björkhem, O. Lantto and L. Svensson, *Clin. Chim. Acta*, 60 (1975) 59.

148 S. Baba, Y. Shinohara and Y. Kasuya, *J. Chromatogr.*, 162 (1979) 529.

149 D. Egert, *J. Chromatogr.*, 135 (1977) 481.

150 J.-T. Lin, E. Heftmann and I.R. Hunter, *J. Chromatogr.*, 190 (1980) 169.

151 R.H. Purdy, C.K. Durocher, P.H. Moore, Jr. and P.N. Rao, in M.P. Kautsky (Editor), *Steroid Analysis by HPLC*, Dekker, New York, 1980, p. 81.

152 F.A. Vandenheuvel, *J. Chromatogr.*, 115 (1975) 161.

153 F.A. Vandenheuvel, *J. Chromatogr.*, 133 (1977) 107.

154 R. Wehner and A. Handke, *J. Chromatogr.*, 177 (1979) 237.

155 K.T. Koshy, *J. Chromatogr.*, 126 (1976) 641.

156 I. Björkhem, R. Blomstrand and O. Lantto, *Clin. Chim. Acta*, 65 (1975) 343.

157 T.A. Baillie, K. Sjövall, J.E. Herz and J. Sjövall, *Advan. Mass Spectrom. Biochem. Med.*, 2 (1977) 161.

158 T.O. Oesterling, *Chromatogr. Sci.*, 9 (1978) 863.

159 V.R. Mattox and R.D. Litwiller, *J. Chromatogr.*, 189 (1980) 33.

160 M.S. Scandrett and E.J. Ross, *Clin. Chim. Acta*, 72 (1976) 165.
161 J.M. Marcos, J. Moreno and J.M. Pla Delfina, *Cienc. Ind. Farm.*, 9 (1977) 92.
162 P.J. van der Merwe, D.G. Müller and E.C. Clark, *J. Chromatogr.*, 171 (1979) 519.
163 D. Apter, O. Jänne and R. Vihko, *Clin. Chim. Acta*, 63 (1975) 139.
164 A. Dyfverman and J. Sjövall, *Anal. Lett.*, B11 (1978) 485.
165 D.R. Boreham, C.W. Vose, R.F. Palmer, C.J.W. Brooks and V. Balasubramanian, *J. Chromatogr.*, 153 (1978) 63.
166 A. Jindra and V. Kučerova, *Čas. Lek. Česk.*, 114 (1975) 927.
167 W.G. Sippell, G. Putz and M. Scheuerecker, *J. Chromatogr.*, 146 (1978) 333.
168 C.D. Kachel and F.A. Mendelsohn, *J. Steroid Biochem.*, 10 (1979) 563.
169 W.G. Sippell, P. Lehmann and G. Hollmann, *J. Chromatogr.*, 108 (1975) 305.
170 W.A. Golder and W.G. Sippell, *J. Chromatogr.*, 123 (1976) 293.
171 A. Wilson, P.A. Mason and R.F. Fraser, *J. Steroid Biochem.*, 7 (1976) 611.
172 C.H. Shackleton and J.W. Honour, *J. Steroid Biochem.*, 8 (1977) 199.
173 P. Bournot, M. Prost and B.F. Maume, *J. Chromatogr.*, 112 (1975) 617.
174 Z. Tomsová, I. Gregorová, K. Horký and J. Dvořáková, *J. Chromatogr.*, 145 (1978) 131.
175 S.J. Gaskell and C.J.W. Brooks, *J. Chromatogr.*, 158 (1978) 331.
176 T. Fehér, L. Bodrogi and A. Váradi, *J. Chromatogr.*, 123 (1976) 460.
177 S.B. Matin and B. Amos, *J. Pharm. Sci.*, 67 (1978) 923.
178 L.P. Romanoff and H.J. Brodie, *J. Steroid Biochem.*, 7 (1976) 289.
179 B.F. Maume, C. Millot, D. Mesnier, D. Patouraux, J. Doumas and E. Tomori, *J. Chromatogr.*, 186 (1979) 581.
180 G. Gavina, G. Moretti, R. Alimenti and B. Gallinella, *J. Chromatogr.*, 175 (1979) 125.
181 G. Schwedt, H.H. Bussemas and C. Lippmann, *J. Chromatogr.*, 143 (1977) 259.
182 C.P. de Vries, C. Popp-Snijders, W. de Kieviet and A.C. Akkerman-Faber, *J. Chromatogr.*, 143 (1977) 624.
183 D. Ishii, K. Hibi, K. Asai, M. Nagaya, K. Mochizuki and Y. Mochida, *J. Chromatogr.*, 156 (1978) 173.
184 J.Q. Rose and W.J. Jusko, *J. Chromatogr.*, 162 (1979) 273.
185 T. Kawasaki, M. Maeda and A. Tsuji, *J. Chromatogr.*, 163 (1979) 143.
186 N.R. Scott and P.F. Dixon, *J. Chromatogr.*, 164 (1979) 29.
187 J. Korpi, D.P. Wittmer, B.J. Sandmann and W.G. Haney, *J. Pharm. Sci.*, 65 (1976) 1087.
188 N.W. Tymes, *J. Chromatogr. Sci.*, 15 (1977) 151.
189 C. Burgess, *J. Chromatogr.*, 149 (1978) 233.
190 E. Smith, *J. Ass. Offic. Anal. Chem.*, 62 (1979) 812.
191 G.E. Reardon, A.M. Caldarella and E. Canalis, *Clin. Chem.*, 25 (1979) 122.
192 S. Gallant, S.M. Bruckheimer and A.C. Brownie, *Anal. Biochem.*, 89 (1978) 196.
193 R. Ballerini, M. Chinol and M. Ghelardoni, *J. Chromatogr.*, 193 (1980) 413.
194 V. Das Gupta and A.G. Ghanekar, *J. Pharm. Sci.*, 67 (1978) 889.
195 J.H. van den Berg, C.R. Mol, R.S. Deelder and J.H. Thijssen, *Clin. Chim. Acta*, 78 (1977) 165.
196 T.H. Chan, M. Moreland, W.T. Hum and M.K. Birmingham, *J. Steroid Biochem.*, 8 (1977) 243.
197 R.L. Farhi and C. Monder, *Anal. Biochem.*, 90 (1978) 58.
198 J.C.K. Loo, A.G. Butterfield, J. Moffatt and N. Jordan, *J. Chromatogr.*, 143 (1977) 275.
199 K.H. Muller and B. Stuber, *Pharm. Acta Helv.*, 53 (1978) 124.
200 D.C. Garg, J.W. Ayres and J.G. Wagner, *Res. Commun. Chem. Pathol. Pharmacol.*, 18 (1977) 137.
201 M.D. Smith and D.J. Hoffman, *J. Chromatogr.*, 168 (1979) 163.
202 J.W. Higgins, *J. Chromatogr.*, 115 (1975) 232.
203 G. Gordon and P.R. Wood, *Proc. Anal. Div. Chem. Soc.*, 14 (1977) 30.
204 A. Ryrfeldt, M. Tonnesson, E. Nilsson and A. Wikby, *J. Steroid Biochem.*, 10 (1979) 317.
205 S.E. Tsuei, J.J. Ashley, R.G. Moore and W.G. McBride, *J. Chromatogr.*, 145 (1978) 213.
206 A. Li Wan Po, W.J. Irwin and Y.W. Yip, *J. Chromatogr.*, 176 (1979) 399.
207 G.B. Neurath and D. Ambrosius, *J. Chromatogr.*, 163 (1979) 230.
208 H.M. Abdou, T.M. Ast and F.J. Cioffi, *J. Pharm. Sci.*, 67 (1978) 1397.

209 S.L. Douglas, *Chromatogr. Sci.*, 9 (1978) 917.

210 F.A. Fitzpatrick, *Advan. Chromatogr.*, 16 (1978) 37.

211 I. Gregorová, Z. Tomsová and K. Macek, *Ergeb. Exp. Med.*, 20 (1976) 51.

212 T. Okumura and A. Azuma, *Bunseki Kagaku*, 28 (1979) 235.

213 L.R. Treiber, *J. Chromatogr.*, 124 (1976) 69.

214 S. Hara and K. Mibe, *Chem. Pharm. Bull.*, 23 (1975) 2850.

215 V.R. Mattox, R.D. Litwiller and P.C. Carpenter, *J. Chromatogr.*, 175 (1979) 243.

216 R.M. Scott and R.T. Sawyer, *Microchem. J.*, 20 (1975) 309.

217 K.L. Bajaj and J.L. Ahuja, *J. Chromatogr.*, 172 (1979) 417.

218 J.C. Kohli and A.K. Arora, *Ann. Chim.*, 2 (1977) 67.

219 J.C. Kohli and N.S. Dhatiwal, *Riechst., Aromen, Kosmet.*, 27 (1977) 132.

220 Y. Yamaguchi, *J. Chromatogr.*, 163 (1979) 253.

221 H.L. Bradlow, *Steroids*, 30 (1977) 581.

222 K.D. Setchell, B. Almé, M. Axelson and J. Sjövall, *J. Steroid Biochem.*, 7 (1976) 615.

223 M. Axelson and J. Sjövall, *J. Steroid Biochem.*, 8 (1977) 683.

224 C.W. Vose, D.R. Boreham and C.J. Lewis, *J. Chromatogr.*, 179 (1979) 187.

225 P. Germeau and J. Duvivier, *J. Chromatogr.*, 129 (1976) 471.

226 H. Gips, K. Korte, B. Meinecke and P. Bailer, *J. Chromatogr.*, 193 (1980) 322.

227 D. Apter, P. Jänne, P. Karvonen and R. Vihko, *Clin. Chem.*, 22 (1976) 32.

228 R.J. Fairclough, M.A. Rabjohns and A.J. Peterson, *J. Chromatogr.*, 133 (1977) 412.

229 S. Hara and S. Hayashi, *J. Chromatogr.*, 142 (1977) 689.

230 S. Hara, M. Hirasawa, S. Miyamoto and A. Ohsawa, *J. Chromatogr.*, 169 (1979) 117.

231 C. Hesse and W. Hövermann, *Chromatographia*, 6 (1973) 345.

232 J.H.M. van den Berg, J. Milley, N. Vonk and R.S. Deelder, *J. Chromatogr.*, 132 (1977) 421.

233 M.J. O'Hare, E.C. Nice, R. Magee-Brown and H. Bullman, *J. Chromatogr.*, 125 (1976) 357.

234 E.C. Nice and M.J. O'Hare, *J. Chromatogr.*, 166 (1978) 263.

235 M.P. Kautsky and D.D. Hagerman, *Chromatogr. Sci.*, 10 (1979) 123.

236 M. Schöneshöfer and H.J. Dulce, *J. Chromatogr.*, 164 (1979) 17.

237 R.C. Cochran and L.L. Ewing, *J. Chromatogr.*, 173 (1979) 175.

238 P.G. Satyaswaroop, E. Lopez de la Osa and E. Gurpide, *Steroids*, 30 (1977) 139.

239 R. Horikawa, T. Tanimura and Z. Tamura, *J. Chromatogr.*, 168 (1979) 526.

240 V. Loppinet and G. Siest, *Pharm. Biol.*, 9 (1975) 437.

241 M. Axelson, *J. Steroid Biochem.*, 8 (1977) 693.

242 M. Axelson, *Anal. Biochem.*, 86 (1978) 133.

243 T. Nambara, T. Iwata and K. Kigasawa, *J. Chromatogr.*, 118 (1976) 127.

244 K. Jacob and W. Vogt, *J. Chromatogr.*, 150 (1978) 339.

245 H. Miyazaki, M. Ishibashi, M. Itoh, K. Yamashita and T. Nambara, *J. Chromatogr.*, 133 (1977) 311.

246 M.A. Quilliam and J.B. Westmore, *Anal. Chem.*, 50 (1978) 59.

247 G. Phillipou, *J. Chromatogr.*, 129 (1976) 384.

248 H. Hosoda, K. Yamashita, H. Sagae and T. Nambara, *Chem. Pharm. Bull.*, 23 (1975) 2118.

249 R.W. Kelly and P.L. Taylor, *Anal. Chem.*, 48 (1976) 465.

250 S.J. Gaskell and C.J.W. Brooks, *Biochem. Soc. Trans.*, 4 (1976) 111.

251 I.A. Blair and G. Phillipou, *J. Chromatogr. Sci.*, 16 (1978) 201.

252 R.W.H. Edwards, *J. Chromatogr.*, 153 (1978) 1.

253 R.W.H. Edwards, *J. Chromatogr.*, 154 (1978) 183.

254 P. Hurter and P. van Ree, *J. Chromatogr.*, 169 (1979) 93.

255 W.L. Zielinski, Jr., K. Johnston and G.M. Muschik, *Anal. Chem.*, 48 (1976) 907.

256 G.M. Janini, W.B. Manning, W.L. Zielinski, Jr. and G.M. Muschik, *J. Chromatogr.*, 193 (1980) 444.

257 B.S. Thomas, *J. High Resolut. Chromatogr. Chromatogr. Commun.*, 3 (1980) 241.

258 U. Matthiesen and W. Staib, *Chromatographia*, 10 (1977) 70.

259 F. Berthou, D. Picart, L. Bardou and H.H. Floch, *J. Chromatogr.*, 118 (1976) 135.

260 C. Madani, E.M. Chambaz, M. Rigaud, J. Durand and P. Chebroux, *J. Chromatogr.*, 126 (1976) 161.

261 J. Krupčík, G.A. Rutten and J.A. Rijks, *Chem. Zvesti*, 30 (1976) 469.

262 P. Sandra, M. Verzele and E. Vanluchene, *Chromatographia*, 8 (1975) 499.
263 M. Novotny, M.P. Maskarinec, A.T. Steverink and R. Farlow, *Anal. Chem.*, 48 (1976) 468.
264 H. Ludwig, J. Reiner and G. Spiteller, *Chem. Ber.*, 110 (1977) 217.
265 C.D. Pfaffenberger and E.C. Horning, *J. Chromatogr.*, 112 (1975) 581.
266 J. Reiner and G. Spiteller, *Monatsh. Chem.*, 106 (1975) 1415.
267 V. Fantl and C.H. Gray, *Clin. Chim. Acta*, 79 (1977) 237.
268 W.J. Leunissen and J.H.H. Thijssen, *J. Chromatogr.*, 146 (1978) 365.
269 G. Phillipou, R.F. Seamark and L.W. Cox, *Aust. N. Z. J. Med.*, 8 (1978) 63.
270 H. Ludwig, G. Spiteller, D. Matthaei and F. Scheler, *J. Chromatogr.*, 146 (1978) 381.
271 C.D. Pfaffenberger, L.R. Malinak and E.C. Horning, *J. Chromatogr.*, 158 (1978) 313.
272 D.S. Millington, *J. Steroid Biochem.*, 6 (1975) 239.
273 L. Siekmann, A. Siekmann and H. Breuer, *Advan. Mass Spectrom. Biochem. Med.*, 2 (1977) 419.
274 G. Sturm and E. Stähler, *Z. Anal. Chem.*, 279 (1976) 164.
275 R.F. Seamark, G. Phillipou and J.E. McIntosh, *J. Steroid Biochem.*, 8 (1977) 885.
276 B.P. Lisboa, H.T. Schneider and H. Breuer, *Acta Endocrinol. (Copenhagen)*, Suppl. 82 (1976) 57.
277 F.M. Menger, J.J. Hopkins, G.S. Cox, M.J. Maloney and F.L. Bayer, *Anal. Chem.*, 50 (1978) 1135.
278 J. Sjövall, in L. Bianchi, W. Gerok and K. Sickinger (Editors), *Liver and Bile*, Falk Symp. No. 23, University Park Press, Baltimore, MD, 1977, p. 67.
279 P. Eneroth, in G.V. Marinetti (Editor), *Lipid Chromatographic Analysis*, Vol. 3, Dekker, New York, 2nd Edn., 1976, p. 819.
280 S. Ikawa and M. Goto, *J. Chromatogr.*, 114 (1975) 237.
281 R. Spears, D. Vukusich, S. Mangat and B.S. Reddy, *J. Chromatogr.*, 116 (1976) 184.
282 S. Ikawa, *J. Chromatogr.*, 117 (1976) 227.
283 W.T. Beher, J. Sanfield, S. Stradnieks and G.J. Lin, *J. Chromatogr.*, 155 (1978) 421.
284 M.N. Chavez, *J. Chromatogr.*, 162 (1979) 71.
285 M.N. Chavez and C.L. Krone, *J. Lipid Res.*, 17 (1976) 545.
286 R. Beke, G.A. de Weerdt, J. Parijs, W. Huybrechts and F. Barbier, *Clin. Chim. Acta*, 70 (1976) 197.
287 S.K. Goswami and C.F. Frey, *Biochem. Med.*, 17 (1977) 20.
288 S.K. Goswami and C.F. Frey, *J. Chromatogr.*, 145 (1978) 147.
289 J.C. Touchstone, R.E. Levitt, R.D. Soloway and S.S. Levin, *J. Chromatogr.*, 178 (1979) 566.
290 A.K. Batta, G. Salen and S. Shefer, *J. Chromatogr.*, 168 (1979) 557.
291 J.C. Touchstone, R.E. Levitt, S.S. Levin and R.D. Soloway, *Lipids*, 15 (1980) 386.
292 G. Parmentier and H. Eyssen, *J. Chromatogr.*, 152 (1978) 285.
293 R. Raedsch, A.F. Hofmann and K. Tserng, *J. Lipid Res.*, 20 (1979) 796.
294 I.A. Macdonald, *J. Chromatogr.*, 136 (1977) 348.
295 M. Jirsa and H. Soldátová, *J. Chromatogr.*, 157 (1978) 449.
296 E. Rocchi and G. Salvioli, *Minerva Gastroenterol.*, 22 (1976) 222.
297 W.A. Taylor, K.G. Blass and C.S. Ho, *J. Chromatogr.*, 168 (1979) 501.
298 H.K. Kim and D. Kritchevsky, *J. Chromatogr.*, 117 (1976) 222.
299 D.T. Belobaba, G.L. Carlson and N.F. La Russo, *J. Chromatogr.*, 172 (1979) 410.
300 R. Beke, G.A. de Weerdt and F. Barbier, *J. Chromatogr.*, 193 (1980) 504.
301 N.A. Parris, *J. Chromatogr.*, 133 (1977) 273.
302 C.A. Bloch and J.B. Watkins, *J. Lipid Res.*, 19 (1978) 510.
303 T. Laatikainen, P. Lehtonen and A. Hesso, *Clin. Chim. Acta*, 85 (1978) 145.
304 D. Baylocq, A. Guffroy, F. Pellerin and J.P. Ferrier, *C. R. Hebd. Seances Acad. Sci. Ser. C*, 286 (1978) 71.
305 R. Shaw and W.H. Elliott, *Lipids*, 13 (1978) 971.
306 J. Goto, H. Kato and T. Nambara, *J. Liquid Chromatogr.*, 3 (1980) 645.
307 N. Parris, *Anal. Biochem.*, 100 (1979) 260.
308 K. Shimada, M. Hasegawa, J. Goto and T. Nambara, *J. Chromatogr.*, 152 (1978) 431.
309 B. Almé, A. Bremmelgaard, J. Sjövall and P. Thomassen, *J. Lipid Res.*, 18 (1977) 339.
310 J. Goto, M. Hasegawa, H. Kato and T. Nambara, *Clin. Chim. Acta*, 87 (1978) 141.
311 S. Miyazaki, H. Tanaka, R. Horikawa, H. Tsuchiya and K. Imai, *J. Chromatogr.*, 181 (1980) 177.

312 J. Goto, H. Kato and T. Nambara, *Lipids*, 13 (1978) 908.
313 S. Okuyama, D. Uemura and Y. Hirata, *Chem. Lett.*, 1976, 679.
314 B. Shaikh, N.J. Pontzer, J.E. Molina and M.I. Kelsey, *Anal. Biochem.*, 85 (1978) 47.
315 F. Stellaard, D.L. Hachey and P.D. Klein, *Anal. Biochem.*, 87 (1978) 359.
316 A. Kuksis, in G.V. Marinetti (Editor), *Lipid Chromatographic Analysis*, Vol. 2, Dekker, New York, 2nd Edn., 1976, p. 479.
317 G.S. Tint, B. Dayal, A.K. Batta, S. Shefer, F.W. Cheng, G. Salen and E.H. Mosbach, *J. Lipid Res.*, 19 (1978) 956.
318 H. Miyazaki, M. Ishibashi and K. Yamashita, *Biomed. Mass Spectrom.*, 5 (1978) 469.
319 G. Karlaganis and G. Paumgartner, *J. High Resolut. Chromatogr. Chromatogr. Commun.*, 1 (1978) 54.
320 R.L. Campbell, J.S. Gantt and N.D. Nigro, *J. Chromatogr.*, 155 (1978) 427.
321 W.G. Brydon, K. Tadesse, D.M. Smith and M.A. Eastwood, *J. Chromatogr.*, 172 (1979) 450.
322 W.S. Harris, L. Marai, J.J. Myher and M.T.R. Subbiah, *J. Chromatogr.*, 131 (1977) 437.
323 P.A. Szczepanik, D.L. Hachey and P.D. Klein, *J. Lipid Res.*, 19 (1978) 280.
324 I.M. Yousef, M.M. Fisher, J.J. Myher and A. Kuksis, *Anal. Biochem.*, 75 (1976) 538.
325 R. Galeazzi, E. Kok and N. Javitt, *J. Lipid Res.*, 17 (1976) 288.
326 K. Imai, Z. Tamura, F. Mashige and T. Osuga, *J. Chromatogr.*, 120 (1976) 181.
327 B.C. Musial and C.N. Williams, *J. Lipid Res.*, 20 (1979) 78.
328 P. Child, A. Kuksis and J.J. Myher, *Can. J. Biochem.*, 57 (1979) 639.
329 C.B. Campbell, C. McGuffie and L.W. Powell, *Clin. Chim. Acta*, 63 (1975) 249.
330 M. Takahashi, R.F. Raicht, A.N. Sarwal, E.H. Mosbach and B.I. Cohen, *Anal. Biochem.*, 87 (1978) 594.
331 T. Kuramoto, B.I. Cohen and E.H. Mosbach, *Anal. Biochem.*, 71 (1976) 481.
332 M.F. Ruh and C. Black, *J. Chromatogr.*, 116 (1976) 480.
333 R. Hardman and T.V. Benjamin, *J. Chromatogr.*, 131 (1977) 468.
334 R.T. Mayer and J.A. Svoboda, *Steroids*, 31 (1978) 139.
335 C.F. Poole, S. Singhawangcha and A. Zlatkis, *J. High Resolut. Chromatogr. Chromatogr. Commun.*, 1 (1978) 96.
336 M. Hashimoto, *Kagaku no Ryoiki, Zokan*, 109 (1976) 127.
337 I.D. Wilson, C.R. Bielby, E.D. Morgan and A.E.M. McLean, *J. Chromatogr.*, 194 (1980) 343.
338 M. Hori, *Steroids*, 14 (1969) 33.
339 M.W. Gilgan, *J. Chromatogr.*, 129 (1976) 447.
340 R. Lafont, G. Martin-Sommé and J.-C. Chambet, *J. Chromatogr.*, 170 (1979) 185.
341 G.M. Holman and R.W. Meola, *Insect Biochem.*, 8 (1978) 275.
342 S. Ogawa, A. Yoshida and R. Kato, *Chem. Pharm. Bull.*, 25 (1977) 904.
343 C.R. Bielby, A.R. Gande, E.D. Morgan and I.D. Wilson, *J. Chromatogr.*, 194 (1980) 43.
344 J.P. Delbecque, M. Prost, B.F. Maume, J. Delachambre, R. Lafont and B. Mauchamp, *C.R. Hebd. Seances Acad. Sci. Ser. D*, 281 (1975) 309.
345 M.E. Gustafson, A.W. Nicholas and J.P. Rosazza, *J. Chromatogr.*, 137 (1977) 465.
346 I.R. Hunter, M.K. Walden, E. Heftmann, E. Glotter and I. Kirson, *J. Chromatogr.*, 170 (1979) 437.
347 Z. Szeleczky, *J. Chromatogr.*, 178 (1979) 453.
348 K. Yoshioka, D.S. Fullerton and D.C. Rohrer, *Steroids*, 32 (1978) 511.
349 C.J. Clarke and P.H. Cobb, *J. Chromatogr.*, 168 (1979) 541.
350 T. Kartnig and P. Kobosil, *J. Chromatogr.*, 138 (1977) 238.
351 D.B. Faber, *J. Chromatogr.*, 142 (1977) 421.
352 D.B. Faber, A. de Kok and U.A.T. Brinkman, *J. Chromatogr.*, 143 (1977) 95.
353 L. Storstein, *J. Chromatogr.*, 117 (1976) 87.
354 W. Lindner and R.W. Frei, *J. Chromatogr.*, 117 (1976) 81.
355 P.H. Cobb, *Analyst (London)*, 101 (1976) 768.
356 F. Erni and R.W. Frei, *J. Chromatogr.*, 130 (1977) 169.
357 M.C. Castle, *J. Chromatogr.*, 115 (1975) 437.
358 K. Shimada, M. Hasegawa, K. Hasebe, Y. Fujii and T. Nambara, *J. Chromatogr.*, 124 (1976) 79.
359 R. Verpoorte, Phan-quôc-Kinh and A. Baerheim Svendsen, *J. Nat. Prod.*, 43 (1980) 347.
360 F. Nachtmann, H. Spitzy and R.W. Frei, *J. Chromatogr.*, 122 (1976) 293.
361 J.C. Gfeller, G. Frey and R.W. Frei, *J. Chromatogr.*, 142 (1977) 271.

Chapter 15

Carbohydrates

SHIRLEY C. CHURMS

CONTENTS

15.1. INTRODUCTION

Since the chromatography of carbohydrates was reviewed in the third edition [1], the field has been revolutionized as a result of the rapid advances in HPLC. Many of the classical techniques once regarded as important in carbohydrate chemistry have now been superseded by HPLC methods, and this remains one of the main areas of growth. Another is seen in the ever-widening application of affinity chromatography to the isolation of polysaccharides and glycoproteins, following the discovery of an increasing number of lectins having appropriate specificity. GLC, particularly when used in conjunction with MS, is of crucial importance in structural studies of carbohydrates, and extensive research continues to produce improvements in derivatization, stationary phases, and other factors determining the degree of resolution in such analyses. The main advance in planar chromatography in recent years has been the advent of HPTLC plates, capable of much higher speed and resolution than was formerly possible with thin-layer methods. In this chaper emphasis will be laid on the use of the newer techniques in the isolation and analysis of carbohydrates and their derivatives, but within this context some of the longer-established methods, such as ion-exchange and gel-permeation chromatography, remain relevant, particularly in view of their compatibility with the various automated detection systems now available for chromatographic analysis of carbohydrates.

15.2. COLUMN CHROMATOGRAPHY

The term "column chromatography" is applied to many different types of system, in which the separation mechanism may depend upon adsorption, partition, ion exchange, or molecular exclusion, according to the nature of the stationary phase. However, it is becoming increasingly apparent that the boundaries between these various mechanisms are vague, as more than one mechanism is frequently involved in a particular separation, especially in the case of some of the packings now used in HPLC. For this reason, the chromatographic systems reviewed in this section have been grouped largely according to the type of column packing used, rather than the mode of sorption involved, except in cases where the latter is self-evident.

15.2.1. Classical methods: use in preparative chromatography

The once important techniques of adsorption chromatography on charcoal–Celite columns [2] and partition chromatography on powdered cellulose [3] are now obsolete as analytical methods, but retain some value for preparative purposes. An example is afforded by the use of preliminary fractionation on a column (58 × 5 cm) packed with charcoal–Celite (dry-wt. ratio 2:3) in the isolation of the oligosaccharides produced on partial hydrolysis of a large sample (50 g) of birch xylan [4]. Stepwise elution with aqueous solutions of ethanol of increasing concentration produced, after four separate runs (each with 25% of the hydrolyzate), fractions from which the oligosaccharides up to the nonasaccharide were subsequently easily isolated by other means (see Chap. 15.2.3).

Chromatography on a cellulose column at ca. 30°C, with 2-butanone–water azeotrope as eluent [5], can be used to isolate individual methylated sugars from the mixtures obtained on hydrolysis of methylated polysaccharides [6]. Stepwise elution with nine mixtures of light petroleum (b.r. 100–120°C) and water-saturated 1-butanol (from 7:3 to 7:50), followed by 1-butanol half-saturated with water, has also proved successful when applied to the preparative-scale separation of methylated sugars on cellulose columns [7].

A method recently recommended for the preparation of macroquantities of D-glucose and its α-$(1 \rightarrow 4)$-linked oligomers from corn syrups involves fractionation of the derived acetate esters on silica gel, with benzene–ethyl acetate (2:1) as eluent [8]. Gram amounts of glucose, maltose, and oligomers up to maltopentaose have been obtained from a syrup (10 g) having dextrose equivalent 43, following acetylation and chromatography of a chloroform extract of the product on a column (100 × 3 cm ID) packed with silica gel (70–230 mesh). With a flowrate of ca. 400 ml/h the separation took 14 h. The sugars were recovered from the fractions collected from this column by deacetylation (with 0.1 M sodium methoxide), deionization with mixed ion-exchange resins and evaporation of the resulting solutions to syrups. It has been suggested that separation of these oligosaccharides as their benzoate esters may improve resolution, owing to the greater differences in molecular weight between oligomers.

15.2.2. High-performance liquid chromatography

Of the sorbents mentioned in Chap. 15.2.1, only silica has the mechanical strength necessary to withstand the high pressures used in modern HPLC. This material is, indeed, an eminently suitable support for HPLC for a number of reasons [9], and currently a high proportion of all such separations are achieved by the use of columns packed with microparticulate silica, either as such or as a support for a chemically bonded phase of appropriate polarity. The chromatography of carbohydrates is no exception. Unmodified silica is too polar to be effective in the chromatography of unsubstituted sugars, but prior derivatization does permit the use of silica columns in HPLC of carbohydrates. The development of packings in which a phase suitable for partition chromatography of sugars is bonded to microparticulate silica has obviated the necessity for derivatization, except where this procedure has certain advantages (see Chap. 15.2.2.1), and such packings are now widely used in carbohydrate analysis.

It must be emphasized that HPLC of carbohydrates is still in the developmental stage and that different chromatographic systems are constantly being evaluated. Of these, the most successful has been that employing a microparticulate cation-exchange resin in the calcium form, which has the signal advantage of permitting elution simply with water. This method is rapidly gaining acceptance as a standard procedure for certain analyses [10]. Applications of each of the three types of column packing mentioned are reviewed below.

References on p. B276

15.2.2.1. Microparticulate silica

The behavior of commonly encountered sugars and alditols in chromatography on microparticulate silica (particle size 5 μm) with eluents consisting of mixtures of ethyl formate, methanol, and water in different proportions has been investigated by Rocca and Rouchouse [11] with the specific objective of finding a rapid method of separating D-fructose, sucrose, lactose, and sorbitol (D-glucitol), which would be of value in the analysis of foodstuffs. Some measure of success was achieved in this regard, and it was found possible to resolve such a mixture in less than 20 min on a stainless-steel column (150 × 4.6 mm ID), packed with this silica, with ethyl formate–methanol–water (6:2:1), at a flowrate of 0.9 ml/min, as the mobile phase. However, while separations of such widely different components are feasible by this method, it is evident, from capacity factors determined by Rocca and Rouchouse [11] and by Churms and Seeman [12], that this system is totally unsuitable for chromatography of mixtures of sugars of the same class.

As has been mentioned, it is in the chromatography of derivatized carbohydrates that column packings consisting of unmodified silica have proved most useful. McGinnis and Fang [13] have published data on the behavior of a variety of partially and completely substituted carbohydrates on a stainless-steel column (250 × 4.6 mm ID) packed with microparticulate (10 μm) silica. The compounds investigated included glycosides, partially methylated sugars, isopropylidene and benzylidene derivatives, and a series of peracetylated carbohydrates. Use of a mobile phase consisting of acetonitrile–water (9:1), at a flowrate of 1.2 ml/min, permitted the separation of a mixture of hexopyranosides in ca. 20 min, with resolution of α- and β-anomers in some cases. For example, the retention times of methyl α- and β-D-mannopyranoside under these conditions were 15.3 and 18.3 min, respectively. The anomeric glycosides were separated from each other and from D-mannose (retention time 22.4 min), and the value of this technique as a means of following glycosidation of D-mannose was thereby demonstrated. Other separations of potential interest in connection with synthetic reactions of carbohydrates included those of 4,6-O-benzylidene- from 1,2-O-isopropylidene-α-D-glucopyranose with acetonitrile–water (18:1) and 1,2:5,6-di-O-isopropylidene-α-D-glucofuranose from 2,3:5,6-di-O-isopropylidene-D-mannofuranose with *n*-hexane–ethyl acetate (1:3), both of which were achieved in less than 15 min under the conditions used.

The retention data obtained by McGinnis and Fang [13] for a series of peracetylated carbohydrates, which included the acetates of α- and β-anomers of cellobiose, D-gluco- and D-galactopyranose, and methyl and phenyl D-glucopyranosides, showed that HPLC on microparticulate silica [with *n*-hexane–ethyl acetate (1:1) as the mobile phase] can be useful as a means of resolving anomers of these derivatives. This application has been further studied by Thiem and coworkers [14,15], using a column (250 × 3 mm ID) of silica (5 μm). After a thorough investigation of the effects of solvent composition and flowrate on the capacity factors for a series of peracetates of 15 monosaccharides and of 12 disaccharides, optimal separations were achieved with *n*-hexane–acetone (10:1) at 0.46 ml/min in the case of the monosaccharide acetates, or *n*-pentane–acetone (7:2) at 0.39 ml/min for the disaccharide derivatives. Resolution of all 15 peracetylated monosaccharides was obtained in 2 h, with excellent separation of the anomeric pairs derived from D-arabino-, D-gluco-, D-galacto-, and D-mannopyranose, and of the derivatives of the epimeric pairs D-gluco- and D-mannopyranose, and D-galacto- and D-talo-pyranose, which are difficult to resolve by other methods. Resolution of the disaccharide derivatives was less satisfactory, but some useful separations (for example, of α- and β-cellobiose and α- and β-mannobiose) were achieved.

A major advantage of derivatization of carbohydrates prior to HPLC is that the introduction of suitable chromophores makes possible the use of a UV detector, which is much more sensitive than the differential refractometers generally used to monitor the emergence of unsubstituted carbohydrates from HPLC columns (see Chap. 15.2.7). For this reason, and also because they are readily isolated in crystalline form and are suitable for NMR and/or MS, benzoates and 4-nitro-benzoates of carbohydrates have recently been thoroughly examined by HPLC. The potential of the use of perbenzoylated derivatives was first demonstrated by Lehrfeld [16], who reported the resolution of a mixture containing the α- and β-anomers of D-xylose, D-mannose, D-glucose, and D-galactose, as well as sucrose, maltose, lactose, and maltotriose. They were chromatographed as the perbenzoates, on a column (2 m $\times$ 2 mm ID) of pellicular silica, with a mobile phase consisting of a linear gradient of diethyl ether in n-hexane (0 to 99% in 110 min). The response of the standard (254 nm) UV detector used was good (3–5 μg of each sugar were injected). This applied also to the benzoates derived from ethylene glycol and glycerol, which were eluted well before the sugar derivatives and were completely resolved by this method. It is evident, therefore, that HPLC of the benzoate esters offers an efficient means of analyzing the complex mixtures of sugars and polyols encountered in degradative studies of polysaccharides.

Application of this method should be expedited following a recent comprehensive study by White et al. [17], who have published retention data for the perbenzoylated derivatives of all the common neutral and amino sugars, the alditols xylitol, D-glucitol, galactitol, and D-mannitol; the disaccharides maltose, lactose, and sucrose; and several methyl glycosides. The HPLC system used consisted of a column (300 $\times$ 3.9 mm) of silica (10 μm), eluted with n-hexane–ethyl acetate (5:1) at 1 ml/min, and a UV detector (254 nm). Excellent separations of anomeric pairs of methyl glycosides were obtained by HPLC of their perbenzoylated derivatives. The simple sugars were found to give multiple peaks, in some cases (e.g. D-fructose, L-rhamnose) as many as five, on HPLC following perbenzoylation. This may facilitate identification of isolated sugars, but complicates the interpretation of chromatograms given by mixtures. For this reason, Thompson [18] advocates oximation prior to benzoylation, in order to obtain a single derivative for each sugar, irrespective of the number of isomers present; the formation of the benzyloxime, which introduces another UV-absorbing chromophore, is particularly recommended.

Incorporation of the chromophoric nitro group into the aryl portion of a benzoylated carbohydrate has been found to shift the absorption maximum from 230 to 260 nm, with a resulting increase in response by a UV detector operating at 254 nm. Nachtmann and Budna [19] have noted a marked increase in detector sensitivity to carbohydrates chromatographed as 4-nitrobenzoates, rather than benzoates, and have investigated the possible application of this method to the analysis of the mixture of D-glucose, D-fructose, D-glucitol, sucrose, and lactose present in certain pharmaceutical syrups. Using a column (150 × 3 mm ID), packed with silica (5 μm), and a mobile phase of n-hexane–chloroform–acetonitrile–tetrahydrofuran (20:10:2:1) at 1.46 ml/min, these workers obtained good resolution of the 4-nitrobenzoates of α- and β-D-glucose, D-glucitol, sucrose, and α- and β-lactose in only 5 min. The use of a spectrophotometric detector at 260 nm permitted the analysis of nanogram quantities of these carbohydrate derivatives. A syrup containing D-fructose showed peaks corresponding to the five isomers of the 4-nitrobenzoylated derivatives, as in the case of the benzoates, all five being resolved in 10 min with n-hexane–chloroform–acetonitrile–water (100:30:19:1), at 1.4 ml/min, as the mobile phase. Another interesting separation acieved by HPLC of 4-nitrobenzoylated derivates was that of the polyols arabinitol, xylitol, mannitol, glucitol, and maltitol, which were resolved by Schwarzenbach [20] as 4-nitrobenzoates on a column (250 × 3 mm ID) of silica (5 μm), eluted with n-hexane–chloroform–acetonitrile (5:2:1) at 0.8 ml/min. Under these conditions this difficult separation was complete in 20 min.

HPLC on microparticulate silica is also proving valuable in the analysis of glycolipids, especially brain gangliosides [21]. Perbenzoylation prior to chromatography is recommended by Bremer et al. [22], who have succeeded in quantitating picomole amounts of monosialogangliosides by HPLC of the derived benzoates on a column (500 × 2.1 mm) of silica (10 μm), with a linear gradient of dioxane in n-hexane (7 to 23% in 18 min) and photometric detection at 230 nm.

The substituted, unsaturated disaccharides produced on enzymatic degradation of chondroitin sulfates A, B and C (chondroitin 4-sulfate, dermatan sulfate, and chondroitin 6-sulfate, respectively) are well separated by HPLC on 5-μm silica: on a commercial pre-packed column (250 × 4.6 mm ID), eluted with dichloromethane–methanol–0.5 M ammonium formate, pH 4.8 (30:17:3) at 2 ml/min, complete separation of the three disaccharide products in 20 min has been reported [23]. The absorbance of these unsaturated disaccharides is sufficiently high to permit detection of ca. 100 ng of each by a standard UV detector (254 nm). This method is potentially valuable as a rapid assay of glycosaminoglycans in urine for the diagnosis of mucopolysaccharidoses such as Hunter syndrome.

15.2.2.2. Bonded phases

The introduction of HPLC packings in which suitable polar phases (amino, nitrile, or a combination of the two) are chemically bonded to microparticulate silica has greatly facilitated the analysis of underivatized carbohydrates. Although some authors have reported a degree of success with bonded-phase packings prepared in their own laboratories [24,25], use of the standardized materials now available commercially, usually in pre-packed stainless-steel columns, is recommended. Of these, the μBondapak column manufactured specifically for carbohydrate analysis by Waters Assoc. has been the most widely used, but good results have been achieved also with several other packings, such as Partisil-10 PAC (Whatman, Reeve

Angel Division), Micropak NH$_2$ (Varian), and more recently [26] Chromosorb LC-9 (Johns-Manville) and Lichrosorb-NH$_2$ (Merck). In all cases, silica having an average particle diameter of 10 μm is used as the support to which the polar phase is bonded.

The solvent system universally employed in HPLC of carbohydrates on bonded-phase columns is acetonitrile–water, the proportions used depending on the nature of the substances to be separated. Linden and Lawhead [27] investigated the conditions necessary for optimal resolution and quantitation of fructose, glucose, sucrose, melibiose, raffinose, and the three isomeric fructosylsucroses 1-, 6- and neokestose, as well as of the α-(1 → 4)-linked D-glucose oligomers (maltodextrins) present in starch hydrolyzates, and concluded that a mobile phase containing 75% of acetonitrile was the best choice for chromatography of oligosaccharides, while higher proportions were necessary for the resolution of monosaccharides. The resolution of

TABLE 15.1

CAPACITY FACTORS FOR HPLC OF SUGARS AND POLYOLS ON WATERS μBONDAPAK CARBOHYDRATE ANALYSIS COLUMN

Compound	Capacity factor (k')		References
	85% aqueous acetonitrile	75% aqueous acetonitrile	
Glycerol	1.3	–	28–30
meso-Erythritol	1.7	–	30
D-Arabinitol	2.8	–	28, 30
D-Glucitol	4.0	–	28, 30
Galactitol	4.3	–	28
D-Mannitol	4.3	–	28, 30
myo-Inositol	11.2	–	28–30
D-Ribose	1.7	–	12, 29
D-Lyxose	2.1	–	12
D-Xylose	2.2	–	12, 29, 30
L-Arabinose	2.8	–	12. 28, 29
D-Mannose	3.8	–	12, 28, 29
D-Glucose	4.2	0.9	12, 28–30
D-Galactose	4.7	–	12, 28, 29
2-Deoxy-D-glucose	1.7	–	12
6-Deoxy-D-glucose	2.0	–	12
L-Rhamnose	1.4	–	12, 29
L-Fucose	1.9	–	12, 29
D-Fructose	3.3	–	12, 28–30
D-*manno*-Heptulose	5.0	–	29
Sucrose	9.0	1.5	12, 28–30
Trehalose	12.4	–	30
Maltose	11.9	2.1	12, 29
Isomaltose	–	2.5	12
Lactose	15.0	2.4	12, 29
Melibiose	19.3	2.7	12, 29
Raffinose	28.2	3.4	12, 29

References on p. B276

glucose and fructose, which were not separated when the acetonitrile content of the eluent was 75%, improved to an increasing extent as the concentration of acetonitrile was increased to 90%. However, long retention times, accompanied by peak broadening, were observed on use of a mobile phase containing 90% of acetonitrile, and 85% was considered optimal for this purpose. In most of the reported HPLC analyses of sugars now in the literature, mobile phases containing 80–88% of acetonitrile have been used to resolve mixtures of monosaccharides, and 65–75% for oligosaccharides.

Under these conditions, and with flowrates of 1–2 ml/min, the common monosaccharides can be separated in ca. 20 min and oligosaccharides up to a degree of polymerization (DP) of 5 in ca. 30–40 min on the standard Waters μBondapak Carbohydrate column (300 × 3.9 mm ID). The application of this column to carbohydrate analyses of importance in the food industry has been reviewed by Conrad and Palmer [28], and further chromatographic data for sugars and polyols will be found in papers dealing with the use of this method in analyses of urine [29] and lichens [30]. Some typical values of capacity factors are listed in Table 15.1.

Resolution of higher oligosaccharides (DP > 5) within a reasonable time is not possible under the standard conditions used for HPLC of carbohydrates, but has been achieved by flow programing (65% aqueous acetonitrile at 2 ml/min for 10 min, then flowrate increased to 4 ml/min over 20 min), which allows the resolution of maltodextrins up to DP 10 in 30 min [28,31]. In an alternative approach, Rabel et al. [32] adjusted the pH of the mobile phase to 5.0 by addition of an acetate buffer and thereby reduced tailing and improved resolution. Using a column (250 × 4.6 mm ID) packed with Partisil-10 PAC (polar amino cyano phase), they resolved maltodextrins up to DP 10 in ca. 40 min by eluting at 1 ml/min with 65% acetonitrile in 0.0025 M sodium acetate, adjusted to pH 5.0 with acetic acid. It is of interest to note that the Partisil-10 PAC column has also proved highly effective in the separation of β-(1 → 4)-linked D-glucose oligomers (cellodextrins) and the corresponding reduced oligosaccharides [33]. Isocratic elution with acetonitrile–water (71:29) at 1.5 ml/min produced good resolution, in only 24 min, of a mixture containing members of both normal and reduced cellodextrin series up to DP 6.

Gradient elution should improve the resolution of higher oligosaccharides on bonded-phase columns, but this is not feasible with the RI detectors generally used in HPLC of unsubstituted carbohydrates. However, this difficulty has recently been obviated by D'Amboise et al. [26] and Noël et al. [34] by employing rapid post-column derivatization (Tetrazolium Blue colorimetric method; see Chap. 15.2.7). The resolution of homologous series of oligosaccharides (from hydrolyzed starch, xylan, and inulin) up to DP 20 has been achieved in 40 min or less on a column (250 × 4.3 mm ID), packed with Chromosorb-NH$_2$ (a bonded amine phase),with linear gradients of acetonitrile in water (70 to 62.5% or 66 to 57%) at 1 ml/min.

Further examples of novel applications of HPLC on bonded-phase packings in the carbohydrate field include the complete separation of the α-, β-, and γ-cyclodextrins (cycloamyloses consisting of 6, 7, and 8 α-(1 → 4)-linked D-glucose units, respectively) in 15 min on the Waters μBondapak Carbohydrate column, maintained at 25°C, with 70% aqueous acetonitrile at 2 ml/min [35], and that of 2-acetamido-2-deoxy-D-glucose and its β-(1 → 4)-linked oligomers (from the hydrolysis of chitin) up to the pentasaccharide in the same time under similar conditions [36]. The disaccharides produced on enzymatic degradation of the clinically important chondroitin sulfates A, B, and C (discussed in Chap. 15.2.2.1) are also rapidly separated on bonded-phase columns [37]. On the standard Partisil-10 PAC column (250 × 4.6 mm ID) the three disaccharides are completely resolved in 10 min with acetonitrile–methanol–0.5 M ammonium formate, pH 4.8 (12:3:5) at 2 ml/min, while on a column of the same dimensions packed with Lichrosorb-NH$_2$, eluted with methanol–0.5 M ammonium formate, pH 4.8 (7:13), also at 2 ml/min, separation is complete in 8 min. These sorbents provide better resolution and higher efficiency than microparticulate silica [23].

Simple separations involving partition chromatography may be achieved on columns packed with unmodified microparticulate silica, if a polyfunctional amine is added to the eluent in small proportion (0.1% before sample injection, 0.01% during chromatography) [38,39]. The resulting amine coating on the surface of the silica is sufficient to produce satisfactory resolution of D-fructose, D-glucose, sucrose, maltose, lactose, and raffinose. The method offers an acceptable alternative to expensive bonded-phase packings for laboratories concerned with analyses involving primarily these sugars. The one major disadvantage of bonded-amine packings is their tendency to deteriorate after only about 4 months of continual use [27,29]; this difficulty is obviated by the use of silica packings coated in situ, since the amine phase is regenerated after each use.

Reversed-phase partition HPLC on bonded-phase packings has hitherto found little application in HPLC of carbohydrates, but two recent papers are likely to stimulate great interest in this method. Wells and Lester [40] have reported excellent separations of oligosaccharides as their peracetylated derivatives on bonded octadecyl hydrocarbon phases (Vydac or Bondapak C$_{18}$/Corasil). Two columns (each 1 m × 3.2 mm ID), packed with these sorbents, were connected in series, maintained at 65°C, and eluted with a gradient (10 to 70%) of aqueous acetonitrile at 2 ml/min, resolving peracetylated oligosaccharides of the maltodextrin series, up to DP 30, in 80 min. Reversed-phase packings based on microparticulate silica (5 μm) can also be used to separate peralkylated oligosaccharides. The highly successful application of this method by Valent et al. [41], who eluted a pre-packed Zorbax ODS (Du Pont) column (250 × 4.6 mm ID) with acetonitrile–water (usually 1:1) at 0.5 ml/min, in the fractionation of complex mixtures of the partially methylated, partially ethylated oligosaccharides obtained during sequencing analysis of polysaccharides, represents a significant advance in the methodology of this field.

15.2.2.3. Ligand exchange on cation-exchange resins

The ability of cation-exchange resins, usually in the Li$^+$, K$^+$, Ca^{2+}, or Ba^{2+} form, to fractionate simple mixtures of monosaccharides and small oligosaccharides with water as the only eluent is well documented [42–45] and has been reviewed [46]. However, separations with the older resins were slow and resolution was, in general, somewhat inadequate, so that little interest was shown in this method until the introduction of microparticulate resins (having particle sizes in the range 10–30 μm)

TABLE 15.2

CHROMATOGRAPHIC DATA FOR SUGARS AND POLYOLS ELUTED WITH WATER FROM Ca^{2+}-FORM RESINS

Compound	Relative retention time *			
Resin	Aminex A-5	Aminex Q15-S	AG 50W-X4	Aminex Q150-S
Column	500 × 2.8 mm ID	600 × 7 mm ID	610 × 8 mm ID	600 × 9.5 mm ID
Flowrate	0.1 ml/min	1 ml/min	0.6 ml/min	1.5 ml/min
Temperature	Ambient	80°C	85°C	50°C
Ref.	47	28	50	30
Glycerol	1.6	1.48	–	1.54
meso-Erythritol	–	–	–	1.49
D-Ribitol	–	–	–	1.44
D-Arabinitol	–	–	–	1.81
D-Glucitol	2.4	1.63	–	2.28
Galactitol	–	–	–	–
D-Mannitol	1.9	1.35	–	1.79
myo-Inositol	–	–	–	1.29
D-Xylose	1.1; 1.2 **	1.19	1.11	1.08
L-Arabinose	–	–	1.26	–
D-Mannose	1.1; 1.3 **	1.12	1.16	–
D-Glucose	1.0 (15 min); 1.2 **	1.00 (15 min)	1.00 (28 min)	1.00 (9.5 min)
D-Galactose	1.2; 1.3 **	1.10	1.13	–
D-Gulose	1.8	–	–	–
D-Talose	3.8	–	–	–
D-Fructose	1.7 (skew peak)	1.18	1.18	1.32
Sucrose	0.9	0.84	0.86	0.88
Lactose	–	–	0.89	–

* Relative to D-glucose.

** Anomers resolved.

made possible a vast increase in column efficiency. Goulding [47] investigated the behavior of a number of sugars and alditols on the resin Aminex A-5 (Bio-Rad Labs.) with water as eluent, testing a wide variety of counterions with the objective of optimizing conditions for rapid separations on this resin (particle size $11 \pm 2 \mu m$). From the capacity factors obtained, it was evident that satisfactory separations were possible with Ag^+, Ca^{2+}, or La^{3+} as counterions; of these, Ca^{2+} is obviously the best choice for practical purposes. The use of microparticulate cation-exchange resins in the Ca^{2+} form, with water as eluent, has now become another standard method for HPLC of carbohydrates [10,28].

Some typical chromatographic data for sugars and polyols separated by this method are shown in Table 15.2. Columns are usually operated at elevated temperatures (50–85°C) to avoid peak distortion due to partial separation of anomers. The time required for any separation varies with the resin used, so that generalization is difficult: an example worth citing is the separation of the polyols glycerol, D-glucitol, and D-mannitol, frequently encountered in the food industry, from one another and from D-glucose, D-fructose, and sucrose in ca. 25 min on a column (600×7 mm ID), at 80°C, packed with Aminex Q 15-S ($22 \pm 3 \mu m$) in the Ca^{2+} form, eluted with water at 1 ml/min [28]. The method is particularly useful for the separation of polyols, which are better resolved under these conditions than by chromatography on bonded-phase packings [28,30]. This is probably related to marked differences in the extent of complex formation with Ca^{2+} among the various polyols. The mechanism involved in this type of chromatography is now generally believed [47,48] to be ligand exchange, the solute molecule exchanging with the water molecules held in the hydration sphere of the counterion on the resin to an extent dependent upon the availability of hydroxyl groups for coordination.

Chromatography on microparticulate resins in the Ca^{2+} form with water as eluent also affords rapid separations of homologous series of oligosaccharides up to DP 6 or 7. The method, which has been recommended as a standard procedure for analysis of fermentable carbohydrates (D-glucose, maltose, and maltotriose) in brewing syrups [10], permits the separation of D-glucose and maltodextrins up to DP 6 in only 20 min [49]; a column (300×7 mm ID), kept at 80°C (by use of an electrically heated column block), packed with Aminex 50W-X4 resin ($20-30 \mu m$) in the Ca^{2+} form, and eluted with water at 0.6 ml/min was the salient feature of this particular chromatographic system. With a longer column (610×8 mm ID), water-jacketed to maintain the temperature at 85°C, under otherwise similar conditions the oligosaccharides of the cellodextrin series up to DP 7 were separated from one another and from D-glucose in 36 min [50].

The members of each homologous series are eluted in order of decreasing molecular size. A linear relationship between their elution volumes (or retention times) and the logarithms of their molecular weights has been noted [49], which suggests that separation of oligosaccharides by this method depends mainly upon a molecular exclusion mechanism. As the larger molecules will be hindered in their approach to the resin counterions, chelation, which is believed to govern the behavior of monosaccharides and polyols in this chromatographic system, is unlikely to influence that of oligosaccharides to any appreciable extent.

References on p. B276

The separation of the α-, β- and γ-cyclodextrins by chromatography on the Ca^{2+} form of Aminex 50W-X4 resin with water at 90°C has recently been reported [51]. The retention time of each cyclodextrin was much longer than that of the corresponding linear analog and it has been suggested that a specific interaction, involving formation of inclusion complexes between the cyclodextrins and the sulfonated polystyrene resin matrix, may be a contributing factor, in addition to differences in molecular size.

The presence of extraneous matter in samples injected into columns packed with microparticulate ion-exchange resins causes deterioration in performance (the average lifetime of a column used continually for the analysis of corn syrups is estimated at about two months [49]). Inorganic salts are particularly undesirable, as are acids, which can, e.g., cause sucrose to be hydrolyzed to D-glucose and D-fructose on contact with hydrogen ions replacing the calcium ions originally on the resin. For this reason, Fitt [52] strongly advocates the use of a precolumn (100 × 3 mm ID), in which the eluent flows first through a cation-exchange resin (20–35 μm) in the H^+ form and then through an equal volume of an anion-exchange resin of similar particle size, in the OH^- form, before entering the analytical column. This has been found to prolong column life without significant loss of resolution.

15.2.3. Partition chromatography on ion-exchange resins

Partition chromatography of sugars and alditols on ion-exchange resins with aqueous ethanol as the eluent has been the subject of extensive research by Samuelson and coworkers [53–57], and some excellent reviews of the technique and its application to carbohydrates have been published [46,58,59]. This type of chromatography is governed by several factors: the partition of the polar carbohydrate molecules between the mobile phase and the resin phase, where the proportion of water is higher, is of major importance, but interactions of these molecules with the counterions on the resin and with the resin matrix also contribute, so that the overall mechanism is complicated and elution order is sometimes difficult to predict. In general, distribution coefficients increase with the number of hydroxyl groups in the sugars, but some exceptions are observed: e.g., D-talose is eluted before D-xylose in chromatography on an anion-exchange resin (sulfate form) in 88% ethanol [57]. In some cases the elution order can be reversed by a change of the counterions on the resin [56]. Distribution coefficients increase with increasing concentration of ethanol and decrease with increasing temperature, though the order of elution from a particular resin is not usually affected by changes in these conditions.

From the results of many comprehensive investigations by Samuelson and his coworkers it is evident that the best separations are achieved by use of an anion-exchange resin in the sulfate form or a cation-exchange resin in the lithium form, at temperatures of 75–90°C and with the concentration of ethanol in the eluent at 86–90% for the separation of mixtures of monosaccharides, or at 65–80% for oligosaccharides. The use of microparticulate resins (8–15 μm) effects excellent resolution of multi-component mixtures of sugars in ca. 3 h with columns 1 m × 2–6 mm, at flowrates of 8–20 ml/cm^2/min. For preparative-scale chromatography with wider columns (12–25 mm ID) and lower flowrates (1–5 ml/cm^2/min), resins of larger particle size may be used. The resins are usually strong-base anion exchangers or strong-acid cation exchangers having crosslinked polystyrene matrices. However, it should be noted that some separations, e.g., of D-fructose and D-tagatose, that are not possible when anion exchangers of this type are used, have been achieved with a quaternary ammonium exchanger based on crosslinked dextran [60]. The alternatives of using anion or cation exchangers may be regarded as complementary, since in some cases pairs of sugars not easily separated on one are well resolved on the other.

This method of chromatography is applicable to a wide variety of sugars, as well as to alditols [55] and sugar derivatives, such as glycosides and ethers [61]. Separation of D-glucuronic and D-galacturonic acid from each other and from neutral monosaccharides has been reported [62], the uronic acids being eluted, with 89% ethanol from a Li$^+$-form cation-exchange resin, well ahead of 2-deoxy-D-ribose and D-erythrose. Excellent separations of homologous series of oligosaccharides, up to DP 8 or 9, are possible: e.g., the members of the β-(1 → 4)-linked D-xylose series having DP from 1 to 9 have been separated by elution with 80% ethanol at 75°C from a column (910 × 4 mm ID), packed with a cation-exchange resin in the Li$^+$ form [4]. The various series of D-glucose oligomers (such as the malto-, cello-, laminari-, and isomaltodextrins) can be separated on cation- or anion-exchange resins, the optimal concentration of ethanol in the eluent varying with the type of glycosidic linkage present [63]. In contrast to their behavior in chromatography on cation-exchange resins with water (see Chap. 15.2.2.3), the oligosaccharides are eluted in order of increasing chain length by aqueous ethanol, provided that the ethanol concentration exceeds a certain critical value (usually 50–60% when a resin in the sulfate form is used, and 60–70% for a Li$^+$-form resin).

Chromatography of carbohydrates on ion-exchange resins in aqueous ethanol is readily adaptable to automation, and such systems were among the first to be coupled to a Technicon AutoAnalyzer (see Chap. 15.2.7). Larsson and Samuelson [54] and Samuelson [59] have made extensive use of the automated orcinol–sulfuric acid method in monitoring the elution of sugars from these columns, while for the detection of alditols an AutoAnalyzer based on the periodate–pentane-2,4-dione method was employed [55]. The latter was combined with the orcinol method for analyses of complex mixtures of sugars and alditols, and of reduced oligosaccharides [64]. Automated colorimetric methods involving the use of noncorrosive reagents, such as Tetrazolium Blue [62,65] and the copper–bicinchoninate reagent developed by Mopper and Gindler [66] and Mopper [67], have also been used in conjunction with chromatographic systems of this type, to which they are particularly suited by virtue of their compatibility with eluents containing ethanol. Use of the moving-wire detector, developed for HPLC (see Chap. 4.7), in chromatography of sugars by this method has been reported [68,69].

15.2.4. Ion-exchange chromatography

Two types of ion-exchange chromatography are of importance in the carbohydrate field. The use of anion-exchange resins in the borate form to separate sugars, as borate complexes, is recognized as a standard analytical procedure, and much recent research has been devoted to the improvement of the efficiency and speed of such chromatographic systems. Ion exchangers based on dextran or cellulose matrices are frequently employed in the fractionation of charged oligosaccharides and polysaccharides. Applications of both types of exchanger are reviewed below, with emphasis on recent developments. A survey of the work in this field published during the period 1962–1970 will be found in a comprehensive review by Jandera and Churáček [46].

15.2.4.1. Ion exchange on resins

The stability of the complexes formed by borate ions with sugars and polyols depends upon various structural factors, which govern the number of adjacent cis-hydroxyl groups in the carbohydrate molecules, and also upon experimental conditions, namely the pH and ionic strength of the medium, and the concentration of borate ions present. This was first exploited in a separation technique thirty years ago by Khym and Zill [70,71], who reported separations of mixtures of sugars on columns packed with the strong-base anion-exchange resin Dowex 1, in the borate form, by stepwise elution with borate buffers of increasing pH (8 to 9) and concentration. The method was subsequently applied also to the separation of alditols [72]. However, under the conditions originally used, resolution was poor and up to 60 h was required for the separation of sugar mixtures. Thus, the method was not generally adopted as an analytical procedure until extensive investigation of the influence of various factors, such as temperature, ionic strength and the particle size of the resin [73], resulted in vast improvements in speed and resolution. By using resins with an average particle size of 20 μm, column temperatures of ca. 50°C, and gradient elution with buffers of increasing concentration of borate ($0.1 \to 0.2\,M$) and chloride ($0 \to 0.2\,M$), ranging in pH from 8.00 to 9.50, it became possible to accomplish the analysis of multi-component mixtures of sugars in ca. 8 h, and a scheme of this type was adopted in the original automated Technicon Carbohydrate Analyzer [74].

In continuation of the search for optimal conditions for anion-exchange chromatography of sugars as borate complexes, Floridi [75] advocated a return to stepwise elution. With a system consisting of only two buffers ($0.025\,M$ potassium tetraborate–$0.125\,M$ boric acid, pH adjusted to 8.40 with M KOH, and $0.11\,M$ potassium tetraborate–$0.17\,M$ boric acid, pH 8.80), the column ($1.1\,m \times 6\,mm$ ID, packed with Dowex 1-X4 resin, 200–400 mesh, temperature 55°C) being eluted at 45 ml/h with the first buffer for 90 min, and thereafter at 60 ml/h with the second, good resolution of a mixture containing twelve sugars (including sucrose, maltose, lactose, melezitose, and raffinose, as well as the common monosaccharides) was achieved in only 6 h.

Another important modification was the use of buffer systems in which the pH was maintained at ca. 7.0, which obviates the problem of the possible isomerization of some sugars in alkaline solution, especially at elevated temperatures. Walborg and coworkers [76,77] recommended the addition of 2,3-butanediol to the borate buffers to increase the amount of ionizable borate at the lower pH. Stepwise elution at a rate of 20 ml/h with a system of two buffers (both containing boric acid and the diol) at pH 7.0, the column temperature being raised from 40 to 60°C on commencement of elution with the second buffer, resulted in the resolution of a mixture of 19 sugars in ca. 12 h on a column (1 m × 4 mm ID), packed with Aminex A-14 (20 μm). However, the diol interferes to some extent with the colorimetric reactions used to detect the sugars emerging from the column, and Hough et al. [78] later demonstrated that its inclusion was unnecessary. A good separation of a mixture containing trehalose and all the neutral monosaccharides commonly occurring in polysaccharides and glycoproteins was achieved in less than 7 h by gradient elution with a system of buffers having borate concentration 0.1–0.4 M, chloride concentration 0–0.2 M, but pH 7.00 throughout, at a flowrate of 37 ml/h, from a column (750 × 6 mm ID), packed with Type-S resin (20 μm, Technicon), which was operated at 53°C. Subsequent changes in column temperature (to 62°C) and flowrate (to 60 ml/h) have permitted the analysis of a larger number of sugars, and the time required has decreased to less than 6 h [79].

The use of the newer microparticulate resins has resulted in further reduction in analysis time. Voelter and Bauer [80], using Durrum DA-X4F resin (11 ± 1 μm), resolved a mixture containing sucrose, D-ribose, D-mannose, L-arabinose, D-galactose, D-xylose, and D-glucose in less than 1 h, by eluting with a single buffer (0.4 M borate, pH 9.2) at a rate of 1.3 ml/min from a column (190 × 6 mm ID) operated at 60°C. Mopper [81] has reported good resolution of a 15-component mixture, containing mono- and disaccharides, in only 2.5 h on a column (300 × 4 mm ID), packed with DA-X4 (20 μm), which had been rinsed with 70% ethanol after conversion to the borate form. This pre-treatment resulted in some shrinkage of the resin, and subsequent packing in water produced a dense, tightly packed column owing to its re-expansion. The separation of the sugar mixture was achieved at a column temperature of 78°C by elution with a 0.5 M borate buffer, pH 8.63, at 24 ml/h.

Promising results have been obtained also by use of the macroporous exchanger DEAE-Spheron (glycol methacrylate matrix) in anion-exchange chromatography of sugars as borate complexes. Separation of a mixture containing trehalose and six common monosaccharides has been achieved in ca. 2 h by isocratic elution with a 0.15 M borate buffer, pH 8.5, at 50°C from a column (500 × 6 mm ID), packed with the borate form of this exchanger [82], while stepwise elution with three buffers having borate concentration 0.03–0.25 M and pH 7.50–8.88, from a column of similar size at 60°C, enabled the resolution of a mixture of 12 sugars, including 4 oligosaccharides, in 4 h [83]. Flowrates of 50 ml/h were maintained in both cases.

The borate anion-exchange method has been applied to the separation and analyses not only of sugars, but also of alditols [84,85]. The chromatographic system used by Hough et al. [85] for this purpose was similar to that developed for sugar chromatography [78], but the column temperature was 75°C and the flowrate 70 ml/h, which allowed separation of a mixture containing ethylene glycol, five alditols, and two aminodeoxyalditols in 4 h. Carbohydrate derivatives separated by anion-exchange chromatography in borate medium include a few methyl glycosides

[80], as well as methyl ethers of D-xylose, D-glucose, and D-mannose [86], some of which are well resolved on Aminex A-14 (20 μm) on elution with 0.11 M potassium tetraborate–0.17 M boric acid (pH 8.8) at 55°C (analysis time ca. 2.5 h with a column 990 × 2 mm ID).

Oligosaccharides up to DP 7 are well resolved by this method, the members of a homologous series being eluted in order of decreasing chain length. This application is exemplified by the separation of the β-(1 → 4)-linked D-glucose oligomers from celloheptaose down to cellobiose in 5 h by Kesler [73], who used gradient elution (borate concentration 0.05 → 0.15 M, pH 7.0 → 9.0) at 53°C, in a system of the Technicon type. Torii and coworkers [87,88] preferred stepwise elution with a three-buffer system (borate 0.13–0.35 M, pH 7.5–9.6) at 65°C, in a Jeol JLC-3BC liquid chromatograph, for the resolution of the members of the α-(1 → 6)-linked isomaltodextrin series from DP 7 down to D-glucose, also in 5 h, and for separating some other di- and trisaccharides containing D-glucose residues (including branched trisaccharides [88]) and their reduced derivatives. The method has proved useful in the analysis of the products yielded by dextrans on acetolysis [87] and enzymatic hydrolysis [88]. Another outstanding example of the value of anion exchange in borate medium in degradative studies of polysaccharides is afforded by the work of Kochetkov and coworkers [89–91], who have successfully used chromatography on the resin DA-X4, eluted isocratically with a 0.5 M borate buffer (pH 8.5) at 55°C, to separate the complex mixture of reduced oligosaccharides obained on sodium borohydride degradation of blood-group substance H.

Although separation of uronic acids, from one another and from neutral sugars, by anion-exchange chromatography in borate medium has been reported [92,93], undesirably high concentrations of borate (0.8–1.0 M) are required to elute the acidic sugars, and the use of resin in the acetate form, with acetate in the eluent [94], is preferable. Mopper [95] has developed an efficient method for the separation of uronic and aldobiouronic acids by chromatography on the microparticulate (11 μm) resin DA-X8F, in the acetate form. With a column (380 × 4 mm ID) at 65°C, eluted with 0.08 M sodium acetate, buffered to pH 8.5, at 30 ml/h a good separation of a mixture containing D-galacturonic acid, D-glucuronic acid and its 4-O-methyl derivative, D-mannuronic acid, L-guluronic acid, L-iduronic acid, and four aldobiouronic acids can be achieved in ca. 2.5 h.

For chromatography of aminodeoxy sugars, cation-exchange resins are used, since these sugars are best separated as their hydrochloride salts [96]. For this purpose an amino acid analyzer, with a citrate buffer system and ninhydrin detection, may be employed: e.g., Donald [97] has achieved good resolution of the sugars 2-amino-2-deoxy-D-glucose, -D-mannose, and -D-galactose, and the corresponding alditols, in ca. 6.5 h by use of an analyzer having a column (40 × 1 cm ID) packed with cation-exchange resin (Na$^+$ form), at 50°C, and eluted at 45 ml/h with a buffer, 0.1 M in Na$^+$, containing both citrate and borate (pH 7.5). Stepwise elution with a two-buffer system (0.1 and 0.2 M in Na$^+$) under the same conditions permitted the separation of reduced and de-N-acetylated derivatives of a number of di- and trisaccharides containing residues of aminodeoxy sugars; the separation time was ca. 10 h. This method is of importance in structural studies of glycoproteins.

Ion-exchange chromatography of carbohydrates is compatible with most of the colorimetric detection methods that have been automated (see Chap. 15.2.7), and several different fully automated systems for analysis of carbohydrates by this method have been described [74,78,98–101].

15.2.4.2. Chromatography on dextran and cellulosic ion exchangers

Ion exchangers based on a crosslinked dextran matrix (Sephadex), which is more porous than that of the crosslinked polystyrene resins, are eminently suitable for chromatography of charged oligosaccharides and polysaccharides. The weakly basic anion exchangers containing diethylaminoethyl (DEAE) groups are the type most used in carbohydrate chromatography; these are available (Pharmacia Fine Chemicals) in two different porosities, as DEAE-Sephadex A-25 and A-50. The former is useful in the fractionation of mixtures of acidic oligosaccharides, as well as the separation of acidic components from neutral ones, in the examination of hydrolyzates from acidic polysaccharides. For example, Aspinall and coworkers [102–104] have isolated the products of partial acid-hydrolysis of certain plant gums by chromatography on DEAE-Sephadex A-25 in the formate form. The neutral sugars present in the hydrolyzates are eluted with water; the acidic components, which are retained by the exchanger in water, are subsequently fractionated by stepwise elution with increasing concentrations (0.05–0.5 M) of formic acid.

DEAE-Sephadex A-50 can be used to fractionate charged polysaccharides: e.g., agars differing in pyruvate and sulfate content have been fractionated on the chloride form of this exchanger [105] by stepwise elution with water and sodium chloride solutions of increasing concentration (0.5 to 3 M). The oligogalacturonic acids (DP 2–9) produced on enzymatic hydrolysis of pectic acid have been isolated on a relatively large scale by chromatography of the hydrolyzate on the same exchanger [106], eluted with water at pH 6.0 and then sodium chloride solutions of increasing concentration (0.05 to 0.275 M, in 9 steps).

For chromatography of larger molecules the highly porous lattice of cellulose affords a suitable matrix, to which various charged groups have been attached in the search for effective ion exchangers. Of these, DEAE-cellulose has been the most extensively applied in fractionation of acidic polysaccharides [107]: the use of this exchanger in the phosphate form, eluted with phosphate buffers (0.1–0.5 M), to fractionate polydisperse plant-gum polysaccharides [103,104] exemplifies this application. Chromatography on the carbonate form with an ammonium carbonate gradient ($0 \rightarrow 0.5 M$) has also been recommended [108].

The introduction of microgranular grades of DEAE-cellulose resulted in improved resolution in chromatography on this material and virtually quantitative recoveries of polysaccharides of the glycosaminoglycan type, which had previously required the use of the weakly basic epichlorohydrin-triethanolamine (ECTEOLA) cellulosic ion exchanger [109,110]. Hallén [111] separated the glycosaminoglycans hyaluronic acid, heparin sulfate, chondroitin 4-sulfate, and heparin by chromatography on a system consisting of a Sephadex G-50 column connected to one packed with the microgranular DEAE-cellulose designated DE-52 (Whatman); the Sephadex column was eluted with 0.15 M sodium chloride and the DE-52 (chloride form) with a lithium chloride gradient (0.2 → 1.2 M), buffered at pH 4.0. The value of DE-52 in the fractionation of acidic carbohydrates is further exemplified by the preparative-scale separation of the mixture of sialyl oligosaccharides occurring in human milk that was achieved by Smith et al. [112] by chromatography on this exchanger, previously equilibrated with a 0.002 M pyridinium acetate buffer, pH 5.4, by stepwise elution with similar buffers (0.012 and 0.060 M). The recent development of DEAE anion exchangers based on crosslinked agarose (Pharmacia and Bio-Rad) should permit the fractionation of much larger polysaccharides by this technique.

15.2.5. Steric exclusion chromatography

When steric exclusion chromatography * (SEC) was introduced over twenty years ago, there was much interest in the application of this technique to the fractionation of dextrans [113,114], which led naturally to its extensive use in the carbohydrate field in general. The method is particularly valuable because it affords not only an effective separation technique but also a means of determining the molecular-weight distribution of polymers fractionated in this way. The principles of SEC are now very well documented, having been the subject of numerous comprehensive reviews (see Chap. 8 and refs. 115–120), and the application of the technique specifically to carbohydrates has also been reviewed [121,122]. In this section, therefore, emphasis will be laid on recent advances in SEC, which have resulted in improvements in speed and efficiency so great that this is now generally regarded as another mode of HPLC.

15.2.5.1. Molecular-weight distribution analysis

Although various theoretical treatments of SEC have resulted in several different equations relating the molecular weights (or molecular-weight averages) of macromolecules to chromatographic parameters, the generally accepted procedure for the estimation of molecular weights by this method is to exploit the simple linear correlation between elution volume, V_e, and log M (where M is the molecular weight or, for polymers, one of the averages $\overline{M}_w$ or $\overline{M}_n$). This empirical relationship has been rationalized on theoretical grounds by Anderson and Stoddart [123], whose treatment applies equally to the correlation between log M and the partition coefficient K_{av} used so successfully in molecular-weight-distribution analysis of dextrans by Granath and coworkers [124,125].

For calibration purposes, a number of substances of known molecular weight or well-characterized polymer fractions of narrow molecular-weight distribution, which should be structurally similar to the substances being examined, are required, and this sometimes presents difficulty in the carbohydrate field. In the case of dextrans,

---

* This term is used in preference to "gel chromatography", since many of the sorbents now used are not gels.

there is no problem, as fractions suitable for use as standards are commercially available (Pharmacia Fine Chemicals), and even rather more polydisperse fractions can be used, provided that suitable corrections, which have been computerized [125,126], are applied to the chromatographic data. In the absence of suitable standards, the Pharmacia dextrans are often used to calibrate columns for molecular -weight-distribution analysis of other polysaccharides, but it must be emphasized that such a calibration is valid only if the relationship between molecular weight and molecular size of the polysaccharide being examined is similar to that of dextran. The "universal" calibration proposed by Grubisic et al. [127], in which V_c is plotted against $\log M \cdot [\eta]$, where $[\eta]$, the intrinsic viscosity, is taken as a measure of the hydrodynamic volume of the molecule in solution, holds only when the polymer chain is sufficiently long for random-coil statistics to be valid, and it has been shown [128] that this calibration does not hold for dextrans having $\overline{M}_w$ below 70,000. Although attempts to find a generally applicable correlation continue [129,130], there is really no substitute for calibration of the column with characterized fractions of the actual polysaccharide under study. The isolation of sharp fractions for characterization (by methods such as ultracentrifugation, light scattering, etc.) requires exclusion chromatography on a preparative scale; this was formerly a slow process, but the introduction of sorbents suitable for use in high-speed liquid chromatographs should expedite such fractionations.

15.2.5.2. Column packings for steric exclusion chromatography

The first packings designed specifically for SEC were the crosslinked dextran gels (Sephadex G series, Pharmacia), of which several types, differing in porosity, are available. Such gels are capable of fractionating polysaccharides only up to molecular weight ca. 200,000. For larger molecules, agarose gels (Sepharose, Pharmacia; or Bio-Gel A series, Bio-Rad) may be used; these are also available in different porosities, and polysaccharides having molecular weights from $< 10^4$ to $> 10^7$ may be fractionated on agarose gels of appropriate pore size. However, although both agarose and dextran gels have been extensively used in chromatography of carbohydrates, they share the disadvantage of having a carbohydrate matrix, so that any material from the gel that may emerge in the column effluent will give rise to spurious peaks when carbohydrates are being detected. The same applies to the gels based on crosslinked cellulose [131] and starch [34] that have been tested in some laboratories. Noncarbohydrate gels are generally considered preferable for chromatography of carbohydrates: of these gels, the polyacrylamide type (Bio-Gel P series, Bio-Rad) has been the most widely used, particularly the tightly crosslinked members of the series (Bio-Gel P-2 and P-4), which are eminently suitable for separating the members of homologous series of oligosaccharides (see Chap. 15.2.5.3). Polysaccharides having $\overline{M}_w$ from 5000 up to ca. 100,000 can be fractionated on Bio-Gel P-300 [132].

Rigid gels like Bio-Gel P-2, which can withstand fairly high pressures, can be used as column packings for HPLC, but softer gels, such as Bio-Gel P-300, lack the necessary mechanical strength. Much recent research has been directed at the

TABLE 15.3

FRACTIONATION OF DEXTRANS ON SEPHADEX AND POROUS GLASS

On Sephadex G-200 + G-100 [124] *		On controlled-pore glass [137] **	
$\overline{M}_w$	K_{av}	$\overline{M}_w$	K_{av}
147,000	0.04	287,000	0.042
96,000	0.064	163,000	0.114
76,000	0.085	74,800	0.235
58,000	0.110	54,100	0.306
48,300	0.148	51,400	0.323
36,000	0.215	30,200	0.425
32,400	0.227	22,400	0.495
27,800	0.266	16,800	0.566
22,400	0.316	13,100	0.615
19,300	0.380	8,500	0.688
13,200	0.476	5,700	0.736
10,000	0.556	3,300	0.806
7,500	0.620	2,600	0.829
6,100	0.671	1,900	0.866
5,400	0.719	1,000	0.917

* G-200 and G-100 in dry-weight ratio 1:2.
** Corning CPG-10: seven porosities (75, 125, 175, 240, 370, 700, and 1250 Å) mixed in inverse proportion to their pore volumes.

application to polysaccharide fractionation of more rigid packings, which are capable of functioning at the pressures used in HPLC. Hydrophilized polystyrene has proved moderately successful [133,134], but inorganic supports such as glass or silica, which are available in a range of precisely controlled pore sizes, are clearly the materials of choice for this application. Porous glass—if the problem of adsorption [135] is overcome by deactivation of polar sites [136]—seems to have the greatest potential, as wide fractionation ranges are made possible by combination of packings differing in pore size [136,137]. This is exemplified by the data presented in Table 15.3, which contrasts the fractionation range for dextrans on controlled-pore glass [137] with that given by a combination of Sephadex gels [124]. Polysaccharides having $\overline{M}_w$ up to ca. 10^7 can be fractionated on glass packings of appropriate pore size [136].

Porous silica shares with porous glass the advantages of rigidity and resistance to thermal, chemical, or bacteriological degradation [138] and—after suitable deactivation of adsorption sites [139,140]—has been found to be an effective packing for SEC of dextrans and glycosaminoglycans [141–143]. Silica packings are available (LiChrospher, Merck; and Porasil, Waters) in a range of porosities, capable of fractionating polysaccharides having $\overline{M}_w$ up to ca. $2 \cdot 10^6$, and the introduction of microparticulate (5 and 10 μm) grades should increase the efficiency of separation to that afforded by other modes of HPLC (see Chap. 15.2.2.1). To obviate the difficulty of adsorption effects in SEC on porous silica, packings in which an inert phase is chemically bonded to the silica have been introduced. For example, in μBondagel (Waters) a polyether phase is bonded to microparticulate (10 μm) Porasil. This packing has recently been reported to be highly effective in SEC of dextrans [144] and of the glycosaminoglycan hyaluronic acid and its degradation products [145]. The development of other bonded-phase packings suitable for rapid fractionation of polysaccharides is continuing [146,147].

Some of the newer "semi-rigid" polymers have found limited application in SEC when only moderate pressures and ambient temperature are used: e.g., the 2-hydroxyethylmethacrylate–ethylenedimethacrylate copolymer, known as Spheron [148], has proved successful when applied to the fractionation of dextrans having $\overline{M}_w$ up to ca. 500,000 [149]. Crosslinked copolymers of vinyl acetate (Merckogel OR-PVA, Merck), used with organic solvents, have found application in SEC of substituted carbohydrates [150], and crosslinked polystyrenes (e.g., Poragel, Waters) can also be used under these conditions. These packings complement the softer gels Sephadex LH-20 and LH-60 (hydroxypropylated dextran gels), which are also useful for separations involving substituted carbohydrates.

15.2.5.3. Eluents and operating conditions

Water may be used as an eluent for SEC of carbohydrates only if the operating conditions are such as to minimize interactions, by electrostatic or Van der Waals forces, of the solute molecules with the column packing or with one another. The role of adsorption in chromatography of oligosaccharides on tightly crosslinked gels, such as Bio-Gel P-2 and Sephadex G-15, has been studied in depth by Brown and coworkers [151–155] and by Dellweg et al. [156], who have demonstrated that solute–gel interaction, which increases with chain length in homologous series, such as the malto-, cello-, xylo-, and mannodextrins, is a major factor, especially in the case of the dextran gel. Interaction decreases with increasing temperature, with resulting decrease in the partition coefficient, K_d, for each oligomer, the effect being greatest for those of greatest chain length, which have the lowest K_d values. There are, therefore, larger differences in K_d among the members of a homologous series of oligosaccharides as the temperature increases, so that better separation can be expected.

It has become standard practice for SEC of neutral oligosaccharides to be performed on tightly crosslinked polyacrylamide gels, with water as eluent, at column temperatures of 55–65°C. Under these conditions, a column packed with Bio-Gel P-2 is capable of resolving the members of the maltodextrin series up to DP 12 or 13 [157–159], and the xylodextrin series up to DP 18 [4]. Use of the slightly more porous Bio-Gel P-4 raises the upper limit of the fractionation range to DP 15 for maltodextrins [160], for the isomalto-oligosaccharides produced on hydrolysis of dextrans [161], and for the reduced oligosaccharides corresponding to these two series [161]. All of these separations have been performed with gels of small particle size (> 400 mesh), in columns 1–2 m long (two 1-m columns in series) and at flowrates of 10–30 ml/h, the conditions recommended by Sabbagh and Fagerson [159] after a detailed investigation of the effects of various operational parameters on resolution of maltodextrins.

References on p. B276

Bio-Gel P-2 and other packings having sufficient mechanical strength can be used in chromatographic systems of the HPLC type, with detection of carbohydrates by an automated colorimetric method [98] or, if the nature of the eluent permits, by a differential refractometer [159,160,162]. Such systems are finding increasing application in SEC of carbohydrates, especially in analyses of mixtures of oligosaccharides.

For carbohydrates containing acidic or basic groups, water may not be a satisfactory eluent, since electrostatic interaction, which cannot be eliminated simply by raising the temperature, is stronger. This does not necessarily present a problem in the case of oligosaccharides containing only aminodeoxy sugar residues, such as those produced on hydrolysis of chitin, which can be separated on a preparative scale by SEC on Bio-Gel P-2 [163] or Sephadex LH-20 [164] in water, but the presence of acidic sugar residues, as in the oligosaccharides from glycosaminoglycans, leads to interactions which totally preclude resolution by SEC if water is used as the eluent [163,165]. However, these oligosaccharides are readily separated (usually on Bio-Gel P-2 or Sephadex G-25) by elution with solutions containing electrolytes. The concentrations required vary according to the ionic strength necessary to eliminate interaction in each specific case: e.g., Flodin et al. [166] used 0.1 M NaCl to resolve the di-, tetra-, hexa-, and octasaccharides produced on enzymatic hydrolysis of hyaluronic acid, but for optimal resolution of the corresponding oligomers from chondroitin 4-sulfate 1 M NaCl was required. Dietrich [167] employed 0.01 M acetic acid as the eluent in the separation of oligosaccharides from heparin on Sephadex G-25, whereas Raftery et al. [163] advocated the use of 25% (ca. 4 M) acetic acid to resolve the hyaluronic acid oligosaccharides on Bio-Gel P-2. For preparative chromatography a solution containing a volatile salt is the best eluent: e.g., the hyaluronic acid oligosaccharides up to the dodecasaccharide have been isolated by chromatography of the digest on Sephadex G-50 with an eluent consisting of 0.25 M pyridinium acetate, pH 6.5 [168], and milk oligosaccharides containing residues of N-acetylneuraminic acid (sialic acid) can be separated on Sephadex G-25 by elution with M pyridinium acetate, pH 4.4 [169].

In SEC of polysaccharides and higher oligosaccharides, an eluent of high ionic strength is essential to counteract molecular association and strong interaction of such molecules with the column packing. This is particularly important in molecular-weight-distribution analysis, for which elution with NaCl solutions of concentration 0.3% [124] to 0.9% [125] has been recommended. The eluent of choice is 1 M NaCl, which has been widely used in determinations of the molecular-weight distributions of a variety of polysaccharides by SEC on polyacrylamide and agarose gels [132,170–173]. The chromatograms obtained are sufficiently reproducible to permit their use in following the course of degradative processes during structural investigations of the polysaccharides [171–173].

SEC of glycosaminoglycans and glycoproteins presents special difficulty because of large variations in molecular shape and size, depending on the pH and ionic strength of the eluent [174–176]. However, even with glycosaminoglycans fractionations have been achieved by elution with NaCl solutions [177–179], although buffering of eluents is generally recommended [145,180], as for glycoproteins [181].

The soft gels, such as Bio-Gel P-300 and Sephadex G-200, that have been extensively used for fractionation of polysaccharides, can be operated only at flowrates of ca. 3 ml/h. The newer, rigid packings still function at flowrates of ca. 1 ml/min [144,182], thus permitting fractionations in less than 1 h, though with some loss in resolution [183].

15.2.6. Affinity chromatography

The important, relatively new technique known as affinity chromatography, which involves the use of column packings containing a ligand with a highly specific affinity for molecules with groups having a particular stereochemistry, is being applied to an increasing extent in carbohydrate chemistry. The ligand, which in this case is a lectin (hemagglutinating glycoprotein), is covalently coupled to an insoluble matrix, usually an agarose or polyacrylamide gel. On chromatography of a mixture on such a packing, polysaccharides or glycoproteins containing the groups with which the lectin binds are retained, and are thereby separated from other components, which rapidly pass through the column. The bound substances are subsequently desorbed, either by displacement by a small carbohydrate molecule containing the group that the lectin binds specifically (hapten inhibition) or as a result of a change in pH or ionic strength of the eluent, such that the conditions necessary for binding are no longer present.

The first lectin to be immobilized for affinity chromatography of carbohydrates was concanavalin A (from the jack bean, *Canavalia ensiformis*). It is a metalloprotein containing Mn^{2+} and Ca^{2+}, the presence of which is essential to the binding activity. This lectin has a strong affinity for α-D-mannopyranosyl and α-D-glucopyranosyl residues [184], and it precipitates polysaccharides containing a sufficient number of terminal groups of these types [185]. It was reported in 1970 [186,187] that, when concanavalin A was covalently bound to an agarose gel, the product adsorbed polysaccharides to an extent governed by the proportion of terminal α-D-mannopyranosyl or α-D-glucopyranosyl groups in the molecule: e.g., a highly branched dextran was strongly bound, whereas one having fewer branches was not, owing to the smaller number of α-D-glucopyranosyl end-groups [186]. Intensive investigation subsequently demonstrated that, although any change at the 3-, 4- or 6-positions of the sugar residue involved in binding interfered with its complexation by concanavalin A, considerable variation at the 2-position was compatible with binding, so that some affinity was shown also by terminal residues of 2-acetamido-2-deoxy-D-glucose and by α-(1 → 2)-linked D-mannopyranosyl chain units in a polysaccharide [188]. Concanavalin A is therefore able to bind a large number of polysaccharides and glycoproteins, and the commercially available Con A-Sepharose (Pharmacia), in which the lectin is covalently attached to the agarose gel Sepharose 4B, is now widely used in the isolation and purification of these biopolymers [189]. Further studies of the binding properties of Con A-Sepharose are in progress [190].

In affinity chromatography on Con A-Sepharose the starting buffer should have a pH in the neutral region and should contain sodium chloride: a recommended buffer [189] is 0.02 M Tris–HCl, pH 7.4, containing 0.5 M NaCl. Bound substances can be quantitatively eluted with a solution containing methyl α-D-mannopyranoside or methyl α-D-glucopyranoside [186,187]; gradient elution $(0 \rightarrow 0.5\,M)$ is advisable where binding is tight, though for many substances a glycoside concentration of $0.1–0.2\,M$ is sufficient [189]. The alternative procedure of using a borate buffer $(0.1\,M$ sodium borate, pH 6.0) for elution [191] has the advantage that the borate, which disrupts the lectin–carbohydrate binding by virtue of its strong complexing action on carbohydrates, is easily removed from the eluate by conversion to volatile methyl borate. A buffer of low pH (not below 3) is another possible eluent [189].

A large number of lectins, having specificities for different sugar residues, are now known. Details of many of these will be found in a recent, very comprehensive review by Goldstein and Hayes [192]; some examples are listed in Table 15.4. Several have been coupled to appropriate supports for use in affinity chromatography, and some of these products are available commercially (e.g., Pharmacia has coupled lentil lectin to Sepharose 4B, and the wheat germ and *Helix pomatia* lectins to Sepharose 6MB). Affinity chromatography is particularly useful in studies of glycoproteins; this application has been reviewed by Kristiansen [193]. However, the technique is also finding application in the isolation of many different types of

TABLE 15.4

CARBOHYDRATE-BINDING SPECIFICITIES OF LECTINS

Source of lectin	Carbohydrate specificity	References
Canavalia ensiformis (jack bean)	α-D-Man$p$ > α-D-Glc$p$ > α-D-Glc$p$NAc	184–192
Lens culinaris (common lentil)	α-D-Man$p$ > α-D-Glc$p$, α-D-Glc$p$NAc	189, 192
Triticum vulgare (wheat germ)	β-D-Glc$p$NAc; β-$(1\rightarrow4)$-linked oligomers (chitin oligosaccharides) have higher affinity	189, 192, 196
Helix pomatia (edible snail)	α-D-Gal$p$NAc $\gg$ α-D-Glc$p$NAc	192, 197
Dolichos biflorus (horse gram)	α-D-Gal$p$NAc $\gg$ α-D-Glc$p$NAc	192, 198
Glycine max (soybean)	α-D-Gal$p$NAc > β-D-Gal$p$NAc $\gg$ α-D-Gal$p$	192, 199
Sophora japonica (Japanese pagoda tree)	β-D-Gal$p$NAc > β-D-Gal$p$ > α-D-Gal$p$	192, 200
Tridacna maxima (giant clam)	β-D-Gal$p$NAc > β-D-Gal$p$	194, 195
Ricinus communis (castor bean)	β-D-Gal$p$ > α-D-Gal$p$	192, 201–203
Crotalaria juncea (sunn hemp)	β-D-Gal$p$ > α-D-Gal$p$	203, 204
Lotus tetragonolobus (asparagus pea)	α-L-Fuc$p$	192, 205
Ulex europaeus (gorse)	α-L-Fuc$p$	192, 206

polysaccharides. The strong affinity of the lectin tridacnin (from the clam *Tridacna maxima*) for polysaccharides containing terminal β-D-galactopyranosyl groups [194], which has recently been exploited in the isolation of an arabinogalactan-protein from *Gladiolus* by chromatography on a packing in which this lectin was covalently coupled to Sepharose 4B [195], demonstrates the potential value of methods of this kind in the investigation of plant polysaccharides. Further examples from the rapidly burgeoning literature of affinity chromatography of carbohydrates are cited in Table 15.4.

15.2.7. Detection methods for column chromatography

The detector most used in HPLC of unsubstituted carbohydrates to date has been the RI detector, though the use of UV detectors will probably increase following the introduction of photometers capable of operating at 190 nm, which can detect sugars [207,208]. The sensitivity of this UV detector for monosaccharides is ca. 12 times greater than that of most refractometers (detection limit ca. 20 μg of sugar), but as the DP of oligosaccharides increases, this sensitivity decreases. Appropriate derivatization of sugars (benzoylation [16] or 4-nitrobenzoylation [19]) permits detection of nanogram amounts by the use of UV detectors operating at 260 nm. For other chromatographic systems, such as those involving ion-exchange or steric exclusion chromatography, use of RI or UV detectors is feasible only if the eluents do not contain salts in high concentration [209,210]. Katz and Thacker [211] have described systems in which UV-absorbing chromophores were produced in the effluent from an anion-exchange column by reaction of the sugars with concentrated sulfuric acid in a dynamic system. An oxidative detector relying upon reduction of Ce^{4+} to the fluorescent Ce^{3+} by the sugars was also proposed [212], but the sensitivity of such detection methods is not high enough for general purposes.

Several of the standard colorimetric methods for the quantitative determination of carbohydrates [213] are adaptable to use in automated analytical systems for the detection of carbohydrates emerging from columns. In the original Technicon AutoAnalyzer for sugars [74] the orcinol–sulfuric acid method was automated [73] by a method involving dynamic mixing of the column effluent with the reagent by a multi-channel peristaltic pump. The liquid stream, segmented by air bubbles, was then passed through a heating coil and, after venting of the air bubbles, through a flow cell in a recording colorimeter (at 420 nm). The detection limit of this system for sugars was ca. 10^{-8} mole. The introduction of pumps constructed of acid-resistant materials, capable of delivering the reagent accurately with low baseline "noise", obviated the necessity for air segmentation and greatly increased the sensitivity of the automated orcinol–sulfuric acid method for sugars; detection limits of $1 \cdot 10^{-10}$ mole for pentoses and $3 \cdot 10^{-10}$ mole for hexoses have been claimed [214]. Such pumps are now widely used in analytical systems of this type [80,98].

The orcinol–sulfuric acid method has the disadvantage that rapid deterioration of the reagent causes precipitation of purple material in the heating coil, with resulting obstruction of the effluent flow and "noisy" baselines. The anthrone–sulfuric acid method [215], which in its automated version [216] has been much used in the analysis of dextran fractions from gel columns [124], also has the disadvantage of reagent instability; a further drawback is that the chloride ions often present in eluents for SEC of polysaccharides interfere with the color response of this reagent. A colorimetric method that has been used extensively in manual analysis of fractions from columns is the phenol–sulfuric acid assay originally developed by Dubois et al. [217]. This simple procedure, which is applicable to a wide variety of carbohydrates and requires only readily available, stable reagents, remains the method of choice for colorimetric detection of carbohydrates in chromatographic systems that are not fully automated, but attempts to automate it [218] have been hampered by the need to control the excessive pulsing and liberation of heat when the required 98% sulfuric acid is mixed with aqueous solutions in a closed system. The L-cysteine–sulfuric acid method, which possesses none of the disadvantages of the other three mentioned, has been successfully automated [219], and affords an excellent detection system for anion-exchange [78] and steric exclusion chromatography [220] of carbohydrates. A more specific analytical method, for the detection of carbohydrates containing uronic acid residues in high proportion (such as the acidic glycosaminoglycans), is the carbazole–sulfuric acid assay [221], which has also proved adaptable to automation [220,222].

The corrosive nature of reagents containing sulfuric acid in high proportion presents a serious problem in automated analytical systems and, therefore, alternative methods for the detection of sugars, based upon their reducing power, have been developed. Mopper and Degens [65] exploited the reduction of Tetrazolium Blue to diformazan in alkaline solution to detect sugars in an automated system based on partition chromatography on an anion-exchange resin in aqueous ethanol (see Chap. 15.2.3), for which high sensitivity (detection limits of the order of 0.1 nmole) was claimed. This method cannot be used unless the eluent contains ethanol, in which the diformazan is soluble, and therefore is not generally applicable. A more versatile procedure is the copper–bicinchoninate colorimetric method, subsequently developed by Mopper and Gindler [66], in which the Cu^+ formed on reduction of Cu^{2+} by sugar is complexed by 2,2'-bicinchoninate anions, with formation of a deep lavender color (absorbance maximum 562 nm). The presence of ethanol increases the color intensity of this complex, so that the reagent is particularly sensitive in automated chromatographic systems employing ethanol eluents [67], but with some modification [99] the copper–bicinchoninate method is now successfully applied in systems involving ion-exchange chromatography of neutral sugars in borate medium [81], acidic sugars in acetate medium [95], and aminodeoxy sugars in citrate buffers [223]. In all cases detection limits below 1 nmole have been claimed. An advantage of this analytical method, when applied to analysis of aminodeoxy sugars in hydrolyzates fom glycoproteins, is that it is not subject to interference fom amino acids, as is the conventional ninhydrin method. The colored complex (absorbance maximum 460 nm) formed by the reagent neocuproin with Cu^+ ions has been exploited in an automated system for the detection of reducing sugars in anion-exchange chromatography by Simatupang and Dietrichs [224], who have developed a method for the simultaneous detection of sugars and alditols [100] by combination of this with an automated periodate–pentane-2,4-dione method for the analysis of alditols, of the type used by Samuelson [55] and others [85,219]. Such combination of analytical methods specific for different classes of carbohydrate has created some very versatile CC systems [101,220].

15.3. GAS–LIQUID CHROMATOGRAPHY

Since McInnes et al. [225] reported the first application of GLC to the separation of carbohydrate derivatives in 1958, developments in this field have been very numerous and varied, and GLC techniques are now among the most important in the methodology of carbohydrate chemistry. The literature on this topic is vast and growing rapidly, necessitating the publication of reviews at frequent intervals. Dutton [226,227] published an excellent review in two parts, covering the period 1963–1972, and specific aspects of the literature have been treated in several important publications that appeared between 1970 and 1972 [228–236], and some more recent papers [237,238]. In the present chapter, emphasis is laid on advances since 1972; it is possible to include only the salient features of the earlier work, which is fully discussed in the reviews cited.

15.3.1. General experimental conditions

The best choice of liquid stationary phase for a particular separation must always be determined by experiment, and it is advisable to investigate the behavior of each new type of volatile derivative on several phases, differing in polarity. In the case of carbohydrate derivatives, it is particularly difficult to generalize regarding choice of liquid phases for GLC, as polarity varies greatly with the molecular structure of the volatile derivative used. For example, although the moderately polar (25% cyanoethyl, 75% methyl) silicone XE-60 has been found [239] to be totally unsuitable for resolution of TMS ethers of partially methylated alditols, which are best resolved on nonpolar phases, such as OV-101 and SE-30 (100% methyl silicones) and SE-52 (5% phenyl, 95% methyl silicone), this phase shows greater potential for separation of TMS derivatives of hexuronic acids and lactones than does SE-30 [240]. In some cases, especially in GLC analysis of mixtures of methylated sugars, it is advisable to use two phases, differing in polarity, for each sample, as compounds that are not separated on one may be resolved on the other and vice versa. For example, Aspinall [241] separated a wide range of fully and partially methylated methyl glycosides by GLC on butanediol succinate polyester and also on a less polar polyphenyl ether phase, and Jansson et al. [242] have used the latter type of phase (OS-138) to resolve certain methylated alditol acetates (notably the acetates of 2,3-di-O-methylrhamnitol and 2,3,4,6-tetra-O-methylglucitol) not separated on the more polar ECNSS-M (a copolymer of ethylene glycol succinate polyester and a cyanoalkyl silicone), which is otherwise the better choice for GLC of these compounds.

In packed columns, the nature of the solid support can have a marked effect upon resolution in GLC. For separation of TMS derivatives of carbohydrates, deactivation of diatomaceous earth supports by acid-washing and silanization is essential

References on p. B276

[243], and such supports (Chromosorb W AW DMCS or Gas-Chrom Q, usually 100–120 mesh) are generally recommended for GLC of carbohydrates. It should be noted, however, that Shaw and Moss [244] have reported that use of untreated Chromosorb W results in better separations of some alditol acetates, and Dutton and Walker [226,227,245] have also found this to hold true with certain methylated alditol acetates. Recently, greatly improved resolution of both acetyl and TMS derivatives of methylated alditols and acetamidodeoxyalditols has been obtained [246] by using a packing consisting of a low proportion (0.3–0.4%) of the liquid phase OV-225 on a Chromosorb W support that had been modified by coating the surface with the polyethylene glycol Carbowax 20M, followed by heat treatment (280°C, under nitrogen) and exhaustive extraction with methanol [247]. The use of glass beads (GLC 110), coated with only 0.05% of ECNSS-M, has proved advantageous when applied to GLC separation of the acetates of methyl ethers of 2-deoxy-2-(N-methyl)acetamido-D-glucitol [248].

While good separations of carbohydrate derivatives can be achieved by use of packed columns, of standard dimensions (ca. 2 m long, 0.3–0.6 mm ID), the higher resolution and speed of separation afforded by capillary columns is obviously preferable, especially in the analysis of complex mixtures, such as those obtained during methylation analyses of polysaccharides. The use of SCOT columns in GLC of methylated sugars as their derived alditol acetates has been strongly advocated by Lindberg and coworkers [228,242], and excellent resolution has been obtained with such columns (15 m × 0.5 mm ID), containing ECNSS-M or the newer phase OV-225. WCOT columns, 25 m × 0.25 mm, containing SP-1000 have proved even more satisfactory [242], and the resolving power of these columns, capable of yielding up to 10^5 theoretical plates, has been strikingly demonstrated by the resolution of the enantiomeric D- and L-forms of certain sugars, as derived chiral glycosides (see Chap. 15.3.3.7), thus achieved by Leontein et al. [249]. Similar resolution of enantiomers on capillary columns (25 m × 0.31 mm ID), wall-coated with SE-30, has been reported by Gerwig et al. [250,251].

Column temperatures required for GLC of carbohydrates range, in general, between 140 and 250°C. In many cases, particularly where the compounds to be separated differ widely in molecular weight, temperature programing is advantageous. An outstanding example of this is afforded by the good separation of TMS derivatives of compounds ranging from a tetritol (erythritol) to a tetrasaccharide (stachyose), achieved in only 23 min with a temperature rise from 160 to 240°C at 10°C/min, 240 to 350°C at 30°C/min, then isothermally at 350°C for 11 min [252]. The use of such high temperatures obviously necessitates a very careful choice of liquid phase; that used in the separation cited was Dexsil 300 GC, a *meta*-carborane that is stable at temperatures up to 500°C. This is an exceptional case, but there is a general tendency to replace the liquid phases originally recommended for GLC of carbohydrate derivatives by others having greater thermal stability, thereby extending the temperature range that can be covered without risking the "bleeding" of liquid phase, which must be avoided particularly when a mass spectrometer is coupled to the gas chromatograph. For example, ECNSS-M, the phase long advocated as the best choice for GLC of the acetates of alditols [253] and their partially methylated derivatives [254], is stable only up to 200°C and tends to "bleed" at temperatures above 180°C. Therefore, the phase OV-225 (25% cyanopropyl, 25% phenyl, 50% methyl silicone), which has similar polarity and characteristics but higher thermal stability (up to 275°C), is now generally preferred [255]. OV-17 (50% phenyl, 50% methyl silicone), which is stable up to 300°C, has proved useful in GLC of carbohydrates not only as alditol acetates [256,257] but also as TMS ethers, especially in the case of oligosaccharides [258].

The carrier gas is normally helium or nitrogen, and the detector the flame-ionization type. If strongly electronegative atoms are introduced on derivatization, as is the case with trifluoroacetylation, use of the more sensitive ECD, which can detect monosaccharide trifluoroacetates at picogram level [259], becomes feasible. Compounds containing nitrogen, such as aminodeoxy sugars and neuraminic acid derivatives, can be differentiated from other sugars by use of a thermionic nitrogen–phosphorus selective detector, for which a selectivity factor of 100 has been claimed [260], the sensitivity to aminodeoxy sugars being ca. 5 times greater than that possible with the conventional FID.

15.3.2. Identification and quantitation of components

In analytical GLC tentative identification of components is possible from their retention times relative to that of a suitable internal standard, but these relative retention times should be verified immediately by GLC of authentic samples of the putative components. Assignment of peaks in a chromatogram to particular compounds by this method alone can never be completely unambiguous, and therefore positive identification necessitates the use of other analytical methods in conjunction with GLC. MS is very extensively used for this purpose, and a considerable body of data on GC–MS of carbohydrates is now available (see Chap. 15.3.4). While use of a system in which the mass spectrometer is coupled to the gas chromatograph is eminently desirable, it is not mandatory: e.g., Choy et al. [261] have reported excellent results even when the two instruments were physically separate. Since carbohydrates are relatively nonvolatile, fairly simple devices [262] suffice for collection of samples at the outlet port of a gas chromatograph. Collection permits characterization of components not only by MS but also by other spectroscopic methods and techniques, such as circular dichroism measurements (which, when applied to peracetylated alditols, indicate the chirality of the parent sugars [263]).

For quantitation of the components of a mixture analyzed by GLC the areas under the corresponding peaks in the chromatogram are measured and their proportions calculated. Peak areas should be determined relative to that of an internal standard, of which a known amount is added to the mixture before chromatography. The internal standard must be chemically similar to the compounds being analyzed and should give a single peak not coincident with any in the sample chromatogram and preferably close to the midpoint. For long runs, the use of more than one internal standard is advisable; this provides additional reference points for calculation of results and serves as a check on any changes in column conditions during the run. Clamp et al. [229] and Holligan and Drew [230,231] recommended the use of various polyols, such as arabinitol, mannitol, and myo-inositol, in appropriately derivatized form, as internal standards in GLC of TMS sugars and glycosides. The molar response factor for a particular compound is obtained by calibration with known quantities of that compound together with the internal standard, and in subsequent analyses the areas of peaks corresponding to the compound are divided by the molar response factor in determining the proportion present. It should be noted that reproducibility of response factors is poor, and therefore these factors should be determined frequently, under the exact experimental conditions to be used in the analysis [226,227]. The molar responses of partially methylated alditol acetates were formerly assumed to be equal, but this has been refuted by Sweet et al. [264]. There is clearly no substitute for careful determination of response factors in quantitative GLC analysis of carbohydrates, irrespective of the method of derivatization employed.

15.3.3. Derivatives

15.3.3.1. Methylated methyl glycosides

The first recorded application of GLC in the carbohydrate field was the separation of the fully methylated methyl glycopyranosides of D-xylose and L-arabinose, and some resolution of the corresponding derivatives of D-glucose, D-galactose, and D-mannose, which were achieved by McInnes et al. [225] with a column, packed with Apiezon M on Celite 545 (1:4), at 170°C. The methyl glycosides, on recovery from the effluent gas stream, were shown to be unchanged by anomerization or hydrolysis under these conditions. With the feasibility of the method thus demonstrated, detailed studies soon followed [265,266], and the technique was rapidly established as a method of analysis of partially as well as fully methylated methyl glycosides. Resolution of anomers [266] and of furanosides from pyranosides [267] demonstrated the potential value of GLC in the analysis of methanolyzates in structural studies of polysaccharides. This application was greatly expedited by the comprehensive studies of Aspinall [241] and Stephen et al. [268], who have published retention data for a large number of methylated methyl glycosides, derived from L-arabinose, D-xylose, L-rhamnose, L-fucose, D-glucose, D-galactose, D-mannose, D-fructose, and the methyl esters of D-glucuronic and D-galacturonic acid. The use of two phases by Aspinall [241] has been mentioned (see Chap. 15.3.1): fully methylated methyl glycosides and those containing only one hydroxyl group were well resolved on butanediol succinate polyester at 175°C, but the less substituted derivatives, such as the glycosides of di-O-methylhexoses, were too strongly retained on this phase, and required the less polar polyphenyl ether (at 200°C). Stephen et al. [268] reported that ethylene glycol succinate polyester (at 155°C) was capable of resolving the highly substituted glycosides without excessively long retention of the di-O-methylhexose derivatives, and this stationary phase (14%, on Chromosorb W, 80–100 mesh) has been much used in analysis of methanolyzates from polysaccharides [269] and acidic oligosaccharides isolated during structural studies of polysaccharides [6,270].

Other investigations of GLC of partially methylated methyl glycosides have been confined mainly to attempts to improve the separations of the various methylated derivatives of a specific sugar. Thus, Ovodov and Pavlenko [271] have published retention data for several methylated derivatives of methyl D-galactopyranoside and -furanoside on neopentyl glycol succinate polyester (5%, on Chromosorb W, 60–80 mesh) at 167°C, and the same phase (2%) at 190°C was used by Bhattacharjee and Gorin [272] in a study aimed at identification of the di-, tri-, and tetra-O-methyl-D-mannose derivatives by GLC. Heyns et al. [273] obtained retention data for 3-O-methyl- and several di-O-methyl-D-glucosides, as well as the tri- and tetra-O-methyl derivatives, by using ethylene glycol succinate polyester (5%) at 200°C. The problem of long retention times of mono- and di-O-methylhexosides can be overcome by acetylation at the unsubstituted positions: e.g., Fournet et al. [274] have reported that the four isomeric mono-O-methyl ethers of methyl α-D-mannopyranoside are easily separated by GLC, on ECNSS-M (3% on Chromosorb W, 60–80 mesh) at 170°C or butanediol succinate polyester (5% on Chromosorb W, 80–100 mesh) at 195°C, following peracetylation of these otherwise highly polar compounds. Partially methylated methyl glycosides derived from pentoses and 6-deoxyhexoses are generally better separated by GLC than are the hexose derivatives. Anderle and coworkers [275,276] have published retention data for all the methyl O-methyl-D-xylofuranosides on several different phases, of which XE-60 (5%, at 160°C) gave optimal separation of this carbohydrate series. The methyl ethers of methyl α-L-rhamnopyranoside have also been studied, by Janeček et al. [277]. In this case, GLC on ECNSS-M (3%) was recommended for resolution of the entire series, with the temperature program 110°C (isothermal) for 8 min, then 110 to 150°C at 2°C/min.

A major advantage of the use of the methyl glycosides as volatile derivatives for GLC of partially methylated sugars is that this method is applicable to methyl ethers derived from uronic acids, which are separated as the methyl glycoside methyl esters. Anderle and Kováč [278] have published retention data for all possible methyl ethers of methyl (methyl-α-D-glucopyranosid)uronate on several phases, of which XE-60 (4% on Gas-Chrom Z, 80–100 mesh) at 175°C was found to be the best choice. There was, however, some irreversible adsorption of those compounds still bearing hydroxyl groups,and therefore acetylation at the unsubstituted positions was recommended. In this case, excellent separations were obtained on butanediol succinate polyester (5% on Gas-Chrom Z) at 190°C. Separation of methylated sugars as their methyl glycosides has the disadvantage that the resulting chromatograms are complex, owing to the production of multiple peaks corresponding to α- and β-anomers and pyranoside and furanoside forms of each sugar. However, in cases where only a few methylated sugars are present in a mixture the method is very useful in that the profiles and retention times corresponding to these multiple peaks can facilitate characterization of the individual sugars.

15.3.3.2. Acetylated alditols

For GLC analysis of complex mixtures of sugars, it is desirable to simplify the chromatogram by eliminating the anomeric center in each sugar, thereby obviating the difficulty of the production of multiple peaks. This may conveniently be achieved by borohydride reduction of the sugars to the corresponding alditols, which on acetylation give derivatives suitable for GLC. The value of this method was first demonstrated in 1960 by Bishop and Cooper [265], who overcame the problem of separating the mono-O-methyl ethers of D-glucose by reduction and acetylation of these compounds prior to GLC on Apiezon M and thus obtained complete separation of all four isomers. Their success stimulated interest in the possible separation of unsubstituted sugars as the acetates of the derived alditols, and in 1961 Gunner et al. [279,280] reported the separation of the peracetylated derivatives of several alditols, ranging from glycerol up to octitols, by GLC at 213°C on packings, consisting of mixtures of the polar phase butanediol succinate polyester with nonpolar phases such as Apiezon M. However, the hexaacetates of galactitol and D-glucitol were not resolved under these conditions, and this fact, together with the difficulty of preparation of the column packings used, delayed general acceptance of this GLC method until the introduction of the silicone–polyester copolymer ECNSS-M. This was reported by Sawardeker et al. [253] to give excellent separations of the peracetylated alditols derived from all common sugars, including glucose and galactose. The conditions used by these workers (3% ECNSS-M on Gas-Chrom Q, 100–120 mesh, at 190°C) soon became standard for GLC of alditol acetates, and their method is now extensively applied in analyses of mixtures of sugars.

Because of limitations of ECNSS-M (see Chap. 15.3.1), this phase is now gradually being replaced by others having greater thermal stability, such as OV-17 [256], and, especially, OV-225 [256,257,281], in the GLC analysis of peracetylated alditols. Metz et al. [281] succeeded in determining both the neutral and the aminodeoxy sugar components in hydrolyzates from glycoproteins by GLC of the derived alditol acetates in a single run on OV-225 (1% on Chromosorb G HP, 80–100 mesh) with a temperature program of 170 to 230°C at 1°C/min, and OV-225 (3%, on Gas-Chrom Q, 100–120 mesh) was also used by Schwind et al. [282] to resolve peracetylated maltitol and lactitol isothermally at 260°C. Another phase stable at higher temperatures than ECNSS-M is neopentyl glycol sebacate polyester, which has been used by Perry and Webb [283] in GLC of all the 2-amino-2-deoxyhexoses as the derived 2-acetamido-1,3,4,5,6-penta-O-acetyl-2-deoxyglycitols, most of which are separated at 240°C on such packings (10%, on Chromosorb W AW, 80–100 mesh). A disadvantage of the alditol acetate method of GLC analysis is that uronic acids cannot be determined directly in this manner. Chemical manipulation, involving reduction of the carboxylic acid groups [284], is required prior to the conversion to alditol acetates in the normal way.

The method shows a major advantage in its application to GLC analysis of mixtures of partially methylated sugars, which results in relatively simple chromatograms, since each methyl ether gives rise to only one peak. The pattern of methylation in alditol acetates may readily be determined by MS, and therefore these are generally considered to be the best derivatives for GLC–MS examination of methylated sugars (see Chap. 15.3.4). This application has been extensively studied by Lindberg and coworkers [228,242,254] and Lönngren and Pilotti [255], who have published retention data for a large number of partially methylated alditol acetates, derived from all the neutral monosaccharides commonly occurring in polysaccharides as well as a few dideoxy sugars found in certain polysaccharides of microbial origin. Again, these scientists [228,254] followed the initial use of ECNSS-M (3%, on Gas-Chrom Q, 100–120 mesh) at 170–180°C by GLC on OV-225 under the same conditions [255], and more recently on SP-1000 capillary columns at 220°C [242]. Another stationary phase that may be used with advantage in GLC of partially methylated alditol acetates is OV-17 (3%, on Chromosorb W AW, 80–100 mesh, at 170°C). It is capable of resolving the derivatives from 2,4,6-tri-O-methyl-D-glucose and 3,4,6-tri-O-methyl-D-mannose [257], which are not separated on ECNSS-M or OV-225. Other phases that have been used to achieve some of the more difficult separations include butanediol succinate [261], Apiezon greases [285], and a mixture of the cyano silicone phases OV-275 and XF-1150 [286]. Temperature programing can also be advantageous [286,287].

Perry and Webb [288] have published GLC retention data for the methyl ethers of 2-amino-2-deoxy-D-glucose as the acetylated 2-acetamido-2-deoxy-D-glucitol derivatives. They used the polyester phase neopentyl glycol sebacate under conditions similar to those employed in their study of the peracetylated acetamidodeoxyalditols [283]. Methyl ethers of 2-deoxy-2-(N-methyl)acetamido-D-glucitol and -galactitol, in the form of acetates, have been examined by Stellner et al. [289] who adopted ECNSS-M as the GLC phase under the standard conditions originally recommended by Björndal et al. [228]. The improvement resulting from the subsequent use of glass beads as a support for this phase [248] has been noted (see Chap. 15.3.1).

A significant recent development in carbohydrate chemistry is the introduction by Sweet et al. [290] of ethylation analysis as a complement to methylation analysis in structural studies of polysaccharides. They have published retention data for a large number of partially ethylated alditol acetates, derived from the sugars (L-arabinose, D-xylose, L-rhamnose, L-fucose, D-glucose, D-galactose, and D-mannose) occurring in the polysaccharides of plant cell-walls; four different GLC columns were used, two of which had been applied previously, under the same conditions, in the examination of the corresponding partially methylated alditol acetates derived from these sugars [287]. This permitted direct comparison, which showed that many polysaccharide components that are not separable as their partially methylated alditol acetates can be resolved on GLC as the partially ethylated derivatives, and vice versa. GLC–MS of the partially methylated, partially ethylated alditol acetates obtained on hydrolysis, reduction and acetylation of peralkylated oligosaccharides isolated by HPLC (see Chap. 15.2.2.2) is an integral part of an important new method for structural analysis of polysaccharides that has recently been described by Valent et al. [41].

15.3.3.3. Permethylated alditols

The GLC separation of fully methylated alditols, derived from the common monosaccharides, has been studied by Ovodov and Evtushenko [291] and by Whyte [292]. The former reported very poor resolution, especially of the arabinitol–xylitol and glucitol–mannitol pairs, on an Apiezon L phase at 138°C, but the latter succeeded in resolving all but the arabinitol and xylitol derivatives on the trifluoropropyl methyl silicone QF-1 (3%, on Gas-Chrom Q) at 110°C. However, insofar as the monosaccharides are concerned, the degree of resolution obtained does not justify the comparatively long time required for the preparation of the permethylated alditols. It is in GLC of oligosaccharides that this method of derivatization has proved most useful, as it affords relatively volatile compounds that are suitable for examination by MS. Kärkkäinen [293] has made extensive use of the technique in GLC–MS studies of disaccharides, including those containing aminodeoxyhexose residues [293,294], and of a large number of trisaccharides [295]. The permethylated reduced disaccharides were best separated on the phenyl methyl silicone phase OV-22 (1%) at 197°C [293]. This phase, which is stable up to 300°C, was also effective in GLC of the trisaccharide derivatives (at 265°C). Permethylated hexopyranosyl-2-acetamido-2-deoxyhexitols were resolved on QF-1 (3%) at 220°C [294].

15.3.3.4. Acetylated aldononitriles

An alternative method of eliminating the anomeric center in a sugar in order to prevent the production of multiple peaks on GLC is to convert it to the aldononitrile acetate. This is readily achieved by heating the sugar with hydroxylamine hydrochloride in pyridine for ca. 30 min and then, after cooling and addition of acetic anhydride, for a further 30 min, so that the initially formed oxime is dehydrated, and simultaneously free hydroxyl groups are acetylated. GLC of sugars as the derived aldononitrile acetates has the advantage that each aldose gives a characteristic derivative; this is not always the case when the sugars are converted to alditols since, e.g., arabinose and lyxose give the same alditol, as do 2- and 4-O-methyl-D-xylose. The value of the aldononitrile acetate derivatives in GLC of the methyl ethers of D-xylose was first demonstrated in 1967 by Lance and Jones [296], who succeeded in resolving the three mono-O-methyl isomers by this method. Dimitriev et al. [297] subsequently investigated the behavior in GLC of aldononitrile acetates, derived from a wide variety of sugars, and their methyl ethers, and showed that these compounds were suitable for examination by GLC–MS. There is a growing interest in the use of acetylated aldononitriles in such studies, particularly those involving partially methylated sugars, and comprehensive GLC retention data for O-acetyl-O-methyl-D-glucononitriles [298,299] and -mannononitriles [300,301] are now available. Such data have been successfully applied by Seymour et al. in GLC–MS analyses for structural studies of dextrans [299] and of a series of α-D-linked mannans [302].

GLC of the peracetylated aldononitrile derivatives of a large number of sugars, ranging from tetrose to heptose and including acetamidodeoxyhexoses, has also been studied by Seymour et al. [303]. Optimal resolution was obtained on neopentyl glycol succinate polyester (3%, on Chromosorb W, 60–80 mesh), with a temperature program of 140 to 250°C at 3°C/min, under which conditions only two pairs of derivatives, those from D-ribose and L-fucose and those from D-mannose and D-talose, were not separated. The stationary phases and conditions recommended by various authors for the GLC of aldononitrile acetate derivatives of sugars and their methyl ethers are, in general, similar to those used for the corresponding alditol acetates: e.g., Morrison [256] has used ECNSS-M (3%, on Gas-Chrom Q, 100–120 mesh) at 185°C, OV-225 (5%, on Chromosorb W AW DMCS, 100–120 mesh) at 210°C and OV-17 (3%, on Supasorb AW DMCS, 100–120 mesh), also at 210°C, to resolve the peracetylated aldononitriles derived from the common sugars. In most cases the peracetylated alditols are retained longer than the corresponding al-dononitrile acetates under the same conditions,and this permits GLC of a mixture of the two types of derivative on a single column. Such a procedure forms the basis of a method of estimating the chain lengths of poly- and oligosaccharides, from the molar ratio of acetylated aldononitrile to acetylated alditol, found by GLC analysis of the products of reduction and hydrolysis of the polymer, followed by conversion of the sugars to aldononitrile acetates [256,304]. Varma and coworkers [305–309] have developed methods for the quantitative analysis of the sugar components of polysaccharides and glycoproteins by GLC of the derived aldononitrile acetates, and these derivatives are now generally accepted as useful alternatives to alditol acetates for such analyses.

15.3.3.5. Trimethylsilylated derivatives

The use of TMS ethers of sugars as volatile derivatives for GLC was first reported in 1960 by Hedgley and Overend [310]. Ferrier [311] subsequently improved resolu-tion on the phases initially employed (Apiezon greases) by operating at a lower temperature (180°C), and observed the separation of anomers under these condi-tions. Further interest in this GLC method was stimulated by the work of Smith and Carlsson [312], who assessed the feasibility of its use in quantitative analysis, but it was the publication in 1963 of the classic study by Sweeley et al. [313] that led to the general acceptance of the technique as a standard method for the analysis of carbohydrates. A major advantage is the rapidity and relative simplicity of the derivatization step which, in the form originally recommended by Sweeley et al. [313] and still widely used, normally produces complete trimethylsilylation of all free hydroxyl groups in the sample within only 5 min. This procedure involves successive treatment of the carbohydrate material, dissolved or suspended in dry pyridine,with hexamethyldisilazane and trimethylchlorosilane followed by shaking or, in the case of sparingly soluble compounds, sonication [226,227]. Following this, the reaction mixture is allowed to stand at room temperature for the requisite time before being injected directly into the gas chromatograph. Modifications proposed by various workers to meet specific needs include the replacement of the pyridine, after the silylation reaction is complete, by another solvent (n-hexane is recommended [314]) in cases where the "tailing" produced by pyridine in GLC is likely to obscure early peaks, use of DMSO instead of pyridine as a solvent for silylation of sparingly soluble compounds [315] and, where the anhydrous conditions demanded by the Sweeley procedure are difficult to achieve (as in the case of corn syrups), replace-ment of the trimethylchlorosilane catalyst by trifluoroacetic acid [316]. An alterna-tive silylating reagent that is relatively tolerant to moisture is N-(trimethylsilyl)imidazole; this can be added to aqueous solutions of sugars or polyols [317]. However, for most purposes the original procedure remains the method of choice for the silylation of carbohydrates.

References on p. B276

Sweeley et al. [313] reported the GLC retention times of a wide variety of sugars and related compounds, as their TMS ethers, on both a polar (15% ethylene glycol succinate polyester on Chromosorb W, 80–100 mesh) and a nonpolar packing (3% SE-52 on silanized Chromosorb W AW, 80–100 mesh). For monosaccharides, especially the aldohexoses, resolution was better on the polar column. This was subsequently corroborated by Sawardeker and Sloneker [318], who used Carbowax 20M, and by Ellis [243], who after studying a number of liquid phases, concluded that XE-60 (1%, on silanized Chromosorb W AW, 60–80 mesh) was the best choice for the resolution of the common monosaccharides as TMS ethers. In all of these studies the column temperature was ca. 140°C. Using somewhat higher temperatures (170–180°C), Cheminat and Brini [319] improved the separation of the monosaccharide ethers on SE-52 and SE-30 (both 10% on silanized Chromosorb W AW, 80–100 mesh), and showed that temperature programing was advantageous in GLC of mixtures of aldoses, ketoses, and deoxy sugars, as their TMS ethers, on SE-52. This is essential if oligosaccharides are also present in the mixture. Sweeley et al. [313] found that a column temperature of 210°C was adequate for GLC of disaccharides as their TMS ethers on SE-52, but 250°C was necessary in the case of tri- and tetrasaccharides. Use of a temperature program (125 to 250°C at 2.3°C/min, then isothermal at 250°C) permitted the separation of a mixture of 20 components, from a tetrose (erythrose) to trisaccharides (raffinose and melezitose) in 75 min. The polyester phase also examined by these workers was, of course, unsuitable for GLC of oligosaccharides as it is stable only up to 200°C. Silicone phases, which are still effective at temperatures up to 300°C (or the even more stable Dexsil 300 GC [252], mentioned in Chap. 15.3.1) are required for this purpose. Of these, OV-17 [258,320], SE-30 [229,321,322], and SE-52 [230,313,316] have been most commonly used, and appropriate temperature programing has permitted the resolution of TMS ethers of a very wide range of carbohydrates, even up to the tetrasaccharide level [316], in 1 h or less.

The use of TMS ethers as volatile derivatives for GLC of carbohydrates has the major advantage over other methods of being an extremely versatile technique, applicable to sugars of all types, including aminodeoxy and acetamidodeoxy sugars [229,313,322–324], and hexuronic acids [240,322,325]. Compounds derived from uronic acids, such as lactones [240,325], aldonic acids and aldonolactones [313,325–328), and aldaric acids [325,328], are all amenable to GLC as their TMS ethers, as are the methyl glycoside methyl esters [322,325]. Neuraminic acid derivatives, such as N-glycolylneuraminic acid and the more common N-acetylneuraminic acid, have been separated as their TMS ethers [260,322], although the trimethylsilylated methyl ester methyl glycosides have the same retention time on SE-30 [260]. The methyl glycosides of aminodeoxy and acetamidodeoxy sugars are separable by GLC as their TMS ethers on SE-30 or OV-17 [260,322–324], as are those of the neutral sugars [322,329], and therefore this method is of great value in the analysis of methanolyzates from glycoproteins, glycosaminoglycans and bacterial cell-wall polysaccharides, since all of the constituents may be determined, as trimethylsilylated methyl glycosides, in a single GLC run [229,260,322,330–332]. Temperature programing, covering the range 80–250°C, is necessary for such separations.

Partially methylated methyl glycosides that are not well separated by GLC, such as the di-O-methylhexosides, are often better resolved after trimethylsilylation of the remaining hydroxyl groups: e.g., Bhattacharjee and Gorin [272] recommend this method for optimal resolution of the di-O-methyl ethers of methyl D-mannoside (on 2% neopentyl glycol succinate on Chromosorb W, 80–100 mesh, at 136°C). Retention data for the TMS ethers of di-, tri-, and tetra-O-methyl-D-mannoses, obtained under similar conditions, have been reported by these authors [272], and the GLC behavior of O-trimethylsilyl-O-methyl ethers of D-glucose [333] and D-xylose [334] has been investigated by various workers. Gorin and Magus [335] have reported retention data for TMS derivatives of the O-methyl ethers expected in methylation analysis of polymers containing 2-acetamido-2-deoxy-D-glucopyranose or -D-galactopyranose units, obtained with polar phases, such as neopentyl glycol sebacate, at 140°C. The application of this GLC method in the methylation analysis of acylneuraminic acids has been investigated by Haverkamp et al. [336], who used SE-30 (4%) at 220°C.

GLC of sugars and glycosides as TMS ethers produces multiple peaks for each component, owing to the resolution of anomers and the separation of pyranoside from furanoside forms. As in the case of the methylated methyl glycosides, the peak profiles can be useful in the characterization of individual components, but if complex mixtures are to be analyzed, a simplification of the chromatogram is desirable. This can be achieved by reduction of the sugars to alditols prior to trimethylsilylation, but this procedure is of little value in analyses involving only pentoses and hexoses, as the TMS ethers of the derived alditols are poorly resolved on GLC [231,313]. The lower polyols, namely ethylene glycol, glycerol, threitol, and erythritol, are separated as their TMS ethers, eluted well ahead of those of pentoses and hexoses and their methyl glycosides. This, therefore, affords a convenient procedure for the simultaneous estimation of polyols and sugars or glycosides in the products of successive periodate oxidation and borohydride reduction, followed by mild hydrolysis (Smith degradation) of polysaccharides [337,338], or methanolysis in the case of glycopeptides [339]. Apart from this application, however, GLC of TMS alditols is useful mainly in the analysis of mixtures of partially methylated sugars, which are in many cases well resolved as the TMS ethers of the derived alditols [239,246,335,340,341]. Reduction of oligosaccharides prior to trimethylsilylation can also prove advantageous, especially in GLC–MS studies [342].

Because of the poor separation of trimethylsilylated pentitols and hexitols, and the ambiguities sometimes introduced when two different sugars produce the same alditol (see Chap. 15.3.3.4), methods other than reduction should be employed to simplify the chromatograms in GLC analysis of mixtures of unsubstituted monosaccharides as their TMS ethers. Morrison and Perry [343] recommended oxidation of aldoses to the corresponding aldonic acids, followed by lactonization. Most aldoses gave only the 1,4-lactone, and in GLC of the TMS ethers a single peak was obtained for each, but D-glucose produced both 1,4- and 1,5-lactones, and hence two peaks. The derivatives most often used are oximes [313,327] or O-methyl oximes [344], prepared by carefully controlled treatment of the sugars with hydroxylamine or methoxylamine hydrochloride in pyridine. Although trimethylsilylated oximes and

TABLE 15.5

GLC OF SUGAR DERIVATIVES

Phases: P1 = 3% ECNSS-M on Gas-Chrom Q, 100–120 mesh; P2 = 1% OV-225 on Chromosorb G HP, 80–100 mesh; P3 = 3% neopentyl glycol succinate on Chromosorb W, 60–80 mesh; P4 = 2% OV-17 on Chromosorb W HP, 80–100 mesh; P5 = 3% SE-30 on Diatoport S, 80–100 mesh; P6 = 0.5% OV-17 on Chromosorb G, 100–120 mesh. Temperature programs: T1 = 170 to 230°C at 1°C/min, 15 min isothermal at 230°C; T2 = 140 to 250°C at 3°C/min; T3 = 130 to 300°C at 5°C/min; T4 = 140 to 200°C at 0.5°C/min.

Parent sugar		Derivatives							
		Peracetyl		Aldononitriles			Trimethylsilyl		
		Alditols		Aldononitriles			Ethers		Oximes
	Phase	P1	P2	P3	P4		P5		P6
	Temp.	190°C	T1	T2	T3		T4		160°C
	Ref.	253	281	303	303		322		327
	RRT*	R1	R1	R2	R2		R3		R4
D-Erythrose		0.26	–	1.00	1.00		–		0.20; 0.23
L-Arabinose		1.00	1.00	2.54	2.35		0.31		0.53
D-Xylose		1.38	1.24	2.79	2.50		0.45; 0.57		0.55
D-Ribose		0.89	–	2.29	2.25		0.36; 0.37		0.60

D-Lyxose	1.00	–	2.43	2.35	0.31; 0.38	–
L-Rhamnose	0.57	–	1.89	2.15	0.33; 0.42	0.68
L-Fucose	0.63	0.76	2.29	2.35	0.35; 0.39; 0.45	0.69
D-Galactose	3.03	2.27	3.89	3.75	0.72; 0.80; 0.90	1.43; 1.57
D-Glucose	3.50	2.43	3.75	3.65	0.86; 1.13	1.55
D-Mannose	2.61	2.11	3.50	3.55	0.66; 0.89	1.45
D-Talose	–	–	3.50	3.55	0.74; 0.78; 0.91	–
D-Allose	–	–	3.39	3.40	0.72; 0.76; 0.82	–
L-Idose	–	–	4.03	3.90	–	–
D-Gulose	–	–	–	–	0.65; 0.71; 1.00	–
D-Altrose	–	–	–	–	0.64; 0.67; 0.82	–
D-Fructose	–	–	–	–	–	1.13
2-Amino-2-deoxy-D-glucose	4.7	–	–	–	0.79; 0.92	–
2-Amino-2-deoxy-D-galactose	5.1	–	–	–	0.73; 0.83	–
2-Acetamido-2-deoxy-D-glucose	–	–	6.11	4.45	1.38	–
2-Acetamido-2-deoxy-D-galactose	–	–	–	4.90	1.30	–
N-Acetylneuraminic acid	–	–	–	–	2.41	–
D-Glucuronic acid	–	–	–	–	1.07; 1.27	–
D-Galacturonic acid	–	–	–	–	0.82; 1.18	–

* RRT = retention time relative to: R1 = arabinitol pentaacetate; R2 = peracetylated erythrononitrile; R3 = trimethylsilylated mannitol; R4 = trimethylsilylated glucitol.

O-methyl oximes derived from certain sugars give two peaks on GLC, owing to the resolution of *syn-* and *anti-*forms, the chromatograms are far less complex than are those given by cyclic forms of the sugars. This derivatization is applicable to both aldoses and ketoses [327], from trioses and tetroses [327,345] to disaccharides [346,347]. The TMS oximes and O-methyl oximes have distinctive mass spectra, and their use in GC–MS analysis of sugars is increasing.

Examples of GLC conditions and retention times of sugar derivatives are presented in Table 15.5.

15.3.3.6. Trifluoroacetate esters

Trifluoroacetylation of carbohydrates, achieved by treatment with trifluoroacetic anhydride [348] or N-methylbis(trifluoroacetamide) [349], in either pyridine or dichloromethane [350], produces highly volatile derivatives that are well suited to use in GLC analysis, particularly if an ECD is available [259]. The application in GLC of trifluoroacetyl (TFA) derivatives of pentoses, hexoses, aminodeoxy sugars, methyl glycosides, and oligosaccharides was first demonstrated in 1966 by Vilkas et al. [351], and subsequently a comprehensive study of the GLC behavior of trifluoroacetates of various hexopyranosides and -furanosides on several different phases was published by Yoshida et al. [329]. Comparison of the separations obtained in GLC of the common monosaccharides and aminodeoxy sugars as the TFA sugars with those of the corresponding methyl glycosides as TFA derivatives under the same conditions showed the latter to be superior, and therefore Zanetta et al. [350] suggested that this method could be applied with advantage in the analysis of the products of methanolysis of glycoproteins and polysaccharides. Using the moderately polar (50% trifluoropropyl, 50% methyl) silicone phase OV-210 (5% on Varaport 30) and the temperature program 90 to 190°C at 1°C/min, they were able to separate, in a single run, most of the naturally occurring pentoses, hexoses, aminodeoxy- and acetamidodeoxyhexoses as TFA O-methyl glycosides, and the methyl ester methyl glycosides of D-glucuronic and D-galacturonic acid were also resolved. The method, which is capable of better resolution of these methyl glycosides than that employing the TMS derivatives, is now being applied to an increasing extent in the examination of methanolyzates from glycoproteins; use of an ECD [352,353] and OV-210 capillary columns [352] permits the analysis of submicrogram quantities of material.

The chromatograms may be simplified by reduction of sugars to alditols before trifluoroacetylation [350,352–354]. Resolution of the TFA alditols derived from pentoses and hexoses is satisfactory, especially with a capillary column [352,354]. Tamura and coworkers [355,356], who recommend the technique as a rapid method for the analysis of mono- and disaccharides in blood and urine, have published GLC–MS data for a number of TFA alditols [357]. This method of derivatization has proved useful in GLC of partially methylated sugars: Anderle and Kováč [348] obtained an excellent separation of all four mono-O-methyl-D-glucoses as the TFA esters of the derived alditols by GLC on XE-60 (1%), with the temperature program 130 to 150°C at 1°C/min, and the methyl ethers of L-rhamnose can also be resolved as TFA alditols [277].

15.3.3.7. Other volatile derivatives

A significant recent development in GLC of carbohydrates has been their conversion to glycosides of chiral alcohols, which has resulted in the separation of enantiomers by GLC of the products on capillary columns. Leontein et al. [249] used (+)-2-octanol to form such glycosides, which were then peracetylated and examined by GLC, isothermally at 230°C, on a WCOT capillary column containing SP-1000. The glycosides gave multiple peaks, with different profiles and retention times for the D- and L-enantiomers, so that it was possible to assign the absolute configurations of the parent sugars by this method. Gerwig et al. [250,251], using TMS (−)-2-butyl glycosides, a capillary column coated with SE-30, and the temperature program 135 to 220°C at 1°C/min, obtained similar results, and the method was shown to be applicable not only to neutral sugars but also to hexuronic acids and acetamidodeoxyhexoses. This new technique for assigning monosaccharide configuration represents an important advance in the methodology of carbohydrate chemistry.

Other types of derivatives that have been used occasionally in GLC of carbohydrates include cyclic butaneboronate esters [358,359] and their TMS ethers [360,361], isopropylidene acetals [362,363] and TMS diethyl dithioacetals [364–366]. The dithioacetal method, which gives a single, sharp peak for each sugar and is applicable to neutral sugars, uronic acids, and the products of nitrous acid deamination of aminodeoxy sugars [365,366], seems the most promising of these derivatization techniques, particularly in view of the highly characteristic mass spectra of sugar dithioacetals.

15.3.4. Use of mass spectrometry

The use of MS in conjunction with GLC of carbohydrates is of crucial importance, especially in the analysis of the complex mixtures encountered in structural studies of polysaccharides and glycoproteins. Most of the carbohydrate derivatives prepared for GLC have been studied by MS, and details of their fragmentation patterns are included in excellent reviews by Kochetkov and Chizhov [367,368], Hanessian [369] and Lönngren and Svensson [370]. Some useful information pertinent to combined GC–MS studies will be found also in the comprehensive review on GLC of carbohydrates by Dutton [226,227].

Derivatives in which the isomerism giving rise to multiple peaks has been eliminated are the most useful for GC–MS studies of carbohydrates. The alditol acetates derived from partially methylated sugars give simple mass spectra which have been extensively studied by Lindberg and coworkers [228,242], with the aid of deuterium labeling [371–373] to obviate the ambiguities sometimes introduced on reduction of sugars to alditols. These studies have been confined to derivatives from neutral sugars; mass spectral data for various O-methylated 2-deoxy-2-(N-methyl)acetamidohexitols are also available [248,289]. Information is now accumulating on the mass spectra of acetylated aldononitriles [297,300,303] and TMS oximes and O-methyl oximes [327,344], which are useful alternatives to alditol acetates for GC–MS analysis of complex mixtures of sugars. The mass spectral data on permethylated reduced oligosaccharides, and the corresponding TMS derivatives, published by Kärkkäinen [293–295,342], have proved invaluable in GC–MS of oligosaccharides.

Electron-impact (EI) MS has been the technique used in most studies of carbohydrates, but chemical ionization (CI) mass spectra, obtained with ammonia, methane, or isobutane as the ionizing gas, are often more easily interpreted [370] and can be used to distinguish compounds not well differentiated by EI spectra [374]. Much valuable information may be obtained by use of both of these complementary techniques in GC–MS studies of carbohydrates [303,374].

15.4. PLANAR CHROMATOGRAPHY

15.4.1. Paper chromatography

PC, first applied to sugars in 1946 by Partridge [375], has been much used in carbohydrate chemistry, but has now been largely superseded by TLC. For this reason, and because PC of carbohydrates is treated very thoroughly in a number of reviews [3, 376–380] and chapters in well-known books [381–384], only a brief discussion is appropriate here.

The relationship between the structure of sugars and their mobility on PC was studied by Isherwood and Jermyn [385]: the R_F values depend upon the number of carbon atoms present and the spatial disposition and degree of substitution of the hydroxyl groups. For oligosaccharides the migration rate decreases with successive addition of sugar residues to the molecule. The linear correlation, first observed by French and Wild [386], between the degree of polymerization (DP) of members of a homologous series and the logarithm of a function α', defined as $R_F/(1 - R_F)$, is characteristic of such a series, the slope of the line depending upon the type of sugar and glycosidic linkage. This relationship is applicable to most oligosaccharides, though an exception has been noted in the case of the oligogalacturonides obtained from pectin [387].

For analytical applications, filter papers such as Whatman No. 1 are used. Thicker sheets (e.g., Whatman No. 3MM or 17) are required for preparative PC. Usually no pretreatment of the paper is necessary, although some separations are greatly improved by the use of papers impregnated with inorganic complexing agents. This is especially true of alditol mixtures and mixtures of alditols with their parent sugars, which are not well separated by conventional PC, but can be resolved on papers impregnated with borate or tungstate. Angus et al. [388], e.g., have studied the behavior of a large number of polyhydroxy compounds, including sugars, alditols and cyclitols, on papers impregnated with sodium tungstate (at pH 6 or 8), and have demonstrated the feasibility of separations, such as those of the four aldopentoses and their three derived alditols, or of D-glucose, D-glucitol, D-galactose, and galactitol, under these conditions. The degree of resolution possible is greater than that obtained by the alternative procedure of including the complexing agent in the chromatographic solvent [389].

In PC of carbohydrates development is usually descending and unidimensional, although there are cases (e.g., in the separation of the sugar phosphates [390]) where two-dimensional chromatography greatly improves separations. Multiple development is normally necessary to resolve oligosaccharides of DP above 6 or 7 [391], but Teichmann [392] has reported good resolution of the members of the isomaltodextrin series, up to DP 12, by a single development on a very long (160-cm) strip in a specially constructed tank. The time required for separations of this type is generaly of the order of 60–120 h, whereas simple separations of monosaccharides can be achieved in 12–24 h, depending upon various factors, such as the porosity of the paper, the viscosity of the solvent, and the temperature. Hough et al. [393] have found that PC at elevated temperature (ca. 37°C) gives more rapid separations and more compact spots.

After development, papers should be allowed to dry at room temperature. At this stage, heating is undesirable as the sugars may undergo modification at temperatures above ca. 60°C. The dried paper chromatograms or—in the case of preparative chromatography—pilot strips cut from the edges and center are then sprayed with an appropriate detecting reagent (see Chap. 15.4.3) to visualize the separated components. The positions of the spots may be recorded as R_F values unless the solvent has run off the end of the paper. When continuous development is used, which is usually necessary if mixtures of closely related sugars are to be resolved, the movement of each component is determined relative to that of a reference sugar on the same paper, and is designated accordingly (for example, R_{Gal} denotes mobility relative to that of galactose). Because these R values are not reproducible, it is essential to chromatograph standard sugars on the same sheet as the sugars under examination to permit direct comparison under identical conditions.

Sugars separated by PC may be quantitated, after clarification of the paper by oiling, by densitometry of the spots: e.g., Menzies [394] has described a technique that permits estimation of sugars in clinical samples of blood or urine at concentrations down to 10 μg/ml, with variation coefficients $\pm 2.0–3.5\%$. In preparative PC, the zones containing the separated components are excised from the paper and eluted with water or 50% ethanol. In any quantitative determination of carbohydrates thus isolated, a correction must be applied for the presence of traces of cellulosic material from the paper, which may be estimated by eluting blank strips of the same size as the excised chromatographic zones.

The solvent systems generally recommended [3,381–383,395,396] for PC of carbohydrates are ternary mixtures containing water, an organic solvent that is completely miscible with water (such as ethanol or pyridine) and an organic solvent that is only partially miscible with water (such as 1-butanol or ethyl acetate). Jermyn and Isherwood [395] advocated the use of solvent systems such as ethyl acetate–pyridine–water (8:2:1) and ethyl acetate–acetic acid–water (3:3:1) for the separation of most sugars. The low viscosity of these solvents permits fairly rapid separations, even of oligosaccharides (which are, however, better resolved with ethyl acetate–pyridine–water (10:4:3) [397]). A disadvantage is that these are rather unstable systems, owing to the tendency of ethyl acetate to undergo hydrolysis. Hough et al. [393] preferred solvent systems containing 1-butanol, such as 1-butanol

TABLE 15.6

PC OF SUGARS AND POLYOLS: SOME RECOMMENDED SYSTEMS

R_{Glc} = mobility relative to that of D-glucose. Papers: W1 = Whatman No. 1; W3 = Whatman No. 3; W1a = Whatman No. 1 impregnated with tungstate at pH 8. Solvents: S1 = 1-butanol–pyridine–water (10:3:3); S2 = ethyl acetate–pyridine–water (8:2:1); S3 = ethyl acetate–pyridine–water (10:4:3); S4 = ethyl acetate–acetic acid–formic acid–water (18:3:1:4); S5 = ethyl acetate–acetic acid–pyridine–water (10:3:3:2); S6 = ethyl acetate–acetic acid–pyridine–water (5:1:5:3); S7 = acetone–1-butanol–water (5:3:2).

Compound	$R_{Glc} \times 100$						
Paper	W1	W1	W1	W1	W1	W3	W1a
Solvent	S1	S2	S3	S4	S5	S6	S7
Ref.	383	383	383	383	399	400	388
L-Arabinose	136	200	114	141	175	–	127
D-Xylose	178	260	130	159	200	–	136
D-Lyxose	–	–	–	–	–	–	104
D-Ribose	214	370	145	186	–	–	80
D-Galactose	80	85	82	94	80	90	67
D-Glucose	100	100	100	100	100	100	100
D-Mannose	143	145	105	119	124	110	94
L-Fucose	196	–	–	184	225	125	187
L-Rhamnose	268	420	159	232	330	–	–
D-Glucuronic acid	4	–	14	98	43	50	–
D-Galacturonic acid	11	–	11	90	32	–	–
2-Amino-2-deoxy-D-glucose	–	–	51	–	–	74	–
2-Amino-2-deoxy-D-galactose	–	–	–	–	–	64	–
2-Acetamido-2-deoxy-D-glucose	–	130	127	–	–	–	–
D-Fructose	137	–	118	133	133	–	–
Sucrose	52	–	72	54	28	–	–
Maltose	–	–	65	–	–	–	47
Maltotriose	–	–	38	–	–	–	–
Raffinose	–	–	27	–	–	–	–
D-Arabinitol	–	–	–	–	–	–	49
Ribitol	–	–	–	–	–	–	63
Xylitol	–	–	–	–	–	–	13
Galactitol	–	–	–	–	–	–	33
D-Glucitol	–	90	91	–	–	–	13
D-Mannitol	–	100	96	–	–	–	33

–pyridine–water (3:1:1), 1-butanol–ethanol–water (40:11:19), and 1-butanol–acetic acid–water (2:1:1). The last one is a good choice for separating acidic components, since ionization of these is suppressed by the presence of acetic acid in the solvent. Formic acid is also effective, and good separations of mixtures containing D-glucuronic and D-galacturonic acid, in addition to neutral monosaccharides, are obtained by PC with ethyl acetate–acetic acid–formic acid–water (18:3:1:4) [383] or 1-butanol–benzene–formic acid–water (100:19:10:25, upper phase) [398]. Basic solvents, such as 1-butanol–pyridine–water systems, retard the movement of acidic sugars, thereby providing a method of separating them from neutral and amino sugars; the latter are retarded by acidic solvents. By suitable variation of the proportions of acetic acid and pyridine in the system ethyl acetate–acetic acid–pyridine–water, all the biologically important sugars, including hexuronic acids and aminodeoxyhexoses, can be resolved by PC [399,400]. Solvent systems recommended for PC of specific classes of carbohydrates include 1-butanol–ethanol–water (4:1:5, upper phase), which is effective in separating methylated sugars [401] and some polyols [402], 2-butanone–water azeotrope, which is, in general, a better solvent system for PC of methylated sugars [5], and t-amyl alcohol–1-propanol–water (8:2:3) for methyl glycosides [403]. Further examples of solvent systems may be found in Table 15.6, together with a list of R_{Glc} values.

15.4.2. Thin-layer chromatography

Since the first application of TLC to sugars in 1961, by Stahl and Kaltenbach [404], the technique has found wide application in the separation and identification of carbohydrates of all types. If offers several advantages over PC, in particular, greater speed and enhanced sensitivity. Numerous publications describing applications of this method in the carbohydrate field have appeared, and an excellent review was published in 1976 by Ghebregzabher et al. [405]. Practical aspects of TLC, as applied to the qualitative and quantitative analysis of carbohydrates, as well as preparative separations, are well presented in three articles by Wing and BeMiller [406].

The sorbents most commonly used in TLC of carbohydrates are cellulose, kieselguhr, and silica. Some of the supports for exclusion chromatography (such as Sephadex or Bio-Gel P) are also available in superfine (>400 mesh) grades for use in thin-layer gel chromatography. It is applied in the rapid fractionation of material of high molecular weight, such as glycosaminoglycans [407], which is, of course, simply a special case of SEC (see Chap. 15.2.5) and will not be considered further here.

Many different solvent systems have been used in TLC of carbohydrates. It is impossible to make any generalizations, as has been done in discussing solvents for PC, which is essentially a form of partition chromatography, since the separation mecanism in TLC varies with the stationary and mobile phases. With cellulose layers, the separation is based on partition, as in cellulose CC (see Chap. 15.2.1) and PC, and in this case ternary solvent mixtures of the type recommended for PC of carbohydrates are effective. In TLC on silica gel and kieselguhr, adsorptive processes often operate together with the partition mechanism. The presence of inorganic salts in the matrix can further influence the separation through complexation and other interactions, so that the optimal choice of solvents depends greatly upon the conditions used and the polarity of the substances being examined. Binary or ternary mixtures of solvents differing in polarity, one of which should be water if unsubstituted sugars are to be separated, are generally used. A useful list of solvent systems recommended for various separations will be found in the review by Ghebregzabher et al. [405]. Because the area available for separation of the substances being examined is restricted in unidimensional TLC, sometimes multiple or continuous development is necessary. Two-dimensional chromatography has been found advantageous in some cases. The recent introduction of HPTLC plates [408], coated with a uniform layer of microparticulate (5–10 μm) silica, slightly thinner than the layers on conventional precoated plates (250 μm thick), has greatly reduced development time and has improved resolution and sensitivity.

After development, the plate is allowed to dry and the separated components are located by spraying it with a suitable detecting reagent (see Chap. 15.4.3). Qualitative analysis by TLC is similar in principle to analysis by PC, and is subject to the same limitations (Chap. 15.4.1). Quantitative analysis by densitometry of the spots on the plate is possible, but presents some difficulties [405,406]. It is generally considered preferable to remove the appropriate zones from the plate and extract the carbohydrate from each for spectrophotometric analysis. As in the case of PC, it is necessary to apply a correction for the effect of any residual sorbent on the analytical results. Similar extraction techniques are used in preparative TLC, for which a layer thickness of ca. 2 mm is recommended [406].

15.4.2.1. TLC on cellulose

Cellulose, available as TLC sorbent in both fibrous and microcrystalline form, has the same chromatographic characteristics as paper, giving good resolution of both mono- and oligosaccharides. More than one development is normally required. Schweiger [409] first demonstrated the usefulness of plates coated with fibrous cellulose (Macherey-Nagel MN 300) by achieving a good separation of D-glucose, D-galactose, D-mannose, D-xylose, D-ribose, and L-rhamnose, and of L-fucose from 2-amino-2-deoxy-D-galactose, on such a plate by two successive developments with ethyl acetate–pyridine–water (2:1:2). Hexuronic acids were separated with 2-propanol–pyridine–acetic acid–water (8:8:1:4). Vomhof and Tucker [410] reported some differentiation of D- and L-arabinose by TLC on cellulose with formic acid–2-butanone–t-butanol–water (3:6:8:3), and Petre et al. [411] successfully applied the same solvent system to TLC analysis of the mixtures of sugars, including hexuronic acids, occurring in plant extracts. The plates, coated with microcrystalline cellulose (CC 41) were developed three times, which required a total time of 3 h. A similar mixture (3:5:7:5), also with triple development, was recommended by Damonte et al. [412] for TLC of the common monosaccharides and small oligosaccharides (up to the tetrasaccharide stachyose) on cellulose (MN 300) layers; the oligosaccharides (DP 2–10) obtained on hydrolysis of inulin were also resolved. Other solvent systems recommended for TLC of monosaccharides and small oligosaccharides on cellulose layers include ethyl acetate–pyridine–water (20:7:5) [413] and ethyl acetate–pyridine–water–acetic acid–propionic acid (10:10:2:1:1) [414].

Spitschan [415] succeeded in resolving the oligosaccharides of the maltodextrin series (up to DP 5) on a cellulose layer by a single development (2 h) with ethyl acetate–acetic acid–pyridine–water (7:1:5:3). The series of oligogalacturonic acids formed on enzymatic hydrolysis of pectic acid was resolved (to DP 10) by Liu and Luh [106], who developed commercial precoated cellulose plates twice (each time for

150 min) with ethyl acetate–acetic acid–water (4:2:3). TLC on cellulose plates has also been applied to the separation of the disaccharides produced on enzymatic degradation of chondroitin sulfates A, B, and C [416]. With the solvent system 1-butanol–acetic acid–1 M ammonium hydroxide (2:3:1) the development time required to achieve this separation was 6 h (cf. HPLC methods [23,37]).

For most purposes, TLC of carbohydrates on cellulose layers does not require prior impregnation of the sorbent with inorganic salts. An exception is the separation of alditols from one another and from their parent sugars. The advantages of the use of paper impregnated with tungstate for such separations have been noted [388], and Briggs et al. [417] have recently reported that the degree of resolution of such mixtures on TLC with 1-butanol–ethanol–water (40:11:19) and acetone–1-butanol–water (5:3:2) was similarly improved if the cellulose layers used were treated with a 5% solution of sodium tungstate dihydrate, at pH 6 or 8, before chromatography.

15.4.2.2. TLC on kieselguhr

The diatomaceous earth kieselguhr is a weaker adsorbent than the cellulose and silica used in TLC, and is effective in separating sugars only when impregnated with a suitable inorganic salt. In the initial application of TLC to carbohydrates, Stahl and Kaltenbach [404], who used layers of Kieselguhr G (containing gypsum as a binder) which had been buffered with 0.02 M sodium acetate, reported that development with ethyl acetate–2-propanol–water (130:47:23) resolved most of the naturally occurring sugars. Bell and Talukder [418] subsequently found that L-fucose could not be separated from the pentoses by this method, but superior resolution of all sugars was obtained when the kieselguhr was buffered with sodium dihydrogen phosphate (0.15 M) instead of sodium acetate [419]. Two developments with ethyl acetate–methanol–1-butanol–water (16:3:3:1) were recommended.

Brown and coworkers [153–155] used TLC on kieselguhr plates, buffered with 0.02 M sodium acetate, and ethyl acetate–2-propanol–water (35:42:23) as the mobile phase, to resolve oligosaccharides of the cello-, manno-, and xylodextrin series, up to DP 6–8. Shannon and Creech [420] showed that kieselguhr was particularly suitable for TLC of oligosaccharides of high DP: good separations of the maltodextrins, up to DP 27, were obtained on layers of unbuffered Kieselguhr G with 1-butanol–pyridine–water mixtures as solvents, and multiple development of large plates with mixtures having different proportions gave optimal resolution of the higher oligosaccharides. The fructans (DP 2–20) from inulin were well resolved on Kieselguhr G layers by a single development (ca. 90 min) with 1-propanol–ethyl acetate–water (3:1:1) [421].

15.4.2.3. TLC on silica

Silica gel, either as such or impregnated with various inorganic salts, has been very extensively used in TLC of sugars and their derivatives. In general, resolution is sharper on silica layers than on either kieselguhr or cellulose. Unmodified silica gel is used mainly in TLC separation of carbohydrate derivatives, such as methylated sugars [422–424] and methyl glycosides [422,423,425,426], which can be resolved by

TABLE 15.7

TLC OF SUGARS: SOME RECOMMENDED SYSTEMS

Layers: L1 = 0.25 mm Cellulose MN 300 (Macherey-Nagel); L2 = 0.10 mm Cellulose F (Merck); L3 = 0.5 mm Kieselguhr G (Merck), impregnated with 0.15 M NaH$_2$PO$_4$; L4 = 0.25 mm Kieselguhr G (Merck); L5 = 0.25 mm Silica Gel G (Merck); L6 = 0.25 mm Silica Gel 60 (Merck); L7 = 0.25 mm Silica Gel 60, impregnated with 0.5 M NaH$_2$PO$_4$; L8 = 0.30-mm layer of Silica Gel G 60 and Syloid 63 (2:1), impregnated with 0.032 M sodium tetraborate and 0.05 M sodium tungstate (3:1). Solvents: S1 = formic acid–butanone-t-butanol–water (3:5:7:5); S2 = ethyl acetate–pyridine–water–acetic acid–propionic acid (10:10:2:1:1); S3 = ethyl acetate–methanol–1-butanol–water (16:3:3:1); S4 = 1-butanol–pyridine–water (13:4:3); S5 = 1-propanol–water (7:1); S6 = 2-propanol–acetone–1 M lactic acid (2:2:1); S7 = 2-propanol–acetone–0.1 M lactic acid (2:2:1); S8 = ethyl acetate–2-propanol–water (2:2:1); S9 = ethyl acetate–methanol–acetic acid–water (12:3:3:2).

Compound	Relative mobility								
Layer	L1	L2	L3	L4	L5	L6	L7	L8	L8
Solvent	S1	S2	S3	S4	S5	S6	S7	S8	S9
Ref.	412	414	419	420	445	431	442	440	440
	hR_{Glc} ***	hR_{Glc} **	hR_F **	hR_F *	hR_F *	hR_F *	hR_F *	hR_F §	hR_F §§
L-Arabinose	112	134	61	–	–	67	47	–	–
D-Xylose	113	154	77	–	–	76	63	33	46
D-Lyxose	114	162	–	–	–	–	65	–	–
D-Ribose	116	176	83	–	–	72	60	20	44
D-Galactose	98	82	28	–	54	58	30	29	27

D-Glucose	100	100	42	87	62	65	41	38	33
D-Mannose	104	115	–	–	–	69	50	41	–
L-Fucose	–	162	91	–	–	75	65	–	42
L-Rhamnose	122	192	98	–	–	85	83	–	–
D-Fructose	109	127	–	–	–	64	46	15	30
L-Sorbose	–	129	–	–	–	69	52	–	–
D-Tagatose	–	–	–	–	–	–	60	–	–
2-Amino-2-deoxy-D-glucose	–	–	–	–	21	34	–	–	–
2-Amino-2-deoxy-D-galactose	–	–	–	–	13	–	–	–	–
2-Acetamido-2-deoxy-D-glucose	–	–	–	–	54	79	–	–	–
2-Acetamido-2-deoxy-D-galactose	–	–	–	–	45	–	–	–	–
N-Acetylneuraminic acid	–	–	–	–	–	3	–	–	–
Sucrose	90	60	–	–	–	66	47	38	20
Maltose	76	48	–	77	–	57	34	29	18
Maltotriose	51	25	–	61	–	46	27	–	–
Maltotetraose	–	–	–	42	–	34	–	–	–
Maltopentaose	–	–	–	24	–	27	–	–	–
Maltohexaose	–	–	–	14	–	23	–	–	–

  * Single development.

  ** Double development.

  *** Triple development.

  § Two-dimensional development, first direction.

  §§ Two-dimensional development, second direction, orthogonal to first.

use of solvent systems such as 2-butanone–water azeotrope [423], 2-butanone saturated with 3% ammonium hydroxide [424], the upper phase of benzene–ethanol–water–conc. ammonium hydroxide (200:47:15:1) [422] or — in the case of the glycosides—1-butanol–acetic acid–diethyl ether–water (9:6:3:1) [422], diethyl ether–toluene (2:1) [425], or benzene–ethanol–water (170:47:15, upper layer) [426]. Anomeric pairs of methyl glycosides are resolved [425,426], as are the various aryl glycosides that have been examined in the form of the peracetylated derivatives by Audichya [427], who used 2-butanone–light petroleum (1:3) as the solvent system. Tate and Bishop [428] achieved good resolution of sugar acetates on silica layers with benzene containing 2 to 10% methanol, depending on the nature of the sugar. Inositol hexaacetates were similarly resolved.

Separation of monosaccharides on layers of unmodified silica is generally poor, except by techniques such as two-dimensional development with two different solvent systems (e.g., chloroform–methanol–water (16:9:2) and ethyl acetate–methanol–acetic acid–water (12:3:3:2) [429]) or continuous-flow TLC (e.g., development with acetone–water (15:1) for 7 h produced good resolution of hexoses on Silica Gel G plates [430]). Hansen [431] achieved greatly improved resolution of the common monosaccharides and of some oligosaccharides (DP 2–5) by a single development with 2-propanol–acetone–1 M lactic acid (2:2:1). Ghebregzabher et al. [432] have studied the effects of adding boric or phenylboronic acids to the solvent system and concluded that the former produced the greater improvement in resolution of the mixtures of sugars occurring in biological fluids: e.g., D-glucuronic acid, D-fructose, D-galactose, D-glucose, L-fucose, and 2-deoxy-D-ribose were well separated on a silica plate after a single development with 2-butanone–2-propanol–acetonitrile–0.5 M boric acid + 0.25 M isopropylamine–acetic acid (200:150:100:75:2).

For TLC of mixtures of closely related sugars the silica gel layer must be impregnated beforehand with an inorganic salt capable of interacting with carbohydrates. Boric acid [433–435], sodium acetate [436] or phosphate [436,437] buffers, and molybdic or phosphotungstic acids [438] have been used for this purpose. Mezzetti et al. [439] demonstrated that two-dimensional development with different solvent systems on coupled layers containing different impregnants produced superior resolution of mixtures of all the monosaccharides and small oligosaccharides commonly encountered. However, the length of time (7–8 h) required for the double development was a disadvantage, and Ghebregzabher et al. [440] have subsequently recommended a simplified technique, involving the use of a uniform layer of silica gel impregnated with a mixture of sodium tetraborate and sodium tungstate. Development with ethyl acetate–2-propanol–water (2:2:1), which separates the sugars mainly according to their capacity to form complexes with borate, is followed by a second development, orthogonal to the first, with the acidic solvent system ethyl acetate–methanol–acetic acid–water (12:3:3:2), which deactivates the borate so that the predominant mechanism becomes a normal partition process. The first development requires 2 h and the second one 50 min, so that, with a ca. 5-min interval between them for drying the plates at 70°C, the total time for a separation by this method is ca. 3 h.

Other methods suggested for improving separations of the common monosaccharides and small oligosaccharides on layers of impregnated silica include the use of vapor programing [441] and the addition of lactic acid to the solvent system [442]. Hansen [442] found 2-propanol–acetone–0.1 M lactic acid (2:2:1) to be more effective on silica plates impregnated with sodium dihydrogen phosphate than on unmodified silica plates [431]. Using 2-propanol–acetone–0.2 M lactic acid (6:3:1) and silica plates similarly impregnated, Kremer [443] has achieved resolution of several alditols, from erythritol to the heptitols, and some differentiation of the pentitols and hexitols from their parent sugars.

Aminodeoxy sugars do not migrate satisfactorily on silica gel impregnated with either phosphate or borate, but can be separated, together with their N-acetylated derivatives, on plates pretreated with cupric sulfate [444]. However, this requires the use of solvent systems containing aqueous ammonia, which are unsuitable for simultaneous TLC of neutral sugars, because they may aminate these sugars. Therefore, TLC on unmodified silica is preferable for the examination of hydrolyzates from glycoproteins and glycolipids, which may contain hexoses, fucose, and neuraminic acid derivatives, in addition to aminodeoxy and acetamidodeoxy sugars. Gal [445] has studied the separation of 2-amino-2-deoxy-D-glucose and -D-galactose, of their N-acetyl derivatives, and of D-glucose, D-galactose, and N-acetylneuraminic acid, on Silica Gel G plates and has reported satisfactory resolution on development with 1-propanol–water (7:1). The method of Hansen [431] was applicable to the separation of L-fucose from 2-amino- and 2-acetamido-2-deoxy-D-glucose, and of these sugars from hexoses.

Nonimpregnated silica is also required for TLC of oligosaccharides having DP > 4. Covacevich and Richards [446] have reported the resolution of the members of the isomaltodextrin series, and of oligosaccharides having a single α-(1 → 3)-linked branch on the α-(1 → 6)-linked D-glucan chain (up to DP 9 or 10) by continuous development for 19 h with 1-propanol–nitromethane–water (5:2:3) at 30°C. Nurok and Zlatkis [447] have used HPTLC plates, developed continuously with various mixtures of acetone, ethanol, 2-propanol, and water at elevated temperatures (40–60°C) to resolve the members of the maltodextrin series (up to DP 16) in 2 h or less. Chromatography on HPTLC plates, which has also been successfully applied to the separation of brain gangliosides [448,449], is now clearly the method of choice for TLC of higher oligosaccharides, superseding the older methods involving kieselguhr or cellulose layers. Some recommended TLC systems are shown in Table 15.7, together with a list of R_{Glc} and R_F values.

15.4.3. Detecting reagents

Many different reagents can be used to visualize carbohydrates separated on paper or thin-layer chromatograms. Those regarded as standard reagents are listed and described in several books and review articles [381–384,405,406,450], and a compilation of information on the preparation, reactions, and sensitivities of these and some newer reagents will be found in a recent handbook [451].

The use of sulfuric acid for charring carbohydrates on a chromatogram and of color-forming reagents containing sulfuric acid is obviously confined to TLC on inorganic layers. With this exception, however, most spray reagents are applicable to both PC and TLC. These may be classified into four types, according to the mechanism upon which color production depends:

(a) The reducing power of sugars. These nonspecific reagents include ammoniacal silver nitrate, alkaline permanganate, 3,5-dinitrosalicylate or 3,4-dinitrobenzoate, and triphenyltetrazolium chloride. Another reagent in this class, recently recommended by Haldorsen [452] for use in TLC, is vanadium pentoxide in sulfuric acid, which permits limited differentiation between different classes of carbohydrates, as some (e.g., ketoses and aldonic acids) react more rapidly than others.

(b) Reaction of the sugars with an acidic component to form furfural derivatives, which condense with phenols or aromatic amines to produce colored compounds, often having a characteristic fluorescence under UV light. Reagents of this type are more specific, giving different colors with the various classes of carbohydrate. Well-known examples are p-anisidine hydrochloride, aniline phosphate or phthalate, and naphthoresorcinol (1,3-dihydroxynaphthalene) in an acid medium. The advantages of the aniline citrate reagent [453] for the differentiation of sugars of various types have been demonstrated by Vitek and Vitek [454], who have described a procedure involving heating paper or thin-layer chromatograms sprayed with this reagent in gradual stages so that, because of the different reaction temperatures and times required by sugars differing in molecular structure and size, the spots formed reach maxima of fluorescence and color intensity at stages characteristic of the corresponding sugars. Another very useful spray for distinguishing carbohydrates having different structures is the diphenylamine–aniline–phosphoric acid reagent, which produces a wide range of colors with the various monosaccharides and also differentiates disaccharides according to the type of glycosidic linkage present [455].

(c) Glycol cleavage, together with tests to reveal the fragments. Spraying with sodium metaperiodate is followed by reaction with pentane-2,4-dione [456], alkaline permanganate [457], or ammoniacal silver nitrate [383]. The alternative of spraying with benzidine to reveal unreacted periodate is no longer recommended, in view of the carcinogenicity of benzidine. Reagents of this type are nonspecific.

(d) Reactions involving certain structural features only. These reagents, specific for a particular class of carbohydrates, include ninhydrin and the Elson–Morgan reagent (pentane-2,4-dione followed by p-dimethylaminobenzaldehyde) for aminodeoxy sugars, Fleury's reagent (mercuric oxide, followed by barium acetate) for polyols, and thiobarbituric acid [458] and dimedone [459] for ketoses.

The various reagents differ greatly in sensitivity [405,460] and the successive application of two or more is often advantageous [405,412].

15.5. ELECTROPHORESIS

15.5.1. Paper electrophoresis

Zone electrophoresis on filter paper can serve as a useful complement to PC in examination of carbohydrates and related compounds. The medium must contain an electrolyte capable of forming charged complexes with carbohydrates. Borate solutions are particularly suitable and have been extensively used for this purpose [461], but electrophoresis has also been performed in media containing arsenite [462], phenylboronate [463], germanate [464], molybdate [465–467], tungstate [467,468], and stannate [467], which form anionic complexes with polyhydroxy compounds, and in the presence of basic lead acetate [462], which forms cationic complexes. The results of these electrophoretic studies have been comprehensively reviewed by Foster [461,469] and Weigel [467], and mobility data for a large number of compounds, including aldoses, ketoses, disaccharides and their reduced forms, alditols, cyclitols, methyl glycosides, and methyl ethers, are listed in these reviews.

Charged compounds can be submitted to electrophoresis without prior complexation, but Haug and Larsen [470] demonstrated that separation of hexuronic acids by this method was much improved by the use of a medium consisting of 0.01 M borax (pH 9.2) containing Ca^{2+} ions (0.005 M).

Detecting reagents used to locate carbohydrates on paper chromatograms are also applicable to paper electrophoresis, unless the presence of ions such as borate interferes with the color-forming reaction. Sensitivities are, in general, lower under these conditions.

15.5.2. Other electrophoretic methods

The electrophoretic mobilities of a number of sugars and polyols, in 0.2 M sodium borate at pH 10, were determined by Bourne et al. [471] using not only paper but also fiber glass sheets. The compounds studied migrated similarly on both supports, the one advantage of fiber glass being that the carbohydrates could be more easily detected, by use of reagents of higher sensitivity which are too corrosive for paper.

Fiber glass electrophoresis has been applied to polysaccharides [472], especially pectic substanes [473], but with limited success, owing to the high rate of endosmotic flow. However, this difficulty can be obviated by silanizing the fiber glass strips. This necessitates addition of a surfactant to the buffer used to wet the strips before electrophoresis, but results in much improved separation of de-esterified pectins [474].

Electrophoresis on cellulose acetate membranes, used in acetate, phosphate, or pyridine–formic acid buffers, has proved useful in separations of acidic glycosaminoglycans [475,476] and in studies of other acidic polysaccharides, such as capsular polysaccharides from bacteria [477]. Neutral polysaccharides can be examined by electrophoresis on cellulose acetate strips in the presence of borate: e.g., Dudman and Bishop [478] have reported electrophoretic studies of various arabinogalactans, galactomannans, and glucans on cellulose acetate, used with a

References on p. B276

0.1 M sodium tetraborate–0.1 M sodium chloride buffer, pH 9.3. In this case, the polysaccharides were dyed before application to the strips to facilitate observation of their behavior, but most experimenters prefer to stain the polysaccharides after electrophoresis with a solution of Alcian blue in acetic acid.

The procedure recommended by Dudman and Bishop [478], of dyeing polysaccharides before examination by electrophoresis, has been applied by Pavlenko and Ovodov [479] in polyacrylamide gel electrophoresis of a number of polysaccharides, both neutral and acidic. Using buffers consisting of mixtures of sodium tetraborate (or boric acid), Tris, and the sodium salt of EDTA, at pH 9.2–9.3, they succeeded in resolving the various components of polydisperse dextrans and amylopectins by disc electrophoresis on polyacrylamide gels (4.6% acrylamide). Polyacrylamide gel electrophoresis has been used in the examination of pectic substances [480] and sulfated polysaccharides from seaweeds [481], but its application in the carbohydrate field has been mainly in studies of glycoproteins [482] and glycosaminoglycans [483,484]. For the latter, use of a buffer containing 0.1 M Na_3PO_4, 0.1 M Na_2HPO_4, and 0.125 M sodium formate, pH 11.5, in a 6% gel has been recommended [483,484]. Alcian blue or toluidine blue are used as staining agents. More recently, electrophoresis on slides of 0.9% agarose gel, in 0.06 M barbital or 0.05 M 1,3-diaminopropane buffers (pH 8.5–8.6), has been applied successfully to the separation and identification of acidic glycosaminoglycans [485]. Studies of gel electrophoresis of glycoproteins [486] have indicated that the use of linear gradient SDS-polyacrylamide gels is advantageous, as it permits the differentiation of groups of compounds not distinguishable by other methods.

REFERENCES

1 S.C. Churms, in E. Heftmann (Editor), *Chromatography*, Van Nostrand-Reinhold, New York, 3rd Edn., 1975, p. 637.
2 R.L. Whistler and J.N. BeMiller, *Methods Carbohydr. Chem.*, 1 (1962) 42.
3 L. Hough, *Methods Biochem. Anal.*, 1 (1954) 205.
4 J. Havlicek and O. Samuelson, *Carbohydr. Res.*, 22 (1972) 307.
5 L. Boggs, L.S. Cuendet, I. Ehrenthal, R. Koch and F. Smith, *Nature (London)*, 166 (1950) 520.
6 E.H. Merrifield and A.M. Stephen, *Carbohydr. Res.*, 74 (1979) 241.
7 A.M. Stephen, *J. Chem. Soc.*, (1962) 2030.
8 S. Dziedzic and M.W. Kearsley, *J. Chromatogr.*, 154 (1978) 295.
9 N.A. Parris, *Instrumental Liquid Chromatography*, J. Chromatogr. Library Series, Vol. 5, Elsevier, Amsterdam, 1976, p. 129.
10 American Society of Brewing Chemists, *Amer. Soc. Brew. Chem. J.*, 35 (1977) 104.
11 J.L. Rocca and A. Rouchouse, *J. Chromatogr.*, 117 (1976) 216.
12 S.C. Churms and U.A. Seeman, in S.C. Churms (Editor), *Handbook of Chromatography Series: Carbohydrates*, CRC Press, Boca Raton, FL, 1982, p. 70.
13 G.D. McGinnis and P. Fang, *J. Chromatogr.*, 153 (1978) 107.
14 J. Thiem, H. Karl, J. Schwentner and J. Reimer, *J. Chromatogr.*, 147 (1978) 491.
15 J. Thiem, J. Schwentner, H. Karl, A. Sievers and J. Reimer, *J. Chromatogr.*, 155 (1978) 107.
16 J. Lehrfeld, *J. Chromatogr.*, 120 (1976) 141.
17 C.A. White, J.F. Kennedy and B.T. Golding, *Carbohydr. Res.*, 76 (1979) 1.
18 R.M. Thompson, *J. Chromatogr.*, 166 (1978) 201.

19 F. Nachtmann and K.W. Budna, *J. Chromatogr.*, 136 (1977) 279.
20 R. Schwarzenbach, *J. Chromatogr.*, 140 (1977) 304.
21 U.R. Tjaden, J.H. Krol, R.P. van Hoeven, E.P.M. Oomen-Meulemans and P. Emmelot, *J. Chromatogr.*, 136 (1977) 233.
22 E.G. Bremer, S.K. Gross and R.H. McCluer, *J. Lipid Res.*, 20 (1979) 1028.
23 G.J.-L. Lee, J.E. Evans and H. Tieckelmann, *J. Chromatogr.*, 146 (1978) 439.
24 R. Schwarzenbach, *J. Chromatogr.*, 117 (1976) 206.
25 A.D. Jones, I.W. Burns, S.G. Sellings and J.A. Cox, *J. Chromatogr.*, 144 (1977) 169.
26 M. D'Amboise, D. Noël and T. Hanai, *Carbohydr. Res.*, 79 (1980) 1.
27 J.C. Linden and C.L. Lawhead, *J. Chromatogr.*, 105 (1975) 125.
28 E.C. Conrad and J.K. Palmer, *Food Technol. Chicago*, 30 (1976) 84.
29 R.B. Meagher and A. Furst, *J. Chromatogr.*, 117 (1976) 211.
30 V. Gordy, J.G. Baust and D.L. Hendrix, *Bryologist*, 81 (1978) 532.
31 J.N. Little and G.J. Fallick, *J. Chromatogr.*, 112 (1975) 389.
32 F.M. Rabel, A.G. Caputo and E.T. Butts, *J. Chromatogr.*, 126 (1976) 731.
33 E.K. Gum, Jr. and R.D. Brown, Jr., *Anal. Biochem.*, 82 (1977) 372.
34 D. Noël, T. Hanai and M. D'Amboise, *J. Liquid Chromatogr.*, 2 (1979) 1325.
35 B. Zsadon, K.H. Otta, F. Tüdös and J. Szejtli, *J. Chromatogr.*, 172 (1979) 490.
36 P. van Eikeren and H. McLaughlin, *Anal. Biochem.*, 77 (1977) 513.
37 G.J.-L. Lee and H. Tieckelmann, *Anal. Biochem.*, 94 (1979) 231.
38 K. Aitzetmüller, *J. Chromatogr.*, 156 (1978) 354.
39 B.B. Wheals and P.C. White, *J. Chromatogr.*, 176 (1979) 421.
40 G.B. Wells and R.L. Lester, *Anal. Biochem.*, 97 (1979) 184.
41 B.S. Valent, A.G. Darvill, M. McNeil, B.K. Robertsen and P. Albersheim, *Carbohydr. Res.*, 79 (1980) 165.
42 J.K.N. Jones, R.A. Wall and A.O. Pittet, *Can. J. Chem.*, 38 (1960) 2285.
43 J.K.N. Jones and R.A. Wall, *Can. J. Chem.*, 38 (1960) 2290.
44 R.M. Saunders, *Carbohydr. Res.*, 7 (1968) 76.
45 S.A. Barker, B.W. Hatt, J.F. Kennedy and P.J. Somers, *Carbohydr. Res.*, 9 (1969) 327.
46 P. Jandera and J. Churáček, *J. Chromatogr.*, 98 (1974) 79.
47 R.W. Goulding, *J. Chromatogr.*, 105 (1975) 229.
48 S.J. Angyal, G.S. Bethell and R.J. Beveridge, *Carbohydr. Res.*, 73 (1979) 9.
49 L.E. Fitt, W. Hassler and D.E. Just, *J. Chromatogr.*, 187 (1980) 381.
50 M.L. Ladisch, A.L. Huebner and G.T. Tsao, *J. Chromatogr.*, 147 (1978) 185.
51 H. Hokse, *J. Chromatogr.*, 189 (1980) 98.
52 L.E. Fitt, *J. Chromatogr.*, 152 (1978) 243.
53 O. Samuelson and B. Swenson, *Acta Chem. Scand.*, 16 (1962) 2056.
54 L.-I. Larsson and O. Samuelson, *Acta Chem. Scand.*, 19 (1965) 1357.
55 O. Samuelson and H. Strömberg, *Carbohydr. Res.*, 3 (1966) 89.
56 P. Jonsson and O. Samuelson, *Anal. Chem.*, 39 (1967) 1156.
57 E. Martinsson and O. Samuelson, *J. Chromatogr.*, 50 (1970) 429.
58 O. Samuelson, in J.A. Marinsky (Editor), *Ion Exchange*, Vol. 2, Dekker, New York, 1969.
59 O. Samuelson, *Methods Carbohydr. Chem.*, 6 (1972) 65.
60 P. Jonsson and O. Samuelson, *J. Chromatogr.*, 26 (1967) 194.
61 L.-I. Larsson, O. Ramnäs and O. Samuelson, *Anal. Chim. Acta*, 34 (1966) 394.
62 R. Nackenhorst and W. Thorn, *Res. Exp. Med.*, 172 (1978) 63.
63 J. Havlicek and O. Samuelson, *Anal. Chem.*, 47 (1975) 1854.
64 J. Havlicek and O. Samuelson, *Chromatographia*, 7 (1974) 361.
65 K. Mopper and E.T. Degens, *Anal. Biochem.*, 45 (1972) 147.
66 K. Mopper and E.M. Gindler, *Anal. Biochem.*, 56 (1973) 440.
67 K. Mopper, *Anal. Biochem.*, 85 (1978) 528.
68 J.S. Hobbs and J.G. Lawrence, *J. Sci. Food Agr.*, 23 (1972) 45.
69 J.S. Hobbs and J.G. Lawrence, *J. Chromatogr.*, 72 (1972) 311.
70 J.X. Khym and L.P. Zill, *J. Amer. Chem. Soc.*, 73 (1951) 2399.

71 J.X. Khym and L.P. Zill, *J. Amer. Chem. Soc.*, 74 (1952) 2090.
72 L.P. Zill, J.X. Khym and G.M. Cheniae, *J. Amer. Chem. Soc.*, 75 (1953) 1339.
73 R.B. Kesler, *Anal. Chem.*, 39 (1967) 1416.
74 *Technicon Sugar Chromatography System Brochure*, Technicon, Ardsley, NY, 1968.
75 A. Floridi, *J. Chromatogr.*, 59 (1971) 61.
76 E.F. Walborg, Jr., D.B. Ray and L.E. Öhrberg, *Anal. Biochem.*, 29 (1969) 433.
77 E.F. Walborg, Jr. and L.E. Kondo, *Anal. Biochem.*, 37 (1970) 320.
78 L. Hough, J.V.S. Jones and P. Wusteman, *Carbohydr. Res.*, 21 (1972) 9.
79 L. Hough and R. Sidebotham, in S.C. Churms (Editor), *Handbook of Chromatography Series: Carbohydrates*, CRC Press, Boca Raton, FL, 1982, p. 89.
80 W. Voelter and H. Bauer, *J. Chromatogr.*, 126 (1976) 693.
81 K. Mopper, *Anal. Biochem.*, 87 (1978) 162.
82 Z. Chytilová, O. Mikeš, J. Farkaš, P. Štrop and P. Vrátný, *J. Chromatogr.*, 153 (1978) 37.
83 P. Vrátný, O. Mikeš, J. Farkaš, P. Štrop, J. Čopíková and K. Nejepínská, *J. Chromatogr.*, 180 (1979) 39.
84 N. Spencer, *J. Chromatogr.*, 30 (1967) 566.
85 L. Hough, A.M.Y. Ko and P. Wusteman, *Carbohydr. Res.*, 44 (1975) 97.
86 M. Sinner, *J. Chromatogr.*, 121 (1976) 122.
87 M. Torii and K. Sakakibara, *J. Chromatogr.*, 96 (1974) 255.
88 M. Torii, K. Sakakibara, A. Misaki and T. Sawai, *Biochem. Biophys. Res. Commun.*, 70 (1976) 459.
89 V.A. Derevitskaya, N.P. Arbatsky and N.K. Kochetkov, *Dokl. Akad. Nauk SSSR*, 223 (1975) 1137.
90 N.K. Kochetkov, V.A. Derevitskaya and N.P. Arbatsky, *Eur. J. Biochem.*, 67 (1976) 129.
91 V.A. Derevitskaya, N.P. Arbatsky and N.K. Kochetkov, *Eur. J. Biochem.*, 86 (1978) 423.
92 M.H. Simatupang, *J. Chromatogr.*, 178 (1979) 588.
93 M.H. Simatupang, *J. Chromatogr.*, 180 (1979) 177.
94 J.X. Khym and D.G. Doherty, *J. Amer. Chem. Soc.*, 74 (1952) 3199.
95 K. Mopper, *Anal. Biochem.*, 86 (1978) 597.
96 S. Gardell, *Acta Chem. Scand.*, 7 (1953) 207.
97 A.S.R. Donald, *J. Chromatogr.*, 134 (1977) 199.
98 J.F. Kennedy and J.E. Fox, *Carbohydr. Res.*, 54 (1977) 13.
99 M. Sinner and J. Puls, *J. Chromatogr.*, 156 (1978) 197.
100 M.H. Simatupang, M. Sinner and H.H. Dietrichs, *J. Chromatogr.*, 155 (1978) 446.
101 M.M. Tikhomirov, A.Y. Khorlin, W. Voelter and H. Bauer, *J. Chromatogr.*, 167 (1978) 197.
102 G.O. Aspinall, J.A. Molloy and C.C. Whitehead, *Carbohydr. Res.*, 12 (1970) 143.
103 G.O. Aspinall and A.K. Bhattacharjee, *J. Chem. Soc. C*, (1970) 361.
104 G.O. Aspinall and G.R. Sanderson, *J. Chem. Soc. C*, (1970) 2256.
105 M. Duckworth, K.C. Hong and W. Yaphe, *Carbohydr. Res.*, 18 (1971) 1.
106 Y.K. Liu and B.S. Luh, *J. Chromatogr.*, 151 (1978) 39.
107 H. Neukom and W. Kuendig, *Methods Carbohydr. Chem.*, 5 (1965) 14.
108 I.R. Siddiqui and P.J. Wood, *Carbohydr. Res.*, 16 (1971) 452.
109 N.R. Ringertz and P. Reichard, *Acta Chem. Scand.*, 13 (1959) 1467.
110 N.R. Ringertz and P. Reichard, *Acta Chem. Scand.*, 14 (1960) 303.
111 A. Hallén, *J. Chromatogr.*, 71 (1972) 83.
112 D.F. Smith, D.A. Zopf and V. Ginsburg, *Anal. Biochem.*, 85 (1978) 602.
113 J. Porath and P. Flodin, *Nature (London)*, 183 (1959) 1657.
114 K.A. Granath and P. Flodin, *Makromol. Chem.*, 48 (1961) 160.
115 K.A. Granath, *Methods Carbohydr. Chem.*, 5 (1965) 20.
116 H. Determann, *Gel Chromatography*, Springer, Berlin, New York, 1968.
117 L. Fischer, in T.S. Work and E. Work (Editors), *Laboratory Techniques in Biochemistry and Molecular Biology*, Vol. 1, Part 2, North-Holland, Amsterdam, 1969.
118 D.M.W. Anderson, I.C.M. Dea and A. Hendrie, *Talanta*, 18 (1971) 365.
119 J. Sherma, in G. Zweig and J. Sherma (Editors), *Handbook of Chromatography*, Vol. 2, CRC Press, Cleveland, OH, 1972, p. 44.

120 N.A. Parris, *Instrumental Liquid Chromatography*, J. Chromatogr. Library Series, Vol. 5, Elsevier, Amsterdam, 1976, p. 191.
121 S.C. Churms, *Advan. Carbohydr. Chem. Biochem.*, 25 (1970) 13.
122 M. Rinaudo, *Bull. Soc. Chim. Fr.*, (1974) 2285.
123 D.M.W. Anderson and J.F. Stoddart, *Anal. Chim. Acta*, 34 (1966) 401.
124 K.A. Granath and B.E. Kvist, *J. Chromatogr.*, 28 (1967) 69.
125 G. Arturson and K. Granath, *Clin. Chim. Acta*, 37 (1972) 309.
126 G. Nilsson and K. Nilsson, *J. Chromatogr.*, 101 (1974) 137.
127 Z. Grubisic, P. Rempp and H. Benoit, *J. Polym. Sci. Part B*, 5 (1967) 753.
128 F.R. Dintzis and R. Tobin, *J. Chromatogr.*, 88 (1974) 77.
129 R.H. Pearce and B.J. Grimmer, *J. Chromatogr.*, 150 (1978) 548.
130 D. Hager, *J. Chromatogr.*, 187 (1980) 285.
131 K. Chitumbo and W. Brown, *J. Chromatogr.*, 87 (1973) 17.
132 D.M.W. Anderson and J.F. Stoddart, *Carbohydr. Res.*, 2 (1966) 104.
133 K.J. Bombaugh, W.A. Dark and R.N. King, *J. Polym. Sci. Part C*, 21 (1968) 131.
134 C.D. Chow, *J. Chromatogr.*, 114 (1975) 486.
135 J.F. Kennedy, *J. Chromatogr.*, 69 (1972) 325.
136 A.C.M. Wu, W.A. Bough, E.C. Conrad and K.E. Alden, Jr., *J. Chromatogr.*, 128 (1976) 87.
137 A.M. Basedow, K.H. Ebert, H. Ederer and H. Hunger, *Makromol. Chem.*, 177 (1976) 1501.
138 A.J. de Vries, M. Le Page, R. Beau and C.C. Guillemin, *Anal. Chem.*, 39 (1967) 935.
139 K.J. Bombaugh, W.A. Dark and J.N. Little, *Anal. Chem.*, 41 (1969) 1337.
140 J.J. Kirkland, *J. Chromatogr.*, 126 (1976) 231.
141 S.A. Barker, B.W. Hatt, J.B. Marsters and P.J. Somers, *Carbohydr. Res.*, 9 (1969) 373.
142 S.A. Barker, B.W. Hatt and P.J. Somers, *Carbohydr. Res.*, 11 (1969) 355.
143 F.A. Buytenhuys and F.P.B. van der Maeden, *J. Chromatogr.*, 149 (1978) 489.
144 T.W. Dreher, D.B. Hawthorne and B.R. Grant, *J. Chromatogr.*, 174 (1979) 443.
145 P.J. Knudsen, P.B. Eriksen, M. Fenger and K. Florentz, *J. Chromatogr.*, 187 (1980) 373.
146 H. Engelhardt and D. Mathes, *J. Chromatogr.*, 142 (1977) 311.
147 H. Engelhardt and D. Mathes, *J. Chromatogr.*, 185 (1979) 305.
148 J. Hradil, *J. Chromatogr.*, 144 (1977) 63.
149 P.E. Barker, B.W. Hatt and S.R. Holding, *J. Chromatogr.*, 174 (1979) 143.
150 G.D. McGinnis and P. Fang, *J. Chromatogr.*, 130 (1977) 181.
151 W. Brown, *J. Chromatogr.*, 52 (1970) 273.
152 W. Brown, *J. Chromatogr.*, 59 (1971) 335.
153 W. Brown and Ö. Andersson, *J. Chromatogr.*, 57 (1971) 255.
154 W. Brown, *J. Chromatogr.*, 67 (1972) 163.
155 W. Brown and K. Chitumbo, *J. Chromatogr.*, 66 (1972) 370.
156 H. Dellweg, M. John and G. Trenel, *J. Chromatogr.*, 57 (1971) 89.
157 M. John, G. Trenel and H. Dellweg, *J. Chromatogr.*, 42 (1969) 476.
158 K. Kainuma, A. Nogami and C. Mercier, *J. Chromatogr.*, 121 (1976) 361.
159 N.K. Sabbagh and I.S. Fagerson, *J. Chromatogr.*, 120 (1976) 55.
160 N.K. Sabbagh and I.S. Fagerson, *J. Chromatogr.*, 86 (1973) 184.
161 E. Schmidt and B.S. Enevoldsen, *Carbohydr. Res.*, 61 (1978) 197.
162 K. Wallenfels, P. Földi, H. Niermann, H. Bender and D. Linder, *Carbohydr. Res.*, 61 (1978) 359.
163 M.A. Raftery, T. Rand-Meir, F.W. Dahlquist, S.M. Parsons, C.L. Borders, Jr., R.G. Wolcott, W. Beranek, Jr. and L. Jao, *Anal. Biochem.*, 30 (1969) 427.
164 B. Capon and R.L. Foster, *J. Chem. Soc. C*, (1970) 1654.
165 P. Flodin and K. Aspberg, in T.W. Goodwin and O. Lindberg (Editors), *Biological Structure and Function*, Vol. 1, Academic Press, New York, 1961, p. 345.
166 P. Flodin, J.D. Gregory and L. Rodén, *Anal. Biochem.*, 8 (1964) 424.
167 C.P. Dietrich, *Biochem. J.*, 108 (1968) 647.
168 V.C. Hascall and D. Heinegard, *J. Biol. Chem.*, 247 (1974) 4242.
169 P. Maury, *Biochim. Biophys. Acta*, 252 (1971) 48.

170 D.M.W. Anderson, I.C.M. Dea and A.C. Munro, *Carbohydr. Res.*, 9 (1969) 363.

171 S.C. Churms and A.M. Stephen, *Carbohydr. Res.*, 15 (1970) 11.

172 S.C. Churms and A.M. Stephen, *Carbohydr. Res.*, 35 (1974) 73.

173 S.C. Churms and A.M. Stephen, *Carbohydr. Res.*, 45 (1975) 291.

174 E.G. Brunngraber and B.D. Brown, *Biochem. J.*, 103 (1967) 65.

175 M. Skalka, *J. Chromatogr.*, 33 (1968) 456.

176 G.B. Sumyk and C.F. Yocum, *J. Chromatogr.*, 35 (1968) 101.

177 A. Wasteson, *Biochem. J.*, 122 (1971) 477.

178 A. Wasteson, *J. Chromatogr.*, 59 (1971) 87.

179 E.A. Johnson and B. Mulloy, *Carbohydr. Res.*, 51 (1976) 119.

180 J.J. Hopwood and H.C. Robinson, *Biochem. J.*, 135 (1973) 631.

181 J.A. Alhadeff, *Biochem. J.*, 173 (1978) 315.

182 G.P. Belue and G.D. McGinnis, *J. Chromatogr.*, 97 (1974) 25.

183 L. Hagel, *J. Chromatogr.*, 160 (1978) 59.

184 I.J. Goldstein, C.E. Hollerman and E.E. Smith, *Biochemistry*, 4 (1965) 876.

185 I.J. Goldstein, C.E. Hollerman and J.M. Merrick, *Biochim. Biophys. Acta*, 97 (1965) 68.

186 K.O. Lloyd, *Arch. Biochem. Biophys.*, 137 (1970) 460.

187 K. Aspberg and J. Porath, *Acta Chem. Scand.*, 24 (1970) 1839.

188 I.J. Goldstein, in H. Bittiger and H.P. Schnebli (Editors), *Concanavalin A as a Tool*, Wiley, London, 1976, p. 55.

189 Pharmacia Fine Chemicals, *Affinity Chromatography: Principles and Methods*, Pharmacia Fine Chemicals AB, Uppsala, 1979.

190 S. Narasimhan, J.R. Wilson, E. Martin and H. Schachter, *Can. J. Biochem.*, 57 (1979) 83.

191 J.F. Kennedy and A. Rosevear, *J. Chem. Soc., Perkin Trans. 1*, (1973) 2041.

192 I.J. Goldstein and C.E. Hayes, *Advan. Carbohydr. Chem. Biochem.*, 35 (1978) 127.

193 T. Kristiansen, *Methods Enzymol.*, 34 (1974) 331.

194 B.A. Baldo, W.H. Sawyer, R.V. Stick and G. Uhlenbruck, *Biochem. J.*, 175 (1978) 467.

195 P.A. Gleeson, M.A. Jermyn and A.E. Clarke, *Anal. Biochem.*, 92 (1979) 41.

196 G. Mintz and L. Glaser, *Anal. Biochem.*, 97 (1979) 423.

197 Pharmacia Fine Chemicals, *Affinity Chromatography Guide*, Pharmacia Fine Chemicals AB, Uppsala, 1979.

198 M.E. Etzler and E.A. Kabat, *Biochemistry*, 9 (1970) 869.

199 M.E.A. Pereira, E.A. Kabat and N. Sharon, *Carbohydr. Res.*, 37 (1974) 89.

200 T. Irimura, T. Kawaguchi, T. Terao and T. Osawa, *Carbohydr. Res.*, 39 (1975) 317.

201 A. Surolia, A. Ahmad and B.K. Bachhawat, *Biochim. Biophys. Acta*, 371 (1974) 491.

202 A. Surolia, A. Ahmad and B.K. Bachhawat, *Biochim. Biophys. Acta*, 404 (1975) 83.

203 T. Majumdar and A. Surolia, *Indian J. Biochem. Biophys.*, 16 (1979) 200.

204 B. Ersson, K. Aspberg and J. Porath, *Biochim. Biophys. Acta*, 310 (1973) 446.

205 M.E.A. Pereira and E.A. Kabat, *Biochemistry*, 13 (1974) 3184.

206 J.-P. Zanetta, A. Reeber, G. Vincendon and G. Gombos, *Brain Res.*, 138 (1977) 317.

207 J. Hettinger and R.E. Majors, *Varian Instrument Applications*, 10 (1976) 6.

208 H. Binder, *J. Chromatogr.*, 189 (1980) 414.

209 J.J. Liljamaa and A.A. Hallén, *J. Chromatogr.*, 57 (1971) 153.

210 L.A.T. Verhaar and J.M.H. Dirkx, *Carbohydr. Res.*, 62 (1978) 197.

211 S. Katz and L.H. Thacker, *J. Chromatogr.*, 64 (1972) 247.

212 S. Katz, W.W. Pitt, Jr., J.E. Mrochek and S. Dinsmore, *J. Chromatogr.*, 101 (1974) 193.

213 J.E. Hodge and B.T. Hofreiter, *Methods Carbohydr. Chem.*, 1 (1962) 380.

214 W.H. Morrison, M.F. Lou and P.B. Hamilton, *Anal. Biochem.*, 71 (1976) 415.

215 T.A. Scott, Jr. and E.H. Melvin, *Anal. Chem.*, 25 (1953) 1656.

216 H. Jenner, in *Proc. European Technicon Symp., Automation in Analytical Chemistry, Brighton*, Technicon, Ardsley, NY, 1967, p. 203.

217 M. Dubois, K.A. Gilles, J.K. Hamilton, P.A. Rebers and F. Smith, *Anal. Chem.*, 28 (1956) 350.

218 S. Katz, S.R. Dinsmore and W.W. Pitt, Jr., *Clin. Chem.*, 17 (1971) 731.

219 S.A. Barker, M.J. How, P.V. Peplow and P.J. Somers, *Anal. Biochem.*, 26 (1968) 219.
220 C.A. White and J.F. Kennedy, *Clin. Chim. Acta*, 95 (1979) 369.
221 T. Bitter and H.M. Muir, *Anal. Biochem.*, 4 (1962) 330.
222 K. von Berlepsch, *Anal. Biochem.*, 27 (1969) 424.
223 R. Dawson and K. Mopper, *Anal. Biochem.*, 84 (1978) 191.
224 M.H. Simatupang and H.H. Dietrichs, *Chromatographia*, 11 (1978) 89.
225 A.G. McInnes, D.H. Ball, F.P. Cooper and C.T. Bishop, *J. Chromatogr.*, 1 (1958) 556.
226 G.G.S. Dutton, *Advan. Carbohydr. Chem. Biochem.*, 28 (1973) 11.
227 G.G.S. Dutton, *Advan. Carbohydr. Chem. Biochem.*, 30 (1974) 9.
228 H. Björndal, C.G. Hellerqvist, B. Lindberg and S. Svensson, *Angew. Chem. Int. Ed. Engl.*, 9 (1970) 610.
229 J.R. Clamp, T. Bhatti and R.E. Chambers, *Methods Biochem. Anal.*, 19 (1971) 229.
230 P.M. Holligan, *New Phytol.*, 70 (1971) 239.
231 P.M. Holligan and E.A. Drew, *New Phytol.*, 70 (1971) 271.
232 K.M. Brobst, *Methods Carbohydr. Chem.*, 6 (1972) 3.
233 C.C. Sweeley and R.V.P. Tao, *Methods Carbohydr. Chem.*, 6 (1972) 8.
234 F. Loewus and R.H. Shaw, *Methods Carbohydr. Chem.*, 6 (1972) 14.
235 J.H. Sloneker, *Methods Carbohydr. Chem.*, 6 (1972) 20.
236 H.G. Jones, *Methods Carbohydr. Chem.*, 6 (1972) 25.
237 F. Drawert and G. Leupold, *Chromatographia*, 9 (1976) 447.
238 J.R. Clamp, *Biochem. Soc. Trans.*, 5 (1977) 1693.
239 B.H. Freeman, A.M. Stephen and P. van der Bijl, *J. Chromatogr.*, 73 (1972) 29.
240 J.F. Kennedy, S.M. Robertson and M. Stacey, *Carbohydr. Res.*, 49 (1976) 243.
241 G.O. Aspinall, *J. Chem. Soc.*, (1963) 1976.
242 P.E. Jansson, L. Kenne, H. Liedgren, B. Lindberg and J. Lönngren, *Chem. Commun. (Univ. Stockholm)*, No. 8 (1976).
243 W.C. Ellis, *J. Chromatogr.*, 41 (1969) 335.
244 D.H. Shaw and G.W. Moss, *J. Chromatogr.*, 41 (1969) 350.
245 G.G.S. Dutton and R.H. Walker, *Cell. Chem. Technol.*, 6 (1972) 295.
246 A.A. Akhrem, G.V. Avvakumov and O.A. Strel'chyonok, *J. Chromatogr.*, 176 (1979) 207.
247 W.A. Aue, C.R. Hastings and S. Kapila, *J. Chromatogr.*, 77 (1973) 299.
248 G.O.H. Schwarzmann and R.W. Jeanloz, *Carbohydr. Res.*, 34 (1974) 161.
249 K. Leontein, B. Lindberg and J. Lönngren, *Carbohydr. Res.*, 62 (1978) 359.
250 G.J. Gerwig, J.P. Kamerling and J.F.G. Vliegenthart, *Carbohydr. Res.*, 62 (1978) 349.
251 G.J. Gerwig, J.P. Kamerling and J.F.G. Vliegenthart, *Carbohydr. Res.*, 77 (1979) 1.
252 G.A. Janauer and P. Englmaier, *J. Chromatogr.*, 153 (1978) 539.
253 J.S. Sawardeker, J.H. Sloneker and A. Jeanes, *Anal. Chem.*, 37 (1965) 1602.
254 H. Björndal, B. Lindberg and S. Svensson, *Acta Chem. Scand.*, 21 (1967) 1801.
255 L. Lönngren and A Pilotti, *Acta Chem. Scand.*, 25 (1971) 1144.
256 I.M. Morrison, *J. Chromatogr.*, 108 (1975) 361.
257 E.H. Merrifield, in S.C. Churms (Editor), *Handbook of Chromatography Series: Carbohydrates*, CRC Press, Boca Raton, FL, 1982, p. 4.
258 J. Haverkamp, J.P. Kamerling and J.F.G. Vliegenthart, *J. Chromatogr.*, 59 (1971) 281.
259 G. Eklund, B. Josefsson and C. Roos, *J. Chromatogr.*, 142 (1977) 575.
260 A. Cahour and L. Hartmann, *J. Chromatogr.*, 152 (1978) 475.
261 Y.-M. Choy, G.G.S. Dutton, K.B. Gibney, S. Kabir and J.N.C. Whyte, *J. Chromatogr.*, 72 (1972) 13.
262 G.G.S. Dutton and K.B. Gibney, *J. Chromatogr.*, 72 (1972) 179.
263 G.M. Bebault, J.M. Berry, Y.-M. Choy, G.G.S. Dutton, N. Funnell, L.D. Hayward and A.M. Stephen, *Can. J. Chem.*, 51 (1973) 324.
264 D.P. Sweet, R.H. Shapiro and P. Albersheim, *Carbohydr. Res.*, 40 (1975) 217.
265 C.T. Bishop and F.P. Cooper, *Can. J. Chem.*, 38 (1960) 388.
266 H.W. Kircher, *Anal. Chem.*, 32 (1960) 1103.
267 G.A. Adams and C.T. Bishop, *Can. J. Chem.*, 38 (1960) 2380.
268 A.M. Stephen, M. Kaplan, G.L. Taylor and E.C. Leisegang, *Tetrahedron*, Suppl. 7 (1966) 233.

269 M. Kaplan and A.M. Stephen, *Tetrahedron*, 23 (1967) 193.
270 S.C. Churms, E.H. Merrifield and A.M. Stephen, *Carbohydr. Res.*, 81 (1980) 49.
271 Yu. S. Ovodov and A.F. Pavlenko, *J. Chromatogr.*, 36 (1968) 531.
272 S.S. Bhattacharjee and P.A.J. Gorin, *Can. J. Chem.*, 47 (1969) 1207.
273 K. Heyns, K.R. Sperling and H.F. Grützmacher, *Carbohydr. Res.*, 9 (1969) 79.
274 B. Fournet, Y. Leroy, J. Montreuil and H. Mayer, *J. Chromatogr.*, 92 (1974) 185.
275 D. Anderle, M. Petriková and P. Kováč, *J. Chromatogr.*, 58 (1971) 209.
276 D. Anderle, P. Kováč and H. Anderlová, *J. Chromatogr.*, 64 (1972) 368.
277 F. Janeček, R. Toman, S. Karácsonyi and D. Anderle, *J. Chromatogr.*, 173 (1979) 408.
278 D. Anderle and P. Kováč, *J. Chromatogr.*, 91 (1974) 463.
279 S.W. Gunner, J.K.N. Jones and M.B. Perry, *Chem. Ind. (London)*, (1961) 255.
280 S.W. Gunner, J.K.N. Jones and M.B. Perry, *Can. J. Chem.*, 39 (1961) 1892.
281 J. Metz, W. Ebert and H. Weicker, *Clin. Chim. Acta*, 34 (1971) 31.
282 H. Schwind, F. Scharbert, R. Schmidt and R. Kattermann, *J. Clin. Chem. Clin. Biochem.*, 16 (1978) 145.
283 M.B. Perry and A.C. Webb, *Can. J. Biochem.*, 46 (1968) 1163.
284 G.G.S. Dutton and S. Kabir, *Anal. Lett.*, 4 (1971) 95.
285 H. Parolis and D. McGarvie, *Carbohydr. Res.*, 62 (1978) 363.
286 A.G. Darvill, D.P. Roberts and M.A. Hall, *J. Chromatogr.*, 115 (1975) 319.
287 K.W. Talmadge, K. Keegstra, W.D. Bauer and P. Albersheim, *Plant Physiol.*, 51 (1973) 158.
288 M.B. Perry and A.C. Webb, *Can. J. Chem.*, 47 (1969) 4091.
289 K. Stellner, H. Saito and S.-I. Hakomori, *Arch. Biochem. Biophys.*, 155 (1973) 464.
290 D.P. Sweet, P. Albersheim and R.H. Shapiro, *Carbohydr. Res.*, 40 (1975) 199.
291 Yu. S. Ovodov and E.V. Evtushenko, *J. Chromatogr.*, 31 (1967) 527.
292 J.N.C. Whyte, *J. Chromatogr.*, 87 (1973) 163.
293 J. Kärkkäinen, *Carbohydr. Res.*, 14 (1970) 27.
294 I. Mononen, J. Finne and J. Kärkkäinen, *Carbohydr. Res.*, 60 (1978) 371.
295 J. Kärkkäinen, *Carbohydr. Res.*, 17 (1971) 11.
296 D.G. Lance and J.K.N. Jones, *Can. J. Chem.*, 45 (1967) 1995.
297 B.A. Dimitriev, L.V. Backinowsky, O.S. Chizhov, B.M. Zolotarev and N.K. Kochetkov, *Carbohydr. Res.*, 19 (1971) 432.
298 D. Anderle and P. Kováč, *Chem. Zvesti*, 30 (1976) 355.
299 F.R. Seymour, M.E. Slodki, R.D. Plattner and A. Jeanes, *Carbohydr. Res.*, 53 (1977) 153.
300 F.R. Seymour, R.D. Plattner and M.E. Slodki, *Carbohydr. Res.*, 44 (1975) 181.
301 G.R. Woolard, E.B. Rathbone and A.W. Dercksen, *S. Afr. J. Chem.*, 30 (1977) 169.
302 F.R. Seymour, M.E. Slodki, R.D. Plattner and R.M. Stodola, *Carbohydr. Res.*, 48 (1976) 225.
303 F.R. Seymour, E.C.M. Chen and S.H. Bishop, *Carbohydr. Res.*, 73 (1979) 19.
304 J.K. Baird, M.J. Holroyde and D.C. Ellwood, *Carbohydr. Res.*, 27 (1973) 464.
305 R. Varma, R.S. Varma and A.H. Wardi, *J. Chromatogr.*, 77 (1973) 222.
306 R. Varma, R.S. Varma, W.S. Allen and A.H. Wardi, *J. Chromatogr.*, 86 (1973) 205.
307 R.S. Varma, R. Varma, W.S. Allen and A.H. Wardi, *J. Chromatogr.*, 93 (1974) 221.
308 R. Varma and R.S. Varma, *J. Chromatogr.*, 128 (1976) 45.
309 R. Varma and R.S. Varma, *J. Chromatogr.*, 139 (1977) 303.
310 E.J. Hedgley and W.G. Overend, *Chem. Ind. (London)*, (1960) 378.
311 R.J. Ferrier, *Tetrahedron*, 18 (1962) 1149.
312 B. Smith and O. Carlsson, *Acta Chem. Scand.*, 17 (1963) 455.
313 C.C. Sweeley, R. Bentley, M. Makita and W.W. Wells, *J. Amer. Chem. Soc.*, 85 (1963) 2497.
314 P.O. Bethge, C. Holmström and S. Juslin, *Svensk Papperstidn.*, 69 (1966) 60.
315 W.C. Ellis, *J. Chromatogr.*, 41 (1969) 325.
316 K.M. Brobst and C.E. Lott, Jr., *Cereal Chem.*, 43 (1966) 35.
317 A.E. Pierce, *Silylation of Organic Compounds*, Pierce Chemical Co., Rockford, IL, 1968.
318 J.S. Sawardeker and J.H. Sloneker, *Anal. Chem.*, 37 (1965) 945.
319 A. Cheminat and M. Brini, *Bull. Soc. Chim. Fr.*, (1966) 80.

320 M.F. Laker, *J. Chromatogr.*, 163 (1979) 9.
321 E. Percival, *Carbohydr. Res.*, 4 (1967) 441.
322 T. Bhatti, R.E. Chambers and J.R. Clamp, *Biochim. Biophys. Acta*, 222 (1970) 339.
323 J. Kärkkäinen and R. Vihko, *Carbohydr. Res.*, 10 (1969) 113.
324 P.L. Coduti and C.A. Bush, *Anal. Biochem.*, 78 (1977) 21.
325 O. Raunhardt, H.W.H. Schmidt and H. Neukom, *Helv. Chim. Acta*, 50 (1967) 1267.
326 M.B. Perry and R.K. Hulyalkar, *Can. J. Biochem.*, 43 (1965) 573.
327 G. Petersson, *Carbohydr. Res.*, 33 (1974) 47.
328 G. Petersson, *J. Chromatogr. Sci.*, 15 (1977) 245.
329 K. Yoshida, N. Honda, N. Iino and K. Kato, *Carbohydr. Res.*, 10 (1969) 333.
330 V.N. Reinhold, *Methods Enzymol.*, 25 (1972) 244.
331 M. Yokota and T. Mori, *Carbohydr. Res.*, 59 (1977) 289.
332 H.A.S. Aluyi and D.B. Drucker, *J. Chromatogr.*, 178 (1979) 209.
333 S. Haworth, J.G. Roberts and B.F. Sagar, *Carbohydr. Res.*, 9 (1969) 491.
334 H.H. Sephton, *J. Org. Chem.*, 29 (1964) 3415.
335 P.A.J. Gorin and R.J. Magus, *Can. J. Chem.*, 49 (1971) 2583.
336 J. Haverkamp, J.P. Kamerling, J.F.G. Vliegenthart, R.W. Veh and R. Schaur, *FEBS Lett.*, 73 (1977) 215.
337 G.G.S. Dutton, K.B. Gibney, G.D. Jensen and P.E. Reid, *J. Chromatogr.* 36 (1968) 152.
338 G.G.S. Dutton and K.B. Gibney, *Carbohydr. Res.*, 25 (1972) 99.
339 K.W. Hughes and J.R. Clamp, *Biochim. Biophys. Acta*, 264 (1972) 418.
340 S. Patel, J. Rivlin, T. Samuelson, O.A. Stamm and H. Zollinger, *Helv. Chim. Acta*, 51 (1968) 169.
341 A.A. Akhrem, G.V. Avvakumov, I.V. Sidorova and O.A. Strel'chyonok, *J. Chromatogr.*, 180 (1979) 69.
342 J. Kärkkäinen, *Carbohydr. Res.*, 11 (1969) 247.
343 I.M. Morrison and M.B. Perry, *Can. J. Biochem.*, 44 (1966) 1115.
344 R.A. Laine and C.C. Sweeley, *Carbohydr. Res.*, 27 (1973) 199.
345 D. Anderle, J. Königstein and V. Kováčik, *Anal. Chem.*, 49 (1977) 137.
346 S. Adam and W.G. Jennings, *J. Chromatogr.*, 115 (1975) 218.
347 T. Toba and S. Adachi, *J. Chromatogr.*, 135 (1977) 411.
348 D. Anderle and P. Kováč, *J. Chromatogr.*, 49 (1970) 419.
349 J.E. Sullivan and L.R. Schewe, *J. Chromatogr. Sci.*, 15 (1977) 196.
350 J.P. Zanetta, W.C. Breckenridge and G. Vincendon, *J. Chromatogr.*, 69 (1972) 291.
351 M. Vilkas, Hiu-I-Jan, G. Boussac and M.-C. Bonnard, *Tetrahedron Lett.*, (1966) 1441.
352 M.M. Wrann and C.W. Todd, *J. Chromatogr.*, 147 (1978) 309.
353 D.G. Pritchard and W. Niedermeier, *J. Chromatogr.*, 153 (1978) 487.
354 J. Shapira, *Nature (London)*, 222 (1969) 792.
355 T. Imanari, Y. Arakawa and Z. Tamura, *Chem. Pharm. Bull.*, 17 (1969) 1967.
356 H. Nakamura and Z. Tamura, *Clin. Chim. Acta*, 39 (1972) 367.
357 H. Haga, T. Imanari, Z. Tamura and A. Momose, *Chem. Pharm. Bull.*, 20 (1972) 1805.
358 F. Eisenberg, Jr., *Carbohydr. Res.*, 19 (1971) 135.
339 F. Eisenberg, Jr., *Anal. Biochem.*, 60 (1974) 181.
360 P.J. Wood and I.R. Siddiqui, *Carbohydr. Res.*, 19 (1971) 283.
361 P.J. Wood, I.R. Siddiqui and J. Weisz, *Carbohydr. Res.*, 42 (1975) 1.
362 S. Morgenlie, *Carbohydr. Res.*, 41 (1975) 285.
363 S. Morgenlie, *Carbohydr. Res.*, 80 (1980) 215.
364 S. Honda, K. Takeda and K. Kakehi, *Carbohydr. Res.*, 73 (1979) 135.
365 S. Honda, N. Yamauchi and K. Kakehi, *J. Chromatogr.*, 169 (1979) 287.
366 S. Honda, K. Kakehi and K. Okada, *J. Chromatogr.* 176 (1979) 367.
367 N.K. Kochetkov and O.S. Chizhov, *Advan. Carbohydr. Chem.*, 21 (1966) 39.
368 N.K. Kochetkov and O.S. Chizhov, *Methods Carbohydr. Chem.*, 6 (1972) 540.
369 S. Hanessian, *Methods Biochem. Anal.*, 19 (1971) 105.
370 J. Lönngren and S. Svensson, *Advan. Carbohydr. Chem. Biochem.*, 29 (1974) 41.

371 H. Björndal, B. Lindberg, A. Pilotti and S. Svensson, *Carbohydr. Res.*, 15 (1970) 339.
372 H.B. Borén, P.J. Garegg, B. Lindberg and S. Svensson, *Acta Chem. Scand.*, 25 (1971) 3299.
373 K. Axberg, H. Björndal, A. Pilotti and S. Svensson, *Acta Chem. Scand.*, 26 (1972) 1319.
374 M. McNeil and P. Albersheim, *Carbohydr. Res.*, 56 (1977) 239.
375 S.M. Partridge, *Nature (London)*, 158 (1946) 270.
376 E.L. Hirst and J.K.N. Jones, *Disc. Faraday Soc.*, 7 (1949) 268.
377 F. Cramer, *Angew Chem.*, 62 (1950) 73.
378 R. Dedonder, *Bull. Soc. Chim. Fr.*, (1952) 874.
379 G.N. Kowkabany, *Advan. Carbohydr. Chem.*, 9 (1954) 303.
380 F.A. Isherwood, *Brit. Med. Bull.*, 10 (1954) 202.
381 E. Lederer and M. Lederer, *Chromatography*, Elsevier, Amsterdam, 2nd Edn., 1957, p. 245.
382 R.J. Block, E.L. Durrum and G. Zweig, *Paper Chromatography and Paper Electrophoresis*, Academic Press, New York, 2nd Edn., 1958, p. 170.
383 L. Hough and J.K.N. Jones, *Methods Carbohyd. Chem.*, 1 (1962) 21.
384 F. Percheron, in E. Heftmann (Editor), *Chromatography*, Van Nostrand-Reinhold, New York, 2nd Edn., 1967, p. 573.
385 F.A. Isherwood and M.A. Jermyn, *Biochem. J.*, 48 (1951) 515.
386 D. French and G.M. Wild, *J. Amer. Chem. Soc.*, 75 (1953) 2612.
387 R. Pressey and R.S. Allen, *J. Chromatogr.*, 16 (1964) 248.
388 H.J.F. Angus, J. Briggs, N.A. Sufi and H. Weigel, *Carbohydr. Res.*, 66 (1978) 25.
389 J.F. Robyt, *Carbohydr. Res.*, 40 (1975) 373.
390 R.S. Bandurski and B. Axelrod, *J. Biol. Chem.*, 193 (1951) 405.
391 K. Umeki and K. Kainuma, *J. Chromatogr.*, 150 (1978) 242.
392 B. Teichmann, *J. Chromatogr.*, 70 (1972) 99.
393 L. Hough, J.K.N. Jones and W.H. Wadman, *J. Chem. Soc.*, (1950) 1702.
394 I.S. Menzies, *J. Chromatogr.*, 81 (1973) 109.
395 M.A. Jermyn and F.A. Isherwood, *Biochem. J.*, 44 (1949) 402.
396 A. Jeanes, C.S. Wise and R.J. Dimler, *Anal. Chem.*, 23 (1951) 415.
397 R.L. Whistler and J.L. Hickson, *Anal. Chem.*, 27 (1955) 1514.
398 T. Fuleki and F.J. Francis, *J. Chromatogr.*, 26 (1967) 404.
399 M.C. Jarvis and H.J. Duncan, *J. Chromatogr.*, 92 (1974) 454.
400 G. Strecker and A. Lemaire-Poitau, *J. Chromatogr.*, 143 (1977) 553.
401 G.O. Aspinall, V.P. Bhavanandan and T.B. Christensen, *J. Chem. Soc.*, (1965) 2677.
402 A.E. Bradfield and A.E. Flood, *Nature (London)*, 166 (1950) 264.
403 J.A. Cifonelli and F. Smith, *Anal. Chem.*, 25 (1954) 1132.
404 E. Stahl and U. Kaltenbach, *J. Chromatogr.*, 5 (1961) 351.
405 M. Ghebregzabher, S. Rufini, B. Monaldi and M. Lato, *J. Chromatogr.*, 127 (1976) 133.
406 R.E. Wing and J.N. BeMiller, *Methods Carbohydr. Chem.*, 6 (1972) 42, 54, 60.
407 N. Taniguchi, *Clin. Chim. Acta*, 30 (1970) 801.
408 A. Zlatkis and R.E. Kaiser (Editors), *HPTLC — High Performance Thin-Layer Chromatography*, J. Chromatogr. Library Series, Vol. 9, Elsevier, Amsterdam, 1977.
409 A. Schweiger, *J. Chromatogr.*, 9 (1962) 374.
410 D.W. Vomhof and T.C. Tucker, *J. Chromatogr.*, 17 (1965) 300.
411 R. Petre, R. Dennis, B.P. Jackson and K.R. Jethwa, *Planta Med.*, 21 (1972) 81.
412 A. Damonte, A. Lombard, M.L. Tourn and M.C. Cassone, *J. Chromatogr.*, 60 (1971) 203.
413 C.W. Raadsveld and H. Klomp, *J. Chromatogr.*, 57 (1971) 99.
414 J.W. Walkley and J. Tillman, *J. Chromatogr.*, 132 (1977) 172.
415 R. Spitschan, *J. Chromatogr.*, 61 (1971) 169.
416 L. Wasserman, A. Ber and D. Allalouf, *J. Chromagr.*, 136 (1977) 342.
417 J. Briggs, I.R. Chambers, P. Finch, I.R. Slaiding and H. Weigel, *Carbohydr. Res.*, 78 (1980) 365.
418 D.J. Bell and M.Q.-K. Talukder, *J. Chromatogr.*, 49 (1970) 469.
419 M.Q.-K. Talukder, *J. Chromatogr.*, 57 (1971) 391.
420 J.C. Shannon and R.G. Creech, *J. Chromatogr.*, 44 (1969) 307.

421 F.W. Collins and K.R. Chandorkar, *J. Chromatogr.*, 56 (1971) 163.
422 G.W. Hay, B.A. Lewis and F. Smith, *J. Chromatogr.*, 11 (1963) 479.
423 S.C. Williams and J.K.N. Jones, *Can. J. Chem.*, 45 (1967) 275.
424 P.A. Mied and Y.C. Lee, *Anal. Biochem.*, 49 (1972) 534.
425 M. Gee, *Anal. Chem.*, 35 (1963) 350.
426 P.J. Brennan, *J. Chromatogr.*, 59 (1971) 231.
427 T.D. Audichya, *J. Chromatogr.*, 57 (1971) 161.
428 M.E. Tate and C.T. Bishop, *Can. J. Chem.*, 40 (1962) 1043.
429 T. Kartnig and O. Wegschaider, *J. Chromatogr.*, 61 (1971) 375.
430 D.S. Bailey, *J. Chromatogr.*, 130 (1977) 431.
431 S.A. Hansen, *J. Chromatogr.*, 105 (1975) 388.
432 M. Ghebregzabher, S. Rufini, G.M. Sapia and M. Lato, *J. Chromatogr.*, 180 (1979) 1.
433 G. Pastuska, *Z. Anal. Chem.*, 179 (1961) 355.
343 V. Prey, H. Berbalk and M. Kausz, *Mikrochim. Acta*, (1961) 968.
435 M. Lato, B. Brunelli, G. Ciuffini and T. Mezzetti, *J. Chromatogr.*, 34 (1968) 26.
436 M. Lato, B. Brunelli, G. Ciuffini and T. Mezzetti, *J. Chromatogr.*, 39 (1969) 407.
437 Yu. S. Ovodov, E.V. Evtushenko, V.E. Vaskovsky, R.G. Ovodova and T.F. Solo'eva, *J. Chromatogr.*, 264 (1967) 111.
438 T. Mezzetti, M. Lato, S. Rufini and G. Ciuffini, *J. Chromatogr.*, 63 (1971) 329.
439 T. Mezzetti, M. Ghebregzhaber, S. Rufini, G. Ciuffini and M. Lato, *J. Chromatogr.*, 74 (1972) 273.
440 M. Ghebregzabher, S. Rufini, G. Ciuffini and M. Lato, *J. Chromatogr.*, 95 (1974) 51.
441 R.A. de Zeeuw and G.G. Dull, *J. Chromatogr.*, 110 (1975) 279.
442 S.A. Hansen, *J. Chromatogr.*, 107 (1975) 224.
443 B.P. Kremer, *J. Chromatogr.*, 166 (1978) 335.
444 M.D. Martz and A.F. Krivis, *Anal. Chem.*, 43 (1971) 790.
445 A.E. Gal, *Anal. Biochem.*, 24 (1968) 452.
446 M.T. Covacevich and G.N. Richards, *J. Chromatogr.*, 129 (1976) 420.
447 D. Nurok and A. Zlatkis, *J. Chromatogr.*, 142 (1977) 449.
448 J.-P. Zanetta, F. Vitiello and J. Robert, *J. Chromatogr.*, 137 (1977) 481.
449 S. Ando, N.-C. Chang and R.K. Yu, *Anal. Biochem.*, 89 (1978) 437.
450 D. Waldi, in E. Stahl (Editor), *Thin-Layer Chromatography*, Academic Press, New York, Springer, Berlin, 1965, p. 483.
451 S.C. Churms (Editor), *Handbook of Chromatography Series: Carbohydrates*, CRC Press, Boca Raton, FL, 1982, p. 187.
452 K.M. Haldorsen, *J. Chromatogr.*, 134 (1977) 467.
453 A.A. White and W.C. Hess, *Arch. Biochem. Biophys.*, 64 (1956) 51.
454 V. Vitek and K. Vitek, *J. Chromatogr.*, 143 (1977) 65.
455 T. Toba and S. Adachi, *J. Chromatogr.*, 154 (1978) 110.
456 J.B. Weiss and I. Smith, *Nature (London)*, 215 (1967) 638.
457 R.U. Lemieux and H.F. Bauer, *Anal. Chem.*, 26 (1954) 920.
458 F. Percheron, *Bull. Soc. Chim. Biol.*, 44 (1962) 1161.
459 S. Adachi, *Anal. Biochem.*, 9 (1964) 224.
460 J. Mes and L. Kamm, *J. Chromatogr.*, 38 (1968) 120.
461 A.B. Foster, *Advan. Carbohydr. Chem.*, 12 (1957) 81.
462 J.L. Frahn and J.A. Mills, *Austr. J. Chem.*, 12 (1959) 65.
463 P.J. Garegg and B. Lindberg, *Acta Chem. Scand.*, 15 (1961) 1913.
464 B. Lindberg and B. Swan, *Acta Chem. Scand.*, 14 (1960) 1043.
465 E.J. Bourne, D.H. Hutson and H. Weigel, *J. Chem. Soc.*, (1960) 4252.
466 E.J. Bourne, D.H. Hutson and H. Weigel, *J. Chem. Soc.*, (1961) 35.
467 H. Weigel, *Advan. Carohydr. Chem.*, 18 (1963) 61.
468 H.J.F. Angus, E.J. Bourne, F. Searle and H. Weigel, *Tetrahedron Lett.*, (1964) 55.
469 A.B. Foster, *Methods Carbohydr. Chem.*, 1 (1962) 51.
470 A. Haug and B. Larsen, *Acta Chem. Scand.*, 15 (1961) 1395.
471 E.J. Bourne, A.B. Foster and P.M. Grant, *J. Chem. Soc.*, (1956) 4311.

472 D.H. Northcote, *Methods Carbohydr. Chem.*, 5 (1965) 49.

473 E. Haaland, *Acta Chem. Scand.*, 23 (1969) 2546.

474 M.C. Jarvis, D.R. Threlfall and J. Friend, *Phytochemistry*, 16 (1977) 849.

475 J.K. Herd, *Anal. Biochem.*, 23 (1968) 117.

476 R. Hata and Y. Nagai, *Anal. Biochem.*, 45 (1972) 462.

477 Y.M. Choy and G.G.S. Dutton, *Can. J. Chem.*, 51 (1973) 198.

478 W.F. Dudman and C.T. Bishop, *Can. J. Chem.*, 46 (1968) 3079.

479 A.F. Pavlenko and Yu.S. Ovodov, *J. Chromatogr.*, 52 (1970) 165.

480 J.Y. Do, J. Ioannou and N.F. Hard, *J. Food Sci.*, 36 (1971) 1137.

481 E.J. Bourne, P.C. Johnson and E. Percival, *J. Chem. Soc. C*, (1970) 1561.

482 J.T. Clarke, *Ann. N.Y. Acad. Sci.*, 121 (1964) 428.

483 J.C. Hilborn and P.A. Anastassiadis, *Anal. Biochem.*, 31 (1969) 51.

484 J.C. Hilborn and P.A. Anastassiadis, *Anal. Biochem.*, 39 (1971) 88.

485 C.P. Dietrich, N.M. McDuffie and L.O. Sampaio, *J. Chromatogr.*, 130 (1977) 299.

486 B.A. Voyles and M. Moskowitz, *Biochim. Biophys. Acta*, 351 (1974) 178.

Chapter 16

Pharmaceuticals

LAWRENCE FISHBEIN

CONTENTS

16.1. INTRODUCTION

The principal objectives of this chapter are to highlight a number of the more recent salient advances in the GC, HPLC, and GC–MS, of a spectrum of important classes of pharmaceuticals with a secondary emphasis on TLC and drugs of abuse and abused drugs. Clearly, the detailed chromatographic elaboration of the thousands of prescribed and over-the-counter (OTC) drugs from myriad classes of pharmaceutical activity is beyond the scope of this review.

Compared to other classes of utility and agents (e.g., industrial chemicals, pesticides) pharmaceuticals represent an area where a variety of chromatographic techniques have been most widely employed and continued advances made, primarily in the utility of HPLC and GC–MS. Chromatography (GC, GC–MS, HPLC, and TLC) is widely employed for the assessment of product purity, homogeneity, and stability, in the analysis of biological fluids and tissues for assessment of proper dosage, metabolic fate, and pharmacokinetics, and in toxicological and forensic applications.

It is generally acknowledged that in the past decade LC has probably experienced a greater development than any other technique, most of the initial applications being focused in the pharmaceutical field. This is so because HPLC is ideally suited to monitoring drug preparations for quality and purity, particularly compounds that are difficult to analyze by GLC due to their thermal instability or poor volatility [1]. In fact, HPLC methods are used more frequently than GLC methods in the quality control sections of most pharmaceutical firms [2]. The sensitivity and specificity of HPLC is reported to be equivalent or superior to corresponding GLC techniques in many instances. Advances in column materials with high selectivity and in the sensitivity of on-line spectrophotometric, spectrofluorimetric, and electrochemical detection should further enhance the utility of HPLC in drug analysis [1]. Additionally, the use of automatic samplers and the application of autoanalyzer–reactor systems in HPLC has significantly advanced the partial automation of LC [3–5] and enhanced its utility in pharmacokinetic investigations [6].

The routine use of HPLC for the identification of drugs in forensic and toxicological laboratories is currently much more limited, although it is being increasingly applied [1]. The major shortcomings of HPLC compared to GLC are:

(a) the precision of relative retention times is lower in HPLC than in GLC,

(b) the corroboration of peak identification by HPLC is more difficult,

(c) there are fewer HPLC data than GLC data on, e.g., narcotics, amphetamines, barbiturates, and tranquillizers in the literature [2]. So far, HPLC has not been significantly employed for large-scale urine screening in drug abuse programs, because it is not sufficiently rapid and sensitive [7].

One of the most promising analytical methods developed in the past decade has been the combination of LC with MS [8]. This technique is of particular interest for two reasons: (a) it could compensate for the lack of versatility and universatility of conventional LC detectors, (b) it could complement the very important functions of GC–MS in organic analysis [8,9]. However, the problems associated with the development of a practical LC–MS system are several times more difficult than those of GC–MS. This is so because all the interfaces so far developed require some compromise in the normal operating modes of either the liquid chromatograph, the mass spectrometer, or both [8]. However, advances have been made in the past decade in regard to technique and method development [10,11] and in the applications to drug analyses [12–14]. There is a strong indication that LC–MS is "not merely of transient interest but rather a new and important analytical technique" [8].

GC has been and continues to be the technique of choice for the separation of thermally stable, volatile organic and inorganic compounds in most laboratories. The unmatched resolving power, speed, precision, efficiency, ease of quantitation, availability of versatile and specific detectors, and the possibility of coupling the gas chromatograph to powerful identification techniques, such as MS, are primarily responsible for its widespread use [15]. However, GLC is not always suitable for highly polar substances, which may be nonvolatile or thermally labile, and for these as well as other reasons HPLC is finding increasing application.

GC–MS is perhaps the most powerful and useful technique for the identification of traces of organic compounds and is becoming the standard method of analysis for a number of classes of pharmaceuticals. Although developed in 1959 [16,17], GC–MS was not utilized to an appreciable extent in toxicological applications until the Seventies [16,18]. By all accounts, GC–MS analysis must be considered to be the ultimate in specificity among all techniques currently available for the detection of drugs of abuse in biological samples [19–22], the certainty of identification approaching 100% [16]. Although a number of GC–MS drug screening procedures have been reported [20–25] and the technique is becoming exceedingly important in forensic and clinical toxicology, its utility for the mass screening of urine samples in treatment programs remains somewhat uncertain, owing to the expense of instrumentation and technical problems [7].

TLC is acknowledged to be an inexpensive technique with reasonable sample capacity and specificity, but it requires sample extraction, does not readily yield quantitative results, and has a limited sensitivity. However, it is well recognized as a useful method for the separation and screening of drugs, primarily in urine screening and toxicological analysis, because the cost is low, operation and maintenance are easy, and the detection reagents are selective. On the other hand, TLC suffers from a lack of specificity and sensitivity, and objective criteria for evaluating its resolving power have often been lacking [26].

Because HPTLC surpasses TLC in speed, sensitivity, and efficiency, and is capable of handling a larger number of samples with the use of less solvents, its use is increasing. Chromatographic plates with chemically-bonded apolar phases, now commercially available, permit the analysis of numerous compounds that could not be directly chromatographed on silica gel layers heretofore. The separations that have been effected on reversed-phase thin-layer plates have been recently surveyed by Brinkman and De Vries [27,28] and their properties and operating conditions have been described by Siouffi et al. [29] and Kaiser and Rieder [30].

16.2. GAS CHROMATOGRAPHY

A number of recent reviews have dealt with the applications of GC in purity tests of drugs in the form of various derivatives and in the analysis of drugs in biological fluids [31–35], with the overall utility of GC in the analysis of pharmaceuticals [36,37] and in urine testing for drug abuse prevention and multimodality treatment programs [7]. In addition, it is important to cite a five-volume series on the analysis of drugs and their metabolites (including the application of GC–MS) by Gudzinowicz and Gudzinowicz [38–42], a two-volume work on the disposition of toxic drugs and chemicals in man by Baselt [43,44], and a short monograph on the analysis of drugs of abuse by Berman [45].

References on p. B320

16.2.1. Anticonvulsants

In recent years, the monitoring of blood levels of anticonvulsant drugs in patients undergoing chronic therapy has become routine practice in many clinical laboratories. For example, this monitoring greatly assists the management of epileptic patients by enabling a level to be maintained between the therapeutic and toxic concentrations of these drugs. It is recognized that individual differences in the speed of absorption, renal clearance, drug conversion rate, and "drug resistance" are factors that have to be taken into account [46]. Because of its sensitivity and specificity in allowing the simultaneous measurement of anticonvulsant drugs, GLC is the current method of choice for monitoring. Several methods have been reported, most of which can determine primidone [5-ethyldihydro-5-phenyl-4,6-(1H,5H)-pyrimidinedione], phenobarbital (phenylethylmalonylurea), diphenylhydantoin (5,5-diphenyl-2,4-imidazolidinedione), ethosuximide (2-ethyl-2-methylsuccinimide) and carbamazepine (5-carbamoyl-5H-dibenz[b,6]azepine simultaneously [46–61].

In GLC packed columns are most frequently used and derivatization is often necessary to improve the peak shape and to prevent irreversible adsorption of the anticonvulsant drugs on the support material [62]. Hill and Latham [57] have reported a quantitative GLC–FID procedure for the determination of blood levels of ethosuximide, phenobarbitone, primidone, and diphenylhydantoin, based on 1 ml of serum and a direct extraction technique. Ethosuximide was analyzed without derivatization, while subsequent flash alkylation with trimethylanilinium hydroxide allowed the simultaneous determination of the remaining three anticonvulsants. Phenobarbitone, primidone, and diphenylhydantoin were analyzed on 3% OV-17 by use of a temperature program, and 4% OV-225 was used for the isothermal analysis of ethosuximide. The technique of flash alkylation "off column" [63] is preferred as a rapid, simple method. It produces complete methylation of barbiturates and hydantoins to single derivatives, which are stable in a subsequent neutral environment for at least 24 h at room temperature [57]. Godolphin and Thoma [60] reported the utility of a new column packing, SP-2510DA (Supelco), for the determination of a wide variety of anticonvulsant drugs without derivatization by GLC–FID. The simple expedient of adding a short "pre-column", containing SP-2250DA, removed any interference from serum cholesterol in the determination of primidone.

Although the FID has been principally employed in the GLC analysis of anticonvulsants, the utility of the alkali flame-ionization detector (AFID) has also been reported. Sengupta and Peat [59] have described the separation of seven anticonvulsants (ethosuximide, phenturide, ethotoin, phenobarbitone, carbamazepine, primidone, and phenytoin) on 1% OV-17 and analysis by the AFID with the simultaneous analysis of sodium valproate on 2% SP-1000 by a FID. The specificity of the AFID allows a considerable saving in extraction time, as there is no need to remove natural interfering constituents, such as cholesterol. Another advantage of the AFID is the enhanced sensitivity for nitrogen-containing compounds, resulting in a lower detection limit than that of the conventional FID and consequently a reduction in sample volume. Since sodium valproate does not contain nitrogen, the conventional FID was employed for it.

Driessen and Emonds [64] introduced micropacked columns of mixed stationary phases (OV-225 and OV-17) for the analysis of a number of anticonvulsant drugs without derivatization. However, the disadvantages of the relatively large area of support material and the short life span of the columns were noted by Cramers et al. [46]. They described the quantitative determination of underivatized anticonvulsant drugs (e.g., phenobarbital, cyheptamide, carbamazepine, primidone, and phenytoin) by high-resolution GC with SCOT columns. A silicious support material (Cab-O-Sil) was first deactivated with benzyltriphenylphosphonium chloride and deposited on the inside wall of a glass capillary column (15 m $\times$ 0.4 mm ID), which was then coated with a polar stationary phase (OV-225). The columns had high plate numbers and did not deteriorate in use. The repeatability of the GC analysis was better than 1.2%, and the minimum detectable quantity of drug was in the order of 10^{-10} g.

The recently increased use of valproic acid (di-*n*-propyl acetate, VPA) in the therapy of certain types of epilepsy has spurred interest in its determination [65]. The degree of protein binding may be of value for the interpretation of plasma levels in individual patients. Small concentrations of VPA may also be encountered particularly in cerebrospinal fluid or in saliva. Jensen and Gugler [65] have reported a GLC method for the determination of VPA in plasma, saliva, spinal fluid, and urine, sensitive to 1 μg or less. It is based on a silanized column, packed with a free fatty acid phase (FFAP, Applied Science) and a FID. All other reported methods have sensitivity limits of ca. 5 μg/ml plasma [66–68].

16.2.2. Barbiturates

Since the first barbituric acid derivatives were introduced into clinical practice in 1913, use and abuse of this group of sedatives has steadily increased. It has been estimated that out of 2500 barbiturates that have been synthesized, over 50 are presently marketed for clinical use throughout the world [69]. Clinical biochemists are increasingly called upon to identify and quantitate a wide variety of sedative drugs in the blood of attempted suicides, and a method to determine these substances that is quick and accurate is essential. Drug abuse involving barbiturates is widespread and the international nature of the illegal markets means that any forensic laboratory may encounter a vast range of these compounds. Additionally, the abused barbiturates often occur in mixtures with other barbiturates, other drugs and/or excipients and hence pose a considerable analytical problem in the isolation and identification of specific barbiturates.

GLC on packed columns with FID has been widely applied to the simultaneous analysis of barbiturates in biological fluids. Some analysts chromatograph the unchanged drugs [47,64,70–73], while others derivatize the acidic hydrogens in the molecules in order to reduce their polarity and, consequently, their adsorption on the column and peak tailing. Several different derivatives have been utilized, including the N-trimethylsilyl (TMS) [74–76] and N-pentafluorobenzyl (PFB) [77], derivatives, and the most widely used N-1,3-dimethyl adducts. Methylation with diazomethane or dimethylsulfate prior to GC [78–82] increases the manipulation steps, but on-column and flash methylation with tetramethylammonium hydroxide [49, 83–87] or trimethylanilinium hydroxide [48,50,56,88,89] are simpler and therefore increas-

References on p. B320

ingly applied. However, by far the most frequently employed method is the on-column methylation of barbiturates with Me_4NOH, despite the decomposition which occurs in this alkaline medium [90–92].

The simultaneous determination of dimethylated barbiturates and other anticonvulsant drugs by high-resolution GC with a solid flash methylation injector and FID was reported by De Graeve and Vanroy [86]. The 2-m Chromosorb R-6470-1 (Supelco) SCOT column, coated with SE-30, was prepared according to the method of Rutten and Luyten [93] and Wetzels [94] to effect the separation of twelve barbiturates and two anticonvulsants after methylation with Me_4NOH. The sensitivity (0.1 $\mu g/ml$) and accuracy ($\pm 5\%$) are acceptable for the routine measurement of therapeutic levels of barbiturates, and the specificity is controlled by the high resolving power of the chromatographic column. The authors suggested that the specificity of the complete procedure makes it competitive and at least comparable in efficiency with others currently in use, which are generally more expensive (e.g., GC–MS, specific nitrogen detection) or more complicated (e.g., mixing of the carrier gas with formic acid vapor). The solid flash methylation injection technique reduces solvent tailing considerably and prevents column overloading. The reproducibility of this mode of injection is equivalent to that of a common septum system and the derivatization yield is not influenced a great deal ($< 10\%$) when precautions are taken [86].

Wu and Pearson [87] described the use of a 15.2-m stainless-steel capillary column, coated with 5% OV-17, and of two methylating agents, Me_4NOH and MeI, in the presence of K_2CO_3 as a condensing agent for the qualitative analysis of ten barbiturates. The technique offers high sensitivity and resolving power so that identification of individual barbiturates is readily accomplished. However, metal capillary columns, though physically strong, generally have the disadvantages of being catalytically active and tending to adsorb sample compounds. Hence, glass capillary columns would be the logical choice for the analysis of these drugs at picogram levels [87].

Although the analysis of barbiturates by GLC has been based almost exclusively on the FID, it is recognized that the FID is nonspecific. This limits its utility in the determination of trace quantities of drugs in complex biological mixtures, such as serum and urine. Additionally, since the FID will respond to any organic compound, clean-up procedures must be used to remove interfering constituents. The lack of specificity can lead to ambiguous results, unless confirmatory techniques are added. Hall and Risk [95] have reported a rapid and specific gas chromatograph for the analysis of free barbiturates in serum and urine at all levels of clinical interest which employs the Hall electrolytic conductivity detector. It can be used for the determination of barbiturates at levels of 0.01 mg/ml without sample clean-up. A column, consisting of 3% OV-17 (or SP-2250) on 100–120-mesh Chromosorb 750, deactivated with 0.2% phosphoric acid, gave the best all-around performance. It displayed excellent stability and could be used for ca. 250 analyses of direct sample extracts.

Sun and Hoffman [96] reported a sensitive GLC method for the determination of a number of barbiturates, which involved an initial extraction from serum, formation of methyl derivatives with methyl iodide by a modification of the procedure of

Dunges and Bergheim-Irps [97], and quantitation by nitrogen-specific detection (with a nitrogen–phosphorus FID). The method has a sensitivity of 0.08 μg/ml for pentobarbital and requires only 0.1 ml of serum. Separation of the barbiturates was accomplished with a 1.2-m column of 2% OV-101. Elahi [98] recently reported a novel encapsulated charcoal extraction technique for the rapid assay of barbiturates in whole blood. The drugs were adsorbed directly from the blood on a premeasured quantity of charcoal within a spherical, porous polypropylene capsule, 1 cm in diameter. Following elution with ether, resolution and quantitation was carried out with a column of 3% SP-2250 and a FID. The response was quantitatively linear between 2.5 μg/ml and 30 μg/ml blood for buta-, amo-, pento-, and secobarbital and between 5 μg/ml and 60 μg/ml for phenobarbital. The lower limit of detection was 0.05 μg/ml, except for phenobarbital (0.1 μg/ml). The extracts of both fresh and postmortem whole blood exhibited very low backgrounds in the range indicated, and the total extraction required only 10 min.

16.2.3. Tricyclic antidepressants

Tricyclic antidepressants are among the most widely prescribed drugs today for the treatment of mental disorders and the therapy of depression. Routine monitoring of plasma concentrations provides useful clinical information, since plasma concentrations in patients treated with these drugs may be as low as 10–20 ng/ml while the concentration after an overdose may reach several milligrams per liter [99]. Numerous analytical techniques for the routine measurement of therapeutic concentrations of these commonly used drugs in serum or plasma have been reviewed by Gupta and Molnar [100] and by Gram [101]. The most frequently used procedures for these measurements include GLC with specific nitrogen detection [100–120], FID [116,121–125], or ECD [126–128].

The majority of GC procedures reported in recent years have focused on the most popular antidepressant, amitriptyline, and its demethylated metabolite, nortriptyline [100,102,104,110,114,121,122,124,126]. The earliest GLC methods utilized the FID [99,121,122,125], which is relatively insensitive and required large volumes of plasma (e.g., 5–10 ml). Although the ECD has greatly increased sensitivity and has provided reliable methods for nortriptyline [127,128], a major limitation has been the extremely time-consuming preparation of fluorinated derivatives [111]. Additionally, since tertiary amine antidepressants cannot be derivatized without prior demethylation, the ECD has only rather limited routine application. GLC with a nitrogen-specific AFID is one of the most practical approaches for the assay of the tricyclic antidepressants in routine clinical laboratories [100,102,111]. However, several of the reported procedures utilizing "nitrogen-specific" detectors are inconvenient for routine use, because they require derivatization [109,110], special pretreatment of glassware [105,109,110], double extraction [107,111,112], or large samples (>4 ml for triplicate analyses) [105,110]. Additionally, a sufficient number of tricyclic antidepressant analyses would appear to be required to justify the purchase of a nitrogen detector, which at present is not common to most laboratories [129].

The stationary phase most commonly employed for the GC determination of the tricyclic antidepressants is 3% OV-17, which is fairly selective and can effect the separation of most members of this class of drugs [100]. However, the relative retention index for many of these drugs is quite similar [130], and this makes quantitation difficult if several of them are prescribed concurrently [100]. Rovei et al. [114] have described the analysis of tricyclic antidepressants by GLC with nitrogen-selective detection by means of either a conventional packed column (3% SP-2250, OV-17) or a capillary column (SE-30). They accurately measured therapeutic concentrations of the parent drugs and their demethylated metabolites down to 5 ng/ml by simple hexane extraction from alkalinized plasma, followed by derivatization with heptafluorobutyric anhydride. The packed column was found to be more suitable for routine purposes, such as monitoring plasma levels during chronic administration of tricyclic antidepressants. It is simpler and less expensive, and the use of a dual gas chromatograph permits the use of two columns, which is not possible with a capillary column and a solid injector system. However, the higher resolution of the GC peaks obtained with a capillary column suggests that this is a superior technique for the determination of these drugs at very low plasma concentrations or in the presence of intefering concomitantly prescribed drugs or endogenous compounds. Moreover, the capillary column appears to be particularly advantageous for pharmacokinetic studies [114].

For the quantitation of tricyclic antidepressants in serum, rapid isothermal GLC with a nitrogen–phosphorus-selective detector was recently reported [120]. The method can be used for the quantitation of amitryptyline, nortriptyline, imipramine, desmethylimipramine, doxepine, desmethyldoxepine, clomipramine, desmethyl-clomipramine, trimipramine, desmethyltrimipramine, and dibenzepine. It requires no derivatization of the metabolites and involves one extraction with hexane–2-propanol and isothermal chromatography on OV-25. Different tricyclic antidepressants are determined within 5–100 min. They give a linear response between 10 and 1500 μg/l and a maximum coefficient of variation of 7.7%.

16.2.4. Benzodiazepines

Since the discovery of chlordiazepoxide (Librium) in 1957, the benzodiazepines have had a tremendous impact on the treatment of disorders of nervous origin and are the most common class of antianxiety drugs prescribed. Indeed, they are the most widely prescribed drugs in the United States [131]. Their potential to produce psychological and physical dependence is of great concern [132]. A broad spectrum of analytical procedures is available for the analysis of the benzodiazepines, their metabolites, or their acid hydrolysis products (e.g., benzophenones) for therapeutic monitoring, pharmacokinetic studies, or screening tests. The analyses of the benzodiazepines have been reviewed through 1974 by Clifford and Smyth [133] and by Hailey [134].

Since drug concentrations during benzodiazepine (e.g., diazepam) therapy are usually less than 1 mg/l, very sensitive methods are required for the simultaneous measurement of the parent drug and its metabolites in biological fluids [133,134]. For example, in clinical practice, single oral doses of diazepam typically range from

2 to 10 mg, administered 2 to 4 times daily. In man, peak concentrations of 221–400 ng/ml of plasma [135] and 137–189 ng/ml of whole blood [136] have been reported following oral administration of 5 mg of diazepam three times daily. This results in steady-state plasma levels of 230–440 ng/ml [137], while chronic oral administration of 10 mg of diazepam once daily gives blood levels of about 200 ng/ml [136]. Concentrations of N-desmethyldiazepam (nordiazepam, the major metabolite of diazepam) can range from 10 to 20 ng/ml of blood when the diazepam levels are at the peak, following single oral administration of 10 mg of diazepam, and may reach steady-state concentrations of 100–150 ng/ml of blood during chronic administration of 10 mg of diazepam daily [136,138].

Diazepam and its metabolites have been determined in body fluids primarily by GLC with specific nitrogen detection [139,140], FID [141], and ECD [133,134,137–156]. Since column adsorption processes (especially with N-desalkyl compounds at the nanogram level) can cause problems, earlier procedures called for conversion of diazepam and desmethyldiazepam to their corresponding O-aminobenzophenone derivatives by acid hydrolysis prior to GLC–ECD analysis [133,134,141,153,154]. The methods involving GLC of the benzophenone hydrolysis products are time-consuming because of the clean-up procedure involved. Moreover, they lack specificity, because metabolites of the parent drug, if present in sufficient amounts, also yield the same benzophenone derivative [133,134]. Several of the more recent procedures permit the determination of diazepam and its metabolites by GLC–ECD without derivatization, by use of OV-17 and a ^{63}Ni ECD [137,145–149].

Diazepam and desmethyldiazepam in whole blood were determined by McCurdy et al. [139], who applied a nitrogen–phosphorus detector (NPD) after derivatization. The problem of quantitating desmethyldiazepam, which tails on certain liquid phases, such as OV-1, was circumvented by derivatization of the drug to its corresponding N-propyl derivative. It had a response to NPD 1.2 times that of diazepam. The lower limit for the detection of diazepam and desmethyldiazepam was 10 ng/ml. A rapid method for the determination of diazepam and desmethyldiazepam in human plasma by GLC–ECD was reported by Weinfeld et al. [138]. The concentration of these drugs was determined by extracting 0.5 ml of plasma with 1.0 ml of benzene, containing 25 ng/ml of methyl nitrazepam as the internal standard. The method has a sensitivity limit of 5 ng diazepam and 10 ng desmethyldiazepam per ml plasma.

The frequency-response ECD can be operated at much slower gas flows than detectors operated at a fixed pulse rate. This allows greater precision than that generally obtained with medazepam as an internal standard [134]. The GLC–ECD method of Linnoila and Dorrity [151] permitted a clinical laboratory to analyze 40 samples of benzodiazepines a day and to report results on the same day. The analysis of about 1000 serum samples over a 5-month period revealed no detrimental effect on the linear dynamic range of the detector over a $5 \cdot 10^2$ concentration range.

The analysis of oxazepam (7-chloro-3-hydroxy-5-phenyl-1,3-dihydro-2H-1,4-benzodiazepin-2-one) is important, because in man it is a major urinary metabolite of diazepam [157] as well as a metabolite of chlordiazepoxide [158] and chlorazepate (Tranxene). It differs from diazepam and nonhydroxylated benzodiazepines in that a substantial proportion of the drug is present in the blood stream as the glucuronide [159]. A variety of GLC procedures have been reported for the determination of oxazepam in biological fluids [143,145,156,159].

Lorazepam is a derivative of oxazepam differing only in having an additional chlorine atom substituted at the 2'-position in the 5-phenyl ring. It is also a metabolite of a new benzodiazepine, 2'-chlordesmethyldiazepam by C-3 hydroxylation. Since lorazepam is a more potent drug than oxazepam, both therapeutic doses and body fluid levels are lower. The analysis of lorazepam has been accomplished primarily by GLC [160–164]. Both the free and conjugated lorazepam in human serum were assayed by the GLC–ECD technique of Knowles et al. [163]. Lorazepam was determined as the 2-amino-2',5'-dichlorobenzophenone on a 3% OV-17 column. The procedure was sufficiently sensitive to measure serum and urine concentrations of the drug, even after 2-mg doses, at a level of 0.01 μg/ml.

Nitrazepam (7-nitro-5-phenyl-1,3-dihydro-2H-1,4-benzodiazepin-2-one) is a widely used hypnotic, which surpasses diazepam in anticonvulsant activity. Various GLC methods have been reported for its determination in body fluids [133,134,149,165,172]. Some GLC methods determine nitrazepam after acid hydrolysis as the benzophenone [165,167,168,172] with a resulting loss of specificity. Other GLC procedures require derivatization, e.g., methylation [149,166] and trimethylsilylation [169]. The GLC–ECD procedure of Beharrell et al. [168] for determining nitrazepam in biological samples involves extraction of the drug and an internal standard, clonazepam, and hydrolysis to the benzophenones, which are analyzed on a 3% OV-17 column. Clonazepam is converted to 2-amino-2'-chloro-5-nitrobenzophenone, which emerges after 2-amino-5-nitrobenzophenone (ANB), the hydrolysis product of nitrazepam. The method is considered specific for nitrazepam in the absence of measurable quantities of ANB and the metabolite, 1,3-dihydro-3-hydroxy-7-nitro-5-phenyl-2H-1,4-benzodiazepin-2-one, which are not normally detected in plasma. Drug recovery from plasma is quantitative ($> 95\%$) and the limit of detection is about 0.1 ng/ml plasma [168].

An assay procedure of underivatized, intact nitrazepam and clonazepam in human plasma, which requires a capillary column coated with 3% OV-17 and an ECD, was described by De Boer et al. [170]. Clonazepam was used as an internal standard in the assay of nitrazepam and vice versa. After a single extraction step, linear calibration curves were obtained in the range of 10–100 ng/ml with standard deviations of less than 4.9%. The sensitivity of the method is about 1 ng/ml plasma for both drugs. Nitrazepam and its main urinary metabolites, 7-aminonitrazepam and 7-acetamidonitrazepam, both free and conjugated, were determined after a single oral dose of 5 mg nitrazepam by utilizing 3% OV-101, an ECD for nitrazepam, and a dual flameless nitrogen-selective detector for the metabolites. The detection limits were about 0.2 ng/ml for nitrazepam and 50 ng/ml for the metabolites [173].

16.2.5. Antiarrhythmic drugs

A variety of antiarrhythmic and beta-adrenergic blocking drugs are currently being used for the treatment of diseases characterized by excess sympathetic nervous activity [174]. The antiarrhythmic drugs, lidocaine, procainamide, propranolol, and quinidine are widely used in clinical practice for the treatment of cardiac disorders. Lidocaine is most extensively employed for treatment of cardiac arrhythmias accompanying acute myocardial infarction. Determination of the plasma concentrations of these drugs enables the achievement of a therapeutic plasma level in a particular patient by dose modification. Therapeutic, toxic and lethal concentrations of these drugs have been reported [175–178]. Monoethylglycinexylidide (MEGX) (II), a dealkylated metabolite formed after the administration of lidocaine (I) has been reported to be approximately 80% as potent an antiarrhythmic agent as the parent drug, while the didesmethylated metabolite glycinexylidiole (GX) had only about 10% of the potency [178].

Lidocaine (I)   MEGX (II)

A number of GLC procedures have been reported for the determination of lidocaine alone [179–183] and of lidocaine as well as its dealkylated metabolites [178,184–188] in biological material. Some of these procedures were found to be wanting. For example, GLC with 10% UCW-98 on Chromosorb W did not completely resolve the lidocaine, MEGX, and GX peaks, and the accuracy of the method was not reported [184]. Kennaghan and Boyes [185] described the quantitative GLC of heptafluorobutyryl derivatives of lidocaine and its metabolites without the use of an internal standard. Strong and coworkers [186,187] have used mass fragmentography for the estimation of lidocaine, MEGX, and GX. The stationary phase was 3% SE-30/OV-17 (6:1) on Chromosorb. The peaks of the primary and secondary amine metabolites were asymmetrical, and the calibration curve for MEGX was nonlinear below about 0.5 μg/ml. A single-column temperature-programing technique was employed to determine lidocaine and its metabolites after conversion to the acetyl derivatives [188]. Nation et al. [178] reported a quantitative method for the isothermal GLC of underivatized lidocaine and MEGX, following extraction from plasma samples. Quantitation of lidocaine and MEGX was accomplished with a 2.5 ft.×1/4-in. column, packed with 2% UCON-75-H-90,000–2% KOH on 80–100-mesh Gas Chrom Q at 175°C and a FID. The KOH-treated solid support minimized adsorption of the amines and obviated the need for derivatization to ensure symmetrical peaks. The response was linear for lidocaine hydrochloride and MEGX hydrochloride in plasma over the range of 0.05–25 μg and 0.05–5 μg, respectively.

Procainamide (III) has been widely prescribed for the prevention or treatment of cardiac arrhythmias. The range of therapeutic plasma concentrations is 4–8 μg/ml [189]. A GLC method for the determination of procainamide (III) in biological fluids was reported by Simons and Levy [190]. By using a dipropyl analog of procainamide (IV) as an internal standard, both compounds could be chromatographed directly, yielding linear calibration responses and a sensitivity that allows quantitative determination of concentrations as low as 0.1 $\mu g/ml$. The extraction procedure was carefully modified to avoid hydrolysis of N-acetylprocainamide (V), a major metabolite of procainamide. The usefulness of the procedure was demonstrated by following the disappearance of procainamide from the plasma and urine of human subjects, treated with the drug.

III Procainamide, $R_1 = H$, $R_2 = -CH_2CH$

IV Internal standard, $R_1 = H$, $R_2 = -CH_2OCH_2CH_3$

V N-Acetylprocainamide, $R_1 = -COCH_3$, $R_2 = -CH_2CH_3$

GLC was carried out with 91×0.63-cm glass columns, packed with 10% OV-7 coated on Gas Chrom Q and maintained at 245°C, and with a FID.

16.2.6. Beta-adrenergic receptor blocking agents

The beta-adrenergic receptor blocking agent propranolol [1-(isopropylamino)-3-(1-naphthoxy)-2-propanol] is a mixture of equal amounts of the D- and L-enantiomers, which have different pharmacological effects [191–193] and different elimination kinetics [194]. It is useful to determine both enantiomers after the administration of racemic propranolol in order to compare the disposition of the enantiomers, because only the L-isomer affects hepatic blood flow and hence decreases hepatic clearance [195]. The simultaneous determination of the propranolol enantiomers in biological samples by GLC was recently accomplished by Caccia et al. [196], who employed a $2 \, m \times 4$-mm column of 3% OV-225 on Chromosorb W HP, operated at 250°C, with an ECD for the analyses of their N-heptafluoro-1-prolyl derivatives and silyl esters. Derivatives of enantiomers with an optically active reagent is known to yield diastereoisomers, which can be separated by GLC [197–201]. Thus, some beta-adrenoceptor antagonists were resolved [202] by the use of N-trifluoroacetyl-1-propyl chloride (TPC) [197] and N-heptafluorobutyryl-1-prolyl chloride [200] as chiral reagents.

Practolol [4-(2-hydroxy-3-isopropylaminopropoxy)acetalinide] is a beta-adrenoceptor blocking drug that has been analyzed in urine and plasma by GLC–ECD [203]. A $3.25 \, m \times 3$-mm glass column, packed with 4% SE-30 on Gas Chrom Q, was operated at 200°C. Practolol and the internal standard propranolol were derivatized in biological samples with trifluoroacetic anhydride prior to the analyses. Oxyprenolol (VI) is a beta-adrenergic blocking drug that has recently been reported to be effective in the control of hypertension in human pregnancy [204] and to be equivalent to alpha-methyldopa in the control of maternal hypertension and improved placental and fetal growth [205]. Several GLC methods have been reported for the quantification of oxprenolol in biological fluids [206–208]. However, most of these methods involve somewhat lengthy and laborious extractions, followed by derivatization procedures. Trifluoroacetic anhydride was used in all the methods reported, and the di-trifluoroacetyl derivative of oxyprenolol was determined by GLC–ECD. The method was sensitive to 10 ng/ml plasma and gave an overall recovery of $80 \pm 15\%$. A 2.7 m $\times$ 4-mm glass column, packed with 3% OV-101 on Gas Chrom Q, was operated at 158°C.

VI   Oxyprenolol R $= -OCH_2CH = CH_2$
VII Alprenolol   R $= -CH_2CH = CH_2$

16.3. HIGH-PERFORMANCE LIQUID CHROMATOGRAPHY

A number of useful books and reviews describing the instrumentation for HPLC [1,209,210] and general reviews of its utility in pharmaceutical analysis [2,211–218] have recently been published.

16.3.1. Anticonvulsants

GLC analysis of anticonvulsant drugs generally requires a relatively large sample and a significant amount of time for sample preparation. Moreover, no single set of analytical conditions in the literature is suitable for the determination of all of these drugs. HPLC offers potential advantages for their determination in that they may be analyzed without derivatization, and eluate fractions of unchanged drugs may be collected for further analysis [218–226].

Kabra et al. [218] have reported the simultaneous determination of phenobarbital, phenytoin, primidone, ethosuximide, and carbamazepine in as little as 25 μl of serum by HPLC. The anticonvulsants were eluted from a reversed-phase column of μBondapak C_{18} with acetonitrile–phosphate buffer and detected by their absorption at 195 nm. Each analysis required ca. 14 min at an optimum column temperature of 50°C. The lower limit of detection for all of these drugs was less than 10 ng, and sensitivities of 1 μg of the drugs per ml of serum were attained routinely. Analytical recoveries for the five drugs varied from 97 to 107% with a day-to-day precision (CV) between 3.9 and 5.9%. Of more than 30 drugs tested, only ethotoin interfered with the analysis of phenobarbital.

A single HPLC procedure for therapeutic monitoring of ethosuximide, theophylline, and acetaminophen in human serum employs a programable auto-injector, and a reversed-phase μBondapak C_{18} column [227]. The mobile phase consists of ion-pairing buffers, such as triethylamine–acetic acid or N-ethylmorpholine–acetic acid at pH 4.8 in aqueous acetonitrile. The procedure permits the separation of the above three drugs from 26 other drugs, metabolites, and related substances.

16.3.2. Barbiturates

The widespread availability of barbiturates, obtained both by prescription and illicit means, in addition to the increasing need for monitoring them in the anticonvulsant therapy of epileptic patients has created a demand for the simultaneous determination of these drugs in serum. The reversed-phase chromatography of barbiturates has been extensively studied [228]. In the free acid form these drugs are easily separated, but little separation occurs when the pH of the mobile phase produces the barbiturate anion as the dominant species. Thus, favorable chromatographic properties require the free acid form and good chromophoric properties require the anionic species [229]. Clark and Chan [230] combined both of these conditions into a single analytical technique. In their HPLC procedure the barbiturates are resolved as the free acids and then ionized with borate buffer (pH 10) after elution for the detection as the anionic chromophores. The pK_a values of the common barbiturates used in therapy range from 7 to 8. Hence at a pH of 10, these weak acids exist almost exclusively in the anionic and thus chromophoric form. The peak area of nonionized barbiturates is increased 20-fold by the ionization.

Gill et al. [69] utilized an effective combination of three HPLC, two GLC and two TLC techniques for the identification of a spectrum of barbiturates. Although the overall correlations observed between pairs of systems were generally low, specific groups of barbiturates showed very high correlations. This determined the approach to the selection of two or more systems to increase chromatographic discrimination among the barbiturate group. CC with lipophilic phases, e.g., HPLC with ODS-silica and GLC with SE-30, proved most suitable for barbiturate identification. Changes of eluent pH in reversed-phase HPLC proved very effective for the separation of barbiturates with closely related structures. For example, a combination of two reversed-phase HPLC systems (one at pH 3.5, the other at pH 8.5) proved very useful for the separation of butobarbitone and secbutobarbitone.

16.3.3. Tricyclic antidepressants

Although therapeutic ranges for plasma concentrations of the tricyclic antidepressants have not been established with the same degree of certainty as for other commonly monitored drugs, such as antiepileptic and cardiac agents, sufficient information is currently available to suggest that measurement of these drugs (in appropriate circumstances) can be helpful in managing the pharmacotherapy of depressed patients. Jatlow [229] has recently reviewed the significance of plasma concentrations of the tricyclic antidepressants and the techniques available for monitoring them. Various HPLC techniques have been reported for this purpose [231–253], in which adsorption, reversed-phase partition, and ion-pair chromatography are applied. While the sensitivity is less than that generally achievable by GLC with nitrogen and mass fragmentography, it is adequate when a good short-wavelength detector is employed. The advantages of HPLC for the determination of this class of drugs are that the technique is simpler and faster than GLC and that the metabolites, including hydroxylated derivatives, behave well without derivatization on reversed-phase columns [229].

TABLE 16.1

RETENTION TIMES OF SOME TRICYCLIC ANTIDEPRESSANTS IN HPLC [246]

Retention times in min; column, LiChrosorb Si 60 (5 μm), 100 × 4.6 mm ID. 1 = Hexane–dichloromethane–methanol (8:1:1), hexane contained 10 ppm methylamine; 2 = 0.05 M NaBr in methanol.

	1	2
Amitriptyline	2.48	5.10
Nortriptyline	11.0	3.70
Clomipramine	2.74	5.06
Desmethylclomipramine	14.5	3.50
Imipramine	3.38	6.16
Desipramine	19.2	3.86
Protriptyline	10.9	3.70
Trimipramine	1.72	4.30
Doxepine	3.26	5.84

A number of HPLC procedures are now available for the analysis of clomipramine or imipramine and their respective demethylated metabolites. Those described for the assay in biological materials involve either ion-pair partition chromatography [231,236–240], adsorption [233,241–248] or reversed-phase chromatography [249–253]. The sensitivity ranges from 1 ng/ml with fluorimetric detection [243,251] to 5–20 ng/ml with UV detection [231,236,239–241,244,245,249,250]. Table 16.1 shows the retention times of nine cyclic antidepressants in HPLC on a silica column [246]. Godbillon and Gauron [235] have devised a HPLC method for the determination of clomipramine or imipramine and their mono-demethylated metabolites in human blood or plasma, based on the separation on a silica gel column with ethanol–hexane–dichloromethane–diethylamine (30:62:8:0.005) as the mobile phase and detection at 254 nm. The limit of sensitivity was 5 ng/ml for clomipramine and imipramine, and 10 ng/ml for the corresponding mono-demethylated metabolites.

16.3.4. Benzodiazepines

The number and frequency of requests for analyses of blood or plasma for benzodiazepines are increasing dramatically [254]. GLC–ECD has become the analytical method of choice for these drugs, because it meets the necessary sensitivity requirements, but the major disadvantage of the GLC–ECD procedure in forensic toxicology is the lack of a suitable confirmatory technique to substantiate the GLC results and provide reliable qualitative identification. The application of HPLC for the determination of this class of drugs is therefore increasing.

Diazepam [254–262], chlordiazepoxide [254,263–268], nitrazepam [269–271], flunitrazepam [272], and their metabolites have been analyzed in biological fluids by HPLC. Cotler et al. [261] have developed a HPLC assay for the determination of

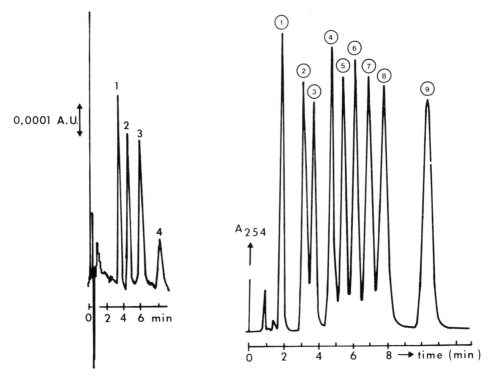

Fig. 16.1. HPLC separation of diazepam and its metabolites. Column, methyl-silica Si 60 (300 × 4.6 mm ID); eluent, methanol–0.05 M phosphate buffer, pH 7.0 (2 : 3); flowrate, 20.3 μl/sec. Peaks: 1 = oxazepam; 2 = hydroxydiazepam; 3 = desmethyldiazepam; 4 = diazepam. (Reproduced from *J. Chromatogr.*, 181 (1980) 227, with permission [260].)

Fig. 16.2. HPLC separation of nine benzodiazepines. Column, methyl-silica Si 60 (100 × 2.8 mm I.D.); eluent, methanol–0.05 M phosphate buffer, pH 6.0 (1 : 1); flowrate: 9.3 μl/sec. Peaks: 1 = 7-aminonitrazepam, 2 = bromazepam; 3 = nitrazepam; 4 = oxazepam; 5 = desmethylchlordiazepoxide; 6 = hydroxydiazepam; 7 = chlordiazepoxide; 8 = desmethyldiazepam; 9 = diazepam. (Reproduced from *J. Chromatogr.*, 181 (1980) 227, with permission [260].)

diazepam and its major metabolites (oxazepam, temazepam, and nordiazepam) in plasma, blood, and urine. The assay involves extraction of the biological fluids, buffered to pH 9.0, with benzene–dichloromethane (9 : 1). The overall recovery of diazepam and its major metabolites from plasma or blood ranged from 60 ± 3.2 to 89 ± 13% (S.D.) and for urine from 79 ± 7.9 to 93 ± 10.5% (S.D.). The sensitivity limits of the assay by UV detection at 254 nm were 50 ng/ml of plasma and blood and 200 ng/ml of human urine (post-Glusulase). The HPLC assay with reversed-phase packing (10-μm Bondapak C$_{18}$) and aqueous methanol was used to monitor the plasma concentration/time profile in man, following a 10-mg oral dose of diazepam, and for the blood concentration/time profile of diazepam and nordiazepam in cats, following a 10 mg/kg intravenous dose of either diazepam or nordiazepam. The HPLC assay data correlated well with the results of GLC–ECD.

The separation and quantitation of benzodiazepines in serum, saliva, and urine at therapeutic levels by HPLC with a highly selective and efficient reversed-phase adsorption system and UV detection was described by Tjaden et al. [260]. A complete separation of nine benzodiazepines within 12 min was achieved by using methyl-silica as the stationary phase and 50% aqueous methanol as the eluent. Fig. 16.1 illustrates the HPLC separation of a test mixture of 1 ng of diazepam and 2 ng of its metabolites. It was estimated that the limit of detection ranges from about 200 to 340 pg. Fig. 16.2 shows the HPLC separation of a test mixture of nine benzodiazepines on a methyl-silica column with methanol–phosphate buffer (0.05 M, pH 6.0) (1:1) as the eluent.

The intrinsic UV absorbance of benzodiazepines and their suitability for HPLC analysis — especially of thermally unstable compounds (e.g., oxazepam) or compounds requiring derivatization prior to GLC–ECD analysis (e.g., temazepam) — have been used to advantage in pharmacokinetic studies [257,259,264]. Reversed-phase HPLC is generally preferred to adsorption chromatography for the analysis of benzodiazepines, since it provides better separation [261,264]. Screening and quantitation of diazepam, flurazepam, chlordiazepoxide, and their metabolites in blood and plasma by both GLC–ECD and HPLC was described by Peat and Kopjak [254]. The screening method involved direct solvent extraction of a buffered sample, followed by GLC–ECD analysis of a small aliquot of the extract on 3% OV-17. Blood or plasma was then extracted with additional solvent and the extract was analyzed by HPLC on a reversed-phase 4-mm ID Bondapak C_{18} column to confirm the presence of the benzodiazepines. Quantitation was accomplished by either

TABLE 16.2

CHROMATOGRAPHIC PROPERTIES OF THE BENZODIAZEPINES [254]

Drug	GLC–ECD Retention time, min *	HPLC Retention volume, ml	
		Eluent a **	Eluent b ***
Chlordiazepoxide	9.95 (major)	12	<5
Desmethylchlordiazepoxide	NA	9.6	<5
Demoxepam	NA	6.7	<5
Diazepam	2.2	16.1	<5
Desmethyldiazepam	3.3	12.7	<5
Flurazepam	4.6	>25	11.0
Desalkylflurazepam	2.5	10.8	<5
Oxazepam	NA	9.8	<5
Prazepam	3.8	>25	8.5
Flunitrazepam	3.8	8.2	<5

   * NA = not analyzed.
  ** Methanol–phosphate buffer (58:42).
*** Methanol–phosphate buffer (73:37).

References on p. B320

method with addition of flunitrazepam as an internal standard. The GLC–ECD method was used for diazepam, desmethyldiazepam, flurazepam, and desalkylflurazepam, while HPLC was used for chlordiazepoxide and desmethylchlordiazepoxide. Although the analytical procedure was described for 1 ml of whole blood or plasma, the GLC–ECD screening procedure could be applied to sample volumes as low as 50 μl, if necessary. Table 16.2 lists the GLC–ECD and HPLC properties of the benzodiazepines examined by Peat and Kopjak [254].

The determination of diazepam, nordiazepam, and clonazepam in plasma by HPLC on a 5-μm porous silica gel (Partisil-5) column with a mobile phase of cyclopentane–chloroform–acetonitrile–methanol (58 : 111 : 30 : 1) and UV detection at 254 nm was described by Perchalski and Wilder [262]. After a single extraction from alkaline plasma with benzene–dichloromethane (9 : 1), the benzodiazepines could be analyzed to a lower limit of 5–10 ng/sample. Relative standard deviations for daily and long-term reproducibility studies were 4% and ⩽ 6%, respectively.

Ascalone [263] described a procedure for the determination of chlordiazepoxide and its metabolites by reversed-phase HPLC that may be used for studying pharmacokinetics in man during long-term chlordiazepoxide therapy. The columns used were: (a) a Hibar column (250 × 4 mm ID), filled with LiChrosorb RP-18 (10 μm); (b) an Altex column (250 × 3.2 mm ID), filled with LiChrosorb RP-18 (10 μm); and

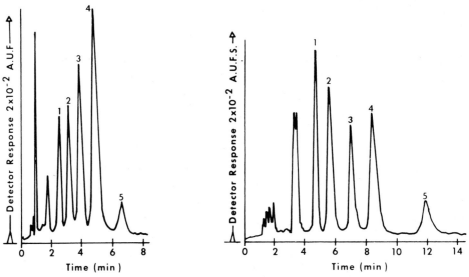

Fig. 16.3. Chromatogram of standards recovered from human plasma: 1 = demoxepam (0.4 μg/ml); 2 = desmethylchlordiazepoxide (0.4 μg/ml); 3 = nitrazepam; 4 = chlordiazepoxide (1 μg/ml); 5 = desmethyldiazepam (0.4 μg/ml). Altex column. (Reproduced from *J. Chromatogr.*, 181 (1980) 141, with permission [263].)

Fig. 16.4. Chromatogram of standards recovered from human plasma: 1 = demoxepam (0.45 μg/ml); 2 = desmethylchlordiazepoxide (0.35 μg/ml); 3 = nitrazepam; 4 = chlordiazepoxide (0.48 μg/ml); 5 = desmethyldiazepam (0.4 μg/ml). Hibar column. (Reproduced from *J. Chromatogr.*, 181 (1980) 141, with permission [263].)

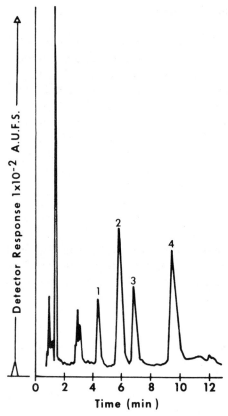

Fig. 16.5. Chromatogram of a plasma extract from a patient 36 h after a single oral administration of 30 mg chlordiazepoxide. Peaks: 1 = demoxepam; 2 = desmethylchlordiazepoxide; 3 = nitrazepam; 4 = chlordiazepoxide. Hibar column. (Reproduced from *J. Chromatogr.*, 181 (1980) 141, with permission [263].)

(c) a precolumn, filled with Corasil (37–50 μm). They were eluted with acetonitrile–0.1% ammonium carbonate (31:69) at a flowrate of 2 ml/min (ca. 1300 p.s.i.). Figs. 16.3 and 16.4 illustrate chromatograms of standards, recovered from human plasma, obtained with Altex and Hibar columns, respectively. The chromatogram shown in Fig. 16.5 was prepared from a plasma extract of a patient 36 h after he received a single oral dose of 30 mg chlordiazepoxide.

Wittwer [273] recently reported the separation of eleven benzodiazepines by HPLC on a 15-cm μPorasil column by isocratic elution and detection at both 254 nm and 280 nm. Powder and tablet dosage forms of ten benzodiazepines were extracted with chloroform. Chlorazepate dipotassium, which is insoluble in chloroform, was decarboxylated to N-desmethyldiazepam prior to analysis. The solvent systems consisted of 9 parts cyclohexane and 1 part of ammonium hydroxide–methanol–chloroform (1:200:800).

References on p. B320

16.3.5. Antiarrhythmic and beta-adrenergic blocking drugs

A variety of methods has been utilized for the determination of lidocaine and procainamide in biological samples, in addition to GLC. These include fluorometry [274], spectrophotometry [275], and HPLC [276,277]. Whatever method is used, it is essential that its precision and accuracy be high, because the therapeutic ranges for the drugs are narrow: e.g., procainamide, 4–8 mg/l [278]; lidocaine, 1–6 mg/l [186,278]. Adams et al. [276] have reported the simultaneous determination of lidocaine and procainamide at therapeutic concentrations in serum by HPLC. The drugs and an added internal standard (procaine) were initially extracted from serum with the aid of charcoal adsorption. The analysis was carried out on a reversed-phase column (0.5 m × 2.6 mm ODS-Sil-X-1) with buffered aqueous acetonitrile as the mobile phase. The drugs were detected by UV at 205 nm. CV values of 10% and 6% for concentrations of 1 mg/l and 20 mg/l, respectively, were attainable routinely. Chromatography was complete in 10 min; the total analysis time was 15 min.

USP Lidocaine injection was assayed by ion-pair HPLC on an octylsilane (RP-8) reversed-phase column with a solution of 464 mg 10-camphorsulfonic acid in 100 ml methanol–acetic acid–water (30:3:67) and detection at 254 nm [277]. Plasma levels of the widely utilized beta-adrenergic blocking drug propranolol, encountered during therapy, may range from 20 to 100 ng/ml [278–280], a concentration range that precludes the use of most common analytical techniques. Nygard et al. [281] have determined propranolol in plasma on a reversed-phase cyanopropyl silane column (μBondapak CN) with acetonitrile–0.02 M acetate buffer, pH 7.0 (7.3). A spectrofluorometric detector with an excitation wavelength of 276 nm and an emission filter with a 340-nm cut-off gave a detectable peak for 0.8 ng propranolol per injection in this system.

Nation et al. [282] have reported a simple and rapid HPLC procedure for the simultaneous quantitative determination of propranolol and its active metabolite, 4-hydroxypropranolol, in plasma. Plasma extracts were chromatographed on a 30-cm reversed-phase column (μBondapak alkyl phenyl) with acetonitrile–0.06% phosphoric acid (27:73) at a flowrate of 2 ml/min and detected by fluorescence monitoring at 205 nm. In 1-ml plasma samples, propranolol and 4-hydroxypropranolol concentrations as low as 1 ng/ml and 5 ng/ml, respectively, could be quantitated. The reproducibility of the procedure was found to be satisfactory, and no interference from endogenous plasma components or other drugs was observed. A single plasma sample could be analyzed in ca. 20 min. The method described was considered suitable for routine clinical monitoring of plasma levels in patients or for use in pharmacokinetic studies.

The quantitation of the beta-adrenergic blocking drug oxprenolol in biological fluids (blood, plasma, urine, or breast milk) by HPLC was recently described by Tsuei et al. [205]. The procedure employed an internal standard [10 μg of alprenolol (VII) hydrochloride per 100 μl] and dichloromethane–diethyl ether (1:4) for extraction. The column (250 × 4.6 mm) was packed with octadecylsilane-bonded silica gel (Whatman PXS10-25 ODS-2, 10 μm) and the mobile phase was 0.005 M 1-octylsulfonic acid in 67% aqueous methanol at a flowrate of 60 ml/h. The sample preparation procedure was relatively simple and required no evaporation of derivatization steps. The total analysis time for a single sample was < 45 min. The method allowed reliable quantification of 30 ng of oxprenolol in plasma and provided a lower sensitivity limit of 10 ng/ml for a 3-ml sample. Other antihypertensive drugs, namely hydralazine, alpha-methyldopa, and the thiazide diuretics, commonly used with oxprenolol, did not interfere with the analysis.

Bretylium (o-bromobenzylethyldimethylammonium) is a quaternary ammonium compound with antiarrhythmic activity, which has recently been released in the United States for treatment of ventricular tachycardia or ventricular fibrillation [283]. Although various methods for its determination, such as colorimetry [285,286], PC [287], and GLC [288], have been reported, they are complex and tedious, and require extraction and chemical manipulations prior to determination. Lai et al. [284] simplified the determination of bretylium in pharmaceutical dosage forms by using HPLC. The method is highly sensitive (50 ng) and reproducible. Exogenous additives from injection, infusion, or tablet dosage forms do not interfere with the assay. The drug is chromatographed on a μBondapak CN column with acetonitrile–buffered aqueous 0.005 M NaH_2PO_4 (3:7) and detected at 254 nm.

The cardiac depressant quinidine is used widely in the treatment of certain cardiac arrhythmias. However, due to the narrow range between its effective and toxic concentrations (3–5 μg/ml) [289], a rapid, accurate, and sensitive procedure is required for monitoring its plasma levels. Quinidine is presented as different salts in a variety of dosage forms, which generally contain 3–10% dihydroquinidine [290,291]. In man, both of these substances undergo transformations, the major urinary metabolites being either 2'-quinidinone and 3-hydroxyquinidine or the dihydro analogs [292]. GLC procedures [293,294] do not differentiate quinidine from dihydroquinidine. A rapid, sensitive, and accurate HPLC method was developed by Sved et al. [295] for the determination of quinidine in plasma, based on ion-pair extraction. The method, which is capable of distinguishing between quinidine and dihydroquinidine, involves the acidification of plasma with perchloric acid, extraction with methyl isobutyl ketone and chromatography of the carbonate-washed extract on a silica gel column with dichloromethane–hexane–methanol–perchloric acid (600:350:55:1), followed by fluorometric detection. The procedure is sensitive to < 50 ng/ml (CV 6.6%) and compares favorably with a standard spectrofluorometric method [296].

HPLC has also been utilized for the determination of other antiarrhythmic and beta-adrenergic blocking drugs, including lorcainide [297], nadolol [298], and atenolol [4-(2-hydroxy)-3-isopropylaminopropoxyphenylacetamide] [299].

16.3.6. Antineoplastic drugs

The folate antagonist methotrexate (MTX) (4-amino-4-deoxy-N^{10}-methyl-pteroylglutamate), followed by citrovorum factor as a biochemical rescue has been widely used in the treatment of a variety of human cancers [300–304]. Because of the inherent risk of toxicity from this regimen, proper patient management necessitates the monitoring of serum MTX to allow identification of patients with high, poten-

References on p. B320

tially toxic, MTX concentrations and/or delayed MTX elimination [305]. Although various analytical methods have been reported for plasma methotrexate, including enzyme inhibition [306,307], protein binding [308], radioimmunoassay [309,310] and fluorimetry [311,312], none of these employ separation steps capable of resolving or quantitating the significant metabolite 7-hydroxymethotrexate (7-OH-MTX). A HPLC assay for MTX, based on fluorescence detection has been reported [313]. It involves the oxidation of MTX to a fluorescent species prior to chromatography and does not differentiate 7-OH-MTX. Watson et al. [314] subsequently reported an HPLC assay by means of a strong ion exchanger, which is capable of quantitating both MTX and 7-OH-MTX after high doses of MTX.

Cohen et al. [305] have utilized a reversed-phase HPLC method for the analysis of methotrexate and 7-OH-MTX in serum from patients receiving MTX therapy in both conventional and high-dose therapeutic regimens. The analyses were performed on a 25 cm × 4.1-mm stainless-steel column, packed with either RP-8 (10 μm) or RP-8 (7 μm). The mobile phase consisted of 0.1 M phosphate buffer (pH 6.8)–methanol (17:3) at a flowrate of 1.5 ml/min. A UV detector at 313 nm gave adequate sensitivity for the measurement of MTX serum levels in most cases [315–317] requiring monitoring until the MTX level drops below $10^{-7} M$. The procedure was ca. 2–3 times as sensitive as the anion-exchange procedure [314], owing mainly to improved peak shape and resolution from interfering components. The 7-μm RP-8 column gave even better peak shapes and resolution than the 10-μm column, resulting in a minimum detectable MTX concentration of 15 ng/ml $(3.3 \cdot 10^{-8} M)$.

MTX in human plasma at a concentration as low as 0.01 μg/ml was assayed by HPLC with fluorescence detection [318]. It was oxidized stoichiometrically to 2,4-diaminopteridine-6-carboxylic acid, a fluorescent product that was separable from other fluorescent materials in plasma on an octadecylsilane column. Two solvent systems were used: 0.1 M Tris, adjusted to pH 6.7 with phosphoric acid; and 0.1 M Tris–PO$_4$ (pH 6.7) in 20% methanol. With the excitation wavelength of the fluorescence detector set at 275 nm, the detector response was linear over the range of 0.01 to 10 μg/ml. Neither folic acid nor citrovorum factor interfered with the analysis. N^4-[2,4-Dihydroxy-(6-pteridyl)methyl]-aminobenzoylglutamic acid could be used as an internal standard, because it can be extracted from plasma and oxidized like methotrexate. The procedure is rapid (ca. 30 min) and potentially useful for monitoring methotrexate plasma concentrations. Lankelma et al. [319] have determined plasma concentrations of 7-OH-MTX by a HPLC method with a detection limit of $2 \cdot 10^{-8} M$. This allowed measurement of 7-OH-MTX after relatively low doses of MTX, and sample clean-up involved only a deproteinization step. The metabolite was chromatographed on a chemically-bonded anion-exchange resin (Partisil SAX) with a mobile phase of phosphate buffer (0.05 M, pH 4.9)–methanol (4:1) and monitored at 306 nm.

Lawson et al. [320] have described a reversed-phase technique allowing direct injection of the supernatant solution, containing MTX and 7-OH-MTX, after protein precipitation from serum. Analyses were performed on a 120 × 4-mm analytical column, connected to a 45 × 4-mm precolumn, packed with Hypersil-octadecylsilane ($5\,\mu$m), with an eluent of 20% methanol in Tris–NaH$_2$PO$_4$ (0.1 M, pH 6.7) and detection at 305 nm. The sensitivity of the assay being 100 ng/ml ($2.2 \cdot 10^{-7} M$) for MTX, toxic levels can be measured up to 48 h after infusion. Samples can be prepared for chromatography in about 20 min. The method has a high degree of specificity, since MTX and 7-OH-MTX are well separated from endogenous compounds, from N^{10}-methylfolic acid and aminopterin (contaminants of commercial MTX preparations and likely to be measurable after high-dose infusions), from folinic acid (citrovorum factor, used as antidote in high-dose regimens), and from the minor metabolite 2,4-diamino-N^{10}-methylpteroic acid (DAMPA).

The pyrimidine analog of 1-beta-D-arabinofuranosylcytosine (Ara-C) is one of the most effective drugs in the treatment of nonlymphocytic leukemia [321]. After administration, Ara-C is rapidly inactivated by deamination to 1-beta-D-arabino-furanosyluracil (Ara-U) and both compounds are eliminated simultaneously in the urine [322,323]. To optimize therapy, suitable methods are required to determine the Ara-C concentration in plasma, and for detailed pharmacokinetic studies, measurement of both Ara-C and Ara-U is required. A variety of procedures have been employed for the determination of Ara-C, including enzymatic assay [324], bioassays [325], radioimmunoassays [326], GLC [327], the radioassay after administration of labeled Ara-C [323], and HPLC [328–331]. Linssen et al. [331] have recently reported the determination of Ara-C and its metabolite Ara-U in human plasma. After deproteinization of the plasma sample, separation was performed by reversed-phase HPLC on a 250×4.6-mm column of Nucleosil 10-C-18, 10 μm. Elution was carried out with 0.2 M KH$_2$PO$_4$, adjusted to pH 2.0 with H$_3$PO$_4$. Ara-C was detected by UV at 280 nm with a detection limit of $2\,\mu$g/l in plasma, while Ara-U was detected at 264 nm with a detection limit ranging from 10 to 100 μg/l in plasma. The CV of the whole procedure was about 6% for Ara-C concentrations above $5\,\mu$g/l, and for Ara-U concentrations above 100 μg/l. For lower concentrations, the CV was about 14%.

Mephalan [4-bis-(2-chloroethyl)amino-1-phenylalanine, L-PAM] is an antineo-plastic alkylating agent used in the clinical treatment of multiple myeloma, ovarian carcinoma, and breast cancer [332,333]. The major hydrolyzate of L-PAM is 4-bis(2-hydroxyethyl)-amino-1-phenylalanine (L-DOH). The measurement of L-PAM itself in various media has been accomplished by GLC [334] and HPLC [335–337]. Ahmed and Hsu [338] have developed a reversed-phase HPLC procedure for the quantitative analysis of both L-PAM and L-DOH in biological samples with a sensitivity of 0.1 ppm. The procedure separates L-PAM (retention time 12 min) from L-DOH (6.5 min) without any interference from background extractives (1.4–3 min). Separations were accomplished with a 25×0.26-cm column, containing 10-μm HCODS/SIL X-C$_{18}$. The eluent was initially 12% acetonitrile in 0.0175 M acetic acid, which was converted by a concave gradient programer to 80% acetonitrile in 0.0175 M acetic acid. L-PAM and its hydrolyzate, L-DOH, were detected by a variable-wavelength UV detector at 263 nm.

Other antineoplastic agents that have been recently determined in blood or plasma by HPLC include: carminomycin (CMM) and a major metabolite carmenomycinol (CMMOH) [339] and 10-chloro-5-(2-dimethylaminoethyl)-7H-in-dolo[2,3-C]quinolin-6(5H)-one [340].

16.3.7. Miscellaneous pharmaceuticals

HPLC has been applied to the analysis of a number of pharmaceutical syrups by using precolumns and adsorption on Amberlite XAD-2 [341]. Muhammad and Bodnar [342] have described an assay method for the quantitative determination of guaifenesin, phenylpropanolamine hydrochloride, sodium benzoate, and codeine phosphate in commercial dosage forms of cough syrups. The assay method was based on paired-ion HPLC with sodium heptanesulfonate as the counter-ion. A fixed-wavelength detector at 254 nm and a μBondapak phenyl column were employed. The total analysis time for all the active ingredients was 22 min.

The selection of preferred systems for the HPLC of basic drugs, with application to the separation of ten antihistamine drugs, was recently described by Massart and Detaevernier [343]. The set of substances used consisted of 100 basic drugs, which had originally been proposed by Moffat and coworkers [344,345]. Table 16.3 lists 15 candidate systems fully investigated. A fixed-wavelength UV detector was used at 254 nm. The drugs were dissolved in the mobile phase, and 1 to 10 μl of 0.1% solutions were injected to determine a suitable detector response. For the three stationary phases, good separation possibilities exist, since among the 5 systems with a discriminating power (DP) $\geqslant 0.900$, there are two adsorption systems (A1, A2),

TABLE 16.3

LIST OF CANDIDATE SYSTEMS FOR HPLC OF BASIC DRUGS AND THEIR DISCRIMINATING POWER (DP) [343]

Systems	Proportions	DP
A. Micro silica gel, 5 μm		
1. n-Heptane–2-propanol–propylamine	100:3:0.5	0.906
2. n-Heptane–dichloromethane–2-propanol–propylamine	75:25:3:0.3	0.900
3. n-Heptane–dichloromethane–2-propanol–propylamine	50:50:3:0.3	0.886
4. n-Heptane–dichloromethane–2-propanol–propylamine	25:75:3:0.3	0.877
5. Dichloromethane–2-propanol–propylamine	100:3:0.3	0.860
B. Nitrile-bonded phase on micro silica gel, 10 μm		
1. Dichloromethane–2-propanol–propylamine	100:3:0.1	0.892
2. n-Heptane–dichloromethane–2-propanol–propylamine	50:50:3:0.2	0.893
3. n-Heptane–dichloromethane–acetonitrile–propylamine	50:50:25:0.1	0.911
4. n-Heptane–2-propanol–propylamine	100:25:0.5	0.867
5. Methanol–water–propylamine	90:10:0.01	0.851
6. Acetonitrile–water–propylamine	90:10:0.01	0.912
C. Amine-bonded phase on micro silica gel, 10 μm		
1. Dichloromethane–2-propanol–propylamine	100:3:0.1	0.880
2. n-Heptane–dichloromethane–2-propanol–propylamine	50:50:3:0.2	0.884
3. n-Heptane–dichloromethane–acetonitrile–propylamine	50:50:25:0.1	0.905
4. n-Heptane–2-propanol–propylamine	100:25:0.5	0.839

TABLE 16.4

k' VALUES OF NINE ANTIHISTAMINE DRUGS [343]

Drugs	k'		
	B3	B6	B5
Bromodiphenhydramine	0.95	1.54	0.82
Chlorphenyramine	8.27	7.51	3.00
Diphenhydramine	1.16	1.90	1.43
Diphenylpyraline	3.08	3.75	3.09
Mepyramine	2.46	2.99	1.69
Methapyrilene	1.98	2.38	1.56
Thenyldiamine	2.62	3.16	1.90
Tripellenamine	2.16	2.87	1.61
Triprolidine	9.09	6.36	2.41

two nitrile systems (B3, B6), and one amine system (C3). Table 16.4 shows a subset of antihistamine drugs and the k' values obtained in Systems B3 and B6 (the preferred systems) and another (less satisfactory) system, B5. Figs. 16.6 and 16.7 show the separation of two groups of antihistamine drugs, the first being a group of seven UV-sensitive ($E_{1\%} > 100$ at 254 nm) and the second a group of five less UV-sensitive substances. In both groups, all the substances were well separated. A chromatogram of the 12 substances together showed that all the substances could be

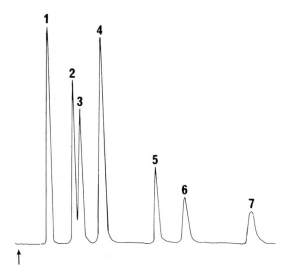

Fig. 16.6. Separation of seven antihistamine drugs on a nitrile-bonded phase (Mikropak CN-10). Mobile phase, *n*-heptane–dichloromethane–acetonitrile–propylamine (50 : 50 : 25 : 0.1); flowrate, 80 ml/h. Peaks: 1 = trimeprazine; 2 = promethazine; 3 = cyproheptadine; 4 = tripellenamine; 5 = dimethindene; 6 = triprolidine; 7 = antazoline. (Reproduced from *J. Chromatogr. Sci.,* 18 (1980) 139, with permission [343].)

References on p. B320

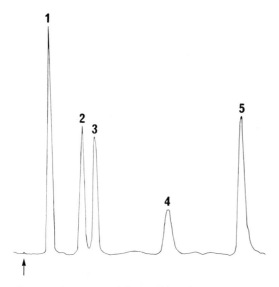

Fig. 16.7. Separation of five antihistamine drugs on a nitrile-bonded phase (Mikropak CN-10). Mobile phase, *n*-heptane–dichloromethane–acetonitrile–propylamine (50 : 50 : 25 : 0.1); flowrate, 30 ml/h. Peaks: 1 = buclizine; 2 = chlorocyclizine; 3 = diphenhydramine; 4 = diphenylpyraline; 5 = chlorphenyramine. (Reproduced from *J. Chromatogr. Sci.*, 18 (1980) 139, with permission [343].)

separated from each other, except promethazine and diphenhydramine. These two substances could be separated with System B6. Hence, the use of both preferred systems, B3 and B6, in combination permits the complete identification and separation of the antihistamine groups investigated. It was also suggested that both systems could possibly be considered as preferred systems for the separation of basic drugs in general.

Analgesic and muscle relaxant–analgesic mixtures are widely prescribed, and many analytical procedures have been reported for salicylates in dosage forms and biological media. For example, an automated HPLC analysis of aspirin, phenacetin and caffeine in dosage forms has been reported by Ascione and Chrekian [346], and Peng et al. [347] described a simple and rapid simultaneous determination of aspirin, salicylic acid, and salicyluric acid in plasma by HPLC. A reversed-phase μBondapak C_{18} column was employed with a mobile phase of 30% acetonitrile in diluted phosphoric acid (0.05%, pH 2.5) and a variable UV detector. With 0.1-ml plasma samples, the HPLC method can accurately and simultaneously measure salicyluric acid, salicylic acid, and aspirin in the same sample in concentrations as low as 0.5 μg/ml. The ability to separate and measure salicyluric acid should make this method also applicable to the assay of salicyluric acid (a major metabolite) in urine samples. The proposed method can also be used in pharmacokinetic and/or plasma level studies of aspirin or salicylates.

Of the numerous methods reported for the determination of aspirin and salicylic acid in pharmaceutical preparations, few permit a simultaneous quantitative determination of both components in thermally degraded samples or have been evaluated as stability-indicating assay procedures. Free salicylic acid [347–349], aspirin [346,350] and aspirin and salicylic acid [347,351–353] have been determined by HPLC. Salicylic acid and aspirin were separated from the other salicylates in thermally degraded multicomponent tablets and determined quantitatively by HPLC analyses of filtered extracts on a silica column with a mobile phase of acetic acid in heptane and UV detection at 300 nm by the procedure of Taguchi et al. [354]. The method was capable of resolving the major thermally induced transformation products in tablet formulations and was sensitive to about 0.1 mg of salicylic acid per tablet. HPLC was used by Honigberg et al. [355] to optimize the resolution of eight widely prescribed therapeutic agents commonly found in muscle relaxant–analgesic mixtures. The compounds were chromatographed on either porous silica or cyanopropylsilane columns with various solvent systems, paired on the basis of Snyder's solvent selectivity scheme [356] to give a polarity index for each system of 33. A carisoprodol, phenacetin, and caffeine mixture was selected to demonstrate the utility of the separation and assay method. The mixture was chromatographed on a porous silica column with tetrahydrofuran–toluene (1:1) as the mobile phase. Each determination could be achieved in approximately 8 min with an accuracy of 3–5%.

Theophylline (1,3-dimethylxanthine), a bronchodilator, has been used in the treatment of asthma for many years. Because the optimum therapeutic effect has been found when plasma concentrations of the drug are in the range 10–20 μg/ml, it is important to monitor drug levels in plasma rapidly [357]. Separation techniques for the drug, based on GLC [358–360] and HPLC [357,361–363] have been developed and used successfully. Eppel et al. [357] compared HPLC and an enzyme-multiplied immunoassay technique (EMIT) for the determination of theophylline in plasma. HPLC utilized 5-μm ODS-Hypersol columns and a mobile phase of 7% acetonitrile in 0.01 M sodium acetate trihydrate solution (adjusted to pH 4.0). Of the three sampling techniques employed (extraction, protein precipitation, and direct injection), protein precipitation was preferred, although the precision for all three methods was acceptable.

For the qualitative and quantitative analysis of standard solutions of sulfonamides ion-exchange [364,365], ion-pairing [366], and reversed-phase [367] HPLC have been used. However, the analytical procedures exclusively used for standard solutions are not satisfactory or reliable for clinical monitoring of therapeutic levels in biological fluids and for the analysis of such compounds in toxicity evaluation. Suber and Edds [368] have recently reported a HPLC procedure for the extraction and quantitation of eight sulfonamides in stock solutions and in vitro plasma samples. The assay consisted of a single, one-step extraction of sulfonamides from plasma and, without additional concentration of the sample, it gave a sensitivity of 10.0 ng/ml at 254 nm. Four sulfonamides (sulfamerazine, sulfamethazine, sulfapyridine, and sulfathiazole) were separated from the plasma matrix by water–methanol (1:1) with acetate buffer at pH 4. The sulfonamides with the highest pK_a values, sulfanilamide (10.5) and sulfaguanidine (11.3), were separated from plasma by water–methanol (1:1) at pH 7.45.

It is common practice in modern swine-rearing to use sulfonamide drugs for prevention and treatment of disease, as well as to promote growth. Hence, it is necessary to monitor human foods for drug residues due to sulfonamide treatment [369]. GLC [370] and HPLC [371] procedures have been proposed for this purpose. Vilim et al. [369] reported an HPLC screening procedure for sulfamethazine residues in pork kidney, liver, and meat. Both the Bratton–Marshall reaction [372] and HPLC were used for the initial screening tests. HPLC was used for quantitative and TLC for qualitative analysis of the derivatized standards and the unknown in the Bratton–Marshall reaction. While the HPLC screening procedure was designed primarily for sulfamethazine residues, it also effects the separation of 14 sulfonamides approved for use in swine in Canada and is specific in combination with a TLC confirmation. A reversed-phase 10-μm C-8 column was used for HPLC with a mobile phase of 25% methanol in 0.01 M ammonium acetate and UV detection.

16.4. GAS CHROMATOGRAPHY–MASS SPECTROMETRY

Recent books and reviews on GC–MS, relating to both general aspects [373–376] and to applications in pharmaceutical and forensic analysis [36,377–380] are available.

16.4.1. Tricyclic antidepressants

GC–MS techniques, based on EI or CI detection, or CC combined with direct-probe field-ionization MS have resulted in more sensitive and specific methods for tricyclic antidepressants [381–394]. However, it has also been noted by Wilson et al. [391] that although mass-fragmentographic methods [382,384,390] can involve impact ionization ions of sufficient intensity, they may often be of low specificity. GC–MS has been employed for monitoring the plasma levels of imipramine [382–386,388,391,394], clomipramine [385,387,388], amitriptyline [389–391], and their metabolites. Garland et al. [389] recently described a method for the determination of amitriptyline and its metabolites nortriptyline, 10-hydroxyamitriptyline, and 10-hydroxynortriptyline in human plasma by stable-isotope dilution and GC–CI–MS. The sensitivity was 0.5 ng/ml for amitriptyline, nortriptyline, and 10-hydroxyamitriptyline, and 1 ng/ml for 1-hydroxynortriptyline; assay precision and accuracy in terms of percent error were both < 5%.

16.4.2. Benzodiazepines

GC–MS procedures for the determination of oxazepam [395–397], chlordesmethyldiazepam and lorazepam [398], clonazepam [399–401], flurazepam [402], and their metabolites in biological fluids have been reported. The determination of clonazepam in human plasma by GC–negative-ion CI–MS was described by Garland and Min [401]. The sensitivity for this technique (< 0.1 ng/ml) was approximately 20 times better than for the positive-ion CI procedure [400] with similar precision.

16.4.3. Phenacetin and acetaminophen

The various methods which have been developed to measure the analgesic phenacetin and its analgesic biotransformation product acetaminophen in human plasma include GLC procedures, having a detection limit of 50 ng/ml [403–405]. While this sensitivity is suitable for measuring acetaminophen, it permits the determination of phenacetin concentrations in normal subjects for only a few hours after the oral administration of a therapeutic phenacetin dose and is insufficient for phenacetin determinations in many subjects with induced drug-metabolizing capabilities. Garland et al. [406] described a GLC–CI–MS assay that requires only 1.1 ml of plasma and is capable of determining 1 ng of phenacetin per ml and 0.1 μg of unconjugated and conjugated acetaminophen per ml. To obtain sufficient sensitivity, selective-ion detection is used to monitor the MH^+ ion of both phenacetin and the methyl derivative of acetaminophen, p-acetanisidine. Deuterated analogs of phenacetin and acetaminophen, phenacetin-d_3 and acetaminophen-d_3, were added to the plasma as internal standards. Phenacetin, acetaminophen (paracetamol), and acetanilide were determined in plasma and urine by mass fragmentography with deuterium-labeled analogs of the above drugs [407]. The plasma levels of phenacetin and acetaminophen were determined down to 20 ng/ml and 1–2 μg/ml, respectively.

16.5. THIN-LAYER CHROMATOGRAPHY

Recent reviews of the applications of TLC in pharmaceutical analysis are available [36,408–412]. Although TLC is well recognized as an ideal technique for the screening of drugs in toxicological analysis, because of its low cost and convenience and the selectivity of detection reagents, objective criteria for the evaluation of the separating ability of TLC have often been lacking [26]. Numerical taxonomy [413] has been used to classify systems according to their similarities, but the measurement of the informing power [414–416] or discriminating power [417–419] is more useful when the selection of optimal systems is required [26].

16.5.1. Basic, neutral, and acidic drugs

The factors involved in assessing discriminating power include:
(a) distribution of chromatographic values over the useful range of the system,
(b) correlation between systems when more than one is used,
(c) speed,
(d) reproducibility,
(e) sensitivity.
These factors were considered by Moffat and Clare [419], who employed discriminating power to select more suitable TLC systems for the analysis of basic drugs. When the better systems for the analyses of basic drugs were chosen, standardization of results from different laboratories became possible, and chromatographic data became transferable from laboratory to laboratory [420].

References on p. B320

Owen et al. [26] utilized a similar selection procedure to choose TLC systems for commonly encountered acidic drugs. The discriminating powers of 15 silica gel TLC systems were measured both individually and in combination. The systems were chosen from those included in standard tests of drug analysis [421–423], those used by British forensic science laboratories and in literature surveys of TLC systems used for the general screening of drugs [424–426], as well as for specific drugs, e.g., barbiturates [427–431] and thiazide diuretics [432,433]. Owen et al. [26] found that the two best mobile phases were ethyl acetate and chloroform–methanol (9:1), and that a combination of ethyl acetate–methanol–ammonia (17:2:1) with either of these gave the best pair of systems.

The efficiencies of 15 silica gel TLC systems for separating commonly encountered neutral drugs were also compared by Owen et al. [434]. The discriminating powers of the systems were measured both individually and in combination. The previously recommended systems selected for either basic or acid drugs [26,418,419] were used in the study, in addition to the TLC systems employed for the benzodiazepines [435,436]. A commonly occurring group of neutral drugs was also considered. Chloroform–acetone (4:1) was found to be the best mobile phase and it yields the best pair of systems in combination with ethyl acetate–methanol–ammonia (17:2:1). Chloroform–acetone (4:1) was recommended as the system of choice when screening for both acidic and neutral drugs.

16.5.2. Barbiturates

A major problem in the determination of barbiturates (as well as other nonvolatile organic drugs) in biological samples is usually the isolation and purification of the active ingredients from the extraneous biological material. Various TLC techniques have been employed for the separation, identification, and determination of barbiturates [427–431,437–443]. Chromatograms on fiber-glass sheets, impregnated with silica gel, and on glass and aluminum foil have been investigated [444,445]. Many reagents have been described for the detection of barbiturates [446–449], the sensitivity of the method varying between 5 and 0.4 μg. Combined TLC and GLC has been employed for the detection and determination of barbiturates [450,451]. Several methods of identification of barbiturates by using one-dimensional [452–454] and two-dimensional [455] TLC have been reported, and the quantitative determination of barbiturates by TLC has also been investigated [456–458]. Abu-Eitlah et al. [459] used some new approaches in the TLC separation and identification of a number of barbiturates, such as preparing silica gel slurries with alkali and/or with cobalt(II) salts. Silica is generally regarded as an acidic adsorbent, and its adsorptive properties depend almost exclusively on the superficial hydroxyl groups. Mixing silica with an alkali results in the inactivation of the acid sites. As the concentration of alkali increases, acidic sites are removed and the sample is strongly adsorbed on the negatively charged oxygen. This results in a decrease in the R_F values of both barbiturates and thiobarbiturate as the pH of the slurry increases. Barbiturates were detected on the chromatograms by spraying with either Co^{2+} or Hg^{2+} solutions. The sensitivity of the determination, which was 10 μg of drug for the cobalt reagent and 1.0 μg for the mercury reagent, increased to 0.5 μg when an alkaline slurry was used.

16.5.3. Urine screening for drugs of abuse

As the illegal use of drugs increases, urine screening for drugs of abuse becomes a necessary adjunct in drug abuse prevention and treatment programs [7,460–464]. A measure of the extent of this abuse can be gleaned from the fact that 15 million urine specimens are tested each year in the United States for the presence of drugs of abuse [461]. Additionally, urine screening for the detection of illicit drug use has become a necessary adjunct to treatment for heroin addiction [7].

In the earlier technique of Dole et al. [465] for the detection of narcotic drugs, quinine, barbiturates, amphetamines, and some tranquilizers in urine, the drugs were first absorbed on ion-exchange paper and then extracted at controlled pH values into an organic solvent. An aliquot of the extract was concentrated and tested by TLC with a series of spray reagents. Although the procedure was satisfactory for drugs such as morphine, codeine, and quinine, such drugs as methadone, phenobarbital, and amphetamines failed to be detected.

Techniques for the analysis of urine samples in a large urine monitoring control program were later refined by Mule and coworkers [464,466]. The procedure involved an initial spectrofluorometric analysis to screen urine samples for the presence of morphine and/or quinine which, if positive, were determined by TLC. It also included an acid hydrolysis of drug conjugates followed by extraction at low pH to remove the acidic drugs, such as the barbiturates, diphenylhydantoin, glutethimide, and an extraction at pH 10–11 for narcotics, tranquilizers, and amphetamine. Detection was accomplished by spray reagents. GLC was recommended for a subsequent positive identification and confirmation. The technique required 60-ml samples and the procedures were designed for the analysis of 500 or more urine samples per day.

Additional screening procedures for drugs were reported by Fujimoto and Wang [467]. They involved an initial isolation of the drugs by passing the urine samples through an Amberlite XAD-2 resin column, elution of the acidic and basic drugs, and separation by TLC. Other urine screening methods which have been utilized for drugs of abuse include those of Wallace and coworkers [462,468] and Kaistha and coworkers [460,469–471]. Kaistha [7] has also published a guide to urine testing in drug abuse prevention and multimodality treatment programs. A single-step extraction method and TLC identification technique detects a wide variety of drugs of abuse [460]. The drugs are initially absorbed on a 6 × 6-cm piece of paper, loaded with SA-2 cation-exchange resin, and then eluted from the paper at pH 10.1 with an ammonium chloride–ammonia buffer. The simultaneous TLC of sedatives, hypnotics, narcotic analgesics, CNS stimulants, and miscellaneous drugs is accomplished by spotting the extract on Gelman precoated silica gel microfiber sheets (ITLC, Type SA), which are subjected to a two-stage development in order to obtain a chromatogram with optimum separation of a wide range of drugs. Different detection reagents are then applied in succession to different marked areas of the developed chromatogram. This method permits the detection of morphine in urine at a sensitivity level of 0.15 μg/ml; amphetamine sulfate, 1.0 μg/ml; methamphetamine hydrochloride, 0.5 μg/ml; phenmetrazine hydrochloride, 0.5 μg/ml; codeine phosphate, 0.5 μg/ml; methadone hydrochloride, 1.0 μg/ml; secobarbital, 0.36 μg/ml; and phenobarbital, 0.5 μg/ml. The minimum volume of urine required at these sensitivities is 20 ml. Of course, confirmation of a positive TLC screening test by GC and/or UV or IR spectrometry is mandatory in most situations, owing to the serious sociological and legal implications [463].

The abuse of anorexigenic drugs is steadily increasing, primarily in the treatment of obesity, where they have CNS stimulatory side-effects [472–476]. The major anorexicants of potential concern in this regard are phenmetrazine (3-methyl-2-phenylmorpholine) (Preludin), diethylpropion (Tenurate, Tepranil), fenfluramine, chlorphentermine, phendimetrazine, mazindol, and phentermine. Winek et al. [477] described the detection and interference of eight CNS stimulant anorexicants in urine drug-screening procedures, by TLC and EMIT. The drugs were extracted from borate-buffered (pH 9.25) urine with dichloromethane–2-propanol (9:1), the extract was spotted on Whatman Linear K Preadsorbent E plates and developed with ethyl acetate–methanol–water–ammonium hydroxide (85:13:5:1). Detection was accomplished with ninhydrin, UV, and Dragendorff's reagent [478]. The limits of detection were 0.1 μg/ml for diethylpropion, phendimetrazine, and mazindol; 0.3 μg/ml for fenfluramine; 0.5 μg/ml for phenmetrazine; and 1.0 μg/ml of phentermine, chlorphentermine, and chlortermine.

16.5.4. Antiarrhythmic and beta-adrenergic blocking drugs

Christiansen [479] reported a micromethod, based on direct TLC for the determination of quinidine and salicylic acid in 10-μl samples of serum or plasma. The samples were applied to the silica gel layer without extraction, and the proteins were precipitated at the site of application by means of ethanol. Chromatograms were evaluated quantitatively by fluorescence scanning with excitation at 366 nm and emission at 456 nm, after development with benzene–diethyl ether–glacial acetic acid–methanol (120:60:18:1).

A quantitative TLC method for the determination of procainamide and its major metabolite, N-acetylprocainamide, in plasma was reported [480]. It involves extraction with dichloromethane at high pH, chromatography with benzene–28% ammonium hydroxide–dioxane (2:3:16), and direct measurement of the absorbance of the compounds on the plate at 275 nm. Quantities as low as 10 ng could be measured, and a linear relationship was observed between peak areas in the scan and amounts of the compound in the spots between 10 and 200 ng. The recovery of both drugs from plasma was between 95.4 and 104.8%. The method was recommended for clinical assays and pharmacokinetic studies.

A sensitive and specific TLC method for the simultaneous determination of the beta-adrenoceptor blocking agent, acebutolol [DL-1-(2-acetyl-4-n-butyramidophenoxy)-2-hydroxy-3-isopropylaminopropane], and its major metabolite, [DL-1-(2-acetyl)-4-acetamido-2-hydroxy-3-isopropylaminopropane], in serum was described by Steyn [481]. A 2-ml sample of serum, containing 350 ng of quinidine as internal standard, was extracted at pH 10, the solvent was evaporated, and the residue was dissolved in 50 μl of methanol. A 10-μl aliquot of this solution was chromatographed on precoated Silica Gel 60 plates with ethyl acetate–methanol–ammonia (15:4:1). The fluorescence was measured at an excitation wavelength of 350 nm and emission wavelength at 450 nm.

Silica-gel-impregnated glass-fiber sheets were used in a rapid, sequential, TLC method for the sensitive and specific analysis of propranolol and seven of its possible metabolites [482]. The first mobile phase was acetonitrile–benzene–*n*-hexane–ammonia (80:40:40:1), and the second one was *n*-hexane–diethyl ether (9:1). The procedure takes only 23 min, is convenient and highly reproducible.

16.5.5. Soap thin-layer chromatography

Soap TLC has given favorable results in the separation of catechol amines [483], primary aromatic amines [484,485], and sulfonamides [485]. The behavior of a number of sulfonamides on layers of silanized silica gel, impregnated with triethanolamine dodecylbenzenesulfonate and N-dodecylpyridinium chloride was investigated by Lepri et al. [485], who studied the influence on the chromatographic behavior of the kind of detergent, the percentage of organic solvent in the eluent, and the apparent pH of the eluent. The soap TLC technique has enabled many separations of sulfonamides and aromatic amines that could not be effected otherwise on ion exchangers.

16.5.6. High-performance thin-layer chromatography

Lee et al. [486] described the simultaneous determination of five antiarrhythmia drugs by HPTLC. Complete separation of all the drugs and clozapine, an internal standard, was achieved by development with two solvent systems of different polarity: benzene–ethyl acetate–methanol (4:4:1) and benzene–ethyl acetate–methanol–pyridine (4:2:3:3). Lidocaine and diphenylhydantoin were scanned at 220 nm after the first development, and procainamide, propranolol, and quinidine were scanned at 290 nm after the second development. The relative standard deviation of the determination varied between 3 and 14%, depending on the nature of the drugs and their concentration.

Gounet and Marichy [487] have recently described the properties and selectivities of the following organic modifiers, methanol, ethanol, 2-propanol, acetonitrile and tetrahydrofuran in reversed-phase HPTLC of a number of pharmaceutical products of toxicological interest. Ethanol appeared to be the most selective organic modifier, but 2-propanol appeared to be the most interesting for HPTLC–HPLC data transfer. In every case, the best resolution was observed with mobile phases containing 40–50% water.

References on p. B320

REFERENCES

1 J.F. Lawrence, *Anal. Chem.*, 52 (1980) 1122A.
2 J.K. Baker, R.E. Skelton and C.-Y. Ma, *J. Chromatogr.*, 168 (1979) 417.
3 R.C. Williams and J.L. Viola, *J. Chromatogr.*, 185 (1979) 505.
4 L.R. Snyder, J.W. Dolan and N. Tanaka, *Clin. Chem.*, 25 (1979) 1117.
5 H. Bethke, *Chromatographia*, 12 (1979) 335.
6 W. Roth, K. Beschke, R. Jauch, A. Zimmer and F.W. Koss, *J. Chromatogr.*, 222 (1981) 13.
7 K.K. Kaistha, *J. Chromatogr.*, 141 (1977) 145.
8 W.H. McFadden, *J. Chromatogr. Sci.*, 18 (1980) 97.
9 W.H. McFadden, *Techniques of Combined Gas Chromatography/Mass Spectrometry*, Wiley-Interscience, New York, 1973.
10 B.L. Karger, D.P. Kirby, P. Vouros, R.L. Foltz and B. Hidy, *Anal. Chem.*, 51 (1979) 2324.
11 P.R. Jones and S.L. Yang, *Anal. Chem.*, 47 (1975) 1000.
12 J.D. Henion, *Anal. Chem.*, 50 (1978) 1687.
13 W.H. McFadden, D.C. Bradford, D.E. Games and J.L. Gower, *Amer. Lab.*, 9 (1977) 55.
14 Y. Hirata, T. Takeuchi, S. Tsuge and Y. Yoshida, *Org. Mass Spectrom.*, 14 (1979) 126.
15 A. Zlatkis and C.F. Poole, *Anal. Chem.*, 52 (1980) 1002A.
16 W.J. Decker, *Clin. Toxicol.*, 10 (1977) 23.
17 R.S. Gohlke, *Anal. Chem.*, 31 (1959) 535.
18 C.E. Costello, H.S. Hertz, T. Sakai and K. Biemann, *Clin. Chem.*, 20 (1974) 255.
19 M.E. Rand, J.D. Hammond and P.J. Moscou, *J. Amer. Coll. Health Assoc.*, 17 (1968) 43.
20 R. Saferstein, J.J. Manura, T.A. Bretell and P.K. De, *J. Anal. Toxicol.*, 2 (1978) 245.
21 P.A. Ullucci, R. Cadoret, P.D. Stasiowski and H.F. Martin, *J. Anal. Toxicol.*, 2 (1978) 33.
22 A. Cailleux, A. Turcant, A. Premel-Cabic and P. Allain, *J. Chromatogr. Sci.*, 19 (1981) 163.
23 R.F. Skinner, F.J. Gallaher, J.B. Knight and E.J. Bonelli, *J. Forensic Sci.*, 17 (1972) 189.
24 B.S. Finkle, D.M. Taylor and E.J. Bonelli, *J. Forensic Sci.*, 17 (1972) 189.
25 B.S. Finkle, R.L. Foltz and D.M. Taylor, *J. Chromatogr. Sci.*, 12 (1974) 304.
26 P. Owen, A. Pendleburry and A.C. Moffat, *J. Chromatogr.*, 164 (1978) 195.
27 U.A.T. Brinkman and G. de Vries, *J. High Resolut. Chromatogr. Chromatogr. Commun.*, 2 (1979) 79.
28 U.A.T. Brinkman and G. de Vries, *J. Chromatogr.*, 192 (1980) 331.
29 A.M. Siouffi, T. Wawrzynowicz, F. Bressolle and G. Guiochon, *J. Chromatogr.*, 186 (1979) 563.
30 R.E. Kaiser and R. Rieder, *J. Chromatogr.*, 142 (1976) 411.
31 J.D. Nicholson, *Analyst (London)*, 103 (1978) 193.
32 T.J. Betts, *Aust. J. Pharm. Sci.*, NS7 (2) (1978) 54.
33 A. Zlatkis and C.F. Poole, *Anal. Chem.*, 52 (1980) 1002A.
34 W.J. Griffin, G.A. Groves, W.A. Harris, B.D. Rawel and P.J. Stewart, *Aust. J. Pharm. Sci.*, NS7 (1978) 2.
35 E. Reid, *Methodological Developments in Biochemistry*, Vol. 5, North-Holland, Amsterdam, 1976.
36 R.K. Gilpin, *Anal. Chem.*, 51 (1979) 275R.
37 S.P. Cram, *Anal. Chem.*, 52 (1980) 324R.
38 B.J. Gudzinowicz and M.J. Gudzinowicz, *Analysis of Drugs and Metabolites by Gas Chromatography-Mass Spectrometry*, Vol. 1, *Respiratory Gases, Volatile Anesthetics, Ethyl Alcohol and Related Toxicological Materials*, Dekker, New York, 1977, p. 233.
39 B.J. Gudzinowicz and M.J. Gudzinowicz, *Analysis of Drugs and Metabolites by Gas Chromatography-Mass Spectrometry*, Vol. 2, *Hypnotics, Anti-Convulsant and Sedatives*, Dekker, New York, 1977, p. 493.
40 B.J. Gudzinowicz and M.J. Gudzinowicz, *Analysis of Drugs and Metabolites by Gas Chromatography-Mass Spectrometry*, Vol. 3, *Antipsychotic, Antiemetic and Antidepressant Drugs*, Dekker, New York, 1977, p. 268.
41 B.J. Gudzinowicz and M.J. Gudzinowicz, *Analysis of Drugs and Metabolites by Gas Chromatography-Mass Spectrometry*, Vol. 4, *Central Nervous System Stimulants*, Dekker, New York, 1977, p. 458.
42 B.J. Gudzinowicz and M.J. Gudzinowicz, *Analysis of Drugs and Metabolites by Gas Chromatography-*

Mass Spectrometry, Vol. 5, *Analgesics, Local Anesthetics and Antibiotics*, Dekker, New York, 1977, p. 541.
43 R.C. Baselt, *Disposition of Toxic Drugs and Chemicals in Man*, Vol. 1, *Centrally Acting Drugs*, Biomed. Publ., Canton, CT, 1978.
44 R.C. Baselt, *Disposition of Toxic Drugs and Chemicals in Man*, Vol. 2, *Peripherally Acting Drugs and Common Toxic Chemicals*, Biomed. Publ., Canton, CT, 1978.
45 E. Berman, *Analysis of Drugs of Abuse*, Heyden, London, 1977.
46 C.A. Cramers, E.A. Vermeer, L.G. van Kuik, J.A. Hulsman and C.A. Meijers, *Clin. Chim. Acta*, 73 (1976) 97.
47 C.I. Pippenger and H.W. Gillen, *Clin. Chem.*, 15 (1969) 582.
48 H.J. Kupfenberg, *Clin. Chim. Acta*, 29 (1970) 283.
49 J. Macgee, *Anal. Chem.*, 42 (1970) 421.
50 J.H. Goudie and D. Burnett, *Clin. Chim. Acta*, 43 (1973) 423.
51 H.L. Davis, K.J. Falk and D.G. Bailey, *J. Chromatogr.*, 107 (1975) 61.
52 D.M. Woodbury, J.K. Penry and R.P. Schmidt, *Antiepileptic Drugs*, Raven, New York, 1972.
53 E.S. Vesell and G.T. Parsananti, *Clin. Chem.*, 17 (1971) 851.
54 C.V. Abraham and H.D. Joslin, *J. Chromatogr.*, 128 (1976) 281.
55 I.P. Baumel, B.B. Gallagher and R.H. Mattson, *Arch. Neurol.*, 27 (1972) 34.
56 G. Kananen, R. Osiewicz and I. Sunshine, *J. Chromatogr. Sci.*, 10 (1972) 283.
57 R.E. Hill and A.N. Latham, *J. Chromatogr.*, 131 (1977) 341.
58 C.V. Abraham and D. Gresham, *J. Chromatogr.*, 136 (1977) 332.
59 A. Sengupta and M.A. Peat, *J. Chromatogr.*, 137 (1977) 206.
60 W. Godolphin and J. Thoma, *Clin. Chem.*, 24 (1978) 483.
61 F. Dorrity, Jr. and M. Linnoila, *Clin. Chem.*, 22 (1976) 860.
62 C.A. Cramers, J.A. Rijks and K. Bocek, *Clin. Chim. Acta*, 34 (1971) 159.
63 R.N. Gupta and P.M. Keane, *Clin. Chem.*, 21 (1975) 1346.
64 O. Driessen and A. Emonds, *Proc. K. Ned. Akad. Wet. Ser. C*, 77 (1974) 171.
65 C.J. Jensen and R. Gugler, *J. Chromatogr.*, 137 (1977) 188.
66 J.W.A. Meijer, *Epilepsia*, 12 (1971) 341.
67 I.C. Dijkhuis and E. Vervloet, *Pharm. Weekbl.*, 109 (1974) 42.
68 F. Schobeen and E. van der Kleijn, *Pharm. Weekbl.*, 109 (1974) 30.
69 R. Gill, A.H. Stead and A.C. Moffat, *J. Chromatogr.*, 204 (1981) 275.
70 P.A. Toseland, J. Grove and D.J. Berry, *Clin. Chim. Acta*, 38 (1972) 321.
71 B. Welton, *Chromatographia*, 3 (1970) 211.
72 D.J. Berry, *J. Chromatogr.*, 86 (1973) 89.
73 R.C. Hall and C.A. Risk, *J. Chromatogr. Sci.*, 13 (1975) 519.
74 T. Chang and A.J. Glazko, *J. Lab. Clin. Med.*, 75 (1968) 145.
75 J.L. Holtzman and D.S. Alberts, *Anal. Biochem.*, 43 (1971) 48.
76 H.V. Street, *J. Chromatogr.*, 41 (1969) 358.
77 T. Walle, *J. Chromatogr.*, 114 (1975) 345.
78 D.H. Sandberg, G.L. Resnick and C.Z. Bacallo, *Anal. Chem.*, 40 (1968) 736.
79 E.M. Baylis, D.E. Fry and V. Marks, *Clin. Chim. Acta*, 30 (1970) 93.
80 J.G.H. Cook, C. Riley, R.F. Nunn and D.E. Budgen, *J. Chromatogr.*, 6 (1961) 182.
81 M.W. Couch, M. Greer and C.M. Williams, *J. Chromatogr.*, 87 (1973) 559.
82 H.F. Martin and J.L. Driscoll, *Anal. Chem.*, 38 (1966) 345.
83 E.W. Robb and J.J. Westbrook, *Anal. Chem.*, 35 (1963) 1644.
84 G.W. Stevenson, *Anal. Chem.*, 38 (1966) 1948.
85 M.J. Barrett, *Clin. Chem. Newsl.*, 3 (1971) 1.
86 J. De Graeve and J. Vanroy, *J. Chromatogr.*, 129 (1976) 171.
87 A. Wu and M.L. Pearson, *Anal. Lett.*, 10 (1977) 381.
88 E. Brochmann-Hanssen and T.O. Oke, *J. Pharm. Sci.*, 58 (1969) 370.
89 R.H. Hammer, B.J. Wilder, R.R. Streiff and A. Mayersdorf, *J. Pharm. Sci.*, 60 (1971) 327.
90 R. Osiewicz, V. Aggarwal, R.M. Young and I. Sunshine, *J. Chromatogr.*, 88 (1974) 157.
91 A. Wu, *Clin. Chem.*, 20 (1974) 630.

92 P.S. Callery and J. Leslie, *Clin. Chem.*, 22 (1976) 22.

93 G.A.F.M. Rutten and J.A. Luyten, *J. Chromatogr.*, 74 (1972) 177.

94 M.L. Wetzels, *Thesis,* Technische Hogeschool, Eindhoven, 1975.

95 R.C. Hall and C.A. Risk, *J. Chromatogr.*, 13 (1975) 519.

96 S.R. Sun and D. Hoffman, *J. Pharm. Sci.*, 68 (1979) 386.

97 W. Dunges and E. Bergheim-Irps, *Anal. Lett.*, 6 (1973) 185.

98 N. Elahi, *J. Anal. Toxicol.*, 3 (1979) 35.

99 P.J. Orsulak and J.T. Schildkraut, *Ther. Drug Monitor.*, 1 (1979) 199.

100 R. Gupta and G. Molnar, *Drug Metab. Rev.*, 9 (1979) 79.

101 L.F. Gram, *Clin. Pharmacokinet.*, 2 (1977) 237.

102 A. Jorgensen, *Acta Pharm. Toxicol.*, 36 (1975) 79.

103 T.B. Cooper, D. Allen and G.M. Simpson, *Psychopharm. Commun.*, 1 (1975) 445.

104 T.B. Cooper, D. Allen and G.M. Simpson, *Psychopharm. Commun.*, 2 (1975) 105.

105 D.N. Bailey and P.I. Jatlow, *Clin. Chem.*, 22 (1976) 777.

106 D.N. Bailey and P.I. Jatlow, *Clin. Chem.*, 22 (1976) 1967.

107 L.A. Gifford, P. Turner and C.M.B. Pare, *J. Chromatogr.*, 105 (1975) 107.

108 M. Bertrand, C. Dupuis, M.A. Gagnon and R. Dugal, *Clin. Biochem.*, 154 (1978) 117.

109 S.F. Reite, *Medd. Nor. Farm. Selsk.*, 37 (1975) 76.

110 R.N. Gupta, G. Molnar, R.E. Hill and M.I. Gupta, *Clin. Biochem.*, 5 (1976) 247.

111 S. Dawling and R.A. Braithwaite, *J. Chromatogr.*, 146 (1978) 449.

112 J. Vasillades and K.C. Bush, *Anal. Chem.*, 48 (1976) 1708.

113 K.K. Midha, C. Charette, J.K. Cooper and I.J. McGilveray, *J. Anal. Toxicol.*, 4 (1980) 237.

114 V. Rovei, M. Sanjuan and P.D. Hrdina, *J. Chromatogr.*, 182 (1980) 349.

115 E. Antal, S. Mercik and P.A. Kramer, *J. Chromatogr.*, 183 (1980) 149.

116 A. Jorgensen, *Acta Pharmacol. Toxicol.*, 36 (1975) 79.

117 W.D. Mitchell, S.F. Webb and G.R. Padmore, *Ann. Clin. Biochem.*, 16 (1979) 47.

118 F. Dorritz, M. Linnoila and R.L. Habig, *Clin. Chem.*, 23 (1977) 1326.

119 A.K. Dhar and H. Kutt, *Ther. Drug Monit.*, 1 (1979) 209.

120 J.E. Bredesen, O.F. Ellingsen and J. Karlsen, *J. Chromatogr.*, 204 (1981) 361.

121 R.A. Braithwaite, B. Widdop, *Clin. Chim. Acta*, 35 (1971) 461.

122 H.B. Hucker and S.C. Stauffer, *J. Pharm. Sci.*, 63 (1974) 296.

123 J.E. O'Brien and O.N. Hinsvark, *J. Pharm. Sci.*, 65 (1976) 1068.

124 G.L. Corona and B. Bonferoni, *J. Chromatogr.*, 124 (1976) 401.

125 G. Nyberg and E. Mårtensson, *J. Chromatogr.*, 143 (1977) 491.

126 J.E. Wallace, H.E. Hamilton, L.K. Goggin and K. Blum, *Anal. Chem.*, 47 (1975) 1516.

127 O. Borgå and M. Garle, *J. Chromatogr.*, 68 (1972) 77.

128 P. Kragh-Sorensen, C.E. Hansen, N.E. Carsen, J. Noestoft and E.F. Hvidberg, *Psychol. Med.*, 4 (1974) 174.

129 J.T. Streator, L.S. Eichmeier and M.E. Caplis, *J. Anal. Toxicol.*, 4 (1980) 58.

130 J.P. Moody, S.F. Whyte and G.J. Naylor, *Clin. Chim. Acta*, 43 (1973) 355.

131 R.I. Shader and D.J. Greenblatt, *Benzodiazepines in Clinical Practice*, Raven, New York, 1974.

132 P. Cushman and D. Benzer, *Drug Alcohol Depend.*, 6 (1980) 365.

133 J.M. Clifford and W.F. Smyth, *Analyst (London)*, 99 (1974) 241.

134 D.M. Hailey, *J. Chromatogr.*, 98 (1974) 527.

135 U. Klotz, G.R. Avent, A. Hoyumpa, S. Schenber and G.R. Wilkinson, *J. Clin. Invest.*, 55 (1974) 347.

136 S.A. Kaplan, M.L. Jack, K. Alexander and R.E. Weinfeld, *J. Pharm. Sci.*, 62 (1973) 1789.

137 A. Berlin, B. Siwers, S. Agurell, A. Hiort, F. Sjöqvist and S. Ström, *Clin. Pharmacol. Ther.*, 13 (1972) 733.

138 R.E. Weinfeld, H.N. Posmanter, K.-C. Khoo and C.V. Puglisi, *J. Chromatogr.*, 143 (1977) 581.

139 H.H. McCurdy, E.L. Slightom and J.C. Harrill, *J. Anal. Toxicol.*, 3 (1979) 195.

140 A.K. Dhar and H. Kutt, *Clin. Chem.*, 25 (1979) 137.

141 J.M. Steyn and H.K.L. Hundt, *J. Chromatogr.*, 107 (1975) 196.

142 J.L. Ferguson and D. Couri, *J. Anal. Toxicol.*, 3 (1979) 171.

143 D.M. Rutherford, *J. Chromatogr.*, 137 (1977) 439.

144 E. Rey, J.M. Turquais and G. Oliver, *Clin. Chem.*, 23 (1977) 1338.
145 J.A.F. de Silva and C.V. Puglisi, *Anal. Chem.*, 42 (1970) 1725.
146 I.A. Zingales, *J. Chromatogr.*, 75 (1973) 55.
147 A.G. Howard, G. Nickless and D.M. Hailey, *J. Chromatogr.*, 90 (1974) 325.
148 E. Arnold, *Acta Pharmacol. Toxicol.*, 36 (1975) 335.
149 J.A.F. de Silva, I. Bekersky, C.V. Puglisi, M.A. Brooks and R.E. Weinfeld, *Anal. Chem.*, 48 (1976) 10.
150 J.J. de Gier and B.J. 't Hart, *J. Chromatogr.*, 163 (1979) 304.
151 M. Linnoila and F. Dorrity, Jr., *Acta Pharmacol. Toxicol.*, 41 (1977) 458.
152 H.P. Gelbke, H.J. Schlicht and G. Schmidt, *Arch. Toxicol.*, 38 (1977) 295.
153 J.A.F. de Silva, M.A. Schwartz, V. Stefanovic, J. Kaplan and L. D'Arconte, *Anal. Chem.*, 36 (1964) 2099.
154 D.J. Hoffman and A.H.C. Chon, *J. Pharm. Sci.*, 64 (1975) 1668.
155 D.J. Greenblatt, *J. Pharm. Sci.*, 67 (1978) 427.
156 R.C. Kelly, R.M. Anthony, L. Krent, W.L. Thompson and I. Sunshine, *Clin. Toxicol.*, 14 (1979) 445.
157 I.A. Zingales, *J. Chromatogr.*, 54 (1971) 15.
158 H.B. Kimmel and S.S. Walkenstein, *J. Pharm. Sci.*, 56 (1967) 538.
159 J.A. Knowles and H.W. Ruelius, *Arzneim.-Forsch.*, 22 (1972) 687.
160 J. Vessman, G. Freij and S. Stronberg, *Acta Pharm. Suecica*, 9 (1972) 447.
161 M. Wretlind, A. Pilbrant, A. Sundwall and J. Vessman, *Acta Pharmacol. Toxicol.*, 40 (Suppl. 1) (1977) 28.
162 J. Lanzoni, L. Airoldi, F. Marcucci and E. Mussini, *J. Chromatogr.*, 168 (1979) 260.
163 J.A. Knowles, W.H. Comer and H.W. Ruelius, *Arzneim.-Forsch.*, 21 (1971) 1055.
164 F. Marcucci, E. Mussini, L. Airoldi, A. Guaitani and S. Garattini, *J. Pharm. Pharmacol.*, 24 (1972) 63.
165 K. Møller Jensen, *J. Chromatogr.*, 111 (1975) 389.
166 H. Ehrsson and A. Tilly, *Anal. Lett.*, 6 (1973) 197.
167 A. Viala, J.P. Cano and A. Angeletti-Philippe, *Eur. J. Toxicol.*, 4 (1971) 109.
168 G.P. Beharrell, D.M. Hailey and M.K. McLaurin, *J. Chromatogr.*, 70 (1972) 45.
169 M.S. Greaves, *Clin. Chem.*, 20 (1974) 141.
170 A.G. de Boer, J. Röst-Kaiser, H. Bracht and D.D. Breimer, *J. Chromatogr.*, 145 (1978) 105.
171 L. Kangas, *J. Chromatogr.*, 172 (1979) 273.
172 Y. Matsuda, *Jap. J. Leg. Med.*, 25 (1971) 445.
173 L. Kangas, *J. Chromatogr.*, 172 (1979) 273.
174 R. Clarkson, H. Tucker and J. Wale, *Annual Reports in Medicinal Chemistry*, Academic Press, New York, 1975, pp. 10 and 51.
175 C.L. Winek, *Clin. Chem.*, 22 (1976) 832.
176 D.E. Jewitt, Y. Kishon and M. Thomas, *Lancet*, i (1968) 266.
177 R. Gianelly, J.O. van der Groeben, A.P. Spivak and D.C. Harrison, *N. Engl. J. Med.*, 277 (1967) 1215.
178 R.L. Nation, E.J. Triggs and M. Selig, *J. Chromatogr.*, 116 (1976) 188.
179 G. Svinhufvud, B. Ortengren and S.E. Jacobsson, *Scand. J. Clin. Lab. Invest.*, 17 (1965) 162.
180 J.B. Kennaghan, *Anesthesiology*, 29 (1968) 110.
181 F. Reynolds and A.H. Beckett, *J. Pharm. Pharmacol.*, 20 (1968) 704.
182 M. Rowland, P.D. Thomson, A. Guichard and K.L. Melman, *Ann. N.Y. Acad. Sci.*, 179 (1971) 383.
183 N. Benowitz and M. Rowland, *Anesthesiology*, 39 (1973) 639.
184 C.A. DiFazio and R.E. Brown, *Anesthesiology*, 34 (1971) 86.
185 J.B. Kennaghan and R.N. Boyes, *Anesthesiology*, 34 (1971) 110.
186 J.M. Strong and A.J. Atkinson, *Anal. Chem.*, 44 (1972) 2287.
187 J.M. Strong, M. Parker and A.J. Atkinson, *Clin. Pharmacol. Ther.*, 14 (1973) 67.
188 K.K. Adjepon-Yamoah and L.F. Prescott, *J. Pharm. Pharmacol.*, 26 (1974) 889.
189 J. Koch-Weser, *N. Engl. J. Med.*, 287 (1972) 227.
190 K.J. Simons and R.H. Levy, *J. Pharm. Sci.*, 64 (1975) 1967.
191 R. Howe and R.G. Shanks, *Nature (London)*, 210 (1966) 1336.

192 A.M. Barrett and V.A. Cullum, *Brit. J. Pharmacol.*, 34 (1968) 43.

193 J.D. Fitzgerald, *Clin. Pharmacol. Ther.*, 10 (1969) 282.

194 C.F. George, T. Fenyvesi, M.E. Conolly and C.T. Dollery, *Eur. J. Clin. Pharmacol.*, 4 (1972) 74.

195 A.S. Nies, G.H. Evans and D.G. Shand, *J. Pharmacol. Exp. Ther.*, 184 (1973) 716.

196 S. Caccia, G. Guiso, M. Ballabio and P. De Ponte, *J. Chromatogr.*, 172 (1979) 457.

197 B. Halpern and J.W. Westley, *Biochem. Biophys. Res. Commun.*, 19 (1965) 361.

198 E. Gordis, *Biochem. Pharmacol.*, 15 (1966) 2124.

199 S. Caccia and A. Jori, *J. Chromatogr.*, 144 (1977) 127.

200 R.W. Souter, *J. Chromatogr.*, 108 (1975) 265.

201 A.H. Beckett and B. Testa, *J. Pharm. Pharmacol.*, 25 (1973) 382.

202 S. Caccia, C. Chiabrando, P. DePonte and R. Fanelli, *J. Chromatogr. Sci.*, 16 (1978) 543.

203 J.P. DeSager and C. Harvengt, *J. Pharm. Pharmacol.*, 27 (1975) 52.

204 E.D.M. Gallery, D.M. Saunders, S.N. Hunyor and A.Z. Gyory, *Med. J. Aust.*, 1 (1978) 540.

205 S.E. Tsuei, J. Thomas and R.G. Moore, *J. Chromatogr.*, 181 (1980) 135.

206 D.B. Jack and W. Riess, *J. Chromatogr.*, 88 (1974) 173.

207 P.H. Degen and W. Riess, *J. Chromatogr.*, 121 (1976) 72.

208 T. Walle, *J. Pharm. Sci.*, 63 (1974) 1885.

209 J.F.K. Huber (Editor), *Instrumentation for High-Performance Liquid Chromatography*, J. Chromatogr. Library Series, Vol. 13, Elsevier, Amsterdam, 1981.

210 L.R. Snyder and J.J. Kirkland, *Introduction to Modern Liquid Chromatography*, Wiley-Interscience, New York, 2nd Edn., 1979.

211 H.F. Walton, *Anal. Chem.*, 52 (1980) 15R.

212 M. Amin and P.W. Schneider, *Analyst (London)*, 103 (1978) 728.

213 M. Amin and P.W. Schneider, *Analyst (London)*, 103 (1978) 1076.

214 W. Roth, K. Beschke, R. Jauch, A. Zimmer and F.W. Koss, *J. Chromatogr.*, 222 (1981) 13.

215 P.J. Twitchett and A.C. Moffat, *J. Chromatogr.*, 111 (1975) 149.

216 Y. Yost, J. Stoveken and W. MacLean, *J. Chromatogr.*, 134 (1977) 73.

217 R.R. Schroeder, P.J. Kudirka and E.C. Toren, Jr., *J. Chromatogr.*, 134 (1977) 83.

218 P.M. Kabra, B.E. Stafford and L.J. Morton, *Clin. Chem.*, 23 (1977) 1284.

219 J.E. Evans, *Anal. Chem.*, 45 (1973) 2428.

220 S. Atwell, V. Green and W. Haney, *J. Pharm. Sci.*, 64 (1975) 806.

221 P.M. Kabra, G. Gotetti, R. Stanfill and L.J. Marton, *Clin. Chem.*, 22 (1976) 824.

222 P.M. Kabra and L.J. Marton, *Clin. Chem.*, 22 (1976) 1070.

223 S. Kitazawa and T. Komura, *Clin. Chim. Acta*, 73 (1976) 31.

224 G. Gauchel, F.D. Gauchel and L. Birkofer, *Z. Klin. Chem. Klin. Biochem.*, 11 (1973) 459.

225 S.J. Soldin and J.G. Hill, *Clin. Chem.*, 22 (1976) 856.

226 A.J. Glazko and A.W. Dill, in D.M. Woodbury, J.K. Penry and R.P. Schmidt (Editors), *Antieleptic Drugs*, Raven, New York, 1972, p. 193.

227 A.J. Quattrone and R.S. Putnam, *Clin. Chem.*, 27 (1981) 129.

228 P.J. Twitchett and A.C. Moffat, *J. Chromatogr.*, 111 (1975) 149.

229 P. Jatlow, *Arch. Pathol. Lab. Med.*, 104 (1980) 341.

230 G.R. Clark and J. Chan, *Anal. Chem.*, 50 (1978) 635.

231 H.F. Proelss, H.J. Lohman and D.G. Miles, *Clin. Chem.*, 24 (1978) 1948.

232 J.C. Kraak and P. Bijster, *J. Chromatogr.*, 143 (1977) 499.

233 F.L. Vandemark, R.F. Adams and G.J. Schmidt, *Clin. Chem.*, 24 (1978) 87.

234 M.R. Detaevernier, L. Dryon and D.L. Massart, *J. Chromatogr.*, 128 (1976) 204.

235 J. Godbillon and S. Gauron, *J. Chromatogr.*, 204 (1981) 303.

236 P.O. Lagerstrom, I. Carlsson and B.A. Persson, *Acta Pharm. Suecica*, 13 (1976) 157.

237 B. Mellström and S. Eksborg, *J. Chromatogr.*, 116 (1976) 475.

238 B. Mellström and G. Tybring, *J. Chromatogr.*, 143 (1977) 597.

239 H.J. Lohmann, H.F. Proelss and D.G. Miles, *Clin. Chem.*, 24 (1978) 1006.

240 B.-A. Persson and P.-O. Lagerström, *J. Chromatogr.*, 122 (1976) 305.

241 H.G.M. Westenberg, B.F.H. Drenth, R.A. de Zeeuw, H. de Cuyper, H.M. van Praag and J. Korf, *J. Chromatogr.*, 142 (1977) 725.

242 J.H.M. van den Berg, H.J.J.M. de Ruwe, R.S. Deelder and T.A. Plomp, *J. Chromatogr.*, 138 (1977) 431.

243 T.A. Sutfin and W.J. Jusko, *J. Pharm. Sci.*, 68 (1979) 703.

244 A. Bonora and P.A. Borea, *Experientia*, 34 (1978) 1486.

245 I.D. Watson and M.J. Stewart, *J. Chromatogr.*, 134 (1977) 182.

246 R.A. de Zeeuw and H.G.M. Westenberg, *J. Anal. Toxicol.*, 2 (1978) 229.

247 H.J. Kuss and M. Nathmann, *Arzeim.-Forsch.*, 28 (1978) 1301.

248 D.R.A. Uges and P. Bouma, *Pharm. Weekbl. Sci. Ed.*, 1 (1979) 417.

249 F.L. Vandemark, R.F. Adams, G.J. Schmidt and W. Salvin, *Clin. Chem.*, 23 (1977) 1139.

250 P.B. Bondo, J.J. Thoma and G.A. Beltz, *Clin. Chem.*, 25 (1979) 1118.

251 P.A. Reece, R. Zacest and C.G. Barrow, *J. Chromatogr.*, 163 (1979) 310.

252 G. Sivorinovsky, *Clin. Chem.*, 25 (1979) 1144.

253 L.P. Hackett and L.J. Dusci, *Clin. Toxicol.*, 15 (1979) 55.

254 M.A. Peat and L. Kopjak, *J. Forensic Sci.*, 24 (1979) 46.

255 K. Harzer and R. Barchet, *J. Chromatogr.*, 132 (1977) 83.

256 T.B. Vree, B. Lenselink, E. van der Kleijn and G.M.M. Nijhuis, *J. Chromatogr.*, 143 (1977) 530.

257 T.B. Vree, A.M. Baars, Y.A. Hekster, E. van der Kleijn and W.J. O'Reilly, *J. Chromatogr.*, 162 (1979) 605.

258 R.R. Brodie, L.F. Chasseaud and T. Taylor, *J. Chromatogr.*, 150 (1978) 361.

259 P.M. Kabra, G.L. Stevens and L.J. Marton, *J. Chromatogr.*, 150 (1978) 355.

260 U.R. Tjaden, M.T.H.A. Meeles, C.P. Thys and M. van der Kaay, *J. Chromatogr.*, 181 (1980) 227.

261 S. Cotler, C.V. Puglisi and J.H. Gustafson, *J. Chromatogr.*, 222 (1981) 95.

262 R.J. Perchalski and B.J. Wilder, *Anal. Chem.*, 50 (1978) 554.

263 V. Ascalone, *J. Chromatogr.*, 181 (1980) 141.

264 N. Strojny, C.V. Puglisi and J.A.F. de Silva, *Anal. Lett.*, B11 (2) (1978) 135.

265 M.A. Peat, B.S. Finkle and M.E. Deyman, *J. Pharm. Sci.*, 68 (1979) 1467.

266 N. Strojny, K. Bratin, M.A. Brooks and J.A.F. de Silva, *J. Chromatogr.*, 143 (1977) 363.

267 H.B. Greizerstein and C. Wojtowicz, *Anal. Chem.*, 49 (1977) 2235.

268 J.B. Zagar, F.J. Vanlenten and G.P. Chrekian, *J. Ass. Offic. Anal. Chem.*, 61 (1978) 678.

269 C.G. Scott and P. Bommer, *J. Chromatogr.*, 8 (1970) 446.

270 D.J. Weber, *J. Pharm. Sci.*, 61 (1972) 1797.

271 B. Moore, G. Nickless, C. Hallet and A.G. Howard, *J. Chromatogr.*, 137 (1977) 215.

272 T.B. Vree, B. Lenselink, E. van der Kleijn and G.M.M. Nijhuis, *J. Chromatogr.*, 143 (1977) 530.

273 J.D. Wittwer, *J. Liquid Chromatogr.*, 3 (1980) 1713.

274 J. Koch-Messer and S.W. Klein, *J. Amer. Med. Ass.*, 215 (1971) 1454.

275 F.E. Barnhardt, in I. Sunshine (Editor), *Methodology for Analytical Toxicology*, CRC Press, Cleveland, OH, 1975.

276 R.F. Adams, F.L. Vandemark and G. Schmidt, *Clin. Chim. Acta*, 69 (1976) 515.

277 D.J. Smith, *J. Chromatogr. Sci.*, 19 (1981) 253.

278 E.S. Vesell and G.T. Passananti, *Clin. Chem.*, 17 (1971) 851.

279 M. Esler, A. Zweifler, O. Rindall and V. DeQuattro, *Clin. Pharmacol. Ther.*, 22 (1977) 299.

280 D.M. Kornhausen, A.J.J. Wood, R.E. Vestal, G.R. Wilkinson, R.A. Branch and D.G. Shend, *Clin. Pharmacol. Ther.*, 23 (1978) 165.

281 G. Nygard, W.H. Shelver and S.K.W. Khalil, *J. Pharm. Sci.*, 68 (1979) 379.

282 R.L. Nation, G.W. Peng and W.L. Chiou, *J. Chromatogr.*, 145 (1978) 429.

283 A.R. Castenada and M.B. Bacaner, *Amer. J. Cardiol.*, 23 (1969) 107.

284 C.M. Lai, Z.M. Look, P.K. Lai and A. Yacobi, *J. Liquid Chromatogr.*, 3 (1980) 93.

285 C.D. Johnson and J.P. Revill, *Acta Pharmacol. Toxicol.*, 22 (1965) 112.

286 W.G. Duncombe and A. McCoulrey, *Brit. J. Pharmacol.*, 15 (1960) 260.

287 S.L. Tompsett, W. Forshall and D.C. Smith, *Acta Pharmacol. Toxicol.*, 18 (1961) 75.

288 R. Kuntzman, I. Tsai, R. Chang and A.H. Conney, *Clin. Pharmacol. Ther.*, 11 (1970) 829.

289 J. Koch-Weser, *Arch. Int. Med.*, 129 (1972) 763.

290 N.J. Pound and R.W. Sears, *Can. J. Pharm. Sci.*, 10 (1975) 122.

291 E. Smith, S. Barkan, B. Ross, M. Maienthal and J. Levine, *J. Pharm. Sci.*, 62 (1973) 1151.

292 F.I. Carroll, A. Philip and M.C. Coleman, *Tetrahedron Lett.*, (1976) 1757.
293 J.L. Valentine, P. Driscoll, E.L. Hamburg and E.D. Thompson, *J. Pharm. Sci.*, 65 (1976) 96.
294 K.K. Midha and C. Charette, *J. Pharm. Sci.*, 63 (1974) 1244.
295 S. Sved, I.J. McGilveray and N. Beaudoin, *J. Chromatogr.*, 145 (1978) 437.
296 J. Armand and A. Badinand, *Ann. Biol. Clin.*, 30 (1972) 599.
297 Y.-G. Yee and R.E. Kates, *J. Chromatogr.*, 223 (1981) 454.
298 B.R. Patel, J.J. Kirschbaum and R.B. Poet, *J. Pharm. Sci.*, 70 (1981) 336.
299 O.H. Weddle, E.N. Anrick and W.D. Mason, *J. Pharm. Sci.*, 67 (1978) 1033.
300 I. Djerassi, S. Farber, E. Abir and W. Neikirk, *Cancer*, 20 (1967) 233.
301 N. Jaffe, E. Frei and D. Traggist, *N. Engl. J. Med.*, 291 (1974) 994.
302 W.A. Bleyer, *Cancer*, 41 (1978) 36.
303 B.A. Chabner, C.E. Myers, C.N. Coleman and D.G. Johns, *N. Engl. J. Med.*, 292 (1975) 1107.
304 D.G. Johns and J.R. Bertino, in J.F. Holland and E. Frei (Editors), *Cancer Medicine*, Lea and Febiger, Philadelphia, PA, 1973, p. 739.
305 J.L. Cohen, G.H. Hisayasu, A.R. Barrientos, M.S.B. Nayar and K.K. Chan, *J. Chromatogr.*, 181 (1980) 478.
306 W.C. Werkheiser, S.F. Zakrzewski and C.A. Nichol, *J. Pharmacol. Exp. Ther.*, 137 (1972) 162.
307 Y.M. Wang, E. Lantin and W.W. Sutow, *Clin. Chem.*, 22 (1976) 1053.
308 C.E. Myers, M.E. Lippman, H.M. Elliot and B.A. Chabner, *Proc. Nat. Acad. Sci. U.S.*, 72 (1975) 3683.
309 V. Raso and R. Schreiber, *Cancer Res.*, 35 (1975) 1407.
310 C. Bohoun, F. Dubrey and C. Bondene, *Clin. Chim. Acta*, 57 (1974) 263.
311 S.G. Chakrabarti and J.A. Bernstein, *Clin. Chem.*, 15 (1969) 1157.
312 M.M. Kincade, W.R. Vogler and P.G. Dayton, *Biochem. Med.*, 10 (1974) 337.
313 J.A. Nelson, B.A. Harris, W.J. Decker and D. Farquhar, *Cancer Res.*, 37 (1977) 3970.
314 E. Watson, J.L. Cohen and K.K. Chan, *Cancer Treat. Rep.*, 62 (1978) 381.
315 W.H. Isacoff, P.F. Morrison, J. Aroesty, K.L. Willis, J.B. Block and T.L. Lincoln, *Cancer Treat. Rep.*, 61 (1977) 1655.
316 D.L. Azarnoff, S.H. Wan and D.H. Huffman, *Clin. Pharmacol. Ther.*, 16 (1974) 884.
317 S.A. Jacobs, R.G. Stoller, B.A. Chabner and D.G. Johns, *Cancer Treat. Rep.*, 61 (1977) 651.
318 J.A. Nelson, B.A. Harris, W.J. Decker and D. Farguha, *Cancer Res.*, 37 (1977) 3970.
319 J. Lankelma, E. van der Kleijn and F. Ramaekers, *Cancer Lett.*, 9 (1980) 133.
320 G.J. Lawson, P.F. Dixon and G.W. Aherne, *J. Chromatogr.*, 223 (1981) 225.
321 R.P. Gale, *N. Engl. J. Med.*, 300 (1979) 1189.
322 D.H.W. Ho and E. Frei, *Clin. Pharmacol. Ther.*, 12 (1971) 944.
323 S.H. Wan, D.H. Huffman, D.L. Azarnoff, B. Hoogstraten and W.E. Larsen, *Cancer Res.*, 34 (1974) 392.
324 R.L. Momparler, A. Labitan and M. Rossi, *Cancer Res.*, 32 (1972) 408.
325 A.L. Harris, C. Potter, C. Bunch, J. Boutagy, D.J. Harvey and D.G. Garham-Smith, *Brit. J. Clin. Pharmacol.*, 8 (1979) 219.
326 E.M. Piall, G.W. Aherne and V.M. Marks, *Brit. J. Cancer*, 40 (1979) 548.
327 J. Boutagy and D.J. Harvey, *J. Chromatogr.*, 146 (1978) 283.
328 Y.M. Rustum, *Anal. Biochem.*, 90 (1978) 289.
329 R.W. Bury and P.J. Keary, *J. Chromatogr.*, 146 (1978) 350.
330 M.G. Pallavicini and J.A. Mazrimas, *J. Chromatogr.*, 183 (1980) 449.
331 P. Linssen, A. Drenthe-Schonk, H. Wessels and C. Haanen, *J. Chromatogr.*, 223 (1981) 371.
332 R.B. Livingston and S.K. Carter, *Single Agents in Cancer Chemotherapy*, Plenum, New York, 1970, p. 99.
333 B. Clarkson, in A.C. Sartorelli and D.G. Johns (Editors), *Antineoplastic and Immunosuppressive Agents*, Vol. I, Springer, New York, 1974, p. 172.
334 J.T. Goras, J.B. Knight, R.H. Iwamoto and P. Lim, *J. Pharm. Sci.*, 59 (1970) 561.
335 S.Y. Chang, D.S. Alberts, L.R. Melnick, P.D. Watsson and S.E. Salmon, *J. Pharm. Sci.*, 67 (1978) 679.

336 S.Y. Chang, T.L. Evans, D.S. Alberts and I.G. Sipes, *Life Sci.*, 23 (1978) 1697.

337 R.L. Furner, L.B. Mellet, R.K. Brown and G. Duncan, *Drug Metab. Dis.*, 4 (1976) 577.

338 A.E. Ahmed and T.-F. Hsu, *J. Chromatogr.*, 222 (1981) 453.

339 S.E. Fandrich and K.A. Pittman, *J. Chromatogr.*, 223 (1981) 155.

340 N. Strojny, L. D'Arconte and J.A.F. de Silva, *J. Chromatogr.*, 223 (1981) 111.

341 Y.H. Mohammed and F.F. Cantwell, *Anal. Chem.*, 50 (1978) 491.

342 N. Muhammad and J.A. Bodnar, *J. Liquid Chromatogr.*, 3 (1980) 113.

343 D.L. Massart and M.R. Detaevernier, *J. Chromatogr. Sci.*, 18 (1980) 139.

344 A.C. Moffat and K.W. Smalldon, *J. Chromatogr.*, 90 (1974) 9.

345 A.C. Moffat and B. Clare, *J. Pharm. Pharmacol.*, 26 (1974) 665.

346 P.P. Ascione and G.P. Chrekian, *J. Pharm. Sci.*, 64 (1975) 1029.

347 G.W. Peng, M.A.F. Gadalla, V. Smith, A. Peng and W.L. Chiou, *J. Pharm. Sci.*, 67 (1978) 710.

348 H. Bundgaard, *Arch. Pharm. Chem. Sci. Ed.*, 4 (1976) 103.

349 S.O. Jansson and I. Andersson, *Acta Pharm. Suecica*, 14 (1977) 161.

350 R.L. Stevenson and C.A. Burtis, *J. Chromatogr.*, 61 (1971) 253.

351 R.G. Baum and F.F. Cantwell, *Anal. Chem.*, 50 (1978) 280.

352 R.G. Baum and F.F. Cantwell, *J. Pharm. Sci.*, 67 (1978) 710.

353 J.F. Reedmeyer and R.D. Kirchhoefer, *J. Pharm. Sci.*, 68 (1979) 1167.

354 V.Y. Taguchi, M.L. Cotton, C.H. Yates and J.F. Millar, *J. Pharm. Sci.*, 70 (1981) 64.

355 I.L. Honigberg, J.T. Stewart and M. Smith, *J. Pharm. Sci.*, 67 (1978) 675.

356 L.R. Snyder, *J. Chromatogr.*, 92 (1974) 223.

357 M.L. Eppel, J.S. Oliver, H. Smith, A. Mackay and L.E. Ramsay, *Analyst (London)*, 103 (1978) 1061.

358 V.P. Shah and S. Riegelman, *J. Pharm. Sci.*, 63 (1974) 1283.

359 G.F. Johnson, W.A. Dechtiaruk and H.M. Solomon, *Clin. Chem.*, 21 (1975) 144.

360 C.J. Least, Jr., G.F. Johnson and H.M. Solomon, *Clin. Chem.*, 22 (1976) 765.

361 R.F. Adams, F.L. Vandemark and G.J. Schmidt, *Clin. Chem.*, 22 (1976) 1903.

362 L.C. Franconi, G.L. Hawk, B.J. Sandmann and W.G. Haney, *Anal. Chem.*, 48 (1976) 372.

363 M.A. Peat and T.A. Jennison, *J. Anal. Toxicol.*, 1 (1977) 204.

364 T.C. Kram, *J. Pharm. Sci.*, 71 (1972) 254.

365 R.B. Poet and H.H. Pu, *J. Pharm. Sci.*, 72 (1973) 809.

366 S.C. Su, A.V. Hartkopf and B.L. Karger, *J. Chromatogr.*, 119 (1976) 523.

367 Waters Associates, Inc., *Antibiotics—Sulfa Drugs*, Application Rept. No. 10, Waters Assoc., Milford, MA, 1979.

368 R.L. Suber and G.I. Edds, *J. Liquid Chromatogr.*, 3 (1980) 257.

369 A.B. Vilim, L. Larocque and A.I. Macintosh, *J. Liquid Chromatogr.*, 3 (1980) 1725.

370 D.P. Goodspead, R.M. Simpson, R.B. Ashwort, J.W. Shafer and H.R. Cook, *J. Ass. Offic. Anal. Chem.*, 61 (1978) 1050.

371 K.L. Johnson, D.T. Jeter and R.C. Claiborne, *J. Pharm. Sci.*, 64 (1975) 1657.

372 A.C. Bratton and E.K. Marshall, *J. Biol. Chem.*, 128 (1939) 537.

373 A.P. De Leenheer, R.R. Roncucci and C. van Peteghem (Editors), *Quantitative Mass Spectrometry in Life Sciences II*, Elsevier, Amsterdam, 1978.

374 A. Frigerio and M. McCamish (Editors), *Recent Development in Mass Spectrometry in Biochemistry and Medicine, 6,* Elsevier, Amsterdam, 1980.

375 B.J. Millard, *Quantitative Mass Spectrometry*, Heyden, London, 1978.

376 M.C. ten Noever de Brauw, *J. Chromatogr.*, 165 (1979) 207.

377 A.L. Burlingame, T.A. Baillie, P.J. Derrick and O.S. Chizhov, *Anal. Chem.*, 52 (1980) 220R.

378 R.M. Smith, *Amer. Lab.*, 10 (1978) 53.

379 E.C. Horning, D.I. Carroll, I. Dzidic, R.N. Stillwell and J.P. Thenot, *J. Ass. Offic. Anal. Chem.*, 61 (1978) 1232.

380 A. Tatematsu, H. Yoshizumi, T. Nadai, T. Kubodera and S. Asai, *Biomed. Mass Spectrom.*, 5 (1978) 192.

381 L.F. Gram, *Clin. Pharmacokinet.*, 2 (1977) 237.

382 A. Frigerio, G. Belvedere, F. De Nadai, R. Fanelli, C. Pantarotto, E. Riva and P.L. Morselli, *J. Chromatogr.*, 74 (1972) 201.

383 G. Belvedere, L. Burti, A. Frigerio and C. Pantarotto, *J. Chromatogr.*, 111 (1975) 313.
384 J.L. Biggs, W.H. Holland, S.S. Chang, P.P. Hipps and W.R. Sherman, *J. Pharm. Sci.*, 65 (1976) 261.
385 J.P. Dubois, W. Kung, W. Theobald and B. Wirz, *Clin. Chem.*, 22 (1976) 892.
386 M. Claeys, G. Muscettola and S.P. Markey, *Biomed. Mass Spectrom.*, 3 (1976) 110.
387 G. Alfredsson, F.A. Wiesel, B. Fryo and G. Sedvall, *Psychopharmacology*, 52 (1977) 25.
388 D. Alkalay, J. Volk and S. Carlsen, *Biomed. Mass Spectrom.*, 6 (1979) 200.
389 W.A. Garland, R.R. Muccino, B.H. Min, J. Cupano and W.E. Fann, *Clin. Pharmacol. Ther.*, 25 (1979) 844.
390 C.G., Hammar, B. Alexanderson, B. Holmstedt and F. Sjöquist, *Clin. Pharmacol. Ther.*, 12 (1971) 496.
391 J.M. Wilson, L.J. Williamson and V.A. Raisys, *Clin. Chem.*, 23 (1977) 1012.
392 V.E. Zeigler and J.R. Taylor, *Clin. Pharmacol. Ther.*, 19 (1976) 795.
393 V.E. Zeigler, P.J. Clayton and J.R. Taylor, *Clin. Pharmacol. Ther.*, 20 (1976) 458.
394 H.A. Heck, N.W. Flynn, S.E. Buttrill, Jr., R.L. Dyer and M. Anbar, *Biomed. Mass Spectrom.*, 5 (1978) 250.
395 M. Wretlind, A. Pilbrant, A. Sundwall and J. Vessman, *Acta Pharmacol. Toxicol.*, 40 (Suppl. 1) (1977) 28.
396 F. Marcucci, R. Bianchi, L. Airoldi, M. Salmona, R. Fanelli, C. Chiabrando, A. Frigerio, E. Mussini and S. Garattini, *J. Chromatogr.*, 107 (1975) 285.
397 W. Sadee and E. van der Kleijn, *J. Pharm. Sci.*, 60 (1971) 135.
398 J. Lanzoni, L. Airoldi, F. Marcucci and E. Mussini, *J. Chromatogr.*, 168 (1979) 260.
399 R.T. Schillings, S.R. Shrader and H.W. Ruelius, *Arzneim.-Forsch.*, 21 (1971) 1059.
400 B.H. Min and W.A. Garland, *J. Chromatogr.*, 139 (1977) 121.
401 W.A. Garland and B.H. Min, *J. Chromatogr.*, 172 (1979) 279.
402 A.J. Clatworthy, L.V. Jones and M.J. Whitehouse, *Biomed. Mass Spectrom.*, 4 (1977) 248.
403 J. Grove, *J. Chromatogr.*, 59 (1971) 289.
404 L.F. Prescott, *J. Pharm. Pharmacol.*, 23 (1971) 111.
405 B.H. Thomas and B.B. Coldwell, *J. Pharm. Pharmacol.*, 24 (1972) 243.
406 W.A. Garland, K.C. Hsiao, E.J. Pantuck and A.H. Conney, *J. Pharm. Sci.*, 66 (1977) 340.
407 J.D. Baty, P.R. Robinson and J. Wharton, *Biomed. Mass Spectrom.*, 3 (1976) 60.
408 G. Zweig and J. Sherma, *Anal. Chem.*, 52 (1980) 276R.
409 C. Gonnet and M. Marichy, *J. Liquid Chromatogr.*, 3 (1980) 1901.
410 A. Aszalos, *J. Liquid Chromatogr.*, 3 (1980) 867.
411 H. Auterhoff, *Thin-Layer Chromatograms and IR Spectra. Identification of Drugs*, Wissenschaftliche Verlagsgesellschaft, Stuttgart, 3rd Edn., 1977.
412 K. Macek (Editor), *Pharmaceutical Applications of Thin-Layer and Paper Chromatography*, Elsevier, Amsterdam, 1972.
413 D.L. Massart and H. de Clercq, *Anal. Chem.*, 46 (1974) 1988.
414 D.L. Massart, *J. Chromatogr.*, 79 (1973) 157.
415 D.L. Massart and R. Sruits, *Anal. Chem.*, 46 (1974) 281.
416 H. de Clercq and D.L. Massart, *J. Chromatogr.*, 115 (1975) 1.
417 A.C. Moffat, K.W. Smalldon and C. Brown, *J. Chromatogr.*, 90 (1974) 1.
418 A.C. Moffat and K.W. Smalldon, *J. Chromatogr.*, 90 (1974) 9.
419 A.C. Moffat and B. Clare, *J. Pharm. Pharmacol.*, 26 (1974) 665.
420 A.C. Moffat, *J. Chromatogr.*, 110 (1975) 341.
421 E.G.C. Clark, *Isolation and Identification of Drugs*, Pharmaceutical Press, London, 1969, p. 43.
422 I. Sunshine, *Handbook of Analytical Toxicology*, CRC, Cleveland, OH, 1969.
423 J.V. Jackson and A.J. Clatworthy, in I. Smith and J.W.T. Seakins (Editors), *Chromatographic and Electrophoretic Techniques*, Vol. 1, Heinemann, London, 1976, p. 380.
424 B. Davidow, N. Li Petri and B. Quame, *Amer. J. Clin. Pathol.*, 50 (1969) 714.
425 S.J. Mulé, *J. Chromatogr.*, 39 (1969) 302.
426 K.K. Kaistha and J.H. Jaffe, *J. Pharm. Sci.*, 61 (1972) 679.
427 J. Cochin, *Psychopharm. Bull.*, 3 (1966) 53.
428 R.A. de Zeeuw and M.T. Feitsma, *Pharm. Weekbl.*, 101 (1966) 957.

429 W.A. Rosenthal, M.M. Kaser and K.N. Milewski, *Clin. Chim. Acta*, 33 (1971) 51.
430 N.C. Jain and R.H. Cravey, *J. Chromatogr. Sci.*, 12 (1974) 228.
431 P.A.F. Pranitis and A. Stolman, *J. Chromatogr.*, 106 (1975) 485.
432 P.J. Smith and T.S. Herman, *Anal. Biochem.*, 22 (1968) 134.
433 D. Sohn, J. Simon, M.A. Hanna, G. Ghali and R. Tolba, *J. Chromatogr.*, 87 (1973) 570.
434 P. Owen, A. Pendlebury and A.C. Moffat, *J. Chromatogr.*, 161 (1978) 187.
435 J.M. Clifford and W.F. Smyth, *Analyst (London)*, 99 (1975) 241.
436 D.M. Hailey, *J. Chromatogr.*, 98 (1974) 527.
437 R.C. Gupta and J. Kofoed, *Nature (London)*, 198 (1963) 384.
438 S. Ebel, H. Holtz and H. Lerche, *Deut. Apoth.-Ztg.*, 108 (1968) 779.
439 M.T. Bush, *Microchem. J.*, 5 (1961) 73.
440 C. Ioannides, J. Chakraborty and D.V. Parke, *Chromatographia*, 7 (1974) 351.
441 S. Larebau and M.C. Saux, *J. Eur. Toxicol.*, 6 (1963) 139.
442 N. Weissman, M. Lowe, J.M. Beattie and J.A. Demetriou, *Clin. Chem.*, 17 (1971) 875.
443 W.A. Rosenthal, M.M. Kaser and K.N. Milewski, *Clin. Chim. Acta*, 33 (1971) 51.
444 K. Itiaba, J.C. Crawhall and C.M. Sin, *Clin. Biochem.*, 3 (1970) 287.
445 G. Kreysing and M. Frahm, *Deut. Apoth.-Ztg.*, 110 (1970) 1133.
446 S. Goenechea, *J. Chromatogr.*, 40 (1969) 182.
447 M. Melzacka and W. Kahl, *Chem. Anal. Warsaw*, 14 (1969) 453.
448 H. Amal, S. Tulus and L. Sanli, *Istanbul Univ. Eczacilik Fak. Mecm.*, 4 (1968) 23.
449 A.S. Curry and R.H. Fox, *Analyst (London)*, 93 (1968) 834.
450 G. Machata, *Mikrochim. Acta*, 1 (1960) 79.
451 G. Machata and H.J. Battista, *Chromatographia*, (1968) 104.
452 D. Sohn, *U.S. Pat.* 3,832,134; *C.A.*, 81 (1974) 164454z.
453 S. Hishida, M. Ueda, T. Tanabe, Y. Mizoi and J. Kobe, *Med. Sci., Tokyo*, 18 (1972) 1.
454 L.T. Kenison, E.L. Loveridge, J.A. Gronlund and A.A. Elmowafi, *J. Chromatogr.*, 71 (1972) 165.
455 R.A. de Zeeuw and J. Wijsbeek, *Pharm. Weekbl.*, 104 (1969) 901.
456 E. Kammerl and E. Mutschler, *Pharm. Ind.*, 35 (1973) 146.
457 R.I.H. Wang and M.A. Mueller, *J. Pharm. Sci.*, 62 (1973) 2047.
458 Y. Garceau, Y. Philopoulos and J. Hasegawa, *J. Pharm. Sci.*, 62 (1973) 2032.
459 R. Abu-Eitlah, A. Osman and A. El-Behare, *Analyst (London)*, 103 (1978) 1083.
460 K.K. Kaistha, R. Tadrus and R. Janda, *J. Chromatogr.*, 107 (1975) 359.
461 G.O. Guerrant and C.T. Hall, *Clin. Toxicol.*, 10 (1977) 209.
462 J.E. Wallace, K. Blum and J.B. Singh, *Clin. Toxicol.*, 7 (1974) 477.
463 W.J. Decker, *Clin. Toxicol.*, 10 (1977) 23.
464 S.J. Mulé, *J. Chromatogr.*, 55 (1971) 255.
465 V.P. Dole, W. Kim and I. Eglitis, *J. Amer. Med. Ass.*, 198 (1966) 349.
466 S.J. Mulé, M.L. Bastos, D. Jukofsky and E. Saffer, *J. Chromatogr.*, 63 (1972) 289.
467 J.M. Fujimoto and R.I.H. Wang, *Toxicol. Appl. Pharmacol.*, 16 (1970) 186.
468 J.E. Wallace, J.D. Biggs, J.H. Merritt, H.E. Hamilton and K. Blum, *J. Chromatogr.*, 71 (1972) 135.
469 K.K. Kaistha, *J. Pharm. Sci.*, 61 (1972) 655.
470 K.K. Kaistha and J.H. Jaffe, *J. Chromatogr.*, 60 (1971) 83.
471 K.K. Kaistha and J.H. Jaffe, *J. Pharm. Sci.*, 61 (1972) 679.
472 J. Anderson, *Practitioner*, 212 (1934) 536.
473 D. Ladewig and R. Battegay, *Int. J. Addict.*, 6 (1971) 167.
474 A. Levin, *Postgrad. Med. J.*, 51, Suppl. 1 (1975) 186.
475 I. Oswald, *Brit. Med. J.*, 3 (1971) 67.
476 F.A. Whitlock and M.I. Nadorfi, *Med. J. Aust.*, 2 (1970) 1097.
477 C.L. Winek, J.J. Kuhlman, Jr. and S.P. Shanor, *Clin. Toxicol.*, 17 (1980) 337.
478 L.B. Rubia and R. Gomez, *J. Pharm. Sci.*, 66 (1977) 1656.
479 J. Christiansen, *J. Chromatogr.*, 123 (1976) 57.
480 B. Wesley-Hadzija and A.M. Mattocks, *J. Chromatogr.*, 143 (1977) 307.
481 J.M. Steyn, *J. Chromatogr.*, 120 (1976) 465.

482 M.B. Abou-Donia, N.M. Bakry and H.C. Strauss, *J. Chromatogr.*, 172 (1979) 463.
483 L. Lepri, P.G. Desideri and D. Heimler, *J. Chromatogr.*, 153 (1978) 77.
484 L. Lepri, P.G. Desideri and D. Heimler, *J. Chromatogr.*, 155 (1978) 119.
485 L. Lepri, P.G. Desideri and D. Heimler, *J. Chromatogr.*, 169 (1979) 271.
486 K.Y. Lee, D. Nurok, A. Zlatkis and A. Karmen, *J. Chromatogr.*, 158 (1978) 403.
487 C. Gonnet and M. Marichy, *J. Liquid Chromatogr.*, 3 (1980) 1901.

Chapter 17

Antibiotics

G.H. WAGMAN and M.J. WEINSTEIN

CONTENTS

17.1. INTRODUCTION

The era which began in the 1940s with the microbiological production and use of penicillin as a chemotherapeutic agent has seen a phenomenal increase in recent years in the production of these microbial metabolites known as antibiotics. Although a number of definitions exist for the term antibiotic, probably the most widely used concept is that of a substance produced by a microorganism which is capable in small concentrations of inhibiting the growth of microorganisms. Antibiotic-producing organisms are widely distributed in nature and their antibiotics differ markedly chemically, in their modes of action and in their spectra of activity. Thus, there is little relationship between the antibiotics per se, although a number of groups of related compounds exist.

In order to isolate and purify these substances, numerous methods for classification and identification of antibiotics have been devised, based mainly on the use of a variety of chromatographic techniques. Initially, simple paper-chromatographic techniques were sufficient, because of the relatively small number of antibiotics known. Today, thousands of antibiotics have already been discovered and the number is increasing due to man's ability to create new, more effective agents from the natural antibiotics by chemical or microbiological manipulations. In the preface to a recent book, Perlman [1] estimated that more than 50,000 semisynthetic antibiotics have been prepared in the last 30 years, most of them in the last eighteen years; the record suggests that only very few of these compounds find a niche in clinical medicine. So, the search continues for natural isolates that can be used in their original form or derivatized to obtain compounds with improved activity, reduced toxicity, or both of these characteristics.

Fortunately, antibiotics can often be identified directly from fermentation broths, and minute quantities may be sufficient for classification and identification by means of chromatographic techniques, unlike many other substances that often require laborious isolation and purification. Many of the recently described HPLC methods are suitable for the isolation of small quantities of the pure antibiotics. Chromatography is an important aid in "Murphy's Law" of antibiotics, i.e., in a complex mixture, the antibiotic produced in the smallest quantity is usually the most interesting one. This chapter will not delve into preparative chromatography, but only cover methodology useful for identification and differentiation of antibiotics.

17.2. CHROMATOGRAPHIC IDENTIFICATION AND CLASSIFICATION

Chromatographic identification and classification of antibiotics can be carried out by a number of techniques. Until recently, the most frequently used method was PC (ascending, descending, circular, centrifugal, one- or two-dimensional PC, chromatography on paper impregnated with buffers [2], ion-exchange paper [3], etc.). However, because it yields results easily and rapidly, TLC, mainly on silica gel or alumina, has made considerable inroads [4–9]. The sorbent can be pretreated, e.g., Nishimoto et al. [10] treated silica gel plates with disodium EDTA for a separation of tetracyclines based on chelating ability. Plates can be coated with sorbent layers such as cellulose [4,11], Sephadex [12] and activated carbon [13].

Electrophoresis [14–17] and countercurrent distribution (CCD) [18–20] can also be used for identification of many antibiotics. Separation and identification of antibiotics by GLC [21–24] has largely been replaced by HPLC in recent years. The HPLC method has the major advantages of rapid and efficient separation without destruction of components, but it also has some disadvantages:

(a) HPLC often requires a specialized system for a specific antibiotic,

(b) unlike PC or TLC, where multiple samples can be chromatographed simultaneously side-by-side, HPLC allows for the separation of only one mixture at the time.

Systematic classification and identification of antibiotics has been devised for but a small portion of the thousands of compounds known. Issaq et al. [25] classified 151 antibiotics exhibiting antitumor properties by the use of TLC and bioautography; Aszalos and coworkers [26,27] used "Instant TLC" (cf. Chap. 5.5.1) for classification of 91 antibiotics; and Aszalos and Issaq [28] have discussed TLC systems for the classification and identification of antibiotics. There are two specific reference books for chromatography of antibiotics [29,30], and several tables of chromatographic data can be found in the *Handbook of Chromatography* [31]. Data on instrumentation for TLC can be found in a comprehensive review by Lott et al. [32].

Generally, detection methods for antibiotics have remained unchanged over the years. They usually comprise bioautography with susceptible organisms, seeded on agar, or visualization with spray reagents and UV light. The agar technique is most useful. The dried papergram, TLC plate, or electropherogram is placed on agar seeded with the appropriate organism and exposed for a given time period. As the plate is incubated, the organisms will grow out as a film with clear areas representing the zones on the chromatogram where antibiotic is present. R_F values are determined from these inhibition zones. One new technique for enhancement of bioautography of *Trichomonas*, in which phenolphthalein monophosphate is used, was described by Meyers and Chang [33]. Methodology for specific antibiotics is generally described in the appropriate references to follow. A section on detection of antibiotics on chromatograms can be found in ref. 30 and in the reviews in two previous editions of this book [34,35].

17.3. SPECIFIC METHODS

This section is divided according to the groups of antibiotics into the following subsections: aminoglycosides, macrolides, penicillins, cephalosporins, tetracyclines, polypeptides, polyenic antifungals, and miscellaneous compounds. It is obviously impossible to cover this enormous body of information in one chapter, but the most recent chromatographic methods for some of the more significant antibiotics are described, as well as older, basic methods that are still useful.

17.3.1. Aminoglycosides

The aminoglycoside group of antibiotics, which is in reality the aminocyclitol group, comprises several major families: gentamicins, neomycins, kanamycins, and streptomycins, as well as numerous lesser known compounds. Many derivatives have been produced chemically or by the biosynthetic route. This section covers only the chromatography of the more important or recently discovered natural and semi-synthetic compounds, along with some newer techniques for characterizing the older antibiotics.

Numerous methods have been published for separation and identification of the gentamicin components. The most generally utilized chromatographic solvent is the lower layer of a mixture of chloroform–methanol–17% aqueous ammonia (2:1:1) to separate the major components consisting of C_1, C_2, and C_{1a} by PC [36,37] or TLC [38,39]. The PC methods are also used as quantitative techniques for measuring the proportions of the gentamicins in a mixture [40,41]. Other, related PC solvents utilize variants of the chloroform–methanol–ammonia system [37,42] or 2-butanone–*t*-butanol–methanol–conc. ammonium hydroxide (16:3:1:6) [37]. The system described in ref. 42 is also useful for separating other, related antibiotics, e.g., minor gentamicin components, gentamines, and sisomicin. A number of TLC systems are based on the use of silica gel with similar solvent systems [38,39,43,44]. The lower phase of chloroform–methanol–conc. ammonium hydroxide (1:1:1), used in ascending development of ChromAR Sheet 500, separated the gentamicin components A, A_1, B, B_1, and X [41]. They were also separable by use of two solvents and TLC media, as noted in ref. 42. Gentamicin C_1 migrates in paper electrophoresis

[45,46], and the C components can be separated by CCD [47] by use of the chloroform–methanol–ammonia solvent system. The gentamicin C components have been differentiated by two HPLC methods [48,49]. Amikacin has been identified by use of TLC on silica gel [50], as have fortimicins A and B [51]. The fortimicins have also been separated on PC with chloroform–methanol–17% ammonia [51], and fortimicin A has been characterized on the basis of four solvent systems [39].

The kanamycins have been resolved by PC [39,52–55], TLC [39,54], electrophoresis [56], GLC [57], and HPLC [58]. Derivatives of kanamycin and biosynthetic products were separated by PC [59,60] and TLC [60–63]. Numerous PC systems have been applied to neomycin and its components and N-acetyl derivatives [39,60,64–75]. Suitable TLC plates are prepared from silica gel [75–79], carbon [80], alumina [81,82], cellulose MN-300 [76,83], and kieselguhr [76]. Electrophoresis of neomycins was carried out on cellulose acetate [84,85], Whatman No. 3 MM paper [46] and acrylamide gel [86]. GC separations were also reported [87,88].

Sisomicin, a more recently introduced antibiotic, has been differentiated from the gentamicins by PC [37, 89–91]. Three solvent systems have been used on silica gel plates to identify sisomicin [51]. It has been detected in HPLC by means of post-column derivatization with o-phthalaldehyde to form fluorescent products [48]. Streptomycin was differentiated by PC from mannisidostreptomycin, dihydroxy-streptomycin, and hydroxystreptomycin [92,93] and by TLC from dihydrostreptomycin and dihydrodesoxystreptomycin [54,94–96]. Electrophoresis has also been performed on Whatman No. 4 paper with six different buffers at 350 V for 1–4 h [56]. Tobramycin, originally identified as nebramycin Factor 6, has been characterized by PC with several solvents [39,51] and by TLC on silica gel with three solvent systems [39].

The aminoglycosides G-418 and G-52 were identified by PC with 2-butanone–t-butanol–methanol–conc. ammonium hydroxide (16:3:1:6) [97]; G-52 and verdamicin were also characterized by means of PC with six solvents [92]. Paromomycins or their derivatives have been chromatographed on a variety of PC systems [92–99] as well as on thin layers of alumina, cellulose [98,99], and silica gel [100,101]. The N-acetyl derivatives of paromomycins I and IIc were characterized by PC with 1-butanol–pyridine–water (6:4:3) [60]. Butirosins were identified by several PC systems [39,102,103] as well as by TLC [39,104]. Butirosin derivatives were differentiated by TLC with five solvent mixtures [105]. Netilmicin, a new semi-synthetic aminoglycoside, was characterized by use of HPLC [48,106]. Two PC methods were used to identify ribostamycin [51,74]; an additional system was used to separate a series of semi-synthetic derivatives [107]. Several TLC methods, including silica gel and cellulose as the migration medium, were also described [51,54,108].

17.3.2. Macrolides

Of the numerous macrolide antibiotics, a selected group has been chosen for discussion. Many other, similar compounds can be separated by the same techniques.

Erythromycin and carbomycin (magnamycin) [109,110] and erythromycin components and their derivatives have been separated by PC [111–113] and by TLC [114–119]. The order of migration in paper electrophoresis with borate buffer was B > A > C [114]. The B component was isolated by CCD [111], and the A and B components were separated by HPLC [120]. Several erythromycins were chromato-graphed by HPLC [121]. Carbomycin was also characterized in a variety of PC and TLC systems [122–127] and by CCD [128].

Megalomicin components A, B, C_1, and C_2 were separated on a thin layer of Silica Gel G with chloroform–methanol (3:2) [129,130] and on aluminum oxide type E (Merck) with ethyl acetate–methanol (9:1) [131]. The latter system also characterized 4′-propionyl megalomicin A. Josamycin was differentiated from its mono- and diacetyl derivatives by TLC on alumina with butyl acetate–methyl ethyl ketone–water (40:9:1) [132]. The junvenimicin group of antibiotics can be resolved by TLC on Spotfilm F or silica gel with a mobile phase of chloroform–methanol–17% aqueous ammonia (10:3:5) [133]; members of the "A" group of components are differentiated on aluminum oxide with the same solvent mixture [133].

Several PC systems were used for the characterization of oleandomycin, including Whatman No. 1 paper [134] and Toyo No. 51 paper [135] with several solvent systems. In TLC systems either bioautography [136,137] or color reagents [138,139] were used for detection. HPLC was successful in separating ten leucomycins with retention times spanning 1.5 to 25.8 min [140].

Niddamycin and desisovaleryl niddamycin were differentiated on silica gel with two solvent systems [141]. A number of solvent mixtures have been used to characterize rosaramicin by PC on Toyo No. 51 paper [127] as well as by TLC on silica gel [142,143]. The quantitative determination of rosaramicin in serum by HPLC was reported by Lin et al. [144]. The mixtures of chloroform–methanol–17% ammonia and butanol–acetic acid–water (3:1:1), in combination with detection by heating the plate at 100°C and spraying with phosphoric acid–methanol (1:1), are useful for differentiating a variety of macrolide antibiotics, including the spiramy-cins, by TLC [142]. Maridomycins were separated by silica gel TLC [145–148] and by PC on paraffin-impregnated paper [145].

17.3.3. Penicillins

The number of semi-synthetic penicillins created in the last few years is literally astronomical. Numerous PC and TLC methods have been devised for separating many of these antibiotics. Recently, HPLC has also become a major technique for the qualitative and quantitative analysis of some of these compounds. A review of some of the more recent procedures follows.

A series of phosphiniminobenzyl penicillins has been separated by PC with butanol–ethanol–water and butanol–acetic acid–water [149]. Benzenesulfonic acid salts of 6-APA were chromatographed with butanol–ethanol–water (1:1:1) and the papergram was sprayed with phenylacetyl chloride prior to detection [150]. Butanol–pyridine–water (1:1:1) was used with Whatman No. 1 paper to distinguish various butyl penicillins [151]. Penicillins have been resolved by use of four solvent systems and detected by bioautography with *Staphylococcus aureus* [54]. TLC on silica gel with different solvent mixtures has been utilized to characterize a variety of penicillins [152–159]. Reversed-phase TLC [160] separates methicillin, penicillin, phenoxymethylpenicillin, phenethicillin, propicillin, oxacillin, doxacillin, and di-doxacillin. Many members of this group can also be characterized by TLC on silica gel with four solvent systems [161]. HPLC was used to characterize 6-APA and penicillins G and V [162] and to differentiate ampicillin from penicillins G and V [58,163]. A number of column packings and mobile phases have been tested for the HPLC separation of some of the penicillins [164].

17.3.4. Cephalosporins

The number of semi-synthetic cephalosporins now in existence is huge, and new ones are being created continuously. Thus, it is virtually impossible to note but a small number of chromatographic systems in this chapter. However, many of these will be found useful for a variety of cephalosporins not specifically mentioned.

Cephalosporins have been separated by chromatography with 1-butanol–acetic acid–water mixtures [165–168], and 1-butanol–ethanol–water mixtures [169–171] in various ratios and with a variety of other solvents on paper with and without buffers [165,169–178]. A series of cephalosporin esters was chromatographed on silica gel plates with nine solvent systems [179]. Zones were detected by ninhydrin in ethanol–2,4,6-collidine–acetic acid, followed by heating. Cephalosporin C was chromatographed on a DEAE-cellulose layer with a mobile phase of 0.05 M sodium acetate buffer [180]. Silica gel plates with mixtures of chloroform and ethyl acetate in different proportions were utilized for identifying a large group of cephalosporins [58,181–183]. N-Acetyl cephalosporins were characterized by means of cellulose-coated plates and four solvent systems [184]. "Selecta" plastic sheets for silica gel TLC and three solvent mixtures separated 23 different cephalosporin derivatives [185]. Deacetyl-7-ACA, 7-ACA, and 7-ACA lactone were separated on silica gel [186], and deacetyl cephalosporin C was characterized on cellulose layers with six solvent mixtures [187]. Reversed-phase TLC was used to separate ceftazole, cefazolin, cephalexin, cephalothin, and cephaloridine [188].

HPLC separated several acetyl derivatives of cephalosporin from the parent compound [184]. Similarly, cephaloridine, cefazolin, cephradine, cephaloglycin, and cephalothin were identified and separated from each other by reversed phase HPLC [189]. 6-APA, 7-ACA, and various mixtures of cephalosporins (e.g., cephalexin, cefazolin, cephradine, cephaloglycin, cephalothin) were separated by HPLC [189]. Specific cephalosporins have also been characterized by a number of investigators. Cephalothin [190,191] and cephardine [185] were identified by HPLC, and cephamycins by PC [168,192] and electrophoresis [168]. Cephaloglycin has been separated from deacetyl cephaloglycin and its lactone by HPLC [193]. Cephaloridine was identified by use of a reversed-phase Corasil column with a mobile phase of methanol–ammonium acetate [194].

17.3.5. Tetracyclines

Voluminous publications on chromatographic methods for the tetracycline family of antibiotics are already available. Identification schemes for only some of the newer numbers of this group and some of the more recent techniques are discussed here.

Tetracycline, oxytetracycline, chlortetracycline, rolitetracycline, demethyl chlortetracycline, doxycycline, and metacycline have been differentiated from each other by their R_F and color reactions in TLC on Polyamide II with five mobile phases [195]. Six tetracyclines have been characterized on silica gel plates by use of two tartaric acid-containing solvents and three color reagents. The antibiotics were identified by R_F and color reaction [196]. Tetracycline was characterized by HPLC on a μBondapak reversed-phase column with the aid of UV detection [197]. This antibiotic has been separated from oxytetracycline, doxycycline, and chlortetracycline by several HPLC methods [58]. The tetracyclines were also separated by another HPLC method [164].

17.3.6. Polypeptides

Bactracin has been chromatographed by TLC on Cellulose MN300 and Kieselgel 60 with several solvent mixtures and detected by bioautography with *Sarcina lutea* [46]. Another TLC method utilizes silica gel, ethanol–water–ammonia (8:8:1), and bioautography with *Micrococcus flavis* [198]. Electrophoresis on acrylamide gel, followed by detection with naphthalene black reagent, resulted in a light blue zone [199]. HPLC has been used to characterize bacitracin peaks with a Bondapak C_{18}/Corasil column [200] and, by eluting it with a convex gradient, 10 components were recognized [201].

Polymyxins S_1 and T_1 [202] as well as B and F [203] have been separated by PC and TLC. These antibiotics have also been chromatographed on Gelman ITLC Type SAF [203]. The electrophoretic system described for bacitracin has also been successful in characterizing polymyxins. A linear gradient in HPLC separated B_1 from B_2 [201]. Numerous polymyxin derivatives were characterized by TLC on silica gel with a solvent mixture of 1-butanol–pyridine–acetic acid–water (30:10:3:12) [204]. The separation and characterization of the peptide antibiotics nisin, subtilin, cinnamycin, duramycin, telomycin, stendomycin, tuberactinomycins, and gramicidins were detailed by Gross [205].

17.3.7. Polyenes

Amphotericin B was differentiated from its various esters, candicidin, nystatin, and their esters on Silica Gel 60 F_{254} (Merck) by the use of 1-butanol–ethanol–acetone–32% ammonium hydroxide (2:5:1:3) [206]. A variety of solvents separated amphotericin A and ascosin on Silica Gel G [207]. CCD separated the A and B components [207,208]. Reversed-phase HPLC was used to identify amphotericin B [209].

References on p. B339

Candicidin, candidin, candehexin, mycoheptin, nystatin, and trichomycin were differentiated by the TLC and CCD methods used for amphotericin. Silica Gel G was used to characterize levorin [206], and its components could also be separated by CCD [207]. Levorin was also identified by HPLC [210]. Antifungal 67-121 was differentiated from most other polyenes on Silica Gel G with 1-butanol–acetic acid–water–dioxane (6:2:2:1) [211] and separated into its four components with chloroform–methanol–water (2:2:1) [212].

17.3.8. Miscellaneous antibiotics

A select number of antibiotics is discussed in this section — those deemed to be of most interest, and/or those which have been characterized by more recent procedures. These are given alphabetically, if possible.

HPLC has been used to differentiate actinomycins C_2, C_3, and D on Bondapak C_{18}/Corasil with a mobile phase of acetonitrile–water (1:1) [213]. A series of actinomycin groups from *Micromonospora floridensis* was separated on Silica Gel GF with a mixture of chloroform–methanol–water (2:1:1) [214]. Resolution of each group into its components was achieved on silica gel with dichloromethane–acetone (7:3) [215]. Apramycin has been chromatographed on paper and silica gel [39]. Bleomycins were separated on Toyo No. 51 paper with 10% ammonium chloride [216], by TLC on silica gel [216,217], and by electrophoresis [216]. Chloramphenicol was chromatographed on paper [218] and on layers of cellulose [46], silica gel [177], silanized Silica Gel HF-254 [219], Avicel [220], and kieselguhr [83]. Several GLC [221,222] as well as HPLC [164,185] systems have been described. A quantitative TLC method was developed for the assay of clindamycin, a variety of its derivatives, and lincomycins [223].

Daunomycin was characterized by HPLC [184]. Diumycin components were separated by TLC [224,225] and CCD [225]. GLC [226] and HPLC [185,227] were used to characterize griseofulvin. Lincomycin 2-monoesters were separated by TLC and by GC on a column of 3% OV-1 on Gas-Chrom Q [228]. A number of TLC systems were used to differentiate mithramycins A, B, and C [229–231]. HPLC also differentiated mitomycins A, B, and C, and it was also used for preparative work [232]. A novel antibiotic mixture produced by *Mycoplasma* species RP III was separated by PC, TLC, and electrophoresis into two factors [233,234].

Nocardicins E and F were differentiated on Eastman Chromagram Sheet No. 6065 with three solvent mixtures [235]. A variety of solvents was used on plates of buffered kieselguhr [236] and silanized silica gel [237] to separate novobiocin and dihydronovobiocin. Novobiocin was also characterized by reversed-phase HPLC [58]. Patulin was characterized by GLC on a column packed with 10% DC-200 plus 15% QF-1 (1:1, v/v) on Gas-Chrom Q [238,239]. Pleuromutilins were separated on Silica Gel G by three solvent mixtures [240]. Rifampicin and two derivatives were chromatographed on buffered Kieselgel H [240]. Various TLC methods were utilized to differentiate rifampin and other rifamycins [241–246]. Rimocidin was characterized by its color reaction and the R_F on Silica Gel G in several solvent systems [247]. The streptothricin family is so large that the review by Khokhlov [248] must be consulted for details about the methods of chromatographic identification for this group. Vancomycin has been identified by HPLC on Bondapak C_{18} [249].

REFERENCES

1 D. Perlman, in D. Perlman (Editor), *Structure–Activity Relationships Among the Semisynthetic Antibiotics*, Academic Press, New York, 1977, p. xi.

2 V. Betina, *J. Chromatogr.*, 15 (1964) 379.

3 E. Addison and R.G. Clark, *J. Pharm. Pharmacol.*, 15 (1963) 286.

4 J. Dobrecky, E.A. Vazquez and R. Amper, *Soc. Arg. Farm. Bioquim. Ind.*, 8 (1968) 204; *C.A.*, 71 (1969) 42362r.

5 A. Aszalos, S. Davis and D. Frost, *J. Chromatogr.*, 37 (1968) 487.

6 J.-P. Schmitt and C. Mathis, *Ann. Pharm. Fr.*, (1970) 205.

7 T. Ikekawa, F. Iwami, E. Akita and H. Umezawa, *J. Antibiot.*, 16 (1963) 56.

8 S. Ochab and B. Borowiecka, *Diss. Pharm. Pharmacol.*, 20 (1968) 449.

9 E. Borowiecka, *Diss. Pharm. Pharmacol.*, 22 (1970) 345.

10 Y. Nishimoto, E. Tsuchida and S. Toyoshima, *Yakugaku Zasshi*, 87 (1967) 516; *C.A.*, 67 (1967) 67635n.

11 Y. Ito, M. Namba, N. Nagahama, T. Yamaguchi and T. Okuda, *J. Antibiot.*, 17 (1964) 218.

12 M.H.J. Zuidweg, J.G. Oostendorp and C.J.K. Bos, *J. Chromatogr.*, 42 (1969) 552.

13 T.F. Brodasky, *Anal. Chem.*, 35 (1963) 343.

14 J. Prath, *Acta Chem. Scand.*, 6 (1952) 1237.

15 S. Nakamura, T. Yajima, M. Hamoda, T. Nishmura, M. Ishizuki, T. Takeuchi, N. Tanaka and H. Umezawa, *J. Antibiot.*, 20 (1967) 210.

16 I. Muramatsu, S. Sofaku and A. Hagitani, *J. Antibiot.*, 25 (1972) 189.

17 M.F. de Albuquerque, F.D. de Andrade Lyra, O.G. de Lima, C.C. de Oliveira, J.S. de Barros Coelho, G.M. Maciel and M. de Salete Barros Cavalcanti, *Recife*, 6 (1966) 35.

18 G.C. Lancini and P. Sensi, *Experientia*, 20 (1964) 83.

19 A.D. Argoudelis, *J. Antibiot.*, 25 (1972) 171.

20 R. Corbaz, L. Ettlinger, E. Gaumann, W. Keller-Schierlein, F. Kradolfer, E. Kyburz, L. Neipp, U. Prelog, A. Wettstein and H. Zahner, *Helv. Chim. Acta*, 39 (1956) 304.

21 S. Iquchi, M. Yamamoto and T. Goromaru, *J. Chromatogr.*, 24 (1966) 182.

22 K. Tsuji and J.H. Robertson, *Anal. Chem.*, 41 (1969) 133.

23 B. van Giessen and K. Tsuji, *J. Pharm. Sci.*, 60 (1971) 1068.

24 R.L. Hamill, H.R. Sullivan and M. Gorman, *Appl. Microbiol.*, 18 (1969) 310.

25 H.J. Issaq, E.W. Bair, T. Wei, C. Meyers and A. Aszalos, *J. Chromatogr.*, 133 (1977) 291.

26 A. Aszalos and D. Frost, *Methods Enzymol.*, 43 (1975) 172.

27 A. Aszalos, S. Davis and D. Frost, *J. Chromatogr.*, 37 (1968) 487.

28 A. Aszalos and H.J. Issaq, *J. Liquid Chromatogr.*, 3 (1980) 867.

29 G.H. Wagman and M.J. Weinstein, *Chromatography of Antibiotics*, J. Chromatogr. Library Series, Vol. 1, Elsevier, Amsterdam, 1973.

30 M.J. Weinstein and G.H. Wagman (Editors), *Antibiotics: Isolation, Separation and Purification*, J. Chromatogr. Library Series, Vol. 15, Elsevier, Amsterdam, 1978.

31 G. Zweig and J. Sherma (Editors), *Handbook of Chromatography*, Vol. 1, CRC Press, Cleveland, OH, 1972, pp. 458 and 616.

32 P.F. Lott, J.R. Dias and S.C. Slahck, *J. Chromatogr. Sci.*, 16 (1978) 571.

33 E. Meyers and C.A. Chang, *J. Antibiot.*, 32 (1979) 846.

34 M. Vondráček, in E. Heftmann (Editor), *Chromatography*, Reinhold, New York, 2nd Edn., p. 718.

35 M. Vondráček, in E. Heftmann (Editor), *Chromatography*, Van Nostrand-Reinhold, New York, 3rd Edn., 1975, p. 818.

36 M.J. Weinstein, G.H. Wagman, E.M. Oden and J.A. Marquez, *J. Bacteriol.*, 94 (1967) 789.

37 M.J. Weinstein, G.H. Wagman and J.A. Marquez, *U.S. Pat.* 3,951,746: Apr. 20, 1976.
38 G.H. Wagman, J.A. Marquez and M.J. Weinstein, *J. Chromatogr.*, 34 (1968) 210.
39 T. Nara, S. Takasawa, R. Okachi, I. Kawamoto, M. Yamamoto, S. Sato and T. Sato, *U.S. Pat.* 3,976,768: Aug. 24, 1976.
40 N. Kantor and G. Selzer, *J. Pharm. Sci.*, 57 (1968) 2170.
41 G.H. Wagman, J.V. Bailey and M.M. Miller, *J. Pharm. Sci.*, 57 (1968) 1319.
42 J. Berdy, J.K. Pauncz, Zs. M. Vajna, Gy. Horváth, J. Gyimesi and I. Koczka, *J. Antibiot.*, 30 (1977) 945.
43 H. Maehr and C.P. Schaffner, *J. Chromatogr.*, 30 (1967) 572.
44 R.S. Egan, R.L. DeVault, S.L. Mueller, M.I. Levenberg, A.V. Sinclair and R.S. Stanascek, *J. Antibiot.*, 28 (1975) 29.
45 H. Umezawa, M. Yagisawa, Y. Matsuhashi, H. Naganawa, H. Yamamoto, S. Kondo and T. Takeuchi, *J. Antibiot.*, 26 (1973) 612.
46 H. Troonen, P. Roelants and B. Boon, *U.S. Pat.* 4,002,530: Jan. 11, 1977.
47 K.M. Byrne, A.S. Kershner, H. Maehr, J.A. Marquez and C.P. Schaffner, *J. Chromatogr.*, 131 (1977) 191.
48 J.P. Anhalt, *Antimicrob. Ag. Chemother.*, 11 (1977) 651.
49 G.W. Peng, M.A.F. Gadalla, A. Peng, V. Smith and W.L. Chiou, *Clin. Chem.*, 23 (1977) 1838.
50 T. Naito, S. Nakagawa, Y. Narita, S. Toda, Y. Abe, M. Oka, H. Yamashita, T. Yamasaki, K. Fujisawa and H. Kawaguchi, *J. Antibiot.*, 27 (1974) 851.
51 T. Nara, S. Takasawa, R. Okachi, I. Kawamoto, M. Yamamoto, S. Sato, T. Sato and A. Morikawa, *U.S. Pat.* 3,939,043: Feb. 17, 1976.
52 J.W. Rothrock, R.T. Goegelman and F.J. Wolf, *Antibiot. Ann.*, (1958–1959) 796.
53 J.W. Rothrock and I. Putter, *U.S. Pat.* 3,032,547; Sep. 12, 1958.
54 M.J. Weinstein, G.M. Ludemann, G.H. Wagman and J.A. Marquez, *U.S. Pat.* 3,819,611; June 25, 1974.
55 N.R. Chatterjee, *Ind. J. Chem.*, 13 (1975) 1182.
56 S. Ochab, *Pol. J. Pharmacol. Pharm.*, 23 (1973) 105.
57 K. Tsuji and J.H. Robertson, *Anal. Chem.*, 42 (1970) 1661.
58 *Varian Aerograph Application Notes*, No. 6, August 19, 1974.
59 A. Fujii, K. Maeda and H. Umezawa, *J. Antibiot.*, 21 (1968) 340.
60 W.J. Shier, K.L. Rinehart, Jr. and D. Gottlieb, *U.S. Pat.* 3,833,556; Sep. 3, 1974.
61 T. Naito, S. Nakagawa and Y. Abe, *U.S. Pat.* 3,904,597; Sep. 9, 1975.
62 H. Umezawa, K. Maeda, S. Kondo and S. Umezawa, *German Pat.* 2,423,591; Dec. 5, 1974.
63 H. Umezawa, K. Maeda, S. Kondo and S. Umezawa, *U.S. Pat.* 3,940,382; Feb. 24, 1976.
64 B.E. Leach and C.M. Teeters, *J. Amer. Chem. Soc.*, 73 (1951) 2794.
65 S.C. Pan and J.D. Dutcher, *Anal. Chem.*, 28 (1956) 836.
66 H. Umezawa, M. Murase and S. Yamazaki, *J. Antibiot.*, 12 (1959) 341.
67 K.L. Rinehart, Jr., A.D. Argoudelis, W.A. Goss, A. Sohler and C.P. Schaffner, *J. Amer. Chem. Soc.*, 82 (1960) 3938.
68 P.G. Kaiser, *Anal. Chem.*, 35 (1963) 552.
69 H. Maehr and C.P. Schaffner, *Anal. Chem.*, 36 (1964) 104.
70 M.K. Majumdar and S.K. Majumdar, *Anal. Chem.*, 39 (1967) 215.
71 M.K. Majumdar and S.K. Majumdar, *Appl. Microbiol.*, 17 (1969) 763.
72 W.T. Shier, K.L. Rinehart, Jr. and D. Gottlieb, *Biochemistry*, 63 (1969) 198.
73 K. Munakata, T. Oda, T. Mori and H. Ito, *U.S. Pat.* 3,870,698; Mar. 11, 1975.
74 H. Baud, A. Betencourt, M. Peyre and L. Penasse, *J. Antibiot.*, 30 (1977) 720.
75 S. Hanessian, R. Masse and M.-L. Capmeau, *J. Antibiot.*, 30 (1977) 893.
76 R. Foppiano and B.P. Brown, *J. Pharm. Sci.*, 54 (1965) 206.
77 B. Borowiecka, *Diss. Pharm. Pharmacol.*, 24 (1972) 210.
78 L.P. Snezknova and L.N. Astanina, *Antibiotiki*, 17 (1972) 263.
79 M. Baudet, *J. Pharm. Belg.*, 31 (1976) 247.
80 T.F. Brodasky, *Anal. Chem.*, 35 (1963) 343.
81 M.J. Weinstein, G.H. Wagman and J.A. Marquez, *U.S. Pat.* 3,956,068; May 11, 1976.

82 K. Munakata, T. Oda, T. Mori and H. Ito, *U.S. Pat.* 3,870,698; Mar. 11, 1975.
83 F. Baldini, G. Frati, G. Paezzani and G. Ambanelli, *Ind. Conserve*, 48 (1973) 135.
84 J.P. Carr, R.J. Stretton and J.W. Watson, *Final Year Study Proj. Theses*, 10 (1969) 17; *Pharm. Anal.*, 73 (1970) 247.
85 R.J. Stretton, J.P. Carr and J. Watson-Walker, *J. Chromatogr.*, 45 (1969) 155.
86 J. Prath, *Acta Chem. Scand.*, 6 (1952) 1237.
87 K. Tsuji and J.H. Robertson, *Anal. Chem.*, 41 (1969) 1332.
88 B. van Giessen and K. Tsuji, *J. Pharm. Sci.*, 60 (1971) 1068.
89 M.J. Weinstein, J.A. Marquez, R.T. Testa, G.H. Wagman, E.M. Oden and J.A. Waitz, *J. Antibiot.*, 23 (1970) 551.
90 G.H. Wagman, R.T. Testa and J.A. Marquez, *J. Antibiot.*, 23 (1970) 555.
91 M.J. Weinstein, G.H. Wagman and J.A. Marquez, *U.S. Pat.* 3,956,068; May 11, 1976.
92 J.N. Pereira, *J. Biochem. Microbiol. Tech. Eng.*, 3 (1961) 79.
93 H. Heding, *Acta Chem. Scand.*, 20 (1966) 1743.
94 P.A. Nussbaumer and M. Schorperet, *Pharm. Acta Helv.*, 40 (1965) 205 and 477.
95 T. Sato and H. Ikeda, *Sci. Papers, Inst. Phys. Chem. Res., Tokyo*, 59 (1965) 159.
96 R. Bossuyt, R. Van Tenterghen and G. Waes, *J. Chromatogr.*, 124 (1975) 37.
97 M.J. Weinstein, G.H. Wagman, R.T. Testa and J.R. Marquez, *U.S. Pat.* 3,997,403; Dec. 14, 1976.
98 M.J. Weinstein, G.M. Ludemann, E.M. Oden and G.H. Wagman, *Antimicrob. Ag. Chemother.*, 1963 (1964) 1.
99 K. Munakata, T. Oda, T. Mori and H. Ito, *U.S. Pat.* 3,891,506; June 24, 1975.
100 T. Naito and S. Nakagawa, *U.S. Pat.* 3,897,412; July 29, 1975.
101 M. Baudet, *J. Pharm. Belg.*, 31 (1976) 247.
102 H. Umezawa, S. Umezawa, K. Maeda, O. Tsuchiya, S. Kondo and S. Fukatsu, *German Pat.* 2,350,169; Apr. 18, 1974.
103 H.W. Dion, P.W.K. Woo, N.E. Willmer, D.L. Kern, J. Onaga and S.A. Fusari, *Antimicrob. Ag. Chemother.*, 2 (1972) 84.
104 H. Kawaguchi, K. Tomita, T. Hoshiya, T. Miyaki, K.-I. Fujisawa, M. Kimeda, K.-I. Numata, M. Konishi, H. Tsukiuara, M. Hatori and H. Koshiyama, *J. Antibiot.*, 27 (1974) 460.
105 P.W.K. Woo, *U.S. Pat.* 3,960,837; June 1, 1976.
106 G.W. Peng, G.G. Jackson and W.L. Chiou, *Antimicrob. Ag. Chemother.*, 12 (1977) 707.
107 E. Akita, T. Tsuchiya, S. Kondo, S. Yasuda, S. Umezawa and H. Umezawa, *German Pat.* 2,342,946; Mar. 14, 1974.
108 T. Fujiwara, T. Tanimoto, K. Matsumoto and E. Kondo, *J. Antibiot.*, 31 (1978) 966.
109 K. Murai, B.A. Sobin, W.D. Celmer and F.W. Tanner, *Antibiot. Chemother.*, 9 (1959) 485.
110 P. Kurath, R.S. Egan and P.H. Jones, *U.S. Pat.* 3,681,323; Aug. 1, 1972.
111 C.W. Pettinga, W.M. Starke and F.R. van Abeele, *J. Amer. Chem. Soc.*, 76 (1954) 569.
112 P.F. Wiley, R. Gayle, C.W. Pettinga and K. Gerzon, *J. Amer. Chem. Soc.*, 79 (1957) 6074.
113 V.C. Stevens, C.T. Pugh, N.E. Davis, M.M. Hoehn, S. Ralston, M.C. Sparks and L. Thompkins, *J. Antibiot.*, 22 (1969) 551.
114 Z. Kotula and A. Kaminska, *Med. Dosw. Microbiol.*, 19 (1967) 381; *C.A.*, 68 (1968) 72292g.
115 A. Banaszek, K. Krowicki and A. Zamojski, *J. Chromatogr.*, 32 (1968) 581.
116 G. Richard, C. Radecka, D.W. Hughes and W.L. Wilson, *J. Chromatogr.*, 67 (1972) 69.
117 M.C. Flickinger and D. Perlman, *J. Antibiot.*, 28 (1975) 307.
118 A. Vilim, M.J. LeBelle, W.L. Wilson and K.C. Graham, *J. Chromatogr.*, 133 (1977) 239.
119 G. Kobrehel, Z. Tamburašev and S. Djokić, *J. Chromatogr.*, 133 (1977) 415.
120 S. Omura, Y. Suzuki, A. Nagakawa and T. Hata, *J. Antibiot.*, 26 (1973) 794.
121 K. Tsuji, *J. Chromatogr.*, 158 (1978) 337.
122 T. Osato, K. Yagishita and H. Umezawa, *J. Antibiot.*, 8 (1955) 161.
123 K. Murai, B.A. Sobin, W.D. Celmer and F.W. Tanner, *Antibiot. Chemother.*, 9 (1959) 485.
124 H. Koshiyama, M. Okanishi, T. Ohmori, T. Miyaki, H. Tsukiura, M. Matsuzaki and H. Kawaguchi, *J. Antibiot.*, 16 (1963) 59.
125 C. DeBoer, A. Dietz, J.R. Wilkins, C.N. Lewis and G.M. Savage, *Antibiot. Ann.*, (1954–1955) 831.

126 I. Kawamoto, R. Okachi, H. Kato, S. Yamamoto, I. Takahashi, S. Takasawa and T. Nara, *J. Antibiot.*, 27 (1974) 493.

127 G.H. Wagman, J.A. Waitz, J. Marquez, A. Murawski, E.M. Oden, R.T. Testa and M.J. Weinstein, *J. Antibiot.*, 25 (1972) 641.

128 F.A. Hochstein and K. Murai, *J. Amer. Chem. Soc.*, 76 (1954) 5080.

129 M.J. Weinstein, G.H. Wagman, J.A. Marquez, E.M. Oden, R.T. Testa and J.A. Waitz, *Antimicrob. Ag. Chemother.*, 1968 (1969) 260.

130 J.A. Marquez, A. Murawski and G.H. Wagman, *J. Antibiot.*, 22 (1969) 259.

131 T. Nara, S. Takasawa, I. Kawamoto and S. Yamamoto, *German Pat.* 2,301,080; July 19, 1973.

132 T. Osono, K. Moriyama, K. Murakami and H. Umezawa, *U.S. Pat.* 3,959,252; May 25, 1976.

133 M. Shibata, K. Hatano, H. Yamana, T. Kishi and E. Higashide, *German Pat.* 2,034,245; Feb. 25, 1971.

134 A.J. Glazko, W.A. Dill and M.C. Rebstock, *J. Biol. Chem.*, 183 (1950) 679.

135 T. Osato, K. Yagishita and H. Umezawa, *J. Antibiot.*, 8 (1955) 161.

136 R.W. Kierstead, R.A. LeMahieu and D. Pruess, *U.S. Pat.* 3,928,387; Dec. 23, 1975.

137 P.A. Nussbauer, *Pharm. Acta. Helv.*, 40 (1965) 210.

138 A.J. Glazko, W.A. Dill and M.C. Rebstock, *J. Biol. Chem.*, 183 (1950) 679.

139 G.H. Wagman, J.A. Waitz, J.A. Marquez, A. Murawski, E.M. Oden, R.T. Testa and M.J. Weinstein, *J. Antibiot.*, 25 (1972) 641.

140 S. Omura, Y. Suzuki, A. Nakagawa and T. Hata, *J. Antibiot.*, 26 (1973) 794.

141 R.J. Theriault, *U.S. Pat.* 3,948,884; Apr. 6, 1976.

142 G.H. Wagman, J.A. Waitz, J. Marquez, A. Murawski, E.M. Oden, R.T. Testa and M.J. Weinstein, *J. Antibiot.*, 25 (1972) 641.

143 M.J. Weinstein, G.H. Wagman and J.A. Marquez, *U.S. Pat.* 4,161,523; July 17, 1979.

144 C. Lin, H. Kim, D. Schuessler, E. Oden and S. Symchowicz, *Antimicrob. Ag. Chemother.*, (1980) 780.

145 E. Higashide, T. Hasegawa, H. Ono, M. Asai, M. Muroi and T. Kishi, *U.S. Pat.* 3,691,280; Sep. 12, 1972.

146 M. Muroi, M. Izawa, M. Asai, T. Kishi and K. Mizuno, *J. Antibiot.*, 26 (1973) 199.

147 N. Kunishige, K. Kawamura, M. Muroi and T. Kishi, *Chem. Pharm. Bull.*, 23 (1975) 3075.

148 K. Kondo, *J. Chromatogr.*, 169 (1979) 337.

149 J.P. Clayton and R. Hubbard, *U.S. Pat.* 3,923,787; Dec. 2, 1975.

150 T.R. Fosker and J.H.C. Naylor, *U.S. Pat.* 3,663,563; May 16, 1972.

151 H. Vanderhaeghe, A. Vlietinck, M. Claesen and G. Parmentier, *J. Antibiot.*, 27 (1974) 169.

152 V.B. Korchagin, L.I. Serova, S.P. Dement'eva, I.N. Navol'neva, I.I. Inozemtseva, D.M. Trakhtenberg and N.I. Kotova, *Antibiotiki*, 16 (1971) 8; *C.A.*, 74 (1971) 218,79535.

153 P.J. O'Neal, *Proc. Ass. Anal. Chem.*, (1974) 39.

154 P.E. Manni, R.A. Lipper, J.M. Blaha and S.L. Hem, *J. Chromatogr.*, 76 (1973) 512.

155 F. Baldini, G. Frati, G. Pezzani and G. Ambanelli, *Ind. Conserve*, 48 (1973) 135.

156 V.B. Korchagin, L.I. Serova, S.P. Eementieva, I.N. Navolenva, I.I. Inozemtseva, D.M. Trakhtenberg and N.I. Kotova, *Antibiotiki*, 16 (1971) 410.

157 M. Pokorny, N. Vitezič and M. Japelj, *J. Chromatogr.*, 77 (1973) 458.

158 M.L. Walash and S.M. Hassan, *J. Drug Res. Egypt*, 5 (1973) 111.

159 J.L. Marin, R.D. Duncombe and W.H.C. Shaw, *Analyst (London)*, 100 (1975) 243.

160 A.E. Byrd and A.C. Marshall, *J. Chromatogr.*, 63 (1971) 313.

161 G.R. Fosker and J.H.C. Nayler, *U.S. Pat.* 3,663,563; May 16, 1972.

162 E.R. White, M.A. Carroll, J.E. Zarembo and A.E. Bender, *J. Antibiot.*, 28 (1975) 205.

163 K. Tsuji and J.H. Robertson, *J. Pharm. Sci.*, 64 (1975) 1542.

164 E.R. White, M.A. Carroll and J.E. Zarembo, *J. Antibiot.*, 30 (1977) 811.

165 J. D'A. Jeffrey, E.P. Abraham and G.G.F. Newton, *Biochem. J.*, 81 (1961) 591.

166 C.W. Hale, G.G.F. Newton and E.P. Abraham, *Biochem. J.*, 79 (1961) 403.

167 B. Fechtig, E. Vischer, H. Bickel, R. Bosshardt and J. Urech, *U.S. Pat.* 3,697,515; Oct. 10, 1972.

168 D.R. Brannon, J.A. Mabe and D.S. Fukuda, *J. Antibiot.*, 29 (1976) 121.

169 B. Loder, G.G.F. Newton and E.P. Abraham, *Biochem. J.*, 79 (1961) 408.

170 N.G. Weir, *U.S. Pat.* 3,929,780; Dec. 30, 1975.

171 N.G. Weir, *U.S. Pat.* 3,974,152; Aug. 10, 1976.

172 G.G.F. Newton and E.P. Abraham, *Biochem. J.,* 62 (1956) 651.

173 J.L. Ott, C.W. Godzeski, D. Pavey, J.D. Farran and D.R. Horton, *Appl. Microbiol.,* 10 (1962) 515.

174 B.H. Olsen, J.C. Jennings and A.J. Junek, *Science,* 117 (1953) 76.

175 H.S. Horton and E.P. Abraham, *Biochem. J.,* 50 (1952) 168.

176 H.R. Sullivan and R.E. McMahon, *Biochem. J.,* (1967) 976.

177 D.R. Brannon, D.S. Fukuda, J.A. Mabe, F.M. Huber and J.G. Whitney, *Antimicrob. Ag. Chemother.,* 1 (1972) 237.

178 E.O. Stapley, M. Jackson, S. Hernandez, S.P. Zimmerman, S.A. Currie, S. Mochales, J.M. Mata, H.B. Woodroof and D. Hendlin, *Antimicrob. Ag. Chemother.,* 2 (1972) 122.

179 B. Fechtig, H. Peter, H. Bickel and E. Vischer, *Helv. Chim. Acta,* 51 (1968) 1108.

180 C.H. Nash and F.M. Huber, *Appl. Microbiol.,* 22 (1971) 6.

181 J. Bouchaudon, P. LeRoy and M.N. Messer, *U.S. Pat.* 3,975,385; Aug. 17, 1976.

182 H. Bickel, J. Muller, R. Bosshardt, H. Peter and B. Fechtig, *U.S. Pat.* RE.29,119; Jan. 18, 1977.

183 H. Nakao, H. Yanagisawa, M. Nagano, B. Shimizu, M. Kanedo and S. Sugawara, *U.S. Pat.* 4,007,177; Feb. 8, 1977.

184 P. Traxler, H.J. Treichler and J. Nuesch, *J. Antibiot.,* 28 (1975) 605.

185 P. Crooij and A. Colinet, *U.S. Pat.* 3,912,728; Oct. 14, 1975.

186 C.J. Budd, *J. Chromatogr.,* 76 (1973) 509.

187 Y. Fujisawa, M. Kikuchi and T. Kanzaki, *J. Antibiot.,* 30 (1977) 775.

188 T. Sawai, K. Matsuba, A. Tamura and S. Yamagishi, *J. Antibiot.,* 32 (1979) 59.

189 E.R. White, M.A. Carroll, J.E. Zarembo and A.D. Bender, *J. Antibiot.,* 28 (1975) 205.

190 T.F. Rolewicz, B.L. Mirkin, M.J. Cooper and M.W. Anders, *Clin. Pharmacol. Ther.,* 22 (1977) 928.

191 I. Nilsson-Ehel, T.T. Yoshikawa, M.C. Schotz and L.B. Guze, *Antimicrob. Ag. Chemother.,* 13 (1978) 221.

192 H. Fukase, T. Hasegawa, K. Hatano, H. Iwasaki and M. Yoneda, *J. Antibiot.,* 29 (1976) 113.

193 J. Haginaka, T. Nakagawa and T. Uno, *J. Antibiot.,* 32 (1979) 462.

194 J.S. Wold and S.A. Turnipseed, *J. Chromatogr.,* 136 (1977) 170.

195 M.S. Santos, M.I. Mamede Santos, L.C. Goncalves and M.T. Morgadinho, *Bol. Fac. Farm., Univ. Coimbra, Ed. Cient.,* 33 (1973) 127.

196 R.T. Wang and Y.H. Tsai, *Tai-Wan K'o Hsueh,* 27 (1973) 36.

197 I. Nilsson-Ehle, T.T. Yoshikawa, M.C. Schotz and L.B. Guze, *Antimicrob. Ag. Chemother.,* 9 (1976) 754.

198 R. Bossuyt, R. Van Tenterghem and G. Waes, *J. Chromatogr.,* 124 (1975) 37.

199 R.G. Coombe, *Aust. J. Pharm. Sci.,* N.S. 1 (1972) 6.

200 K. Tsuji, J.H. Robertson and J.A. Bach, *J. Chromatogr.,* 99 (1974) 597.

201 K. Tsuji and J.H. Robertson, *J. Chromatogr.,* 112 (1975) 663.

202 J. Shoji, H. Hinoo, Y. Wakisaka, K. Koizumi, M. Mayama and S. Matsuura, *J. Antibiot.,* 30 (1977) 1029.

203 W.L. Parker, M.L. Rathnum, L.D. Dean, M.W. Nimeck, W.E. Brown and E. Meyers, *J. Antibiot.,* 30 (1977) 767.

204 M. Igloy and A. Mizsei, *J. Chromatogr.,* 28 (1967) 458.

205 E. Gross, in M.J. Weinstein and G.H. Wagman (Editors), *Antibiotics. Isolation, Separation and Purification,* J. Chromatogr. Library Series, Vol. 15, Elsevier, Amsterdam, 1978, p. 415.

206 T. Bruzzese, M. Cambieri and F. Recusani, *J. Pharm. Sci.,* 64 (1975) 462.

207 A.H. Thomas, *Analyst (London),* 101 (1976) 321.

208 H.A.P. Linke, W. Mechlinski and C.P. Schaffner, *J. Antibiot.,* 27 (1974) 155.

209 I. Nilsson-Ehle, T.T. Woshikawa, J.E. Edwards, M.C. Schotz and L.B. Guze, *J. Infec. Dis.,* 135 (1977) 414.

210 S.H. Hansen and M. Thomsen, *J. Chromatogr.,* 123 (1976) 205.

211 G.H. Wagman, R.T. Testa, M. Patel, J.A. Marquez, E.M. Oden, J.A. Waitz and M.J. Weinstein, *Antimicrob. Ag. Chemother.,* 7 (1975) 457.

212 M.J. Weinstein, G.H. Wagman, J.A. Marquez and M.G. Patel, *U.S. Pat.* 4,027,015; May 31, 1977.

213 W.J. Rzeszotarski and A.B. Mauger, *J. Chromatogr.*, 86 (1973) 246.

214 G.H. Wagman, J.A. Marquez, P.D. Watkins, F. Gentile, A. Murawski, M. Patel and M.J. Weinstein, *Antimicrob. Ag. Chemother.*, 9 (1976) 465.

215 M.J. Weinstein, G.H. Wagman, J.A. Marquez and P.D. Watkins, *U.S. Pat.* 3,954,970; May 4, 1976.

216 A. Fuji, T. Takita, K. Maeda and H. Umezawa, *J. Antibiot.*, 26 (1973) 396.

217 T. Nara, S. Takasawa, R. Okachi, I. Kawamoto, M. Kohagura and I. Takahashi, *U.S. Pat.* 3,922,343; Nov. 25, 1975.

218 A.J. Glazko, W.A. Dill and M.C. Rebstock, *J. Biol. Chem.*, 183 (1950) 679.

219 R. Rangone and C. Ambrosio, *Pharmaco, Ed. Prat.*, 26 (1971) 237.

220 G.S. Chung and R.T. Wang, *Tai-Wan K'o Hsueh*, 27 (1973) 27.

221 T. Nakagawa, M. Masada and T. Uno, *J. Chromatogr.*, 111 (1975) 355.

222 G. Janssen and H. Vanderhaeghe, *J. Chromatogr.*, 82 (1973) 297.

223 L.W. Brown, *J. Pharm. Sci.*, 67 (1978) 1254.

224 A. Sattler and F. Kreuzig, *J. Antibiot.*, 28 (1975) 200.

225 W.A. Slusarchyk, J.L. Bouchard-Ewing and F.L. Weisenborn, *J. Antibiot.*, 26 (1973) 391.

226 W.L. Epstein, V.P. Shah and S. Riegelman, *Arch. Dermatol.*, 106 (1972) 344.

227 L.P. Hackett and L.J. Dusci, *J. Chromatogr.*, 155 (1978) 206.

228 W. Morozowich, A.A. Sinkula, F.A. Mackellar and C. Lewis, *J. Pharm. Sci.*, 62 (1973) 1102.

229 K. Stajner, M. Blumauerova, D.A.S. Callieri and Z. Vanek, *Folia Microbiol. (Prague)*, 19 (1974) 498.

230 B.A. Soben, J.B. Routlen, K.V. Rao, W.S. March and A.L. Garretson, *U.S. Pat.* 3,906,093; Sep. 16, 1975.

231 B.A. Soben, J.B. Routlen, K.V. Rao, W.S. March and A.L. Garretson, *U.S. Pat.* 3,646,194; Feb. 29, 1972.

232 S.C. Srivastava and L. Hornemann, *J. Chromatogr.*, 161 (1978) 393.

233 T. Sakai and D. Perlman, *J. Antibiot.*, 28 (1975) 749.

234 M. Sylvestre and D. Perlman, *J. Antibiot.*, 28 (1975) 73.

235 J. Hosoda, H. Aoki and H. Imanake, *U.S. Pat.* 4,146,536; Mar. 27, 1979.

236 T. Murakawa, Y. Kono and M. Nishida, *J. Antibiot.*, 25 (1972) 243.

237 R. Rangone and C. Ambrosio, *Farmaco, Ed. Prat.*, 26 (1971) 237.

238 T. Suzuki, Y. Fujimoto, Y. Hoshino and A. Tanaka, *J. Chromatogr.*, 105 (1975) 95.

239 Y. Fujimoto, T. Suzuki and Y. Hoshino, *J. Chromatogr.*, 105 (1975) 99.

240 F. Knauseder and E. Brandl, *J. Antibiot.*, 29 (1976) 125.

241 K. Winsel, H. Iwainsky, E. Werner and H. Eule, *Pharmazie*, 31 (1976) 95.

242 J. Birner, P.R. Hodgson, W.R. Lane and E.H. Baxter, *J. Antibiot.*, 25 (1972) 356.

243 W.L. Wilson, K.C. Graham and M.J. LeBelle, *J. Chromatogr.*, 144 (1977) 270.

244 W.D. Celmer, F.C. Sciavolino, J.B. Routien and T.C. Cullen, *German Pat.* 2,500,898; July 17, 1975.

245 L. Marsili, V. Rossetti and C. Pasqualucci, *U.S. Pat.* 4,007,169; Feb. 8, 1977.

246 R. White, G. Lancini and P. Antonini, *Belgian Pat.* 832,921; Dec. 17, 1975.

247 R.C. Pandey and K.L. Rinehart, Jr., *J. Antibiot.*, 30 (1977) 146.

248 A.S. Khokhlov, in M.J. Weinstein and G.H. Wagman (Editors), *Antibiotics, Isolation, Separation, and Purification*, Elsevier, Amsterdam, 1978, p. 617.

249 R.L. Kirchmeier and R.P. Opton, *Anal. Chem.*, 50 (1978) 349.

Chapter 18

Nucleic acids

GRAHAM J. COWLING

CONTENTS

18.1. INTRODUCTION

This chapter deals with thin-layer, ion-exchange, and affinity chromatography, as well as paper and gel electrophoresis of nucleic acids and their monomeric constituents. General texts covering the biological roles and chemical structures of these molecules include *The Biochemistry of Nucleic Acids* [1] and *Molecular Biology of the Gene* [2]. The practical and theoretical aspects of this subject may be found in either the references cited or volumes of *Laboratory Techniques in Biochemistry and Molecular Biology* [3–6].

This essay follows the example of other reviews on the subject in that it covers the separation of progressively larger molecules. These include bases, nucleosides, nucleotides, oligonucleotides, and various classes of macromolecular RNA and DNA species. Final sections highlight the use of chromatography and electrophoresis in sequence analysis and nucleoprotein separation. The descriptions of chromatographic applications usually precede those of electrophoresis. This review is not

References on p. B372

intended to be comprehensive but rather an evaluation of the contemporary status of the field with emphasis on the developments of the last few years. In many instances the use of a particular technique is shown to illustrate its possibilities.

18.2. BASES, NUCLEOSIDES, AND NUCLEOTIDES

18.2.1. Base composition

The base composition of an unknown RNA or DNA molecule is obtained by first degrading the polymer to purine and pyrimidine bases, nucleosides, or nucleotides, separating the individual species and estimating the proportion of each. Before surveying separation techniques, it may be useful to review briefly the ways in which RNA and DNA can be degraded.

The products of chemical hydrolysis vary with the conditions used. Heating RNA or DNA at 100°C in 70% perchloric acid generates bases [7], whereas hydrolysis of RNA in HCl produces purine bases and pyrimidine nucleotides [8]. Alkaline hydrolysis of RNA gives a mixture of 3'- and 5'-ribonucleotides [9]. The products of nuclease digestion also depend on the enzyme used. RNase T_2 degrades RNA to 3'-ribonucleotides [10] and, after further digestion with snake venom phosphodiesterase, to nucleosides [11]. Likewise, DNA can be sequentially treated with pancreatic DNase I and snake venom phosphodiesterase to produce 5'-deoxyribonucleotides [11] or with hog spleen phosphodiesterase to give 3'-deoxyribonucleotides [12]. The advantage of nuclease degradation is that products can usually be applied directly to ion-exchange columns or thin layers without neutralization or removal of the reagent. Chemical degradation tends to destroy many of the modified sugar residues found in tRNA species [13].

18.2.2. Column chromatography

The common purine and pyrimidine bases possess ring nitrogen atoms and, with the exception of thymine (DNA) and uracil (RNA), exocyclic amino groups, which accept protons. Hence, by changing the pH of the solvent, the degree of protonation on each type of base can be controlled. Using this principle of charge control, Cohn [14] demonstrated that with $2 N$ HCl the common bases are eluted from strong cation-exchange columns in an order (U, C, G, A) related to the pK_a value of each base. This relationship between charge and order of elution soon breaks down at higher pH values. Nonionic interactions which are, in turn, influenced by the shape, size, and aromatic character of the solute, begin to play a major role in controlling the order of elution. Crampton et al. [15] have shown that bases are eluted from the same cation exchanger (Dowex-50) in a changed order (U, G, C, A) at higher pH (sodium acetate buffer, pH 4).

In general, the polystyrene sulfonic acid exchangers, such as Dowex-50, have limited use for nucleoside separation, as the acidity of the support is detrimental to the acid-labile purine–glycosyl linkages. An alternative approach has been to use anion-exchange resins, again based on polystyrene beads of high porosity, but containing quaternary ammonium groups. At low pH (≤ 4), bases and nucleosides are selectively excluded from these resins [16] – a process called "cation-exclusion chromatography". At higher pH, the same resins fractionate bases and nucleosides by anion exchange [16,17].

Recently, microparticulate reversed-phase supports, in which long-chain hydrocarbons are covalently bonded to silicate beads, have been successfully used to separate mixtures of bases and nucleosides derived from DNA and tRNA [18]. Base or nucleoside mixtures (1.5 μg digested or degraded nucleic acid) are applied to the octadecylsilyl columns in 50 mM KH_2PO_4 and resolved by applying gradient elution of the same buffer to methanol. Sensitive UV absorbance monitoring of the eluate allows as little as 2–3 nmoles of each substance to be detected. Each chromatographic analysis takes less than 1 h, and complex mixtures of bases and nucleosides are well resolved. Future developments in HPLC may include the use of radioactive substrates, allowing even greater detection sensitivity, to study the biosynthesis of modified bases and nucleosides found in low cellular abundance.

The simple nucleotides are phosphoric esters of purine or pyrimidine nucleosides and, thus, display inherent amphoteric behavior. Several isomeric forms exist, including 2'-, 3'-, and 5'-nucleoside monophosphates and 2', 3'- and 3', 5'-cyclic phosphates [19]. The phosphate group in nucleotides provides a convenient "anionic handle" on these molecules by which to separate them on ion-exchange columns. The order of elution is governed by nonionic interactions between solute and support as well as the pH value of the solvent. Cohn and Volkin [20] showed that 2'-, 3'-, and 5'-ribonucleotides are separated on Dowex-1 columns (formate form) by stepwise elution with increasing concentrations of formic acid and sodium formate. Gradient elution by sodium acetate buffer, pH 4, was later shown to improve resolution [21].

Nucleotides are also selectively excluded from cation-exchange columns. Blattner and Erickson [22] described two useful systems for the separation of mixtures of 2'- and 3'-ribonucleotides and 5'-deoxyribonucleotides on Dowex-50 columns (Fig. 18.1). HPLC on Aminex A-25 anion exchanger gives comprehensive resolution of most bases, nucleosides and nucleotides [23]. This rapid method is ideal for studying the metabolic fate of nucleoside analogs administered in cancer chemotherapy.

18.2.3. Thin-layer chromatography

Partition chromatography on paper [24] or thin layers of cellulose [25–27] has been used extensively to analyze the modified nucleoside composition of nucleic acids. Modified bases found in DNA are usually analyzed by one-dimensional ascending PC [28]. However, complex mixtures of modified ribonucleosides found in tRNA species [13] are more successfully resolved by using two-dimensional cellulose TLC. Nucleoside mixtures are developed in one direction in solvents of low pH, followed by a second development in a perpendicular direction at higher pH. When the chromatogram is viewed under UV light, each nucleoside appears as a single spot with characterstic R_F [25]. Randerath and coworkers [26,27] have introduced a more

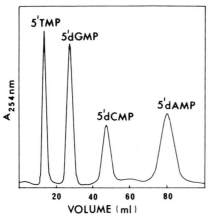

Fig. 18.1. Anion-exclusion chromatography of 5'-deoxyribonucleotides. Chicken erythrocyte DNA was extensively digested with DNase I and snake venom phosphodiesterase [14]. The reaction mixture was applied directly to a Bio-Rad AG 50W-X4 column (400-mesh, H^+ form, 30×1.1 cm ID, flowrate 55 ml/h). The 5'-deoxyribonucleotides were eluted in 0.1 M ammonium formate buffer, pH 3.2, and the absorbance at 254 nm was continuously monitored with an ISCO absorbance monitor. (G.J. Cowling, unpublished results based on the method of Blattner and Erikson [22]).

sensitive radiochemical method of nucleoside analysis. It involves oxidizing the *cis*-OH groups of the ribose sugar with sodium periodate to give a mixture of nucleoside dialdehydes. Sodium or potassium [^{3}H]borohydride reduces these to tritiated nucleoside di- and trialcohols. After two-dimensional TLC, products are visualized by low-temperature fluorography [27]. Fig. 18.2A shows the pattern of [^{3}H]ribonucleoside derivatives found in two serine-accepting species of *Drosophila* tRNA [29].

18.2.4. Electrophoresis

High-voltage paper ionophoresis in ammonium formate buffer, pH 3.5, has been commonly used to separate ribonucleotides in one dimension [30]. They migrate towards the anode but, with the exception of Gp, the 2'- and 3'-isomers do not separate. Conditions used for the two-dimensional TLC of deoxyribonucleotides and ribonucleotides are similar to those of nucleoside analysis [31]. A comparison of nucleotide and nucleoside analysis is shown in Fig. 18.2.

Thin layers of cellulose, treated with polyethyleneimine (PEI-cellulose), exhibit anion-exchange properties and can be used to separate nucleotides by development with increasing concentrations of a counteranion [e.g., with LiCl or $(NH_4)_2SO_4$] [25]. Nucleoside mono-, di-, and tri-phosphates are resolved by PEI-cellulose TLC, their mobilities being inversely related to the number of phosphate groups. Usually, 0.75 M NaH_2PO_4 is used as the developer [5].

18.2.5. Gel chromatography

Attempts to utilize the secondary binding forces of Sephadex, a crosslinked

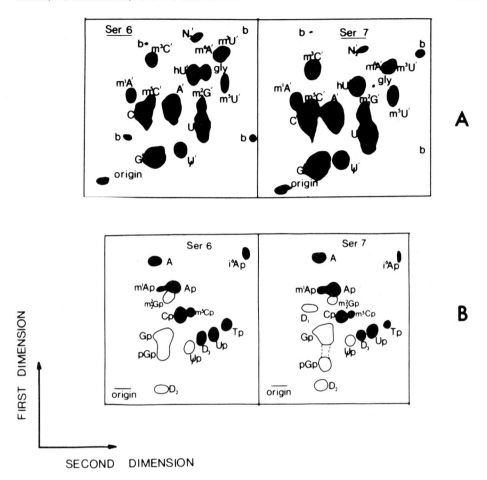

Fig. 18.2. Cellulose TLC of ^{3}H-labeled nucleoside derivatives from *Drosophila* tRNA$_6^{Ser}$ and tRNA$_7^{Ser}$ (A) and nucleotides from the same tRNA species (B). (A) Maps of ^{3}H-labeled nucleoside trialcohol derivatives, prepared by the method of Randerath et al. [26] and separated on cellulose thin-layer plates by development in the first direction with acetonitrile–ethyl acetate–1-butanol–2-propanol–6 N aqueous ammonia (40:30:10:20:27) and in the second direction with *t*-amyl alcohol–methyl ethyl ketone–water–formic acid (20:20:10:1). The figure is a representation of the fluorogram; background spots are labeled as "b" and glycerol as "gly". (B) Two-dimensional separation of RNase T$_2$ digest of serine tRNAs from *Drosophila* by cellulose TLC, developed in the first direction with isobutyric acid–0.5 M NH$_4$OH (5:3) and in the second direction with *t*-butyl alcohol–HCl–water (14:3:3). Under UV illumination the open areas were fluorescent in acid. N$_1$p has the same chromatographic mobility and spectral properties as mt^{6}Ap. D$_1$, D$_2$, and D$_3$ are probably dinucleoside diphosphates, resulting from the presence of 2′-O-methyl modifications. (Reproduced from *J. Biol. Chem.*, 250 (1975) 515, with permission [29]).

dextran, and Biogel, a beaded polyacrylamide gel, for the separation of nucleic acid constituents have been reported [32,33]. However, the major use of gel-exclusion chromatography is the fractionation of nucleic acids or their constituents by size. Nucleosides or nucleotides are excluded from Sephadex G-10 columns, whereas

References on p. B372

smaller inorganic molecules are included [34]. Alternatively, larger oligonucleotides (> 10–15 nucleotides) are excluded from Sephadex G-25 or G-50 columns, and mononucleotides included [3]. This is a commonly used method for removing excess nucleoside triphosphates from biosynthetic reactions [35].

18.3. OLIGONUCLEOTIDES

Polymers of nucleoside monophosphates, in which the phosphoric group connects the 3'-carbon of one ribosyl residue to the 5'-carbon of another, form the chemical basis of the nucleic acids. Molecules containing a smaller number of nucleotide residues (< 50) are usually referred to as oligonucleotides. Oligonucleotides containing a single type of purine or pyrimidine are called homopolymers. They arise either by synthesis from smaller molecules or from the breakdown of larger polymers. This section deals with the CC of these molecules; fractionation of oligonucleotides by paper and gel electrophoresis is treated in Chap. 18.7.

18.3.1. Anion-exchange chromatography

Early attempts to use polystyrene-based anion-exchange columns completely failed to resolve molecules larger than triribonucleotides, because of strong nonionic interactions between solute and column. Weaker anion exchangers, based on cellulose and first used for protein separations, have overcome this problem. Their reduced charge and high porosity allows much larger molecules to be separated. DEAE-cellulose columns fractionate di-, tri-, and tetraribonucleotides, regardless of base composition [36]. Similar experiments with linear salt gradient elution showed that deoxyribonucleotide homopolymers up to octamers can be resolved according to size [37]. If 7 M urea is included in the gradient buffer, DEAE-cellulose columns separate oligonucleotides containing up to 15 residues [38,39]. Urea abolishes the secondary nonionic interactions between the solute and the column at neutral pH and allows separation based solely on charge difference [38]. At lower pH, urea promotes the protonation of adenosine and cytidine residues, and so, oligonucleotides of the same length but different base composition can be separated by DEAE-cellulose chromatography with 7 M urea at pH 2.5–4 [40]. DEAE-Sephadex, which exhibits far less nonionic interactions with oligonucleotides even in the absence of urea, allows good resolution of oligonucleotides according to size [40].

HPLC on RPC5, devised by Egan and Kelmers [41], has been successfully applied to oligonucleotide fractionation [41–43]. The sorbent consists of minute polytrifluorochloroethylene beads, covered by a film of quaternary ammonium halides, containing three long alkyl residues (C_8–C_{12}). Columns of RPC5 behave, in most respects, like weak anion exchangers. Oligonucleotides of increasing size are eluted by progressively higher counteranion concentrations. We have observed that the plastic beads without coating display an affinity for purine but not pyrimidine nucleosides. Because the anion-exchange processes occur at the surface of the beads, equilibrium between adsorbed solute and displacing counteranion is rapidly established. This

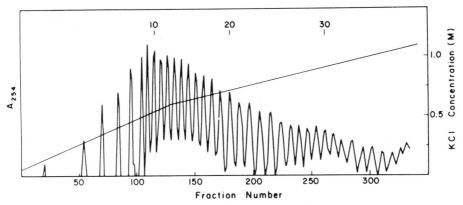

Fig. 18.3. RPC5 column chromatography of rA$_n$ oligonucleotides. Oligoribonucleotides, generated by alkali treatment of poly rA, were loaded onto the RPC5 column (100×2.5 cm ID) in 10 mM Tris–HCl, 0.1 mM EDTA, pH 8, and eluted with a linear KCl gradient (concentration as indicated). Oligonucleotide sizes (as indicated) were determined by analytical 20% polyacrylamide gel electorphoresis. (Reproduced from *Methods Enzymol.*, 65 (1980) 327, with permission [43]).

probably accounts for the high resolution seen in oligonucleotide separations. Size and base composition and – in some cases – sequence can affect the order of elution [43]. Unlike cellulose anion exchangers, which have a fibrous structure able to trap large nucleic acids, RPC5 columns fractionate oligonucleotides over a wide range of molecular weights. Examples include oligoribonucleotides [41], tRNAs (Chap. 18.4), DNA fragments (Chap. 18.6.3), and homopolymers of deoxyribonucleotides and ribonucleotides [42,43] (Fig. 18.3). Polymers of Gp or dGp, which have strong nonionic affinities for RPC5 supports, can be fractionated at alkaline pH [43].

18.4. TRANSFER RNA

Transfer RNA (tRNA) plays a major role in protein biosynthesis. There are between 40 and 150 different species of tRNA in each living cell. In all tRNA molecules that have been sequenced, the 3'-end of the polynucleotide chain has a final base sequence CCA$_{OH}$. The aminoacyl-tRNA-synthetases esterify the 3'-hydroxyl group of this terminal adenosine unit with an amino acid which corresponds specifically to a triplet of nucleotides (anticodon) in one loop of the tRNA molecule. Species which accept the same amino acid are called "isoacceptors". The discovery that tRNA is composed of distinct molecular species resulted in an intensive search for methods for their fractionation.

The small differences in physical and chemical properties between individual tRNA molecules have been used to achieve separation. tRNA can be partitioned between two immiscible phases and, as the partition coefficient of each tRNA species is slightly different, countercurrent distribution allows their separation. Doctor et al. [44] had early successes in purifying yeast tRNA[Ala] by partitioning unfractionated tRNA between 1.25 M sodium phosphate, pH 6.0, formamide, and 2-propanol. Other countercurrent distribution methods are reviewed elsewhere [45].

Anion-exchange chromatography is widely used for the fractionation of tRNA. At low ionic strength, the polyanionic phosphate backbone of tRNA binds easily to the cationic groups of DEAE-cellulose or DEAE-Sephadex. tRNA can then be eluted from the column with salt solutions of high ionic strength [46] or with salt concentration gradients [47]. The former method is commonly used for concentrating nucleic acids. The interaction of tRNA with DEAE-cellulose depends not only on ion exchange but also on weaker forces of attraction, such as multiple hydrogen bonding interactions between tRNA and the cellulose matrix. These secondary interactions are influenced by the pH and the temperature used for chromatography [47].

The characteristics of DEAE-cellulose are changed enormously if a portion of the hydroxyl groups in the cellulose matrix is benzoylated or naphthoylated. There are two effects of such substitutions: one is to shield some of the ionized groups and, hence, change the affinity of tRNA for the exchanger, the other is to introduce selective weak association between the bases and the aromatic groups introduced into the cellulose [48]. Gillam and coworkers [48–50] first prepared benzoylated DEAE-cellulose (BD-cellulose) and demonstrated that yeast tRNA could be fractionated with an increasing salt concentration gradient. tRNA species, containing lipophilic nucleosides, bind tightly to BD-cellulose columns and require for elution the addition of 10% ethanol or formamide to 1.0 M NaCl. The high resolution of tRNA species achieved by BD-cellulose has led to its extensive use in the fractionation of tRNAs [48–55] and in the analysis of isoaccepting species [51] from many biological systems. A typical opical density profile of *Drosophila* tRNA, fractionated by BD-cellulose chromatography, is shown in Fig. 18.4.

By using the property of BD-cellulose columns of preferentially retaining tRNA molecules containing residues with aromatic character, Tener and coworkers [53,54] have separated isoaccepting species by chemical derivatization. This involves the reaction of 2-napthoxyacetyl or phenoxyacetyl esters of N-hydroxysuccinimide with the NH$_2$ group of the amino acid in aminoacyl-tRNA. Nonderivatized tRNAs are removed by washing the BD-cellulose column with 1.2 M NaCl, and the reacted species are eluted by the addition of 10% ethanol to the NaCl solution. The reaction of aminoacyl-tRNA with the active ester of N-3-(4-hydroxyphenyl) propionate and subsequent purification by BD-cellulose chromatography, provide a convenient method for radiochemically labeling specific tRNAs with [125]I [54].

The so-called reversed-phase chromatography of tRNA is actually a combination of partition and ion-exchange methods. By using stationary phase, adsorbed on an inert matrix, and a salt concentration gradient as the moving phase, the partition coefficients of the tRNA species are continuously changed and, thus, another parameter is introduced into the experiment. Muench and Berg [56] used Sephadex G-25 as an inert support, K$^+$ in the aqueous stationary phase, and an increasing concentration of organic amine in the mobile phase. Although the fractionation of tRNA was successful, this method was superseded by one in which a water-immiscible organic extractant, coating an inert support, was used as the stationary phase

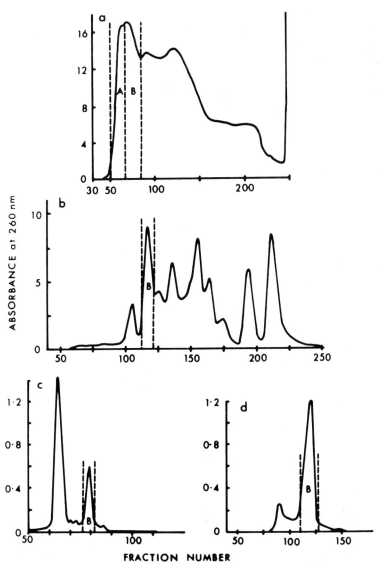

Fig. 18.4. Chromatographic purification of *Drosophila* initiator tRNA^Met. (a) tRNA (8700 $A_{260\ nm}$) was applied to a column of BD-cellulose (90 × 1.2 cm ID) in 0.5 M NaCl, 10 mM MgCl$_2$, and eluted at 20°C with a gradient of the same buffer to 1.0 M NaCl, 10 mM MgCl$_2$. Zones A and B contained tRNA^Met, as assayed by aminoacylation. (b) Pooled fractions from zone B (1854 $A_{260\ nm}$), containing initiator tRNA_i^Met, were further fractionated on a Sepharose 6B column (95 × 1.6 cm ID) with a decreasing ammonium sulfate gradient 1.5 M → 0. Zone B contains tRNA_i^Met. (c) Part of the pooled fractions from zone bB (90 $A_{260\ nm}$) was fractioned on a RPC5 column (36 × 2.5 cm ID) at 37°C by gradient elution from 0.45 M NaCl, 10 mM MgCl$_2$, 10 mM Tris–HCl, pH 8, to 1.0 M NaCl in the same buffer. The peak marked B contained tRNA_i^Met. (d) Further purification of fraction cB was obtained by fractionation on the same RPC5 column but by elution with a 0.45 M → 0.65 M NaCl gradient, containing 10 mM MgCl$_2$, 50 mM sodium formate buffer, pH 3.5. Peak B contained pure initiator tRNA^Met. (Methionine acceptance 1510–1720 pmoles amino acid/$A_{260\ nm}$) (Reproduced from *Nucleic Acids Res.*, 6 (1979) 421, with permission [55].)

References on p. B372

[57–60]. Individual tRNA species were separated by passing aqueous solutions through the column. The support consisted of acid-washed kieselguhr, which had been treated with dimethyldichlorosilane to eliminate adsorption effects. More recently, polychlorotrifluoroethylene resin has been used as an inert support for reversed-phase chromatography (see Chap. 18.3.1). These supports are coated with a water-insoluble quaternary alkylammonium salt of high molecular weight, which functions as an active extractant and effective anion exchanger.

A simple model of the mechanism of tRNA mobility on these RPC columns is mass-action controlled anion exchange. When tRNAs are applied to the column in a dilute NaCl solution, chloride ions bound to the quaternary ammonium extractant exchange for tRNA phosphate anionic sites. The tRNAs are thus retained on the column with essentially zero mobility. At higher sodium chloride concentrations, mass action then favors chloride binding with the quaternary ammonium extractant; the tRNAs are thus released from the support to the aqueous phase and eluted from the column. The resolution of tRNA species can be drastically altered by changes in (a) temperature, (b) presence of divalent metal ions, (c) pH (d) amount of tRNA used, and (e) flowrate [59]. These systems have been used extensively for the purification of tRNA species from bacterial and mammalian sources [59–61]. The improved resolution of tRNA species on the newer RPC5 chromatographic system has led to its widespread use for analysis of isoaccepting tRNA species. RPC5 chromatography of tRNA is shown in Figs. 18.4 and 18.5.

Chromatography on hydroxyapatite columns, eluted with increasing concentrations of sodium phosphate buffer, has been successfully used for the fractionation of tRNA species [62–65]. According to Bernardi's theory [66], Ca^{2+} ions on the crystal surface of hydroxyapatite bind specific phosphate residues in the RNA backbone. Separation therefore depends on the spatial distribution of phosphate groups in RNA which, in turn, is governed by the secondary and tertiary structure of the molecule. RNA molecules lacking secondary structure are desorbed at distinctly lower phosphate concentrations [66] and, in at least one case, hydroxyapatite chromatography can distinguish between uncharged tRNA^{Trp} from yeast and its corresponding aminoacyl-tRNA [67]. Disadvantages of hydroxyapatite column chromatography for tRNA fractionation arise from the weak physical properties of the crystalline support and the variability of its chromatographic behavior. Spencer and coworkers [68–70] have suggested a preparation method which overcomes these problems.

Sepharose columns bind unfractionated tRNA in 1.5 M $(NH_4)_2SO_4$. Fractionation is achieved by applying a linear gradient, ranging from 1.5 M $(NH_4)_2SO_4$ to water [71]. The binding forces are thought to arise from the nonpolar aliphatic chains of the gel, binding to similar sites on the tRNA, as well as hydrogen bonding interactions between tRNA molecules and the Sepharose. As the sulfate ion concentration decreases, tRNA species with increased nonpolar character are eluted (Fig. 18.4).

Dihydroxyboryl-cellulose contains borate groups, capable of binding to the 2′- and 3′-hydroxyls of the 3′-terminal adenosine of noncharged tRNAs. Aminoacyl-tRNAs pass through the column [72]. However, in practice, the high pH of the

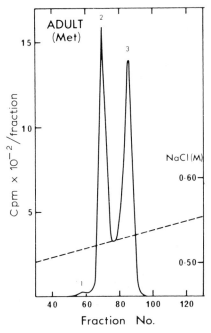

Fig. 18.5. Chromatography of ^{14}C-labeled methionyl-tRNAs from *Drosophila* on a RPC5 column. Crude *Drosophila* tRNA, aminoacylated with [^{14}C]methionine, was eluted from a RPC5 column (14×0.9 cm ID) at 22°C with a linear gradient of NaCl (concentration as indicated), containing 10 mM MgCl$_2$, 10 mM sodium acetate buffer, pH 4.5, 1 mM 2-mercaptoethanol. Fractions were counted after the addition of Aquasol. Methionyl-tRNA$_3^{Met}$ could be aminoacylated with *E. coli* aminoacyl-tRNA synthetase, indicating that it was the initiator species (see Fig. 18.4). (Reproduced from *Dev. Biol.*, 33 (1973) 185, with permission [61].)

eluent, required for sugar residues to complex with the support, has a tendency to destabilize the aminoacyl bonds of many isoaccepting species of tRNA. tRNAs containing the nucleoside Q, which itself possesses *cis*-hydroxyl groups, may be isolated by dihydroxyboryl-cellulose chromatography [72].

The purification of a single tRNA species usually requires several chromatographic steps. After each fractionation, the presence of a particular tRNA is measured by radioactive aminoacylation and the appropriate fractions are pooled and applied to a second chromatographic column. This process is continued until the amino acid acceptance of the tRNA approaches that of a pure species. An example of this strategy is shown in Fig. 18.4. In general, the order of elution of tRNA on anion-exchange columns (i.e. BD-cellulose, DEAE-Sephadex, and RPC5) is similar, but by varying the conditions most tRNAs can be separated. The high capacity of BD-cellulose and DEAE-Sephadex makes them obvious choices for the initial fractionations of tRNA. The advantages of hydroxyapatite and Sepharose CC become apparent when one is dealing with mixtures of tRNA that cannot be resolved by other forms of chromatography. These systems separate solely on the basis of nonionic interactions and, hence, the order of elution of tRNAs differs considerably from that seen with anion exchangers. Aminoacyl-tRNAs, especially those containing aromatic amino acids, behave differently from noncharged species when fractionated on anion-exchange columns. The presence of an amino acid often tends to retard the elution of a particular tRNA, and this may often be used to great advantage in certain purifications.

Polyacrylamide gel electrophoresis systems have been described for separating unfractionated tRNAs. *Escherichia coli* tRNA is resolved into as many as 45 distinct spots after being subjected to two-dimensional electrophoresis on a 16% poly-acrylamide gel [73]. Continuing development in high-resolution gel electrophoresis may allow rapid fractionations of tRNAs, comparable with those by chromatography.

18.5. HIGH-MOLECULAR-WEIGHT RNA

A great impetus to the study of RNA metabolism came from the discovery of messenger RNA (mRNA), which acts as the direct bearer of genetic information between a DNA sequence and the protein-synthesizing apparatus. The mRNAs are a heterogeneous collection of molecules of defined length and sequence. The ribosome, site of protein biosynthesis, also contains distinct species of ribosomal RNA, complexed with ribosomal proteins. These RNAs fall into three size classes: 5S, 16S, and 23S in prokaryotic cells; and 5S, 18S, and 28S in most eukaryotic cells.

Conventional anion-exchange chromatography has proved to be unsatisfactory for the purification of RNAs containing more than 100–150 nucleotide residues. Ribosomal and messenger RNAs are therefore usually separated by polyacrylamide or agarose gel electrophoresis, which offer the advantages of speed and high resolution of single-stranded species differing in size. Affinity chromatography relies on the high degree of specificity that can be achieved by molecular hybridization or chemical interactions between RNAs in solution and agents covalently attached to solid supports.

18.5.1. Affinity chromatography

Most eukaryotic messenger RNAs have 3′-polyadenylate tails, containing between 20–250 adenosine residues. At high ionic strength (1.0 M NaCl), the 3′-poly A sequences in mRNA can form base pairs with small stretches of poly dT or poly U, covalently attached to a column matrix. After the removal of RNAs containing no poly A by washing the column with concentrated salt solutions, mRNA can be recovered by decreasing the ionic strength of the eluent (0.1 M NaCl) [35]. Oligo dT-cellulose [74,75] and poly U-Sepharose [76,77] columns have been used for mRNA isolation. After having undergone continuous recycling through these sorbents, highly purified poly A-containing mRNA is recovered, free of ribosomal RNA contamination [78].

The chemical interaction between mercuric ions and thiol groups forms the basis of several affinity methods for the purification of specific RNAs. Mercurated RNA can be synthesized by direct treatment with mercuric acetate [34] or produced by in vitro DNA transcription in the presence of mercurated CTP or UTP [35]. Whereas RNA fails to bind to agarose columns containing sulfydryl groups, mercurated species bind tightly at low ionic strength. The inclusion of 10–100 mM 2-mercaptoethanol in the eluent allows the elution of mercurated species [34,35]. Although care is needed to eliminate nonspecific interactions between mercurated and nonmercurated RNAs during separation [35], this method has proved useful in the removal of endogenous RNAs from transcripts synthesized in vitro with mercurated nucleoside triphosphates [35,80].

A recent report suggested a further use of this affinity method for the purification of specific mRNAs [81]. Mercurated bacterial plasmid DNA, containing an inserted sequence of the messenger RNA to be isolated, is reassociated with total cellular RNA. When the products of this reaction are passed through a thiol–agarose column, only RNA sequences which have formed hybrids with the mercurated plasmid DNA bind to the column. The purified RNA species are released from the column by denaturing the hybrids, either by adding formamide to the eluate or by increasing the temperature. Finally, mercurated plasmid DNA is recovered by the addition of 2-mercaptoethanol to the eluent. Reeve et al. [82] have demonstrated that RNA chains possessing a γ-thiol group on the phosphate residue of the 5′-terminal nucleotide bind specifically to Hg–agarose columns. This provides a simple method for isolating RNA chains initiated in vitro.

18.5.2. Gel electrophoresis

The electrophoretic mobility of a substance is determined by the net electrostatic force acting on the particle, which depends on the potential gradient, the effective charge (taking account of counterion binding and shielded charges), and the frictional resistance. For nucleic acids, the constancy of the charge/mass ratio limits the possibility of separation by purely electrophoretic means, for each added unit of charge (nucleotide) is accompanied by an additional unit of frictional resistance. The frictional properties of double- and single-stranded species of RNA (and DNA) vary greatly and, yet, in practice, both native and denatured forms migrate in polyacrylamide or agarose gels with a rate proportional to the logarithm of their molecular weights. This relationship only holds for low-molecular-weight single- or double-stranded species (MW 10^7 in the weakest polyacrylamide gel). With increasing MW, the pore size of the support begins to limit separation by physically trapping the molecules.

In the case of RNA molecules possessing rigid secondary structure (composed of single- and double-stranded regions), migration during gel electrophoresis depends not only on size but also on base composition and sequence. This can be exploited, especially when the separation of RNA species of similar size but different base composition is attempted. In this case, the amount of denaturant used in the gel can be varied, or the denaturant may even be left out. On the other hand, the MW determination of RNA species by gel electrophoresis depends on the abolition of RNA secondary structure prior to separation.

References on p. B372

The gel electrophoresis of nucleic acids necessitates a compromise between high resolution (high gel concentration) and high-molecular-weight limits (low gel concentration) for any particular set of molecules. The upper MW limit may be extended by decreasing the concentration of crosslinker in the gel and/or the total concentration of the support. This has proved to be an effective method for the preparation and analysis of messenger and ribosomal RNAs [83,84]. It allows extremely accurate MW determinations of many RNA species. The problem of completely denaturing RNA species, especially those with high GC content, has been solved by two approaches:

(a) the addition of denaturants, such as urea or formamide, to RNA, gel, and running buffer;

(b) the reaction of RNA with reagents known to prevent the formation of secondary structure.

In the first approach, urea has been commonly used as a denaturant in the gel electrophoresis of RNA species. RNAs between 12 and 150 nucleotides long can be separated on high-percentage polyacrylamide gels, containing $7 M$ urea [85], and larger RNAs (1500–4500 nucleotides long) on low-percentage polyacrylamide or agarose gels, containing $6 M$ urea [86,87].

Complex mixtures of RNAs having the same MW but different sequence can be separated by exploiting the differences in migration rate as molecules encounter increasing concentrations of denaturant during electrophoretic fractionation. This principle was elegantly demonstrated by Gross et al. [88], who fractionated sea urchin histone mRNAs by electrophoresis in gels containing urea concentration gradients (Fig. 18.6). Urea has the effect of destabilizing the internal hydrogen bonding of RNA molecules, but under acid conditions it also promotes the protonation of adenosine and cytosine residues. Therefore, as molecules are subjected to increasing concentrations of urea, their charge begins to depend on base composition as well as size, and this is reflected in the electrophoretic separation by changes in the mobility of RNA species.

Another effective RNA denaturant is deionized formamide. When RNA, dissolved in 98% formamide, is subjected to electrophoresis in low-percentage polyacrylamide gels, also containing 98% formamide, the migration of RNA species is again proportional to their size [89,90]. Both analytical and preparative separations of many messenger and ribosomal RNAs have been accomplished by this method [78] (Fig. 18.7).

The secondary structure of RNA can be abolished prior to gel electrophoresis by several methods, including the reaction of RNA with $1.1 M$ formaldehyde [91–93], methyl mercury [94] or $1 M$ glyoxal in 50% aqueous DMSO at 50°C [95,96] (Fig. 18.8). These reagents react reversibly with purine and pyrimidine bases in RNA, destroying their ability to form internal hydrogen bonds.

Gel electrophoresis has seen widespread use in the purification of biologically active mRNAs [78] and in the accurate MW determination of RNA molecules [97]. Recently, it has been used to analyze RNA species by molecular hybridization. Bands of RNA, resolved by slab gel electrophoresis, are allowed to diffuse vertically onto strips of diazobenzyloxymethyl-cellulose (DBM-cellulose) paper with which

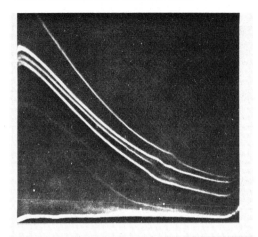

9% 4%
POLYACRYLAMIDE GRADIENT

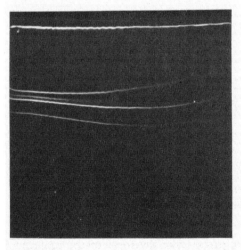

5M 0M
UREA GRADIENT

Fig. 18.6. Fluorograms of polyacrylamide slab gel separations of histone mRNAs of *Psammechinus*, Top, [^{3}H]uridine-labeled polyribosomal RNA (ca. $3 \cdot 10^5$ cpm) from *Psammechinus* was subjected to electrophoresis, from top to bottom, in a gel, containing a transverse gradient of acrylamide (4–9%) at 34°C, and detected by fluorography. Bottom, as top, except electrophoresis in a 6% polyacrylamide gel, containing a transverse gradient of 0–5 M urea, at 22°C. (Reproduced from *Cell*, 8 (1976) 455, with permission [88].)

References on p. B372

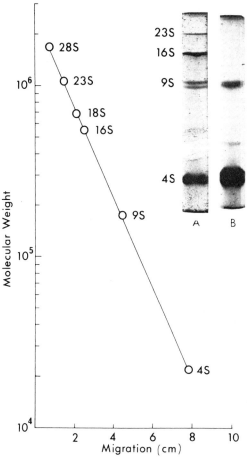

Fig. 18.7. Electrophoretic separation of RNAs on 6% polyacrylamide gels in 99% formamide. Plot of the logarithm of MW against the distance migrated by various ribosomal RNAs (28S, 23S, 18S, and 16S), globin mRNA (9S), and tRNA (4S). A shows the separation of 23S, 16S, rabbit globin mRNAs, and 4S RNA on 6% polyacrylamide gel in 99% formamide. Note separation of α- and β-globin mRNAs. B shows a comparative cylindrical gel separation of total mRNA and 4S RNA from chicken reticulocytes. Both gels were stained with 0.02% methylene blue. (D. Maryanka and G.J. Cowling, unpublished results.)

they bond covalently [98]. Such "RNA transfers" take the form of the replicated pattern of separated RNA species found in the gel and can be used in subsequent molecular hybridization experiments with radioactively labeled "probe" RNA or DNA sequences. These "probes" are usually a single RNA or DNA species. Hybrids formed between RNA species, attached to the activated cellulose and ^{32}P-labeled "probe" molecules can be detected by autoradiography [98]. Fig. 18.8 shows an example of this technique, used to measure the level of mRNAs during the developmental stages of the slime mold [99]. Analogous experiments with nitrocellulose strips to which purified DNA sequences are bound, have been used to isolate pure mRNAs found in gels after electrophoretic separation [100].

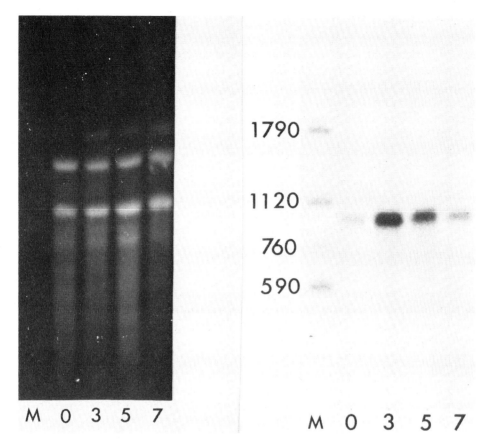

Fig. 18.8. Agarose gel electrophoretic separation of glyoxated RNA and autoradiogram of subsequent "RNA transfer". Left, Poly A$^+$ cytoplasmic RNA, isolated from the developing cells of *Dictyostelium discoideum* at the indicated stage of development (0, 3, 5, 7 h) was denatured with glyoxal [95] and subjected to electrophoresis on 1.5% agarose slab gel. After removal of glyoxal by alkali treatment, the gel was stained with ethidium bromide and photographed under UV illumination. The major bands are 27S rRNA (top) and 17S rRNA (bottom). Right, RNA in the gel was transferred to DBM-paper ("RNA transfer") and hybridized with ^{32}P-labeled pDd 812 DNA (a plasmid, containing sequences of a mRNA, observed late in the development of the slime mold). After hybridization, autoradiography was performed with preflashed X-ray film. Lanes M are Hinf 1 digests of SV$_{40}$ DNA, end-labeled with ^{32}P by means of T4 polynucleotide kinase. The size is indicated in base pairs of DNA. (Reproduced from *Cell,* 17 (1979) 903, with permission [99].)

18.6. DEOXYRIBONUCLEIC ACID

18.6.1. Hydroxyapatite chromatography

Native and denatured DNA molecules bind to hydroxyapatite columns at low phosphate concentrations (0.01–0.03 M sodium phosphate buffer, pH 6–8). At intermediate concentrations (0.12–0.14 M), single-stranded species are eluted, and at high concentrations (0.4–0.5 M), double-stranded molecules are released [101–104]. The presence of other monovalent anions does not seem to be important but urea and/or a detergent, such as sodium dodecyl sulfate improves the recovery of high-molecular-weight DNA. The phosphate ion concentration at which native or denatured DNA molecules are eluted from hydroxyapatite columns does not depend on the temperature. Fractionation on hydroxyapatite columns can therefore exploit either the process of DNA denaturation (separation of two complementary strands of DNA by heat or denaturing agents, such as formamide) or DNA reassociation (formation of a complementary double helix from single strands of DNA). Reassociation depends mainly on the concentrations of complementary strands and, hence, on the repetition frequency of the sequence involved [105]. By allowing controlled reassociation of genomic DNA with "probe" RNA or DNA sequences and then separating the hybrids on hydroxyapatite columns, the number of repetitive or single-copy sequences in genomic DNA can be determined [105].

18.6.2. Gel electrophoresis

The sheer size of genomic DNA restricts the chromatographic and electrophoretic methods that can be used for its separation. Even viral DNAs are many times larger than naturally occurring RNA species. Most bacterial chromosomes consist of a single DNA molecule with a MW of about $3 \cdot 10^9$ (10^6 base pairs), whereas animal cells contain three orders of magnitude more nuclear DNA (10^9 base pairs). The increased negative charge and nonionic interactions between high-molecular-weight DNA and the support make anion-exchange chromatography virtually impossible. Problems are also encountered in gel electrophoresis. Although the electrophoretic mobility of DNA molecules through gel matrices is, more or less, dependent on charge, the pore size of the gel limits the separation process by physically excluding large molecules [106,107]. Even the lowest percentage agarose gel is unable to resolve DNA molecules larger than $1 \cdot 10^4$ to $2 \cdot 10^4$ base pairs.

The analysis of genomic DNA sequences by gel electrophoresis did not begin until the discovery of restriction endonucleases. Hundreds of these enzymes have been isolated by now, and each cuts native DNA at specific sequences, usually between 2 and 8 base pairs long [108]. Cuts may be blunt-ended or staggered on double-stranded DNA. Because of the frequency and specificity of cutting sites, which vary with the restriction endonuclease used, it is possible to cleave high-molecular DNA into discrete fragments, many of which are in the size range that can be separated by polyacrylamide or agarose gel electrophoresis [107]. In this way, unique fragments of genomic DNA have been prepared for insertion into recombinant DNA vectors [109]. This section deals with the separation of such fragments.

Double-stranded DNAs migrate during polyacrylamide or agarose gel electrophoresis at a rate proportional to the logarithm of their MW [107]. This relationship is only true for low-molecular-weight species, for the reasons explained previously. The buffers for these electrophoresis experiments usually have a low ionic strength and contain Tris, adjusted to pH 7–8 with NaH_2PO_4, boric acid, or acetic acid [107]. The choice of gel system depends on the size of fragments to be separated. Species

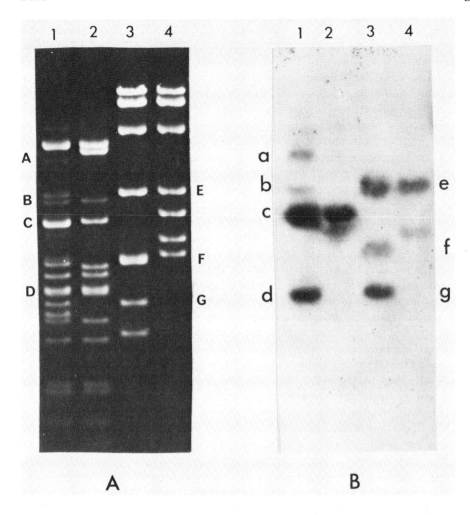

Fig. 18.9. Separation of cloned human DNA fragments by agarose gel electrophoresis and identification of fragments containing β-globin-related sequences. (A) Electrophoretic separation of restriction endonuclease digests of λ$_{Charon\ 4a}$DNAs, containing 15 kilobase inserted sequences around the human β-globin gene [113]. Lanes 1 and 3 are HβG1, digested with EcoR1 and Pst 1, respectively. Lanes 2 and 4 are the same digests of HβG3. The slab gel contains 0.8% agarose, the buffer 40 mM Tris, 5 mM sodium acetate, 1 mM EDTA, adjusted to pH 7.4 with acetic acid. Bands were visualized under UV light after the gel was stained with ethidium bromide (0.5 μg/ml). (B) DNA fragments in the gel were transferred to DBM-paper and hybridized with ^{32}P-labeled JW102 plasmid, containing human β-globin cDNA sequence. After hybridization, autoradiography was performed with preflashed X-ray film. Bands a–d (Lane 1) and e–g (Lane 3) contain human β-globin sequences and correspond to fragments A–D (Lane 1) and E–G (Lane 3) on the gel. Similar patterns of bands relate to fragments in Lanes 2 and 4. (Cloned λ bacteriophages HβG1 and HβG3 were gifts from T. Maniatis, Plasmid JW102 was a gift from B. Forget.)

References on p. B372

containing less than 400 base pairs are separated on 2–10% polyacrylamide gels, 400–1000 base pairs on composite polyacrylamide–agarose gels, and larger fragments on 0.5–1.0% agarose gels [107]. A wide range of sizes can be resolved on gels containing an acrylamide concentration gradient [106].

The analysis of gene sequences in DNA became possible through the technique introduced by Southern [110] for transferring DNA fragments, resolved by gel electrophoresis, to nitrocellulose strips ("Southern blotting"). This method preceded the "RNA transfer" technique, described in Chap. 18.5.2. DNA species diffusing out of the gel are trapped on nitrocellulose filters. Heating the filter covalently binds the DNA molecules to the nitrocellulose. This allows it to be incubated with ^{32}P-labeled RNA or DNA hybridization probes. Sequences on the filter which hybrize with the "probe" sequences can then be detected by autoradiography [110]. DNA species can also be transferred to DBM-cellulose [111,112]. The advantages of this system are that papers are reusable and bind small DNA fragments with greater efficiency than nitrocellulose. Many laboratories are using these techniques to locate specific gene sequences in nuclear DNA [113] or to probe the chromatin structure surrounding active gene regions [114]. Fig. 18.9 shows the agarose gel electrophoresis of DNA fragments and subsequent molecular hybridization analysis.

One serious limitation of gel electrophoresis for the isolation of specific DNA fragments is that molecules of similar size but different sequence tend to remain unresolved. One possible solution to this problem comes from the work of Fischer and Lerman [115] on two-dimensional gel separations of DNA fragments. Mixtures of DNA molecules are separated according to size by conventional gel electrophoresis, followed by further electrophoresis in a second dimension on a 4% polyacrylamide slab gel, containing a descending gradient of 4–30% formamide and $0.7–5.25\ M$ urea. The separation is performed at elevated temperature. This technique utilizes the abrupt decrease in electrophoretic mobility when part of a native DNA molecule undergoes denaturation or melts. Electrophoresis in the second dimension subjects DNA fragments to progressively stronger denaturing conditions, and distinct transitions in their mobility occur when part of a molecule melts. The complex relationship between mobility and sequence is, as yet, not understood [115], but this powerful technique, already able to resolve 10^3 discrete DNA fragments, has many potential uses.

Two-dimensional polyacrylamide gel electrophoresis can also be used to localize the 5′-ends, internal "splice" points and 3′-ends of nuclear or viral RNA molecules with respect to the template DNA sequence [116]. Hybrids are formed between RNA and DNA, digested with single-strand-specific endonuclease S1, and the resulting duplexes are sized in one dimension by gel electrophoresis. Electrophoresis in a second dimension under denaturing conditions can be used to determine the size of the single-stranded DNA molecules produced. Specific sequences can be detected by "Southern blotting" and subsequent molecular hybridization analysis [116]. Single-stranded DNA molecules are easily separated by alkaline polyacrylamide gel electrophoresis. Denatured samples are subjected to gel electrophoresis in 30 mM NaOH, 2 mM EDTA buffer [117]. Their electrophoretic behavior resembles that of single-stranded RNA species (see Chap. 18.5.2).

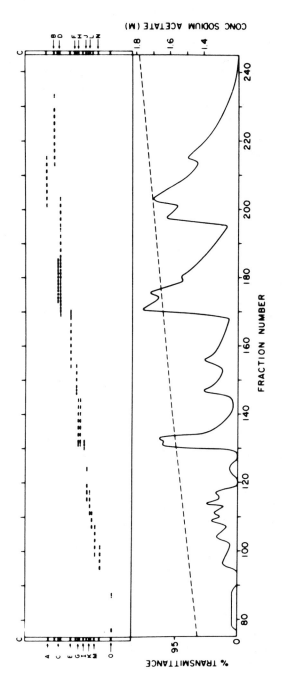

Fig. 18.10. Fractionation of a Hae III digest of pRZ2 DNA by RPC5 chromatography. The lower panel shows the transmittance profile. The upper panel is a drawing of the polyacrylamide gel electropherograms on certain fractions across the sodium acetate gradient (concentration as indicated, electrophoresis from top to bottom). A continuous line represents a strong band, a dotted line represents a weak one. On either end of the upper panel, the control digests (C) are shown and the bands are labeled in alphabetical order of decreasing size. (Reproduced from R.K. Patient, S.C. Hardies and R.D. Wells, *J. Biol. Chem.*, 254 (1976) 5542, with permission.)

18.6.3. RPC5 chromatography

Wells et al. [43] pioneered the use of RPC5 column chromatography for the fractionation of genomic DNA, cleaved by restriction endonucleases [118], and for the isolation of specific DNA fragments [119] (see, e.g., Fig. 18.10). A linear salt concentration gradient elutes DNA fragments in an order related to base composition, sequence, and MW. This technique is proving useful in separating DNA fragments of similar size that are unresolved by gel electrophoresis. Recovery from RPC5 chromatography is high ($>80\%$), and fractionated DNA species do not contain impurities – often found after gel electrophoresis – which inhibit enzyme reactions. When 12 mM NaOH is included in the eluate, single-stranded DNA molecules can be fractionated on RPC5 columns [120].

18.7. SEQUENCE ANALYSIS

Progress in sequence analysis depends on the progress made in oligonucleotide fractionation. The availability of purified RNA species and known ribonucleases which cut single-stranded RNA at particular bases helped to accelerate this research. Many RNAs are now completely sequenced, including numerous tRNAs, 5S, 5.8S, and messenger and ribosomal RNAs. DNA sequencing started later, but as a result of the introduction of rapid electrophoretic methods, it has rapidly overtaken older methods of RNA sequencing. Many excellent reviews have been written to explain the experimental strategies behind nucleic acid sequencing [5,121–123]. This section concentrates on the paper- and gel-electrophoretic methods used for the separation of ribonucleotides and deoxyribonucleotides.

In early sequencing experiments, oligoribonucleotides, produced by ribonuclease digestion of purified tRNAs, were fractionated by DEAE-cellulose chromatography [124]. The base composition and sequence of two overlapping sets of oligoribonucleotides were compared to deduce the full sequence. This laborious procedure allowed only small RNAs to be sequenced and relied on the presence of modified nucleosides to act as structural markers. The development of separation methods capable of handling much larger oligoribonucleotides, the products of partial ribonuclease digestion, that would simplify sequence deduction, began when Sanger [121] introduced the technique of high-voltage ionophoresis on modified cellulose papers. The earlier fractionations of oligoribonucleotides by paper ionophoresis, which had been extensively used for peptide separation, proved far from satisfactory, since the larger oligomers tended to streak and failed to resolve. Ionophoresis on cellulose acetate at pH 3.5 in one direction, followed by a second ionophoresis on DEAE-cellulose paper at pH 1.9 in a perpendicular direction gave clear resolution of di-, tri-, and tetraribonucleotides [125] (Fig. 18.11).

The problem of resolving larger oligoribonucleotides was solved by Brownlee and Sanger [126], who introduced a method known as "homochromatography". After an initial fractionation of oligoribuonucleotides by cellulose acetate ionophoresis, ^{32}P-labeled oligomers are transferred from the cellulose strip to a sheet of DEAE-cellulose paper. The radioactive oligomers are carried along by a mixture of nonlabeled oligonucleotides, usually a crude hydrolyzate of RNA, which act as counteranions in a chromatographic process where molecules of the same or smaller size displace the larger molecules. Homochromatography on DEAE-cellulose paper allows resolution of molecules up to 25 ribonucleotides long, but mixed DEAE-cellulose/cellulose TLC was found to extend this separation to oligomers up to 50 nucleotides long [126].

Two-dimensional gel electrophoresis can also be used to fractionate oligonucleotides [127]. The first electrophoretic step in $0.025\,M$ citric acid, $6\,M$ urea buffer separates oligomers according to base composition as well as length. The second step involves conventional electrophoresis at pH 8 [127]. This technique is capable of resolving oligomers up to 80 nucleotides in length. By comparing the pattern of spots ("fingerprints") obtained when oligoribonucleotides from known and unknown RNAs are resolved by two-dimensional electrophoresis (and/or homochromatography), areas of sequence homology can be established. Only oligomers which differ in sequence then need further analysis (Fig. 18.11).

Having separated the oligoribonucleotides produced by either complete or partial ribonuclease digestion, most RNA sequences are built up by comparing the order of bases in overlapping fragments. Each individual oligoribonucleotide is partially

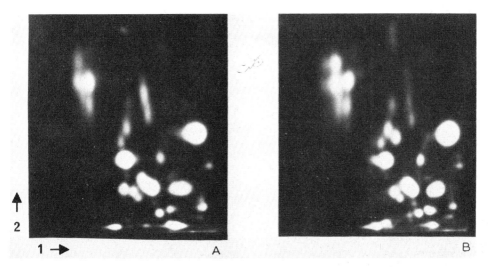

Fig. 18.11. Autoradiograms of two-dimensional separations of T_1-RNase digests of initiator tRNA^{Met} from *Drosophila* (A) and trout liver (B). Oligoribonucleotides, obtained after T_1-RNase digestion of isolated tRNA species, were 5′-end labeled with ^{32}P by means of T4-polynucleotide kinase. The labeled digests were applied to strips of cellulose acetate paper and fractionated in the first dimension by ionophoresis at pH 3.5. After transfer to DEAE-cellulose paper, they were subjected to further ionophoresis at pH 1.9 in a secod dimension. The pattern of spots was obtained by autoradiography with preflashed X-ray film. The similarity of the two "fingerprints" shows the strong conservation of nucleotide sequence among initiator tRNAs in general. Only 6 out of 75 positions are different in *Drosophila* from the highly conserved sequence of vertebrate cytoplasmic initiator tRNAs, trout liver being an example [55]. (R.J. Dunn and G.J. Cowling, unpublished results.)

References on p. B372

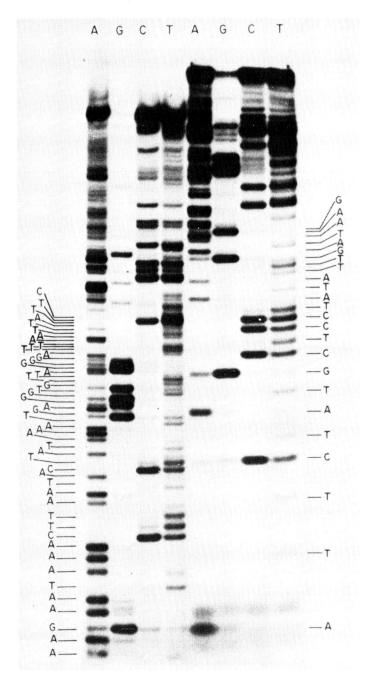

Fig. 18.12. Direct readout DNA sequencing. Colicin regions of Col E 1, sequenced by the method of Maxam and Gilbert [132]. The actual sequence is indicated on the left and right for each region. The reactions (A, G, C, T) loaded in each slot are noted at the top (see text). (Reproduced from *Nucleic Acids Res.*, 6 (1979) 2647, with permission [149].)

digested by a less specific ribonuclease (for example, RNase T_2) which cuts at every residue. The reaction is controlled so that a series of molecules is produced, each differing by one residue. Two-dimensional homochromatography on DEAE-cellulose plates resolves these products, and the characteristic change in mobility between adjacent molecules (differing by one nucleotide) indicates the type of base at that position. Thus, by reading across the chromatogram, a complete sequence can be deduced [5,128].

DNA and the newer RNA sequencing procedures also involve the preparation of a series of single-stranded fragments differing in length by one nucleotide. Four series of fragments are produced, each terminating at a particular base. They are prepared either by copying a single-stranded DNA molecule in the presence of low concentrations of chain terminators, such as dideoxyribonucleoside triphosphates [129–131], or by controlled chemical reactions that destroy selected bases in a single-stranded DNA molecule [132]. In both types of experiment, the reaction is controlled so that, on the average, only one site per chain is affected. Similar methods for generating series of oligoribonucleotides involve the aqueous hydrolysis of RNA at 80°C [27] or the digestion of RNA by ribonucleases that cleave at specific bases [133,134].

These oligonucleotide fragments are separated by high-resolution polyacrylamide slab gel electrophoresis [129,135] (see Fig. 18.12). Both the loading buffer and the gel contain urea as a denaturant to prevent the formation of internal secondary structure, especially in large oligomers. Early in the development of these gel systems, it was noticed that the degree of resolution governs the length of sequence that can be read and that this, in turn, depends on the thickness of the gel. Thick gels (1–2 mm) cause band spreading in the autoradiogram, as the source of radiation (DNA fragments) is further away from the film. Thinner gels (0.1–0.4 mm) give extremely high resolution of oligonucleotides and have the added advantage that the heat generated during electrophoresis causes a complete denaturation of GC-rich loops [129,135].

18.8. NUCLEOPROTEINS

Many proteins recognize and specifically interact with nucleic acid sequences. Such interactions play decisive roles in DNA replication, gene expression, RNA maturation, and protein biosynthesis. For most of their lifetime in the cell, DNA and RNA molecules are complexed with proteins which organize their functional sequences. Early studies of nucleic acid metabolism were concerned with the chemical and biological characterization of the molecules themselves, but more recently, molecular biologists have turned their attention to elucidating the structure and function of various nucleoproteins in order to solve the complex problems of replication and gene expression. Existing chromatographic and electrophoretic methods for separating nucleic acids have proved invaluable in solving nucleoprotein structures. Other separation methods have, in turn, allowed complex nucleoprotein mixtures to be analyzed with respect to both protein and nucleic acid sequences.

References on p. B372

18.8.1. Chromatin structure

DNA in the nuclei of all eukaryotic cells is packaged, by association with histone proteins, into chromatin fibers consisting of a linear repeating array of particles called nucleosomes. Each nucleosome contains 140 base pairs of DNA, supercoiled around an octamer of histones (two molecules of each histone H2A, H2B, H3, and H4). The nucleosomes are connected by "linker" DNA varying in length between 0 (yeast) and 100 (sea urchin sperm) base pairs, depending on organism and tissue. Staphylococcal nuclease cleaves in the linker DNA between nucleosomes, generating fragments which are multiples of the nucleosomal DNA length [136]. Deproteinized DNA fragments separate in agarose gel electrophoresis, thus allowing differences in the repeating unit of chromatin to be examined (see Fig. 18.13A). When chromatin is digested with DNase I, cleavage of the nucleosomal DNA is internal and results in the production of multiples of 10.4 (average) base pair DNA fragments [137]. These smaller DNA fragments separate in polyacrylamide gel electrophoresis (see Fig. 18.13B).

Studies on the kinetics of nuclease digestion of chromatin and analysis of the products by gel electrophoresis continue to be major approaches in solving the nucleoprotein architecture of the eukaryotic nucleus and the structural requirements for active gene expression [138].

Garrard and coworkers [139–142] have developed gel electrophoresis methods for probing the differences in mononucleosome populations. At low ionic strength, whole nucleosome particles are separated by polyacrylamide tube gel electrophoresis. The DNA and/or protein components of different nucleosome fractions can then be analyzed by two-dimensional slab gel electrophoresis in the presence of a detergent, usually SDS [139,140]. In the first dimension, mono- or oligonucleosomes migrate as dispersed protein–DNA complexes, while in the second dimension protein and DNA migrate independently. In order to trace precursor–product relationships, separated mono- or polynucleosomes are digested by nuclease in situ prior to electrophoresis in the second dimension. This is accomplished by incorporation in the first gel matrix and buffer of reversibly inactivated micrococcal nuclease or DNase I, which can be reactivated after electrophoresis by the addition of the appropriate divalent metal ion. This technique can distinguish between different levels of nucleosome organization with respect to core or spacer DNA length. Distinct classes of mononucleosome [139] as well as those containing histone H1 [140–142] have been examined by this method.

The sites at which an individual histone binds to the DNA in a nucleosome have been investigated by a second type of two-dimensional gel electrophoresis method [143,144]. Lysine groups of different histones are crosslinked to partially depurinated nucleosomal DNA at neutral pH. Splitting the DNA strand at the points of crosslinking leaves the 5'-terminal DNA fragments bound to the histone. In order to measure the position at which a particular histone binds to the strand of "core" DNA (relative to its 5'-end), the crosslinked nucleoproteins are separated by 7% polyacrylamide slab gel electrophoresis according to size. Before electrophoresis in a perpendicular direction, DNA or protein is removed in situ. Hydrolysis of DNA

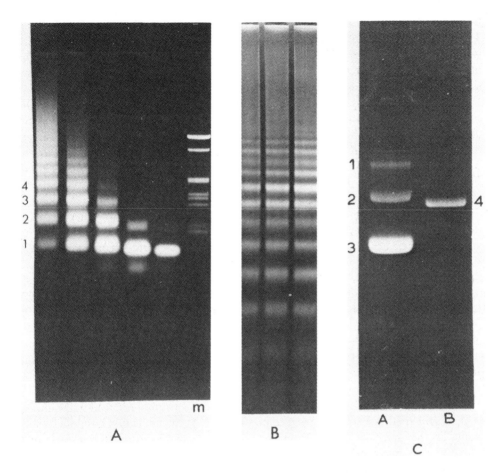

Fig. 18.13. Gel electrophoresis as a tool in chromatin research. (A) Chicken erythrocyte chromatin was digested with micrococcal nuclease (2 units/ml) for 0.5, 1, 2, 5, and 10 min at 37°C (lanes left to right). The DNA fragments produced were isolated by phenol extraction and subjected to electrophoresis in a 1% agarose slab gel containing 30 mM Tris, 36 mM NaH$_2$PO$_4$, 1 mM EDTA buffer, pH 7.5. DNA was visualized under UV light after staining with ethidium bromide. Bands corresponding to mono-, di-, tri-, and tetranucleosome DNA are labeled 1–4. Lane m contained a Hae III restriction endonuclease digest of PM2 DNA. Sizes are in base pairs (top to bottom) 1860, 1760, 1410, 890, 845, 672, 615, 525, 333, 295, 272, 167, 152, 120, and 95. (J. Allan, unpublished results.) (B) The DNA fragments obtained after chicken erythrocyte chromatin was digested with DNase I were subjected to electrophoresis in 10% polyacrylamide slab gel, containing 40 mM Tris, 20 mM sodium acetate, 2 mM EDTA, 7 M urea buffer, pH 7.8. The step-ladder pattern is the result of DNase I cutting the DNA within a nucleosome as 10 base pair intervals. (J. Allan, unpublished results.) (C) $E.\ coli$ PBr 322 plasmid DNA, subjected to electrophoresis in a 0.8% agarose slab gel. Lane A shows the three forms of plasmid DNA, as isolated. Band 1, relaxed circular, Band 2, linear; Band 3, supercoiled. Lane B shows the result of making one double-stranded break in the DNA with restriction endonuclease Bam H1 (Band 4). (D. Sullivan, unpublished results.)

References on p. B372

leaves ^{125}I-labeled histones, while [^{32}P]DNA remains after proteinase K digestion. Two types of autoradiograms are obtained. One indicates the size of the DNA fragment binding to each histone and hence, the position of the histone, while the other demonstrates the specific histone. In this way, a map of each histone binding site on the nucleosomal DNA is built up.

Agarose gel electrophoresis is proving a valuable method for establishing the conformational states of circular DNA molecules. Fig. 18.13C shows the distinct differences in electrophoretic mobility between linear, relaxed circular, and super-coiled forms of plasmid DNA. Some laboratories have already used this method to correlate the extent of DNA supercoiling with replication [145] and transcription [146]. In both of these processes, it is becoming clear that DNA conformation may play an important role.

Recent experiments have shown that nucleoproteins [147] and other proteins [148], separated by high-resolution gel electrophoresis, can be "transferred" to nitrocellulose or DBM-cellulose filters. Furthermore, transferred proteins retain the ability to bind DNA [148]. This type of experiment may allow the isolation and characterization of DNA and RNA binding proteins. *Lac* repressor protein from *E. coli* has been shown to bind specifically to DBM-cellulose columns to which DNA sequences, containing the *E. coli lac* operator site, have been covalently bound [149]. Unlike electrophoretic separations, which require rather specific ionic conditions, this type of affinity chromatography is ideal for investigating the effect of ionic strength, pH, and temperature on nucleic acid–protein interactions. It may also be possible to purify the "putative" proteins which control gene expression by DNA affinity chromatography.

The rapid proliferation of chromatographic and electrophoretic techniques in the nucleic acid field, accelerated by recombinant DNA "technology", continues especially in the areas of DNA fragment, mRNA, and nucleoprotein separations. Already, vast areas of the genome in many animals have been sequenced, and continuing progress in high-resolution gel electrophoresis, coupled with the use of computers to determine full sequences, will allow further progress to be made. Early researchers in this field adapted chromatographic and electrophoretic methods from other areas of science to achieve separation of mono- and oligonucleotides. Thirty years later, nucleic acid biochemistry is itself supplying the sciences of chromatography and electrophoresis with new concepts in macromolecular separation.

REFERENCES

1 J.N. Davidson, *The Biochemistry of the Nucleic Acids,* Chapman & Hall, London, 7th Edn., 1972.
2 J.D. Watson, *Molecular Biology of the Gene,* Benjamin, London, 3rd Edn., 1976.
3 L. Fischer, in T.S. Work and E. Work (Editors), *Laboratory Techniques in Biochemistry and Molecular Biology,* Vol. 1, Part II, *An Introduction to Gel Chromatography,* Elsevier, Amsterdam, 1969.
4 E.A. Peterson, in T.S. Work and E. Work (Editors), *Laboratory Techniques in Biochemistry and Molecular Biology,* Vol. 2, Part II, *Cellulosic Ion Exchangers,* Elsevier, Amsterdam, 1970.
5 G.G. Brownlee, in T.S. Work and E. Work (Editors), *Laboratory Techniques in Biochemistry and Molecular Biology,* Vol. 3, Part I, *Determination of Sequences in RNA,* Elsevier, Amsterdam, 1972.

6 H. Gould and H.R. Matthews, in T.S. Work and E. Work (Editors), *Laboratory Techniques in Biochemistry and Molecular Biology*, Vol. 4, Part II, *Separation Methods of Nucleic Acids and Oligonucleotides*, Elsevier, Amsterdam, 1976.

7 A. Marshak and H.J. Vogel, *J. Biol. Chem.*, 189 (1951) 597.

8 G. Bjork and I. Svensson, *Biochim. Biophys. Acta*, 138 (1967) 430.

9 A. Zamir, R.W. Holley and M. Marquisee, *J. Biol. Chem.*, 240 (1965) 1267.

10 P. Leder and M. Nirenberg, *Proc. Nat. Acad. Sci. U.S.*, 52 (1964) 420.

11 Y. Miyazawa and C.A. Thomas, *J. Mol. Biol.*, 11 (1965) 223.

12 J. Josse, A.D. Kaiser and A. Kornberg, *J. Biol. Chem.*, 236 (1961) 864.

13 B.G. Barrell and B.F.C. Clark (Editors), *Handbook of Nucleic Acid Sequence*, Joynson-Bruvvers, Oxford, 1974.

14 W.E. Cohn, *Science*, 109 (1949) 377.

15 C.F. Crampton, F.R. Frankel, A.M. Benson and A. Wade, *Anal. Biochem.*, 1 (1960) 249.

16 R.P. Singhal and W.E. Cohn, *Biochemistry*, 12 (1973) 1532.

17 R.P. Singhal and W.E. Cohn, *Anal. Biochem.*, 45 (1972) 585.

18 C.F. Mischke and E. Wickstrom, *Anal. Biochem.*, 105 (1980) 181.

19 D. Lipkin, R. Markham and W.H. Cook, *J. Amer. Chem. Soc.*, 81 (1959) 6075.

20 W.E. Cohn and E. Volkin, *Nature (London)*, 167 (1951) 483.

21 N.G. Anderson, J.G. Green, M.L. Barber, Sr. and F.C. Ladd, *Anal. Biochem.*, 6 (1963) 153.

22 F.R. Blattner and H.P. Erickson, *Anal. Biochem.*, 18 (1967) 220.

23 E. Nissinen, *Anal. Biochem.*, 106 (1980) 497.

24 B.G. Lane, *Biochim. Biophys. Acta*, 72 (1963) 110.

25 K. Randerath and H. Stuck, *J. Chromatogr.*, 6 (1961) 365.

26 E. Randerath, C.-T. Yu and K. Randerath, *Anal. Biochem.*, 48 (1972) 172.

27 K. Randerath, R.C. Gupta and E. Randerath, *Methods Enzymol.*, 65 (1980) 638.

28 B.F. Vanyushin, S.G. Tkacheve and A.N. Belozerky, *Nature (London)*, 225 (1970) 948.

29 B.N. White, R. Dunn, I. Gillam, G.M. Tener, D.J. Armstrong, F. Skoog, C.R. Frihart and N.J. Leonard, *J. Biol. Chem.*, 250 (1975) 515.

30 J.D. Smith, in E. Chargaff and J.N. Davidson (Editors), *The Nucleic Acids*, Vol. 1, Academic Press, New York, 1955, p. 267.

31 S. Nishimura, in J.N. Davidson and W.E. Cohn (Editors), *Progress in Nucleic Acid Research and Molecular Biology*, Vol. 12, Academic Press, New York, 1972, p. 50.

32 R. Braun, *Biochim. Biophys. Acta*, 149 (1967) 601.

33 M. Carrara and G. Bernardi, *Biochim. Biophys. Acta*, 155 (1968) 1.

34 R.M.K. Dale, E. Martin, D.C. Livingson and D.C. Ward, *Biochemistry*, 14 (1975) 2447.

35 H.J. Gould, D. Maryanka, S.J. Fey, G.J. Cowling and J. Allan, in G. Stein, J. Stein and L.J. Kleinsmith (Editors), *Methods in Cell Biology*, Vol. 19, *Chromatin and Chromosomal Protein Research, IV*, Academic Press, New York, 1978, p. 387.

36 M. Staehelin, *Biochim. Biophys. Acta*, 49 (1961) 11.

37 G.M. Tener, H.G. Khorana, R. Markham and E.H. Pol, *J. Amer. Chem. Soc.*, 80 (1958) 6223.

38 R.V. Tomlinson and G.M. Tener, *Biochemistry*, 2 (1963) 697.

39 H.G. Khorana and W.J. Conners, *Biochem. Prep.*, 11 (1966) 113.

40 G.M. Tener, *Methods Enzymol.*, 12 (1967) 398.

41 B.Z. Egan and A.D. Kelmers, *Methods Enzymol.*, 29 (1974) 469.

42 J.B. Dodgson and R.D. Wells, *Biochemistry*, 16 (1977) 2367.

43 R.D. Wells, S.C. Hardies, G.T. Horn, B. Klein, J.E. Larson, S.K. Neuendorf, N. Panayotatos, R.K. Patient and E. Selsing, *Methods Enzymol.*, 65 (1980) 327.

44 B.P. Doctor, J. Apgar and R.W. Holley, *J. Biol. Chem.*, 236 (1961) 1117.

45 S.R. Ayad, *Techniques of Nucleic Acid Fractionation*, Wiley-Interscience, London, 1972, p. 85.

46 Y. Kawade, T. Okamoto and Y. Yamamoto, *Biochem. Biophys. Res. Commun.*, 10 (1963) 200.

47 J.D. Cherayil and R.M. Bock, *Biochemistry*, 4 (1965) 1174.

48 I. Gillam, S. Millward, D. Blew, M. von Tigerstrom, E. Wimmer and G.M. Tener, *Biochemistry*, 6 (1967) 3043.

49 I. Gillam, D. Blew, R.C. Warrington, M. von Tigerstrom and G.M. Tener, *Biochemistry*, 7 (1968) 3459.

50 I.C. Gillam and G.M. Tener, *Methods Enzymol.*, 20 (1971) 55.

51 B.N. White and G.M. Tener, *Anal. Biochem.*, 55 (1973) 394.

52 B.N. White and G.M. Tener, *Biochim. Biophys. Acta*, 312 (1973) 267.

53 B.N. White and G.M. Tener, *Can. J. Biochem.*, 51 (1973) 896.

54 G.M. Tener, A.D. Delaney, T.A. Grigliatti, G.J. Cowling and I.C. Gillam, *Biochemistry*, 17 (1978) 741.

55 S. Silverman, J. Heckman, G.J. Cowling, A.D. Delaney, R.J. Dunn, I.C. Gillam, G.M. Tener, D. Söll and U.L. RajBhandary, *Nucleic Acids Res.*, 6 (1979) 421.

56 K.H. Muench and P. Berg, *Biochemistry*, 5 (1966) 970.

57 A.D. Kelmers, G.D. Novelli and M.P. Stulberg, *J. Biol. Chem.*, 240 (1965) 3979.

58 A.D. Kelmers, *J. Biol. Chem.*, 241 (1966) 3540.

59 A.D. Kelmers, H.O. Weeren, J.F. Weiss, R.L. Pearson, M.P. Stulberg and G.D. Novelli, *Methods Enzymol.*, 20 (1971) 9.

60 L.C. Waters and G.D. Novelli, *Methods Enzymol.*, 20 (1971) 39.

61 B.N. White, G.M. Tener, J. Holden and D.T. Suzuki, *Dev. Biol.*, 33 (1973) 185.

62 R.M. Kothari and V. Shankar, *J. Chromatogr.*, 98 (1974) 449.

63 R.L. Pearson and A.D. Kelmers, *J. Biol. Chem.*, 241 (1966) 767.

64 P. Schofield, *Biochemistry*, 9 (1970) 1694.

65 K.H. Muench, in J.N. Davidson and W.E. Cohn (Editors), *Procedures in Nucleic Acid Research*, Vol. 2, Academic Press, New York, 1971, p. 515.

66 G. Bernardi, *Nature (London)*, 206 (1965) 779.

67 K.H. Muench, *Biochemistry*, 8 (1969) 4880.

68 M. Spencer and M. Grynpas, *J. Chromatogr.*, 166 (1978) 423.

69 M. Spencer, *J. Chromatogr.*, 166 (1978) 435.

70 M. Spencer, E.J. Neave and N.L. Webb, *J. Chromatogr.*, 166 (1978) 447.

71 W.M. Holmes, R.E. Hurd, B.R. Reid, R.A. Rimerman and G.W. Hatfield, *Proc. Nat. Acad. Sci. U.S.*, 72 (1975) 1068.

72 T.F. McCutchan, P.T. Gilham and D. Söll, *Nucleic Acids Res.*, 2 (1975) 853.

73 F. Varricchio and H.J. Ernst, *Anal. Biochem.*, 68 (1975) 485.

74 C. Astell and M. Smith, *J. Biol. Chem.*, 246 (1971) 1944.

75 H. Aviv and P. Leder, *Proc. Nat. Acad. Sci. U.S.*, 69 (1972) 1408.

76 U. Lindberg and T. Persson, *Eur. J. Biochem.*, 31 (1972) 246.

77 M. Adesnik, M. Salditt, W. Thomas and J.E. Darnell, *J. Mol. Biol.*, 71 (1972) 21.

78 J.M. Taylor, *Ann. Rev. Biochem.*, 48 (1979) 681.

79 R.M.K. Dale and D.C. Ward, *Biochemistry*, 14 (1975) 2458.

80 R. Weinmann and L.O. Aiello, *Proc. Nat. Acad. Sci. U.S.*, 75 (1978) 1662.

81 S. Longacre and B. Mach, *Methods Enzymol.*, 69 (1979) 192.

82 A.E. Reeve, M.M. Smith, V. Pigiet and R.C.C. Huang, *Biochemistry*, 16 (1977) 4464.

83 A.C. Peacock and C.W. Dingman, *Biochemistry*, 7 (1968) 668.

84 U.E. Leoning, *Biochem. J.*, 113 (1969) 131.

85 T. Maniatis and A. Efstratiadis, *Methods Enzymol.*, 65 (1980) 299.

86 L. Reijnders, P. Sloof, J. Sival and P. Borst, *Biochim. Biophys. Acta*, 324 (1973) 320.

87 J.M. Rosen, S.L.C. Woo, J.W. Holder, A.R. Means and B.W. O'Malley, *Biochemistry*, 14 (1975) 69.

88 K. Gross, E. Probst, W. Schaffner and M. Birnstiel, *Cell*, 8 (1976) 455.

89 J.C. Pinder, D.Z. Staynov and W.B. Gratzer, *Biochemistry*, 13 (1974) 5373.

90 D.Z. Staynov, J.C. Pinder and W.B. Gratzer, *Nature (New Biol.)*, 235 (1972) 108.

91 H. Boedtker, *Biochim. Biophys. Acta*, 240 (1971) 448.

92 M.W. Schwinghamer and R.J. Shepherd, *Anal. Biochem.*, 103 (1980) 426.

93 N. Rave, R. Crkvenjakov and H. Boedtker, *Nucleic Acids Res.*, 6 (1979) 3559.

94 J.M. Bailey and N. Davidson, *Anal. Biochem.*, 70 (1976) 75.

95 G.K. McMaster and G.C. Carmichael, *Proc. Nat. Acad. Sci. U.S.*, 74 (1977) 4835.

96 G.K. McMaster and G.C. Carmichael, *Methods Enzymol.*, 65 (1980) 380.

97 H. Lehrach, D. Diamond, J.M. Wozney and H. Boedtker, *Biochemistry*, 16 (1977) 4743.
98 J.C. Alwine, D.J. Kemp and G.R. Stark, *Proc. Nat. Acad. Sci. U.S.*, 74 (1977) 5350.
99 J.G. Williams, M.M. Lloyd and J.M. Devine, *Cell*, 17 (1979) 903.
100 J. Burckhardt, J. Telford and M.L. Birnstiel, *Nucleic Acids Res.*, 6 (1979) 2963.
101 G. Bernardi, *Methods Enzymol.*, 21 (1971) 95.
102 G. Bernardi, *Biochim. Biophys. Acta*, 174 (1969) 423, 435.
103 H.G. Martinson, *Biochemistry*, 12 (1973) 139, 145.
104 H.G. Martinson, *Biochemistry*, 12 (1973) 2731.
105 R.J. Britten, D.E. Graham and B.R. Neufeld, *Methods Enzymol.*, 29 (1974) 363.
106 P.G.N. Jeppesen, *Methods Enzymol.*, 65 (1980) 305.
107 E. Southern, *Methods Enzymol.*, 65 (1980) 152.
108 R.J. Roberts, *Methods Enzymol.*, 65 (1980) 1.
109 A.M. Chakrabarty (Editor), *Genetic Engineering*, CRC Press, Boca Raton, FL, 1978.
110 E. Southern, *J. Mol. Biol.*, 98 (1975) 503.
111 J. Reiser, J. Renart and G.R. Stark, *Biochem. Biophys. Res. Commun.*, 85 (1978) 1104.
112 G.M. Wahl, M. Stern and G.R. Stark, *Proc. Nat. Acad. Sci. U.S.*, 76 (1979) 3683.
113 R.M. Lawn, E.F. Fritsch, R.C. Parker, G. Blake and T. Maniatis, *Cell*, 15 (1978) 1157.
114 J. Stalder, A. Larsen, J.D. Engel, M. Dolan, M. Groudine and H. Weintraub, *Cell*, 20 (1980) 451.
115 S.G. Fischer and L.S. Lerman, *Methods Enzymol.*, 68 (1979) 183.
116 J. Favaloro, R. Treisman and R. Kamen, *Methods Enzymol.*, 68 (1979) 719.
117 M.W. McDonnel, M.N. Siman and W.F. Studier, *J. Mol. Biol.*, 110 (1977) 119.
118 S.C. Hardies and R.D. Wells, *Proc. Nat. Acad. Sci. U.S.*, 73 (1976) 3117.
119 J.E. Larson, S.C. Hardies, R.K. Patient and R.D. Wells, *J. Biol. Chem.*, 254 (1979) 5535.
120 H. Eshaghpour and D.M. Crothers, *Nucleic Acids Res.*, 5 (1978) 13.
121 F. Sanger, *Biochem. J.*, 124 (1971) 833.
122 S.M. Weissman, *Anal. Biochem.*, 98 (1979) 243.
123 R. Wu, *Ann. Rev. Biochem.*, 47 (1978) 607.
124 R.W. Holley, J. Apgar, G.A. Everett, J.T. Madison, M. Marquisse, S.H. Merrill, J.R. Penswick and A. Zamir, *Science*, 147 (1965) 1462.
125 F. Sanger, G.G. Brownlee and B.G. Barrell, *J. Mol. Biol.*, 13 (1965) 373.
126 G.G. Brownlee and F. Sanger, *Eur. J. Biochem.*, 11 (1969) 395.
127 R. DeWachter and W. Fiers, *Anal. Biochem.*, 49 (1972) 184.
128 M. Silberklang, A.M. Gillum and U.L. RajBhandary, *Nucleic Acids Res.*, 4 (1977) 4091.
129 F. Sanger and A.R. Coulson, *J. Mol. Biol.*, 94 (1975) 441.
130 F. Sanger, S. Nicklen and A.R. Coulson, *Proc. Nat. Acad. Sci. U.S.*, 74 (1977) 5463.
131 P.H. Schreier and R. Cortese, *J. Mol. Biol.*, 129 (1979) 169.
132 A.M. Maxam and W. Gilbert, *Proc. Nat. Acad. Sci. U.S.*, 74 (1977) 560.
133 H. Donis-Keller, A.M. Maxam and W. Gilbert, *Nucleic Acids Res.*, 8 (1977) 2527.
134 A. Simonesits, G.G. Brownlee, R.S. Brown, J.R. Rubin and H. Guilley, *Nature (London)*, 269 (1977) 833.
135 F. Sanger and A.R. Coulson, *FEBS Lett.*, 87 (1978) 107.
136 R. Axel, W. Melchior, B. Sollner-Webb and G. Felsenfeld, *Proc. Nat. Acad. Sci. U.S.*, 71 (1974) 4101.
137 M. Noll, *Nucleic Acids Res.*, 1 (1974) 1573.
138 G. Felsenfeld, *Nature (London)*, 271 (1978) 115.
139 R.D. Todd and W.T. Garrard, *J. Biol. Chem.*, 252 (1977) 4729.
140 R.D. Todd and W.T. Garrard, *J. Biol. Chem.*, 254 (1979) 3074.
141 P.P. Nelson, S.C. Albright, J.M. Wiseman and W.T. Garrard, *J. Biol. Chem.*, 254 (1979) 11751.
142 J. Boulikas, J.M. Wiseman and W.J. Garrard, *Proc. Nat. Acad. Sci. U.S.*, 77 (1980) 127.
143 V.V. Shick, A.V. Belyavsky, S.G. Bavykin and A.D. Mirabekov, *J. Mol. Biol.*, 138 (1980) 491.
144 A.V. Belyavsky, S.G. Bavykin, E.G. Goguadze and A.D. Mirabekov, *J. Mol. Biol.*, 138 (1980) 519.
145 D.M.J. Lilley and M. Houghton, *Nucleic Acids Res.*, 6 (1979) 507.
146 B. Wasylyk and P. Chambon, *Eur. J. Biochem.*, 103 (1979) 219.
147 B. Bowen, J. Steinberg, U.K. Laemmli and H. Weintraub, *Nucleic Acids Res.*, 8 (1980) 1.
148 G. Herrick, *Nucleic Acids Res.*, 8 (1980) 3721.
149 R.K. Patient, *Nucleic Acids Res.*, 6 (1979) 2647.

Chapter 19

Porphyrins and related tetrapyrrolic substances

DAVID DOLPHIN

CONTENTS

19.1. STRUCTURES AND CHEMICAL PROPERTIES OF TETRAPYRROLES

The porphyrins, e.g., uroporphyrin III (19.1), heme [iron protoporphyrin (19.2)] and their closely related reduced partners, the photosynthetic pigments chlorophyll *a* (19.3) and bacteriochlorophyll *a* (19.4), as well as the catabolic products of the cyclic systems, which are open-chain bile pigments, e.g., bilirubin IXα (19.5), have various and important biochemical functions in nature. When the normal biosynthetic

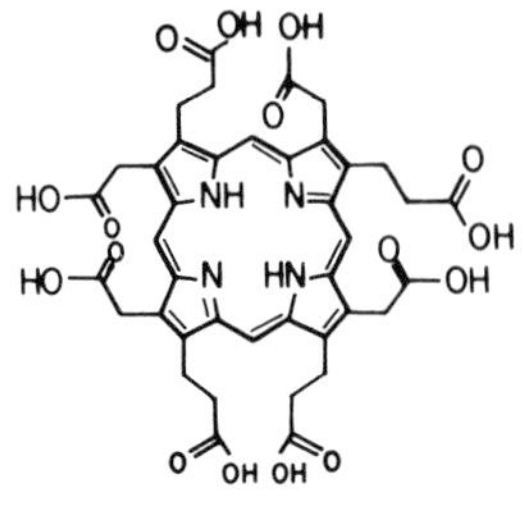

UROPORPHYRIN III
19.1

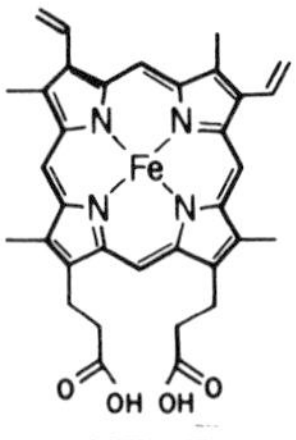

HEME
19.2

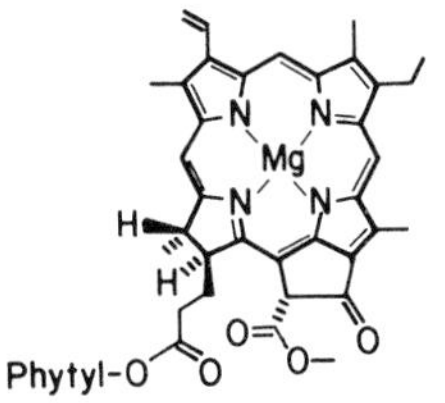

CHLOROPHYLL a
19.3

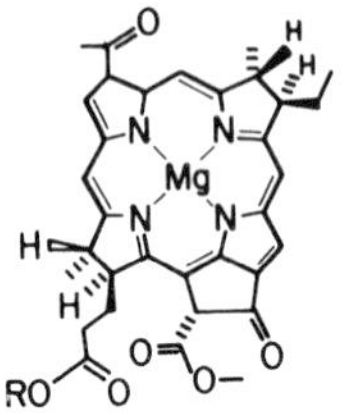

BACTERIOCHLOROPHYLL a
19.4

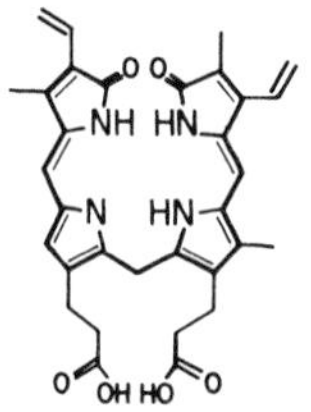

BILIRUBIN IX α
19.5

pathways for the production of the tetrapyrroles in man malfunction, a variety of pathological disorders, termed porphyrias, ensue. These disease states are frequently accompanied by excessive production, storage, and excretion of porphyrins. Because the tetrapyrroles, in addition to their biochemical roles, prove to be of great interest to other branches of the physical sciences, a wealth of information on various aspects of these subjects is available [1]. The relatively high molecular weights of the compounds, the dominating effect of the large, flat aromatic ring in the cyclic systems, and their generally low volatility makes separation and purification by classical techniques, such as distillation, crystallization, and sublimation very difficult. On the other hand, all of the tetrapyrrolic compounds described here are highly colored and many are fluorescent, and this makes their detection, both visually and spectrophotometrically, a relatively simple matter. It is perhaps not surprising, when all of the above factors are taken into account, that the first experiments in chromatography, by Tsvet [2], were used to separate leaf pigments, and that at the present time the method of choice for the isolation, separation, purification, and analysis of tetrapyrroles is chromatography.

The diverse nature of the peripheral substituents determines both the stability of the molecule and the chromatographic method best suited for its purification. Additional considerations are involved when dealing with the metallated systems, since the mode of axial ligation and the lability of the axial ligands will affect the net charge on the molecule, its solubility in solvent systems, and, consequently, its stability and interactions with chromatographic media. This diversity of chemical and physical properties does not allow for any simple generalization about the chromatographic method of choice nor for simple classifications of the compounds themselves. Consequently, we have chosen to divide the compounds into two main categories (hydrophobic and hydrophilic) and then to discuss the different methods separately according to the structure of the chromophores. Nevertheless, some general comments can be made, particularly about the stability and detection of tetrapyrrolic substances.

19.1.1. Stability

Many of the tetrapyrrolic compounds can act as photosensitizers for the production of singlet oxygen from ground-state triplet oxygen. Since the double bonds of the conjugated aromatic systems as well as the unsaturated peripheral substituents can react with singlet oxygen, it is advisable — at least until one is familiar with the chemical properties of the test material — to perform the chromatographic experiments in the absence of light (aluminum foil around the column or tank will usually suffice) and to protect the material before and after chromatographic separation from exposure to light. Furthermore, the aromaticity of the tetrapyrroles facilitates both the one-electron oxidation of the ring and the autoxidation of peripheral substituents via "benzylic-like" radicals. When they are adsorbed on a large surface area, such oxidations can be accelerated by both light and oxidants, e.g. peroxides in the solvents. Thus, as in the majority of other chromatographic experiments, only solvents of acceptable analytical purity should be used. Since metal-free tetrapyrroles exhibit great avidity for metal ions, care must always be exercised to ensure that the compounds being chromatographed are not metallated by impurities in the chromatographic solvents or media. In practice, this does not usually pose a problem at the preparative level. However, at the analytical level, particularly when compounds are extracted from natural sources, be they biological tissues or geological samples, one must be aware of the ready formation of artifacts generated by facile metallation. There are no easy or general methods for avoiding this problem other than ensuring that the sample during extraction, preparation, and chromatography has minimal contact with metal ions and metals (even a nickel spatula can cause contamination). Since porphyrins often exhibit binding constants for metal ions which are comparable to those of sequestering agents such as EDTA, the use of sequestering agents in low concentrations may prove ineffective and will most likely interfere with the chromatographic system in high concentrations.

At the other extreme, demetallation of metal complexes must also be guarded against. Most metalloporphyrins do not readily demetallate, but this is not always so. For instance, magnesium porphyrins are demetallated even by the traces of hydrochloric acid found in chloroform or by the anhydrous magnesium sulfate sometimes used as a drying agent. Facile demetallation of the magnesium-containing chlorophylls and bacteriochlorophylls to the corresponding pheophytins poses one of the major problems in the isolation and chromatographic separation of the photosynthetic pigments. In addition, being di- and tetrahydroporphyrins, they are subject to oxidation to porphyrins as well as to isomerization and other facile reactions producing altered chlorophylls. In fact, it is their delicate nature which has led to the development of the very gentle isolation and chromatographic techniques described below.

References on p. B403

19.1.2. Detection

As noted above, all of the tetrapyrrolic compounds described in this chapter are highly colored, showing intense absorption throughout the whole UV and visible region of the spectrum. This makes both the qualitative and quantitative analysis fairly easy. Visual observation of components during both CC and TLC is feasible, amounts of less than a microgram being detectable. Even greater sensitivity is possible in cases where the tetrapyrrolic compound fluoresces, i.e. for most metal-free porphyrins, bile pigments, and chlorophylls. The fluorescence on thin-layer plates is often much more easily observed when it is still wet with the developing solvent. On paper chromatograms, the intensity of fluorescence can be similarly enhanced by spraying with isooctane [3]. In both cases nanogram quantities can thus be visualized. Excitation at the strong Soret absorption bands around 400 nm gives rise to the most intense fluorescence but excitation at 366 nm (where many commercially available UV lamps emit) also produces intense fluorescence. Since the intensity of absorption and emission for most porphyrins is high, the use of reagent sprays, practiced in TLC of other organic compounds, does not usually increase the sensitivity of detection. However, for the many metalloporphyrins that do not fluoresce, layers with added fluorophores or spraying the developed chromatogram with a fluorochrome, such as fluoranthene [3], are recommended.

A useful spray for the detection of hemins has been reported [4], which detects nanogram quantities when a fresh solution is used. The original recipe calls for benzidine, but because it is highly carcinogenic, we recommend that o-dianisidine [5] be used instead. A saturated solution of o-dianisidine hydrochloride in methanol (50 ml) is diluted with water (25 ml) and then glacial acetic acid (10 ml), 3% hydrogen peroxide (5 ml), and pyridine (1 ml) are added. However, since all aromatic amines are likely carcinogens, great care should be taken in handling this spray.

Fluoroscanning of thin-layer plates, particularly for the clinical detection of naturally occurring porphyrins, has been employed in semi-quantitative analysis [6]. However, with the advent of HPLC, which is adaptable to all of the chromatographic systems described below, the quantitation of tetrapyrrolic compounds is now greatly simplified. Thus, with a fixed-wavelength detector, or by scanning the whole visible region [7], nanogram quantities of porphyrins are readily quantitated; and with a fluorescence detector as little as 30 pg of porphyrins can be detected in urine [8].

19.2. WATER-SOLUBLE COMPOUNDS

19.2.1. Porphyrin carboxylic acids

The close structural similarity among many of the naturally occurring porphyrins, such as the uroporphyrins I (19.6) and III (19.1) or the coproporphyrins I (19.7) and III (19.8), and the similarity in the pK_a values of the carboxyl groups attached to porphyrins pose some very special problems in chromatographic separation. The isolation of water-soluble, carboxylic group-containing porphyrins from other water-soluble metabolites is often best achieved by esterifying the carboxyl groups to produce compounds that are water-insoluble but soluble in a variety of organic solvents. Since the naturally occurring porphyrins are often isolated after esterification, this has the added advantage that numerous methods for the separation and analysis of the porphyrin esters are already available (cf. Chap. 19.3.1.1.). We recommend that porphyrins containing carboxyl groups be separated and analyzed as their methyl esters, except under circumstances where, e.g., the compounds are being used in reconstitution studies with heme proteins. In addition, the analysis of native porphyrin carboxylic acids plays an important role in the direct analysis — particularly by HPLC — of biological samples without prior isolation or derivatization of the porphyrins.

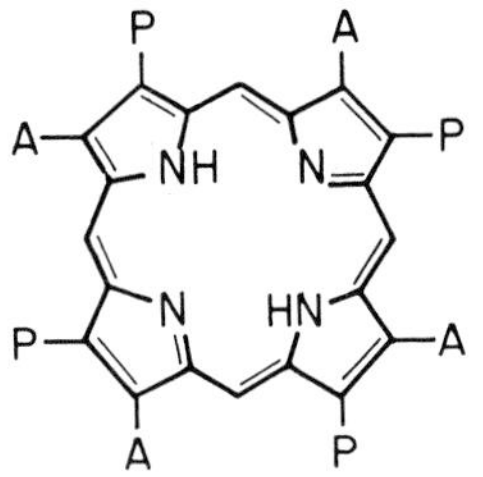

UROPORPHYRIN I
19.6

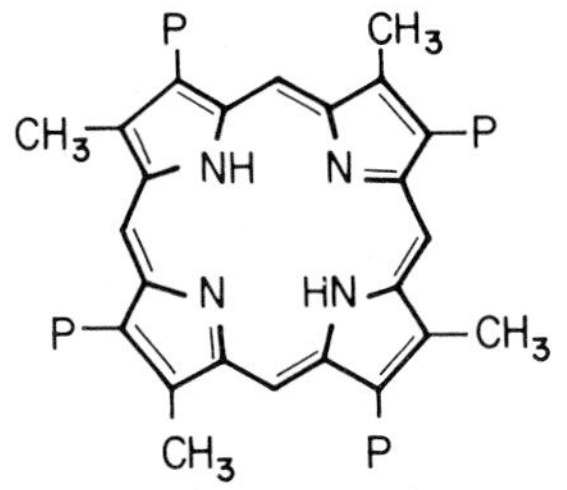

COPROPORPHYRIN I
19.7

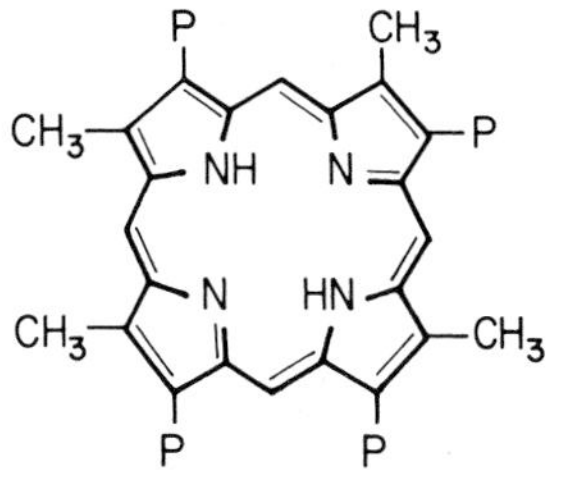

COPROPORPHYRIN III
19.8

$A = -CH_2CO_2H : P = -CH_2CH_2CO_2H$

19.2.1.1. Paper chromatography

Because development takes so much longer in PC than in cellulose TLC and HPLC, PC is now mainly of historical interest. The original method of Nicholas and Rimington [9], who used a mixture of water-saturated 2,4- and 2,5-lutidines, was modified by Kehl and Stich [10], who recommended the use of 2,6-lutidine, and has seen further modification by Eriksen [11] and Mauzerall [12].

A typical chromatogram may be developed as follows. The porphyrins, dissolved in $2N$ NH$_4$OH, are applied to the paper (Whatman No. 1 or its equivalent) as a spot containing 1–10 μg, and the solvent is allowed to evaporate. The chromatogram is developed with a 10:7 mixture of 2,6-lutidine and $0.7N$ NH$_4$OH, the solvent being allowed to ascend 20–30 cm (this may take overnight). With this system, the R_F increases as the number of carboxyl groups decreases. As the absolute R_F values depend on a number of variables, the use of standards (available from Porphyrin Products, P.O. Box 31, Logan, UT, USA) is advisable. Although this system effectively separates porphyrins according to the number of carboxyl groups, it is less effective in resolving porphyrins containing the same number of carboxyl groups. Thus, while this system separates coproporphyrins I (19.7) and III (19.8), neither uroporphyrins I (19.6) and III (19.1) nor a number of important or naturally occurring porphyrin dicarboxylic acids are resolved. Hematoporphyrin (19.9), protoporphyrin IX (19.10), mesoporphyrin IX (19.11, $R_1 = R_2 = Et$), and deuteroporphyrin (19.11, $R_1 = R_2 = H$) can be separated by the method of Belcher et al. [13], where pyridine–0.2 M sodium borate (pH 8.6) (1:9) is the developing solvent.

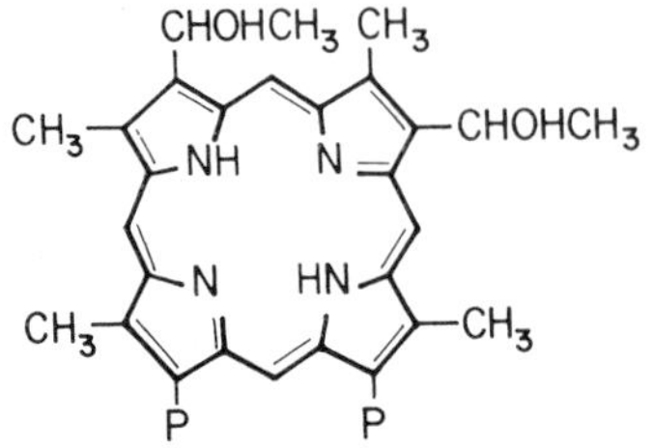

HEMATOPORPHYRIN
19.9

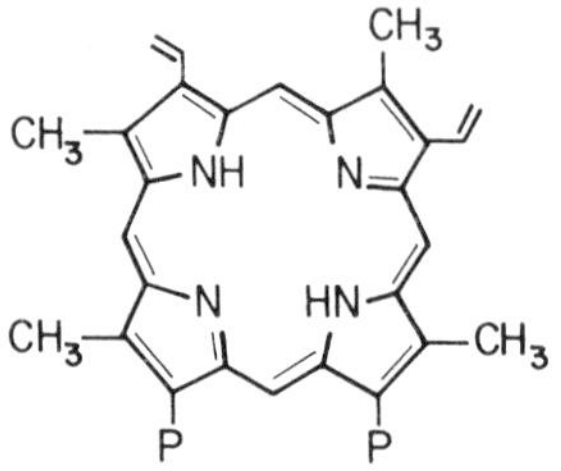

PROTOPORPHYRIN IX
19.10

For faster development of paper chromatograms, With [14] recommended a 0.1 M LiCl solution as the mobile phase in an atmosphere of ammonia. In this sytem the R_F values are in the same order as the number of carboxyl groups on the porphyrin.

19.2.1.2. Thin-layer chromatography

TLC of porphyrin carboxylic acids has been carried out on cellulose, talc, and silica gel layers. Cellulose and talc are included here more for their historical significance than practical usefulness. In our experience, silica gel gives superior resolution in less time. Application of TLC to the analysis of porphyrins from the body fluids, tissues, and feces of porphyric patients has revolutionized the diagnosis of these diseases [15].

The results of cellulose TLC [16–19] parallel those described above for PC. Separations of uroporphyrin, coproporphyrin and protoporphyrin [17] were accomplished with Jensen's [20] solvent system of 2,6-lutidine–water (10:3) in an ammonia atmosphere. The method was further improved by Yuan and Russell [18] who, using a mixture of 2,6-lutidine–ammonia–water–0.1 M EDTA (500:210:140:1), were able to separate coproporphyrins I and III and deutero- and mesoporphyrins. However, optimum development, at 25°C in the dark, took ca. 24 h and, although the resolution on the cellulose layer was somewhat better than on paper, the time required for development was much greater than that on silica gel plates, described below.

With [21,22] described the chromatography of porphyrin carboxylic acids on thin layers of talc, and Okuda et al. [23] and Belcher et al. [13] published improvements in these systems. A mobile phase of ethanol–2,6-lutidine–water (30:3:67) in an atmosphere saturated with ammonia vapor, separated porphyrin S-411 (19.12) from coproporphyrin. Hemato-, deutero-, meso-, pempto-, and protoporphyrins were similarly chromatographed on talc, but the separations achieved were inferior to those with PC, reported by the same author [14].

19.11

PORPHYRIN S-4II

19.12

Mundschenk [19] has described the TLC of porphyrin carboxylic acids on silica gel, but we prefer the method of Ellfolk and Sievers [24]. The solvent system chloroform–methanol (17:3), containing formic acid in 0.3 M concentration, separates porphyrin carboxylic acids, the porphyrins containing the lowest number of carboxyl groups having the highest mobility. In this system, protoporphyrin, mesoporphyrin, deuteroporphyrin, hematoporphyrin, and the half-hydrated forms of protoporphyrin, 2-hydroxyethyl-4-vinylporphyrin (19.11, R_1 = hydroxyethyl, R_2 = vinyl) and 2-vinyl-4-hydroxyethylporphyrin (19.11, R_1 = vinyl, R_2 = hydroxyethyl), can be separated.

19.2.1.3. Column chromatography

There are still few reports of the separation of porphyrin carboxylic acids by HPLC, but the high resolution, speed, and convenience which it offers suggest that much activity can be expected in this area in the near future, particularly if the porphyrins do not have to be isolated from contaminants or transformed into derivatives (see Fig. 19.1). Evans et al. [25] showed that porphyrin di-, tetra-, and pentacarboxylic acids could be separated in less than 10 min on a Pellionex SA ion-exchange column eluted with a gradient of methanol and acetic acid. More recently, Bonnett et al. [26] have studied the separation of porphyrin carboxylic acids using a reversed-phase column (μBondapak C_{18}, Waters Assoc.) and solvent systems of aqueous methanol or ethanol in the presence of tetra-n-butylammonium dihydrogen phosphate as a phase transfer agent. As shown in Fig. 19.2, the hexa- through tetra-porphyrin carboxylic acids are readily separated, but the type I and type III isomers of neither copro- nor uroporphyrins are resolved. Nevertheless, the potential for clinical analyses is shown in Fig. 19.3, where a urine sample from a patient suffering from congenital porphyria, injected directly without preliminary work-up, gave a clean separation of the variously carboxylated porphyrins. This technique will have reached maturity when resolution of the type I and III isomers of uroporphyrin can be achieved with the free carboxylic acids. Recently, the corresponding octamethyl esters were separated by HPLC [27].

The chromatography of porphyrin carboxylic acids on cellulose columns has been described [28–30]. The method of Eriksen [28] parallels that described for PC of porphyrins (cf. Chap. 19.2.1.1). The mixture of porphyrins to be separated is first adsorbed on cellulose powder, which is dried, and this dried powder is then added to the top of a column of dry cellulose. 2,6-Lutidine–water (3:1) first elutes the dicarboxylic acids, followed by the tetracarboxylic acids, which migrate at about half the rate. Under these conditions, uroporphyrin remains at the top of the column. As

References on p. B403

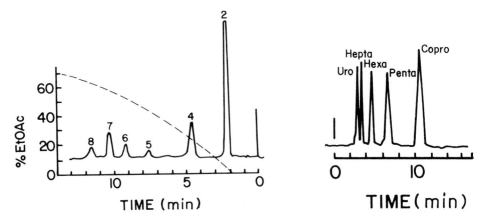

Fig. 19.1. HPLC of porphyrin methyl esters with 2 to 8 carboxyl groups. Column, 24×1/8 in. OD, Merckosorb SI 60 (10 μm silica). Flowrate 3.0 ml/min with a gradient of hexane to ethyl acetate. Detected at 400 nm. (Reproduced from *J. Chromatogr.*, 115 (1975) 325 [25].)

Fig. 19.2. HPLC of porphyrins containing 4 to 8 carboxyl groups. Column, 30×0.4 cm ID, μBondapak C$_{18}$ (Waters Assoc., Milford, MA, USA); eluent, methanol–water (4:1), containing 1 mM tetrabutylammonium dihydrogen phosphate; flowrate, 1.0–1.5 ml/min; detected at 400 nm. (Reproduced from *Biochem. J.*, 173 (1978) 693, with permission [26].)

the water concentration in the eluent is increased (3:2), the tetracarboxylic acids migrate faster, and addition of a trace of concentrated ammonia finally removes the uroporphyrin. While this is an efficient and reasonably rapid method, we prefer the following method, based on silica gel column chromatography, because it gives a better resolution of porphyrin carboxylic acids.

Fig. 19.3. HPLC of a urine sample from a patient suffering from congenital porphyria. Column, 30×0.4 cm μBondapak C$_{18}$ (Waters); eluent, methanol–water (4:1), containing 1 mM tetrabutylammonium dihydrogen phosphate; flowrate, 1.0–1.5 ml/min; detected at 400 nm. (Reproduced from *Biochem. J.*, 173 (1978) 693, with permission [26].)

Slurry 100 g silica gel (particle size 0.2–0.5 mm, "Suitable for Column Chromatography") in benzene–methanol–formic acid (110 : 30 : 1) and pour the slurry into a column, 150 cm × 2.5 cm ID, leaving a 5-cm layer of solvent at the top of the column. Dissolve 600 mg of the porphyrin mixture in a minimum of pyridine and mix it with 5 g silica gel. Remove the solvent under vacuum and finely divide the residue. Slurry it in a small volume of the solvent mixture used to prepare the column, and carefully apply the slurry to the top of the column. During development, maintain a flowrate of at least 1 ml/min by using slight nitrogen pressure, if necessary. If the porphyrins remain in the column too long, esterification with methanol, catalyzed by the formic acid, may occur.

The chromatography of free porphyrin acids on columns of talc [31–33], Celite [34–38], and Al$_2$O$_3$, CaCO$_3$, MgO, and MgCO$_3$ [39] has been described. Schwartz et al. [40] have separated urinary porphyrins into octa-, tetra-, and dicarboxylic acids directly by chromatography on Florisil (magnesium trisilicate). Polytrifluorochloroethylene [41,42], polyethylene [43], and polyamide [44] columns have also been used for separating porphyrin carboxylic acids. Although the use of ion-exchange or gel chromatography would be anticipated for resolving porphyrin carboxylic acids, the literature actually contains only a few reports. Dowex-2 [45] has been used to separate urinary porphyrins, and gel permeation on Sephadex G-25 [46] has also been reported.

19.2.2. Metalloporphyrin carboxylic acids

The chromatography of metalloporphyrin carboxylic esters is more complicated than that of the metal-free compounds, owing to the stability of the metal–porphyrin complex and the lability of the axial ligands. In many instances it is, in fact, advisable to purify the metal-free porphyrin by the methods described above and, after insertion of the metals — which in most cases can be quantitative — to remove excess metal ions by ion exchange or dialysis [47].

Metalloporphyrins containing carboxylic acids are often prepared for reconstitution studies with apoheme proteins. Thus, a considerable body of experience on the chromatography of metalloporphyrin dicarboxylic acids of protoporphyrin, mesoporphyrin, hematoporphyrin, and derivatives of deuteroporphyrin is available. This is particularly true of the ferric complexes of these porphyrins, which are called hemins. It should be noted here that most of the chromatographic methods described below will bring about exchange of the axial ligands of the original porphyrin for those encountered during the chromatography. For other than analytical purposes, it is important that after chromatography a homogeneous metalloporphyrin be prepared by religating the metalloporphyrin with the desired ligands.

19.2.2.1. Paper chromatography

Application of acid-washed filter paper [48] and paper impregnated with silicic acid [49] has been described. In analogy to the method outlined above, 2-6,lutidine–water was used to separate hemins according to their number of carboxylic acid groups [50]. The same article described a reversed-phase separation of hemins on silicone-impregnated paper.

A sample of the hemins, dissolved in pyridine, dilute ammonia, or freshly prepared caustic soda, is applied to the silicone-treated paper, which is developed in the inner chamber of a dual chromatographic tank. The developing solvent, water–1-propanol–pyridine (55:1:4), is allowed to ascend ca. 10 cm above the origin, which takes ca. 1 h. The outer chamber of the chromatographic tank is lined with moist paper, and pyridine (0.4 ml/1 of chamber volume) is placed at the bottom. If the pyridine concentration in the vapor phase is too high, R_F values will be high and resolution will be poor, while too low a concentration will cause the protohemin to streak. In this reversed-phase system the hemins with the larger number of carboxyl groups have the higher R_F values. For hemins with high R_F values water–1-propanol–pyridine (60:1:1) is a better solvent system. In the dried chromatogram as little as 50 ng can be visualized, and even lower concentrations can be detected with o-dianisidine, as described in Chap. 19.1.2.

19.2.2.2. Thin-layer chromatography

There are several reports on TLC of metalloporphyrins [20,51,52]. The two most useful techniques are the separation of hemins on silica gel [53], and the TLC of a wider range of metalloporphyrins on polyamide [54]. An evaporative TLC technique for the separation of hemins has been reported [55] in which benzene–methanol–formic acid (92:8:1) and silica gel are used. However, this technique is time-consuming and does not always produce adequate separations. Better separations of the iron complexes of protoporphyrin (19.10), hematoprophyrin (19.9), deuteroporphyrin (19.11, $R_1 = R_2 = H$) and diacetyldeuteroporphyrin (19.11, $R_1 = R_2 = CH_3CO-$) have been achieved [53] on silica gel with hexane–1-propanol–acetic acid (20:10:3). Best results are obtained when the plates are exposed to the room atmosphere for at least 12 h before they are used. Despite the fact that the R_F values of the hemins are low in this system (diacetyldeuterohemin R_F 0.08, hematohemin R_F 0.13, deuterohemin R_F 0.20, protohemin R_F 0.28, mesohemin R_F 0.32), the resolution is adequate and cannot be improved by an increase in the amount of propanol, which merely causes the protohemin to move faster and to streak badly. A modification of this system, 1-butanol–water–acetic acid (500:15:14), resolved the trivalent manganese, cobalt, and iron complexes of mesoporphyrin [56].

Polyamide TLC is based on the interaction of the porphyrin carboxyl groups of metalloporphyrin carboxylic acids with the amide groups of polyamide. In our experience polyamide thin-layer sheets from Cheng Chin, Taiwan (available from Gallard Schellinger, Cante Place, NY, USA), give the best and most consistent results. Lamson et al. [56] have separated protohemin (R_F 0.31), deuterohemin (R_F 0.43), mesohemin (R_F 0.49), and hematohemin (R_F 0.81) on polyamide using 2.5% glacial acetic acid in methanol. Srivastava and Yonetani [54] have shown that a variety of metalloporphyrins, including cobalt, ruthenium, rhodium, zinc, and magnesium porphyrins, can be purified on this layer. While no single solvent system was suitable for the resolution of all of these compounds, they found that pyridine, mixed with less polar solvents, gave the best results for these metalloporphyrins. Typical examples of solvent systems are pyridine–1-propanol–hexane (1:1:2) for cobalt protoporphyrins and pyridine–hexane mixtures for ruthenium and rhodium porphyrins.

19.2.2.3. Column chromatography

As yet no report on the separation of metalloporphyrin carboxylic acids by HPLC techniques has appeared. However, since adequate separation by silica TLC is reported (cf. Chap. 19.2.2.2), adaptation of TLC systems to HPLC should be a routine matter.

The chromatography of metalloporphyrin carboxylic acids on columns of alumina, silica gel, cellulose, and silicone-treated cellulose [49,56–62] has been reported, but the most useful sorbent is polyamide [54]. No single solvent system is suitable for all of the various metalloporphyrins, but systems analogous to those described above for TLC on polyamide sheets are applicable to CC. Thus, hematohemin was purified on a polyamide column with 10% acetic acid in benzene, cobalt protoporphyrin with pyridine–1-propanol–hexane (1:1:2), which removes a minor fraction, followed by the same solvents in a ratio of 2:1:2, and ruthenium mesoporphyrin and its carbon monoxide complex were separated with pyridine–hexane (3:2). This method was extended in our laboratories [44] to the purification of iron porphyrins (hemins). The low solubility of most metalloporphyrin carboxylic acids does not usually allow their direct application to the column and necessitates a preabsorption of the material on some suitable support. The crude hemin (600 mg) was dissolved in a minimal amount of pyridine and applied as a thin band to a 2-mm silica gel layer. The plate was dried under a stream of warm air, the band was scraped off the plate, and the silica gel was thoroughly powdered. Polyamide powder (100 g, Macherey-Nagel, 70 μm, available from Brinkmann, Westbury, NY, USA) was slurried in benzene–methanol–formic acid (110:30:1) and poured into a column 125 × 2.5 cm ID. The silica gel (containing the hemin) was slurried in the eluent and carefully added to the top of the column without disturbing the polyamide layer. The column was then eluted with the same solvent mixture at a flowrate of 6–9 ml/min by applying a slight nitrogen pressure. The gravity flowrate of about 1 ml/min allows sufficient time for transesterification of the carboxylic acid groups to occur but when it is rapidly eluted, hemin behaves as it does in polyamide TLC (Chap. 19.2.2.2). The main hemin fraction was collected and concentrated under vacuum. During chromatography, the initial axial ligands of the hemin may exchange for various ligands in the solvent system, but crystallization from a mixture of pyridine, chloroform, acetic acid, and HCl [63] gives the hemin chloride.

19.2.3. Water-soluble porphyrins lacking carboxyl groups

meso-Tetra(4-sulfonatophenyl)porphyrin (19.13) and meso-tetra(N-methylpyridyl)porphyrin (19.14) are the principal water-soluble synthetic porphyrins. Both are soluble in acids and bases. There are no reports on PC of water-soluble synthetic porphyrins, but the method described above for the water-soluble porphyrin carboxylic acids could be applied.

The separation of *meso*-tetra(4-sulfonatophenyl)porphyrin and the mono-, di-, and tri-sulfonated precursors by TLC on silica gel has been accomplished with the organic layer of a mixture of pyridine–chloroform–water (2:1:1) as the developing solvent [64].

The separation of sulfonated tetraphenylporphyrins on a column of acid-washed

19.13   19.14

diatomaceous earth by elution with pyridine–chloroform–water (2:1:1) was re- ported [65], but Pasternak et al. [66] found the method unsatisfactory and recom- mended instead the chromatography on Dowex 50W-X8. Additional reports on the separation of the synthetic water-soluble porphyrins are scarce, but the methods described above for the water-soluble porphyrin carboxylic acids should be applica- ble to them.

19.3. WATER-INSOLUBLE COMPOUNDS

19.3.1. Porphyrins

19.3.1.1. Porphyrin carboxylic esters

The conversion of water-soluble porphyrin carboxylic acids to esters, which are insoluble in water but soluble in organic solvents facilitates their isolation from other water-soluble materials. Several convenient methods for the esterification are availa- ble, including methanol–sulfuric acid, and methanol–boron trifluoride [15]. Di- azomethane can also be used conveniently for esterifying the metal-free porphyrin carboxylic acids but not the iron porphyrins, since it produces intractable green side-products.

19.3.1.1.1. Paper chromatography

Chu et al. [67] separated the methyl esters of uroporphyrin I (19.6), copro- porphyrin I (19.7), coproporphyrin III (19.8), protoporphyrin IX (19.10), and mesoporphyrin (19.11, $R_1 = R_2 = Et$) by first developing the chromatogram with kerosene–chloroform (20:13) in a chloroform atmosphere. The chromatogram was then dried and developed in the same direction with kerosene–1-propanol (5:1) in a kerosene atmosphere. This method was somewhat improved [68] by carrying out the second development at right angle to the first. It was also modified by Barrett [58,69] for porphyrins containing hydroxyl groups in the side chain.

While the above methods are suitable for separating porphyrins with a different number of carboxylic ester groups, they are unsatisfactory for porphyrin dicarboxylic esters. In order to separate them, Chu and Chu [70] devised a complicated two-dimensional system.

The outer chamber of a dual chromatography tank was lined with paper, soaked in kerosene, and tetrahydropyran (0.2 ml/l of chamber volume) was placed at the bottom of the outer chamber. The chromatogram was developed with kerosene–tetrahydropyran–methyl benzoate (100:28:7). After the solvent had risen about 15 cm, the chromatogram was dried at 105–110°C for 10 min. The paper was then treated with light petroleum (b.r. 65–110°C) and dipped into a 12.5% (w/v) solution of silicone oil (Dow-Corning 550) in light petroleum and dried at 105–110°C. The second reversed-phase partition was carried out at right angle to the first with a mixture of water–acetonitrile–1-propanol–pyridine (38:10:20:5) as the developer in the inner chamber and the paper liner in the outer chamber saturated with water. This system is capable of separating the methyl esters of protoporphyrin IX, hematoporphyrin, deuteroporphyrin, and mesoporphyrin.

Falk and coworkers [71,72] reported that the methyl esters of uroporphyrin I (19.6) and uroporphyrin III (19.1) could be separated by PC with dioxane in a modification of the method of Chu et al., described above [67]. However, although two distinct bands are seen, Bogorad and Marks [73] showed, using radioactively labeled porphyrins, that both bands contained both the I and III isomers. Further studies on this method [74,75] allowed a quantitative estimate of the ratio of the uroporphyrins I and III to be made, but it is plagued with difficulties and has now been replaced by the HPLC method of Bommer et al. [27] (cf. Chap. 19.3.1.1.3).

19.3.1.1.2. Thin-layer chromatography

The versatility, speed, and resolution achieved by TLC accounts for the numerous reports on separation of porphyrin carboxylic esters in the literature. This topic has been recently reviewed [62,76], as have the applications of TLC to the clinical identification of porphyrins and their relationship to the porphyrias [15,77]. Various sorbents, including cellulose, talc, polyamide, alumina, and silica gel [16,17,23,75,78–103] have been used, but in our experience the majority of separations can be achieved with silica gel and a suitable solvent system. Doss [77] recommended benzene–ethyl acetate–methanol (170:27:3) for analytical and the same solvent system with a slightly higher concentration of methanol for preparative work. For consistent and reproducible results it is, of course, necessary to use plates of constant activity (cf. Chap. 5.4.2). After the sample has been applied to the plate, the spots may be concentrated into a very narrow band by allowing chloroform–methanol (5:1) to rise 5 mm above the initial line of application or by using a commercially available plate (e.g., Whatman Linear-K plates) with a pre-adsorbent zone (cf. Chap. 5.3). Solvent systems for TLC include chloroform–light petroleum (b.r. 30–60°C) (5:1) in an atmosphere of ammonia, provided by 10% aqueous ammonia in a separate container [92] and other hydrocarbon–chloroform mixtures in an ammonia atmosphere.

For the separation of the methyl esters of the dicarboxylic acid porphyrins protoporphyrin (19.10), hematoporphyrin (19.9), and deuteroporphyrin (19.11, $R_1 = R_2 = H$) silica plates were prepared from a slurry containing 3% (w/v) aqueous $FeSO_4 \cdot 7H_2O$, or preformed plates were sprayed with ferrous sulfate solution and then developed with benzene–methanol (20:1) [101]. Protoporphyrin (19.10) dimethyl ester was separated from mesoporphyrin (19.11, $R_1 = R_2 = Et$) dimethyl ester on silica gel by elution with kerosene–2,4-pentanedione–methyl benzoate (12:7:1) and protoporphyrin (19.10) dimethyl ester from deuteroporphyrin (19.11, $R_1 = R_2 = H$) dimethyl ester by reversed-phase partition chromatography with silicone oil as the stationary phase and dioxane–acetonitrile–water (1:7:2) as the mobile phase [102]. Additional solvent systems for resolving porphyrin carboxylic esters according to the number of ester groups are: kerosene–chloroform–propanol (60:35:2) [104], benzene–ethyl acetate–methanol (85:13:3) [100], and carbon tetrachloride–dichloromethane–ethyl acetate–ethyl propionate (2:2:1:1) [81]. It is apparent that a variety of comparable solvent systems can be used for these separations.

More difficult separations, such as the resolution of the esters of porphyrin dicarboxylic acids and the separation of isomeric coproporphyrins and closely related compounds, require more specialized conditions. No adequate methods are available for the complete resolution of the tetramethyl esters of uroporphyrins I, II, III, and IV. The solvent system of Cardinal et al. [92], consisting of chloroform–light petroleum (5:1) in an ammonia atmosphere, separates coproporphyrin I (R_F 0.69) from coproporphyrin IV (R_F 0.66), but leaves the II and III isomers unresolved (R_F 0.55), while the system of Doss [100] cleanly separates the II, III, and IV isomers (R_F 0.67, 0.69, 0.72) from each other but hardly separates the isomer I from II (R_F 0.67). No TLC techniques for the separation of the octamethyl esters of uroporphyrins I and III are available, but these isomers are separable by HPLC [27]. If HPLC equipment is unavailable, the uroporphyrins can be decarboxylated to the corresponding coproporphyrins [105] and the resulting coproporphyrins can then be characterized either as free acids (Chap. 19.2.1.1) or as their tetramethyl esters (see above).

19.3.1.1.3. High-performance liquid chromatography

Because porphyrin carboxylic esters are easily separated and quantitated by HPLC, this technique is especially useful for the analysis of biologically important porphyrins. The analyses of porphyrins from body tissues and fluids accounts for the majority of publications on this topic. Although the separation of the naturally occurring porphyrin carboxylic acids can, in some cases, be achieved by reversed-phase HPLC (Chap. 19.2.1.3), the resolutions described so far are inferior to those obtainable with the corresponding esters. Numerous methods for esterification have been reported, but when they are applied to clinical samples, some preliminary separation from interfering material is required. Various techniques are available [25,106–108], but will not be discussed here.

Using a Porasil T (Waters Assoc.) column and isocratic elution with light petroleum–dichloromethane (5:4), Carlson and Dolphin [7] readily achieved base-line separations of the porphyrin methyl esters containing 2 to 8 carboxylic acid groups (Fig. 19.4). In this example, the column was flushed with 1-propanol–trieth-ylamine–dichloromethane (5:2:193) before injection of the sample in order to reduce peak broadening and to eliminating tailing. Similar results have been reported for a silica column eluted with a gradient of hexane to ethyl acetate or isocratically with ethyl acetate–light petroleum (b.r. 60–80°C) with flow programing [25]. Similar resolutions are also achieved with other eluents, including ethyl acetate–cyclohexane [108,109], benzene–ethyl acetate–chloroform (7:1:2) and *n*-heptane–methyl acetate (3:2) [110]. By isocratic elution with acetonitrile–water (7:3) from a reversed-phase (μPorasil C_{18}) column, Battersby et al. [111] readily separated the ethyl esters of coproporphyrin I and II from III and IV. They achieved the more difficult separation of the III and IV isomers by recycling the methyl esters on μPorasil ten times with ether–*n*-heptane (2:3), 90% saturated with water.

One of the most difficult chromatographic problems relating to the analysis of naturally occurring porphyrins, namely the separation of uroporphyrins I and III, has been recently achieved. These two compounds (19.6 and 19.1) differ only in the arrangement of their acetic and propionic acid groups on ring D. In 1979, Bommer et al. [27] showed that the two isomers could be separated as their methyl esters on a μPorasil (Waters Assoc.) column by recycling with *n*-heptane–acetic acid–acetone–water (1800:1200:600:1). After five cycles, the resolution of the isomers was greater

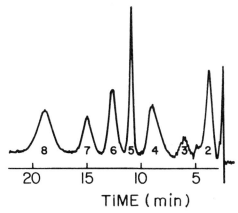

TiME (min)

Fig. 19.4. Reference chromatogram of porphyrin methyl esters with 2 to 8 carboxyl groups. Column, 24×1/8 in. OD, Porasil T (Waters). Column initially flushed with 1-propanol–triethylamine–dichloromethane (5:2:193). Separation achieved by isocratic elution with dichloromethane–light petroleum (5:2) with flow programing. Detected at 403 nm. (Reproduced from M. Doss (Editor), *Porphyrins in Human Diseases*, Karger, Basel, 1976, p. 465, with permission [7].)

References on p. B403

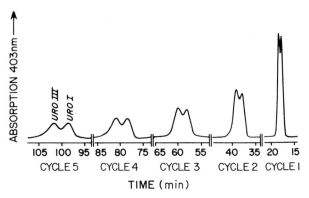

Fig. 19.5. HPLC of approximately equal quantities (1 μg) of uroporphyrins I and III as their methyl esters. Two μPorasil columns (Waters), each 300×4 mm ID; eluent, *n*-heptane–glacial acetic acid–acetone–water (1800 : 1200 : 600 : 1) in the recycling mode; flowrate, 1.5 ml/min; detected at 403 nm. (Reproduced from *Anal. Biochem.,* 95 (1979) 444, with permission [27].)

than 92% (Fig. 19.5). The same chromatographic system was used for the separation of the methyl esters of porphyrins containing 2 to 8 carboxylic acid groups (Fig. 19.6).

19.3.1.1.4. Conventional column chromatography

The relatively high solubility and stability of porphyrin carboxylic esters makes it convenient to purify them by CC. Columns of calcium carbonate, magnesium carbonate, magnesium oxide, alumina, Celite (diatomaceous earth), silica gel, and Sephadex have been used for this purpose.

Calcium carbonate was one of the first adsorbents employed for porphyrin esters [112,113], e.g., for the separation of protoporphyrin from deuteroporphyrin esters [39], but irreproducible results were frequently encountered. White et al. [76]

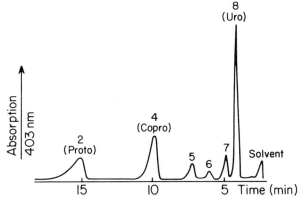

Fig. 19.6. Separation of porphyrin methyl esters (2 to 8 carboxyl groups). Two μPorasil columns (Waters), each 300×4 mm ID; eluent, *n*-heptane–glacial acetic acid–acetone–water (1800 : 1200 : 600 : 1); flowrate, 1.5 ml/min; detected at 403 nm. (Reproduced from *Anal. Biochem.,* 95 (1979) 444, with permission [27].)

observed that changing the activity of the calcium carbonate does not solve this problem. Since it is the crystal structure of the calcium carbonate which is critical for proper separation, they noted that a mixture containing 60% of the aragonite and 40% of the calcite crystal modification gives optimal resolution and suggested that mixtures of calcium carbonate in the calcite form and barium carbonate, which always occurs in the aragonite form, give satisfactory results. Fuhrhop and Smith [114] suggested selecting an inactive sample of calcium carbonate and activating it by heating at 100°C for 12 h. The calcium carbonate should be packed into the column dry and then wetted with the eluent by applying either a slight pressure at the top or a slight vacuum at the bottom of the column. Benzene–chloroform (10:1) or light petroleum–dichloromethane (1:1) are suitable solvent systems. A slow increase in the polar component of the eluent during chromatography will remove the slower bands at an acceptable rate.

On both magnesium oxide and magnesium carbonate columns the porphyrins with the largest number of ester groups are eluted first. Thus, the order of elution is different from that on all other adsorbents discussed here. The separation of the dimethyl esters of protoporphyrin, mesoporphyrin, and deuteroporphyrin on magnesium oxide has been reported [5].

Chu and Chu [34–36] have extensively studied the separation of porphyrin carboxylic esters on diatomaceous earth (Celite) and found that the mobility of the esters increases as the number of ester groups decreases. Celite is packed dry into the column, the porphyrin sample is dissolved in a suitable solvent, such as dichloromethane, adsorbed on a small volume of Celite, and then the solvent is removed. After the dried material is placed on top of the column, it is eluted with chloroform–light petroleum (1:2). The development of the chromatogram is slow, and suction is usually required to increase the flowrate. It is recommended that the porphyrins not be eluted from the column, but rather that it be dried by suction. The Celite is then extruded from the column and cut into appropriate sections. The porphyrins are recovered by extraction with dichloromethane containing 5% methanol. If development with chloroform–light petroleum is too slow, dichloromethane with increasing amounts of methanol is used to develop the column until the colored front reaches the bottom.

Bachmann and Burnham [115] reported the separation of the esters of uroporphyrin, coproporphyrin, mesoporphyrin, and deuteroporphyrin on two columns (90×2.5 cm) of Sephadex LH-20, connected in series, by elution with chloroform–methanol (1:1).

Nicholas [39] was first to describe the separation of porphyrin carboxylic esters on alumina. Since that time, a wide variety of porphyrin carboxylic esters has been successfully purified, including the protoporphyrin ester, with alumina, Grade V, and benzene [116]. A similar purification of protoporphyrin dimethyl ester was reported by O'Keefe [117], who chromatographed 20 g of the crude porphyrin on 450 g of alumina with rapid elution of the column by chloroform. Under these conditions, there is minimal damage to the vinyl groups of the protoporphyrin. The great versatility of aluminum oxide was demonstrated by Caughey et al. [63], who successfully purified several derivatives of the dimethyl ester of deuteroporphyrin

References on p. B403

with diverse R_1 and R_2 groups (of 19.11), such as acetyl, propionyl, bromo, nitro, and cyano. In each case, chromatography of the porphyrin on alumina (ca. 35 g per 1 g of porphyrin) and elution with chloroform, which probably contained ethanol as preservative, gave adequate purification. They also purified the 2-formyl-4-vinyl (19.11, R_1 = formyl, R_2 = vinyl) and the 2-vinyl-4-formyl (19.11, R_1 = vinyl, R_2 = formyl) porphyrins on alumina with chloroform–1,2-dichloroethane (1:2). Inhoffen et al. [118] later successfully separated these two isomers on alumina. Similar separations of the dimethyl esters of protoporphyrin, 2-formyl-4-vinyldeuteroporphyrin, and 2,4-diformyldeuteroporphyrins as well as of the 2- and 4-monoformyldeuteroporphyrins on alumina with chloroform and ether as eluents have been reported [119].

For large-scale preparations, involving the separation of a major porphyrin component from small amounts of impurities, 1 g of material can be readily separated on only 50 g of alumina, and frequently half this amount will suffice. However, for more difficult separations up to 10 times this amount of alumina should be used. The great solubility of porphyrin carboxylic esters allows them to be applied to the column in a small volume of benzene or chloroform. The appropriate eluents can frequently be deduced from the chromatographic behavior of the mixture on alumina thin-layer plates. Mixtures of benzene, chloroform, or dichloromethane with methanol are usually suitable, and all porphyrin carboxylic esters can be eluted by gradually increasing the polarity of the mixture, first by increasing the concentration of chloroform and then that of methanol. Some de-esterification may occur, causing porphyrin carboxylic acids to remain on the column, but they can later be removed with higher methanol concentrations. When alcohol is used in the eluent mixture, trans-esterification may occur. This can be avoided by using the esterifying alcohol in the eluent.

Silica gel is as versatile as alumina for the chromatography of porphyrin carboxylic esters, but the capacity of silica gel is lower than that of alumina. Lucas and Orten [120] chromatographed the methyl esters of uroporphyrin I, coproporphyrin I, mesoporphyrin, and protoporphyrin on silica gel with light petroleum–chloroform mixtures. By gradually increasing the concentration of chloroform, they achieved separation, the protoporphyrin being eluted first and the remaining porphyrins being eluted according to the number of carboxylic esters per molecule. White et al. [76], using the dry-column technique, adapted the solvent systems of Cardinal et al. [92] [light petroleum (b.r. 30–60°C)–chloroform (1:5)] but omitted the ammonia. In fact, the solvent systems suitable for silica TLC are generally also suitable for silica CC.

19.3.1.2. Octaalkyl- and meso-tetraarylporphyrins

Octaalkylporphyrins are widely distributed in petroleum and other geological samples, where they frequently occur as either nickel or vanadyl complexes. In addition, octaalkylporphyrins, notably octaethylporphyrin are extensively used as stable, symmetric analogs of the natural systems, as are the *meso*-tetraarylporphyrins, e.g., *meso*-tetraphenylporphyrin. Both classes of compounds show similar chromatographic behavior and, in general, the systems developed for one can, with but small modifications, be applied to the other. Furthermore, these systems are only slightly less polar than those of the porphyrin dicarboxylic esters, so that the techniques suitable for them can usually be applied to these compounds.

Porphyrins exhibit little volatility below 250°C and, therefore, the techniques of classical gas chromatography are not applicable. Karayannis and Corwin [121] have used fluorinated hydrocarbons above their critical temperatures as carrier gases at pressures up to 1400 p.s.i. and temperatures of ca. 170°C to separate a variety of octaalkylporphyrins and porphyrin carboxylic esters. The advent of HPLC, coupled with the experimental difficulties of supercritical gas chromatography have reduced it to historical value only, at least for the porphyrin chemist.

PC has found little application to the hydrophobic alkyl- and aryl-substituted porphyrins. One reported example [122] is the separation of geological porphyrins, but this method has nothing to recommend it.

In our experience, the solvent systems and supports described for TLC of porphyrin carboxylic esters can be used directly for the present class of compounds, but at lower solvent strength. The low solubility of almost all porphyrins may cause overloading of the plates. For analytical and preparative purposes, both alumina and silica gel plates may be developed with binary systems, e.g., dichloromethane, chloroform, or benzene, in combination with light petroleum, ethyl acetate, methanol, or tetrahydrofuran.

There are not many examples where the high resolution obtainable with HPLC has been required for the study of alkyl- and aryl-substituted porphyrins, but there is no doubt that the systems described for the metal-free porphyrin carboxylic esters (Chap. 19.3.1.1.3) may be adapted to these compounds. In fact, Evans et al. [25] have separated octamethyl- from octaethylporphyrin on Corasil II (Waters) by elution with ethyl acetate–light petroleum mixtures. In the area of petroporphyrins, where complex mixtures of closely related porphyrins are found, this method is expected to find wide application. Hajibrahim et al. [123], using a variety of silica supports and gradients of hexane to toluene, have obtained partial resolutions of demetallated porphyrins from crude oil. Similar separations were achieved with a reversed-phase system, in which a μBondapak C_{18} (Waters) column was isocratically eluted with 85% aqueous acetonitrile.

Solvent systems developed for TLC are directly applicable to CC, particularly to alumina and silica columns. Some care is required in the application of samples to the column. The low solubility of porphyrins often requires that they be applied in a fairly large volume. In general, chloroform and dichloromethane are the best solvents for this purpose. If the sample does not stay at the top of the column as a tight band, benzene or toluene should be used for packing the column and for dissolving the sample. If this also proves to be unsatisfactory the use of a dry column may solve this problem. Dry-column chromatography [43] was first introduced into porphyrin chemistry by Adler et al. [124], where the method is especially useful for separating the chlorins that contaminate *meso*-tetraarylporphyrins after their synthesis from pyrrole and benzaldehydes.

References on p. B403

19.3.2. Metalloporphyrins

The same precautions must be used here as those detailed above for water-soluble metalloporphyrins (Chap. 19.2.2). In particular, attention must be paid to ligand-exchange processes, which frequently occur during chromatography. Moreover, many metalloporphyrins do not fluoresce and may consequently escape detection at low concentrations (Chap. 19.1.2).

19.3.2.1. Metalloporphyrin carboxylic esters

Because the iron porphyrins (hemins) are biologically important, a large body of literature has accrued on their chromatographic properties. While the information on other metalloporphyrins is not extensive, the experience gained with the hemins is directly transferable to the chromatography of other trivalent, substitution-labile, metal complexes that show some "ionic character". Furthermore, it has been our experience that neutral divalent metalloporphyrins exhibit chromatographic properties similar to their parent porphyrins and that the conditions employed for the metal-free systems are also applicable to these metal complexes.

Hemin methyl esters of uro-, copro-, hemato-, deutero-, meso- and proto-porphyrins have been chromatographed on a reversed-phase system of paper impregnated with silicone oil [50]. Best resolution was achieved using kerosene–chloroform, followed by kerosene–1-propanol. While this sytem easily separated hemins differing in the number of ester side chains, it was incapable of separating the four dicarboxylic ester-containing porphyrins.

Chu and Chu [102] separated the copper complexes of uro-, copro- and deutero-porphyrins, as well as the copper and zinc complexes of uroporphyrin, by silica gel TLC with n-decane–chloroform (1:9) as the mobile phase. As noted above, the migration of these complexes was similar to that of their metal-free precursors. An excellent analytical technique for hemin esters is TLC on polyamide plates with benzene–acetic acid (9:1) [56]. Cobalt and rhodium complexes have also been separated on polyamide [54], which presumably would also work well for various other metalloporphyrins.

The acceptable separations that are already achieved with TLC ensure that excellent results will be forthcoming with HPLC. So far, Miller and Malina [125] have resolved the copper chelates of the porphyrins with 2 to 8 carboxylic acid groups in the form of their methyl esters on a Micro Pak CN (10 μm, Varian) column by isocratic elution with ethyl acetate–n-heptane–isopropanol (80:120:1). Similarly, Carlson et al. [126] have separated the copper complexes of proto- and coproporphyrins from their free bases in the form of their methyl esters by using n-heptane–methyl acetate (3:1) on a μPorasil (Waters) column.

Hemin methyl esters, particularly the diesters, can be conveniently separated on a polyamide column with the systems of Lamson et al. [56]. A polyamide column was also used for cobalt and rhodium porphyrins, and it is undoubtedly applicable to a wide range of other metalloporphyrins. Columns of alumina have been employed with benzene–chloroform [127], but it converts the majority of ferric porphyrins to their μ-oxo dimers [128]. When the purified hemin has been recovered after chromatography, the μ-oxo dimer can then be religated by treating the organic phase with the appropriate acid [127]. Numerous other metal complexes of porphyrin esters have been purified on alumina and, in general, this is the method of choice [129–131].

Most metalloporphyrins do not demetallate under the chromatographic conditions described above, but the magnesium complexes are an exception. Because even the smallest traces of acid will cause demetallation, the chromatography of magnesium complexes should be carried out under basic conditions, and chloroform should be avoided. Baum and Ellsworth [128] have resolved the zinc and magnesium protoporphyrin dimethyl esters on a sucrose–cornstarch column with light petroleum–pyridine (199:1).

19.3.2.2. Metallo-octaalkyl- and -meso-tetraarylporphyrins

Few examples of the analysis of complex mixtures of these metalloporphyrins have been reported. Usually, chromatography was used to rid a principal component of minor porphyrin impurities. For the sake of completeness, we note that work on supercritical gas chromatography in Corwin's laboratory, described in Chap. 19.3.1.2, was extended to the study of metalloporphyrins [121,132], but the poor resolution and complexity of the apparatus do not speak for this technique.

Using silica gel TLC with a variety of solvent systems (all of which contained acetic acid and chloroform, toluene, and methanol) Hui et al. [133] resolved the iron, cobalt, manganese, zinc, and rhodium complexes of meso-tetraphenylporphyrin. Using this technique, coupled with integrated ion-current mass spectroscopy, they could detect as little as 10^{-14} moles of metal, but the mercury, cadmium, and lead complexes had the identical mobility. Both silica [134] and alumina [135] thin layers are suitable for the metal complexes of octaethylporphyrin. Mixtures of toluene, chloroform, or dichloromethane with methanol or light petroleum make eluents of appropriate solvent strength.

As part of the study on the chromatography of porphyrins from geological sources, described in Chap. 19.3.1.2, Hajibrahim et al. [123] reported on the analysis of nickel and vanadyl octaethyl and etioporphyrin I and of complex mixtures of petroporphyrins. While the separations were not spectacular, they suggest that HPLC holds much promise for the future.

Both silica and alumina have found considerable use in conventional CC. Buchler and coworkers [134,135] have used both wet and dry chromatography on alumina with dichloromethane–methanol for separating metalloporphyrins. Adler et al. [136] described the purification of copper meso-tetraphenylporphyrin on a dry alumina column with chloroform, and they emphasized that better results are usually achieved with dry columns. Hanson et al. [137] reported the use of Sephadex LH-20 (which is compatible with organic solvents) and chloroform–methanol mixtures for the purification of metalloporphyrins, particularly when demetallation problems are encountered with silica or alumina. Because magnesium complexes are especially subject to demetallation, magnesium octamethylporphyrin was purified on magnesol–cellulose with chloroform–pyridine [138].

References on p. B403

19.3.3. Chlorophylls and related photosynthetic pigments

A thorough and critical review (including practical details) on the isolation, preparation, and chromatographic purification of chlorophylls and bacterio-chlorophylls has been published [139]. We shall consider here the general problems encountered in the chromatography of photosynthetic pigments and bring the literature in this area up to date.

The photosynthetic pigments occur in nature always along with a variety of other pigments, including tetrapyrroles, carotenes and xanthophylls, and any preliminary isolation that can be achieved will be beneficial. Also, because the photosynthetic pigments are especially labile (Chap. 19.1.1), special precautions must be taken during the isolation and chromatographic procedures. Numerous systems for the separation of chlorophylls and other plant pigments by PC have been described, but, with the advent of superior techniques, such as TLC and HPLC, they are now obsolete.

19.3.3.1. Thin-layer chromatography

Both one- and two-dimensional TLC of chlorophylls on cellulose have been reported [140–145]. While a variety of solvent systems have been described, light petroleum or hexane, in combination with a more polar component, such as chloroform, ethanol, or acetone will effect satisfactory separations. Recently, the use of two systems containing either light petroleum (b.r. 60–80°C) or *n*-heptane and pyridine have been shown [145] to give excellent results with the chlorophylls, pheophytins, and pheophorbides, as well as with chlorin, rhodins, and their esters. This method is rapid, and no chemical alterations during chromatography have been observed.

Sucrose is an excellent packing material for CC of chlorophyll derivatives, and

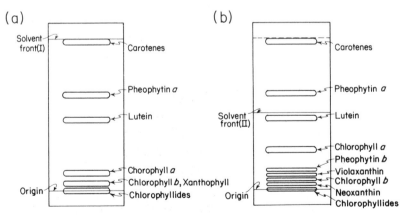

Fig. 19.7. Typical sucrose TLC of leaf pigment extract. (a) Crude pigment extract developed initially with 20% diethyl ether in light petroleum. (b) Pigment distribution on the sucrose thin-layer plate from (a) after a second development with 0.5% 2-propanol in light petroleum. (Reproduced from *J. Chromatogr.*, 47 (1970) 395, with permission [147].)

thin layers of sucrose have also been used for analytical work [142,146,147]. As with cellulose plates, light petroleum in combination with diethyl ether, 1- or 2-propanol, acetone, or methanol provides suitable systems (Fig. 19.7), where degradation is minimal. Sucrose plates must not be exposed to moisture, as this may cause caking and irregular development. Both one- and two-dimensional TLC methods have been described for silica gel [146,148–151]. However, results should be interpreted with care, as alterations and decompositions were reported to occur on these layers [148].

Silica gel coated with Wesson oil [152], cellulose coated with unsaturated tri-glycerides, and Kieselguhr G impregnated with either triolein, castor oil, or paraffin oil [153] have served supports for reversed-phase TLC systems. Impregnated kiesel-

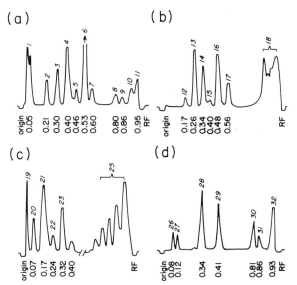

Fig. 19.8. Scanning fluorimeter tracings of reversed-phase thin-layer chromatograms of chlorophyll derivatives. The chromatograms were scanned (1 cm/min) in a Turner filter fluorometer with a far-red sensitive photomultiplier (R136, Hitachi TV Corp.) and a Turner No. II TLC scanner door. The primary filter was a Kodak Wratten narrow-pass filter No. 47B, and the secondary filter was a Corning sharp-cut No. 2-64. (a) Chromatogram on a 10-μm reversed-phase Kieselguhr G layer; plate dipped into 4% triolein in light petroleum and then air-dried for 12 h. Developing solvent, methanol–acetone–2-propanol–water–benzene (30:10:5:5:1). 1 = Pheophytins, 2 = chlorophyll a', 3 = chlorophyll a, 4 = allomerized chlorophyll a, 5 = chlorophyll b', 6 = chlorophyll b, 7 = allomerized chlorophyll b, 8 = pheophytin c, 9 = chlorophyll c, 10 = pheophorbides a and b, 11 = chlorophyllides a and b. (b) Chromatogram prepared as in (a), but plate dipped into 8% paraffin oil (N.F., viscosity 125/135) in light petroleum. Developing solvent, methanol–acetone–propanol–water–benzene (35:50:10:10:2). 12 = Pheophytin a', 13 = pheophytin a, 14 = allomerized pheophytin a, 15 = pheophytin b', 16 = pheophytin b, 17 = allomerized pheophytin b, 18 = chlorophyllides, pheophorbides and chlorophylls. (c) Chromatogram prepared as in (a), but plate dipped into 10% castor oil in methanol. Developing solvent, methanol–acetone–isopropanol–water–benzene (160:5:5:30:4). Composite diagram of two scans: 19 = chlorophylls and pheophytins, 20 = pheophorbide a', 21 = pheophorbide a, 22 = allomerized pheophorbide a, 23 = pheophorbide b, 24 = allomerized pheophorbide b, 25 = unknown chlorophylls. (d) Plate prepared as in (a), but developing solvent was methanol–acetone–benzene–water (6:2:1:1). 26 = Pheophytin a, 27 = pheophytin b, 28 = chlorophyll a, 29 = chlorophyll b, 30 = pheophorbide a, 31 = pheophorbide b, 32 = chlorophyllides. (Reproduced from *J. Chromatogr.*, 76 (1973) 175, with permission [153].)

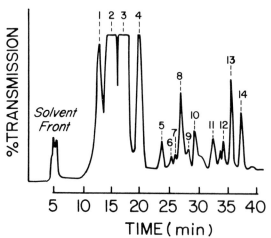

Fig. 19.9. Reversed-phase HPLC of 150 mg of bacteriochlorophylls *c*, obtained from *Chlorobium limicola thiosulfatophilum* (Tassajara). Column, 9 ft.×3.5 in., Microethene FN 500 polyethylene, 3–45 μm (US Industrial Chem. Co., New York, USA); gradient elution with 65–75% aqueous acetone; flowrate, 14 ml/min. Detected in a 1-cm flow cell with a red-light emitting diode and a phototransistor having a response maximum in the IR region. (Reproduced from *J. Chromatogr.*, 151 (1978) 357, with permission [154].)

Peak No.	R_1	R_2	R_3
1	Et	Me	Farnesol
2	Et	Et	Farnesol
3	*n*-Pr	Et	Farnesol
4	*i*-Bu	Et	Farnesol
5	Et	Et	4-Undecyl-2-furanmethanol
6	Et	Et	Geranylgeraniol
7	*n*-Pr	Et	4-Undecyl-2-furanmethanol
8	Et	Et	*cis*-9-Hexadecen-1-ol
9	*n*-Pr	Et	Geranylgeraniol
10	*n*-Pr	Et	*cis*-9-Hexadecen-1-ol
11	Et	Et	10,14-Tetrahydrogeranylgeraniol
12	*n*-Pr	Et	10,14-Tetrahydrogeranylgeraniol
13	Et	Et	Phytol
14	*n*-Pr	Et	Phytol

guhr, in combination with methanol–acetone–isopropanol–water mixtures, enabled the rapid separation and identification of eighteen derivatives of chlorophylls *a*, *b*, and *c*, as well as numerous derivatives of bacteriochlorophyll, as shown in Fig. 19.8. Reversed-phase TLC, as described here, is a rapid yet mild method, providing excellent resolution. Several copper and zinc complexes of pheophytins and pheophorbides have been separated on kieselguhr coated with peanut oil with a methanol–water–acetone system [152].

19.3.3.2. Column chromatography

The complexity of the natural plant pigment mixtures, coupled with their susceptibility to degradation and sensitivity to light, makes HPLC appear as the ideal technique for their analysis and purification. Although only few reports on the use of this method have appeared so far, initial experiments suggest that this chromatographic method promises to be a powerful tool in the future.

Using a reversed-phase column of polyethylene powder, Chow et al. [154] were able to separate a complex mixture of bacteriochlorophylls, isolated from *Chlorobium limicola thiosulfatophilum* (Tassajara), with a 67 to 75% aqueous acetone gradient. This separation is notable in that complex molecules differing by only a methylene group or one double bond were resolved (Fig. 19.9). Svec [139] also gave details of the rapid purification of chlorophyll *a* from spinach by HPLC on polyethylene with aqueous acetone.

The systems described for TLC (Chap. 19.3.3.1) can be used directly for the column chromatography of chlorophylls. Sucrose [146,155–158], cellulose [140,146,148,159–163], starch [140,146,148], polyamide [164], polyethylene [153,158],

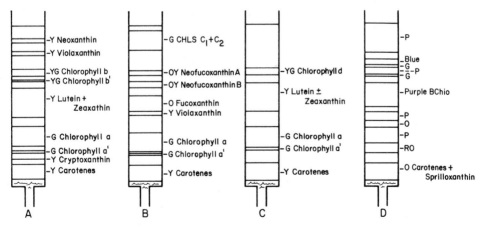

Fig. 19.10. Typical adsorption chromatograms on a column of powdered sugar, developed with light petroleum plus 0.5 to 2% 1-propanol. (A) Green plants; (B) brown algae; (C) red algae; (D) purple sulfur bacterium (G, green; YG, yellow green; Y, yellow; O, orange; P, pink). (Reproduced from D. Dolphin (Editor), *The Porphyrins*, Vol. 5, Academic Press, New York, 1979, p. 341, with permission [139].)

References on p. B403

and polypropylene [160] have been successfully employed as column packing materials. In our experience, best results are obtained with columns packed dry with powdered commercial sucrose (containing 3% starch to prevent caking) and developed with light petroleum and 1-propanol diethyl or ether. When the column is developed as shown in Fig. 19.10, the solvent is removed by suction. The adsorbent is extruded from the column and is then cut into appropriate sections. The products are eluted with ether or, if necessary, dichloromethane (to which a trace of pyridine may be added) to remove polar compounds.

Ion-exchange resins must be used with caution, since extremes of pH cause alterations and degradations. However, the large-scale preparation of chlorophylls *a*, *b*, and *c* on a DEAE-cellulose column appears to offer distinct advantages of speed over sucrose columns [165]. The separation of chlorophylls from carotenoid pigments by chromatography on a polystyrene gel column has been reported [116,167]. The more recent [168] purification of chlorophylls *a* and *b* on Sephasorb HP Ultrafine C (Pharmacia, Bromma, Sweden), which is similar in structure to Sephadex LH-20, with diethyl ether–hexane (1:9), followed by the addition of acetone (0.5%) to the eluent is an improvement over the earlier Sephadex LH-20 methods [146,169].

19.3.4. Bile pigments

The bile pigments are linear tetrapyrroles derived from the breakdown and further interconversions of porphyrins. Some thirty different silica gel TLC procedures for the separation of bilirubin and biliverdin, as both the free carboxylic acids and as their esters, have been described [170]. For the free acids, TLC on silica gel with chloroform, containing 1% acetic acid, or on polyamide with methanol, containing 1% concentrated NH_4OH, have been proposed [170]. The separation of the dimethyl esters of the biliverdins can be achieved on silica gel, but it is important to note that the calcium sulfate binder cannot be omitted [171], for it plays an active role in the separation. On silica Gel G, the α-, β-, γ-, and δ-isomers of both the proto- and mesobiliverdins were separated [120] and it was also found suitable for the chromatography of bilirubin esters with solvent systems such as benzene–ethanol [172–174], benzene–chloroform, or chloroform–ethyl acetate [161,162].

Only a few examples of the application of HPLC to bile pigments have yet appeared, but the separation of a complex mixture of bile pigments [163], as their dimethyl esters, on Corasil II by methyl acetate–isooctane (1:2) with flowrate programing, indicates the strength of the method.

19.4. ACKNOWLEDGEMENTS

The preparation of this review was aided by the United States National Institutes of Health (AM 17989) and the Canadian National Science and Engineering Research Council. The author wishes to express his thanks to the Harvard Chemistry Department for its assistance in the preparation of this manuscript during his stay as a Visiting Professor.

REFERENCES

1 D. Dolphin (Editor), *The Porphyrins*, Vols. 1–7, Academic Press, New York, 1978–1979.
2 M. Tsvet, *Ber. Deut. Bot. Ges.*, 24 (1906) 384.
3 M. Blumer, *Anal. Chem.*, 28 (1956) 1640.
4 J.L. Connelly, M. Morrison and E. Stotz, *J. Biol. Chem.*, 233 (1958) 743.
5 J.A. Owen, H.J. Silberman and C. Got, *Nature (London)*, 182 (1958) 1373.
6 M. Doss, *Z. Klin. Chem. Klin. Biochem.*, 8 (1970) 197.
7 R.E. Carlson and D. Dolphin, in M. Doss (Editor), *Porphyrins in Human Diseases*, Karger, Basel, 1976, p. 465.
8 W. Slavin, A.T. Rhys Williams and A.F. Adams, *J. Chromatogr.*, 134 (1977) 121.
9 R.E.H. Nicholas and C. Rimington, *Biochem. J.*, 48 (1951) 306.
10 R. Kehl and W. Stich, *Hoppe-Seylers' Z. Physiol. Chem.*, 289 (1951) 6.
11 L. Eriksen, *Scand. J. Clin. Lab. Invest.*, 5 (1953) 155.
12 D. Mauzerall, *J. Amer. Chem. Soc.*, 82 (1960) 2601.
13 R.V. Belcher, S.G. Smith, R. Mahler and J. Campbell, *J. Chromatogr.*, 53 (1970) 279.
14 T.K. With, *Scand. J. Clin. Lab. Invest.*, 9 (1957) 395.
15 L. Eales, in D. Dolphin (Editor), *The Porphyrins*, Vol. 6, Academic Press, New York, 1979, p. 663.
16 C.S. Russell, *J. Chromatogr.*, 25 (1966) 163.
17 M. Gajdos-Török, *Bull. Soc. Chim. Biol.*, 50 (1968) 925.
18 M. Yuan and C.S. Russell, *J. Chromatogr.*, 87 (1973) 562.
19 M. Mundschenk, *J. Chromatogr.*, 25 (1966) 380.
20 J. Jensen, *J. Chromatogr.*, 10 (1963) 236.
21 T.K. With, *Clin. Biochem.*, 1 (1967) 30.
22 T.K. With, *J. Chromatogr.*, 42 (1969) 389.
23 T. Okuda, H. Nakajima, K. Yatsuki, M. Amano and G. Umeda, *Brit. J. Ind. Med.*, 35 (1978) 61.
24 N. Ellfolk and G. Sievers, *J. Chromatogr.*, 25 (1966) 373.
25 N. Evans, D.E. Games, A.H. Jackson and S.A. Matlin, *J. Chromatogr.*, 115 (1975) 325.
26 R. Bonnett, A.A. Charalambides, K. Jones, I.A. Magnus and R.J. Ridge, *Biochem. J.*, 173 (1978) 693.
27 J.C. Bommer, B.F. Burnham, R.E. Carlson and D. Dolphin, *Anal. Biochem.*, 95 (1979) 444.
28 L. Eriksen, *Scand. J. Clin. Lab. Invest.*, 9 (1957) 97.
29 R. Lemberg and M. Steward, *Aust. J. Exp. Biol. Med. Sci.*, 33 (1955) 451.
30 D.B. Morell, J. Barrett and P.S. Clezy, *Biochem. J.*, 78 (1961) 793.
31 H. Fischer and H. Hofmann, *Hoppe-Seyler's Z. Physiol. Chem.*, 246 (1937) 15.
32 A. Comfort, *Biochem. J.*, 44 (1949) 111.
33 T.K. With, *S. Afr. Med. J.*, special issue *Lab Clin. Med.*, 17 (1971) 227.
34 T.C. Chu and E.-J. Chu, *J. Biol. Chem.*, 227 (1957) 505.
35 T.C. Chu and E.-J. Chu, *Anal. Chem.*, 30 (1958) 1678.
36 T.C. Chu and E.-J. Chu, *J. Biol. Chem.*, 234 (1959) 2741.
37 T.C. Chu and E.-J. Chu, *Biochem. J.*, 83 (1962) 318.
38 T.L. Popper and H. Tuppy, *Acta Chem. Scand.*, 17 (1963) 547.
39 R.E.H. Nicholas, *Biochem. J.*, 48 (1951) 309.
40 S. Schwartz, B. Stephenson and D. Sarkar, *Clin. Chem.*, 22 (1976) 1057.
41 H. Mundschenk, *J. Chromatogr.*, 38 (1968) 106.
42 H. Mundschenk, *J. Chromatogr.*, 37 (1968) 431.
43 B. Loev and M.M. Goodman, *Progr. Separ. Purif.*, 3 (1970) 73.
44 R.K. DiNello and C.K. Chang, in D. Dolphin (Editor), *The Porphyrins*, Vol. 1, Academic Press, New York, 1978, p. 289.
45 T. Heikel, *Scand. J. Clin. Lab. Invest.*, 10 (1958) 193.
46 C. Rimington and R.V. Belcher, *J. Chromatogr.*, 28 (1967) 112.
47 C.A. Busby, R.K. DiNello and D. Dolphin, *Can. J. Chem.*, 53 (1975) 1554.
48 W.N. Ramsay, *Anal. Biochem.*, 7 (1964) 366.
49 M. Yoshida and N. Shimazono, *J. Biochem.*, 58 (1965) 39.

50 T.C. Chu and E.-J. Chu, *J. Biol. Chem.*, 212 (1955) 1.
51 P.K. Warme and L. Hager, *Biochemistry*, 9 (1970) 1599.
52 C. Miyauchi, *Proc. Jap. Acad.*, 45 (1969) 72.
53 R.K. DiNello and D. Dolphin, *Anal. Biochem.*, 64 (1975) 444.
54 T.S. Srivastava and T. Yonetani, *Chromatographia*, 8 (1975) 124.
55 D. Lamson and T. Yonetani, *Anal. Biochem.*, 52 (1973) 470.
56 D.W. Lamson, A.F.W. Coulson and T. Yonetani, *Anal. Chem.*, 45 (1973) 2273.
57 M. Kiese and H. Kurz, *Biochem. Z.*, 325 (1954) 299.
58 J. Barrett, *Biochem. J.*, 64 (1956) 626.
59 M. Morrison and E. Stotz, *J. Biol. Chem.*, 213 (1955) 373.
60 W.S. Caughey and J.L. York, *J. Biol. Chem.*, 237 (1962) PC2414.
61 M. Wilchek, *Anal. Biochem.*, 49 (1972) 572.
62 K.M. Smith (Editor), *Porphyrins and Metalloporphyrins*, Elsevier, Amsterdam, 1975.
63 W.S. Caughey, J.O. Alben, W.Y. Fujimoto and J.L. York, *J. Org. Chem.*, 31 (1966) 2631.
64 J. Winkelman, G. Slater and J. Grossman, *Cancer Res.*, 27 (1967) 2060.
65 J. Winkelman, *Cancer Res.*, 22 (1962) 589.
66 R.F. Pasternak, P.R. Huber, P. Boyd, G. Engasser, L. Francesconi, E. Gibbs, P. Fasella, G.C. Venturo and L. deC. Hinds, *J. Amer. Chem. Soc.*, 94 (1972) 4511.
67 T.C. Chu, A.A. Green and E.J.-H. Chu, *J. Biol. Chem.*, 190 (1951) 643.
68 L. Bogorad and S. Granick, *J. Biol. Chem.*, 202 (1953) 793.
69 J. Barrett, *Nature (London)*, 183 (1959) 1185.
70 T.C. Chu and E.J.-H. Chu, *J. Biol. Chem.*, 208 (1954) 537.
71 J.E. Falk, E.I.B. Dresel, A. Benson and B.C. Knight, *Biochem. J.*, 63 (1956) 87.
72 J.E. Falk and A. Benson, *Biochem. J.*, 55 (1953) 101.
73 L. Bogorad and G.S. Marks, *Biochim. Biophys. Acta*, 41 (1960) 356.
74 P.A.D. Cornford and A. Benson, *J. Chromatogr.*, 10 (1963) 141.
75 A.M. del C. Batlle and A. Benson, *J. Chromatogr.*, 25 (1966) 117.
76 W.I. White, R.C. Bachmann and B.F. Burnham, in D. Dolphin (Editor), *The Porphyrins*, Vol. 1, Academic Press, New York, 1978, p. 553.
77 M. Doss, in A. Niederwieser and G. Pataki (Editors), *Progress in Thin-Layer Chromatography and Related Methods*, Vol. 3, Ann Arbor Sci. Publ., Ann Arbor, MI, 1972, p. 145.
78 J. Pinol Aguadé, C. Herrero, J. Almeida, S.G. Smith and R.V. Belcher, *Brit. J. Dermatol.*, 93 (1975) 277.
79 T.K. With, *Dan. Med. Bull.*, 22 (1975) 74.
80 S.G. Smith, *Brit. J. Dermatol.*, 93 (1975) 291.
81 W.G. Sears, T. Darocha and L. Eales, *Enzyme*, 17 (1974) 69.
82 G.H. Elder, *S. Afr. J. Lab. Clin. Med.*, 17 (1971) 45.
83 G.H. Elder, *J. Chromatogr.*, 59 (1971) 234.
84 S.R. Heller, S.F. Labbé and J. Nutter, *Clin. Chem.*, 17 (1971) 525.
85 M. Doss, *Z. Klin. Chem. Klin. Biochem.*, 8 (1970) 197.
86 M. Doss, *Z. Klin. Chem. Klin. Biochem.*, 8 (1970) 208.
87 G.H. Elder and J.R. Chapman, *Biochim. Biophys. Acta*, 208 (1970) 535.
88 M. Doss, *Z. Klin. Chem. Klin. Biochem.*, 7 (1969) 133.
89 M. Doss, *Hoppe-Seyler's Z. Physiol. Chem.*, 350 (1969) 499.
90 M. Doss, *J. Chromatogr.*, 30 (1971) 234.
91 M. Doss, *Klin. Wochenschr.*, 47 (1969) 1281.
92 R.A. Cardinal, I. Bossenmaier, Z.J. Petryka, L. Johnson and C.J. Watson, *J. Chromatogr.*, 38 (1968) 100.
93 M. Doss and U. Bode, *J. Chromatogr.*, 35 (1968) 248.
94 M. Doss, *Deut. Med. Wochenschr.*, 93 (1968) 2223.
95 M. Doss and U. Bode, *Z. Klin. Chem. Klin. Biochem.*, 6 (1968) 383.
96 M. Doss, *Klin. Wochenschr.*, 46 (1968) 731.
97 M. Doss, *Z. Klin. Chem. Klin. Biochem.*, 6 (1968) 498.
98 P.A. Burbridge, C.L. Collier, A.H. Jackson and G.W. Kenner, *J. Chem. Soc. B*, (1967) 930.

99 T.C. Chu and E.J.-H. Chu, *J. Chromatogr.*, 28 (1967) 475.

100 M. Doss, *J. Chromatogr.*, 30 (1967) 265.

101 R.W. Henderson and T.C. Morton, *J. Chromatogr.*, 27 (1967) 180.

102 T.C. Chu and E.J.-H. Chu, *J. Chromatogr.*, 21 (1966) 46.

103 E. Demole, *J. Chromatogr.*, 1 (1958) 24.

104 Y. Grosser, G.D. Sweeney and L. Eales, *S. Afr. Med. J.*, 41 (1967) 460.

105 P.R. Edmondson and S. Schwartz, *J. Biol. Chem.*, 205 (1953) 605.

106 J.H.P. Wilson, J.W.O. van den Berg, A. Edixhoven and L.H.M. van Gastel-Quist, *Clin. Chim. Acta*, 89 (1978) 165.

107 R.E. Carlson and D. Dolphin, in P.F. Dixon, C.H. Gray, C.K. Lim and M.S. Stoll (Editors), *High Pressure Liquid Chromatography in Clinical Chemistry*, Academic Press, London, 1976, p. 87.

108 N. Evans, A.H. Jackson, S.A. Matlin and R. Towill, *J. Chromatogr.*, 125 (1976) 345.

109 N. Evans, A.H. Jackson, S.A. Matlin and R. Towill, in P.F. Dixon, C.H. Gray, C.K. Lim and M.S. Stoll (Editors), *High Pressure Liquid Chromatography in Clinical Chemistry*, Academic Press, London, 1976, p. 71.

110 C.H. Gray, C.K. Lim and D.C. Nicholson, in P.F. Dixon, C.H. Gray, C.K. Lim and M.S. Stoll (Editors), *High Pressure Liquid Chromatography in Clinical Chemistry*, Academic Press, London, 1976, p. 79.

111 A.R. Battersby, D.G. Buckley, G.L. Hodgson, R.E. Markwell and E. McDonald, in P.F. Dixon, C.H. Gray, C.K. Lim and M.S. Stoll (Editors), *High Pressure Liquid Chromatography in Clinical Chemistry*, Academic Press, London, 1976, p. 63.

112 C.J. Watson, S. Schwartz and V. Hawkinson, *J. Biol. Chem.*, 157 (1945) 345.

113 R. Hill, *Biochem. J.*, 19 (1925) 341.

114 J.-H. Fuhrhop and K.M. Smith, in K.M. Smith (Editor), *Porphyrins and Metalloporphyrins*, Elsevier, Amsterdam, 1975, p. 757.

115 R.C. Bachmann and B.F. Burnham, cited in ref. 77.

116 M. Grinstein, *J. Biol. Chem.*, 167 (1947) 515.

117 D.H. O'Keefe, *Ph.D. Thesis*, Arizona State University, Tempe, AZ, 1974.

118 H.H. Inhoffen, K. Bliesener and H. Brockmann, Jr., *Tetrahedron Lett.*, (1967) 727.

119 R. Lemberg and J. Parker, *Aust. J. Exp. Biol. Med. Sci.*, 30 (1952) 163.

120 J. Lucas and J.M. Orten, *J. Biol. Chem.*, 191 (1951) 287.

121 N.M. Karayannis and A.H. Corwin, *Anal. Biochem.*, 26 (1968) 34.

122 H.N. Dunning and J.K. Carlton, *Anal. Chem.*, 28 (1956) 1362.

123 S.K. Hajibrahim, P.J.C. Tibbetts, C.D. Watts, J.R. Maxwell and G. Eglinton, *Anal. Chem.*, 50 (1978) 549.

124 A.D. Adler, F.R. Longo, F. Kampas and J. Kim, *J. Inorg. Nucl. Chem.*, 32 (1970) 2443.

125 V. Miller and L. Malina, *J. Chromatogr.*, 145 (1978) 290.

126 R.E. Carlson, D. Dolphin and M. Bernstein, *Clin. Chem.*, 24 (1978) 2009.

127 C.K. Chang, R.K. DiNello and D. Dolphin, in D.H. Bush (Editor), *Inorganic Syntheses*, Vol. 20, Wiley, New York, 1980, p. 147.

128 S.J. Baum and R.K. Ellsworth, *J. Chromatogr.*, 47 (1970) 503.

129 J.O. Alben, W.H. Fuchsman, C.A. Beaudreau and W.S. Caughey, *Biochemistry*, 7 (1968) 624.

130 B.D. McLees and W.S. Caughey, *Biochemistry*, 7 (1968) 642.

131 L.J. Boucher and J.J. Katz, *J. Amer. Chem. Soc.*, 89 (1967) 1340.

132 N.M. Karayannis and A.H. Corwin, *J. Chromatogr.*, 47 (1970) 247.

133 K.S. Hui, B.A. Davis and A.A. Boulton, *J. Chromatogr.*, 115 (1975) 581.

134 J.W. Buchler, L. Puppe, K. Rohbock and H.H. Schneehage, *Chem. Ber.*, 106 (1973) 2710.

135 J.W. Buchler and K.L. Lay, *Inorg. Nucl. Chem. Lett.*, 10 (1974) 297.

136 A.D. Adler, F.R. Longo and V. Váradi, in F. Basolo (Editor), *Inorganic Syntheses*, Vol. 16, McGraw Hill, New York, 1976, p. 213.

137 L.K. Hanson, M. Gouterman and J.C. Hanson, *J. Amer. Chem. Soc.*, 95 (1973) 4822.

138 P.E. Wei, A.H. Corwin and R. Arellana, *J. Org. Chem.*, 27 (1962) 3344.

139 W.A. Svec, in D. Dolphin (Editor), *The Porphyrins*, Vo. 5, Academic Press, New York, 1979, p. 341.

140 K.R. Mattox and J.P. Williams, *J. Phycol.*, 1 (1965) 191.

141 J. Sherma, *Anal. Lett.*, 3 (1970) 35.

142 S.W. Jeffrey, *Biochim. Biophys. Acta*, 162 (1968) 271.

143 M.F. Bacon, *J. Chromatogr.*, 17 (1965) 322.

144 D.Y.C. Lynn and S.H. Schanderl, *J. Chromatogr.*, 26 (1967) 442.

145 G. Sievers and P.H. Hynninen, *J. Chromatogr.*, 134 (1977) 359.

146 H.H. Strain and W.A. Svec, *Advan. Chromatogr.*, 8 (1969) 119.

147 A.S.K. Chan, R.K. Ellsworth, H.J. Perkins and S.E. Snow, *J. Chromatogr.*, 47 (1970) 395.

148 H.H. Strain and J. Sherma, *J. Chem. Educ.*, 46 (1969) 476.

149 K. Schaltegger, *J. Chromatogr.*, 19 (1965) 75.

150 J. Sherma, *J. Chromatogr.*, 52 (1970) 177.

151 S.H. Schanderl and D.Y.C. Lynn, *J. Food Sci.*, 31 (1966) 141.

152 I.D. Jones, L.S. Butler, E. Gibbs and R.C. White, *J. Chromatogr.*, 70 (1972) 87.

153 R.J. Daley, C.B.J. Gray and S.R. Brown, *J. Chromatogr.*, 76 (1973) 175.

154 H.-C. Chow, M.B. Caple and C.E. Strouse, *J. Chromatogr.*, 151 (1978) 357.

155 H.H. Strain and W.A. Svec, in E. Heftman (Editor), *Chromatography*, Van Nostrand-Reinhold, New York, 3rd, Edn., 1975, p. 744.

156 H.H. Strain, B.T. Cope and W.A. Svec, *Methods Enzymol.*, 23 (1971) 452.

157 H.H. Strain and W.A. Svec, in L.P. Vernon and G.R. Seely (Editors), *The Chlorophylls*, Academic Press, New York, 1966, p. 21.

158 A.F.H. Anderson and M. Calvin, *Nature (London)*, 194 (1962) 285.

159 W.F. Smith, Jr. and K.L. Eddy, *J. Chromatogr.*, 22 (1966) 296.

160 H.H. Strain, J. Sherma and M. Grandolfo, *Anal. Biochem.*, 24 (1968) 54.

161 C.C. Kuenzle, M.H. Wieber and R. Pelloni, *Biochem. J.*, 133 (1973) 357.

162 C.C. Kuenzle, *Biochem. J.*, 119 (1970) 395.

163 M.S. Stoll, C.K. Lim and C.H. Gray, in P.F. Dixon, C.H. Gray, C.K. Lim and M.S. Stoll (Editors), *High Pressure Liquid Chromatography in Clinical Chemistry*, Academic Press, London, 1976, p. 97.

164 F. Frič and E. Haspel-Horvatovič, *J. Chromatogr.*, 68 (1972) 264.

165 N. Sato and N. Murata, *Biochim. Biophys. Acta*, 501 (1978) 103.

166 J.J. Zwolenik, *U.S. Pat.*, 3,514,467 (Cl. 260-314; CO96) (1970).

167 J.H. Argyroudi-Akoyunoglou and G. Akoyunoglou, *Photosynthetica*, 5 (1971) 153.

168 K. Iriyama, M. Yoshiura and M. Shiraki, *J. Chem. Soc. Chem. Commun.*, (1979) 406.

169 S. Shimizu, *J. Chromatogr.*, 59 (1971) 440.

170 A.F. McDonagh, in D. Dolphin (Editor), *The Porphyrins*, Vol. 6, Academic Press, New York, 1979, p. 293.

171 P. O'Carra and E. Colleran, *J. Chromatogr.*, 50 (1970) 458.

172 F. Eivazi, M.F. Hudson and K.M. Smith, *Tetrahedron Lett.*, (1976) 3837.

173 A.J. Fatiadi and R. Schaffer, *Experientia*, 27 (1971) 1139.

174 W. Rudiger, *Hoppe-Seylers' Z. Physiol. Chem.*, 350 (1969) 1291.

Chapter 20

Phenolic compounds

JEFFREY B. HARBORNE

CONTENTS

20.1. INTRODUCTION

Two early pioneers who developed chromatographic techniques for separating phenolic compounds were Karrer and Strong [1], who in 1930 resolved anthocyanin mixtures on columns of calcium sulfate and of alumina. It was not, however, until paper partition chromatography was applied to the separation of these pigments by Bate-Smith [2] in 1948 that chromatography became an established procedure in this field. Apart from amino acids and sugars, possibly no other group of natural compounds is so easily separated and identified on paper chromatograms. This is partly because polyphenols have just the right range of solubility characteristics for ease of separation, and partly because most of them can be seen on chromatograms without the use of chromogenic reagents. In recent years, other chromatographic

References on p. B430

techniques have been introduced, but PC remains the most important method for these substances.

Of the newer techniques, TLC is undoubtedly the most versatile and has proved especially valuable for separating the classes of phenols (e.g. phenolic acids and methylated flavones) that do not lend themselves readily to separation on paper. The application of GLC to these compounds has been frequently explored [3,4], and it is clearly the best technique for separating mixtures of the simple, more volatile phenols. HPLC has been extensively applied to the flavonoids [5] and promises to be of especial value in the quantitative analysis of involatile plant polyphenols generally. Finally, paper electrophoresis [6] and conventional CC [7,8] both have limited but important applications in this field.

The term "phenolic compounds" embraces a wide range of naturally occurring substances [9], but there are two main groups: simple phenolics and flavonoids. Simple phenolics are phenols, such as catechol and resorcinol; phenolic acids, such as protocatechuic and syringic acids; and cinnamic acids (e.g. caffeic acid) and their lactone derivatives, the coumarins. The flavonoids comprise the widely occurring water-soluble plant pigments, the anthocyanins and flavones, and a number of related substances (e.g., isoflavones, catechins, tannins, and biflavonyls) [10]. Because the plant and animal quinone pigments [11] contain phenolic hydroxyl groups, their separation will be described here, as well as that of the related plant xanthones. A small but significant number of phenolics also contain a nitrogen function; in general, these will not be considered here in any detail.

20.2. PAPER CHROMATOGRAPHY

20.2.1. Solvent systems

The solvent systems most frequently used for separating phenols on paper are shown in Table 20.1. The exact proportions of solvents used in the mixtures are not usually critical, and, indeed, may be varied on occasion to give improved separations. For example, a single-phase mixture of n-butanol–acetic acid–water (BAW) (6:1:2) is as effective as the more usual BAW 4:1:5 systems. However, R_F values of phenols are generally lower than in the latter mixture. A single-phase mixture of t-butanol–acetic acid–water (3:1:1) can be used in place of BAW [12], but it has the disadvantage of a longer developing time. There is much to be said for using well-known standardized systems when determining R_F values of new compounds so that they can be directly compared with published data. The most important systems from this point of view are BAW (4:1:5), acetic acid–conc. HCl–water (30:3:10) (Forestal), phenol, and water (either alone, containing 2–6% acetic acid, or containing 1% HCl).

A few practical precautions are appropriate here. BAW mixtures may be used either immediately or 1–3 days after preparation, but R_F values vary with the time of equilibration. Equilibration of the upper with the lower layer of the solvent mixture and of the paper in the tank are essential when using the systems involving

TABLE 20.1

SOLVENT SYSTEMS FOR PAPER CHROMATOGRAPHY OF PHENOLS

Composition * and abbreviation	Proportions (v/v)	For
n-Butanol–acetic acid–water (BAW)	4 : 1 : 5 (upper layer)	all classes of phenols
n-Butanol–ethanol–water (BEW)	20 : 5 : 11 (miscible)	most flavonoid glycosides
n-Butanol–2 N HCl (BuHCl)	1 : 1 (upper layer)	anthocyanins **
n-Butanol–2 N ammonia (BN)	1 : 1 (upper layer)	cinnamic acids, coumarins, xanthones, biflavonyls, and simple phenols
Phenol–water (PhOH)	4 : 1 (miscible)	many flavonoids, especially partially methylated derivatives
Benzene–propionic acid–water (BPA)	2 : 2 : 1 (upper layer)	phenolic acids
Benzene–acetic acid–water	125 : 72 : 3 (miscible)	isoflavones and methylated flavones
2-Propanol–ammonia–water (PAW)	8 : 1 : 1 (miscible)	phenolic acids
Acetic acid–conc. HCl–water (Forestal) ***	30 : 3 : 10 (miscible)	most flavonoid aglycones
Water or (5–15%) aqueous acetic acid	–	most flavonoid glycosides, simple phenols, and flavanones
Chloroform–acetic acid–water	13 : 6 : 1 (miscible)	flavonols

* All solvent mixtures can safely be used for overnight separation (18–24 h) except water (2–4 h) and TAW (6–8 h).

** Anthocyanins fade, unless mixtures containing HCl or acetic acid are used.

*** Various mixtures of acetic acid and water may be used for flavonoid aglycones other than anthocyanidins.

n-butanol and water, i.e. n-butanol–2 N ammonia (1 : 1) (BN) and n-butanol–2 N HCl (1 : 1) (BuHCl). However, equilibration is undesirable with the benzene–acetic acid–water system, because substances streak unless the solvent mixture is fresh [13,14].

Although the systems outlined in Table 20.1 work very well with most classes of phenols, they give poor separations with simple phenols. Compounds such as vanillin and eugenol can be separated on Schleicher & Schüll 2043b paper, with dimethylformamide–acetone (3 : 1) as stationary phase and cyclohexane–ethyl acetate (5 : 1) as mobile phase [15]. Alternatively, BN may be used on Whatman No. 7 paper, impregnated with boric acid–1 N NaOH [16]. Another approach is to separate phenols on paper as their p-nitrophenylazobenzene derivatives [17]. There is, however, much to be said for using TLC (Chap. 20.4) or GLC (Chap. 20.6) for such separations.

References on p. B430

TABLE 20.2
COLOR REACTIONS OF VARIOUS PHENOLS ON PAPER

Reagent	Light source	Color	Class of phenols responding
None	Visible	Orange, magenta, mauve	Anthocyanins, betacyanins
		Bright yellow	Chalcones, aurones
		Faint yellow	Flavones, xanthones
	UV 253 nm	Absorbing black on a fluorescent background	Simple phenols, isoflavones, flavone itself
	UV 300 nm	Dull brown	Flavonol 3-glycosides, flavones
			Xanthones
		Apricot	Flavonol with free 3-OH group
		Bright yellow	Flavonol without a 5-OH group
		Yellow-green fluorescence	Cichoriin, etc.
		Yellow fluorescence	3,5-Methoxylated flavonols, coumarins, and cinnamic acids
		Blue fluorescence	
NH$_3$ vapor or 2 N Na$_2$CO$_3$ spray	UV 300 nm	Color changes	Most phenols
		Appearance of blue color	p-Coumaric acid, phloroglucinol, etc.
		Loss of blue color	Methylated cinnamic acids
		Fluorescence	coumarins, etc.
FeCl$_3$/K$_3$Fe(CN)$_6$ spray	Visible	Blue	All phenolic compounds
Diazotized p-nitroaniline spray	Visible	Yellow, brown, or red	Most phenolic compounds
Ammoniacal AgNO$_3$	Visible	Reduction in the cold	o-Dihydroxy compounds
Sodium molybdate	Visible	Yellow	o-Dihydroxy compounds
Vanillin-p-toluenesulfonic acid	Visible	Violet–red	Phloroglucinol and derivatives
		Pale pink (after heating)	Catechol and derivatives
Sodium borohydride–ethanolic AlCl$_3$	Visible	Red	Flavanones only
p-Toluenesulfonic acid	Visible	Red or orange (after heating at 105°C for 10 min)	Flavan-3,4-diols only
Bromphenol blue	Visible	Yellow on a blue background	Phenolic acids
2,4,5-Trinitrophenol in EtOH/ethanolic KOH	Visible	Red	Flavan-3-ols (catechins) only

20.2.2. Color reactions

Table 20.2 indicates briefly the more useful reagents for characterizing phenolic compounds on paper. Examination in UV light, with and without ammonia vapor, provides a highly satisfactory way of locating and provisionally identifying these compounds. Their colors fade only slowly, and chromatograms can be kept for reference purposes almost indefinitely. Selective spray reagents may be employed on duplicate chromatograms for confirmation and are also available for showing the presence or absence of particular structural groups in the substances being examined. Details of the preparation of the spray reagents and of the color reactions of individual phenols are given in refs. 18–24.

20.2.3. Flavonoids

As already mentioned, PC is ideally suited to the separation of flavonoid pigments. Also, flavonoids are so regular in their behavior on paper that it is possible to discern relationships between chromatographic behavior and chemical structure. R_M values can be calculated for known substances and used to predict the behavior of new compounds. The numbers of hydroxyl, methyl, and glycosyl substituents a flavonoid possesses are the most important factors in determining its mobility; for example, increasing hydroxylation decreases mobility in all solvent systems.

The relationship between hydroxylation and R_F in benzene–acetic acid–water is so regular that R_F measurements can be used to estimate the number of free phenolic groups in an unknown flavonoid [25]. Other structural features are also significant, such as planarity. Planar flavonoids (flavonol aglycones, chalcones, aurones, etc.) can be separated from nonplanar compounds (flavonol 3-glycosides, flavanones, catechins, etc.) by their immobility in aqueous solvents.

In the following discussion, a few examples are selected to illustrate (a) the power of paper chromatography in resolving mixtures of closely related flavonoids, and (b) the relationship between chromatographic behavior and chemical structure. This subject is considered in more detail elsewhere [20].

20.2.4. Anthocyanins

There are over 200 known natural anthocyanins [10,19], and because these pigments do not have sharp melting points and cannot always be distinguished by spectral measurements, R_F values are of particular importance in their characterization. The main source of structural variation is in the nature and number of sugar residues and their position of attachment (through phenolic hydroxyl groups) to the basic anthocyanidin chromophore. Some typical data for a series of cyanidin glucosides are illustrated in Table 20.3. It will be seen at once that all these pigments are variously separated from each other on the basis of R_F values in the four solvent systems shown.

It is apparent from the data in Table 20.3 that glycosylation lowers the R_F in

References on p. B430

TABLE 20.3

GLYCOSYLATION AND BEHAVIOR OF THE CYANIDIN SERIES IN PC

AAH = acetic acid–conc. HCl–water (15 : 3 : 82). For key to other solvents, see Table 20.1.

Cyanidin glycosides	hR_F values in			
	BAW	BuHCl	1% HCl	AAH
Aglycone	68	72	0	3
3-Glucoside	23	23	6	24
5-Glucoside	32	43	6	24
7-Glucoside	34	27	4	20
3'-Glucoside	–	44	2	15
3,5-Diglucoside	14	4	13	40
3,7-Diglucoside	11	3	20	49
3,3'-Diglucoside	16	7	20	49
3-Sophoroside-5-glucoside	15	6	53	75
3,7,3'-Triglucoside	8	1	47	67

BAW and BuHCl, while increasing it in aqueous systems. A regular decrease in R_F values in BAW and an increase in 1% HCl occurs, largely irrespective of the position of glucose attachment (compare mono-, di-, and triglucosides in Table 20.3). However, variation from this regular behavior occurs, if the nature of the sugar or the type of linkage in the case of biosides is altered. The substitution of rhamnose for glucose, e.g., causes a marked increase in R_F values in all solvents. Again, the presence of sugars at the 3-, 7- and 3'-hydroxyls of cyanidin produces a distinctly lower R_F in BAW and BuHCl than if only two hydroxyls are substituted, as in the 3-sophoroside-5-glucoside (Table 20.3). Data similar to those in Table 20.3 are available for the glycosides of the other five common anthocyanidins of plants.

A range of other sugars besides those illustrated in Table 20.3 may be associated with anthocyanidins. Pairs of closely related glycosides are separated without much difficulty: the 3-galactoside and the 3-glucoside, the 3-sambubioside (xylosyl-β1 → 2-glucoside) and the 3-lathyroside (xylosyl-β1 → 2-galactoside), the 3-gentiobioside and the 3-sophoroside. The only pigments that are more difficult to separate are those having acylated sugars. Acylation with p-coumaric, caffeic, or ferulic acid lowers the R_F in aqueous solvents, but increases it in BAW. More than one sugar in the anthocyanin may carry an acyl group, and complex pigments do occur in some plants. Members of the *Commelinaceae*, for example, contain cyanidin 3,7,3'-triglucoside, which appears to have an aromatic acyl substituent linked to each of the three glucose residues [25,26].

20.2.5. Flavonols and flavones

PC has been widely used for characterizing these compounds and their derivatives, and there does not seem to be any mixture of flavones which has not been successfully resolved by one or another of the solvent systems listed in Table 20.1. A few examples, selected at random, illustrate the type of separations that have been achieved.

20.2.5.1. Flavonol methyl ethers

Quercetin and its isomeric monomethyl ethers are readily separated in BAW: quercetin, 0.64; 3-OMe, 0.93; 5-OMe, 0.48; 7-OMe, 0.72; 3'-OMe, 0.74. The two methyl ethers most similar in BAW, i.e. the 7-methyl ether (rhamnetin) and the 3'-methyl ether (isorhamnetin), separate better in propanol–acetic acid–water (1:1:1). The unexpectedly low R_F value of the 5-methyl ether (azaleatin), when compared with quercetin, is due partly to the fact that in this compound the 5-hydroxyl group can no longer chelate with the adjacent carbonyl group. Some difficulty may be experienced in distinguishing between methylated and unmethylated flavonols on paper. For example, 7-O-methylmyricetin (europetin) has the same R_F (0.40) as quercetin in Forestal. It does, however, separate from it in BAW (quercetin 0.64, europetin 0.53) or in phenol (quercetin 0.38, europetin 0.55). Thus, in identifying flavonols on paper, one should use several different solvent systems.

20.2.5.2. Flavonol glycosides

Closely similar glycosides can be separated without trouble. For example, the six known 3-monoglycosides of quercetin have the following R_F values in BAW: rhamnoside, 0.72; arabinoside, 0.70; xyloside, 0.65; glucoside, 0.58; galactoside, 0.55; and glucuronide, 0.40. The rhamnoside and arabinoside, which run rather closely to each other in BAW, have quite different mobilities in water (0.19 and 0.07, respectively). A change in the position of substitution of the sugar residues in the flavonol also alters the R_F values appreciably, and the 3-, 7-, and 4'-glucosides of quercetin are easily separated, as are the 3,4'-, 3,7-, and 7,4'-diglucosides (R_F values in BAW: 0.58, 0.37, and 0.48; 0.35, 0.30 and 0.23, respectively).

20.2.5.3. C-Glycosylflavones

Much work has been devoted to distinguishing the many different C-glycosides of flavones which occur in plants [27]. PC is most valuable for such separations. In the first instance, two-dimensional separations in BAW and 15% acetic acid can be used to distinguish C-glycosides from the many related O-glycosides that are also known; C-glycosides have higher R_F values in 15% acetic acid and lower values in butanolic solvents. Again, such separations usefully distinguish between different C-glycosides, 6-substituted glycosides occupying a different area on the chromatogram from the corresponding 8-substituted glycosides (Fig. 20.1). In the case of the many possible 6,8-disubstituted apigenin derivatives illustrated in Fig. 20.1, there is some degree of overlap, and TLC (cf. Chap. 20.4) is a useful complementary technique for resolving these phenolic derivatives.

References on p. B430

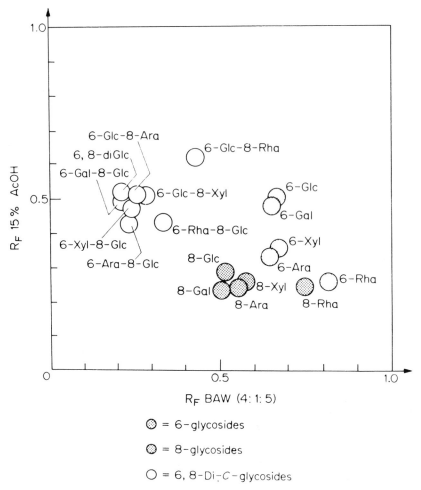

Fig. 20.1. Two-dimensional PC of C-glycopyranosylapigenins on Whatman No. 1 paper. Glc = β-D-glucopyranosyl; Gal = β-D-galactopyranosyl; Xyl = β-D-xylopyranosyl; Ara = α-L-arabinopyranosyl; Rha = α-L-rhamnopyranosyl. (Reproduced from *The Flavonoids,* Chapman & Hall, London, 1975, p. 680, with permission [27].)

20.3. PAPER ELECTROPHORESIS

Paper electrophoresis (PE) (cf. Chap. 9) has no general advantage over PC for phenols, but should be tried when other chromatographic methods fail. It does provide a diagnostic test for catechol groupings in unknown phenols and it is important for separating betacyanins, the only group of phenolic compounds for which its use is superior to PC.

The general techniques of PE of phenols are no different from those used with other substances. Phenolic compounds must, in general, be ionized or complexed with a metal ion before becoming mobile in an electric field. Potential gradients ranging from 20–130 V/cm will produce separations in 15–45 min. Mobility is customarily related to a reference compound (e.g., M_{SA}, mobility relative to salicyclic acid; M_{HB}, mobility relative to *p*-hydroxybenzoic acid; etc.). The main variables in the electrophoresis of phenols are the pH and the complexing nature of the electrolyte used.

Noncomplexing electrolytes will separate simple phenols from phenolic amines or from phenols having additional functional groups. For example, in acetate buffer (pH 5.2) or phosphate buffer (pH 7.2) phenols are immobile, whereas phenolic amines and acids move. The only other phenols that migrate under these conditions are anthocyanins and betacyanins. Anthocyanins carry a positive charge in acid solution, but require a longer time for separation. Thus, Markakis [28] found that cyanidin 3-rutinoside and cyanidin 3-sophoroside required 5 h for separation at pH 2 and 7 V/cm. The betacyanins are more mobile than other phenols, because they possess three carboxyl groups in their structures, and thus electrophoresis has become an essential procedure for their separation and characterization. Linstedt [29] used citrate buffers (pH 2.7), but aqueous formic acid (pH 2.4) and pyridine–formate (pH 4.5) with gradients of 16 V/cm are used more routinely [30,31].

Alkaline buffers are less useful than acidic buffers, because polyphenols oxidize in alkaline solution, and serious streaking may occur in the electropherogram. Glycine and borate buffers at pH 10 have been used most frequently. Although simple phenols are fairly mobile at this pH (e.g., catechol M_{SA} 0.67), flavonols show low mobility and do not separate particularly well (e.g., quercetin M_{SA} 0.21 and myricetin M_{SA} 0.18; compared with PC in Forestal, R_F 0.40 and 0.24, respectively). Electrophoresis is less valuable than chromatography for the structural analysis of

TABLE 20.4

ELECTROPHORETIC MOBILITIES OF SOME FLAVONE AND FLAVONOL SULFATES

Compound	Mobility*	Compound	Mobility*
8-Hydroxyluteolin 8-sulfate	0.52	Quercetin 7-sulfate	0.18
Luteolin 7-sulfate	0.56	Isorhamnetin 7-sulfate	0.25
Apigenin 7-sulfate	0.70	Kaempferol 7-sulfate	0.34
Luteolin 7-rutinoside sulfate	0.72	Quercetin 3'-sulfate	0.66
Luteolin 4'-sulfate	0.75	Quercetin 3-sulfate	1.00
Apigenin 4'-sulfate	0.80	Isorhamnetin 3-sulfate	1.15
Apigenin 7,4'-disulfate **	2.20	Quercetin 7-sulfate 3-glucuronide	1.22
Luteolin 7-sulfate 3'-glucoside	2.50	Isorhamnetin 7-sulfate 3-glucuronide	1.28
Luteolin 7,3'-disulfate	3.0	Kaempferol 7-sulfate 3-glucuronide	1.44
Luteolin 7,4'-disulfate **	3.5		

* Mobility relative to quercetin 3-sulfate on Whatman No. 3 paper in formate–acetate buffer (pH 2.2) at 400 V/cm for 2.5 h.
** Synthetic compounds; all others have been obtained by isolation and most of them also by synthesis.

References on p. B430

most phenolics, because it is difficult to relate mobility to structure in any but the simplest systems [6].

Complexing electrolytes separate catechol and its derivatives, which complex with metal ions and increase in mobility, from related monophenols. Alkaline borate has been utilized, but good separations can also be obtained with sodium molybdate at pH 5.2, which has the added advantage that catechols form visible brown complexes in it [32]. As a diagnostic test for catechols, the only drawback is that phenolic acids are also mobile (e.g., catechol M_{HB} 1.15, salicylic acid M_{HB} 1.18), but this difficulty can be overcome by preparing a second electropherogram in acetate buffer, pH 5.2, in which catechols are immobile.

Recently, PE in formate–acetate buffer (pH 2) has been employed for screening plant extracts for flavone and flavonol potassium sulfate salts [33–35]. Having an overall negative charge, the sulfate derivatives migrate to the anode (Table 20.4), whereas all normal flavones and flavonols are immobile at this pH. Not surprisingly, disulfates migrate further than monosulfates, although the position of attachment of the sulfate (compare luteolin 7,3'- and 7,4'-disulfates in Table 20.4) has a modifying effect on mobility [35]. Another unusual type of flavone, this time with malonic acid as a covalently bound acyl group, which has recently been discovered, is immobile at pH 2, but migrates towards the anode at pH 4 [36]. Similarly, flavone O-glucuronides carry a net negative charge and, like the malonates, are immobile at pH 2 but move at pH 4.

20.4. THIN-LAYER CHROMATOGRAPHY

20.4.1. Introduction

TLC has been applied to the separation of nearly all classes of phenols, and the method is so versatile and successful that it can be used instead of PC for much of that work. However, in the author's view, it is best considered as a complementary technique to PC. The water-soluble flavonoids, for example, are so well adapted to separations on paper, that there is no advantage to be gained in sacrificing the greater convenience of paper and the more consistent R_F values it produces, if the resolving power is the same. TLC is undoubtedly faster, but the rates of separation on cellulose plates, for example, differ only by a factor of 4–5 from those on paper sheets.

A distinct disadvantage of TLC is that flavonoids are not so easily located as on paper. Flavonols display a range of characteristic fluorescences on paper, but all appear as dark, absorbing spots on silica gel. In addition, the fluorescences and colors of catechol derivatives and anthocyanins fade and disappear more quickly on thin-layer adsorbents than on paper, and have to be recorded immediately after development. A further disadvantage of TLC is that no single system will separate several different classes of phenols on the same plate (compare the use of BAW on paper).

Despite these disadvantages, TLC is a most important technique for phenol separations, and it is the method of choice when (a) the time factor is crucial, (b) separation by PC is difficult to achieve (e.g. as in the case of nonpolar flavonoids), (c) only minute amounts of substances are available, because it is the more sensitive procedure. The problem of locating flavonoids on polyamide thin-layer plates has diminished somewhat with the recent widespread use of "Naturstoffreagenz A" (diphenylboric acid β-aminoethyl ester) spray [37], which gives a range of fluorescent colors according to the structure [38]. Another technique applicable to silica gel plates is to incorporate anilinium chloride in the adsorbent. After development, a simple spray with aqueous sodium nitrite will produce a stable colored diazonium salt with any phenol in the sample [39].

20.4.2. Sorbents and solvents

Silica gel is by far the most popular adsorbent, in spite of the fact that polyamide powder was introduced specifically for the separation of flavones. Cellulose powder,

TABLE 20.5

SOLVENT SYSTEMS FOR TLC OF PHENOLS

Class of phenols	Adsorbent	Solvent systems
Simple phenols and phenolic acids	Silica gel	Chloroform–acetic acid (9:1), ethyl acetate–benzene (9:11), ethyl acetate–light petroleum (3:1), diethyl ether–light petroleum (1:1)
Coumarins	Silica gel	Benzene–acetone (9:1), benzene–ethyl acetate (9:1), chloroform–ethyl acetate–formic acid (2:1:1)
Hydroxycinnamic acids	Silica gel	Dichloromethane–toluene–formic acid (5:4:1), benzene–ethyl acetate–formic acid (8:2:1)
Phenylpropenes	Silica gel	Benzene–chloroform (9:1), n-hexane–chloroform (3:2)
Anthocyanidins	Silica gel	Formic acid–conc. HCl–water (85:9:6)
Anthocyanins	Silica gel	Ethyl acetate–formic acid–water (14:3:3)
Flavones and flavonols	Polyamide	Ethanol–water (3:2), methanol–acetic acid–water (19:1:1), benzene–methyl ethyl ketone–methanol (4:3:3)
Isoflavones	Silica gel	Methanol–chloroform (11:89)
Biflavonyls	Silica gel	Toluene–ethyl formate–formic acid (5:4:1)
Aurones	Silica gel (buffered with sodium acetate)	Benzene–ethyl acetate–formic acid (9:7:4)
Hydroxyquinones	Polyamide Silica gel	Methanol–water (6:4), light petroleum–ethyl acetate (7:3)
Xanthones	Silica gel	Benzene–chloroform (3:7)
Chlorinated phenols	Silica gel	Light petroleum satd. with formic acid; xylene, satd. with formamide; benzene–acetic acid–water (2:2:1)

References on p. B430

as might be expected, gives results very similar to filter paper, but it is reported to produce better resolutions of pairs of closely similar substances [40]. Mixtures of cellulose with silica gel, which have been used for anthocyanins [41] and for phenolic acids [42], appear to combine many of the advantages of the two sorbents. Cellulose, impregnated with 10% polyamide, has been employed by Bark and Graham [43] for separating series of simple alkylated and halogenated phenols.

A much greater range of solvent systems has been used in TLC than in PC. Many of the solvents used for PC were tried in the first experiments, but other mixtures, especially those containing chloroform, ethyl acetate, and light petroleum, were devised for TLC in particular. A selection of recommended systems is given in Table 20.5. It should be noted that aqueous mixtures, although excellent on polyamide, are to be avoided on silica gel.

For extensive lists of R_F values the reader is referred to one of the various monographs on TLC [44]. Some typical values for phenols and phenolic acids, given in Table 20.6, illustrate the kind of separations that can be achieved [45].

20.4.3. Simple phenolic compounds

The TLC of simple phenols has attracted a great deal of attention, because these substances are difficult to resolve by PC. A range of systems has been used, usually with Silica Gel G, but no one system can be used universally. Thus, diethyl ether–light petroleum (1:1) will separate the closely similar 2,3-dimethylquinol (R_F 0.20) and 2,5-dimethylquinol (R_F 0.50), but it will not separate phenol from p-cresol (both R_F 0.44) [46]. However phenol and p-cresol can be separated as their 3,5-dinitrobenzoates, which have R_B values[*] of 0.76 and 1.05, respectively, with benzene–light petroleum (1:1) [47]. Mixtures of phenolic aldehydes and ketones are often difficult to resolve, and a procedure involving pretreatment of silica gel plates with chlorobenzene was found to be required, followed by development in benzene–acetone (24:1) containing 0.4% acetic acid [48]. Systems for separating the common phenols and phenolic acids of plants on silica gel and cellulose plates are indicated in Table 20.6.

Phenols, such as phloroglucinol, catechol, and protocatechuic acid, can be detected either on silica gel layers containing fluorescent indicators or by spraying the chromatogram with one of the usual phenol reagents (cf. Table 20.2). For spraying, the Folin–Ciocalteu reagent is preferred to the ferric chloride–ferricyanide reagent. Sodium cobaltinitrite in aqueous acetic acid has been suggested as an alternative general reagent for detecting phenolic compounds on thin-layer plates; it gives brown to yellow colors and will detect as little as 2 μg [49]. If titanium chloride in concentrated HCl is used as a spray reagent, monophenols (yellow) can be distinguished from diphenols (brown); the limit of detection is 4 μg [50].

TLC also has much to offer for the separation of cinnamic acids and coumarins. Strohl and Seikel [51] devised a two-dimensional system [chloroform–acetone (2:1) and ethyl acetate–dimethylformamide (50:1) on silica gel] for the separation of

---

[*] R_B is the R_F relative to butter yellow.

TABLE 20.6

TLC OF SIMPLE PHENOLS AND PHENOLIC ACIDS [45]

Phenolic compounds	hR_F values *				Color and detection methods **
	1	2	3	4	
Simple phenols					
Orcinol	19	62	46	67	Bluish pink (V)
4-Methylresorcinol	25	63	59	65	Brick red (V)
2-Methylresorcinol	40	64	58	73	Bluish pink (V)
Resorcinol	17	59	48	74	Red (V)
Catechol	35	66	58	72	
Hydroquinone	18	58	34	69	Blue (FC + NH$_3$)
Pyrogallol	8	15	19	72	
Phloroglucinol	5	47	9	62	
Phenolic acids					
Gallic	5	40	5	40	Blue (FC)
Protocatechuic	19	44	19	52	
Gentisic	33	44	41	61	
p-Hydroxybenzoic	55	80	60	62	
Syringic	79	58	74	52	Blue (FC + NH$_3$)
Vanillic	82	73	70	57	
Salicylic	91	82	86	66	

* Systems: 1, Silica Gel G/acetic acid–chloroform (1:9); 2, Silica Gel G/ethyl acetate–benzene (9:11); 3, Cellulose MN 300/benzene–methanol–acetic acid (45:8:4); 4, Cellulose MN 300/6% aqueous acetic acid.
** V = vanillin–HCl; FC = Folin–Ciocalteu reagent.

p-coumarate esters in pine pollen. The system chloroform–ethyl acetate–formic acid (2:1:1) can be used for resolving caffeic acid esters; it separates caffeylglucose from orobanchin, two substances which had similar R_F values on paper in all systems tried [45]. For the free hydroxycinnamic acids, a general analytical procedure has been described involving purification of extracts on polyamide columns, followed by TLC on silica gel in the lower phase of dichloromethane–water–acetic acid (2:1:1) [52]. A battery of other techniques, including the use of steamed silica gel plates, of "multiple-elimination" TLC (i.e. removal of impurities from successive chromatograms [53]), and of cellulose–silica gel (1:1) mixtures has also been applied to the determination of these substances [53].

Thin-layer systems have also been reported for coumarins [54]. Whereas simple hydroxycoumarins are probably more easily separated on paper, complex coumarins, particularly those with isoprenoid side chains, are ideal subjects for TLC. Suitable solvents on silica gel include: chloroform, diethyl ether–benzene (1:1), and diethyl ether–benzene–10% acetic acid (1:1:1).

The hydroxyphenylpropenes, found in the essential oil fraction of many plant extracts, also separate well on thin layers of silica gel with solvents such as benzene, benzene–chloroform (9:1), and n-hexane–chloroform (3:2). They are detected by

References on p. B430

spraying the chromatograms with vanillin–sulfuric acid [55] or with Gibbs reagent [56]. Phenylpropenes are sometimes accompanied by their allylic isomers, and such pairs of compounds may be difficult to resolve. Guven and Bayer [57] recommended hexane–ethyl acetate (17:3) for separating one such pair, eugenol (R_F 0.25) and isoeugenol (R_F 0.18).

Most of the naturally occurring depsides found in lichens have phenolic as well as carboxylic acid substituents. Such compounds are conveniently separated on silica gel layers with benzene–dioxane–acetic acid (36:9:1), hexane–diethyl ether–formic acid (13:8:2), or toluene–acetic acid (20:3). A standardized method for the identification of depsides and their phenolic degradation products, based on R_F values in these three systems, has been devised by Culberson [58], who listed data for 220 such compounds. Further data on lichen compounds are available in refs. 59 and 60.

20.4.4. Flavonoids

TLC has been applied with considerable success to the anthocyanin pigments, but the resolving power of TLC on silica gel is not significantly better than that of PC. The limitation in TLC as well as PC is that solvent systems must contain 10–30% of formic, acetic, or hydrochloric acid, in order to stabilize the pigments. Nybom [40] has elaborated a method of two-dimensional TLC on Cellulose MN 300 with formic acid–conc. HCl–water (10:1:3) and amyl alcohol–acetic acid–water (2:1:1). Its distinct advantage over two-dimensional PC is that the pigment spots are much more compact. A good one-dimensional solvent system for TLC on silica gel is ethyl acetate–formic acid–2 N HCl (85:6:9); it clearly separates malvidin from peonidin, the two common anthocyanidins which are most difficult to distinguish by PC when present together. The system works well on standard grades of Silica Gel G. However, on purified silica gel separations were much inferior, indicating that traces of metal ion are useful for reducing the mobility of the anthocyanidins with a catechol nucleus, such as cyanidin. Hess and Meyer [61] have separated the corresponding glycosides, the anthocyanins, on silica gel; Birkofer et al. [62] have employed polyacrylonitrile–polyamide (7:2), buffered with 0.05 M potassium phosphate; and Wrolstad [63] prepared thin layers of polyvinylpyrrolidone–cellulose mixtures. In spite of the availability of these procedures, anthocyanins are most frequently separated on microcrystalline cellulose by the use of similar solvents as for PC.

For the separation of flavones and flavonols with up to three free hydroxyl groups (i.e. mono-, di-, or trihydroxyflavones and polyhydroxyflavones with extensive O-methylation), TLC is an ideal procedure. Systems that work well on silica gel are 10% acetic acid in chloroform and 45% ethyl acetate in benzene, while on polyamide, one may use benzene–light petroleum (b.r. 100–140°C)–methyl ethyl ketone–methanol (60:26:7:7) [64,65]. For more highly hydroxylated flavonoids, polyamide is the adsorbent of choice, silica gel being less suitable. In such cases, chromatography on thin layers of polyamide is particularly valuable when used in conjunction with PC. By combining these procedures, it is possible to distinguish clearly between many closely related structures. Typical R_F data for various querce-

TABLE 20.7

R_F VALUES OF QUERCETIN METHYL ETHERS ON PAPER AND ON THIN LAYERS OF POLYAMIDE [66]

BMM = benzene–methyl ethyl ketone–methanol (4:3:3); BLMM = benzene–light petroleum (b.r. 100–140°C)–methyl ethyl ketone–methanol (60:26:7:7). For key to other solvents, see Table 20.1.

Quercetin derivatives	hR_F values on				
	paper			polyamide *	
	60% HOAc	Forestal	BAW	BMM	BLMM
Quercetin	33	42	73	20	0
3-methyl ether	63	76	87	40	0
5-methyl ether	40	55	67	20	0
7-methyl ether	55	55	77	45	0
3'-methyl ether	37	53	73	40	0
4'-methyl ether	42	58	79	45	0
3,7-dimethyl ether	74	86	90	65	10
7,4'-dimethyl ether	60	76	86	65	25
3,7,4'-trimethyl ether	82	91	93	80	55
7,3',4'-trimethyl ether	64	84	88	85	65
3,7,3',4'-tetramethyl ether	85	93	51	90	80

* Polyamide grade DC 11, Macherey-Nagel.

tin methyl ethers are illustrated in Table 20.7 [66].

If it is desired to separate the commonly occurring O-glycosides of the flavones and flavonols, it is advisable to use microcrystalline cellulose and the same solvents as for PC. Silica gel and polyamide systems have been explored [67,68] but, in general, the resolutions achieved are inferior. By contrast, silica gel plates have much to offer for the separation of C-glycosylflavones. One recommended solvent system is ethyl acetate–pyridine–water–methanol (16:4:2:1). It is remarkable for producing excellent separations of the closely similar C-glucosyl- and C-galactosyl-apigenin derivatives (Table 20.8) [27], although PC will achieve the same end [69]. The type of

TABLE 20.8

TLC OF C-GLYCOSYLAPIGENINS

hR_F values on activated Silica Gel G in ethyl acetate–pyridine–water–methanol (16:4:2:1).

6-Glycosylapigenins	hR_F	8-Glycosylapigenins	hR_F
6-C-Glucopyranosyl	59	8-C-Glucopyranosyl	71
6-C-Galactopyranosyl	37	8-C-Galactopyranosyl	59
6-C-Xylopyranosyl	72	8-C-Xylopyranosyl	73
6-C-Arabinopyranosyl	61	8-C-Arabinopyranosyl	68
6-C-Rhamnopyranosyl	78	8-C-Rhamnopyranosyl	72

References on p. B430

separation that can be achieved for glycoflavones in several medicinal plant extracts is nicely illustrated in color in ref. 70.

For separating other classes of flavonoids, such as the biflavonyls and the isoflavonoids, TLC is now a routine method. For purifying biflavonyls from crude plant extracts, one recommended procedure depends on using n-butanol–$2N$ NH_4OH (1:1) on paper and toluene–ethyl formate–formic acid (5:4:1), followed by benzene–pyridine–formic acid (36:9:5) on silica gel [45]. R_F values for 28 biflavonyls in five solvents on silica gel plates have been collated by Chexal et al. [71]. Isoflavones have been chromatographed on both silica gel and alumina plates; one suitable solvent on both adsorbents is chloroform–methanol–water (65:25:4) [72,73]. Pterocarpans and isoflavans are typically separated on silica gel (Merck 60) with benzene–ethyl acetate–methanol–light petroleum (b.r. 60–80°C) (6:4:1:6) or chloroform–methanol mixtures [74].

20.4.5. Hydroxyquinones

TLC has been conspicuously successful in separating quinonoid compounds. Most of the naturally occurring quinones are also phenolic, so that their separation is briefly dealt with here. p-Benzoquinones can be separated in many systems [75,76], e.g., on acetylated polyamide with aqueous methanol or aqueous acetone as solvents. Naphthoquinones separate well on silica gel, in light petroleum (b.r. 60–80°C)–ethyl acetate mixtures (e.g., in 7:3 ratio), or in benzene–light petroleum (b.r. 30–50°C) (2:1) [77].

Separation of anthraquinone mixtures is a little more difficult; different techniques are needed, depending on the type of pigment being studied. Anthraquinone glycosides can be chromatographed on silica gel plates with ethyl acetate–methanol–water (200:33:27) [78]. Mixtures of free anthraquinones are better separated on tartaric acid-treated silica gel plates by chloroform–methanol (99:1) or on polyamide by methanol–benzene (4:1) [79]. Separation of anthraquinones from their reduced forms, the anthrones and dianthrones, which accompany them in some plant extracts, is best achieved on plates of Silica Gel G–Kieselguhr G (1:6) in light petroleum (b.r. 40–60°C)–ethyl formate–formic acid (90:4:1) [80].

Finally, ubiquinones and plastoquinones can be resolved on paraffin-treated Kieselguhr G plates with acetone–water (9:1) [81] or on Silica Gel G with benzene–light petroleum (b.r. 40–60°C (2:3) [82].

20.5. LARGE-SCALE COLUMN CHROMATOGRAPHY

20.5.1. Comparison of methods

Large-scale separations of phenolics require rather different techniques from those described thus far. Some form of CC is often applied for the preliminary purification of the phenolic fraction in plants, for the separation of different classes of phenols, and for actually separating the individual components of a mixture on a milligram scale. For example, a convenient means of cleaning up anthocyanin extracts before PC or TLC is to transfer the pigments from an aqueous HCl solution to a weak anion-exchange column (e.g., Zeokarb 226), which is washed with water and then eluted with methanolic acetic acid [83]. A range of adsorbents has been applied to phenolic compounds in general, including: cellulose [84], cellulose acetate [85], Celite [86], ion-exchange resins [87], magnesol–Celite [88], silica gel [89], and polyamide [7]. An important recent advance has been the use of Sephadex columns, particularly for separating the various oligomers and polymers of proanthocyanidins [90].

Chromatography on cellulose or silica gel columns has not often been employed in practice for large-scale separations of phenolic compounds, because the procedures required are cumbersome and lack reproducibility. Moreover, they can be circumvented by the substitution of thick sheets of Whatman No. 3 paper [91,92] for cellulose columns, or thick (1-mm) layers of silica on glass plates [44] for silica gel columns. The present discussion of CC is therefore restricted to polyamide and Sephadex, which are the two most widely used sorbents at the present time.

20.5.2. Polyamide

Polyamide was first introduced as a chromatographic sorbent in 1957 for the separation of tannins [93], and has since become popular for separating most other groups of phenols. Its advantages over other sorbents, as listed by Hörhammer [7], are that it has a great capacity and that simple solvents can be used with it; water is used first, then alcohol–water mixtures, and finally pure alcohol. It can also be regenerated easily, which is a significant factor, because it is fairly expensive. Its excellent resolving power can be gauged from the fact that it separates eight flavonol glycosides present in flowers of the lime, *Tilia argentea,* and also the two closely related glycosides, linarin and pectolinarin, present in *Cirsium oleraceum* leaves. Polyamide columns also provide a simple procedure for separating different classes of phenols, particularly the coumarins from flavones and the chalcones from flavanones. For example, after adsorption of rutin and aesculin on polyamide, the flavone, rutin, can be eluted with water and the coumarin, aesculin, with methanol [7]. Similarly, chalcone–flavanone mixtures can be resolved by elution of the flavanone with water, followed by elution of the chalcone with methanol [94].

20.5.3. Sephadex

The structural elucidation of proanthocyanidins (also known as condensed tannins or leucoanthocyanidins) has been delayed by the many difficulties in purifying them. The application of chromatography on Sephadex has undoubtedly greatly aided research in this field [90,95]. Sephadex chromatography provides the first satisfactory means of separating the various oligomers according to their molecular weight and of purifying polymers for further analysis.

Tannin oligomers are eluted from Sephadex G-25 with aqueous acetone (1:1). This separates flavonoids in the MW range 300–900 and excludes polymers over

1100 [96]. The procedure for MW determination is based on the straight-line relationship between the elution volume and log MW, where catechin (MW 290), procyanidin dimer (MW 578), and procyanidin trimer (MW 834) are used as markers. Proanthocyanidin polymers may be purified on Sephadex G-50 or G-100, with aqueous alcohols as eluents [97]. In order to obtain pure polymers from plant extracts, Jones et al. [98] conducted a preliminary purification on Sephadex G-50, which was eluted with acetone–water (1 : 1) containing 0.1% ascorbate. The eluate was then applied to a Sephadex LH-20 column with aqueous methanol (1 : 1), the proanthocyanidins being eventually recovered with acetone–water (3 : 7).

For the phenols of lower MW, Sephadex G-25 and LH-20 have been proposed as sorbents. Adsorption as well as molecular sieving are involved in such separations. This is a rather useful procedure for separating 5 to 250-mg samples of simple mixtures of phenolic compounds. Phenols are eluted from G-25 with water or 0.1 M acetic acid [99], and isoflavones with 0.1 M NH_4OH [100]. Flavonoids are rather strongly adsorbed on Sephadex and are better separated as their molybdate complexes. Elution with water, followed by 1 M molybdate, separates mixtures of common flavone and flavonol glycosides on Sephadex G-25 [99]. Methanol resolves a range of flavones and flavanones on Sephadex LH-20, flavones (e.g., luteolin) being eluted before flavonols (e.g., quercetin) and methylated quercetins before quercetin itself [101].

20.6. GAS–LIQUID CHROMATOGRAPHY

Only within the last few years has GC been widely exploited for separating phenolic compounds. The technique is now clearly the method of choice for separating mixtures of naturally occurring simple phenols and is also useful for particular applications with flavonoids and other more complex phenolic compounds.

In order to increase volatility, phenols are generally chromatographed as their trimethylsilyl ethers. Even so, fairly high temperatures may have to be employed to effect resolution. Some indication of the types of stationary phases, supports, and column temperatures employed for separating different classes of phenols is given in Table 20.9 [102–109].

For the separation of variously substituted simple phenols, it is imperative to employ several types of column, because closely related structures (e.g., *o*-, *m*-, and *p*-cresols) may well have identical retention times on one or even two columns. Irvine and Saxby [102] found it necessary to use three stationary phases (cf. Table 20.9) in order to resolve and identify the 38 phenols present in cured tobacco leaf. Dihydric phenols can be conveniently distinguished from monohydric phenols by GC on Porapak P, coated with 3% Carbowax 20M at 230°C [110]; relative retention times of phenol, catechol, hydroquinone, and resorcinol are 0.33, 1.00, 1.66, and 1.83 respectively. Phenols present in the bark of members of the Salicaceae were chromatographed as their trimethylsilyl ethers on columns coated with 3% OV-1, 2% OV-17, 2% OV-25, and 4% OV-25 [111]. Similar procedures were employed for

TABLE 20.9

GLC OF PHENOLS

Phenol classes	Derivatives	Supports	Column temperatures, °C
Anthocyanidins	TMS	3% OV-225 on Chromosorb W HP (100–200 mesh)	270
Anthraquinones	TMS	1.5% SE-30 on Chromosorb W (60–80 mesh)	240
Cannabinoids	None	3% OV-17 on Gas Chrom Q (100–120 mesh)	235
Furanocoumarins	None	QF-1	174
Mono- and di-hydroxyflavones	TMS or methyl ethers	2% OV-17 Gas Chrom Q	250–270
Hydroxyanthones	TMS or methyl ethers	2% silicone rubber on silanized Chromosorb W (80–100 mesh)	
Phenolic acids	TMS	3% SE-30 on Chromosorb W AW DMCS (60–80 mesh) 3% OV-1 or 3% UCW-98 on Chromosorb W HP (100–120 mesh)	175
Phenols	Acetates	Polyphenyl ether OS-124 Trixylenyl phosphate Diethylene glycol succinate	130–160 150–180 150–180
Phenylpropenes	None	20% Reoplex 400 on Gas Chrom Q	80–200

fractionating the phenolic glycosides present in the same plants [112,113].

One clear advantage of GC over most other separation techniques is its great sensitivity. The phenolic coumarin, scopolin, was determined in plant extracts in microgram amounts by an ECD [114]. The sensitivity of detection of phenol or cresol using that detector can be increased a 100-fold by brominating phenols prior to GLC [115]. One situation where good sensitivity is important is in the detection of traces of phenolic acids in Recent sediments from the soil, where contents are in the order of 110 $\mu g/g$ dry wt. Successful analyses have been conducted on columns of 1% silicone OV-1 on Chromosorb W AW [116].

Phenols containing isoprenoid residues are often fairly volatile and may be separated without conversion to TMS ethers. For instance, the cannabinoids of marijuana are isoprenylated resorcinols, and GLC is the method of choice for their analysis. Most of the major cannabinoids can be adequately resolved on any of several types of stationary phases, e.g., SE-30, QF-1, OV-7, OV-17, Carbowax 20M, or XE-60. The two isomers of tetrahydrocannabiol, Δ^1-THC and Δ^6-THC, separate with some difficulty, but because Δ^1-THC is nearly always present as the major

References on p. B430

isomer, this is not often a serious handicap. If necessary, they can be resolved as the TMS ethers on the columns mentioned above. In fresh *Cannabis* plants or resin, Δ^1-THC, the hallucinogenic principle, occurs as a mixture of related acids, and these are decarboxylated during GLC. For routine analysis, this may be an advantage, because this reaction parallels the chemical changes that occur during the smoking of marijuana. Decarboxylation can be prevented by silylation or methylation of the cannabinoids prior to GLC [109,117].

GLC is of limited use for the commonly occurring flavones and flavonols, because of their low volatility. It does, however, have significant value for separating mixtures of highly methylated flavonoids or simple mono- and dihydroxyflavones when they are converted to the TMS ethers [108]. Flavonoids may also be identified by chromatographing the more volatile substituted acetophenones formed by al-kaline degradation of the methylated flavones [118]. Even the GLC of the very involatile anthocyanidins has been accomplished [119], although the compounds have to be first silylated. Subsequent mass spectral identification is complicated by the fact that during silylation the heterocyclic oxygen of the anthocyanidin nucleus is partly replaced by nitrogen.

20.7. HIGH-PERFORMANCE LIQUID CHROMATOGRAPHY

The modern development of HPLC as a highly sensitive and rapid procedure for separating and analyzing involatile organic constituents can be dated from about 1967 [120]. The method was soon applied to phenolic compounds, e.g., to those present in lichen extracts [121], but much of the early enthusiasm for the method was dampened by two practical problems. First, most of the column packings available at that time were unsuitable for the majority of phenolic compounds, which are very polar, and second, even when nonpolar phenolic derivatives (e.g. the methylated flavones present in citrus extracts) must be separated, an extensive clean-up of plant extracts is necessary before the compounds can be applied to the column.

The first problem largely disappeared following the introduction of reversed-phase HPLC in 1976 by Wulf and Nagel [122], who used a chemically bonded silica gel (μBondapak C$_{18}$) column for separating flavonoids and phenolic acids, the eluent being water–methanol–acetic acid (13:6:1). The second problem remains, since it is still not possible, except in favorable circumstances, to monitor the presence or absence of particular groups of phenolic derivatives in crude plant extracts by HPLC. Precolumn treatment is usually essential and, in addition, some previous fractionation by some other chromatographic procedure (e.g. polyamide CC or TLC) is often advisable.

As a result of this limitation, the most successful applications of the technique have been in situations where extensive purification is not required. For example, the crystalline deposits of flavonoids on the under surface of certain fern fronds are relatively free from contamination. In such cases, it is possible to take a small, 1-cm segment of a pigment-bearing pinnule, dissolve it in tetrahydrofuran and inject the solution directly into a reversed-phase column of Ultrasphere-ODS (Altex, Berkeley,

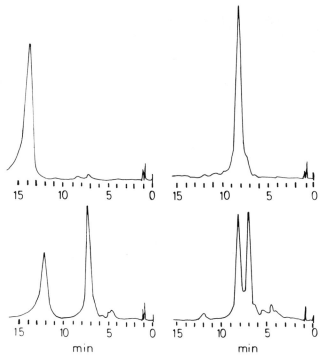

Fig. 20.2. High-performance liquid chromatograms of fern flavonoid exudates (*Pityrogramma triangularis*). Top left, galangin 7-methyl ether chemotype; top right, galangin chemotype; bottom left, kaempferol 4'-methyl ether and 7,4'-dimethyl ether chemotype; bottom right, hybrid of bottom left and top right chemotypes. (Reproduced from *Bull. Torrey Botan. Club,* 107 (1980) 134, with permission [123].)

CA, USA). The flavonoids are then separated with 10% aqueous acetic acid–tetrahydrofuran (2 : 1) [123]. By monitoring the eluate spectrophotometrically at 367 nm, excellent separations of the various kaempferol methyl ethers, chalcones, and flavanones present can be demonstrated (Fig. 20.2). A very ready distinction can be drawn between different chemical races within a given taxon, in this case the *Pityrogramma triangularis* fern complex. Furthermore, the analyses give better resolutions than TLC and provide a more accurate quantitative picture of chemical variation in these plants [123].

Another successful direct application of HPLC to flavonoid separations has been with *Poinsettia* cultivars, where the deeply colored sepal extracts have been analyzed in order to distinguish the various cultivated forms by quantitative differences in the five anthocyanins present. Good separations have been achieved on columns of μBondapak C_{18} with isocratic elution [124] or of LiChrosorb RP-18 with gradient elution [125]. Similar analyses on the same cultivars have been applied to the 8 to 10 flavonol glycosides which are present in these tissues, together with the anthocyanins. For the flavonol glycosides, a 300×3.9 mm column of μBondapak C_{18} was

References on p. B430

employed, the elution gradients being made from 2% acetic acid and acetonitrile or 2% acetic acid and tetrahydrofuran [126].

These successes with sepal pigments suggest that the direct injection of crude petal extracts to columns might be feasible, especially since the flavonoid content of

TABLE 20.10

COLUMN SUPPORTS AND SOLVENTS FOR HPLC OF PHENOLIC COMPOUNDS

Class	Column materials	Solvents	Ref.
Acylphloroglucinols	μBondapak C_{18} *	Tetrahydrofuran–phosphoric acid–water (650:1:350)	128
Anthocyanidins	μBondapak C_{18} *	Water–acetic acid–methanol (71:10:19)	129
Anthocyanins	LiChrosorb RP-18 *	1.5% Phosphoric acid, 20% acetic acid and 25% acetonitrile in water	125
Anthraquinones	Micropak CH-5	Methanol–water (1:1), acidified to pH 3	130
Benzo- and naphtho-quinones	Micropak Si-10	1% 2-Propanol in light petroleum	130
Biflavonoids	Merckosorb S-160	8% Methanol in isopropyl ether	131
Flavones and flavonols	μBondapak alkylphenyl	Ethanol–water–acetic acid (650:1:350)	132
Flavonol glycosides	μBondapak C_{18} *	2% Acetic acid and tetrahydrofuran (gradient)	126
Furanocoumarins	Corasil I (37–50 μm)	Chloroform–cyclo-hexane mixtures	133
Glycosylflavones **	LiChrosorb NH_2 *	Acetonitrile–water mixtures	134
Hydroxycinnamate esters	ECTEOLA-cellulose	Dichloromethane–acetic acid (99:1) and 1-propanol	135
Isoflavonoids	Porasil	3 Parts dichloromethane–ethanol–acetic acid (485:15:1) and 22 parts hexane	136
Methylated flavones	LiChrosorb Si-60	Heptane–ethanol (3:1)	137
Phenolic acids	μBondapak C_{18} *	Water–methanol–acetic acid (12:6:1)	122
Phenolic aldehydes	Spherisorb C_{18} *	Water–acetic acid–n-butanol (342:1:14)	138
Proanthocyanidins	LiChrosorb RP-8	Water–methanol mixtures	139
Xanthones	Micropak CN	n-Hexane–chloroform (13:7)	140

* Reversed-phase separations.
** A variety of other systems have also been used for these compounds, see ref. 5.

floral tissues is often very high, while the amount of possible column contaminants is likely to be low. With leaf tissues, however, considerable clean-up of crude extracts is almost inevitable. One procedure which would avoid extensive precolumn purification is to acetylate the flavones present in the crude dried plant extracts and separate the crystalline acetate mixture. This could then be placed on a silica gel column and eluted with benzene–acetone (18:3) or isooctane–ethanol–acetonitrile (140:32:11) solvent systems [127].

The mild conditions of HPLC mean that labile compounds can be easily handled without loss of material. Among the classes of naturally occurring phenolic compounds, there are some which are unstable, especially in the presence of traces of alkali. This is true, e.g., of the acylphloroglucinols present in various *Dryopteris* species. HPLC would, therefore, seem to be especially appropriate for their separation. Indeed, satisfactory results have been obtained with a reversed-phase Bondapak C_{18} column and the solvent mixture tetrahydrofuran–phosphoric acid–water (650:1:350) [128].

Suitable columns and solvent systems have been devised for almost every class of natural phenols. A selection of these procedures is shown in Table 20.10, and further references are available in reviews [5]. It is apparent from Table 20.10 that HPLC is already recognized as an important technique to be added to the repertoire of chromatographic procedures for phenolic constituents. Perhaps its most important advantage over other widely used procedures is that it gives precise quantitative data. In the case of flavonoids, this was not so easy before, but now the opportunity of determining the flavonoid constituents in various tissues will undoubtedly stimulate further improvements in HPLC aimed at overcoming the present disadvantages. This will result in a much wider application of this powerful analytical method to research on phenolic compounds.

20.8. CONCLUSION

While HPLC is the latest technique to be widely used in phenol analysis, other procedures are continually evolving, and some of these may eventually find an important place in phenol research. Reversed-phase HPTLC on silica gel has recently been applied to a number of phenolic acids and simple flavonoids. The separations it produces are quite different from those by ordinary TLC, although the selectivity is not as good as in the conventional procedures [132]. Droplet countercurrent (DCC) chromatography has been suggested as an efficient and rapid method for separating flavonoid glycosides on a preparative scale (mg to g amounts) [141]. For example, a crude fraction (130 mg) from the plant *Tecoma stans* gave 27 mg of pure isoquercitrin in ca. 6 h by DCC separation with the upper phase of chloroform–methanol–water (7:13:8).

In choosing a chromatographic system for phenol separations, the cost/effectiveness ratio must now be considered in view of limited budgets and increased costs of organic solvents. While making use of every new technique that appears, it is wise not to discard existing methods as being necessarily inferior, especially if they are

References on p. B430

cheaper. Very recently, phenolic lignans, long known as plant products, have been identified in animal tissues [142,143]. It may be noted here that, although HPLC procedures have been described for lignans [144], the authors relied on GLC of the TMS ethers (with a column of OV-1 on Gas Chrom Q) for the key separations. The link-up with MS was probably an important determinant in this decision. They also employed chromatography on Sephadex LH-20 and on thin layers during the purification process. This leads to the truism that no single method can be relied upon to produce a perfect separation because a great variety of different phenolic substances may be encountered in living tissues. For the foreseeable future at least, a number of different chromatographic systems will be needed to solve problems in this particular research area.

REFERENCES

1 P. Karrer and F.M. Strong, *Helv. Chim. Acta*, 19 (1936) 25.
2 E.C. Bate-Smith, *Nature (London)*, 161 (1948) 835.
3 R.O.C. Norman, J.R.L. Smith and G.K. Radda, in J.B. Pridham (Editor), *Methods in Polyphenol Chemistry*, Pergamon, Oxford, 1964, p. 125.
4 S.T. Preston, Jr., *A Guide to the Analysis of Phenols by Gas Chromatography*, Polyscience, Evanston, IL., 1966.
5 C.F. Van Sumere, W. Van Brussel, K. Vande Casteele and L. Van Rompaey, in T. Swain, J.B. Harborne and C.F. Van Sumere (Editors), *Biochemistry of Plant Phenolics*, Plenum, New York, 1979, p. 1.
6 J.B. Pridham, in J.B. Pridham (Editor), *Methods in Polyphenol Chemistry*, Pergamon, Oxford, 1964, p. 111.
7 L. Hörhammer, in J.B. Pridham (Editor), *Methods in Polyphenol Chemistry*, Pergamon, Oxford, 1964, p. 89.
8 J.B. Woof and J.S. Pierce, *J. Chromatogr.*, 28 (1967) 94.
9 J.B. Harborne, in E.A. Bell and B.V. Charlwood (Editors), *Encyclopedia of Plant Physiology, New Series*, Vol. 8, Springer, Berlin, 1980, p. 329.
10 J.B. Harborne, T.J. Mabry and H. Mabry (Editors), *The Flavonoids*, Chapman & Hall, London, 1975.
11 R.H. Thomson, *Naturally Occurring Quinones*, Academic Press, London, 1971.
12 T.J. Mabry, K.R. Markham and M.B. Thomas, *The Systematic Identification of Flavonoids*, Springer, Berlin, 1970.
13 R.K. Ibrahim and G.H.N. Towers, *Arch. Biochem. Biophys.*, 87 (1960) 125.
14 E. Wong and A.O. Taylor, *J. Chromatogr.*, 9 (1962) 449.
15 E. Sundt, *J. Chromatogr.*, 6 (1961) 475.
16 P. Colombo, B. Corbetta, A. Pirotta and G. Ruffini, *J. Chromatogr.*, 6 (1961) 467.
17 G.B. Crump, *J. Chromatogr.*, 10 (1963) 21.
18 V. Jiraček and Z. Procházka, in I.M. Hais and K. Macek (Editors), *Paper Chromatography*, Academic Press, New York, 1963, p. 254.
19 J.B. Harborne, *Comparative Biochemistry of the Flavonoids*, Academic Press, London, 1967.
20 J.B. Harborne, *J. Chromatogr.*, 2 (1959) 581.
21 E. Lederer and M. Lederer, *Chromatography*, Elsevier, Amsterdam, 1957.
22 R.J. Bloch, E.L. Durrum and G. Zweig, *A Manual of Paper Chromatography and Paper Electrophoresis*, Academic Press, New York, 1958.
23 L. Reio, *J. Chromatogr.*, 4 (1960) 458.
24 R. Hansel, in F. Linskens (Editor), *Paper-chromatographie in der Botanik*, Springer, Berlin, 2nd Edn., 1959, p. 228.
25 K. Yoshitama, *Bot. Mag.*, 91 (1978) 207.

26 J.Z. Stirton and J.B. Harborne, *Biochem. Syst. Ecol.*, 8 (1980) 285.
27 J. Chopin and M.L. Bouillant, in J.B. Harborne, T.J. Mabry and H. Mabry (Editors), *The Flavonoids*, Chapman & Hall, London, 1975, p. 632.
28 P. Markakis, *Nature (London)*, 187 (1960) 1092.
29 G. Lindstedt, *Acta Chem. Scand.*, 10 (1956) 698.
30 A.S. Dreiding, *Helv. Chim. Acta*, 45 (1962) 640.
31 M. Piatelli and L. Minale, *Phytochemistry*, 3 (1964) 547.
32 J.B. Pridham, *J. Chromatogr.*, 2 (1959) 605.
33 J.B. Harborne and C.A. Williams, *Z. Naturforsch.*, 26b (1971) 490.
34 C.A. Williams, J.B. Harborne and H.T. Clifford, *Phytochemistry*, 12 (1973) 2417.
35 J.B. Harborne, *Progr. Phytochem.*, 4 (1978) 189.
36 F. Kreuzaler and K.H. Hahlbrock, *Phytochemistry*, 12 (1973) 1149.
37 R. Neu, *Naturwissenschaften*, 44 (1957) 181.
38 H. Homberg and H. Geiger, *Phytochemistry*, 19 (1980) 2443.
39 B.R. Chhabra, R.S. Gulati, R.S. Dhillon and P.S. Kalsi, *J. Chromatogr.*, 135 (1977) 521.
40 N. Nybom, *Physiol. Plant.*, 17 (1964) 157.
41 S. Asen, *J. Chromatogr.*, 18 (1965) 602.
42 C.F. Van Sumere, G. Wolf, H. Tenchy and J. Kint, *J. Chromatogr.*, 20 (1965) 48.
43 L.S. Bark and R.J.T. Graham, *J. Chromatogr.*, 27 (1967) 131.
44 E. Stahl, *Thin Layer Chromatography*, Allen & Unwin, London, English Edn., 1965.
45 J.B. Harborne, *Phytochemical Methods*, Chapman & Hall, London, 1973.
46 D. McHale, *Thin-Layer Chromatography*, United Trade Press, London, 1964, p. 47.
47 J.H. Dhont and C. de Rhooy, *Analyst (London)*, 86 (1961) 527.
48 H.M. Chawla, I. Gambhir and L. Kathuria, *J. Chromatogr.*, 188 (1980) 289.
49 I.S. Bhatia, K.L. Bajaj, A.K. Verma and J. Singh, *J. Chromatogr.*, 62 (1971) 471.
50 N.A.M. Eskin and C. Frenkel, *J. Chromatogr.*, 150 (1978) 293.
51 M.J. Strohl and M.K. Seikel, *Phytochemistry*, 4 (1964) 383.
52 J.M. Schulz and K. Herrmann, *J. Chromatogr.*, 195 (1980) 95.
53 C.F. van Sumere, *Rev. Ferment. Ind. Aliment.*, 24 (1969) 91.
54 F.M. Abdul Ahy, E.A. Abu-Mustafa, B.A.H. El-Tawil and M.B.E. Fayez, *Planta Med.*, 13 (1965) 91.
55 J.B. Harborne, V.H. Heywood and C.A. Williams, *Phytochemistry*, 8 (1969) 2223.
56 J.K. Forrest, R. Richard and R.A. Heacock, *J. Chromatogr.*, 65 (1972) 439.
57 K.C. Guven and P. Bayer, *Eczacilik Bul.*, 12 (1970) 185.
58 C.F. Culberson, *J. Chromatogr.*, 72 (1972) 113.
59 C.F. Culberson, *J. Chromatogr.*, 97 (1974) 107.
60 C.F. Culberson and A. Johnson, *J. Chromatogr.*, 128 (1976) 253.
61 D. Hess and C. Meyer, *Z. Naturforsch.*, 17b (1962) 853.
62 L. Birkofer, C. Kaiser, H.A.M. Stoll and F. Suppam, *Z. Naturforsch.*, 17b (1962) 352.
63 R.E. Wrolstad, *J. Chromatogr.*, 37 (1968) 542.
64 J.B. Harborne, *Phytochemistry*, 7 (1968) 1215.
65 F. Wollenweber and K. Egger, *Phytochemistry*, 10 (1971) 225.
66 M. Jay, J.F. Gonnett, E. Wollenweber and B. Voirin, *Phytochemistry*, 14 (1975) 1605.
67 Y.F. Chia, *Yao Hsuek Hsuek Pao*, 11 (1964) 485.
68 J. Davidek and Z. Procházka, *Collect. Czech. Chem. Commun.*, 26 (1961) 2947.
69 R.M. Castledine and J.B. Harborne, *Phytochemistry*, 15 (1976) 803.
70 G.P. Forni, *Fitoterapia*, 51 (1980) 13.
71 K.K. Chexal, B.K. Handa and W. Rahman, *J. Chromatogr.*, 48 (1970) 484.
72 L.C. Wang, *Anal. Biochem.*, 42 (1971) 296.
73 J. Sachse, *J. Chromatogr.*, 58 (1971) 297.
74 J.L. Ingham and P.M. Dewick, *Phytochemistry*, 19 (1980) 1767.
75 M. Barbier, *J. Chromatogr.*, 2 (1958) 649.
76 G. Pettersson, *J. Chromatogr.*, 12 (1963) 352.
77 M.H. Zenk, M. Fürbringer and W. Steglich, *Phytochemistry*, 8 (1969) 2199.

78 L. Hörhammer and H. Wagner, *Deut. Apoth.-Ztg.*, 105 (1965) 827.
79 J.W. Fairbairn and K. El-Muhtadi, *Phytochemistry*, 11 (1972) 263.
80 R.P. Labadie, *Pharm. Weekbl.*, 104 (1969) 257.
81 H. Wagner, L. Hörhammer and B. Dengler, *J. Chromatogr.*, 7 (1962) 211.
82 G.R. Whistance and D.R. Threlfall, *Phytochemistry*, 9 (1970) 737.
83 S.J. Jarman and R.K. Crowden, *Phytochemistry*, 12 (1973) 171.
84 B.V. Chandler and K.A. Harper, *Nature (London)*, 181 (1958) 131.
85 E.A.H. Roberts and M. Myers, *J. Sci. Food Agr.*, 10 (1959) 176.
86 G.S. Pope, P.V. Elcoate, S.A. Simpson and D.G. Andrews, *Chem. Ind. (London)*, (1953) 1092.
87 M. Hori, *Bull. Chem. Soc. Jap.*, 42 (1969) 2333.
88 C.H. Ice and S.H. Wender, *Anal. Chem.*, 24 (1952) 1616.
89 K.C. Li and A.C. Wagenknecht, *J. Amer. Chem. Soc.*, 78 (1956) 979.
90 T.C. Somers, *Nature (London)*, 209 (1966) 368.
91 B.H. Koeppen, C.J.B. Smit and D.G. Roux, *Biochem. J.*, 83 (1962) 507.
92 M.K. Seikel, J.H.S. Chow and L. Feldman, *Phytochemistry*, 5 (1966) 439.
93 W. Grassmann, H. Endres, W. Pauckner and H. Mathes, *Chem. Ber.*, 90 (1957) 1125.
94 R. Neu, *Arch. Pharm.*, 293 (1959) 169.
95 T.C. Somers, *J. Sci. Food Agr.*, 18 (1967) 193.
96 L.J. Porter and R.D. Wilson, *J. Chromatogr.*, 71 (1972) 570.
97 T.C. Somers, *Phytochemistry*, 10 (1971) 2175.
98 W.T. Jones, R.B. Broadhurst and J.W. Lyttleton, *Phytochemistry*, 15 (1976) 1407.
99 J.B. Woof and J.S. Pierce, *J. Chromatogr.*, 28 (1967) 94.
100 J. Porath, *Nature (London)*, 218 (1968) 2044.
101 K.M. Johnston, D.J. Stern and A.C. Waiss, *J. Chromatogr.*, 33 (1968) 539.
102 W.J. Irvine and M.J. Saxby, *Phytochemistry*, 8 (1969) 2067.
103 E.R. Blakley, *Anal. Biochem.*, 15 (1966) 350.
104 H. Morita, *J. Chromatogr.*, 71 (1972) 149.
105 H. Wagner and J. Hölzl, *Deut. Apoth.-Ztg.*, 42 (1968) 1620.
106 R.E. Reyes and A.G. Gonzalez, *Phytochemistry*, 9 (1970) 833.
107 A. Jefferson, C.I. Stacey and F. Scheinmann, *J. Chromatogr.*, 57 (1971) 247.
108 C.G. Nordstrom and T. Kroneld, *Acta Chem. Scand.*, 26 (1972) 2237.
109 R. Mechoulam, *Science*, 168 (1970) 1159.
110 V. Kusy, *J. Chromatogr.*, 57 (1971) 132.
111 J.W. Steele and M. Bolan, *J. Chromatogr.*, 71 (1972) 427.
112 M. Bolan and J.W. Steele, *J. Chromatogr.*, 36 (1968) 22.
113 J.W. Steele, M. Bolan and R.C.S. Audette, *J. Chromatogr.*, 40 (1969) 370.
114 R.A. Andersen and T.H. Vaughn, *Phytochemistry*, 11 (1972) 2593.
115 Y. Hoshika and G. Muto, *J. Chromatogr.*, 179 (1979) 105.
116 G. Matsumoto and T. Hanya, *J. Chromatogr.*, 193 (1980) 89.
117 M.D. Willinsky, in R. Mechoulam (Editor), *Marijuana. Chemistry, Metabolism, Pharmacology and Clinical Aspects*, Academic Press, New York, 1973, p. 137.
118 S. Bekassy and M. Nogradi, *Acta Chim. Acad. Sci. Hung.*, 59 (1969) 423.
119 E. Bombardelli, A. Bonati, B. Gabetta, E.M. Martinelli and G. Mustich, *J. Chromatogr.*, 139 (1977) 111.
120 J.C. Giddings, in E. Heftmann (Editor), *Chromatography*, Van Nostrand-Reinhold, New York, 1975, p. 27.
121 C.F. Culberson, *Bryologist*, 75 (1972) 54.
122 L.W. Wulf and C.W. Nagel, *J. Chromatogr.*, 116 (1976) 271.
123 D.M. Smith, *Bull. Torrey Botan. Club*, 107 (1980) 134.
124 R.N. Stewart, S. Asen, D.R. Massie and K.H. Norris, *Biochem. Syst. Ecol.*, 7 (1979) 281.
125 D. Strack, N. Akavia and H. Reznik, *Z. Naturforsch.*, 35c (1980) 533.
126 R.N. Stewart, S. Asen, D.R. Massie and K.H. Norris, *Biochem. Syst. Ecol.*, 8 (1980) 119.
127 R. Galensa and K. Herrmann, *J. Chromatogr.*, 189 (1980) 217.

128 C.J. Widen, H. Pyysalo and P. Salovaara, *J. Chromatogr.*, 188 (1980) 213.
129 M. Wilkinson, J.G. Sweeney and G.A. Jacobucci, *J. Chromatogr.*, 132 (1977) 349.
130 B. Rittich and M. Krska, *J. Chromatogr.*, 130 (1977) 189.
131 R.S. Ward and A. Pelter, *J. Chromatogr. Sci.*, 12 (1974) 570.
132 M. Vanhaelen and R. Vanhaelen-Fastré, *J. Chromatogr.*, 187 (1980) 255.
133 F.R. Stermitz and R.D. Thomas, *J. Chromatogr.*, 77 (1973) 431.
134 H. Becker, G. Wilking and K. Hostettmann, *J. Chromatogr.*, 136 (1977) 174.
135 L. Nagels, W. Van Dongen, J. De Brucker and H. De Pooter, *J. Chromatogr.*, 187 (1980) 181.
136 R.E. Carlson and D. Dolphin, *J. Chromatogr.*, 198 (1980) 193.
137 J.P. Bianchini and E.M. Gaydon, *J. Chromatogr.*, 190 (1980) 233.
138 R.D. Hartley and H. Buchan, *J. Chromatogr.*, 180 (1979) 139.
139 A.G.H. Lea, *J. Chromatogr.*, 194 (1980) 62.
140 K. Hostettmann and A. Jacot-Guillarmod, *J. Chromatogr.*, 124 (1976) 381.
141 K. Hostettmann, M. Hostettmann-Kaldas and K. Nakanishi, *J. Chromatogr.*, 170 (1979) 355.
142 S.R. Stitch, J.K. Toumba, M.B. Groen, C.W. Funke, J. Leemhuis, J. Vink and G.F. Woods, *Nature (London)*, 287 (1980) 738.
143 K.D.R. Setchell, A.M. Lawson, F.L. Mitchell, H. Adlercreutz, D.N. Kirk and M. Axelson, *Nature (London)*, 287 (1980) 740.
144 R. Andersson, T. Popoff and O. Theander, *Acta Chem. Scand.*, 29B (1975) 835.

Chapter 21

Pesticides

LAWRENCE FISHBEIN

CONTENTS

21.1. INTRODUCTION

The principal objectives of this chapter are to highlight the salient advances in the HPLC, GC, GC–MS, and TLC of a spectrum of selected important classes of pesticides with secondary focus on the determination of important trace contaminants, such as chlorinated dibenzodioxins and nitrosated pesticides and nitrosamines.

21.2. GAS CHROMATOGRAPHY

Gas chromatographic techniques, because of their superior resolving power, reasonably high sensitivity, and universal acceptance, continue to be widely employed for the separation and determination of pesticides, primarily in residue analysis, biological tissues, and in environmental samples [1].

21.2.1. Carbamates

GC techniques have been widely used for the determination of carbamates, which are extensively employed as insecticides, fungicides, herbicides, nematocides, miti-

References on p. B453

cides, and molluscicides [2,3]. However, it is recognized that a number of available GC techniques for the determination of carbamates present difficulties. Although direct GC analysis of the parent carbamate would be the obvious method of choice, most N-methyl carbamates are either retained on the chromatographic column or are decomposed to the corresponding phenols [4–8]. In general, utilization of glass columns, low-polarity methyl and phenyl silicone stationary phases, special column conditioning techniques, and moderate column temperatures have been employed with some success [9–11]. The limits of detection for direct GC analysis of under-ivatized carbamates [4,5,12], for which specially treated column supports are used, generally exceeds 10 ng [4,5]. GC analysis of carbamates in the picogram range can be accomplished via initial formation of heat-stable derivatives to minimize degrada-tion [2,4,13–21]. For example, N-perfluoroacylated derivatives of carbamates have been prepared using anhydrides of differing fluoride content [16]. Various tech-niques have been utilized to catalyze the reaction, including heat [16] and reaction with pyridine [19,20] or trimethylamine [21]. The trimethylamine method required only a 30-min reaction time for derivatization, and 14 out of 16 carbamate compounds could be detected with a Coulson detector [21].

However, derivatization procedures for aromatic carbamates, which involve initial hydrolysis to the corresponding phenols or amines and reaction with halogen-rich reagents for the ECD, often suffer several limitations that can reduce their sensitivity and versatility: the presence of contaminants from side reactions, unknown identity of the derivative in some procedures, and limited number of pesticides that can be analyzed by a given method [4,5,22]. An excellent general review of derivatization techniques for the ECD was published by Zlatkis and Poole [23].

The analysis of animal tissues for carbamate insecticides is conceded to be more difficult than procedures for plant materials, soil, and water, since hydrolysis of the parent compound is believed to occur through the activity of tissue enzymes [22]. Mount and Oehme [24] reported a procedure for the microanalysis of carbaryl (1-naphthylmethylcarbamate) in blood and tissue of animals poisoned with this insecticide. Ball-mill extraction with dichloromethane and acetone, removal of lipid material by freezing, and Florisil microcolumn chromatography with 20% ethyl acetate in hexane were employed. Derivatization of carbaryl with heptafluorobutyric anhydride in the presence of trimethylamine permitted rapid processing for GLC analysis on 3% OV-17 with an ECD. The limits of carbaryl detection were 0.02 ppm for blood and 0.1 ppm for tissue. This procedure allowed the simultaneous process-ing and sensitive detection of carbaryl in tissues and blood.

Propoxur (2-isopropoxyphenyl N-methylcarbamate) and its metabolite, 2-isopro-poxyphenol have been determined in blood, urine and tissues, after prefractionation by TLC, by GLC on 10% DC-200 with an ECD [25], according to the procedure of Cohen et al. [26]. Hall and Harris [5] have investigated the direct GC determination of 32 carbamate pesticides on six different Carbowax 20M-modified supports with the aid of an electrolytic-conductivity detector. The following columns were tested: Ultra-Bond, 3% OV-101 on Ultra-Bond, 1% OV-17 on Ultra-Bond, 1% OV-210 on Ultra-Bond, 1% Carbowax 20M on Ultra-Bond, and 0.5% OV-210 plus 0.65% OV-17 on Ultra-Bond. Chemical-ionization GC–MS was used to verify that the carbamates

TABLE 21.1

RELATIVE RETENTION INDICES FOR CARBAMATE PESTICIDES [5]

Column temperature 170°C.

Compound [*]	Purity	Ultra-bond	3% OV-101	1% OV-17	1% Carbowax 20M	1% OV-210	0.5% OV-210 + 0.65% OV-17
EPTC	99.5	–	0.20	0.08	0.07	–	–
Butylate	99.5	–	0.25	0.09	0.07	–	–
Pebulate	99.0	–	0.25	0.12	0.09	–	–
Vernolate	99.0	–	0.28	0.12	0.08	–	–
Propham	100.0	0.19 [**]	0.31	0.19	0.22	–	0.22
Diallate	99.0	0.20	0.67	0.31	0.21	0.32	0.28
Meobal	99.0	0.33	0.59	0.42	0.52	0.56	0.50
CDEC	99.5	0.34	0.66	0.40	0.30	0.40	0.37
Pyramat	98.0	0.35	0.62	0.43	0.29	0.39	0.36
Triallate	99.5	0.53	1.01	0.48	0.26	0.39	0.38
Propoxur	98/99	0.55	0.55	0.48	0.53	0.63	0.52
2,3,5-Landrin	98.0	0.60	0.69	0.51	0.58	0.65	0.58
Chlorpropham	99.5	0.61	0.66	0.45	0.59	0.56	0.55
Bux	98.0	0.78	1.04	0.72	0.71	0.75	0.71
Terbutol	98.0	0.82	1.47	0.91	0.66	0.82	0.78
3,4,5-landrin	98.0	0.85	0.94	0.78	0.85	0.88	0.85
Benthiocarb	98.0	0.85	1.80	1.26	0.82	0.75	1.02
Aminocarb	98.0	0.93	1.07	0.89	0.95	1.02	0.94
Mexacarbate	99.0	0.98	1.32	0.98	0.94	1.02	0.96
Carbofuran	99.5	1.00	1.00	1.00	1.00	1.00	1.00
SWEP	98.0	1.36	1.19	0.97	1.47	1.19	1.28
Dimetilan	98.0	1.37	1.79	1.93	1.38	1.86	1.64
Methiocarb	99.0	2.10	2.25	2.13	2.28	1.96	2.20
Carbaryl	99.5	2.75	2.48	2.41	3.10	2.81	2.82

[*] Common names.
[**] Column temperature 150°C.

were chromatographed intact. Although a wide variety of carbamate pesticides (Table 21.1) could be chromatographed intact on Carbowax 20M-modified supports with or without additional liquid phase coatings, it was stressed that moderate column temperatures ($< 185°C$) and relatively short analysis times be employed. Carbaryl was the only carbamate pesticide that exhibited significant degradation (about 50%). Eleven carbamate pesticides could be separated on a OV-201/OV-17 mixed-phase column under isothermal conditions, and a total of 15 carbamates could be separated with baseline resolution on the same column with temperature programing. Soil samples fortified with carbamates were readily analyzed by electrolytic detection down to levels of 0.1 ppm. Excellent recoveries were obtained from a slightly basic water-acetone-sodium bicarbonate-sodium chloride solution.

References on p. B453

21.2.2. s-Triazines

GLC is the most popular method for the analysis of triazine herbicides, which are among the most abundantly used class of pesticidal compounds. The analysis of the triazines has been reported in reviews [27–29] and comparison studies [30–35]. The most frequently used stationary phases for the analysis of s-triazines in various substrates include: silicones (OV-101, SE-30, DC-200, OV-17, QF-1), polyesters (DEGS, Reoplex 400), polyethylene glycol (Carbowax 20M), and polaymides (Versamid 900). The detector found most useful for triazines has been the nitrogen-selective electrolytic-conductivity detector. Owing to the high nitrogen content by weight of the s-triazines, it often permits detection at subnanogram levels, whereas with the ECD, the sensitivity can vary from subnanogram to microgram levels, depending on the character of the triazine and its substituents. The ECD is most sensitive to chloro-s-triazines and least to methoxy-s-triazines [36]. However, while most pesticide-residue laboratories are equipped with an ECD, many may not have either nitrogen-selective electrolytic-conductivity or alkali-flame ionization detectors (AFID) [36].

Bailey et al. [36] reported the GLC of 10 triazine herbicides as their heptafluoro-butyryl (HFB) derivatives with some applications to their determination in foods. The HFB derivatives were prepared by allowing the s-triazines to react with

TABLE 21.2

KOVÁTS RETENTION INDICES (I) AND THEIR TEMPERATURE DEPENDENCES ($\delta I/\delta T$) FOR s-TRIAZINES, OBTAINED ON A GLASS CAPILLARY COLUMN WITH A CARBOWAX 20M STATIONARY PHASE [39]

No.	Common name	Substituent positions			$I(463°K)$	$\delta I(20°)$
		2-	4-	6-		
1	Ipazine	Cl	$N(C_2H_5)_2$	$NHCH(CH_3)_2$	2475.1	18.6
2	–	Cl	$NHC(CH_3)_3$	$NHCH(CH_3)_2$	2518.4	14.6
3	Trietazine	Cl	$N(C_2H_5)_2$	NHC_2H_5	2559.3	20.7
7	Propazine	Cl	$NHCH(CH_3)_2$	$NHCH(CH_3)_2$	2659.1	15.1
8	Terbutylazine	Cl	NHC_2H_3	$NHC(CH_3)_3$	2686.1	18.5
11	Atrazine	Cl	NHC_2H_5	$NHCH(CH_3)_2$	2747.3	18.6
14	Simazine	Cl	NHC_2H_5	NHC_2H_5	2833.9	21.4
4	Prometon	OCH_3	$NHCH(CH_3)_2$	$NHCH(CH_3)_2$	2570.3	12.1
5	Terbuton	OCH_3	NHC_2H_5	$NHC(CH_3)_3$	2597.9	15.3
6	Atraton	OCH_3	NHC_2H_5	$NHCH(CH_3)_2$	2633.4	14.4
9	Secbumeton	OCH_3	NHC_2H_5	$NHCH(CH_3)C_2H_5$	2707.8	15.1
10	Simeton	OCH_3	NHC_2H_5	NHC_2H_5	2714.1	17.9
12	Prometryn	SCH_3	$NHCH(CH_3)_2$	$NHCH(CH_3)_2$	2779.6	19.1
13	Terbutryn	SCH_3	NHC_2H_5	$NHC(CH_3)_3$	2812.1	22.8
15	Ametryn	SCH_3	NHC_2H_5	$NHCH(CH_3)_2$	2859.3	22.5
16	Desmetryn	SCH_3	$NHCH_3$	$NHCH(CH_3)_2$	2888.9	25.9
17	Simetryn	SCH_3	NHC_2H_5	NHC_2H_5	2937.1	26.1

heptafluorobutyric anhydride in benzene, in the presence of trimethylamine or pyridine as catalyst. The reactions produced mainly the mono-HFB products, but small quantities of the di-HFB derivatives were observed with some of the herbicides. The derivatives were 300 to several thousand times more sensitive to the ECD than the underivatized triazines. They were also 5–10 times more sensitive than their parent compounds to electrolytic conductivity detection in the halogen mode but showed similar sensitivity to the same detector in the nitrogen mode. The HFB derivatives emerged in the same general order as the parent triazines from stationary phases of OV-1, OV-101, OV-101/QG-1, and OV-210. This procedure was successfully applied to the analysis of potatoes, peas, and tomatoes, spiked with various triazines (e.g., simazine, atrazine, propazine, terbutylazine, ametryn, prometryn, terbutryn, atratone, and prometone) at levels of 0.13–0.86 ppm. Although earlier studies have suggested that the best separation of several s-triazine mixtures (underivatized) can be achieved on DEGS, Versamid 900, or Carbowax 20M (2–10%) as the stationary phase [27,29,37], it has proved impossible, even on Carbowax, to effect the separation of complex mixtures of chloro-, methoxy-, and methylthio-s-triazines, either isothermally or with temperature programing [38].

Matisová et al. [39] have described the successful separation of a complex mixture of 17 chloro-, methoxy-, and methylthio-s-triazines by high-resolution GLC in capillary columns, $1.3\,m \times 3\,mm$ ID, coated with 3% Carbowax 20M on 80–100-mesh Chromosorb W AW, at a column temperature of 487°K with a FID. The optimal temperature for the analysis of s-triazines was determined from the dependence of log t_R on $1/T$. For the characterization of s-triazines in environmental samples, Kováts retention indices were determined on glass capillary columns with dynamically coated nonpolar (OV-101) and polar (Carbowax 20M) stationary liquids. Table 21.2 lists the Kováts retention indices measured on Carbowax 20M glass capillary columns at 463°K as well as the $\delta I(20°)$ values. The standard deviation of the Kováts retention indices obtained from four measurements for s-triazines on the Carbowax 20M column was 0.1–0.4 index unit. The Kováts retention indices decrease depending of the alkyl substituent and on the shielding of the amino group. Higher values of $\delta I(20°)$ were found for trietazine, simazine, simeton, and simetryn (Compounds 3, 14, 10 and 17, respectively), all of which have an ethyl group bonded to amino groups at the 4- and 6-positions.

21.2.3. Organochlorine pesticides

Although efforts have increased to curtail the use of certain chlorinated pesticides, principally the persistent and ubiquitous chlorophenoxy acids and cyclodienes, their metabolites and photodegradation products remain in the ecosystem and in many cases as part of the human body burden. Hence, there is a continued need for their determination. The determination of chlorinated pesticides (as well as polychlorinated biphenyl contaminants) in the environment as well as in biological samples is usually carried out by GLC [40–42]. GLC has largely replaced TLC, which is useless for the detection of picogram quantities. Thus far, HPLC has not been widely used for the determination of trace amounts of chlorinated compounds because the

References on p. B453

sensitivity of HPLC detectors in general is inadequate [43–45].

The separation and/or detection of chlorophenoxyalkyl acids, as the free acids or after derivatization under a variety of experimental conditions, by GLC combined with the microcoulometer, FID, or ECD, has been achieved in herbicide formulations [46], water [47,48], plant tissues and crops [49–56], food [57–64], soil [47,52,56,65,66], and animal and human samples [67–69]. In most of the GLC methods, the chlorophenoxy acids are determined as their methyl esters [70–73] or PFB esters [73–76] after direct derivatization or after transesterification [77,78]. The utility of the PFB bromide reagent is of added interest because of the increased electron-capture sensitivity of the derivatives formed [79,80] and the fact that the PFB ester derivatives of 10 chlorophenoxyalkyl acids could be completely separated by GLC on DC-200, unlike their methyl esters [73]. De Beer et al. [73] have examined the GLC behavior of the PFB and methyl esters of 10 structurally related chlorophenoxy alkyl acids (e.g., 2,4-D, 2,4-DB, 2,4-DP, MCPA, MCPB, MCPD, 2,4,5-T, 2,4,5-TB, and 2,4,5-TP) by comparing their retention indices on 9 frequently used liquid phases with increasing McReynolds constants. The resolution of each liquid phase was different for the PFB and methyl ester derivative. However, a plot of the retention index (I) on an apolar phase against the difference in retention indices (δI) between a polar and nonpolar phase, showed a good correlation between structure and retention in GLC.

Residue levels of a number of chlorinated insecticides in human adipose tissue and human milk samples have been quantitated by GLC [81–85]. Christ and Moseman [85] described the utility of the Hall detector and a surface-bonded Carbowax 20M column for the determination of oxychlordane, *trans*-nonachlor, dieldrin, β-HCH, heptachlor epoxide, p,p'-DDE and p,p'-DDT at sub-ppm concentration levels in human adipose tissue and human milk samples. The sample extracts were also analyzed with GLC–ECD for comparison purposes. Gel permeation chromatography (GPC) was utilized to remove additional lipid material from certain extract samples having an adverse effect on the performance of the Hall detector, and this proved to be valuable as an adjunct clean-up technique [86–88]. A 5% Carbowax 20M column was used for the determination of oxychlordane, *trans*-nonachlor, p,p'-DDE, p,p'-DDT and dieldrin with the Hall detector, and, because of an interference from heptachlor epoxide on the Carbowax 20M column, a 5% OV-1 column was substituted in the determination of hexachlorocyclohexane (HCH). A 1.5% OV-17/1.95% QF-1 column served for the determination of the pesticides by GLC–ECD. Saxena et al. [89] determined BHC, lindane, DDT, DDE, DDD, and aldrin in human placenta and accompanying fluid by GLC–ECD on a 1.5% OV-17/1.95% OV-210 column.

Hexachlorocyclopentadiene is an extensively used precursor in the manufacture of many important organochloride pesticides (as well as flame retardants and other industrial organics). Octachlorocyclopentene is an important intermediate in the manufacture of hexachloropentadiene and has been found as a minor contaminant in raw and technical-grade hexachlorocyclopentadiene [90]. DeLeon et al. [90] have described a rapid and sensitive method for the analysis of these chloro derivatives in blood and urine. The procedure requires 5.0 ml of blood or urine and involves the isolation of the compounds from the sample by liquid–liquid extraction, followed by screening and assay by capillary column GLC–ECD and confirmation by GC–MS. The assays were suitable for the detection and identification of nanogram quantities of these compounds in body fluids (lower detection limit, 50 ng/ml for blood and 10 ng/ml for urine).

The GLC determination of chlorinated pesticides in the presence of polychlorinated biphenyls (PCB) has long been recognized to be difficult since the chlorinated pesticides and PCB are extracted together in routine residue analysis and the GLC retention times of several PCB peaks are almost identical with those of a number of peaks of chlorinated pesticides, most notably of the DDT group [91]. Moreover, the PCB interference may vary, since PCB mixtures may have different chlorine contents. It is common for PCB to be very similar to many chlorinated insecticides. Hence, the complete separation of chlorinated pesticides from PCB is extremely difficult and rarely possible by GLC alone [92–96]. Luckas et al. [91] reported a novel determination of chlorinated pesticides and PCB in environmental samples via the simultaneous use of GLC and derivatization gas chromatography (both with ECD). The method was based on the different stabilities of chlorinated pesticides and PCB towards MgO in a microreactor. Extracts of samples were injected twice, first into a regular gas chromatograph and then into a gas chromatograph equipped with a microreactor for derivatization. A "basic" chromatogram and a "derivatization" chromatogram were obtained, and the combination of the two chromatograms provided a satisfactory solution.

The GLC determination of several cyclodiene insecticides in the presence of PCB was accomplished by Mansour and Parlar [97] by utilizing photoisomerization reactions. The photoisomers of cyclodiene insecticides give longer retention times than their parent compounds. Separation of aldrin, dieldrin, Mirex, heptachlor epoxide, and lindane from PCB and polychloronaphthalene (PCN) matrices (Aroclor 1242 and Halowax 1013) was reported by Paramasigamani et al. [98], who used two chromatographic separations. The sample was first injected into a nonpolar column containing 3% Apiezon L on Carbowax 20M-modified Chromosorb W. The appropriate fraction was transferred by valve switching to a second polar column, containing 10% OV-275 on Chromosorb W AW, where the insecticide separated in most cases cleanly from the PCB or PCN matrix.

21.2.4. Organophosphorus pesticides

The application of GC procedures to the analysis of organophosphorus pesticides has been reviewed by Cram et al. [1], Thornburg [40], and Smart et al. [99]. The flame-photometric phosphorus- and sulfur-specific detectors and the phosphorus-specific thermionic detectors continue to be preferred for the quantitative analysis of these compounds [40]. Considerable efforts have recently been made in the United Kingdom and other European countries to establish by collaborative study multi-residue methods for the determination of organophosphorus (and other) pesticides in fruits and vegetables for use by analysts, particularly in ascertaining whether samples conform with maximum residue limits [99]. Methods developed by Abbott et al. [100] by scientists at the National Food Institute in Denmark [101], Watts et al. [102] and Sissons and Telling [103,104] were compared in the U.K. [101]. Other countries have examined the Becker method [105], a European Economic Community (EEC) method [106] and sweep co-distillation techniques [107,108]. Smart et al. [99] have reported a comparison of the methods of Abbott et al. [100], Sissons and Telling [103,104], and sweep co-distillation [107] for the determination of some residues of malathion, dimethoate, parathion, α-chlorfenvinphos, β-chlorfenvinphos, and omethoate in lettuce, tomatoes, and cabbage. Purified extracts were examined by GLC, both with a flame-photometric detector and an AFID, and the mean was reported. The chromatographic columns used were: 5% Carbowax 20M and 80–100-mesh Gas Chrom Q for omethoate and 5% OV-17 with 0.02% Epon 100 on 80–100-mesh Gas Chrom Q for all other pesticides. No significant differences between results obtained by use of the two detectors were found when five replicate determinations were carried out for each crop–pesticide combination. For regulatory purposes, the above methods [100,103,104,107] appeared to be roughly equivalent for the limited number of crop–pesticide combinations tested.

Miles and Dale [109] described the GLC determination of a number of organophosphate, phosphorothioate, and phosphorodithioate pesticides by in-block methylation. Microgram quantities of the organophosphorus pesticides were mixed with methanolic solutions of trimethylanilinium hydroxide (TMAH) and injected into a gas chromatograph equipped with a phosphorus-sensitive flame-photometric detector. The efficiency of the reaction of TMAH with the various pesticides was determined by measurement of the quantity of trialkyl phosphates formed. The efficiency of the in-block methylation in 0.01 M TMAH varied from 61% for phoxim to 100% for azinphosmethyl. The rate of reaction of the pesticides with TMAH at ambient temperatures varied from 0% per day for malathion to 75% per day for chlorphoxim. The columns used were 5% OV-225 and 3% OV-275, both on 100–120-mesh Chromosorb W HP. The derivatization technique, in which short-chain trialkylphosphates are formed and easily separated, was found to be useful for the qualitative and quantitative analysis of many organophosphorus pesticides.

Ionic alkyl and aryl phosphates, phosphonates, and thioanalogs are metabolites of organophosphorus pesticides and of several other categories of economically important chemicals [110]. No single method is available for the routine analysis of these compounds, especially diprotic alkyl phosphonates, in biological or abiotic samples. Most of the chromatographic methods, which involve esterification with diazoalkanes and determination of the alkyl esters by GLC [111], have been designed for dialkyl phosphates and thiophosphates. Daughton et al. [112] reported the GLC determination of a variety of phosphorus-containing pesticide metabolites after benzylation. Mono- and di-protic alkyl and aryl phosphates, phosphonates, and thio-analogs were determined in spiked samples, bacterial growth media and human urines by this procedure with a detection limit of less than 2 pmole of each. The acids were fully protonated by passing the aqueous samples through a BioRad Ag 50W-X8(H$^+$) resin and then thoroughly dried before they were refluxed with 3-benzyl-1-p-tolyltriazene (BTT) in acetone. The benzyl esters that formed were quantitatively partitioned into cyclohexane and then analyzed by GLC with a flame-photometric detector (phosphorus mode) and a glass column of 5% OV-210 and 80–100-mesh Gas Chrom Q. The benzylation of the phosphoric acids by BTT proceeds according to the equation

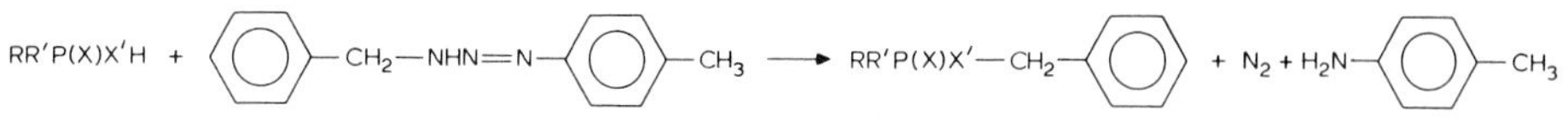

$$RR'P(X)X'H \; + \; \langle\bigcirc\rangle - CH_2 - NHN{=}N - \langle\bigcirc\rangle - CH_3 \; \longrightarrow \; RR'P(X)X' - CH_2 - \langle\bigcirc\rangle \; + \; N_2 + H_2N - \langle\bigcirc\rangle - CH_3$$

where R = alkoxy or hydroxy, R′ = alkoxy, alkyl, or aryl, and X and X′ = sulfur or oxygen. The procedure is so simple that more than 20 samples per day can be analyzed. The benzyl esters give a better response than alkyl esters in FID and MS. More significantly, their aromaticity greatly enhances the partition coefficients, permitting the use of a variety of extraction solvents.

21.3. HIGH-PERFORMANCE LIQUID CHROMATOGRAPHY

HPLC has probably become more useful than any other analytical technique for the analysis of low levels of chemicals in biological and environmental samples. General reviews on HPLC [114–116] and more specific reviews on the application of HPLC to pesticide residue analysis [40,117–119] have recently been published. HPLC represents an alternative to GLC for the residue analysis of pesticides lacking sufficient volatility or thermal stability for GLC [113,114]. The prime requirement

TABLE 21.3

UV ABSORPTION SENSITIVITIES OF PESTICIDES [117]

Pesticide class	Sensitivity, μg *	Wavelength, nm
Phenyl ureas	0.03 – 0.06	254
Phenyl carbamates	0.05 – 0.08	254
Phenyl carbamates	0.005– 0.01	207
Methyl carbamates	0.5 –20	254
Methyl carbamates	0.2 – 2	200–206
Triazines	0.1 –10	254
Triazines	0.01 – 0.02	220
Organophosphates	0.5 –16	254
Phenoxy acids	0.05 – 0.35	280
Phenoxy esters	15	254
Organochlorines (DDT type)	1 –15	254
Anilides	0.05	254
Nitrophenols	0.02 – 0.05	254
Uracils	0.2	254
Uracils	0.06 – 0.07	270–280
Thiocarbamates	0.07 – 0.60	205
Thiocarbamates	0.3 –10	254

* Amount injected to produce a response equivalent to 0.01 absorbance units.

References on p. B453

for LC detectors that are to be used for trace analysis (ppb concentrations) is, of course, high sensitivity [114]. For most applications, detectors sensitive to 1–50 ng of material are required.

The utility of HPLC would be further extended if element-selective detectors were available, comparable to those in current use for GLC. Among the commercially available detectors, only the UV and fluorescence detectors have provided the sensitivity needed for practical applications in this field [113,114,118–121]. Table 21.3 shows the range of UV sensitivities for various classes of pesticides [117]. Of course, the choice of wavelengths is very important for maximum sensitivity. As is evident from Table 21.3, wavelengths below 254 nm often increase the sensitivity 10 to 100-fold. Other detectors, offering some selectivity that may be useful in pesticide residue analysis include: electrochemical [122–127], spray impact [128], AFID [129], ECD [130], thermal ionic [131]. Coulson electrolytic conductivity (CECD) [132], and chemiluminescence [133] detectors. Electrochemical detectors appear to be especially promising, as they provide a very sensitive, simple, inexpensive, and somewhat selective alternative to conventional UV and fluorescence detectors for CC [127]. Since many electroactive organic compounds have minimum detection limits ranging from 10^{-10} to 10^{-14} mole per sample [126], LC with electrochemical detectors (LC–EC) would appear quite useful for trace analysis. LC–EC detection limits often are in the picogram range, which is attainable with GLC–ECD only after derivatization [126]. The applicability of LC–EC to a wide range of analyses has been reviewed by Kissinger [134].

21.3.1. Carbamates

Lawrence and Turton [117] have reported HPLC data for 166 pesticides, including 37 carbamate pesticides, giving the chromatographic conditions (e.g., type of packing material, column dimensions, and mobile phase composition), elution volumes, type of sample, and detectors used for analysis. A major potential advantage of HPLC is the feasibility of direct carbamate analysis without derivatization, since relatively gentle operating conditions are applied. HPLC procedures have been reported for a wide range of carbamates [4,113,117,135–143], both in uncombined form by UV detection [4,135–137] and, after derivatization, by fluorescence detection [138–143]. Detection limits in the low nanogram range were reported for most of the carbamates when monitored with commercial UV detectors at wavelengths of optimum response (e.g., 190–210 nm) [135]. By employing fluorimetric derivatization and detection, detection limits were obtained which typically ranged between 1 and 10 ng for dansyl derivatives formed prior to injection [138,139] and between 500 pg [143] and 1–10 ng [138,141] for post-column derivatization with o-phthalaldehyde.

A reversed-phase HPLC method with ECD, reported by Anderson and Chesney [127], is applicable to the trace analysis in water of selected examples of the important and ubiquitously applied carbamate pesticides. The thin-layer, Kel-F-graphite ECD was operated in the oxidative, constant-potential amperometric mode at +1.1 V vs. Ag/AgCl/3.5 M KCl. A column (250 × 42 mm ID) packed with

Bondapak C_{18} Corasil, was employed. Calibration curves were linear over a wide range of sample concentration—typically at least 3 orders of magnitude—with relative standard deviations for repetitive determinations of 1–2% through most of the range. Detection limits in the range of 40–140 pg (signal-to-noise ratio 2:1) were obtained for eight carbamates corresponding to sample concentrations of 2–7 ppb for 20-μl injections. Since the response in LC is not appreciably degraded by increases in sample injection volume [144], detection limits of less than ppb of most of the carbamates studied could be achieved by increasing injection volumes to 100 μl. Moye et al. [140] have published a HPLC procedure for the separation and detection of seven carbamate insecticides based on dynamic fluorogenic labeling with o-phthalaldehyde. An attractive feature of the procedure is that the internal fluorescence of the fluorophore formed in the reaction enables detection of nanogram quantities of the carbamate insecticides. Moreover, the fluorescence detection system responds only to the fluorophore and to other compounds that fluoresce at the same excitation and emission wavelengths.

Further refinement of a post-column fluorometric labeling technique for the determination of five carbamate insecticides (carbaryl, carbofuran, methiocarb, methomyl, and propoxur) and two metabolites (3-hydroxycarbofuran and β-naphthol) were described by Krause [141]. The compounds were separated on a μBondapak C_{18} HPLC column by gradient elution with aqueous methanol at a flowrate of 1.5 ml/min. A solution of 0.05 N NaOH was added to the column effluent at 0.5 ml/min to hydrolyze the carbamates to methylamine by a 16-sec reaction at 100°C. A solution containing 0.5 mg o-phthalaldehyde and 1 μl 2-mercaptoethanol per ml was added at 0.5 ml/min to react with the methylamine.

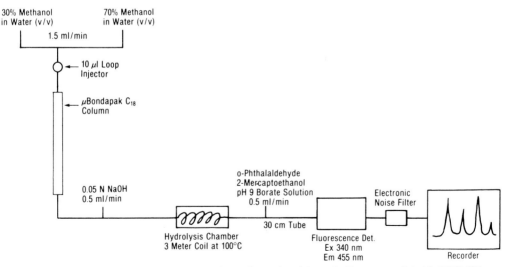

Fig. 21.1. HPLC-fluorometric carbamate analyzer (Reproduced from *J. Chromatogr. Sci.*, 16 (1978) 281, with permission [141].)

References on p. B453

The resulting fluorophore was monitored with a fluorescence detector containing a 10-μl cell. In a seven-day system stability study, relative standard deviations in peak height responses and peak retention times for the seven compounds were 3.5 and 1%,

Table 21.4

CECD RESPONSE AND RELATIVE CECD/UV$_{254}$ RESPONSE FOR PESTICIDES AND RE-LATED CHEMICAL STANDARDS [132]

Chemical	Retention relative to p,p'-DDT	Hetero-atoms	Response	
			CECD (area/ picomoles)	CECD/ UV$_{254}$
Mirex	2.39	Cl_{12}	207	–
Heptachlor	0.94	Cl_7	113	26.3
Heptachlor epoxide	0.72	Cl_7	105	25.0
Hexachlorobenzene	2.15	Cl_6	101	0.46
Aldrin	1.21	Cl_6	84	6.98
Dieldrin	0.84	Cl_6	77	10.4
Lindane	0.62	Cl_6	71	–
Endrin	0.84	Cl_6	66	4.06
o,p'-DDT	1.08	Cl_5	56	1.23
p,p'-ODDD	0.76	Cl_4	41	1.33
Pentachloronitrobenzene	1.01	Cl_5	41	0.24
Pentachloroanisole	0.95	Cl_5	40	0.64
p,p'-DDT	1.00	Cl_5	40	0.58
p,p'-DDE	1.15	Cl_4	40	0.03
Methoxychlor	0.73	Cl_3	40	0.10
Trifluralin	0.78	F_3N_3	36	0.008
Dursban	0.79	Cl_3NPS	36	1.04
Kelthane	0.81	Cl_5	35	1.16
Ronnel	0.78	Cl_3PS	31	3.66
Pentachlorophenol	0.37	Cl_5	30	0.04
Captan	0.53	Cl_3NS	30	2.41
Perthane	1.04	Cl_2	20	0.51
Dichlorobenzophenone	0.71	Cl_2	15	0.01
Paraoxon	0.50	NP	11	0.16
Disyston	0.63	PS_3	11	1.33
p,p'-DDA	0.47	Cl_2	10	0.33
2,4-D	0.49	Cl_2	10	0.73
Parathion	0.59	NPS	9.9	0.02
Abate	0.72	P_2S_3	8.3	0.01
Guthion	0.55	N_3PS_2	7.1	0.02
Phosdrin	0.49	P	7.1	–
Meta-Systox-R	0.47	PS_2	6.5	–
Azodrin	0.47	NP	5.1	–
Methyl parathion	0.53	NPS	4.1	0.01
Omite	0.76	S	3.8	0.65
Methomyl	0.49	N_2S	2.3	–
Carbaryl	0.50	N	0.08	0.004

respectively. Fig. 21.1 illustrates the HPLC–fluorometric carbamate analyzer of Krause [141]. This system is claimed to have the sensitivity, linearity, and stability required for routine pesticide residue analysis.

21.3.2. s-Triazines

So far, HPLC has not been extensively employed for determining residue levels of the s-triazines. Lawrence and Turton [117], in their review of HPLC data for 166 pesticides, tabulate operating parameters for the analysis of fifteen s-triazines, chromatographed primarily on LiChrosorb Si 60 (5 μm) with 2% 2-propanol in trimethylpentane as the mobile phase and detected at 254 and 220 nm with a sensitivity of 0.1–10 μg and 0.01–0.02 μg, respectively.

21.3.3. Organochlorine pesticides

In contrast to the well-developed GLC methodology for the determination of organochlorine pesticides, the HPLC analysis of this class has not yet been extensively explored. This may reflect, in part, the lack of generally available selective detection methods for organochlorine pesticides resolved by reversed-phase chromatography. The UV absorption sensitivity at 254 nm for the organochlorines (DDT type) and phenoxyesters is 1–15 μg and 15 μg, respectively (Table 21.3) [117].

In an attempt to improve the HPLC detection sensitivity for organochlorine pesticides, Dolan and Sieber [132] designed and developed a CECD. This detector was high selective for organochlorine compounds relative to hydrocarbons, and had a linear range of 10^5 and minimum detectability of 5–50 ng for lindane. The CECD detectability was better than that of the UV_{254} detector and its selectivity greater than that of the UV_{220} detector when chlorine-containing aliphatic pesticides were examined. The utility of the CECD analytical system was demonstrated in the analysis of crude extracts of lettuce and river water, fortified at sub-ppm levels with aldrin and dieldrin. The UV detector at either 254 or 220 nm was useless in this analysis.

To evaluate selectivity, the LC–CECD system response was determined for 37 pesticides and related compounds, covering a wide range of chemical classes (Table 21.4). In general, compounds containing Cl or F responded better than non-halogenated chemicals, and among related compounds the response increased with halogen content. The chlorine response was also somewhat dependent on structure (cf., e.g., endrin and hexachlorobenzene), indicating that a combustion temperature even higher than 975°C would be needed to eliminate inter-halogen selectivity. Moreover, the nature of other functional groups in organochlorine compounds might also influence response.

21.3.4. Organophosphorus pesticides

As in the determination of organochlorine pesticides, HPLC has not yet been extensively employed in the analysis of organophosphorus pesticide residues

References on p. B453

[117,122,145–147]. Detection methods have included UV absorption at 254 nm [145,147], cholinesterase inhibition [146], and polarography [122]. The UV absorption sensitivity at 254 nm for organophosphates ranges from 0.5 to 16 μg [117,145,147]. Ramsteiner and Hörmann [146] have explored the analysis of organophosphate and carbamate insecticide residues by coupling HPLC with an autoanalyzer for cholinesterase inhibitors. The organophosphates examined were: diazoxon, dichlorovos, dicrotophos, monocrotophos, phosphamidon, dimetilan, dioxacarb, and CGA 18809. The approximate limits of detection ranged from 10 pg for diazoxon to 200 ng for dioxacarb. In most instances, these detection limits were satisfactory for residue analysis, considering the fact that there was no interference from other cholinesterase inhibitors in the extracts. The cholinesterase inhibition technique has an added advantage over existing methods for the determination of carbamate residues in requiring no preliminary hydrolysis and/or derivatization.

Compton and Purdy [148] have demonstrated the utility of phosphorus-sensitive detectors in HPLC for selective detection of organophosphorus compounds. The modification they described involved conversion of a FID to a thermionic detector (TD) by addition of a rubidium silicate glass bead above the FID. Enhanced sensitivity as well as selectivity for phosphorus over hydrocarbon responses can be expected for any detector of the TD class and, hence, it should be of value for qualitative and quantitative analysis of organophosphorus compounds in complex mixtures. Compared to the FID, the TD showed up to 500 times greater sensitivity for certain organophosphorus pesticides and is presumably capable of even greater sensitivity.

21.3.5. Ureas

A broad spectrum of substituted ureas are among the most widely used herbicides. HPLC is being increasingly employed for determining ureas in herbicide residues [117,149–152], because it has certain advantages over GLC. For example, under the conditions normally used for GLC, phenylureas are degraded to isocyanates [153]. Since this does not occur in HPLC, urea herbicides can be separated from some of their metabolites. For detection, UV absorption has been used; the sensitivity for phenylureas at 254 nm is between 0.03 and 0.06 μg [117].

21.4. GAS CHROMATOGRAPHY–MASS SPECTROMETRY

It is well recognized that analytical power can be increased by combining several analytical techniques [154]. GC–MS is currently the most powerful and useful technique for the identification of trace levels of organic compounds. It can provide qualitative information on nanogram quantities of compounds in a sample by providing a mass spectrum of each peak as it emerges from the gas chromatograph. Fundamental aspects of GC–MS [155–160] and representative examples of its usage may be found in the literature [156,157,159,160].

21.4.1. Organochlorine pesticides

GC–MS is being increasingly employed to elucidate the extent of toxic environmental exposure. Since the original disclosure of the environmental contamination of Kepone in 1975 in the vicinity of the sole U.S. producer, this pesticide has been under intense methodological and toxicological investigation [161]. Harless et al. [161] demonstrated the utility of combining GLC with CI and high-resolution MS in the detection, identification, and confirmation of the presence of Kepone, Kepone photoproducts, and a reduction product of Kepone in environmental and human samples. Selected environmental samples were subjected to GC–CI–MS analyses for Kepone, and the selected-ion monitoring technique was used to monitor the four largest ions (i.e., m/z 488.7, 490.7, 492.7, and 494.7) in the quasimolecular ion (M + H) region. The minimum detectable amount of a standard solution of Kepone was approximately 700 pg. Table 21.5 illustrates the confirmed presence of Kepone in a number of environmental and human samples by GC–CI–MS.

Onuska et al. [162] have reported the quantitative determination of Mirex and its degradation products in fish and sediment by high-resolution capillary GC–MS. The widespread contamination of human populations in the United States [163,164], Western Europe [165], and Japan [166] with pentachlorophenol (PCP) at concentrations roughly 10^{-3} of the lethal dose for PCP in man has been the subject of much recent concern [167]. It has been suggested that PCP contamination may also be due, in part, to the metabolism of hexachlorobenzene by mammals [168].

Recently, Kuehl and Dougherty [169] obtained evidence for the origin of PCP in the environment by means of GC–negative ion CI–MS. Commercial PCP contains significant quantities of tetrachlorophenol (TCP). The occurrence of TCP in environmental samples provided a chemical marker for PCP originating from commercial formulations. Commercial PCP formulations and a series of environmental and human samples were analyzed for TCP by the ion current at m/z 229, and for

Table 21.5

CONFIRMED PRESENCE OF KEPONE IN ENVIRONMENTAL AND HUMAN SAMPLES BY GC–CI–MS [161]

Type of sample	Concentration range	Number of samples
Water	0.15–3.38 ppm	3
Air	0.21–54.8 μg m^{-3}	18
Soil, sediment, sludge	1.31–13,250 ppm	3
Fish with fins	1.04–14.4 ppm	10
Shellfish	0.18–0.91 ppm	10
Human sebum	0.05–3.06 μg per forehead wipe	10
Human blood *	2–15 ppm	2
Human feces *	2–5 ppm	2
Human bile *	0.1–1.5 ppm	3

* Unpublished data.

References on p. B453

tetrachlorophenoxide and PCP by the ion current at m/z 267 (pentachlorophenoxide). Pentachlorophenol residues in human adipose tissues of Japanese subjects were determined by Ohe [170] with GLC and GC–MS.

21.4.2. Chlorinated dibenzo-*p*-dioxins

The formation, occurrence, persistence, transport, and toxicity of polychlorinated dibenzo-*p*-dioxins (CDD) are of continued interest. The discovery that combustion, as well as the use of certain chlorinated phenols and phenoxy herbicides can result in the transmission of CDD in the environment has necessitated the development of very sophisticated technology for their determination. The problem is compounded by the existence of 75 positional CDD isomers, ranging from the mono- to the octachloro compounds. There are 22 tetrachlorodibenzodioxins (TCDD) alone. Positional isomers of the various CDD vary greatly in acute toxicity and biological activity, the most toxic isomer identified so far being 2,3,7,8-TCDD. This compound was involved in several industrial accidents and has recently caused severe environmental contamination at Seveso, Italy [171].

For investigation of the bioaccumulation of CDD in biological and environmental samples, analytical methodology was developed to determine dioxin residues at picogram levels. (Much of the work in this area reported prior to 1974 is summarized in ref. 172.) More recently, a number of techniques based on GC–MS have been developed and applied to the determination of chlorinated dioxins at parts-per-trillion (ppt, 10^{-12}) levels in environmental and biological samples [173–192].

The MS determination of TCDD at the ppt level in many biological and environmental samples depends upon the concentration of contaminants such as DDT, PCB, and a minor component of toxaphene, which can interfere with the analyses [193,194]. Hence, it is vital that the sample clean-up portion of the analytical procedure for TCDD be as selective as possible [184,186,187]. Methods for determining ppt levels of TCDD in extracts have been reported that are based on high-resolution MS [194,195], packed-column GC in conjunction with low- and high-resolution MS [176,191,192], high-resolution GC with low-resolution MS [183], GC–negative ion CI–MS [185], capillary GC–atmospheric negative ion CI–MS [187] and capillary GC–high-resolution MS [190].

Harless et al. [190] gave a description of the extraction, clean-up, and capillary GC–high-resolution MS methods for determining low concentrations (pg/g levels) of TCDD in human milk, beef liver, fish, water, and sediment. This TCDD method gave a mean recovery of 80% for 2.5 to 10 ng [^{37}Cl$_4$]TCDD (internal standard) and a mean accuracy of ±23% for 1 to 1250 pg TCDD in quality assurance samples. The precision of the GC–high-resolution MS technique during daily operations was ±20% for 2- to 10-pg TCDD quantitation standards. A combined capillary GC–atmospheric negative ion CI–MS technique was devised by Mitchum et al. [187] for determining TCDD in fish and wildlife tissues. TCDD was isolated from tissue samples by multistep HPLC, including stable isotope dilution. PCB were found not to interfere with TCDD analysis at the low ptt levels.

A method, developed and validated by Langhorst and Shadoff [180] for the

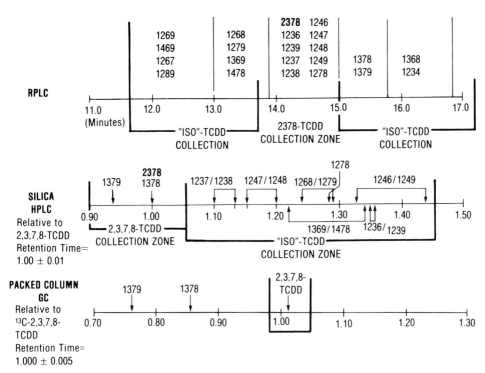

Fig. 21.2. Fractionation scheme for isolating 2,3,7,8-TCDD from the other 21 tetrachlorodioxin isomers by reversed-phase chromatography, silica adsorption HPLC, and packed-column GC. (Reproduced from *Anal. Chem.*, 52 (1980) 2037, with permission [180].)

determination of ppt levels of tetra-, hexa-, hepta-, and octachlorodibenzo-*p*-dioxins in human milk, involved sample digestion with acid, extraction with an organic solvent, multiple clean-up by adsorbents, two HPLC steps, and final detection by multiple-ion-mode GC–MS. The procedure effectively isolated 2,3,7,8-TCDD for a separate isomer-specific determination and also specified conditions for determining 18 other TCDD isomers, as well as hexa-(HCDD), hepta-(H_7CDD)-, and octa-(OCDD) chlorodibenzo-*p*-dioxins. ^{13}C-Labeled dioxins were used as internal standards and carriers. The procedure was validated at levels between 1 and 12 ppt for 2,3,7,8-TCDD and at higher ppt levels for TCDD, H_7CDD, and OCDD. Fig. 21.2 illustrates the fractionation scheme for isolating 2,3,7,8-TCDD from the other 21 TCDD by reversed-phase chromatography, silica adsorption HPLC, and packed-column GC. The procedure effectively removed potential interferences by various chromatographic steps. The removal efficiencies of the basic alumina and reversed-phase LC steps for PCB, DDE, and polychlorinated benzylphenyl ethers from fish samples has been demonstrated [196]. In all cases, it is expected that a 10^6-fold excess of these compounds would not interfere with the dioxin determination.

The separation of the 22 tetrachlorodibenzo-*p*-dioxin isomers on high-resolution glass capillary columns with different stationary phases (Silar 10C, OV-17, OV-101) and use of EI–MS detection was reported by Buser and Rappe [189]. Silar 10C allowed the highest number of isomers to be unequivocally assigned, including 2,3,7,8-TCDD. Some additional isomers could be identified on other columns, e.g., 1,2,7,8-, 1,2,7,9-, and 1,2,3,9-TCDD on OV-101, and 1,2,3,6-TCDD on OV-17. The determination of the various TCDD isomers was illustrated in the analysis of samples from known contaminated areas in Seveso, Italy, and in eastern Missouri, and the method was also applied to the analysis of fish from the Tittabawassee River in Michigan and fly ash samples from municipal incinerators in Switzerland.

A procedure has been developed by Nestrick et al. [197] which permits the acquisition of photo-decomposition data for the 22 isomers of TCDD and requires only 1 ng of material of each compound. Under equivalent light exposure conditions, photolytic half-lives were determined for each of the individual TCDD isomers in dilute hydrocarbon solution and in a diffuse molecular dispersion on a clean soft-glass surface. As a means of defining the molecular structure of each TCDD isomer, pattern recognition techniques were used to correlate specific rate data with chlorine substituent locations. The application of this system should permit the acquisition of fundamental property data for a large number of compounds not currently available in macro amounts. Additionally, photodecomposition data acquired at very low concentration may ultimately prove valuable in predicting the fate of TCDD in the ecosphere.

21.5. THIN-LAYER CHROMATOGRAPHY

In addition to classical TLC methods, employing precoated plates or sheets, there has been a significant increase in the use in pesticide residue analysis of HPTLC, reversed-phase TLC on chemically-bonded layers, and quantitations by in situ densitometry [198]. The advantages of HPTLC over TLC include: higher speed, greater sensitivity and efficiency, saving or solvent, and a larger number of samples per plate [198–200]. Densitometry is gradually replacing visual estimation and zone elution techniques in the quantitative analysis by TLC and in nanogram analysis by HPTLC [198], e.g., in the determination of triazine herbicides [201,202].

Jork and Roth [202] have compared the applicability of GC, HPLC, TLC, and HPTLC for residue analysis of s-triazine herbicides. The detection limit for fourteen s-triazines were as follows: 0.02–0.03 ng in GC with AFID; 0.8 ng in GC with FID; 1 ng in HPLC with UV; 3–5 ng in HPTLC with UV; and 8–13 ng in TLC with UV. The linear range in GC–AFID was 10^4, whereas in the three CC methods it was 10^2. The reproducibility of CC with a confidence limit of $P = 95\%$ had a coefficient of variation of $\pm 5\%$. In routine analysis requiring a large number of separations, HPTLC has distinct advantages, since the time for a single separation is only 40 sec.

21.6. NITROSATED PESTICIDES AND NITROSAMINE TRACE CONTAMINANTS

An area of increasing concern is the possibility that a variety of pesticides which are applied to soil and plants may contain nitroso compounds as impurities [203]. These impurities may arise from the three most probable routes of N-nitroso contamination: (a) formation in the manufacturing process, (b) formation during storage, and (c) contamination of amines used in the manufacturing process [204]. The second focal point of concern is that certain pesticides as residues in soil, water, and plants may be nitrosated in situ [203,204]. The pesticides that are most likely to undergo this reaction are the amides, triazines, ureas, carbamates, and dinitroaniline herbicides [203–206]. The most important products thus far shown to contain nitrosamines are the dinitroaniline herbicides, the major herbicides trifluralin (α,α,α-trifluoro-2,7-dinitro-N,N-dipropyl-p-toluidine) and the s-triazine, atrazine, being of particular concern [203–206].

The principal analytical techniques employed for the analysis of volatile N-nitrosamines and nitrosated pesticides have been GC–MS [207–209] and HPLC with chemiluminescent detection (thermal energy analyzer, TEA) [208–213]. The technique accepted as the most reliable for the confirmation of N-nitrosamines is based on MS [208,209]. Low-resolution MS is satisfactory for the analysis of relatively simple mixtures and in those instances in which extensive clean-up of samples has been performed. However, complex samples require more sophisticated GC and MS procedures (e.g., high-resolution MS). Chemiluminescence detectors have considerable selectivity for nitrosamines, because the light emitted from the NO/O_3 reaction is in the near IR region, whereas other known chemiluminescent reactions with O_3 emit light in the visible near-UV region. The TEA is sensitive to picogram quantities of N-nitroso compounds [208,210,212], the linear response extending over five orders of magnitude. In the routine analysis of N-nitroso compounds, possible TEA responses to compounds other than N-nitroso derivatives normally do not present a problem, since the identity of a compound can be readily established by admixture of known standards in GC–TEA and/or HPLC–TEA systems [214,215]. Additional confirmation is provided by analyzing the sample by both GC–TEA and HPLC–TEA [214].

REFERENCES

1 S.P. Cram, T.H. Risby, L.R. Field and W.L. Yu, *Anal. Chem.*, 52 (1980) 324R.
2 J.H. Ruzicka, *Proc. Soc. Anal. Chem.*, 10 (1973) 32.
3 E.D. Magallona, *Residue Rev.*, 56 (1975) 1.
4 H.W. Dorough and J.H. Thorstenson, *J. Chromatogr. Sci.*, 13 (1975) 212.
5 R.C. Hall and D.E. Harris, *J. Chromatogr.*, 169 (1979) 245.
6 W.L. Zielinski and L. Fishbein, *J. Gas Chromatogr.*, 3 (1965) 333.
7 S.R. Lowry and C.R. Gray, in J. Harvey, Jr. and G. Zweig (Editors), *Pesticide Analytical Methodology*, Amer. Chem. Soc. Symp. Series 136, Amer. Chem. Soc., Washington, DC, 1980, p. 17.
8 L. Wheeler and A. Strother, *J. Gas Chromatogr.*, 6 (1968) 110.
9 R.F. Cook, R.P. Stanovick and C.C. Cassil, *J. Agr. Food Chem.*, 17 (1969) 277.

10 C.C. Cassil, R.P. Stanovick and R.F. Cook, *Residue Rev.*, 26 (1969) 63.

11 M. Riva and A. Carisano, *J. Chromatogr.*, 42 (1969) 464.

12 M. Oda, N. Shida and T. Kashiwa, *Noyaku Kensasho Hokoku*, 16 (1976) 60.

13 J.F. Lawrence, D.A. Lewis and H.A. McLeod, *J. Chromatogr.*, 138 (1977) 143.

14 H.G. Loebering, L. Weil and K.E. Quentin, *Vom Wasser*, 51 (1978) 265.

15 G.H. Tjan and J.T.A. Jansen, *J. Ass. Offic. Anal. Chem.*, 62 (1979) 269.

16 J.N. Seiber, *J. Agr. Food Chem.*, 20 (1972) 443.

17 J. Sherma and T.M. Shafik, *Arch. Environ. Contam. Toxicol.*, 3 (1975) 55.

18 E.R. Holden, *J. Ass. Offic. Anal. Chem.*, 56 (1973) 713.

19 L.J. Sullivan, J.M. Eldridge and J.B. Knaak, *J. Agr. Food Chem.*, 15 (1967) 927.

20 R.A. Chapman and J.R. Robinson, *J. Chromatogr.*, 140 (1977) 209.

21 J.F. Lawrence, *J. Chromatogr.*, 123 (1976) 287.

22 R.J. Kuhr and H.W. Dorough, *Carbamate Pesticides: Chemistry, Biochemistry and Toxicology*, CRC Press, Cleveland, OH, 1976, p. 201.

23 A. Zlatkis and C.F. Poole, *Anal. Chem.*, 52 (1980) 1002A.

24 M.E. Mount and F.W. Oehme, *J. Anal. Toxicol.*, 4 (1980) 286.

25 J. Krechniak and W. Foss, *Bull. Environ. Contam. Toxicol.*, 23 (1979) 531.

26 I.C. Cohen, J. Norcup, J.H.A. Ruzicka and B.B. Wheals, *J. Chromatogr.*, 49 (1970) 215.

27 L. Fishbein, *Chromatogr. Rev.*, 12 (1970) 167.

28 A.M. Mattson, R.A. Kahrs and R.T. Murphy, *Residue Rev.*, 32 (1970) 32.

29 W.P. Cochrane and R. Purkayastha, *Toxicol. Environ. Chem. Rev.*, 1 (1973) 137.

30 J.F. Lawrence, *J. Agr. Food Chem.*, 22 (1974) 137.

31 C.E. McKone, T.H. Byast and R.J. Hance, *Analyst (London)*, 97 (1972) 653.

32 R. Delley, K. Friedrich, B. Karlhuber, G. Szekely and K. Stammbach, *Z. Anal. Chem.*, 228 (1967) 23.

33 R. Purkayastha and W.P. Cochrane, *J. Agr. Food Chem.*, 21 (1972) 93.

34 R. Greenhalgh and W.P. Cochrane, *J. Chromatogr.*, 70 (1972) 37.

35 H.Y. Young and A. Chu, *J. Agr. Food Chem.*, 21 (1973) 711.

36 R. Bailey, G. LeBel and J.F. Lawrence, *J. Chromatogr.*, 161 (1978) 251.

37 L. Fishbein, *Chromatography of Environmental Hazards*, Vol. 3, *Pesticides*, Elsevier, Amsterdam, 1975, p. 733.

38 K. Ramsteiner, W.D. Hörmann and D.O. Eberle, *J. Ass. Offic. Anal. Chem.*, 57 (1974) 192.

39 E. Matisová, J. Krupčík, O. Liška and N. Szentiványi, *J. Chromatogr.*, 169 (1979) 261.

40 W. Thornburg, *Anal. Chem.*, 51 (1979) 198R.

41 W.A. Aue, *J. Chromatogr. Sci.*, 13 (1975) 329.

42 T.A. Gosink, *Environ. Sci. Technol.*, 9 (1975) 630.

43 K. Aitzetmüller, *J. Chromatogr.*, 107 (1975) 411.

44 B. Zimmerli and B. Marek, *Mitt. Geb. Lebensmittelunters. Hyg.*, 66 (1975) 362.

45 H. Rohleder, M. Staudacher and W. Sümmermann, *Z. Anal. Chem.*, 279 (1976) 152.

46 P.L. Pursley and E.D. Schall, *J. Ass. Offic. Agr. Chem.*, 48 (1965) 327.

47 R. Purkayastha, *J. Agr. Food Chem.*, 22 (1974) 453.

48 K.L. Choi, S.S. Quee Hee and R.G. Sutherland, *J. Environ. Sci. Health*, B11 (1976) 175.

49 G. Yip, *J. Ass. Offic. Agr. Chem.*, 45 (1962) 367.

50 C. Chow, M.L. Montgomery and T.C. Yu, *Bull. Environ. Contam. Toxicol.*, 6 (1971) 576.

51 H.E. Munro, *Pestic. Sci.*, 3 (1972) 371.

52 W.P. Cochrane and J.B. Russell, *Can. J. Plant Sci.*, 55 (1975) 323.

53 A.E. Dupuy, T.J. Forehand and Han Tai, *J. Agr. Food Chem.*, 23 (1975) 827.

54 J.E. Allebone and R.J. Hamilton, *J. Chromatogr.*, 108 (1975) 188.

55 C.J. Soderquist and D.G. Crosby, *Pestic. Sci.*, 6 (1975) 188.

56 S.U. Khan, *J. Ass. Offic. Anal. Chem.*, 58 (1975) 1027.

57 A. Bevenue, G. Zweig and N. Nash, *J. Ass. Offic. Agr. Chem.*, 46 (1963) 881.

58 G. Yip, *J. Ass. Offic. Agr. Chem.*, 47 (1964) 1116.

59 D.L. Klingman, C.H. Gordon, G. Yip and H.P. Burchfield, *Weeds*, 14 (1966) 164.

60 D.G. Crosby and J.B. Bowers, *Bull. Environ. Contam. Toxicol.*, 1 (1966) 104.

61 T.R. Duffy and P. Shelfoon, *J. Ass. Offic. Agr. Chem.*, 50 (1967) 1098.

62 G. Yip, *J. Ass. Offic. Agr. Chem.*, 54 (1971) 966.

63 W.P. Cochrane, R. Greenhalgh and N.E. Looney, *J. Ass. Offic. Agr. Chem.*, 59 (1976) 617.

64 W.P. Cochrane, R. Greenhalgh and N.E. Looney, *Can. J. Plant Sci.*, 56 (1976) 207.

65 W.H. Gutenmann and D.J. Lisk, *J. Ass. Offic. Agr. Chem.*, 47 (1964) 353.

66 D.W. Woodham, W.G. Mitchell, C.D. Loftis and C.W. Collier, *J. Agr. Food Chem.*, 19 (1971) 186.

67 J.B. Rivers, W.L. Yauger, Jr. and H.W. Klemmer, *J. Chromatogr.*, 50 (1970) 334.

68 L. Renberg, *Anal. Chem.*, 46 (1974) 459.

69 C.R. Nony, M.C. Bowman, C.L. Holder, J.F. Young and W.L. Oller, *J. Pharm. Sci.*, 64 (1976) 1810.

70 J.E. Scoggins and C.H. Fitzgerald, *J. Agr. Food Chem.*, 17 (1969) 156.

71 S.F. Howard and G. Yip, *J. Ass. Offic. Anal. Chem.*, 54 (1971) 970.

72 J. Horner, S.S.Q. Hee and R.G. Sutherland, *Anal. Chem.*, 46 (1974) 110.

73 J. De Beer, C. Van Peteghem and A. Heyndrickx, *J. Chromatogr.*, 157 (1978) 97.

74 A.S.Y. Chau and K. Terry, *J. Ass. Offic. Agr. Chem.*, 59 (1976) 633.

75 H. Agemian and A.S.Y. Chau, *Analyst (London)*, 101 (1976) 732.

76 J. De Beer, C. Van Peteghem and A. Heyndrickx, *Meded. Rijksfac. Landbouwwet. Gent*, 42 (1977) 1739.

77 G. Yip, *J. Ass. Offic. Agr. Chem.*, 54 (1971) 343.

78 C. van Peteghem and A. Heyndrickx, *Meded. Rijksfac. Landbouwwet. Gent*, 38 (1973) 857.

79 F.K. Kawahara, *Anal. Chem.*, 40 (1968) 1009.

80 F.K. Kawahara, *Anal. Chem.*, 40 (1968) 2073.

81 J. Robinson and C.G. Hunter, *Arch. Environ. Health*, 13 (1966) 558.

82 A. Curley and R. Kimbrough, *Arch. Environ. Health*, 18 (1969) 156.

83 W.E. Dale, M.F. Copeland and W.J. Hayes, Jr., *WHO Bull.*, 33 (1965) 471.

84 E.M. Brevik, *Bull. Environ. Contam. Toxicol.*, 19 (1978) 281.

85 H.L. Christ and R.F. Moseman, *J. Chromatogr.*, 160 (1978) 49.

86 D.L. Stalling, R.C. Tindle and J.L. Johnson, *J. Ass. Offic. Agr. Chem.*, 55 (1972) 32.

87 L.D. Johnson, R.H. Waltz, J.P. Ussary and F.E. Kaiser, *J. Ass. Offic. Agr. Chem.*, 59 (1976) 174.

88 R.C. Tindle and D.L. Stalling, *Anal. Chem.*, 44 (1972) 1768.

89 M.C. Saxena, T.D. Seth and P.L. Mahajan, *Int. J. Environ. Anal. Chem.*, 7 (1980) 245.

90 I.R. DeLeon, N.J. Brown, J.P. Cocchiara, S.K. Miles, J.L. Laseter, E.H. Kremer and L. Makk, *J. Anal. Toxicol.*, 4 (1980) 314.

91 B. Luckas, H. Pscheidl and D. Haberland, *J. Chromatogr.*, 147 (1978) 41.

92 W.L. Oller and M.F. Cramner, *J. Chromatogr. Sci.*, 13 (1975) 296.

93 R. Edwards, *Chem. Ind.*, (1970) 1340.

94 L. Fishbein, *J. Chromatogr.*, 68 (1972) 345.

95 E. Schulte, H.P. Thier and L. Acker, *Deut. Lebensm. Rundsch.*, 72 (1976) 229.

96 G. Göke, *Deut. Lebensm. Rundsch.*, 71 (1975) 309.

97 M. Mansour and P. Parlar, *J. Agr. Food Chem.*, 25 (1977) 201.

98 V. Paramasigamani, S. Kapila and W.A. Aue, *J. Chromatogr. Sci.*, 18 (1980) 191.

99 N.A. Smart, A.R.C. Hill and P.A. Roughan, *Analyst (London)*, 103 (1978) 770.

100 D.C. Abbott, S. Crisp, K.R. Tarrant and J.O.G. Tatton, *Pestic. Sci.*, 1 (1970) 10.

101 Report by the Panel on Determination of Residues of Certain Organophosphorus Pesticides in Fruits and Vegetables, *Analyst (London)*, 102 (1977) 858.

102 R.R. Watts, R.W. Storherr, J.R. Pardue and T. Osgood, *J. Ass. Offic. Anal. Chem.*, 52 (1969) 522.

103 D.J. Sissons and G.M. Telling, *J. Chromatogr.*, 47 (1970) 328.

104 D.J. Sissons and G.M. Telling, *J. Chromatogr.*, 48 (1970) 468.

105 G. Becker, *Deut. Lebensm. Rundsch.*, 67 (1971) 125.

106 Dormal van den Bruel, personal communication (see ref. 99).

107 R.W. Storherr and R.R. Watts, *J. Ass. Offic. Anal. Chem.*, 51 (1968) 662.

108 W. Ebing, personal communication (see ref. 99).

109 J.W. Miles and W.E. Dale, *J. Agr. Food Chem.*, 26 (1978) 480.

110 A.M. Coor, C.G. Daughton and M. Alexander, *Appl. Environ. Microbiol.*, 36 (1978) 668.

111 C.G. Daughton, D.G. Crosby, R.L. Garnas and D.P.H. Hsieh, *J. Agr. Food Chem.*, 24 (1976) 236.
112 C.G. Daughton, A.M. Coor and M. Alexander, *Anal. Chem.*, 51 (1979) 1949.
113 H.A. Moye, *J. Chromatogr. Sci.*, 13 (1975) 268.
114 J.F. Lawrence, *Anal. Chem.*, 52 (1980) 1122A.
115 H.F. Walton, *Anal. Chem.*, 52 (1980) 15R.
116 R.E. Majors, *J. Ass. Offic. Anal. Chem.*, 60 (1977) 186.
117 J.F. Lawrence and D. Turton, *J. Chromatogr.*, 159 (1978) 207.
118 *American Chemical Society, Pesticide Analytical Methodology*, ACS Symp. Ser. No. 136, Amer. Chem. Soc. Washington, DC, 1980.
119 K. Ivie, in G. Zweig (Editor), *Analytical Methods for Pesticides and Plant Growth Regulators*, Vol. 11, Academic Press, New York, 1980.
120 F. Eisenbeiss and H. Sieper, *J. Chromatogr.*, 83 (1973) 439.
121 D.F. Horgan, Jr., in J. Sherma and G. Zweig (Editors), *Analytical Methods for Pesticides and Plant Growth Regulators*, Vol. 7, Academic Press, New York, 1973.
122 J.G. Koen, J.F.K. Huber, H. Poppe and G. den Boef, *J. Chromatogr. Sci.*, 8 (1970) 192.
123 P.T. Kissinger, C. Refshange, R. Dreiling and R.N. Adams, *Anal. Lett.*, 6 (1973) 465.
124 B. Fleet and C.J. Little, *J. Chromatogr. Sci.*, 12 (1974) 747.
125 R.C. Buchta and L.J. Papa, *J. Chromatogr. Sci.*, 14 (1976) 213.
126 C. Bollet, P. Oliva and M. Caude, *J. Chromatogr.*, 149 (1977) 625.
127 J.L. Anderson and D.J. Chesney, *Anal. Chem.*, 52 (1980) 2156.
128 R.A. Mowery, Jr. and R.S. Juvet, Jr., *J. Chromatogr. Sci.*, 12 (1974) 687.
129 K. Šlais and M. Krejči, *J. Chromatogr.*, 91 (1974) 181.
130 F.W. Willmott and R.J. Dolphin, *J. Chromatogr. Sci.*, 12 (1974) 695.
131 B.J. Compton and W.C. Purdy, *J. Chromatogr.*, 169 (1979) 39.
132 J.W. Dolan and J.N. Sieber, *Anal. Chem.*, 49 (1977) 326.
133 A.S. Bhown, J.E. Mole, A. Weissinger and J.C. Bennett, *J. Chromatogr.*, 148 (1978) 532.
134 P.T. Kissinger, *Anal. Chem.*, 49 (1977) 447A.
135 C.M. Sparacino and J.W. Hines, *J. Chromatogr. Sci.*, 14 (1976) 549.
136 C.F. Aten and J.B. Bourke, *J. Agr. Food Chem.*, 25 (1976) 549.
137 I. Stoeber and R. Reupert, *Vom Wasser*, 51 (1978) 273.
138 R.W. Frei and J.F. Lawrence, *J. Chromatogr.*, 83 (1973) 321.
139 R.W. Frei, J.F. Lawrence, J. Hope and R.M. Cassidy, *J. Chromatogr. Sci.*, 12 (1974) 40.
140 H.A. Moye, S.J. Scherer and P.A. St. John, *Anal. Lett.*, 10 (1977) 1049.
141 R.T. Krause, *J. Chromatogr. Sci.*, 16 (1978) 281.
142 W.P. Cochrane, *J. Chromatogr. Sci.*, 17 (1979) 124.
143 J. Muth and J. Giles, *Altex Chromatogram*, 3 (1980) 5.
144 B.L. Karger, M. Martin and G. Guiochon, *Anal. Chem.*, 46 (1974) 1640.
145 J.N. Seiber, *J. Chromatogr.*, 94 (1974) 151.
146 K.A. Ramsteiner and W.D. Hörmann, *J. Chromatogr.*, 104 (1975) 438.
147 A.A. Carolstrom, *J. Ass. Offic. Anal. Chem.*, 60 (1977) 1157.
148 B.J. Compton and W.C. Purdy, *J. Chromatogr.*, 169 (1979) 39.
149 J.A. Sidwell and J.H.A. Ruzicka, *Analyst (London)*, 101 (1976) 111.
150 D.S. Farrington, R.G. Hopkins and J.H.A. Ruzicka, *Analyst (London)*, 102 (1977) 377.
151 J.F. Lawrence, *J. Ass. Offic. Anal. Chem.*, 59 (1976) 1066.
152 T.H. Byast, *J. Chromatogr.*, 134 (1977) 216.
153 D. Spengler and B. Hamroll, *J. Chromatogr.*, 49 (1970) 205.
154 T. Hirschfeld, *Anal. Chem.*, 52 (1980) 297A.
155 C.J.W. Brooks and B.S. Middleditch, in R.A.W. Johnstone (Editor), *Mass Spectrometry*, Vol. 5, Chemical Society, London, 1979.
156 C.J.W. Brooks, C.G. Edmonds, S.J. Gaskell and A.G. Smith, *Chem. Phys. Lipids*, 21 (1978) 403.
157 M.C. ten Noever de Brauw, *J. Chromatogr.*, 156 (1979) 207.
158 W.H. McFadden, *J. Chromatogr. Sci.*, 17 (1979) 2.
159 B.J. Millard, *Quantitative Mass Spectrometry*, Heyden, London, 1978, p. 171.
160 A.L. Burlingame, T.A. Baillie, P.J. Derrick and O.S. Chizhov, *Anal. Chem.*, 52 (1980) 220R.

161 R.L. Harless, D.E. Harris, G.W. Sovocool, R.D. Zehr, N.K. Wilson and E.O. Oswald, *Biomed. Mass Spectrom.*, 5 (1978) 232.

162 F.I. Onuska, M.E. Comba and J.A. Coburn, *Anal. Chem.*, 52 (1980) 2272.

163 A. Bevenue and H. Beckman, *Bull. Environ. Contam. Toxicol.*, 2 (1967) 319.

164 R.C. Dougherty and K. Piotrowska, *Proc. Nat. Acad. Sci. U.S.*, 73 (1976) 1777.

165 M.F. Carnmer and J. Freal, *Life Sci.*, 9 (1970) 121.

166 T. Akisada, *Bunseki Kagaku*, 14 (1965) 101.

167 R.C. Dougherty, in K.P. Raol (Editor), *Pentachlorophenol: Chemistry, Pharmacology, and Environmental Toxicology*, Plenum, New York, 1978, p. 351.

168 R.S. Yang, H.F. Coulston and L. Goldberg, *J. Ass. Offic. Anal. Chem.*, 58 (1975) 1197.

169 D.W. Kuehl and R.C. Dougherty, *Environ. Sci. Technol.*, 14 (1980) 447.

170 T. Ohe, *Bull. Environ. Contam. Toxicol.*, 22 (1979) 287.

171 E. Homberger, G. Reggiani, J. Sambeth and H.K. Wipf, *Ann. Occup. Hyg.*, 22 (1979) 327.

172 G. Lucier (Editor), *Environ. Health Perspect.*, 5 (1973).

173 R.A. Hummel, *J. Agr. Food Chem.*, 25 (1977) 1049.

174 C.D. Pfeiffer, *J. Chromatogr. Sci.*, 14 (1976) 386.

175 C.D. Pfeiffer, T.J. Nestrick and C.W. Kocher, *Anal. Chem.*, 50 (1978) 800.

176 L.A. Shadoff and R.A. Hummel, *Biomed. Mass Spectrom.*, 5 (1978) 7.

177 N.H. Mahle, H.S. Higgins and M.E. Getzandaner, *Bull. Environ. Contam. Toxicol.*, 18 (1977) 123.

178 R.H. Stehl and L.L. Lamparski, *Science*, 197 (1977) 1008.

179 L.L. Lamparski, N.H. Mahle and L.A. Shadoff, *J. Agr. Food Chem.*, 26 (1978) 1113.

180 M.L. Langhorst and L.A. Shadoff, *Anal. Chem.*, 52 (1980) 2037.

181 L.L. Lamparski and T.J. Nestrick, *Anal. Chem.*, 52 (1980) 2045.

182 E.K. Chess and M.L. Gross, *Anal. Chem.*, 52 (1980) 2057.

183 H.R. Buser, *Anal. Chem.*, 49 (1977) 918.

184 T.J. Nestrick, L.L. Lamparski and R.H. Stehl, *Anal. Chem.*, 51 (1978) 2273.

185 J.R. Hass, M.D. Friesen, D.J. Harvan and C.E. Parker, *Anal. Chem.*, 50 (1978) 1474.

186 L.L. Lamparski, T.J. Nestrick and R.H. Stehl, *Anal. Chem.*, 51 (1979) 1453.

187 R.K. Mitchum, G.F. Moler and W.A. Korfmacher, *Anal. Chem.*, 52 (1980) 2278.

188 A. Cavallaro, G. Bartolozzi, D. Carreri, G. Bandi, L. Luciani, G. Viala, A. Giorni and G. Invernizzi, *Chemosphere*, 9 (1980) 623.

189 H.R. Buser and C. Rappe, *Anal. Chem.*, 52 (1980) 2257.

190 R.L. Harless, E.O. Oswald, M.K. Wilkinson, A.E. Dupuy, Jr., D.D. McDaniel and H. Tal, *Anal. Chem.*, 52 (1980) 1239.

191 L.A. Shadoff, L.L. Lamparski and J.H. Davidson, *Bull. Environ. Contam. Toxicol.*, 18 (1977) 478.

192 R.L. Harless and E.O. Oswald, in F. Cattabini, A. Cavallaro and G. Galli (Editors), *Dioxin, Toxicological and Chemical Aspects*, Spectrum Publications, New York, 1978, p. 51.

193 R. Baughman and M. Meselson, *Advan. Chem. Ser.*, 120 (1973) 92.

194 P.W. O'Keefe, M.S. Meselson and R.W. Baughman, *J. Ass. Offic. Anal. Chem.*, 61 (1978) 621.

195 R.W. Baughman and M.S. Meselson, *Environ. Health Perspect.*, 5 (1973) 27.

196 L.L. Lamparski, T.J. Nestrick and R.H. Stehl, *Anal. Chem.*, in press.

197 T.J. Nestrick, L.L. Lamparski and D.I. Townsend, *Anal. Chem.*, 52 (1980) 1865.

198 G. Zweig and J. Sherma, *Anal. Chem.*, 52 (1980) 276R.

199 H. Halpaap and J. Ripphahn, *Chromatographia*, 10 (1977) 613.

200 P.P. Ilinov, *Lipids*, 14 (1979) 598.

201 J. Sherma and N.T. Miller, *J. Liquid Chromatogr.*, 3 (1980) 901.

202 H. Jork and B. Roth, *J. Chromatogr.*, 144 (1977) 39.

203 L. Fishbein, *Nitrosamines in Pesticides and Agricultural Residues*, *Proc. Toxicol. Forum*, Toxicology Forum Assoc., Washington, DC, 1980.

204 J.E. Oliver, *Chem. Tech.*, 9 (1979) 366.

205 E.W. Day, Jr., S.D. West, D.K. Koenig and F.L. Powers, *J. Agr. Food Chem.*, 27 (1979) 1081.

206 S.D. West and S.N. Day, Jr., *J. Agr. Food Chem.*, 27 (1979) 1075.

207 R. Fanelli, C. Chiabrando and L. Airoldi, *Anal. Lett.*, A 11 (1978) 845.

208 A. Preussman, E.A. Walker, A.E. Wasserman and M. Castegnaro (Editors), *Environmental Carcinogens. Selected Methods of Analysis*, Vol. 1, *Nitrosamines*, IARC Scient. Publ. No. 18, Intern. Agency Res. Cancer, Lyon, 1978.

209 T.A. Gough, *Analyst (London)*, 103 (1978) 785.

210 D.H. Fine, F. Rufeh and B. Gunther, *Anal. Lett.*, 6 (1973) 731.

211 D.H. Fine, F. Rufeh, D. Lieb and D. Rounbehler, *Anal. Chem.*, 47 (1975) 1188.

212 P.E. Oettinger, F. Huffman, D.H. Fine and D. Lieb, *Anal. Lett.*, 8 (1975) 411.

213 I.S. Krull, K. Mills, G. Hoffmann and D.H. Fine, *J. Anal. Toxicol.*, 4 (1980) 260.

214 I.S. Krull, E.U. Goff, G.G. Hoffman and D.H. Fine, *Anal. Chem.*, 51 (1979) 1706.

215 M. Castegnaro and E.A. Walker, in E.A. Walker, M. Castegnaro, L. Griciuite and R.E. Lyle (Editors), *Environmental Aspects of N-Nitroso Compounds*, IARC Scient. Publ. No. 79, Intern. Agency Res. Cancer, Lyon, 1978, p. 53.

Chapter 22

Inorganic compounds

M. LEDERER

CONTENTS

22.1. INTRODUCTION

The purpose of this chapter is to give a perspective of inorganic chromatography today. It is not intended as a complete survey of the field, which would require more space and would duplicate numerous recent monographs on the various branches of inorganic chromatography. Both chromatographic and electrophoretic techniques will be discussed, as they are often used together in the same investigation and are essential for a complete picture. Inorganic chromatography has two main purposes:

(a) The isolation and/or determination of an element. Unlike organic chromatography, inorganic chromatography has a wide range of competing methods available for this purpose, such as polarography, atomic absorption spectrometry, spectrographic methods in general, and radioactivation techniques. Although many chromatographic methods were proposed earlier, very few of them have survived competition with other techniques, except for instances where their relative cheapness and the convenience in carrying out a large number of analyses is advantageous.

(b) The separation and determination of ionic or molecular species in solution. It is in this field that chromatography has fostered considerable advances in inorganic chemical research. Some fields were developed almost entirely by the application of chromatographic methods, such as the chemistry of polyphosphates and of astatine compounds and the solution chemistry of the platinum group metals.

References on p. B488

22.2. REVERSIBLE COMPLEX EQUILIBRIA

The general idea of the solution chemistry of metal ions is usually based on elementary electrochemistry, which leaves most chemists with the impression that in the usual concentrations encountered in "wet analysis" the ions are essentially in the "free" form. For instance, an "ionic mobility" for the "Fe^{3+}" ion obtained by conductivity measurements at infinite dilution is even quoted in textbooks on electrochemistry. Thus, let us take Fe^{3+} as an example to illustrate the solution chemistry of metal ions.

(a) When pink ferric nitrate hydrate is dissolved in distilled water, a pale-brown solution is obtained, which contains members of the following hydrolysis species:

$$Fe(H_2O)_6^{3+} \rightarrow Fe(H_2O)_5OH^{2+} \rightarrow Fe(H_2O)_4(OH)_2^{+} \rightarrow Fe(OH)_3^{0}$$

unstable   polymers

In fresh solutions, the preponderant species is $Fe(H_2O)_5OH^{2+}$. Dimers between various species are also possible, linked by OH bridges. They, as well as the main species, vary with the age, concentration, and temperature of the solution. These reactions are not entirely reversible, and some aged solutions or old samples of the solid give a number of zones in various chromatographic systems.

(b) When pink ferric nitrate is dissolved in HCl, a yellow to brown solution is obtained, containing mainly the following species:

$$Fe(H_2O)_6^{3+} \rightleftharpoons FeCl^{2+} \rightleftharpoons FeCl_2^{+} \rightleftharpoons FeCl_3^{0} \rightleftharpoons H^{+}FeCl_4^{-}$$

The neutral and the anionic species ($H^{+}FeCl_4^{-}$ seems to be a rather strong ion pair) are preponderant in ca. $6N$ HCl. In none of the solutions can the Fe^{3+} (i.e. the $Fe(H_2O)_6^{3+}$ ion) be found to any extent.

As a simple example of a chromatographic separation of inorganic ions, let us consider the separation Ni^{2+}/Fe^{3+}. Ni^{2+} does not form complexes beyond possibly $NiCl^{+}$ in HCl of a concentration of up to $12N$. Fe^{3+}, on the other hand, exists in HCl of a concentration of $6N$ or above mainly as the ion pair $H^{+}FeCl_4^{-}$, which is strongly hydrophobic and behaves as an anion. Thus, as long as the concentration of HCl is above $6N$, it is possible to separate these two metals by virtually any chromatographic method, e.g.:

(a) Solvent extraction with ether: Ni^{2+}, 0% extracted; Fe^{3+}, 99% extracted.

(b) Anion-exchange chromatography on a Dowex-1 column with $8N$ HCl: Fe^{3+}, adsorbed as a brown band; Ni^{2+}, eluted.

(c) Partition chromatography on paper with 1-butanol–$8N$ HCl $(1:1)$: Ni^{2+}, R_F 0.36; Fe^{3+}, R_F 0.98; the hydrophobic $H^{+}FeCl_4^{-}$ is well extracted into the butanol phase.

Another separation, $In^{3+}/Ga^{3+}/Tl^{3+}$, is illustrated in Fig. 22.1. It is obvious that Tl^{3+} and In^{3+} can be separated by either partition chromatography or ion-exchange chromatography or even by solvent extraction at a wide range of HCl concentrations, but Ga^{3+} will separate from In^{3+} only above 2–$3N$ HCl and from Tl^{3+} only below $5N$ HCl.

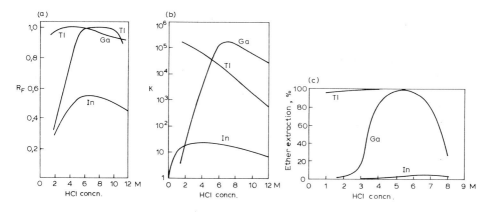

Fig. 22.1. Dependence of the R_F value in n-butanol (a), the ion-exchange equilibrium on Dowex 1 (b), and the partition equilibrium in ether/water (c) on the concentration of HCl in the solvent for Tl^{3+}, In^{3+}, and Ga^{3+}. (Reproduced from *Metodi di Separazione nella Chimica Inorganica*, Vol. 1, 1963, p. 199, with permission [64].)

The chromatographic data for most metal ions (and for many other ions) are usually presented in the form of the periodic table. As shown for a few different systems in Figs. 22.2–22.5, the separations depend much more on the HCl concentration in the aqueous solution than on the chromatographic system.

We can see here an important feature of inorganic chromatography: numerous metals form complexes with HCl at various HCl concentrations, and this makes it possible to change the separation factors over a wide range and, as shown in Fig. 22.6, to elute metals by suitably varying the concentrations of HCl. In this kind of chromatography there is thus no need for a large number of theoretical plates. For most of the chromatographic work on anion exchange the columns were cut-off 10-ml pipettes with a plug of glass wool at the bottom to hold the resin. As many of the metal ion halide complexes are colored (e.g., those of Co, Cu, Fe), there is often no need for a fraction collector either, as one can collect each colored fraction in a separate beaker. On the other hand, a suitable combination of an efficient column with (in this case) suitable HCl concentrations and organic solvents permits the elution of a dozen heavy metals in a matter of approximately 30 min, as is shown, e.g., in Fig. 22.7.

In summary, reversible complexes, especially those of the heavy metals, have been separated on cation exchangers, anion exchangers, liquid ion exchangers supported by cellulose or polymers, and by partition chromatography, reversed-phase chromatography (e.g., on acetylated cellulose) and by paper electrophoresis. The data on planar methods have been collected [1,2], and these collections make an excellent starting point for newcomers in the field.

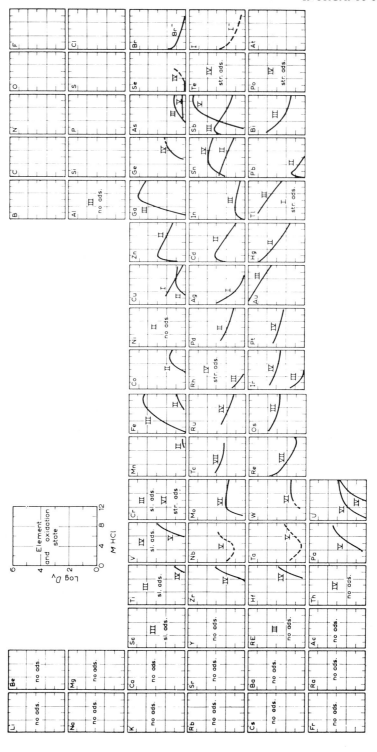

Fig. 22.2. Ion-exchange data for Dowex-1 with HCl. no ads. = no adsorption in 0.1–12 M HCl; sl. ads. = slight adsorption in 12 M HCl; str. ads. = strong adsorption; distribution coefficient, $D_v \gg 1$. (From *Proc. Int. Conf. Peaceful Uses At. Energy, Geneva*, Vol. 7, 1955, as reproduced in ref. 65, with permission.)

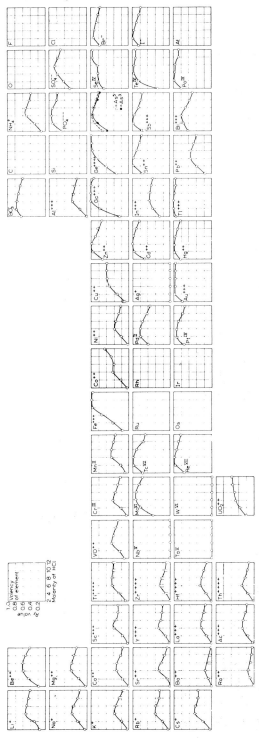

Fig. 22.3. R_F values of ions in butanol–HCl mixtures. All solvents were prepared by shaking equal volumes of aqueous acid and butanol; with 2 M HCl only one phase is formed. (From *Anal. Chim. Acta*, 16 (1957) 555, as reproduced in ref. 66, with permission.)

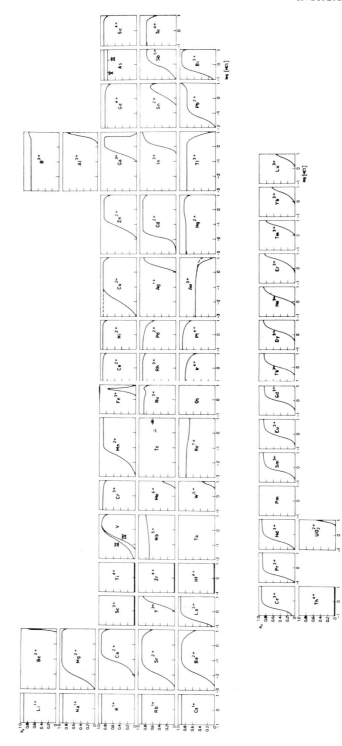

Fig. 22.4. Chromatography on paper, treated with 0.1 M di-(2-ethylhexyl)orthophosphoric acid in cyclohexane. Plot of R_F values of the elements as a function of the logarithm of the molarity of the solvent (HCl). (Reproduced from *J. Chromatogr.*, 24 (1966) 383, with permission [67].)

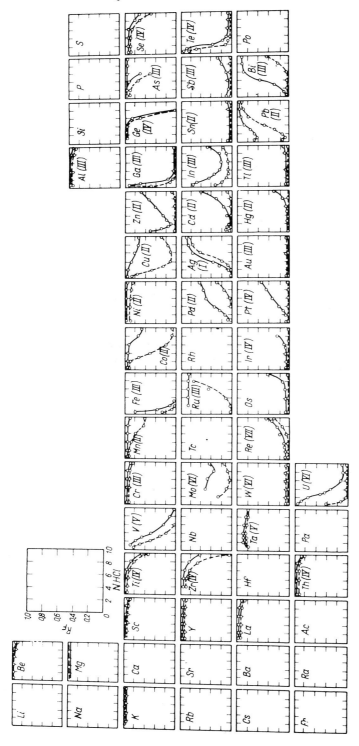

Fig. 22.5. R_F values of metal ions in HCl on papers, impregnated with liquid ion exchangers. Paper: Whatman No. 1, impregnated with a 0.1 M solution of the exchanger in benzene; solvent: HCl: ———, tri-n-octylamine. (From *Chem. Anal. (Warsaw)*, 12 (1967) 1071, as reproduced in ref. 68., with permission.)

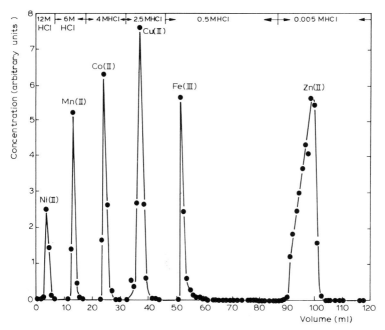

Fig. 22.6. Separation of transition elements Mn to Zn by ion-exchange chromatography. Column, 26 cm×0.29 cm², Dowex-1; flowrate, 0.5 cm/min. (From *J. Amer. Chem. Soc.*, 75 (1953) 1460, as reproduced in ref. 69, with permission.)

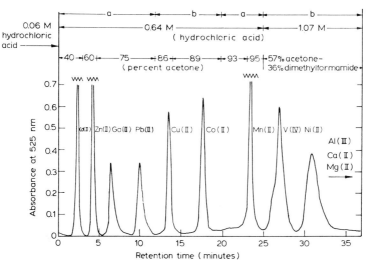

Fig. 22.7. Separation of metal ions on Amberlite 200 (25–30 μm). Column, 120 × 5 mm ID; flowrate, 22 cm/min; temperature, 40°C; column inlet pressure, 16–25 atm; color-forming reagent solution, 0.02% 4-(2-pyridylazo)resorcinol in ammonia, concentration (a) 0.7 M and (b) 1.2 M; sample volume, 60 ml. Amount of metal (×10⁻⁸ mole); Cd^{2+}, 9.6; Zn^{2+}, 4.8; Ga^{3+}, 4.5; Pb^{2+}, 12; Cu^{2+}, 4.8; Co^{2+}, 7.2; Mn^{2+}, 12; V^{4+}, 57; Ni^{2+}, 12; Al^{3+}, 9.6; Ca^{2+}, $1.2 \cdot 10^3$; Mg^{2+}, $6 \cdot 10^2$. (Reproduced from *J. Chromatogr.*, 137 (1977) 381, with permission [70].)

22.3. COMPLEXING EQUILIBRIA THAT ARE NOT INSTANTLY REVERSIBLE

Looking at Figs. 22.2–22.5, one may think that the separation of any mixture of metal ions is merely a matter of selecting the optimum conditions for obtaining a difference in complexation and then applying them in the most suitable separation technique. However, there are a number of metal ions, notably Rh^{3+}, Ru^{3+} and Cr^{3+}, that form complexes, but at rather slow rates, so that the variation of complexing concentrations usually yields a mixture of complexes. For example, $RhCl_6^{3-}$ in aqueous HCl will first hydrolyze quickly to $Rh(H_2O)Cl_5^{2-}$ and then more slowly to form a mixture of chloro–aquo complexes, as is best illustrated in Figs. 22.8 and 22.9. Such metal ions, therefore, do not yield one band, but a number of bands in all chromatographic systems.

Although many authors include such metal ions in the "periodic table" type of representation of the chromatographic data, this usually refers only to a predominant species, and application of the data will give erroneous results in most actual analytical separations. Special emphasis should be given here to Ru^{3+}, which forms extremely stable nitrosyl and nitro- as well as nitrato-complexes. A solution of ruthenium in HNO_3 will always be a mixture of such complexes (sometimes also

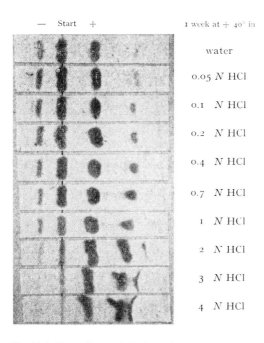

— Start +	1 week at + 40° in
	water
	0.05 N HCl
	0.1 N HCl
	0.2 N HCl
	0.4 N HCl
	0.7 N HCl
	1 N HCl
	2 N HCl
	3 N HCl
	4 N HCl

Fig. 22.8. Dependence of the hydrolysis of rhodium on HCl concentration. Rhodium chloro complexes (0.1 N); buffer, 0.3 N acetic acid–0.2 N sodium acetate; paper, Schleicher & Schüll 2043 b mgl; electrophoresis for 30 min at 3 kV. (Reproduced from *Chromatogr. Rev.*, 6 (1964) 191, with permission [3].)

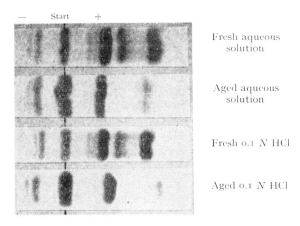

—   Start   +

Fresh aqueous
solution

Aged aqueous
solution

Fresh 0.1 N HCl

Aged 0.1 N HCl

Fig. 22.9. Electrophoretic separations of fresh and aged rhodium solutions. Rhodium chloro complexes (0.1 N); buffer, 0.3 N acetic acid–0.2 N sodium acetate; paper, Schleicher & Schüll 2043 b mgl; electrophoresis for 30 min at 3 kV. (Reproduced from *Chromatogr. Rev.*, 6 (1964) 191, with permission [3].)

containing polymeric species) in slow evolution. This is still one of the major problems in analyzing solutions of the common fission products. Chromatography and electrophoresis have for the first time permitted the separation of such mixtures and thus also the isolation of individual species as well as the calculation of complexing (stability) constants. An excellent review by Blasius and Preetz [3] indicates the possibilities of chromatographic and electrophoretic methods in this field.

22.4. HYDROPHOBIC ADSORPTION AND THE "PERCHLORATE EFFECT"

A number of chloro-complexes of metal ions, especially those of the last line, viz. Au^{3+}, Hg^{2+}, Po^{4+}, but also Sb^{5+}, Fe^{3+}, Ga^{3+}, have extremely high K_d values (order of 10^6) when chromatographed in HCl solutions on anion-exchange resins. They also move near the solvent front in partition chromatography with butanol–HCl and are rather strongly adsorbed on neutral or sulfonic acid resins. Some are even adsorbed on cellulose paper from HCl. This is obviously not due to an "ionic attraction" between an anion such as $AuCl_4^-$ and an ionized substituted ammonium group of the resin, but has the character of a hydrophobic adsorption. On neutral supports, there is also increased adsorption by salting-out (usually with LiCl) and desorption with organic solvents. Typical results for cellulose paper are shown in Figs. 22.10a and b. These ions usually cannot be eluted from anion-exchange resins by HCl at any concentration, but they are rapidly eluted with $HClO_4$. This drastic desorption was first termed the "perchlorate effect". However, it was shown later that not only perchlorate but also nitrate and other oxygenated anions may have the same effect, which is best explained by the competition of hydrophobic anions for the same adsorption site.

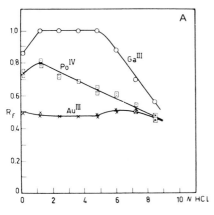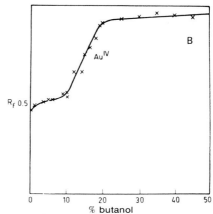

Fig. 22.10. (A) Variation of the R_F values of Ga^{3+}, Po^{4+} and Au^{3+} with the concentration of HCl used as eluent. (B) Variation of the R_F values of Au^{3+} with the concentration of butanol in butanol–6 N HCl mixtures. (Reproduced from *Metodi di Separazione nella Chimica Inorganica*, Vol. 1, 1963, p. 109, with permission [71].)

Adsorption of hydrophobic nature is also responsible for the separation of numerous mixtures of neutral coordination complexes. Table 22.1 lists the R_F values of *cis*-dihalodiammineplatinum(II) complexes. Evidently, as the size of the halogen increases, the R_F value decreases.

TABLE 22.1

R_F VALUES OF *cis*-DIHALODIAMMINEPLATINUM(II) COMPLEXES

Solvent, water; paper, Whatman No. 3MM; temperature, 5°C [71].

Complex	R_F value
$Pt(NH_3)_2Cl_2$	0.65
$Pt(NH_3)_2ClBr$	0.58
$Pt(NH_3)_2Br_2$	0.53
$Pt(NH_3)_2ClI$	0.56
$Pt(NH_3)_2I_2$	0.44

22.5. ION-PAIR OR OUTER-SPHERE COMPLEX FORMATION

Metal ions with a completely filled first coordination sphere may still form a second sphere of coordination. This was first discovered by Werner, who noted that the color (and spectrum) of the very stable $Co(NH_3)_6^{3-}$ changed in the presence of sulfate ions. In the various systems used for separation, the most striking outer-sphere effects were noted in the paper electrophoresis of Co(III) complexes, as best illustrated in Fig. 22.11.

References on p. B488

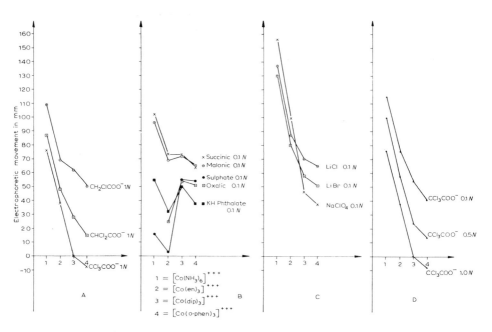

Fig. 22.11. Graphic representation of the electrophoretic movement of Co(III) complexes in various electrolytes. (A) Comparison of mono-, di-, and trichloroacetate (1 N). (B) Comparison of several divalent anions (0.1 N). (C) Comparison of LiCl, LiBr, and NaClO₄ (0.1 N). (D) Comparison of 0.1, 0.5, and 1 N trichloroacetate. Complexes: 1 = hexamminecobalt(III); 2 = tris(ethylenediamine)cobalt(III); 3 = tris(dipyridyl)cobalt(III); 4 = tris(o-phenanthroline)cobalt(III). (Reproduced from *J. Chromatogr.*, 35 (1968) 201 [72].)

Chloride and other monovalent hydrated anions, such as acetate and nitrate, will form a nonspecific anion cloud, which will retard all complexes considerably but to a similar degree. Sulfate, chromate, and most divalent anions will form hydrogen bridges to coordinated NH_3 or aquo groups, but not with groups such as dipyridyl or o-phenanthroline, which have no possibility for hydrogen bonding. This effect can be of such magnitude as to make trivalent cations behave like neutral species. Hydrophobic anions, such as perchlorate or trichloroacetate, will form "hydrophobic ion pairs", especially with voluminous hydrophobic coordination groups, such as dipyridyl or o-phenanthroline. Again, the effect can be very strong.

In ion exchange, outer-sphere complexation between the sulfonic groups of a resin and the aquo or ammine groups of a complex has been shown to exist. It should be stressed here that inner-complex formation has also been observed. There is also evidence that outer-sphere complexation between, e.g., aluminate groups on the surface of alumina or silanol groups on the surface of silica and coordination complexes does play a part in the chromatography of such complexes on inorganic exchangers.

22.6. ION CHROMATOGRAPHY

A fascinating recent advance in analytical ion-exchange chromatography was introduced by Small et al. [4] (cf. Chap. 7.5.4). Because in ion-exchange chromatography a strong electrolyte is usually employed as eluent, continuous effluent monitoring by means of conductivity measurements is not possible, except in a few cases.

In the commercial "Ion Chromatograph" (Dionex Corp.), weak acid salts are used as eluents for anions. The eluents pass over the hydrogen form of a cation exchanger ("suppressor column"), which removes the cation, leaving only a weak acid as background. The zones of strong electrolytes can then be recorded by their conductivity. Typical eluents are, e.g., carbonate–bicarbonate mixtures, and as the resins used have a very low capacity (0.007 meq/g) the concentrations of eluents can also be rather low ($2 \cdot 10^{-4} M$). Extensive work on a wide range of anions was published by Gjerde et al. [5]. Figs. 22.12–22.14 show some of the excellent separations obtained. They also demonstrate that separations can be obtained in as little as 10 min. Thus, ion chromatography is quite competitive with HPLC techniques.

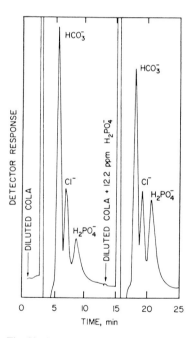

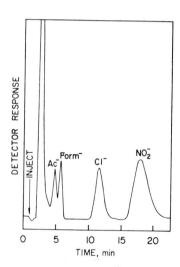

Fig. 22.12. Separation of phosphate from all other anions in a diluted cola drink by ion chromatography with $1 \cdot 10^{-4} M$ potassium benzoate, pH 6.25, on a column of XAD-2 (0.007 meq/g). (Reproduced from *J. Chromatogr.*, 187 (1980) 35, with permission [5].)

Fig. 22.13. Ion chromatography of 27.1 ppm acetate, 6.8 ppm formate, 5.1 ppm chloride, and 15.8 ppm nitrite on Vydac SC anion-exchange resin with $5 \cdot 10^{-4} M$ potassium benzoate, pH 6.25. (Reproduced from *J. Chromatogr.*, 187 (1980) 35, with permission [5].)

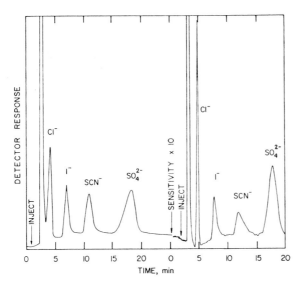

Fig. 22.14. Ion chromatography of 7.7 ppm chloride, 29.5 ppm iodide, 28.5 ppm thiocyanate and 16.5 ppm sulfate and of the same mixture, diluted by a factor of 10, on XAD-1 (0.007 meq/g) with $5 \cdot 10^{-5} M$ potassium phthalate, pH 6.25. (Reproduced from *J. Chromatogr.*, 187 (1980) 35, with permission [5].)

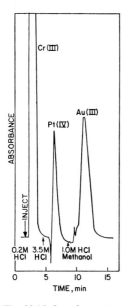

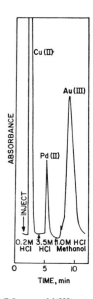

Fig. 22.15. Ion chromatography of 2.0 mg chromium(III), 39 μg platinum(IV), and 7.8 μg gold(III) on XAD-4 (0.21 meq/g) with eluents shown. Detection at 225 nm. (Reproduced from *J. Chromatogr.*, 188 (1980) 391, with permission [6].)

Fig. 22.16. Ion chromatography of 1.3 mg copper(II), 2.1 μg palladium(II), and 9.8 μg gold(III). For conditions, see Fig. 22.15. (Reproduced from *J. Chromatogr.*, 188 (1980) 391, with permission [6].)

Separations of metal ions are also possible on low-capacity resins, but Gjerde and Fritz [6] had to develop all-glass and plastic apparatus and use spectrophotometric detection for this purpose. Some of the separations obtained are shown in Figs. 22.15 and 22.16. These separations took as little as 10–15 min.

22.7. HIGH-PERFORMANCE LIQUID CHROMATOGRAPHY

Commercial instrumentation presently available is rather ill suited for the separation of metal ions. Stainless-steel columns cannot be used for eluents containing chloride ions, especially in acid solutions, and, because the various fitting and detector parts are usually also made of stainless steel or other metals, the choice of eluents is rather drastically limited. Early optimism prompted numerous applications of HPLC to metal chelates, but the methods either proved to be inapplicable to actual analytical problems or failed to improve the separations already in use. Most of these methods are listed in a recent review [7].

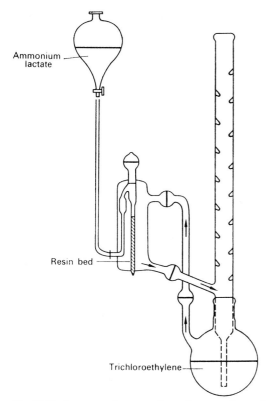

Fig. 22.17. Apparatus for separations of rare-earth metals by ion-exchange chromatography with lactate. (Reproduced from *J. Amer. Chem. Soc.* 76 (1954) 6229, with permission [8].)

References on p. B488

High-speed separations would be very valuable for radioactive isotopes, especially those with short half-lives. A forerunner of HPLC (in general) may be found in the fundamental work on the transuranic elements [8], some ten years ahead of the invention of HPLC (cf. Preface). Fig. 22.17 shows the apparatus used, including a short column, 2 mm in diameter and packed with rather fine resin particles. Further, Fig. 22.18 is evidence that Thompson et al. [8] already considered the plate number at various flowrates. Their results (Fig. 22.19) show that a separation was effected in ca. 200 min, (100 drops at a rate of 2 min/drop), but the separation of Er–Ho was already complete in 30 min, which is clearly in the range of some modern HPLC separations.

In order to effect HPLC separations of radioactive tracers, Horwitz et al. [9] built special apparatus in which only glass or PTFE parts were in contact with the eluent. The stationary phase, 25–30% di-(2-ethylhexyl)orthophosphoric acid in dodecane, was held on a support of porous silica microspheres. The columns, which were only 1 cm high, were operated at a maximum pressure of 500 psi. Fig. 22.20 shows a typical separation of ^{225}Ac and its daughter nuclides, which was effected in only 74 sec. Horwitz et al. also reported isotope enrichment of ^{48}Ca/^{40}Ca by use of a column 75 cm in length (ca. 30,000 theoretical plates) at 9°C, which indicates a considerable improvement over the previous attempts with "slow" chromatography.

While there are difficulties in the use of commercial HPLC equipment for the separation of metal ions, this is not so for anions. Reeve [10] separated anions on a

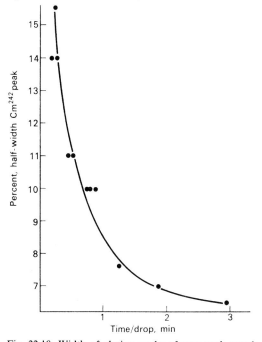

Fig. 22.18. Width of elution peaks of rare-earth metals vs. drop rate of ammonium lactate eluent in ion-exchange chromatography. (Reproduced from *J. Amer. Chem. Soc.*, 76 (1954) 6229, with permission [8].)

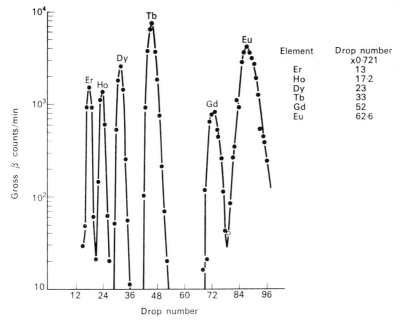

Fig. 22.19. Elution of homologous lanthanides with ammonium lactate eluent (ca. 2 min/drop = 35 μl); column, Dowex 50. (Reproduced from *J. Amer. Chem. Soc.*, 76 (1954) 6229, with permission [8].)

column of cyano-bonded silica, Sil 60-D 10-CN (250 × 4.6 mm); eluted with 0.1 *M* Na_2HPO_4, 0.1 *M* KH_2PO_4, and 0.1% cetrimide, mixed with methanol (3:2); and detected the anions at 210–220 nm. Typical separations are shown in Figs. 22.21 and 22.22 (cf. also Chap. 22.13).

22.8. GEL CHROMATOGRAPHY

Elementary chemistry textbooks still like to present the formation of an insoluble hydroxide, such as $Fe(OH)_3$, as a reaction obeying the law of mass action (solubility product), although it was realized as long as ca. 50 years ago during the work on "radiocolloids" that this picture is incorrect. A whole range of unstable and stable, soluble and colloidal polymers can be formed, depending on the concentration of the metal ion, the rate of addition of the OH^- ion, temperature, etc. Very few of these hydrolysis polymers have been well characterized; e.g., zirconium(IV) occurs in dilute HCl as a polymer of ionic weight of about 20,000. This was, perhaps, the first polymeric substance separated by "gel filtration", i.e. it was eluted from sulfonic ion-exchange resins [11], although, owing to its high charge, it should be strongly held if it were a monomer.

At first sight, gel chromatography would be the ideal method for the isolation and characterization of such hydrolysis polymers. However, most of the present gel filtration media are unsuitable for such work. Sephadex gels have on their surface vicinal hydroxyl groups which complex readily with most metal ions at neutral pH (a

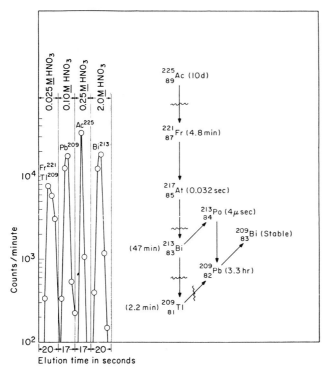

Fig. 22.20. High-speed separation of ^{225}Ac and daughter nuclides. Column, Zorbax-SIL (5 μm), 10×2 mm ID with 30% (w/w) of di-(2-ethylhexyl)orthophosphoric acid in dodecane as stationary phase; eluent, HNO$_3$; flowrate, 17 cm/min; temperature, 50°C. (Reproduced from *J. Chromatogr.*, 125 (1976) 203, with permission [9].)

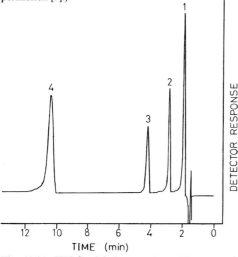

Fig. 22.21. HPLC of halogen anions. Eluent, methanol–water (2:3), being 0.1 M Na$_2$HPO$_4$, 0.1 M KH$_2$PO$_4$ and 0.1% (w/v) cetrimide with respect to the aqueous component; flowrate, 2 ml/min. For other conditions, see text. Peaks: 1 = iodate; 2 = bromate; 3 = bromide; 4 = iodide. (Reproduced from *J. Chromatogr.*, 177 (1979) 393, with permission [10].)

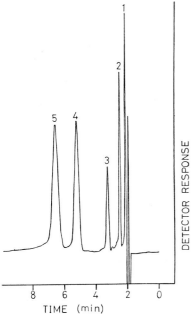

Fig. 22.22. HPLC of iodide and polythionates. Eluent, methanol–water (11:9). For other conditions, see Fig. 22.21. Peaks: 1 = thiosulfate; 2 = iodide; 3 = trithionate; 4 = tetrathionate; 5 = pentathionate. (Reproduced from *J. Chromatogr.*, 177 (1979) 393, with permission [10].)

typical compound of this type is the "glycerate of iron" of the British Pharmacopoeia). Hence, these ions leave a trail when applied in high concentrations, and they are adsorbed altogether in low concentrations. Porous glasses and porous silica (and their derivatives) usually have on their surfaces a high concentration of silanol groups which also interact strongly with monomeric and polymeric metal ions.

In spite of these shortcomings, quite a few polymeric hydrolytic species could be separated on gel-filtration media, e.g.: polymeric ruthenium [12], polymeric rhodium(III) [13], and soluble ferrocyanides [14–16]. The movement of small monomeric ions inside gel-filtration media is largely governed by ion exchange with residual carboxyl groups, ion exclusion due to these groups, or hydrophobic adsorption, which is rather strong on Sephadex LH-20, where also salting-out effects can be observed [17].

Gel chromatography has been used to study the interaction between metal ions and large molecules, such as proteins, dextrans, and polyphosphates. Most of the early work discussed in an excellent review by Yoza [18]. Another fascinating area that can be studied by gel chromatography deals with the interactions between inorganic polymers and small ions. For instance, the zirconium hydrolysis polymer is excluded from Sephadex, while small ions, such as chromate, are not. When a mixture of polymeric zirconium and chromate is chromatographed on Sephadex G-10, there is a slow-moving yellow band as well as a yellow excluded band, showing clearly that the polymer binds some of the chromate [19].

References on p. B488

22.9. GAS CHROMATOGRAPHY

The early literature on GC of inorganic compounds was reviewed by Tadmor [20], and there are also monographs by Guiochon and Pommier [21] and by Moshier and Sievers [22] and a review by Uden and Henderson [23]. We shall mention here only some of the main topics of research.

GC separation of metals presupposes that a volatile derivative can be prepared having properties suitable for the available apparatus. Trifluoroacetylacetone complexes and similar derivatives fulfill these requirements, and some useful methods for some metal ions, notably for beryllium and chromium, have been reported. They seem to be competitive with other methods, such as atomic absorption spectrometry or radioactivation analysis [24–27].

Metal organic compounds often have suitable properties for GC. For instance, inorganic mercury can be derivatized and detected at the 0.0025-ppm level [28]. Also, the organic compounds of astatine were characterized by GC (see Fig. 22.23) [29].

Volatile metal chlorides have been either quantitatively determined by GC (e.g., germanium or arsenic [30]) or GC was used for studying their thermodynamic properties [31]. Special apparatus has to be used for the rather corrosive volatile halides.

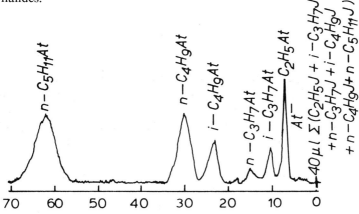

Fig. 22.23. GLC separation of alkyl astatides, produced as a result of exchanging astatine, adsorbed at the column inlet in the form of astatines, with iodine in alkyl iodides. Column, 2 m × 4 mm ID, 10% dinonyl phthalate on Chromosorb G; carrier gas, helium; flowrate, 30 ml/min; column temperature, 95°C. (Reproduced from *J. Chromatogr.*, 60 (1971) 414, with permission [29].)

22.10. RESOLUTION OF OPTICAL ISOMERS

Enantiomeric pairs of chelate complexes of the type $[Co(en)_3]^{3+}$ or $[Co(ox)_3]^{3-}$ have been separated by using both optically active adsorbents and optically active eluents. For example, the enantiomers of $[Co(en)_3]^{3+}$ were separated on CM-Sephadex with disodium D-tartrate as eluent [32]. In this kind of separation, ap-

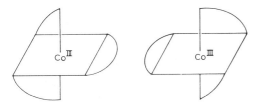

parently ion-pair formation between an optically active anion (such as tartrate) and the two forms of the chelate complex is responsible for the separation [33].

Similar results were also obtained by paper electrophoresis with optically active electrolytes (usually tartrate or antimonyl tartrate for complex cations; and strychnine, cinchonine, etc. for complex anions). It is usually considered necessary for the separation of optical isomers that a "three-point" adsorption or complexation is operating. The three points are usually hydrogen bonding and/or ionic attraction. However, there is evidence from paper electrophoresis experiments that separations are possible even when the conditions for a "three-point" interaction are nonexistent or unlikely, as is the case in the resolution of the optical isomers of $[Fe(II)(dip)_3]^{2+}$ in tartrate [34–36]. Neutral tris(aminoacidato)cobalt(III) complexes were also separated on ion exchangers (e.g., the strong cation exchanger TSK 211), eluted with tartrate or antimonyl tartrate. Fig. 22.24 shows that the optical activity on one ligand (D- or L-serine) yields a separation with antimonyl tartrate (or with tartrate), but not with sulfate as eluent [37].

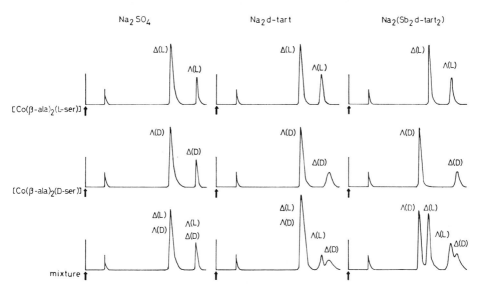

Fig. 22.24. Elution curves of fac-[Co(L-ser)₂(β-ala)] and fac-[Co(D-ser)₂(β-ala)] by ion-exchange chromatography. Column, 250×4 mm ID, stainless steel, packed with TSK 211 strongly acidic cation exchanger; eluent, $0.05\ M$ aqueous Na_2SO_4, $0.1\ M$ aqueous Na_2D-tart or $0.1\ M\ Na_2[Sb_2(D\text{-}tart)_2]$. Δ(L), Λ(L), Δ(D) and Λ(D) refer to the complexes with Δ or Λ configuration containing L- or D-aminoacidato anions. (Reproduced from *J. Chromatogr.*, 175 (1979) 317, with permission [37].)

References on p. B488

22.11. SEPARATION OF COORDINATION COMPLEXES

Unlike the reversible complexes of the type $FeCl^{2+}$, many metal ions, especially Co^{3+}, Rh^{3+}, Ru^{3+}, and Cr^{3+}, can form extremely strong complexes with ammonia, amines, substitued phosphines, and substituted arsines and stibines. For example, one of the simplest Co(III) complexes, namely $Co(NH_3)_6^{3+}$, can be heated to several hundred degrees or treated with KOH without undergoing decomposition. It forms a hydroxide which is a strong base and, in solution, absorbs CO_2 from the atmosphere.

Such complexes can be subjected to chromatography or electrophoresis in a wide range of acids or bases without any risk of decomposition. Most methods for the preparation of such stable complexes involve, e.g., the oxidation of Co(II) (by air or H_2O_2) in the presence of the ligands, e.g., NH_4OH or ethylenediamine. Usually, an array of compounds forms, from which the desired compound is isolated by crystallization (by evaporation or by addition of ethanol or acetone). One of the uses of chromatographic techniques is to show that in numerous syntheses the isolated complex is actually a mixture of as many as six compounds (e.g., $[Co(NH_3)_3(NO_2)_3]^0$ or $Co(tn)_3Cl_3$). A pure complex can then be isolated by preparative chromatography for the first time. This field has been reviewed [38], and we can therefore restrict ourselves to mentioning only some of the interesting features.

Paper chromatography has maintained its popularity as a convenient small-scale preparative technique [39,40]. Quantities up to 30 mg can be applied as a line to a sheet of Whatman No. 3MM paper, which is developed overnight. The zones, which are usually colored, can be excised and eluted chromatographically. The amount so

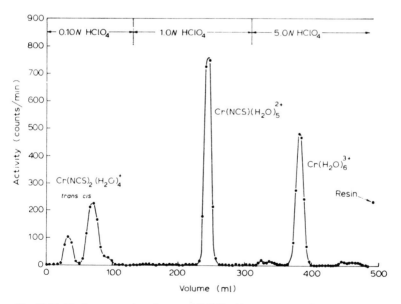

Fig. 22.25. Elution curve of a mixture of Cr(III)–thiocyanate complexes from Dowex-50 cation-exchange resin. (From *J. Amer. Chem. Soc.*, 82 (1960) 2963, as reproduced in ref. 38.)

obtained is usually sufficient for further reactions, analyses, and spectroscopic studies.

Ion-exchange column chromatography is employed in practically all synthetic work, usually for purification or for the separation of several complexes from a reaction mixture. The amounts obtained are again of the order of 10–100 mg. A typical example is shown in Fig. 22.25.

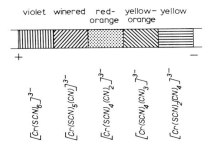

Fig. 22.26. Moving-boundary ionophoresis of mixed ligand complexes $[Cr(SCN)_x(CN)_{6-x}]^{3-}$, where $x = 2...6$, in acetonitrile. (Reproduced from *J. Chromatogr.*, 50 (1970) 319, with permission [43].)

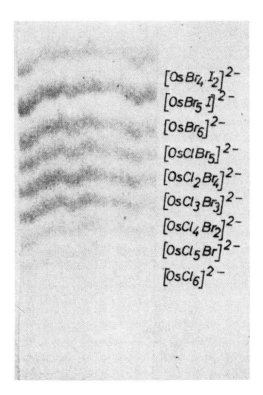

Fig. 22.27. Paper electrophoresis of the chlorobromo and bromoiodo complexes of osmium(IV). (Reproduced from *Chromatogr. Rev.*, 6 (1964) 191, with permission [3].)

References on p. B488

Alumina was advocated for the separation of coordination complexes by Jensen et al. [41] mainly because colored complexes could be easily distinguished on the white alumina but not on the brown ion-exchange resin. In these separations, which were carried out with aqueous eluents, alumina functions essentially as an ion exchanger (surface aluminate groups) with possibly an additional outer-sphere complexing effect.

HPLC has not been applied extensively so far. Its possibilities are indicated by preliminary work [42], which demonstrated one of the attractive features of the technique, i.e. that spectra can be obtained on the eluate by stop-flow methods.

Electrophoretic methods, TLC, and HPTLC have been applied widely, mainly in the study of reactions of coordination complexes. A beautiful displacement electrophoretic separation of a series of Cr(III) thiocyanato–cyano complexes is shown in Fig. 22.26 [43]. A paper-electrophoretic separation of chlorobromo and bromoiodo complexes of osmium(IV) is shown in Fig. 22.27 [3].

22.12. SCANDIUM, YTTRIUM, THE RARE EARTHS, AND ACTINIUM

Together with Sc, Y, and Ac, the rare earths constitute 18% of the elements of the periodic table. There was no effective method for separating these elements before the advent of chromatography. The first successful analytical separations date from the Second World War [44,45] and are among the best examples of how separations

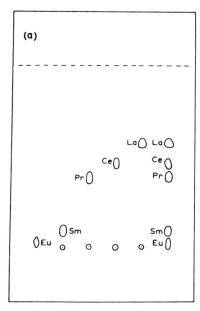
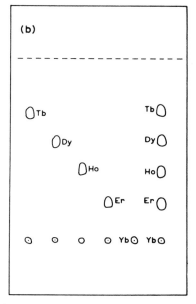

Fig. 22.28. TLC separation of (a) cerium earths and (b) ytterbium earths. Eluent: (a) 0.5 N HCl; (b) 4 N HNO₃. Layers (500 μm) were prepared from a mixture of 30 g silica gel, 12 ml di-(2-ethylhexyl)phosphoric acid and 34 ml 1-butanol. (Reproduced from *J. Chromatogr.*, 24 (1966) 153, with permission [48].)

can be achieved by using a suitable complexing agent on an ion exchanger. Since then, separations have been devised by numerous column systems, including HPLC [46,47], by TLC (Fig. 22.28) [48,49], paper electrophoresis (Fig. 22.29) [50], and isotachophoresis (Fig. 22.30) [51].

As chromatography appeared to be the best method yet for preparing gram quantities of the rare-earth elements, several laboratories have devised large-scale

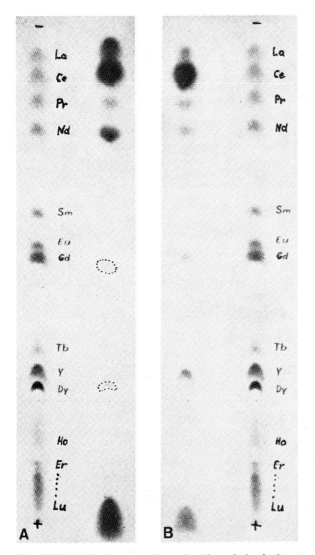

Fig. 22.29. Qualitative paper-electrophoretic analysis of mixtures of rare-earth metal ions, carried out in the ligand buffer system $Zn^{2+}-ZnL^{2-}$ (pH 2). (A) Flint stones; (B) rare-earth alloy. Paper, Whatman No. 2; potential drop, 28–30 V/cm; concn. of free ligand, $10^{-17.5}$ M. (Reproduced from *J. Chromatogr.*, 74 (1972) 325, with permission [50].)

References on p. B488

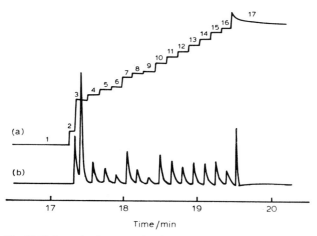

Fig. 22.30. Isotachopherogram for the simultaneous separation of lanthanides. (a) Potential gradient; (b) differential gradient. Leading electrolyte: 0.027 M KOH, 0.015 M 2-hydroxyisobutyric acid, acetic acid, and 0.0025% poly(vinyl alcohol), pH 4.92. Migration current, 225 μA; chart speed, 40 mm/min; sample, 5.0 μl of a $10^{-3}M$ solution of mixed lanthanides. $1 = K^+$, $2 = Na^+$, $3 = La^{3+}$, $4 = Ce^{3+}$, $5 = Pr^{3+}$, $6 = Nd^{3+}$, $7 = Sm^{3+}$, $8 = Eu^{3+}$, $9 = Gd^{3+}$, $10 = Tb^{3+}$, $11 = Dy^{3+}$, $12 = Ho^{3+}$, $13 = Er^{3+}$, $14 = Tm^{3+}$, $15 = Yb^{3+}$, $16 = Lu^{3+}$, $17 = \beta$-Ala. (Reproduced from *J. Chromatogr.*, 205 (1981) 95, with permission [51].)

preparative separations (for a typical example, see Fig. 22.31). Spedding's group [52–54] laid the foundations for this technology with their demonstration that displacement development is the most efficient preparative method. For most of these methods organic complexing agents were used; first citric acid and later α-hydroxybutyric acid as well as lactic acid and EDTA, which were found to yield better separation factors (i.e., larger differences in stability constants). Nitric acid can also be used for the separation of the lighter rare-earth elements [55].

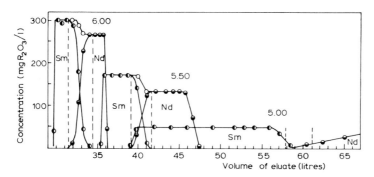

Fig. 22.31. Preparative ion-exchange chromatography of rare-earth elements. Elution of 1.713 g equimolar mixtures of Sm_2O_3 and Nd_2O_3 from a 120×2.2-cm column of Amberlite IR-100 (30–40 mesh), with 0.1% citrate at pH 5.0, 5.5, and 6.0; at a flowrate of 0.5 cm/min. ○, Total R_2O_3; ◐, Sm_2O_3; ●, Nd_2O_3; broken vertical lines indicate amount of overlap between Sm and Nd bands. (From *Discuss. Faraday Soc.*, 7 (1949) 214, as reproduced in ref. 73.)

22.13. CONDENSED PHOSPHATES

The chemistry of condensed phosphates could not be investigated before the advent of chromatographic methods. There are several good reviews and even a book on the main developments in this field [56–58]. Chromatography is now the classical (if not the only) method for characterizing new phosphates or mixed condensed arseno-phosphates.

After it was realized that the Ca^{2+} and Mg^{2+} ions present in Whatman papers (due to their manufacture in hard water) must be removed by suitable acid washes, PC was used in all pioneering work. Subsequently, excellent TLC, ion-exchange, gel filtration, and HPLC methods have been described. In retrospect, these methods certainly improved quantitative analysis and speeded up the separations, but they did not materially improve the rather adequate separations obtained by PC. Figs. 22.32–22.34 show some typical examples.

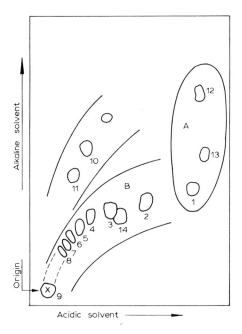

Fig. 22.32. Separation of a mixture of phosphorus oxoacids by two-dimensional PC. Alkaline solvent, 2-propanol–2-butanol–water–20% ammonia (40:20:39:1); acidic solvent, 2-propanol–water–20% tri-chloroacetic acid–25% ammonia (700:100:200:3). Compounds: 1 = orthophosphate; 2 = pyrophosphate; 3 = triphosphate; 4 = tetraphosphate; 5 = pentaphosphate; 6 = hexaphosphate; 7 = heptaphosphate; 8 = octaphosphate; 9 = Graham salt; 10 = trimetaphosphate; 11 = tetrametaphosphate; 12 = hypophosphite; 13 = phosphite; 14 = hypophosphate. (Reproduced from *Metodi di Separazione nella Chimica Inorganica*, Vol. 1, 1963, p. 199, with permission [58].)

References on p. B488

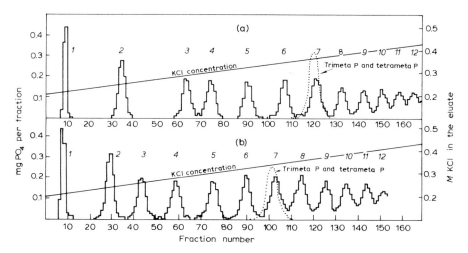

Fig. 22.33. Ion-exchange chromatography of a polyphosphate mixture. Ion exchanger, Dowex 1-X4; gradient elution. (a) Borate buffer, pH 8.0; (b) ammoniacal buffer, pH, 9.3. (From *J. Biochem. (Tokyo)*, 44 (1957) 65, as reproduced in ref. 58.)

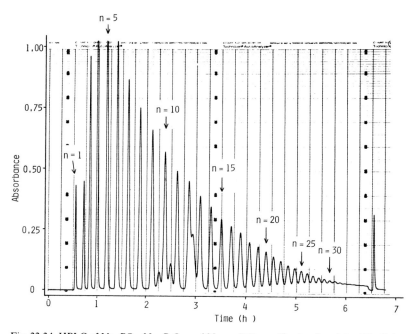

Fig. 22.34. HPLC of Na_3PO_4, $Na_4P_2O_7$ and $Na_{n+2}P_nO_{3n+1}$ (for $\bar{n} = 5$ and $\bar{n} = 10$). Column, 100×9 mm ID, Hitachi 2630 anion-exchange resin (4% crosslinked). Eluent, NaCl–5 mM Na_4EDTA, pH 10.0; gradient, convex from 0.22 M NaCl ($t = 0$) to 0.53 M NaCl ($t = 6$ h). (Reproduced from *J. Chromatogr.*, 172 (1979) 131, with permission [74].)

22.14. SILICATES

The early literature on PC contains claims of the separation of polysilicates, which, however, could not be verified. A gel-chromatographic separation of monosilicic acid from polysilicic acid on a Sephadex G-25 column was reported by Tarutani [59]. Intermediate species seem to be too unstable to be isolated by this method. TMS derivatives of condensed silicates were separated by GC [60,61], but the method of preparation of the derivatives seems to have had a large influence on the species isolated. Decomposition during derivatization seemed excessive. More recently, Shimono et al. [62] have reported both GC and gel-filtration separations of TMS derivatives of polysilicates. Typical separations are shown in Figs. 22.35 and 22.36. Even colloidal silica can be studied by gel filtration [63].

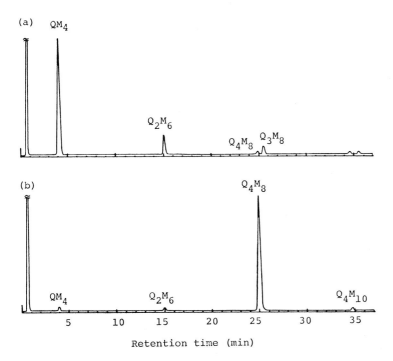

Fig. 22.35. Gas chromatograms of (a) trimethylsilyl olivine and (b) trimethylsilyl laumontite. Column, 3.0 m × 3.0 mm ID, packed with 2% OV-1 on Chromosorb W AW DMCS (60–80 mesh); carrier gas, helium; temperature, programed from 100 to 300°C at 5°C/min; detector, flame ionization. Q is one tetrahedron corresponding to one SiO_2 unit in a condensed silicate system; M is the $(CH_3)_3SiO_{1/2}$ atomic group resulting from silylation. (Reproduced from *J. Chromatogr.*, 197 (1980) 59, with permission [62].)

References on p. B488

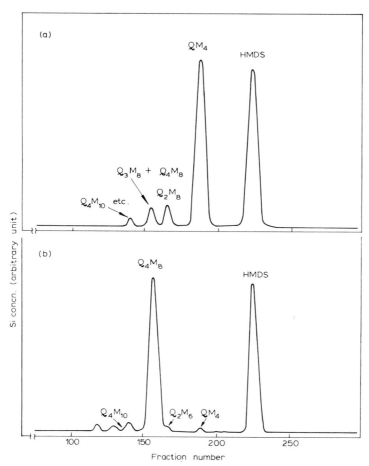

Fig. 22.36. Gel chromatogram of (a) trimethylsilyl olivine and (b) trimethylsilyl laumontite. Two 90×1.5-cm ID columns, total volume ca. 318 ml. Gel, Bio-Beads S-X1; eluent, 2-propanol–chloroform (2:3). Q is one tetrahedron corresponding to one SiO_2 unit in a condensed silicate system; M is the $(CH_3)_3SiO_{1/2}$ atomic group resulting from silylation. (Reproduced from *J. Chromatogr.*, 197 (1980) 59, with permission [62].)

REFERENCES

1 M. Lederer and C. Majani, *Chromatogr. Rev.*, 12 (1970) 239.
2 U.A.Th. Brinkman, G. de Vries and R. Kuroda, *J. Chromatogr.*, 85 (1973) 187.
3 E. Blasius and W. Preetz, *Chromatogr. Rev.*, 6 (1964) 191.
4 H. Small, T.S. Stevens and W.C. Bauman, *Anal. Chem.*, 47 (1975) 1801.
5 D.T. Gjerde, G. Schmuckler and J.S. Fritz, *J. Chromatogr.*, 187 (1980) 35.
6 D.T. Gjerde and J.S. Fritz, *J. Chromatogr.*, 188 (1980) 391.
7 G. Schwedt, *Chromatographia*, 12 (1979) 613.
8 S.G. Thompson, B.G. Harvey, G.R. Choppin and G.T. Seaborg, *J. Amer. Chem. Soc.*, 76 (1954) 6229.
9 E.P. Horwitz, W.H. Delphin, C.A.A. Bloomquist and G.F. Vandegrift, *J. Chromatogr.*, 125 (1976) 203.
10 R.N. Reeve, *J. Chromatogr.*, 177 (1979) 393.

11 J.A. Ayres, *J. Amer. Chem. Soc.*, 69 (1947) 2879.
12 I. Kitayevitch, M. Rona and G. Schmuckler, *Anal. Chim. Acta*, 61 (1972) 277.
13 M. Sinibaldi and A. Braconi, *J. Chromatogr.*, 94 (1974) 338.
14 H. Saito and Y. Matsumoto, *J. Chromatogr.*, 168 (1979) 227.
15 Y. Matsumoto, M. Shirai and H. Saito, *Bull. Chem. Soc. Jap.*, 41 (1975) 210.
16 D. Corradini and M. Sinibaldi, *J. Chromatogr.*, 187 (1980) 458.
17 V. Di Gregorio and M. Sinibaldi, *J. Chromatogr.*, 129 (1976) 407.
18 N. Yoza, *J. Chromatogr.*, 86 (1973) 325.
19 M. Sinibaldi, G. Matricini and M. Lederer, *J. Chromatogr.*, 129 (1976) 412.
20 J. Tadmor, *Chromatogr. Rev.*, 5 (1963) 223.
21 G. Guiochon and C. Pommier, *Gas Chromatography in Inorganics and Organometallics*, Ann Arbor Sci. Publ., Ann. Arbor, MI, 1973.
22 R.W. Moshier and R.E. Sievers, *Gas Chromatography of Metal Chelates*, Pergamon, New York, 1965.
23 P.C. Uden and D.E. Henderson, *Analyst (London)*, 102 (1977) 889.
24 L.C. Hansen and W.G. Scribner, *Anal. Chem.*, 43 (1971) 349.
25 W.R. Wolf, M.L. Taylor, B.M. Hughes, T.O. Tiernan and R.E. Sievers, *Anal. Chem.*, 44 (1972) 616.
26 M.S. Black and R.E. Sievers, *Anal. Chem.*, 48 (1976) 1872.
27 J.G. Lo and S.J. Yeh, *J. Chromatogr. Sci.*, 18 (1980) 359.
28 P. Jones and G. Nickless, *J. Chromatogr.*, 89 (1974) 201.
29 M. Gesheva, A. Kolachkovsky and Yu. Norseyev, *J. Chromatogr.*, 60 (1971) 414.
30 B. Iatridis and G. Parissakis, *J. Chromatogr.*, 122 (1976) 505.
31 J. Rudolph and K. Bächmann, *J. Chromatogr.*, 187 (1980) 319.
32 Y. Yoshikawa and K. Yamasaki, *Inorg. Nucl. Chem. Lett.*, 6 (1976) 523.
33 H. Nakazawa and H. Yoneda, *J. Chromatogr.*, 160 (1978) 89.
34 L. Ossicini and C. Celli, *J. Chromatogr.*, 115 (1975) 655.
35 V. Cardaci, L. Ossicini and T. Prosperi, *Ann. Chim. (Rome)*, 68 (1978) 713.
36 V. Cardaci and L. Ossicini, *J. Chromatogr.*, 198 (1980) 76.
37 S. Yamazaki, T. Yukimoto and H. Yoneda, *J. Chromatogr.*, 175 (1979) 317.
38 V. Carunchio and G. Grassini Strazza, *Chromatogr. Rev.*, 8 (1966) 260.
39 O. Bang, A. Engberg, K. Rasmussen and F. Woldbye, *Acta Chem. Scand.*, 29 (1975) 749.
40 F.P. Dwyer, T.E. MacDermott and A.M. Sargeson, *J. Amer. Chem. Soc.*, 85 (1963) 2913.
41 A. Jensen, J. Bjerrum and F. Woldbye, *Acta Chem. Scand.*, 12 (1958) 1202.
42 G. Grassini Strazza and C.M. Polcaro, *J. Chromatogr.*, 147 (1978) 516.
43 E. Blasius, H. Augustin and U. Wenzel, *J. Chromatogr.*, 50 (1970) 319.
44 B.H. Ketelle and G.E. Boyd, *J. Amer. Chem. Soc.*, 69 (1947) 2800.
45 E.R. Tompkins, J.X. Khym and W.E. Cohn, *J. Amer. Chem. Soc.*, 69 (1947) 2769, and the other papers in this issue.
46 F. Schoebrechts, E. Merciny and G. Duyckaerts, *J. Chromatogr.*, 174 (1979) 351.
47 F. Schoebrechts, E. Merciny and G. Duyckaerts, *J. Chromatogr.*, 179 (1979) 63.
48 H. Holzapfel, Le Viet Lan and G. Werner, *J. Chromatogr.*, 24 (1966) 153.
49 H. Holzapfel, Le Viet Lan and G. Werner, *J. Chromatogr.*, 20 (1965) 580.
50 V. Jokl and Z. Pikulíková, *J. Chromatogr.*, 74 (1972) 325.
51 I. Nukatsuka, M. Taga and H. Yoshida, *J. Chromatogr.*, 205 (1981) 95.
52 J.E. Powell and F.H. Spedding, *Chem. Eng. Progr., Symp. Ser.*, 55 (1959) 101.
53 D.B. James, J.E. Powell and H.R. Burkholder, *J. Chromatogr.*, 35 (1968) 423.
54 F.H. Spedding, *Discuss. Faraday Soc.*, 7 (1949) 214.
55 K. Ishida, *Bunseki Kagaku (Jap. Anal.)*, 19 (1970) 1250.
56 H. Hettler, *Chromatogr. Rev.*, 1 (1959) 225.
57 H. Grunze and E. Thilo, *Die Papierchromatographie der kondensierten Phosphate*, Akademie Verlag, Berlin, 1954/1955.
58 J.P. Ebel, in *Metodi di Separazione nella Chimica Inorganica*, Vol. 1, CNR, Rome, 1963, p. 199.
59 T. Tarutani, *J. Chromatogr.*, 50 (1970) 523.
60 G. Garzó, D. Hoebbel, Z.J. Ecsery and K. Ujszászi, *J. Chromatogr.*, 167 (1978) 321.
61 G. Garzó and D. Hoebbel, *J. Chromatogr.*, 119 (1976) 173.

62 T. Shimono, H. Takagi, T. Isobe and T. Tarutani, *J. Chromatogr.*, 197 (1980) 59.
63 J.J. Kirkland, *J. Chromatogr.*, 185 (1979) 273.
64 E. Blasius, in *Metodi di Separazione nella Chimica Inorganica*, Vol. 1, CNR, Rome, 1963, p. 141.
65 K.A. Kraus and F. Nelson, *J. Chromatogr.*, 1 (1958) X.
66 R.A. Guedes de Carvalho, *J. Chromatogr.*, 1 (1958) XII.
67 E. Cerrai and G. Ghersini, *J. Chromatogr.*, 24 (1966) 383.
68 S. Przeslakowski, *Chromatogr. Rev.*, 12 (1970) 383.
69 E. Lederer and M. Lederer, *Chromatography — A Review of Principles and Applications*, Elsevier, Amsterdam, 2nd Edn., 1957, p. 468.
70 K. Kawazu, *J. Chromatogr.*, 137 (1977) 381.
71 M. Lederer, in *Metodi di Separazione nella Chimica Inorganica*, Vol. 1, CNR, Rome, 1963, p. 109.
72 M. Lederer and M. Mazzei, *J. Chromatogr.*, 35 (1968) 201.
73 E. Lederer and M. Lederer, *Chromatography — A Review of Principles and Applications*, Elsevier, Amsterdam, 2nd Edn., 1957, p. 448.
74 H. Yamaguchi, T. Nakamura, Y. Hirai and S. Ohashi, *J. Chromatogr.*, 172 (1979) 131.

Chapter 23

Nonhydrocarbon gases

JAROSLAV JANÁK

CONTENTS

23.1. INTRODUCTION

Modern gas analysis has developed from the classical to instrumental methods, characterized by measuring physical values, and from macro- to microtechniques, characterized by improved apparatus for measurement. Although this development can be traced from the beginning of this century, the greatest advances began in the Fifties when chromatographic principles and instrumentation were accepted as a new basis of gas analysis. This was characterized by a shift from the separation of simple to complex mixtures and to trace analysis. Interest in analyzing permanent and hydrocarbon gases by separation methods was, of course, supported by practical industrial needs. GSC in the elution mode is the most convenient method for this purpose, but frontal and sometimes displacement techniques as well as GLC have found applications in the solution of some problems (e.g., enrichment procedures, determination of the composition of gases containing higher-boiling vapors or vice versa).

Not only have changes in methodology and instrumentation taken place in industrial laboratories, but new fields of research and application have also opened up in the last decade, such as trace analysis in environmental sciences and pollution control and in clinical practice as a diagnostic means. The object of contemporary

GC is no longer the separation and analysis of gases as such, but important possibilities have arisen from the analysis of by-products, mixed with or transported by gases. Fig. 23.1 gives the boiling points of important gases and provides the information about vapor pressures, which is, of course, relevant to the chromatography of gases. Interest has now shifted mainly to the analysis of trace components in pure gases, the composition of which reflects industrial processes, natural sources, the role in nature, etc. This trend is reflected by the literature, where many thousands of chromatographic analyses of gases are now reported. Books [1–9] and reviews [10,11] will serve for the critical selection of techniques used in this field, and special applications may be found in the original literature (see the bibliographic section of the *Journal of Chromatography*). The aim of this chapter is to collect recent material that illustrates contemporary research and industrial activity on the

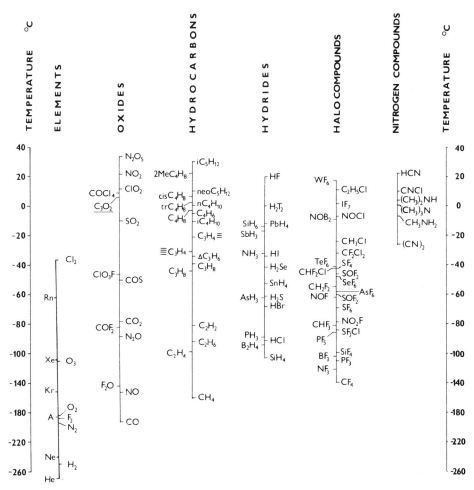

Fig. 23.1. Boiling points of the main representatives of various classes of gases.

topic, but the list of references published in the previous edition of this book [5] has lost little of its value and can still serve as a source of information supplementing this new collection, since earlier references are only repeated in exceptional circumstances.

In any GC separation, we must consider whether a sufficiently large difference in vapor pressures and separation factors is followed by an acceptable value of the column capacity ratio, related to absolute retention volumes of about 10–100 ml/g at a relevant temperature. By changing these parameters, we can shift the capacity ratio of any gas or vapor in either direction to fall within the favorable range of retention volumes. However, there are some specific features of gas analysis by chromatography. The handling of gases requires special techniques, due to their temperature-dependent parameters, such as volume or pressure. The difference between the molecular weights of the carrier gas and the gas to be analyzed is relatively small, which makes it necessary to calibrate detectors for all the components. With some exceptions, such as molecular sieves, solid sorbents rarely have a homogeneous and uniform porous structure, and this causes a considerable nonlinearity of the sorption isotherm at the concentrations relevant to GC separations. The nature of the carrier gas exerts an influence on retention and peak shape, due to a considerable modification of the solid surface, physically as well as chemically, by self-adsorption.

23.2. SEPARATION OF GASES AND ITS APPLICATIONS

23.2.1. General considerations

The separation of permanent gases can be performed on any material having sufficient sorption capacity. This capacity is given by a sufficiently large specific surface of the solid (having two-dimensional character only) or by the solvating ability of the liquid (having a two-dimensional surface and a three-dimensional bulk structure). This condition is fulfilled very well by solids with microporous structure, serving as adsorbents or as supports for the liquid phase distributed in a thin film on their surface. Obviously, molecules of the carrier gas will occupy the surface of the solid to a certain degree as well as saturate the liquid phase. This has two main consequences:

(a) In the case of adsorbents, the physical nature of the solid is changed, the initial surface area is dimished, and the average pore diameter available for the separation tends toward a higher value owing to the blocking of the most narrow pores. As a result, a decrease in retention times and an improvement of the linearity of the sorption isotherm is expected, but a simultaneous competition among the gases for adsorption on the surface exerts certain displacement effects.

(b) The chemical nature of the solid surface is also changed, and this can cause considerable changes in polarity and may influence the retention sequence of some gases. In the case of a liquid, the solution represented by the liquid saturated with the carrier gas will sometimes have a chemical potential that is quite different from that of the pure liquid, and this will influence to some degree the adsorption on the surface, the mass transfer phenomena at the boundary between the gaseous and the liquid phases, and the solubility of the gas sample in the liquid phase.

We can illustrate these phenomena by practical examples. When carbon dioxide (ammonia, water, steam, etc.) and active charcoal are used as carrier gas and adsorbent, respectively, less adsorbed gases will have symmetrical peaks (e.g., Ar) or peaks with a leading front (e.g., H_2), whereas gases sorbed more strongly than CO_2 (e.g., propane) will tail considerably. The same effect can be observed with the same adsorbent when H_2 is used as a carrier gas at subambient temperature ($-196°C$). Under these conditions, Ne, which is sorbed less strongly, forms a peak with a leading front and Ar, which is sorbed more strongly, tails. When silica gel, alumina, aluminosilicates, etc. are used as adsorbents, CO_2 is sorbed more strongly than propane, and only hydrocarbons higher than butanes will tail and retention times of all peaks will generally be shorter. On the other hand, the most narrow pores present in active charcoal with a large specific surface are blocked by CO_2 so that the $Ar-O_2-N_2$ group may not separate, but with H_2 as carrier gas the separation of $Ar + O_2$ from N_2 is feasible, especially at subambient temperature. In the case of molecular sieves having crystal lattices with geometrically uniform pores of molecular dimensions, the separation of the $O_2 + N_2$ pair is very easy when H_2, He, or Ne is used as a carrier gas, and the separation of the whole group of gases is feasible. When CO_2 or N_2O is used as a carrier gas, its adsorbed molecules fill the inner pore spaces, and none of the permanent gases are entirely separated from one another.

The chemical nature of the sorbents influences the separation selectivity. Carbonaceous sorbents (active charcoal, powdered carbon deposited on supports) show a weak selectivity toward carbon-containing gases, such as CO, CH_4, CO_2, COS, and CS_2. Solids with hydroxy groups and/or oxygen bridges, such as silica gel, alumina, and molecular sieves, exhibit a strong selectivity toward substances that

TABLE 23.1

INFLUENCE OF SORBENT POLARITY AND TEMPERATURE ON RETENTION RATIOS OF SOME PERMANENT GASES

Data calculated from: J. Janák, M. Krejčí and H.E. Dubský; *Collect. Czech. Chem. Commun.*, 24 (1959) 1080 [12].

Gas	Molecular sieve 5A wetted with water, in percent (w/w) (18°C)			Temperature, °C water content, 3.5% (w/w)			
	2.07	4.18	9.43	20	50	100	135
Oxygen	0.19	0.18	0.15	0.20	0.24	0.30	0.33
Nitrogen	0.45	0.36	0.28	0.45	0.45	0.47	0.45
Methane	1.00	1.00	1.00	1.00	1.00	1.00	1.00
Carbon monoxide	2.32	1.18	0.68	2.16	1.40	1.00	0.75

can participate in donor–acceptor interactions, e.g., unsaturated hydrocarbons and polar or polarizable molecules. An example of the relation between the surface polarity and the retention sequence of some gases is given in Table 23.1 [12]. When Molecular Sieve 5A is wetted with water vapor, not only is its specific surface area diminished (making the retention times of all gases shorter) but also the retention sequence of the $CH_4 + CO$ pair is affected by the change of polarity. A similar effect is caused by an increase in temperature. Some specific selectivities can be achieved by incorporating metal ions, such as Ag, Zn, Cd, Cu, and Ni, into molecular sieves of aluminosilicate character, sometimes causing stronger chemisorption effects. Porous solids having an inorganic surface covered by chemically bonded organic functional groups can be used for the separation of gases. Currently, synthetic, highly porous organic polymers are employed for the separation of gases with great success. They combine the best properties of solids (regular porosity) with those of liquids (linear partition), but liquids coated on supports, i.e. GLC, can also be used in the separation of specific gas mixtures.

23.2.2. Permanent gases

The best separation of H_2, O_2, N_2, CH_4, and CO at room temperature has been achieved [14] on Molecular Sieves 5A or 13X. However, the practical analysis of most gas mixtures of industrial interest requires that these gases be separated one from another in the presence of CO_2 and water vapor, very often in the presence of gaseous hydrocarbons and sometimes in the presence of nitrogen- or sulfur-containing gases. From this point of view, the separation on synthetic organic solids, such as styrene–ethylvinylbenzene–divinylbenzene copolymers with defined porous structure has great potential [15]. For example, the separation of permanent gases is easily accomplished at $-70°C$, but not at room temperature. Further potentialities of porous polymers can be derived by inspecting the retention data given in Table 23.2 [16].

The detection systems must be selected to suit the nature of the gases involved and must meet the demands of selectivity, sensitivity, and linearity of response. The detector used most frequently for routine gas analysis is the TCD with H_2 or He as carrier gas [17]. Unfortunately, this detector requires calibration for each gas. Its sensitivity has been improved for trace analysis [18]. The FID is used with success only for the analysis of hydrocarbon gases. However, CO and CO_2 can easily be determined as CH_4 by reduction with H_2 in a Ni catalytic reactor prior to entering the FID [19]. It is accepted that an ordinary FID gives no response to permanent and related gases [20], but recently, a hydrogen-rich FID has been described for analyzing permanent and other inorganic gases with adequate sensitivity [21]. Ionization detectors, particularly Ar and He detectors [22–24], ECD, and mass spectrometers [25–27] have very frequently been used for special gas analysis. Some less frequent, but possibly also useful detector systems are based on, e.g., the measurements of the modulation of ultrasound [28] or the electrochemical activity of some gases (CO, nitrogen and sulfur oxides) [29,30]. For additional information, see ref. 31.

References on p. B508

TABLE 23.2

RETENTION DATA FOR SOME GASES ON SYNTHETIC POLYMERS OF THE PORAPAK SERIES

Values measured with He as carrier gas (not corrected to dead volume) are calculated from data in O.L. Hollis and W.V. Hayes, in A.B. Littlewood (Editor), *Gas Chromatography 1966,* Institute of Petroleum, London, 1966, p. 57 [16].

Gas	Porapak Q *			Porapak R *			Porapak T *		
	Temperature, °C								
	−78	+26	+65	−78	+26	+65	−78	+26	+65
H_2	–	0.16	0.32	0.29	0.19	0.34	0.31	0.07	0.25
N_2	0.72	0.23	0.42	0.78	0.23	0.40	0.81	0.10	0.29
O_2	0.90	0.23	0.42	0.91	0.23	0.40	0.94	0.10	0.29
Ar	1.00	0.23	0.42	1.00	0.23	0.40	1.00	0.10	0.29
CO	1.23	0.26	–	1.19	0.25	–	1.30	0.11	–
NO	1.65	0.26	0.48	1.40	0.26	0.43	1.86	0.12	0.39
CO_2	–	1.00	1.00	–	1.00	1.00	–	1.00	1.00
N_2O	–	1.42	1.24	–	1.24	1.18	–	–	–
H_2O	–	5.46	3.25	–	30.9	13.6	–	30.9	22.6
H_2S	–	5.26	3.15	–	6.1	3.47	–	5.06	3.40
COS	–	11.1	5.21	–	9.2	4.94	–	6.52	4.00
SO_2	–	22.6	6.85	–	–	19.6	–	–	22.4
CS_2	–	–	72.2	–	–	–	–	–	48.0
HCN	–	17.8	5.17	–	–	31.3	–	–	34.0
CH_4	–	0.42	0.57	5.92	0.39	0.56	6.57	0.18	0.41
CH_3CH_3	–	3.27	2.24	–	2.66	2.01	–	1.40	1.18
$CH_2=CH_2$	–	–	1.66	–	1.84	1.57	–	1.18	1.07
$CH\equiv CH$	–	–	1.66	–	3.03	2.01	–	3.19	2.49
$CH_3CH_2CH_3$	–	20.0	8.22	–	14.0	7.77	–	6.75	5.47
$CH_2=CHCH_3$	–	15.3	7.18	–	11.6	6.93	–	7.85	5.85

* Trade names of Waters Assoc., Framingham, MA, USA.

The instrumentation is influenced by the necessity of separating gaseous constituents with large differences in vapor pressure or chemical nature, e.g., H_2, CO_2, and water vapor. Dual- or multi-column systems for splitting, cutting out, or backflushing of components have been developed [32,33], and special packings have been examined for improving the column performance [34]. Typically, permanent gases will break through the first (usually GLC) column unresolved, while the more strongly retained components are well separated. The permanent gas fraction then enters the second (usually GSC) column and is either separated with a stream of a second carrier gas or is stored there for a certain time. A dual-phase column has recently been described [35], where one column is inside another one and each of them is packed with a different packing (e.g., also GLC and GSC). The carrier gas flows from a single inlet through both columns simultaneously, a proper ratio of the flowrates through both sections being achieved by the selection of packing mesh sizes and by controlling the porosity of the fritted discs at the ends of both columns. The shortcoming of both methods is the requirement for keeping the split flowrates constant, although they are strongly influenced by any physical (shaking down, changes in the temperature) or chemical (aging, irreversible sorption) changes in the column packings.

23.2.3. Industrial gases

Methods for the separation of gases from air processing, production of town and fuel gases, production of ammonia, methanol, basic petrochemical products, high-grade gases, exhaust gases, etc., have been standarized. They require a single column with multiple detectors or multiple columns with single detectors and sometimes a combination of both. New variations are often published without introducing any fundamentally new approaches, such as the rapid and isothermal analysis of gases with a single detector, described in refs. 33 and 36–38. A simultaneous determination of gases in the presence of N_2O, SO_2, and gaseous hydrocarbons has also been achieved [28,39].

A simple portable gas chromatograph has recently been described for the analysis of stack and related gases [40] and a simple monitor, developed for measuring the exposure to CO in the workplaces is now available [41]. Computing methods [42], particularly those using digital techniques [43] and microprocessors [44], will profoundly influence future instrumentation.

Interesting special methods have been developed for the determination of O_3 [45] and H_2 with the use of H_2 as a carrier gas [46]. Problems connected with the analysis of pure gaseous [47,48] and liquid O_2 [49]. O_2 in NH_3 [50], in H_2O [51], or in nonaqueous solvents used for electrochemical purposes [52] call for special methods. Similarly, constituents of technical-grade H_2 [53] and of H_2 used for cooling large electrical generators have been successfully analyzed by GC [54]. H_2, CO, and ethylene in air have been detected with a TCD specially adapted to measure the thermochemical reaction on the surface of the wire [55]. The effusion rate of H_2 serves for the proof of the welding of metals [56]. GC separation of common gases in natural minerals has been developed for geological and mine services [57].

23.2.4. Gases of biological origin

The basal metabolism is often followed by analyzing O_2, CO_2, N_2, and water vapor [58,59]. Sometimes, reactions producing permanent gases are used for applications, such as the determination of enzyme activity [60] of for following the progress of fermentation [61]. The chromatographic separation of traces of volatile compounds in gases, mainly in the air, is being studied far more frequently now [62]. This type of research, based preferably on GC–MS, is proving to be very useful as a diagnostic means for many types of metabolic disorders in human medicine [63,64] and in veterinary science [65].

References on p. B508

23.2.5. Rare gases

Molecular sieves have proved to be the most suitable sorbents for the separation of rare gases [5,66,67]. As in the case of other categories of gases, there is now greater interest in the determination of traces of rare gases in other gases or vice versa. Determinations of low levels of Kr [68], of H_2 and water vapor [69,70], and of permanent gases [71,72] in He or of traces of radioactive rare gases in the atmosphere have been published [73].

23.2.6. Gaseous nitrogen compounds

The use of GC for the separation and analysis of nitrogen oxides is very effective only in the case of N_2O and NO, because of the chemical reactivity of other oxides, particularly NO_2. Careful selection of sorbents is necessary for the successful isolation of NO_2. PTFE, treated with NaCl at 320°C, has been found [74] to be the best sorbent, if NO_2 is present in the gas mixture. The analysis of such mixtures, based on boiling point differences, has recently been described [75]. A determination of these gases at subatmospheric pressures is of industrial interest [76]. The analysis of nitrogen oxides in the presence of N_2, O_2, CO, CO_2, SO_2, and water vapor has been achieved by programing the column temperature [77].

Much work has been done on the trace analysis of N_2O since the discovery of its high electron-capturing ability at temperatures of 300–350°C with [63]Ni [78] by using its strong adsorption on molecular sieves [79]. N_2O has been determined in the gases emanating from soils [80], ground water [81,82], or sea water [83]. Precise results have been achieved in the field analysis of air [79,84], as well as in the analysis of Freons in the troposphere [85] and stratosphere [86].

NO [87] and NH_3 [88] have been determined in cigarette and cigar smoke. Trimethylamine in air was studied by means of a Tenax-tube concentrator and a chemiluminscent nitrogen detector [89]. O_2 has been analyzed in anhydrous NH_3 [90], and moisture has been determined in liquid NH_3 [91].

HCN is often isolated by GC. Procedures for its determination in combustion effluents [92], exhaust gases [93], and in blood [94] have recently been developed. The cyanide ion can be derivatized and determined by GC in the gaseous phase [95]. Similarly, cyanides [96] and cyanide in the presence of thiocyanates have been analyzed in biological fluids [97].

23.2.7. Gaseous sulfur compounds

Only a few new methods have been published for the analysis of sulfur compounds in industrial gases, such as town gas [98]. For a complete analysis of H_2S, COS, CO_2, and SO_2 in gases and hydrocarbon streams it is necessary to use three columns and flame-photometric detection [99,100]. A specially treated Porapak QS can also be used as a column packing for gas mixtures of this type [101]. Because of the chemical activity of gaseous sulfur compounds, the electrolytic conductivity detector permits the analysis of H_2S, organic sulfides, mercaptans, and CO_2 in hydrocarbon streams [102]. The sensitivity of the electrochemical detector is very convenient, as it permits the estimation of picogram quantities of sulfur [103]. Thermionic [104], piezoelectric [105], and chemiluminescence [106] detectors have been adopted for the determination of sulfur gases. Very often, the total sulfur is determined only as H_2S [107] or as S by flame photometry [108].

Much effort has been devoted to the analysis of traces of H_2S or other sulfur gases in pure gases, such as ethylene [109], H_2 [110], CO_2 [111], N_2 [112], and, of course, in the air [113–116], frequently at the ppb level. Traces of H_2S and CH_3SH were analyzed in the atmosphere of a cellulose plant [117] and in expired air [118]. Trace analysis requires special techniques for the calibration of the analytes and for the storage of test gases. A calibration system for measuring total reduced sulfur and SO_2 at ppb concentrations has been developed for air pollution studies and control [119]. The stability of low-level sulfur gases in a gaseous matrix stored in high-pressure aluminum cylinders has been studied [120].

A concentration method is required for studying the transport and fate of sulfur gases, particularly of SO_2 in air masses, in order that a sufficiently precise analysis may be obtained. This problem has been studied successfully with tracers, SF_6 being used as a suitable model compound [121], and with the ECD as the most effective sensor [122], permitting the analysis of concentrations in the range of 10^{-13} parts of SF_6 in the air [123,124]. Analysis of the mixture originating during the production of SF_6 (e.g., SF_4, SOF_2, S_2F_{10}, $SFCl$, CHF and C_2F_2) was described earlier [125]. Oxidation processes yielding gas mixtures with higher concentrations of SO_2 and some [126] or much [127] SO_3 prompted the development of suitable GC methods for the analysis.

23.2.8. Halocompounds

There are only few new reports on the separation of free halogens. They are difficult to handle, owing to their corrosive nature. The large number of possible interhalogen compounds has led to an attempt at forecasting retention volumes by a semi-empirical method [128]. The separation of F_2, ClF, Cl_2, ClF_3, ClF_5, and BrF_3 on Fluorolube-4 and PTFE is relatively easy at ambient temperatures [129–132]. Pure electrolytic Cl_2 [133,134] and free F_2 mixed with HF [135] have thus been analyzed. Table 23.3 gives the retention ratios for some halocompounds on Kel-F [136] and Silicone 702 [137] columns.

Complex mixtures, originating from the chlorination of metal oxides and containing CO, CO_2, $COCl_2$, HCl, and Cl_2 are frequently separated by GC [138,139]. In such cases, a nickel-plated gas density balance is recommended. Nevertheless, the problem of detection is not an easy one. Although the ECD seems to be the most sensitive for halogens and halocompounds, the halogens corrode it very quickly. Coulometric [140], aerosol ionization [141], and He microwave [142] detectors all give measurable signals at mass flows of about 10 pg/sec of halogens. The GC determination of traces of phosgene in air has been carried out by pulse-flow coulometry [143], FID and IR spectroscopy [144]. Traces of HCl in air have also been determined [145], and the presence of ClO_2 in gases from chlorination processes

References on p. B508

TABLE 23.3

RETENTION RATIOS OF SOME HALOCOMPOUNDS

Gas	Column	
	Kel-F, 60°C [136]	Silicone 702, 48°C [137]
Chlorine monofluoride	0.48	0.60
Hydrogen fluoride	0.60	0.70
Chlorine	1.00	1.00
Chlorine trifluoride	1.45	1.40
Chlorine pentafluoride	0.85	–
Bromide	2.60	2.28
Bromide pentafluoride	–	3.05
Chlorine monoxide	1.76	–
Chlorine dioxide	1.45	–
Chlorine oxidifluoride	1.28	–
Fluorine monoxide	0.40	–
Uranium hexafluoride	6.00	5.25

has been demonstrated [146]. A COF_2 and CO_2 mixture was successfully separated under similar conditions [147]. Determination of trace impurities in hydrogen halides is of industrial importance. CO_2 in HCl [148] has been determined by headspace GC analysis. Inorganic and organic trace compounds at a level of $1 \cdot 10^{-4} - 5 \cdot 10^{-4}$ vol.% in HCl [149] and anhydrous HF [150] have been analyzed with good results.

It seems relevant at this point to mention some papers dealing with the determination of gaseous halogenated hydrocarbons in connection with air and water pollution. Trace analysis of vinyl chloride in air is of great interest, due to the carcinogenicity of this compound and to its large industrial use for the production of plastics [151,152]. A selective separation was achieved on a Porapak S-T column [152]. A good list of halogenated organic pollutants in ambient air will serve as a guide to these analyses [153]. Low-molecular-weight chlorinated hydrocarbons are analyzed most frequently [154,155]. GC–MS has been found to be the best method for the identification of complicated mixtures, permitting the detection of individual compounds at the ppt level. Because they are claimed to have dangerous decomposition effects on the ozone layer in the stratosphere, the trace analysis of low-molecular-weight fluorocarbons in air is a very active field. This analysis has also been carried out with a sensitivity in the ppt range [156,157]. Pollution of the environment with volatile hydrocarbons is growing to such an extent that practically all water resources are now polluted with gaseous hydrocarbons at the ppb level, albeit without toxic or organoleptic consequences. However, disinfective chlorination of drinking water gives rise to halogenated hydrocarbons with potentially toxic effects. Therefore, GC analysis of very volatile halogenated organic compounds in water has been developed, with the aid of headspace techniques [158] and liquid–liquid extraction [159].

23.2.9. Gaseous hydrides and miscellaneous gases

This section includes miscellaneous gases and volatile compounds which can be separated to advantage by GC. The relatively large range of boiling points and the strong chemical reactivity of some of these compounds require special arrangements of GC equipment, column conditioning, and optimization of the separation [160]. Although the paper by Rudolph and Baechmann [160] deals with the separation of the chlorides of Nb, Te, Mo, Sb, and Zr, their approach is applicable to a wider range.

The present interest in this category of gases is focused on the separation of hydrides, such as PH_3, AsH_3, H_2Se, GeH_4 and SnH_4, and on the application of new, sensitive detectors. A sequential determination of the As, Se, Ge, and Sn hydrides has been performed on one sample by using an atomic absorption detector [161]. Amounts as low as 3 ppb of As or 13 ppb of Ge were analyzed in 5 min. Plasma emission spectroscopy, coupled with GC, gives satisfactory results in the determination of the above-mentioned hydrides, including SbH_3 [162]. A gold gas-porous electrode sensor [163], measuring the respective electro-oxidative currents caused by reaction with the As, Sn, and Sb hydrides eluted from a Porapak Q column, has been applied to air and water pollution studies. As little as 0.2 ppb of As, 0.8 ppb of Sn, and 0.2 ppb of Sb were determined in 5-ml samples. The flame-photometric detector has been found [164,165] suitable for the estimation of traces (10^{-4}–10^{-6} vol.%) of impurities, such as B_2H_2, SiH_4, GeH_4, AsH_3, PH_3, H_2S, and H_2Se in inorganic hydrides of high purity. The determination of O_2, N_2, and CO_2 in the same type of hydrides [166], water in germanium hydrides [167], and inorganic and organic impurities in H_2Se [168] are examples of the ability of GC to solve analytical problems associated with such diverse gaseous mixtures. Mixtures of GeH_4 with H_2, CO, CO_2, N_2, and the Si, S, As, and P hydrides [169] and of SiH_4 with NH_3, HCl, H_2B_2, and the S, P, As, Ge hydrides [170] were analyzed in the concentration range of 10^{-2} to 10^{-7} vol.%. Much attention has been given to the determination of phosphine [171], particularly in ambient air. A modification of the inlet of the gas chromatograph has been found to be necessary for satisfactory results [172]. Air pollution by PH_3 in chemical plants [173] and by PH_3 and CH_2Br in corn silos [174] was examined.

Air pollution control for lead caused by alkyl-lead emission is a very active field, in which the gas chromatograph with an atomic-adsorption detector is used [175]. Absolute detection limits were reported to be 0.04–0.09 ng of alkyl-lead compounds, i.e. limits of 0.1–0.3 ng/m^3 in a 1-h sampling procedure. The detection limit for mercury emissions was reported [176] to be 45 pg of Hg^{2+}, when it was volatilized as phenylmercury(II) chloride and separated on OV-17 or EGS.

We will now deal with some representative papers concerned with the GC separation of volatile inorganic chlorides or bromides, although they are not true gases. Good separations of Zr, Nb, Mo, Te, In, Sn, Sb, Bi, or Te bromides on silica, treated with KBr and CsBr, as a column packing and with bromine-doped nitrogen as a carrier gas were achieved by programed-temperature GC [177]. The separating power of this method is indicated by the retention ratios of Nb/Zr: 0.58 on silica and 2.66 on KBr. Otozai and Tohyama [178] have earlier reported the separation of many other chlorides and bromides. High-purity BCl_3 was analyzed for total carbon content by GC [179] with a sensitivity limit of 0.0011% of carbon. As [180], Ge [181], and Sb [182] in ores and alloys of different kinds and silicon in alloys of Si and Al [183] were analyzed after conversion to the chlorides. Chlorosilanes are easily separated [184].

The determination of water in a great variety of gases and liquids represents a problem of great industrial importance. Examples are the analysis, in the ppm range, of natural gas [185] and 1,3-butadiene [186] and the analysis of organic solvents containing active chlorine and hydrogen chloride [187] and of liquid fuels [188]. To obtain a greater sensitivity, reaction GC [188] or the ECD [189] have been applied.

23.3. SEPARATION OF ISOTOPIC AND SPIN SPECIES

The research interest in this topic seems to be decreasing. A good separation of p- and o-H_2 and o- and p-D_2 was achieved on γ-alumina at 77°K in less than 3 min with Ne as a carrier gas and a katharometer as the sensor [190]. Adsorption and molecular sieve properties of some polymeric Schiff's-base complexes were used for a very rapid separation of hydrogen spin isomers. On a Cr(II) complex, the pairs p- and o-H_2 and o- and p-D_2 separated completely at room temperature in less than 20 sec with He as the carrier gas [191]. However, no separation was achieved for the pairs p-H_2 and o-D_2, and o-H_2 and p-D_2. Some special features of the thermistor-equipped TCD for the determination of o- and p-H_2 have been defined theoretically and proved experimentally [192].

The mass differences of isotopic gases having relatively low molecular weights play a significant role in the contributions of compounds that are similar in chemical structure or activity (e.g., NO–HNO_3, SO_2–H_2SO_4, CO–Ni surface) to adsorption enthalpy and to enthalpies of exchange equilibria between the gaseous and liquid or solid phases. Enrichment has been obtained in the case of ^{12}C and ^{13}C in CF_4 on Porapak Q at −25 to 200°C [193] and of CO on finely powdered Ni (ratio 1.037) [194]. Heavier isotopes tend to concentrate in the leading parts of the peaks. The same tendency has been observed in the case of ^{34}S and ^{36}S in SF_6 [193,195]. Partial separation of deuterated methanes in etched glass capillaries at 83°K with a He–N_2 mixture as the carrier gas has been described [196], but no baseline separation was obtained. Tailing occurred, even when N_2 was doped as a surface modifier, but retention times were reduced from 320 min to 35 min [197].

23.4. TRACE ANALYSIS

Contemporary developments in environmental protection, chemical diagnosis of diseases, space research, and the manufacture of plastics and microelectronics have created a considerable interest in trace analysis [2]. Books [3,198–200] and review articles [201,202] on trace analysis by GC are now available.

TABLE 23.4

SURVEY OF BASIC APPROACHES IN TRACE ANALYSIS

Method	Applicability	Detection limit, ppm	References
Concentration of traces			
Static sorption	General	0.01–0.1	151, 159, 205, 208, 211, 222, 223
Dynamic sorption	General	0.001–0.1	81, 114, 204–207, 214–216
Condensation	Special	1	209, 226, 227
Elimination of matrix			
Chemical	Special	10–1	209
Physicochemical	General	0.01–0.1	81, 82, 158, 203, 224, 228, 229, 231
Increase of detection sensitivity			
More sensitive detectors	General	0.001–0.1	78, 81, 89, 94, 99, 102, 103, 106, 111, 114, 124
Derivatization	Special	0.001–0.1	95, 119, 145, 232, 233
Sampling	Special	25	209

References on p. B508

GC with its highly developed detection systems and concentration techniques is a convenient tool for this purpose. Indeed, it lowers the limits of classical trace analysis from concentrations of, say, 0.1 vol.% to the ppb and, in some cases, to the ppt range. In the case of gases, it is necessary to examine every reference carefully for a comparable expression of quantity of the compound under consideration, as there are many ways of expressing the gas content of gaseous, liquid, or solid samples, e.g., v/v (vol.%), w/v (mg/ml or mg/m^3), v/w (ml/g), or w/w (g/g), and sometimes detection limits are expressed in terms of the quantity of the compounds detected per unit time, e.g., w/sec or v/sec.

In principle, there are still only three basic ways of separating and analyzing trace components with precision and sufficiently high sensitivity:

(a) concentration of the trace component(s) to a level adequate for detection,

(b) removal of most or all of the matrix,

(c) increase in the sensitivity of the detector toward the trace substances.

Table 23.4 represents a survey of methods for trace analysis and the approximate limits of sensitivity that are usually obtainable. In the earlier edition of this book [5], it was suggested that best results are achieved by combining the most sensitive analytical methods, such as, e.g., chromatographic trapping procedures, with the most sensitive detectors. The efficacy of this approach is borne out the more recent applications cited here.

23.4.1. Concentration of trace constituents of gases

Basically, there are three ways to achieve enrichment:

(a) Concentration is accomplished by a static sorption, which may be performed by a liquid, solid, or gas. If a liquid is used, it will extract the trace material from the gaseous mixture. In the case of a solid, adsorption will concentrate the traces. If a gas serves for the extraction from a liquid or a solid into the gas headspace, the traces will then be analyzed by GC.

(b) Dynamic concentration is effected by passing a continuous stream of the gas mixture through a column of sorbent. This is a chromatographic process (frontal analysis). Dynamic development can be effected by simply using one open column or by using a closed loop in a coupled system (Fig. 23.2).

(c) Concentration of traces by condensation is a third possibility,

It is clear that for the successful concentration of a trace component a sufficient difference in the vapor pressure of the trace material and the gas is necessary, so that this type of procedure is most convenient for trace compounds of lower volatility than that of the gas under analysis.

During dynamic concentration, a frontal chromatographic development of all constituents of the gas mixture takes place (Fig. 23.2). A condition for complete retention (e.g., of Substance D) is given to a first approximation by the volume of analyzed gas $V \leq V_R(1 - 2\sqrt{N})$, where V_R is the retention volume of Substance D in the system and N is the plate number. In other words, the leading front of the component being concentrated must remain completely in the column. If the gas is continuously passed through the column, Substance D will break through when

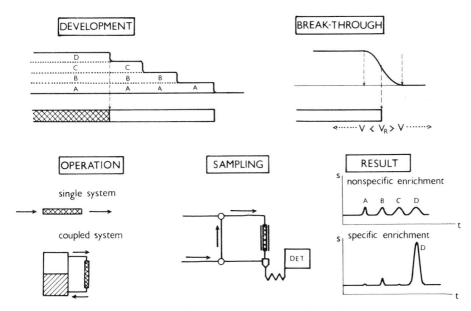

Fig. 23.2. Chromatographic concentration of trace constituents of gas mixtures.

$V \leqslant V_R(1 + 2\sqrt{N})$. The amount of D sorbed on the column is given by the equilibrium, defined by the ratio between the concentration in the sorbed and gaseous phases, i.e., it is related to the partition coefficient, since $V_R = V_G + KV_S$, where V_G and V_S are the volumes of gas and sorptive material, respectively, in the column [203]. The greatest advantage of dynamic coupled systems is that only a knowledge of the temperature of the system, and not of the volume of gas streaming through the column is necessary, if the mentioned condition is fulfilled. A selective enrichment of trace components is possible by a suitable choice of packing material [204].

The estimation of other gases in He [205] or of organic traces in air [206] are good examples of static sorption techniques. Similarly, precolumn enrichment of toxic traces in air and injection into a gas chromatograph by a rapid heating of the trap was applied to ppb analysis [207–209]. Because this procedure is frequently used in air pollution studies, different column packings have been compared [210]. For trapping organic contaminants in gases, porous polymers have been found to be the best column material and were studied thoroughly [206,211–213]. Traces of CO in air were measured with a ca. 2000-fold enrichment [214]. After preconcentration, sulfur gases are very often separated and determined by GC [100,215]. Noble gases [216] as well as N_2 in He [217] have been thus determined. The problem of representative sampling of an enclosed inhabited environment has been studied statistically [209]. A dryer for field use in the determination of trace atmospheric gases [218] and passive monitoring means for the estimation of personnel exposure to toxic gases [219] have been described. GC–MS is the best tool for the analysis of unknown organic compounds in air, and software improvement of a computer-as-

References on p. B508

sisted unit may be of interest in this connection [220]. The problem of standard reference gases is a very serious one in those cases where GC–MS is not available. Use of reference gases in the trace analysis of turbine exhaust gas measurements [221] has a broader interest for environmental studies.

In cases where trace components are less sorbed than the major gas components adsorbents make more suitable packing materials than liquids. The enrichment tends to be a displacement process, because the adsorbent will become blocked by large quantities of strongly adsorbed gas. Chromatography has been advocated for such an approach [222]. Thus, 10^{-4} to 10^{-6} vol.% of He and Ne were determined in about 100–150 ml of air with high precision [223].

Headspace gas analysis is now being applied very successfully to the determination of volatile environmental pollutants in biological matrices [224]. Dissolved gases, such as N_2O, have been estimated in aquatic systems [81,82], and CO in the workplace has also been measured by this technique [41]. Closed-loop trapping techniques, using the principle of headspace gas analysis, gave very good results in the determination of traces of organic compounds in water pollution studies [225]. Cryogenic enrichment [226] and particularly co-condensation with light hydrocarbons, like n-pentane [227], are novel concentration methods in environmental trace analysis.

23.4.2. Quantitation in trace analysis with unknown matrix composition

The analysis of gaseous trace materials in a gas mixture is relatively easy when they are collected by one of the methods described above. If the matrix is a liquid or solid, quantitation becomes more complicated, because the matrix is generally

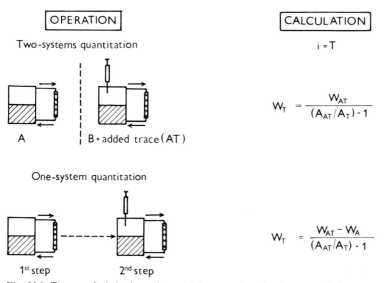

Fig. 23.3. Trace analysis in dynamic coupled system (e.g., headspace analysis).

complex and of unknown composition. The headspace gas analysis is simple, but it may lead to erroneous quantitative results.

In normal stripping procedures a stream of gas is employed to carry the components to the trapping column up to the moment of total transfer of the trace from the condensed phase into the trap (Fig. 23.3). The quantity of the trace compound i at time t, W_{it}, is then given by an exponential function [203]:

$$W_{it} = W_i \left[1 - \exp\left(- \frac{F \cdot t}{V_G + K_{LG}V_L} \right) \right]$$

where F is the volumetric flowrate of the stripping gas, V_G and V_L are the volumes of gaseous and liquid phase, respectively, in the system, and K_{GL} is the distribution coefficient. The equation is valid if at any time the concentration of trace compound i in the stripping gas entering the trapping column is in equilibrium with the condensed phase.

If a closed loop technique is used, the same reasoning applies. However, if the circulation is continued beyond the moment the trace compound breaks through the trapping column (see Fig. 23.2), an equilibrium is reached between the condensed phase and the trapping column. In such a case, the following equation is valid [203]:

$$W_{it} = W_i \left[\frac{K_{SG}V_S}{(V_{Gt} + K_{SG}V_S)} \cdot \frac{V_G}{K_{SG}V_S} + \frac{K_{LG}V_L}{K_{SG}V_S} + 1 \right]^{-1}$$

where V_{Gt} is the void volume in the trapping column, V_S is the volume of sorbent in the trapping column, K_{SG} is the distribution coefficient (ratio of the equilibrium constant of trace compound in the sorbent and in the circulating gas at the temperature of the trapping column), and all the other terms are the same as before. In most cases, it can be assumed that $K_{SG}V_S$ is much greater than V_{Gt} so that the factor $K_{SG}V_S (V_{Gt} + K_{SG}V_S)^{-1}$ tends to unity. Unfortunately, the term K_{LG} is not known, because of the unknown composition of the matrix, a fact which is often not sufficiently taken into the consideration.

The standard-addition method [228,229] has proved to be a very effective way to eliminate, or at least reduce matrix effects. It is acceptable to assume that, if a known quantity of the trace compound to be analyzed is added to the system, it will follow the same distribution constants (particularly K_{LG}) as the trace material originally present. In this case, no experimental parameters need to be known, but at least two analyses for the trace compound must be performed (Fig. 23.3) under the same conditions. The method has been successfully applied in water pollution studies; quantitation was possible at the ppb level [230], and precision was adequate. The quantitative determination of hydrophilic substances dissolved in water has been performed by a standard addition method in Grob's close-loop strip/trap procedure with good results [225].

23.4.3. Increase of detection sensitivity

The earlier survey in ref. 5 is still valid. Generally, detectors based on ECD [78,111,124], MS [26,156], electrochemical processes [30,103], and photo-ionization

References on p. B508

phenomena [234,235] have found increasing acceptance. A FID has been developed which is sensitive to inorganic gases [22]. Frequently, chemical reactions are used that change the original gas to some other compound giving a greater response. As examples, we may mention reactions of many acidic gases with sodium butylate [232] and detection of the stoichiometrically evolved butane or of methane from the reduction of carbon-containing gases by FID [233]. Similarly, the reaction of HCl with epibromohydrin, producing chlorobromopropanol, yields three carbon atoms to give a signal with a FID [145]. Many other possibilities for chemical conversion can be thought of.

REFERENCES

1 P.G. Jeffery and P.J. Kipping, *Gas Analysis by Gas Chromatography*, Pergamon, Oxford, 2nd Edn., 1972.
2 S. Hertz and S.N. Chesler, *Trace Organic Analysis. A New Frontier in Analytical Chemistry*, Nat. Bur. Stand. Spec. Publ. No. 519, Washington, DC, 1979.
3 H. Hachenberg and A.P. Schmidt, *Gas Chromatographic Headspace Analysis*, Heyden, London, 1977.
4 B.I. Anvaer and Yu.S. Drugov, *Gazovaya Khromatografiya Neorganichskikh Veshchestv (Gas Chromatography of Inorganic Substances)*, Izd. Khimiya, Moscow, 1976.
5 J. Janák, in E. Heftmann (Editor), *Chromatography*, Van Nostrand-Reinhold, New York, 3rd Edn., 1975.
6 G.G. Devyatykh and A.D. Zorin, *Letuchie Neorganicheskie Gidridy Osoboi Chistoty (Volatile Inorganic Hydrides of High Purity)*, Izd. Nauka, Moscow, 1974.
7 G. Guiochon and C. Pommier, *Gas Chromatography in Inorganics and Organometallics*, Ann Arbor Sci. Publ., Ann Arbor, MI, 1973.
8 H. Rüssel and G. Tölg, *Anwendung der Gaschromatographie zur Trennung und Bestimmung anorganischer Stoffe*, Springer, Berlin, 1972.
9 V.A. Kiselev and Ya.I. Yashin, *Gas-Adsorption Chromatography*, Plenum, New York, 1969.
10 Yu.S. Drugov, *Zh. Anal. Khim.*, 35 (1980) 559.
11 J. Janák, *Ann. N.Y. Acad. Sci.*, 72 (1959) 606.
12 J. Janák, M. Krejčí and H.E. Dubský, *Collect. Czech. Chem. Commun.*, 24 (1959) 1080.
13 T.J. Nestrick, R.H. Stehl, J.N. Driscoll, L.F. Jaramillo and E.C. Atwood, *Ind. Res. Dev.*, 22 (1980) 126.
14 G. Kyryacos and C.E. Boord, *Anal. Chem.*, 29 (1957) 787.
15 W.F. Wilhite and O.L. Hollis, *J. Gas Chromatogr.*, 6 (1968) 84.
16 O.L. Hollis and W.V. Hayes, in A.B. Littlewood (Editor), *Gas Chromatography 1966*, Institute of Petroleum, London, 1967, p. 57.
17 K. Tadesse, A. Smith, W.G. Brydon and M. Eastwood, *J. Chromatogr.*, 171 (1979) 416.
18 H. Kern and M. Elser, *Microchim. Acta*, (1978) 319.
19 A.L. Pertsiovskii, V.M. Belyakov and L.M. Kremko, *Gig. Sanit.*, 7 (1978) 64.
20 J.C. Sternberg, W.S. Dallaway and D.T.L. Jones, in N. Brenner, J.E. Callen and M.D. Weiss (Editors), *Proc. 3rd Internat. Symp. GC*, Academic Press, New York, 1962, p. 231.
21 P. Russev, T.A. Gough and C.J. Woollam, *J. Chromatogr.*, 119 (1976) 461.
22 F.J. Bourke, R.W. Dawson and W.H. Denton, *J. Chromatogr.*, 19 (1965) 425.
23 F.F. Adrews and E.K. Gibson, *Anal. Chem.*, 50 (1978) 1146.
24 P. Popp and G. Oppermann, *J. Chromatogr.*, 148 (1978) 265.
25 P.G. Simmonds, *Int. Lab.*, No. 1/2 (1981) 8.
26 P.G. Simmonds, *J. Chromatogr.*, 166 (1978) 593.
27 H. Nozoye, Y. Yamazaki, H. Tomita and K. Someno, *Bunseki Kagaku*, 29 (1980) T54.
28 H.W. Dürbeck and R. Niehaus, *Chromatographia*, 11 (1978) 14.

29 E. Punfor, K. Tóth, Z. Fehér, G. Nagy and M. Varadi, *Anal. Lett.*, 8 (1975) 9.

30 J.R. Stetter, D.R. Rutt and K.F. Blurton, *Anal. Chem.*, 48 (1976) 924.

31 D. Jentzsch and E. Otte, *Detektoren in der Gas-Chromatographie*, Akademie Verlag, Frankfurt/M., 1970.

32 J.O. Terry and J.H. Futrell, *Anal. Chem.*, 37 (1965) 1165.

33 F.F. Andrawes and E.K. Gibson, Jr., *Anal. Chem.*, 51 (1979) 462.

34 R. Dandanea and S. Hawkes, *Chromatographia*, 13 (1980) 686.

35 T.W. Rendl, J.M. Anderson and R.A. Dolan, *Amer. Lab.*, 12 (1980) 60.

36 M.A. Hossain, M. Forissier and Y. Trambouze, *Chromatographia*, 9 (1976) 471.

37 A. Kumar, R.D. Dua and M.K. Sarkar, *J. Chromatogr.*, 107 (1975) 190.

38 J. Barcicki and A. Machowski, *Chem. Anal. (Warsaw)*, 20 (1975) 421.

39 S.D. Sarotova and V.I. Ostroushko, *Zavod. Lab.*, 46 (1980) 702.

40 D.L. Miller, J.S. Woods, K.W. Grubauch and L.M. Jordan, *Environ. Sci. Technol.*, 14 (1980) 97.

41 D. Bauer, in B. Kolb (Editor), *Applied Headspace Gas Chromatography*, Heyden, London, 1980, p. 41.

42 E.M. Emery, *J. Chromatogr. Sci.*, 14 (1976) 261.

43 F.W. Karasek, *Ind. Res. Dev.*, 25 (1974) 28.

44 J.D. Mitchell, F.C. Duranza and P.B. Fuhrman, *Int. Lab.*, No. 11/12 (1981) 51.

45 R.V. Holland and P.W. Board, *Analyst (London)*, 101 (1976) 887.

46 V.E. Stepanenko and Z.N. Golovnya, *Zavod. Lab.*, 41 (1975) 670.

47 J. Karlsson, *Chromatographia*, 8 (1975) 155.

48 A.A. Datskievich, D.E. Bobrina and N.V. Menshikova, *Zh. Anal. Khim.*, 34 (1979) 550.

49 E.S. Orlova, R.V. Pilipenko and L.N. Chudovich, *Zavod. Lab.*, 42 (1976) 1178.

50 U.D. Choubey, G. Mohan, D.S. Mathur and E.K. Ghosh, *Fertil. Technol.*, 14 (1977) 125.

51 K.S. Mall, *J. Chromatogr. Sci.*, 16 (1978) 311.

52 J.M. Achord and C.L. Hussey, *Anal. Chem.*, 52 (1980) 601.

53 A.N. Timofeeva, T.E. Alymova, N.I. Lulova and A.V. Kuz'mina, *Zh. Anal. Khim.*, 33 (1978) 2249.

54 D.J.A., Dear, A.F. Dillon and A.N. Freedman, *J. Chromatogr.*, 137 (1977) 315.

55 V.G. Guglya, *Zh. Anal. Khim.*, 34 (1979) 405.

56 I.K. Pokhodnya and A.P. Pal'tsevich, *Avtom Svarka*, No. 1 (1980) 37.

57 O.F. Mironova, *Zh. Anal. Khim.*, 33 (1978) 981.

58 U. Smidt, in G.R. Waller and O.C. Dermer (Editors), *Biochemical Applications of Mass Spectrometry*, Wiley, New York, 1980, p. 703.

59 J. Janák, in P. Porter (Editor), *Gas Chromatography in Biology and Medicine*, Churchill, London, 1969, p. 86.

60 E. Noszticzius, D. Kalmar and Z. Noszticzius, *Acta Chim. (Budapest)*, 103 (1980) 225.

61 J.W. Czerkawski and G. Breckenridge, *Anal. Biochem.*, 67 (1975) 476.

62 S.D. Sastry, K.T. Buck, J. Janák, M. Dressler and G. Preti, in G.R. Waller and O.C. Dermer (Editors), *Biological Application of Mass Spectroscopy*, Suppl. Vol., Wiley, New York, 1980, p. 1085.

63 E. Jellum, *Trends Anal. Chem.*, 1 (1981) 12.

64 E. Jellum, *J. Chromatogr.*, 143 (1977) 427.

65 P. Hradecký, *Thesis*, Veterinary University, Brno, 1978.

66 F. Thouzeau, *Chromatographia*, 9 (1976) 506.

67 L. Wilhelmova and F. Cejnar, *J. Radioanal. Chem.*, 31 (1976) 383.

68 D.L. Evans and A.K. Mukherji, *Anal. Chem.*, 52 (1980) 579.

69 V.G. Reznikov, T.S. Kuznetsova and A.D. Zorin, *Zh. Anal. Khim.*, 32 (1977) 60.

70 V.S. Mirzayakov, *Zavod. Lab.*, 43 (1977) 664.

71 H. Aoyagi and M. Takahashi, *Bunseki Kagaku*, 24 (1975) 144.

72 V.M. Nemets, A.A. Petrov and A.A. Solov'ev, *Zh. Anal. Khim.*, 35 (1980) 1751.

73 A. Botlino, G. Lembo, F. Scacco, G. Sciocchetti and J. Lasa, in *Environmental Surveillance around Nuclear Installations*, Vol. 1, Intern. Atomic Energy Agency, Vienna, 1974, p. 191.

74 J. Amouroux, D. Rapakoulias and A. Saint-Yrieix, *Analusis*, 5 (1977) 372.

75 B.D. Timofeev, *Zh. Anal. Khim.*, 35 (1980) 2029.

76 K.A. Pearsall and F.T. Bonner, *J. Chromatogr.*, 200 (1980) 224.

77 T. Fujii, *Bunseki Kagaku*, 28 (1979) 388.

78 P.G. Simmonds, *J. Chromatogr.*, 166 (1978) 593.
79 T. Punpeng, J.O. Frohliger and N.A. Esmen, *Anal. Chem.*, 51 (1979) 159.
80 C.J. Smith and P.M. Chalk, *Analyst (London)*, 104 (1979) 538.
81 J.W. Elkins, *Anal. Chem.*, 52 (1980) 263.
82 K.C. Hall, *J. Chromatogr. Sci.*, 18 (1980) 22.
83 Y. Cohen, *Anal. Chem.*, 49 (1977) 1238.
84 M. Hirota, *Bull. Chem. Soc. Jap.*, 51 (1978) 3075.
85 B.L. Tyson, J.C. Arwesen and N. O'Hara, *Geophys. Res. Lett.*, 5 (1978) 535.
86 B.L. Tyson, J.F. Vedder, J.C. Arwesen and R.B. Brever, *Geophys. Res. Lett.*, 5 (1978) 369.
87 A.D. Horton, J.R. Stokely and M.R. Guerin, *Anal. Lett.*, 7 (1974) 177.
88 K.D. Brunneman and D. Hoffmann, *J. Chromatogr. Sci.*, 13 (1975) 159.
89 N. Kashihira, K. Kirita, Y. Watanabe and K. Tanaka, *Bunseki Kagaku*, 29 (1980) 853.
90 A.D. Hunt, *Analyst (London)*, 102 (1977) 846.
91 V.M. Davydova, E.A. Solomentseva and V.I. Anisimov, *Zavod. Lab.*, 43 (1977) 246.
92 A.L. Myerson and J.J. Chludzinski, Jr., *J. Chromatogr. Sci.*, 13 (1975) 554.
93 N. Fukuda, H. Itoh, A. Tsukamoto and H. Tamari, *Bunseki Kagaku*, 28 (1979) 569.
94 R.W. Darr, T.L. Capson and F.D. Hileman, *Anal. Chem.*, 52 (1980) 1379.
95 K. Funazo, K. Kusano, M. Tanaka and T. Shono, *Anal. Lett.*, 13 (1980) 751.
96 H. Honna, K. Suzuki, M. Yoshida and H. Yanashima, *Bunseki Kagaku*, 28 (1979) 756.
97 I. Thomson and R.A. Anderson, *J. Chromatogr.*, 188 (1980) 357.
98 Y. Hoshika and Y. Iida, *J. Chromatogr.*, 134 (1977) 423.
99 C.D. Pearson and W.J. Hines, *Anal. Chem.*, 49 (1977) 123.
100 M.S. Black, R.P. Herbst and D.R. Hitchcock, *Anal. Chem.*, 50 (1978) 848.
101 T.L.C. De Souza, D.C. Lane and S.P. Bhatia, *Anal. Chem.*, 47 (1975) 543.
102 R.G. Schiller and R.B. Bronsky, *J. Chromatogr. Sci.*, 15 (1977) 541.
103 J.R. Stetter, J.M. Sedlak and K.F. Blurton, *J. Chromatogr. Sci.*, 15 (1977) 125.
104 J. Novotný and A. Müller, *J. Chromatogr.*, 148 (1978) 211.
105 L.M. Webber, K.H. Karmarkar and G.G. Guilbaut, *Anal. Chim. Acta*, 97 (1978) 29.
106 H. Kawana, K. Nidaira, Y. Takada and F. Nakajima, *J. Chem. Soc. Jap., Chem. Ind. Sec. B*, (1978) 691.
107 M.I. Afanas'ev, L.V. Denisova, T.I. Lomonovtseva, B.P. Okhotnikov, V.A. Rotin and V.S. Yusfin, *Zavod. Lab.*, 41 (1975) 1203.
108 A.I. Butuzova, G.S. Beskova, L.B. Preobrazhenskaya and T.I. Korotkova, *Zavod. Lab.*, 45 (1979) 24.
109 F.Kh. Kudasheva, M.A. Plan, M.A. Mirgaleeva and R.A. Istambaeva, *Zh. Anal. Khim.*, 33 (1978) 1225.
110 H.J. Rath and J. Wimmer, *Chromatographia*, 13 (1980) 513.
111 M.E. Pick, *J. Chromatogr.*, 171 (1979) 305.
112 L. Giry, M. Chaigneau and L.P. Ricard, *Analusis*, 6 (1978) 203.
113 A.F. Fatkullina, N.V. Zakharova, G.S. Akhmetzyanova and L.L. Bondarenko, *Khim. Tekhnol. Topl. Masel*, No. 4 (1979) 15.
114 J. Godin, J.L. Cluet and C. Boudene, *Anal. Chem.*, 51 (1979) 2100.
115 F. Bruner, P. Ciccioli and G. Bertoni, *J. Chromatogr.*, 120 (1976) 200.
116 F. Bruner, P. Ciccioli and F. Di Nardo, *Anal. Chem.*, 47 (1975) 141.
117 A.G. Vitenberg, L.M. Kuznetsova, I.L. Butaeva and M.D. Inshakov, *Anal. Chem.*, 49 (1977) 128.
118 A.R. Blanchette and A.D. Cooper, *Anal. Chem.*, 48 (1976) 729.
119 T.L.C. De Souza and S.P. Bhatia, *Anal. Chem.*, 48 (1976) 2234.
120 F.J. Kramer and S.G. Wechter, *J. Chromatogr. Sci.*, 18 (1980) 674.
121 I. Zoccolillo and A. Liberti, *J. Chromatogr.*, 108 (1975) 219.
122 C.M. Derks, *J. Chromatogr.*, 108 (1975) 222.
123 L.E. Thomsen, *Atmos. Environ.*, 10 (1976) 917.
124 R.N. Dietz, E.A. Cote and R.W. Goodrich, in *Measuring, Detection and Control of Environmental Pollutants*, Intern. Atomic Energy Agency, Vienna, 1976, p. 277.
125 R.S. Juvet and F. Zado, in J.C. Giddings and R.A. Keller (Editors), *Advances in Chromatography*, Vol. 1, Dekker, New York, 1965, p. 249.

126 M.S. Wainwright and D.W.B. Westerman, *Chromatographia*, 10 (1977) 665.

127 J.P. Briggs, R.R. Hudgins and P.L. Silveston, *J. Chromatogr. Sci.*, 14 (1976) 335.

128 K. Otozai and I. Tohyama, *Z. Anal. Chem.*, 279 (1976) 195.

129 V.S. Pervov, V.F. Sukhoverkhov and L.G. Podzolko, *Zh. Anal. Khim.*, 34 (1979) 2369.

130 V.F. Sukhoverkhov, L.G. Podzolko and V.F. Garanin, *Zh. Anal. Khim.*, 30 (1975) 330 and 33 (1978) 1360.

131 D.A. Donokh'yu, T.D. Nevitt and A.V. Zletts, in I.I. Monseev (Editor), (*Contemporary Chemistry of Rocket Fuels*), Atomizdat, Moscow, 1972, p. 204.

132 A.G. Linch, N.R. McQuacker and M. Gurney, *Environ. Sci. Technol.*, 12 (1978) 169.

133 S.K. Dang, P.D. Grover and S.L. Chawla, *J. Chromatogr.*, 139 (1977) 207.

134 I.P. Ogloblina, I.I. Novozhilova, and I.A. Il'icheva, *Zavod. Lab.*, 44 (1978) 540.

135 R. Foon and G.P. Reid, *J. Chromatogr. Sci.*, 14 (1976) 421.

136 A.G. Hamlin, G. Iveson and T.R. Phillips, *Anal. Chem.*, 35 (1963) 2037.

137 J.F. Ellis and G. Iveson, in R.P.W. Scott (Editor), *Gas Chromatography 1960*, Butterworths, London, 1960, p. 308.

138 A. Baiker, H. Geisser and W. Richarz, *J. Chromatogr.*, 147 (1978) 453.

139 G. Alexander and G. Garzó, *Acta Chim. (Budapest)*, 88 (1976) 329.

140 S.I. Krichmar, Zh.B. Levchenko and A.E. Semenchenko, *Zh. Anal. Khim.*, 32 (1977) 1703.

141 P. Popp, H.-J. Grosse and G. Opperman, *J. Chromatogr.*, 147 (1978) 47.

142 B.S. Quimbly, P.C. Uden and R.M. Barnes, *Anal. Chem.*, 50 (1978) 2112.

143 H.B. Singh, D. Lillian and A. Appleby, *Anal. Chem.*, 47 (1975) 860.

144 G.G. Esposito, D. Lillian, G.E. Podolak and R.M. Tuggle, *Anal. Chem.*, 49 (1977) 1774.

145 K. Bächmann, K. Goldbach and B. Vierkorn-Rudolph, *Mikrochim. Acta*, (1979) 17.

146 M. Ziołkowska, T. Goldowski and Z. Figura, *Przem. Chem.*, 56 (1977) 480.

147 L.A. Agas, T.F. Dobrovol'skikh and V.S. Shaidurov, *Zavod. Lab.*, 43 (1977) 427.

148 H.J. Rath, D. Schmidt and J. Wimmer, *Chromatographia*, 12 (1979) 567.

149 V.I. Maiorov, A.D. Molodyk, L.N. Morozova and G.V. Bondar, *Zavod. Lab.*, 46 (1980) 113.

150 W.P. Cottom and D.E. Stelz, *Anal. Chem.*, 52 (1980) 2073.

151 B. Miller, P.O. Kane, D.B. Robinson and P.J. Whittingham, *Analyst (London)*, 103 (1978) 1165.

152 D. Krockenberger, H. Lorkowski and L. Rohrschneider, *Chromatographia*, 12 (1979) 787.

153 D.E. Harsch, D.R. Cronn and W.R. Slater, *J. Air Pollut. Control Assoc.*, 29 (1979) 975.

154 E.P. Grimsrud and D.A. Miller, in H.S. Hertz and S.N. Chesler (Editors), *Trace Organic Analysis: A New Frontier in Analytical Chemistry*, Nat. Bur. Stand. Spec. Publ. No. 519, Washington, DC, 1979, p. 143.

155 R. Tatsukawa, T. Okamoto and T. Wakimoto, *Bunseki Kagaku*, 27 (1978) 164.

156 D.R. Cronn and D.E. Harsch, *Anal. Lett.*, 12 (1979) 1489.

157 E. Heil, H. Oeser, R. Hatz and H. Kelker, *Z. Anal. Chem.*, 297 (1979) 357.

158 G.J. Piet, P. Slingerland, F.E. de Grunt, M.P.M. van den Heuvel and B.C.J. Zoeteman, *Anal. Lett.*, 11 (1978) 437.

159 H.J. Brass, *Int. Lab.*, 10, No. 6 (1980) 17.

160 J. Rudolph and K., Baechmann, *Chromatographia*, 10 (1977) 731.

161 M.H., Hahn, K.J. Mulligan, M.E. Jackson and J.A. Caruso, *Anal. Chim. Acta*, 118 (1980) 115.

162 W.B. Robbins and J.A. Caruso, *J. Chromatogr. Sci.*, 17 (1979) 360.

163 P.R. Gifford and S. Bruckenstein, *Anal. Chem.*, 52 (1980) 1028.

164 A.E. Ezheleva, M.F. Churbanov, G.E. Snopatin and B.G. Vtorov, *Zh. Anal. Khim.*, 33 (1978) 317.

165 A.E. Ezheleva, G.E. Snopatin and L.S. Malygina, *Zh. Anal. Khim.*, 34 (1979) 2308.

166 N.T. Ivanova, N.A. Vislykh, V.V. Voevodina, L.A. Protasova and L.A. Frangulyan, *Zavod. Lab.*, 44 (1978) 649.

167 G.G. Devyatykh, V.V. Balabanov, A.V. Gusev, E.N. Karataev and N.Kh. Agliulov, *Zh. Anal. Khim.*, 30 (1975) 1630.

168 R.L. Belford and J.R. Marguart, *Anal. Chem.*, 50 (1978) 656.

169 A.E. Ezheleva, L.S. Malygina and M.F. Churbanov, *Zavod. Lab.*, 45 (1979) 978.

170 A.E. Ezheleva, L.S. Malygina and M.F. Churbanov, *Zh. Anal. Khim.*, 35 (1980) 1972.

171 M. Braunová, G. Minarovič and B. Blahovec, *Chem. Prům.*, 28 (1978) 568.

172 J.R. Bean and R.E. White, *Anal. Chem.*, 49 (1977) 1468.

173 A. Vinsjansen and K.E. Thrane, *Analyst (London)*, 103 (1978) 1195.

174 S. Noak, C. Reichmuth and P. El-Lakwah, *Z. Anal. Chem.*, 291 (1978) 121.

175 W. De Jonche, D. Chakradorti and F. Adams, *Anal. Chem.*, 52 (1980) 1974.

176 V. Luckow and H.A. Rüssel, *J. Chromatogr.*, 150 (1978) 187.

177 S. Tsalas and K. Baechmann, *Talanta*, 27 (1980) 201.

178 K. Otozai and I. Tohyama, *Z. Anal. Chem.*, 274 (1975) 353.

179 N.T. Ivanova, L.D. Prigozhina, L.A. Frangulyan and Yu.V. Malakhovskii, *Zh. Anal. Khim.*, 34 (1979) 2343.

180 B. Iatridis and G. Parissakis, *J. Chromatogr.*, 122 (1976) 505.

181 B. Iatridis and G. Parissakis, *Anal. Chem.*, 49 (1977) 909.

182 B. Iatridis and G. Parissakis, *Anal. Chim. Acta*, 89 (1977) 347.

183 G. Parissakis and B. Iatridis, *Chromatographia*, 10 (1977) 37.

184 H.J. Rath and D. Schmidt, *Z. Anal. Chem.*, 290 (1978) 316.

185 V.S. Yusfin, B.P. Okhotnikov, V.A. Rotin, V.N. Khokhlov, M.I. Afanas'ev, G.V. Pogosbekyan and Z.I. Filatova, *Zavod. Lab.*, 41 (1975) 542.

186 W.-T., Wang, X.-D. Ding and X.-J. Wu, *J. Chromatogr.*, 199 (1976) 149.

187 T. Sakano, Y. Hori and Y. Tamari, *J. Chromatogr. Sci.*, 14 (1976) 501.

188 A. Konopczyński and A. Siedlecki, *Chem. Anal. (Warsaw)*, 25 (1980) 777.

189 B. Scholz and K. Ballschmiter, *Z. Anal. Chem.*, 302 (1980) 264.

190 J. Dericbourg, *J. Chromatogr.*, 123 (1976) 405.

191 M. Riederer and W. Sawodny, *J. Chromatogr.*, 179 (1979) 337.

192 D.N. Mitchell and D.J. Le Roy, *Can. J. Chem.*, 56 (1978) 1817.

193 E. Bayer, G. Nicholson and R.E. Sievers, *J. Chromatogr. Sci.*, 8 (1970) 467.

194 J.C. Fetzer, P.A. Bloxham and L.B. Rogers, *Separ. Sci. Technol.*, 15 (1980) 49.

195 N. Moiseev and I. Platzner, *J. Chromatogr. Sci.*, 14 (1976) 143.

196 S. Lukáč, *J. Chromatogr.*, 166 (1978) 287.

197 F. Bruner, G.P. Cartoni and M. Possanzini, *Anal. Chem.*, 41 (1969) 1122.

198 J. Novák, *Quantitative Analysis by Gas Chromatography*, Dekker, New York, 1975, p. 107.

199 H. Hachenberg, *Industrial Gas Chromatographic Trace Analysis*, Heyden, London, 1973.

200 V.G. Berezkin and V.S. Tatarinskii, *Gas-Chromatographic Analysis of Trace Impurities*, Plenum, New York, 1973.

201 Yu.S. Drugov and V.G. Berezkin, *Usp. Khim.*, 48 (1979) 1884.

202 Y. Hoshika and G. Muto, *Bunseki Kagaku*, 29 (1980) T10.

203 J. Novák, J. Janák and J. Goliáš, in H.S. Hertz and S.N. Chesler (Editors), *Trace Organic Analysis: A New Frontier in Analytical Chemistry*, Nat. Bur. Stand. Spec. Publ. 519, Washington, DC, 1979, p. 739.

204 J. Novák, V. Vašák and J. Janák, *Anal. Chem.*, 37 (1965) 660.

205 H. Aoyagi and M. Takahashi, *Bunseki Kagaku*, 24 (1975) 144.

206 R.H. Brown and C.J. Purnell, *J. Chromatogr.*, 178 (1979) 79.

207 R.G. Melcher and V.J. Caldecourt, *Anal. Chem.*, 52 (1980) 875.

208 R.B. Denyszyn, J.M. Harden, D.L. Hardison, J.F. McGaughey and A.L. Sykes, in H.S. Hertz and S.N. Chesler (Editors), *Trace Organic Analysis: A New Frontier in Analytical Chemistry*, Nat. Bur. Stand. Spec. Publ. 519, Washington, DC, 1979, p. 153.

209 H.G. Eaton, F.W. Williams, J.R. Wyatt, J.J. DeCorpo, F.E. Saalfeld, D.E. Smith and T.L. King, in H.S. Hertz and S.N. Chesler (Editors), *Trace Organic Analysis: A New Frontier in Analytical Chemistry*, Nat. Bur. Stand. Spec. Publ. 519, Washington, DC, 1979, p. 213.

210 C. Vidal-Madjar, M.F. Gonnord, F. Benchan and G. Guiochon, *J. Chromatogr. Sci.*, 16 (1978) 190.

211 H. Peterson, G.A. Eiceman, L.R. Field and R.E. Sievers, *Anal. Chem.*, 50 (1978) 2152.

212 R. Sydor and D.J. Pietrzyk, *Anal. Chem.*, 50 (1978) 1842.

213 T.N. Gvozdovich, A.V. Kiselev and Ya.I. Yashin, *Chromatographia*, 11 (1978) 596.

214 A.K. Ghosh, D.P. Rajwar, P.K. Bandyopadhyay and S.K. Ghosh, *J. Chromatogr.*, 117 (1976) 29.

215 D.S. Walker, *Analyst (London)*, 103 (1978) 397.

216 A.A. Datskevich, D.E. Bodrina and N.V. Men'shykova, *Zh. Anal. Khim.*, 34 (1979) 550.

217 V.M. Nemets, A.A. Petrov and A.A. Solov'eva, *Zh. Anal. Khim.*, 35 (1980) 1741.

218 B.E. Foulger and P.G. Simmonds, *Anal. Chem.*, 51 (1979) 1089.

219 P.W. West, *Int. Lab.*, 10 (1980) 39.

220 L.C. Dickson, R.E. Clement, K.R. Betty and F.W. Karasek, *J. Chromatogr.*, 190 (1980) 311.

221 J.H. Elwood, D.J. Robertson, D.G. Gardner and R.N. Groth, *J. Air Pollut. Control Assoc.*, 26 (1976) 1158.

222 A.A. Zhukhovitskii, L.G. Novikova, S.M. Yanovskii, E.V. Kanunnikova and V.P. Shvartsman, *Zh. Anal. Khim.*, 30 (1975) 2251.

223 M.L. Sazonov, M.Kh. Lunskii and S.N. Morozova, *Zavod. Lab.*, 46 (1980) 698.

224 L.C. Michael, M.D. Erickson, S.P. Parks and E.D. Pellizzari, *Anal. Chem.*, 52 (1980) 1836.

225 K. Grob and F. Zürcher, *J. Chromatogr.*, 117 (1976) 285.

226 S.O. Farwell, S.J. Gluck, W.L. Bamesberger, T.M. Shutte and D.F. Adams, *Anal. Chem.*, 51 (1979) 609.

227 D.J. Freed, in H.S. Hertz and S.N. Chesler (Editors), *Trace Organic Analysis: A New Frontier in Analytical Chemistry*, Nat. Bur. Stand. Spec. Publ., 519, Washington, DC, 1979, p. 95.

228 J. Drozd and J. Novák, *J. Chromatogr.*, 136 (1977) 37.

229 J. Drozd and J. Novák, *J. Chromatogr.*, 152 (1978) 55.

230 J. Drozd, J. Novák and J.A. Rijks, *J. Chromatogr.*, 158 (1978) 471.

231 J. Novák, J. Goliáš and J. Drozd, *J. Chromatogr.*, 204 (1981) 421.

232 V.M. Sakharov, G.S. Beskova and A.I. Butuskova, *Zh. Anal. Khim.*, 31 (1976) 250.

233 L.Ya. Ruvinskii, V.V. Cherezova, F.G. Akchurin and G.A. Seichasova, *Khim. Technol. Topl. Masel*, No. 5 (1979) 43.

234 D.B. Smith and L.A. Krause, *Amer. Ind. Hyg. Assoc.*, 39 (1978) 939.

235 J.N. Driscoll, *Ind. Hyg. News*, 3, No. 1 (Mar. 1980).

Chapter 24

Hydrocarbons

E.R. ADLARD

CONTENTS

24.1. INTRODUCTION

It is fitting that a book on chromatography should contain a chapter on the chromatography of hydrocarbons, since the hydrocarbon chemist has probably contributed more to the advancement of GC in the last thirty years than any other group of workers. The vast amount of development work carried out by the hydrocarbon chemist stems from the great commercial importance of the oil industry, the chemical complexity of crude oil and product streams, and the relative chemical inertness of hydrocarbon mixtures, which makes them difficult to separate by other means. However, it is the very inertness of hydrocarbon mixtures that makes them ideal for separation by chromatography, and anyone who has witnessed the separation of a complex hydrocarbon mixture, such as a gasoline, on a capillary column, yielding more than 200 peaks in an hour, would agree that this must surely represent the apogee of the chromatographic art.

24.2. GAS CHROMATOGRAPHY

Gas–solid chromatography (GSC) has been used from the earliest days, at one time rivalling gas–liquid chromatography (GLC) in importance. Over the years, however, the importance of the former has decreased and that of the latter has increased. This is due to a number of reasons, such as the limited range of adsorbents available, the difficulty in reproducing their properties from batch to

References on p. B542

batch, and the relatively high temperatures (compared with the boiling points of the compounds to be separated) that must be used in order to obtain reasonably symmetrical peaks. In spite of these drawbacks, GSC is still used for specific applications, such as trace analysis, where the high amplifier gain required to obtain the necessary detector sensitivity precludes the use of liquid stationary phases [1].

In contrast to GSC, the range of stationary liquids potentially available for GLC is very large. In fact, it has been argued that there are far too many available, and attempts have been made to draw up lists of "preferred" phases. For example, there can be little virtue in one laboratory using dinonyl phthalate and another didecyl phthalate for the separation of similar mixtures, but the American Society for Testing Materials (ASTM) GC methods frequently list several stationary liquids that may be used. It has also been argued that with the increasing use of capillary columns of high separating power, the number of stationary liquids required can be reduced to a minimum. Although there is considerable justice in these arguments, the fact remains that selective liquids enable low-efficiency columns to effect separations that would be much more difficult to obtain if nonselective liquids were used. It is likely, therefore, that many uncommon stationary liquids that happen to fill a particular practitioner's needs will be retained, to the chagrin of the rationalizers.

With one class of packing, the porous polymer materials introduced by Hollis [2], it is not easy to be sure whether the process of separation is GSC or GLC, or a mixture of both. In view of the high operating temperatures required, it is probable that adsorption is the more important process. These materials are useful for the separation of lower-boiling hydrocarbons and have the advantage that they also allow the permanent gases, such as O_2, N_2, and CO_2 to be separated on the same column [3].

Another class of column packing is one that absorbs or reacts with a specific type of compound and removes it from a mixture. This process has been called subtractive chromatography and is exemplified by the removal of n-paraffins with 5A Molecular Sieves and by the removal of olefins and aromatics with mercuric perchlorate. In the latter case, the process is an irreversible chemical reaction and is, therefore, not a chromatographic technique at all, but in conjunction with chromatography, it can be used to simplify a complex mixture either before a chromatographic separation or between the outlet of the chromatographic column and the detector. Picker and Sievers [4] have recently described the application of lanthanide metal chelates as selective complexing agents in a precolumn. The chelating agent removes aldehydes, ketones, esters, ethers, and alcohols, but alkanes, alkenes, and aromatics pass through and are analyzed by the gas chromatograph in a normal fashion. Subsequent heating of the precolumn dissociates the complexes, which are analyzed as a separate fraction.

24.3. APPARATUS

The relative inertness of hydrocarbons makes them less demanding on apparatus design and materials of construction than more reactive and polar mixtures. For example, in the early days of GC, on-column injection of the sample was universal, but in the late 1950s and early 1960s various instrument manufacturers introduced precolumn vaporization chambers with large surface areas and high thermal mass. It is probable that these designs were successful only because the majority of customers at that time worked in the petroleum field, and the hydrocarbon mixtures to be separated were tolerant of such conditions. The last decade has witnessed a gradual realization of the advantages of on-column injection and a return to its use. A somewhat similar story applies to columns. Most of the early, home-made equipment contained straight or U-tube glass columns, but the commercial manufacturers introduced coiled metal columns which required much smaller ovens. In this case, there was a less radical departure from early practice in that coiled glass columns were often used, since the advantage of being able to see the packing outweighed the disadvantage of fragility. Glass capillary columns have now achieved a place in the analyst's repertoire after two decades of slow development, first by Desty [5] and more recently by Grob and coworkers [6–11] and many others.

In discussing the detectors used for the GC analysis of hydrocarbons, there can be few who would disagree about the importance of the FID. When properly constructed and used, this detector has many virtues and hardly any vices. Among the virtues are a high sensitivity, a large linear range, and a response proportional to the carbon number of the hydrocarbon eluted in the flame. The only drawback of any significance is the absence of response to permanent gases. This precludes the sole use of the FID for mixtures of lower-boiling hydrocarbons and permanent gases, in particular, air. Such mixtures are of considerable practical importance and must be tackled either with a TCD alone or with an FID in series with a TCD. Most of the major manufacturers now sell instruments with the two detectors in series, but they are still regarded as somewhat "nonstandard", a factor reflected in prices and delivery times. Enormous advances have been made in the design and performance of TCDs, but they still need more calibration than FIDs.

In the petroleum industry, hydrocarbon mixtures frequently act as a matrix for small (1% and less) amounts of impurities such as sulfur and nitrogen compounds, and at the low concentrations involved, it is often impossible to detect the presence of the compounds containing hetero-atoms. One solution to this problem is to use selective GC detectors, which have a much greater response to the hetero-compounds than to hydrocarbons [12–14]. The flame-photometric detector, which is selective for sulfur and phosphorus compounds; the flame-thermionic detector, which is selective for nitrogen compounds; and the electron-capture detector, which is especially sensitive to some polycyclic aromatic hydrocarbons, halogenated compounds, and lead alkyls, all enjoy considerable usage in the petroleum industry for the analysis of products of essentially hydrocarbon composition.

Spectroscopic methods, such as UV and IR spectrometry play a relatively unimportant role as detectors for the GC separation of hydrocarbon mixtures, since the amounts of material required by these methods are large by GC standards. GC followed by MS (GC–MS), on the other hand, is probably the most powerful combination available for the identification of the individual compounds separated from a complex hydrocarbon mixture. Without GC–MS and a computer for data

References on p. B542

interpretation, the labor of identification would be so great as to be impractical in many cases. An important form of GC–MS is that in which the mass spectrometer is used in the selective-ion mode, i.e. the spectrometer is set in such a way that it only detects a few (usually from one to about six) ions of specifically designated masses that are characteristic for certain types of compounds. For example, mass 71 is characteristic for paraffins, mass 91 for mono-aromatics (excluding benzene), mass 141 for methyl naphthalenes, and so on. Use of the instrument in this way can give an increase in sensitivity of up to 10^3 times, and this increase is particularly desirable in the analysis of environmental samples which often consist of a complex mixture in dilute solution [13,14].

The early work on GC was carried out with isothermal column conditions, but it was soon found that this limited the range of the mixture to be separated to about five or six carbon numbers. Beyond this range, the later peaks broadened so much that quantitative work, especially at low concentrations, became very difficult. It was realized that the solution to this problem was to raise the temperature of the column gradually as the separation proceeded. In modern equipment, this temperature programing is usually carried out with a linear increase of temperature with time. The rate of temperature rise depends on the mixture to be separated, but normally lies in the range of about 2–10°C/min, although both lower and higher rates are used for special purposes. For example, a high rate can be used at the end of an analysis to remove high-boiling material that is not of interest. Most instrument manufacturers now sell gas chromatographs with sub-ambient temperature programing capabilities. Such facilities make the separation of low-boiling compounds much easier. With temperature programing the temperature of a GC column is often raised during the separation of a wide-boiling-range hydrocarbon mixture, such as a crude oil, by 200–300°C, thus allowing a carbon number range of 40–50 to be analyzed on one column.

24.4. GAS ANALYSIS

Hydrocarbon gas mixtures generally range from methane to n-pentane, but, on occasions, gas mixtures may be analyzed for small amounts of components up to C_8. Al-Thamir et al. [15] state that of the 32 possible C_1 to C_4 hydrocarbons, six are thermally unstable and six more can be obtained only by synthesis. By using alumina, coated with 7.25% (w/w) squalane in a 33-ft. column at 50°C, these workers were able to separate nearly all the remaining hydrocarbons. In a later paper [16], the same authors demonstrated that mixing the squalane-treated alumina with an inert diluent (Chromosorb P) improved the permeability of the column and gave the same separation in a much shorter time. This work, an extension of that first described by Eggertsen et al. [17] and by McKenna and Idleman [18], is an elegant piece of chromatographic virtuosity, but most of the hydrocarbon gas mixtures encountered in practice are relatively simple in composition, and difficulties arise mainly from the presence of permanent gases rather than in the separation of the hydrocarbons themselves.

The modern approach to this type of work takes one of two forms:

(a) A multi-column technique, which at its simplest may require several separate analyses. In its more sophisticated form, only one injection is required, and column switching is carried out manually or automatically at the appropriate time.

(b) A single-column method, making use of extreme temperature programing conditions from, say, $-50°C$ to $+200°C$.

These alternatives are exemplified by two methods recommended by the UK Institute of Petroleum (IP). The first method, IP 337/78, is intended for the analysis of the permanent gases, O_2, N_2, He, and CO_2, and paraffin hydrocarbons from C_2 to C_5 in natural gas (methane). It utilizes two columns and requires three separate analyses. The first column, 2 m × 4 mm ID, packed with 80–100-mesh 13X Molecular Sieve, effects the separation of He, O_2, and N_2 at 50°C. The second column, of the same dimensions, packed with 80–100-mesh Porapak Q (a porous polymer bead

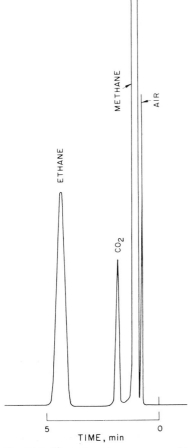

Fig. 24.1. Separation of permanent and hydrocarbon gases on a porous polymer bead column at 50°C. For other conditions, see text. (From IP 337, by courtesy of the Institute of Petroleum.)

References on p. B542

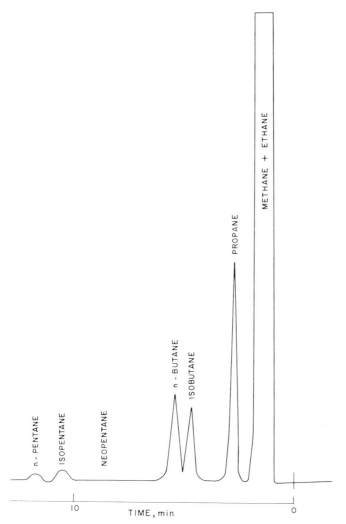

Fig. 24.2. Separation of hydrocarbon gases on a porous polymer bead column at 140°C. For other conditions, see text. (From IP 337, by courtesy of the Institute of Petroleum.)

packing) is operated at 50°C and at 140°C to give the separations shown in Figs. 24.1 and 24.2. A TCD is used, and the method requires the use of Ar carrier gas for the first column and He for the second. It is apparent that the complete analysis is tedious and slow, but it has the virtue of requiring very simple GC equipment. With an apparatus capable of temperature programing, the two analyses on the Porapak column could be combined, and with column-switching valves the whole analysis might be carried out on one gas sample, resulting in a drastic reduction in the time required at the expense of more complex equipment. There are several instrument

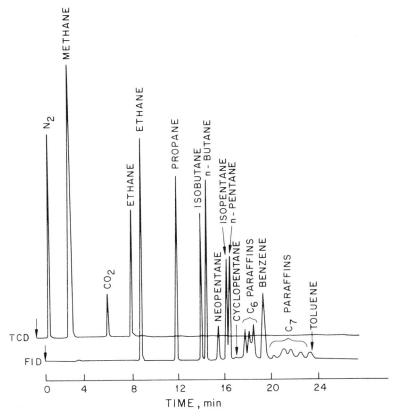

Fig. 24.3. Typical natural gas chromatogram. For conditions, see text. (Reproduced from *Anal. Chem.*, 47 (1975) 383, with permission [3].)

manufacturers who will make a custom-built chromatograph with multiple columns and automatic valve switching to cope with a given combination of permanent gases and hydrocarbon gases.

The second method, currently in the process of evaluation by the IP, is based on the work of Stufkens and Bogaard [3]. The method is intended to separate O_2, N_2, CO_2, individual C_1 to C_5 paraffins, and total C_6, C_7, and C_8 hydrocarbons as carbon number groups (Fig. 24.3). This gas mixture is obtained when crude oil, emerging from a well at high pressure, is reduced to a lower pressure to facilitate transport. By analyzing the gas and the low-boiling compounds left in the liquid phase, the composition of the front-end of the original crude oil at high pressure can be calculated. The need for determining low concentrations of the C_6 to C_8 hydrocarbons demands the high sensitivity of an FID, while the determination of the permanent gases requires a TCD. The solution to the problem is to use both detectors in series at the column exit—the TCD first, followed by the FID—and to link the response of the two detectors by a compound, such as ethane or propane, which gives a substantial peak with both detectors. The separation is effected on a 3 m × 2.3-mm ID stainless-steel column, packed with Porapak R, temperature-programed from −50°C to +150°C, with He as the carrier gas. This method is simpler than multi-column techniques, which depend on the perfect operation of the switching valves. These must be leak-tight and be in synchronization with the chromatographic separations taking place on different columns.

It will be noted that both of the methods involve GSC, with molecular sieves and porous polymers as the adsorbents. Experience has shown that these types of adsorbent are the most reproducible in their properties, although even these materials do sometimes vary from batch to batch. Many separations of hydrocarbon gases on other adsorbents have been published, e.g., the excellent separation achieved by McTaggart et al. [19], who used hydrated alumina, as shown in Fig. 24.4. However, GSC methods tend to founder in that they require considerable skill and experience in order to give satisfactory and reproducible results. A number of early methods for the separation of hydrocarbon gases were based on GLC, in which the properties of the stationary phase are easier to reproduce than in GSC. An example of this approach is the separation shown in Fig. 24.5 [20]. Although the separation is satisfactory, the time of four hours required for the separation is not acceptable. However, with the advances made in GC since this publication appeared, especially the ready availability of temperature programing (both normal and sub-ambient), it is probable that the time requirements could be drastically reduced, and the analysis of hydrocarbon gases may yet see a return to GLC in preference to GSC.

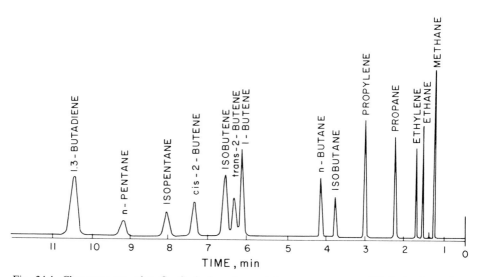

Fig. 24.4. Chromatogram of a C_1–C_5 hydrocarbon mixture on a water-modified alumina column. Chromatograph, Perkin-Elmer 452; column, 40 ft.×0.012-in. ID glass capillary, packed with water-modified γ-alumina; temperature, 80°C; carrier gas, H_2, 35 psi; presaturator, $CuSO_4 \cdot 5 \, H_2O$ at 30°C; splitter ratio, 1:500. (Reproduced from *J. Inst. Petrol.*, 54 (1968) 265, with permission [19].)

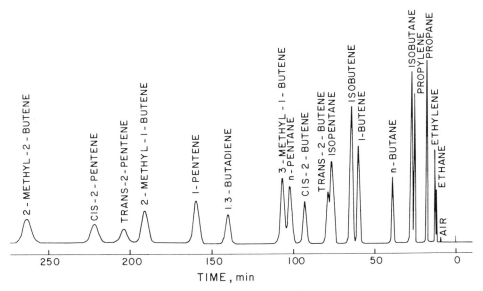

Fig. 24.5. Chromatogram of a C_2–C_5 hydrocarbon mixture on a 50-ft. dimethylsulfolane column at $0°C$. (Reproduced from *Anal. Chem.*, 28 (1956) 297, with permission [20].)

The most difficult task encountered in the analysis of hydrocarbon gas mixtures is not generally the actual separation, but the calibration of the detectors. In order to calibrate the detector, gas blends of known composition must be prepared, either gravimetrically or based on partial pressures. Pressure blending is the easier of the two, but gas law imperfections must be allowed for when calculating composition. Gravimetric blending is the more accurate method, but is tedious to carry out. Great care is required, because the container invariably weighs about three orders of magnitude more than the mass of each gas to be added. Whichever way the blends are prepared, they should be allowed several days to equilibrate, preferably with mechanical mixing, since—contrary to popular belief—gases take an appreciable time to mix. When gas blends containing parts per million of various compounds are required, static blends made up in containers are suspect owing to adsorption on the walls of the vessel. At these concentrations, blends made by dynamic dilution are more reliable, although, as with static blending, the setting up of the equipment is tedious and time-consuming. A minimum of two dilution stages is required, and three stages are more commonly used.

24.5. VOLATILE LIQUID HYDROCARBON MIXTURES

Gasoline, which ranges in boiling point from $-42°C$ (propane) to about $216°C$ (*n*-dodecane) and consists of a complex mixture of saturated hydrocarbons, olefins, and aromatics, may be taken to be typical of this class. Other important examples are naphtha and olefin feedstocks for petrochemical manufacture, saturated hydrocarbon solvent mixtures, and individual, commercially pure compounds, such as toluene and xylene. It is in this boiling range that GC is at its most versatile, and the analysis may be tackled in two alternative ways—either by using nonselective high-resolution columns or by using columns of lower resolution containing highly selective stationary phases. Variations on these two themes include hybrid systems, such as a low-resolution packed column in series with a capillary column or several different packed columns, connected by valves to give the appropriate elution pattern at the correct time. This latter approach may be necessary for the analysis of mixtures of hydrocarbons and oxygenated compounds, such as alcohols and ethers, now being used on an experimental basis in a number of countries as a fuel for internal combustion engines.

The routine application of long (50 m and longer) high-resolution capillary columns for the analysis of gasoline-range mixtures was pioneered by Sanders and Maynard [21] and has been developed more recently by Whittemore [22], Adlard et al. [23] and Schulz et al. [24]. All these workers used squalane-coated columns, since this stationary liquid gives the best separation of low-boiling hydrocarbons and also because much information is available on the identification of complex mixtures

TABLE 24.1

PRECISION DATA FOR PEAK AREA MEASUREMENTS FOR SOME OF THE COMPONENTS IN A TYPICAL FULL-RANGE GASOLINE, ANALYZED 42 TIMES [23]

$V = 100\,\sigma/\bar{x}.$

Peak identification	Mean % peak area, $\bar{x}$	Standard deviation, σ	% Standard deviation, V
Isobutane	1.56	0.110	7.05
Isobutene + 1-butene	0.555	0.028	5.12
Isopentane	1.889	0.046	2.44
2,2-Dimethylbutane	0.382	0.010	2.62
3-Methylpentane + ethyl-1-butene	2.89	0.042	1.42
cis-3-Methyl-2-pentene + cis-2-hexene	4.84	0.080	1.65
Methylcyclopentane	2.63	0.091	3.46
Benzene	7.90	0.100	1.27
3-Methylhexane	3.29	0.035	1.07
cis-1,2-Dimethylcyclopentane	0.223	0.006	2.65
2,5-Dimethylhexane	0.812	0.020	2.36
Toluene	10.72	0.170	1.58
2,2-Dimethyl-3-ethylpentane + 2-methyl-4-ethylhexane	0.149	0.010	6.37
1,3,5-Trimethylbenzene	0.296	0.011	3.83
n-Propylbenzene	0.410	0.020	4.87
1,2,3-Trimethylbenzene	0.771	0.024	3.15
1,3-Dimethyl-5-ethylbenzene	0.330	0.018	5.44
1,4-Dimethyl-2-ethylbenzene	0.160	0.007	4.37
s-Buthylbenzene	0.116	0.005	4.38

TABLE 24.2

PRECISION DATA FOR PEAK RETENTION TIMES FOR SOME OF THE COMPONENTS IN A TYPICAL FULL-RANGE GASOLINE, ANALYZED 42 TIMES [23]

Peak identification	Mean retention time, sec	Standard deviation, σ	% Standard deviation, V
Isobutane	462.5	3.47	0.75
Isobutene + 1-butene	485.2	3.45	0.71
Isopentane	673.9	4.96	0.74
2,2-Dimethylbutane	899.4	5.11	0.57
3-Methylpentane + ethyl-1-butene	1140.1	5.92	0.52
cis-3-Methyl-2-pentene + cis-2-hexene	1236.9	5.43	0.43
Methylcyclopentane	1370.3	5.75	0.42
Benzene	1429.6	6.08	0.43
3-Methylhexane	1685.2	5.82	0.35
cis-1,2-Dimethylcyclopentane	1948.1	5.66	0.29
2,5-Dimethylhexane	2032.6	8.58	0.42
Toluene	2115.7	6.48	0.30
2,2-Dimethyl-3-ethylpentane + 2-methyl-4-ethylhexane	2487.7	6.20	0.25
1,3,5-Trimethylbenzene	2612.4	6.88	0.26
n-Propylbenzene	3261.0	11.04	0.33
1,2,3-Trimethylbenzene	3533.6	10.56	0.30
1,3-Dimethyl-5-ethylbenzene	3952.5	12.85	0.33
1,4-Dimethyl-2-ethylbenzene	4431.4	16.44	0.37
s-Butylbenzene	5470.4	23.64	0.43

separated on such columns. However, squalane suffers the serious limitation of having a temperature maximum of about 100°C and, as a consequence, flow programing has been employed in addition to temperature programing.

Attempts by the ASTM in the USA and the IP in the UK to standardize capillary column methods for the quantitative analysis of gasolines have so far been unsuccessful because of the poor reproducibility obtained. There is no doubt, however, that a given laboratory can, with practice, achieve acceptable precision and accuracy. Precision and accuracy data obtained by Adlard et al. [23] for the analysis of gasoline range blends are shown in Tables 24.1–24.3. The compounds showing poorer area repeatability are either very volatile, such as isobutane, or are present in small amount, e.g., 2,2-dimethyl-3-ethylpentane. The recent introduction of capillary columns made of pure silica, as opposed to stainless steel or glass, may lead to improved reproducibility, but the main source of variation probably lies in the injector, followed to a lesser extent by the detector and data-handling systems. Less demanding analyses with capillary columns have been standardized (e.g., ASTM method D2268 for the analysis of high-purity n-heptane and isooctane).

The main disadvantage of the analysis of complex hydrocarbon mixtures by squalane-coated capillary columns is the long analysis time required, which is

References on p. B542

TABLE 24.3

COMPARISON OF EXPERIMENTAL AND KNOWN COMPOSITION FOR A SYNTHETIC MIXTURE [23]

Hydrocarbon	Time, sec	% Area	%w
1. 2-Methylbutane	656	10.59	10.41
2. 1-Pentene	676	0.64	0.71
3. *n*-Pentane	739	1.23	0.92
4. *trans*-2-Pentene	744	0.63	0.99
5. *cis*-2-Pentene	756	0.42	
6. 3,3-Dimethyl-1-butene			0.16
7. 2-Methyl-2-butene	790	0.05	0.03
8. 2,2-Dimethylbutane	867	3.06	3.05
9. 4-Methyl-1-pentene	926	0.17	0.20
10. Cyclopentane	984	1.01	1.09
11. 2,3-Dimethylbutane	1005	3.94	3.77
12. 2-Methylpentane	1028	7.15	6.38
13. 2-Methyl-1-pentene	1076	0.31	0.42
14. 1-Hexene	1087	1.12	1.10
15. 2-Ethyl-1-butene	1139	0.46	0.53
16. *trans*-2-Hexene	1165	0.14	0.98 c/t
17. 2-Methyl-2-pentene	1171	0.26	0.37
18. *n*-Hexane	1181	1.25	1.26
19. 4,4-Dimethyl-1-pentene			0.17
20. *cis*-3-Methyl-2-pentene	1195	0.58	0.38 c/t
21. *cis*-2-Hexene			
22. *trans*-3-Methyl-2-pentene	1246	0.10	–
23. Methylcyclopentane	1305	0.38	0.22
24. 2,3-Dimethyl-2-butene	1309		0.25
25. 2,4-Dimethylpentane	1335	0.52	0.60
26. Benzene	1358	1.35	1.42
27. 2,2,3-Trimethylbutane	1374	1.10	0.49
28. 2,4-Dimethyl-1-pentene			0.58
29. 2,4-Dimethyl-2-pentene	1404	0.11	0.11
30. 3-Methyl-1-hexene	1412	0.48	0.51
31. *trans*-2-Methyl-3-hexene	1437	0.13	0.18
32. 5-Methyl-1-hexene	1446	0.38	0.43
33. 4-Methyl-1-hexene	1494	1.43	0.13
34. Cyclohexane			1.33
35. 2,3-Dimethylpentane	1563	0.44	0.27
36. 1,1-Dimethylcyclopentane			0.45
37. 3-Methylhexane	1597	0.31	0.37
38. 2-Methyl-1-hexene	1609	0.15	0.16
39. *cis*-1,3-Dimethylcyclopentane	1617	0.16	0.42 c/t
40. 1-Heptene	1632	0.38	0.45
41. *trans*-1,3-Dimethylcyclopentane	1645	0.39	0.18
42. *cis*-2,5-Dimethyl-3-hexene			
43. 3-Ethylpentane	1649		0.13
44. *trans*-1,2-Dimethylcyclopentane	1657	0.38	0.43 c/t
45. *trans*-3-Heptene	1665	0.34	0.06
46. 2,2,4-Trimethylpentane			0.06

TABLE 24.3 (continued)

Hydrocarbon	Time, sec	% Area	%w
47. *cis*-3-Heptene	1679	} 0.21	0.37 c/t
48. 2-Methyl-2-hexene	1686		0.11
49. *trans*-2,5-Dimethyl-3-hexene	1712	0.17	0.20
50. *trans*-2-Heptene	1730	} 1.91	0.03
51. *n*-Heptane	1739		1.78
52. *cis*-2-Heptene	1753	0.28	0.96 c/t
53. *cis*-1,2-Dimethylcyclopentane	1840	0.02	–
54. Methylcyclohexane	1867	0.78	0.83
55. 2,5-Dimethylhexane	1899	0.29	0.41
56. 2,4-Dimethylhexane	} 1920	0.44	0.06
57. Ethylcyclopentane			0.42
58. 2,3-Dimethyl-1-hexene	1960	0.23	0.30
59. *trans*-2-Methyl-3-heptene	1976	0.10	0.11
60. Toluene	1993	5.77	4.90
61. 2,5-Dimethyl-2-hexene	2025	0.17	0.20
62. 2,3,4-Trimethylpentane	2030	0.15	0.14
63. 2,3-Dimethylhexane	2081	0.16	0.19
64. 2-Methylheptane	2110	0.26	0.32
65. 4-Methylheptane	2123	0.12	0.13
66. 3,4-Dimethylhexane	2141	0.20	0.24
67. 3-Methylheptane	} 2153	0.21	0.18
68. 3-Ethylhexane			0.06
69. 2-Methyl-1-heptene	} 2176	0.26	0.06
70. 2,2,5-Trimethylhexane			0.25
71. 1-Octene	2205	0.30	0.38
72. *trans*-1,4-Dimethylcyclohexene	} 2228	} 1.04	0.41
73. *cis*-1,3-Dimethylcyclohexene			0.61 c/t
74 1,1-Dimethylcyclohexane	2236		0.37
75. 2-Methyl-2-heptene	2256	0.10	0.10
76. *trans*-2-Octene	2301	0.13	0.64 c/t
77. *n*-Octane	} 2308	} 4.82	3.77
78. *trans*-1,2-Dimethylcyclohexene			0.42
79. *cis*-2-Octene	2325		–
80. *trans*-1,3-Dimethylcyclohexene	} 2344	} 0.56	–
81. *cis*-1,4-Dimethylcyclohexene			0.43
82. 2,4-Dimethylheptane	2430	1.04	1.02
83. 2,6-Dimethylheptane	2465	0.44	0.63
84. *cis*-1,2-Dimethylcyclohexene	2486	0.74	0.70
85. 2,5-Dimethylheptane	2497		0.16
86. Ethylbenzene	2526	7.52	6.53
87. 1,4-Dimethylbenzene	2608	2.15	2.11
88. 1,3-Dimethylbenzene	} 2622	} 3.49	2.43
89. 2,3-Dimethylheptane			0.70
90. 3,4-Dimethylheptane	2641	0.25	0.25
91. 4-Methyloctane	2668	} 0.93	0.13
92. 2-Methyloctane	2678		0.81
93. 3-Methyloctane	2710	0.19	0.19

References on p. B542

TABLE 24.3 (continued)

Hydrocarbon	Time, sec	% Area	%w
94. 1,2-Dimethylbenzene	2733	3.64	3.21
95. 1-Nonene	2773	0.55	0.62
96. Isopropylbenzene	⎱ 2877	⎱ 4.47	0.65
97. n-Nonane	⎰	⎰	3.48
98. n-Propylbenzene	3046	2.56	2.52
99. 3-Ethyltoluene	3133	0.10	0.08
100. 4-Ethyltoluene	3147	0.16	0.14
101. 2-Ethyltoluene	⎱ 3240	⎱ 0.21	0.15
102. 5-Methylnonane	⎰	⎰	0.07
103. 4-Methylnonane	3251	0.15	0.13
104. 2-Methylnonane	⎱ 3275	⎱ 1.21	0.49
105. 1,3,5-Trimethylbenzene	⎰	⎰	0.94
106. t-Butylbenzene	3305	0.29	0.31
107. 3-Methylnonane	3323	0.22	0.19
108. 1,2,4-Trimethylbenzene	3410	2.52	1.78
109. 1-Decene	⎱	⎱	0.25
110. Isobutylbenzene	⎱ 3461	⎱ 0.41	0.22
111. s-Butylbenzene	⎰	⎰	0.23
112. n-Decane	3567	1.32	1.39
113. 1,2,3-Trimethylbenzene	⎱ 3625	⎱ 1.21	1.23
114. 4-Isopropyltoluene	⎰	⎰	0.11
115. Indan	3647	0.49	0.52
116. 1,3-Diethylbenzene	3801	0.29	0.28
117. n-Butylbenzene	3868	0.59	0.65
118. 4-n-Propyltoluene	3906	0.27	0.21
119. 4-Methyldecane	4178	0.16	0.14
120. 2-Methyldecane	4233	0.27	0.25
121. 3-Methyldecane	4303	0.05	0.07
122. n-Undecane	4729	0.76	0.91
123. 1,2,4,5-Tetramethylbenzene	4780	0,51	0.51
124. 1,2,3,5-Tetramethylbenzene	4869	0.26	0.24
125. 1,2,3,4-Tetramethylbenzene	5284	0.08	0.08
126. Naphthalene	5578	0.09	0.10

dictated by the temperature limitation of squalane. Fig. 24.6, showing the analysis of a gasoline on such a column, demonstrates this point: most of the sample has emerged after 60 min, but a further 40 min is required to elute the (mainly aromatic) higher-boiling compounds. This temperature limitation of squalane has prompted the search for a stationary liquid giving as good a separation but a higher temperature limit [25–28], but so far, squalane still appears to be the best material available.

If it is required to separate and measure only the aromatics, then either a packed or a capillary column, with a polar stationary liquid, such as 1,2,3-tris(2-

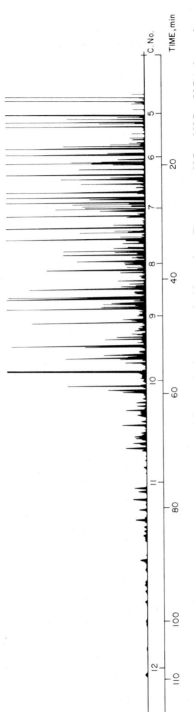

Fig. 24.6. Chromatogram of a gasoline on a 70 m × 0.25-mm stainless-steel column, coated with squalane. Temperature, 0°C to 95°C at 2°C/min; carrier gas, He, 15 to 65 psi; sample size, 0.5 μl, 50:1 split.

cyanoethoxy)propane (TCEP) or polyethylene glycol 400 (PEG 400), may be used. These stationary liquids permit the rapid elution of saturates and olefins with minimal separation, followed by the aromatics, which are well resolved from each other. Fig. 24.7 shows the separation of the aromatics in a gasoline, obtained on a 50 m × 0.2-mm ID silica capillary column, coated with PEG 400.

By combining a selective stationary phase (thiodipropionitrile) to separate the aromatics from the saturates and olefins, and mercuric perchlorate–perchloric acid to remove the olefins, Martin [29] was able to separate a gasoline into the three main classes, saturates, olefins, and aromatics, to give results equivalent to those from LC with fluorescent indicators (FIA) (see Chap. 24.9). The method has been improved by Robinson et al. [30] and by Stavinola [31], but it has not gained great popularity, parly because of the somewhat intractable properties of mercuric perchlorate. This compound absorbs water readily and turns into a hard, cement-like mass with low

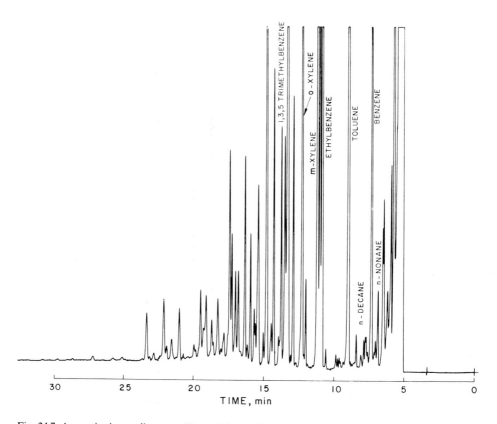

Fig. 24.7. Aromatics in gasoline on a 50 m × 0.2-mm silica column, coated with PEG 400. Temperature, 50°C for 6 min, then 10°C/min to 110°C; carrier gas, He at 30 psi; sample size, 0.5 μl, 100:1 split.

permeability. A further problem is surface adsorption on the solid support used for the perchlorate.

Another technique that has gained greater acceptance, especially for the analysis of petrochemical feedstocks, separates naphtha samples boiling up to about 200°C into paraffins, napthenes, and aromatics. In addition to a polar column for separating the aromatics from the saturates, a column containing 13X Molecular Sieve is utilized. This adsorbent has the property of separating paraffins and naphthenes from each other to give two discrete peaks for each carbon number throughout the range. The overall method is complicated, with trapping-out, backflushing, and column switching at the appropriate times during the analysis. Further information on this technique can be found in the IP method 321/77; a specialized GC apparatus for this analysis is commercially available.

The separation of olefins from saturated hydrocarbons by the use of stationary phases containing silver salts dates back to the earliest days of GC [32]. Silver nitrate, dissolved in glycerol, is the most common stationary phase, but Wasik and Tsang [33] pointed out that the active agent in these separations is the Ag$^+$ ion. Therefore, aqueous solutions are more effective, although there are obviously severe temperature limitations to such systems and hence to the maximum molecular weight of the mixtures that can be handled, which would appear to be about 80–100, i.e. C_6–C_7.

An interesting feature of stationary phases incorporating silver nitrate is their ability to separate hydrocarbons and deuterated hydrocarbons [33,34]. Olefin separations using other metal complexes have been reported [35,36] and Guha and Janák [37] have published a comprehensive review of the subject. More recently Soják and coworkers [38,39] have used high-resolution capillary columns for the separation of complex olefin mixtures. A paper by Schurig [40] describes the application of optically active rhodium complexes for the separation of deuterated ethylenes and also for the separation of the enantiomers of cyclic olefins with optically active sites.

24.6. HYDROCARBON MIXTURES IN THE KEROSINE AND GAS OIL RANGE

Kerosine and gas oils cover the boiling range from about 100°C to about 380°C, i.e. from about C_7 to C_{24}; in this boiling range, selective separations of the kind described in the preceding section are no longer feasible by GC. Techniques such as the separation of n-paraffins with 5A Molecular Sieves are still applicable, but cannot be considered to be truly chromatographic in nature. Although type separation is not possible, excellent boiling point separations can still be obtained, especially if capillary columns coated with methylsilicone stationary liquids are used. Fig. 24.8 shows the separation of a < 343°C fraction of a crude oil obtained on a 50 m × 0.2-mm ID silica capillary column coated with methylsilicone SP-2100. The small peak just appearing between n-C_{17} and pristane and the two peaks between n-C_{18} and phytane are particularly noteworthy. Pristane and phytane (2,6,10,14-tetramethylpentadecane and 2,6,10,14-tetramethylhexadecane, respectively) are two important members of the isoprenoid hydrocarbon series; their presence in crude oil

References on p. B542

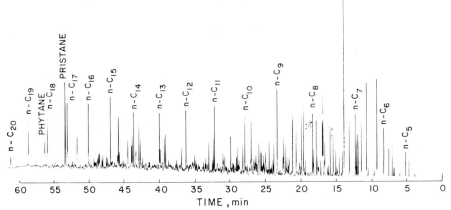

Fig. 24.8. $< 343°C$ fraction of Light Nigerian crude oil on a 50 m $\times$ 0.2-mm silica column, coated with SP-2100. Temperature, 0°C for 3 min, then 4°C/min to 275°C; carrier gas, He at 0.8 ml/min; sample size, 1 μl, 200:1 split.

is of importance not only to organic geochemists and to biologists but also as a method of oil spill identification, since their relative amounts vary widely from oil to oil [41].

Although it is not possible to separate nearly every component of a mixture in the gas oil boiling range, as it is in the gasoline range, the analyst is frequently in the position of being able to give the applications chemist or engineer far more information than can be used to solve a particular problem. With modern computer facilities it is a simple matter to recombine the information produced by the GC analysis in any one of a number of ways so as to present it in the best form for interpretation. For example, it is possible to group all the peaks with a given carbon number and present the results as a carbon number distribution, or to give the approximate n-paraffin content (approximate, since no matter how good the column, it is impossible to guarantee that there is no other compound under a n-paraffin peak). The results may also be presented in the form of a simulated distillation curve; this method of presentation is discussed in Chap. 24.8.

24.7. POLYCYCLIC AROMATIC HYDROCARBONS

In the last ten years the polycyclic aromatic hydrocarbons (PCA) have attracted considerable analytical effort because of the potent carcinogenic properties of some of the members of this group. Considerable care is needed when reading the literature on PCA, since several systems of nomenclature have been used, and historically hallowed names die hard. The names by which the compounds are commonly known usually follow the International Union of Pure and Applied Chemistry (IUPAC) rules [42], but there are exceptions, such as benzo[a]pyrene which, according to IUPAC rules, should be called benzo[def]chrysene; this is an example of a historically accepted name resisting attempts at systematization.

Before the PCA mixture is analyzed by GC, the original starting material is subjected to a preconcentration and separation stage of a varying degree of complexity, depending to some extent on the nature of the starting material. Exhaust gases from automobile engines [43,44], lubricating oils [45–47], gasoline [44,46], tobacco smoke [48,49], particulate matter [50,51], marine organisms and sediments [52,53], foodstuffs [54], and industrial wastewater [55] are just some of the materials examined. The references quoted are given merely as typical examples and in no way constitute a comprehensive list.

In spite of the wide range of starting materials, there are considerable similarities between clean-up procedures in many of the papers. After solvent extraction, the aromatics are separated from other hydrocarbons by LC on silica gel, followed by gel permeation chromatography (GPC) to isolate the PCA fractions. A typical scheme of this sort is shown in Fig. 24.9 [45]. Sometimes, more selective solvents, such as nitromethane, acetonitrile, and dimethyl sulfoxide, are used to simplify extraction. Because of the complexity of the extraction/concentration procedure, losses are inevitable, and it is common practice to add a labeled PCA compound, such as [^{14}C]benzo[a]pyrene, to the starting material to measure the recovery [49]. An inactive compound, such as picene, is also introduced prior to GC as an internal marker.

Brown and Searl [56] used a 3 m × 2.2-mm ID stainless-steel column, packed with 2% (w/w) silicone gum SE-30 on 80–100-mesh Chromosorb G. Presumably because this column gave relatively poor separation, they split the effluent between an FID and a succession of traps. The trapped fractions were then analyzed, both qualitatively and quantitatively by UV spectroscopy. Similarly, John and Nickless [50], using a 1 m × 4-mm ID column, packed with Dexsil-300, trapped fractions for UV fluorescence analysis. Grimmer and Böhnke [45] have made extensive use of high-ef-

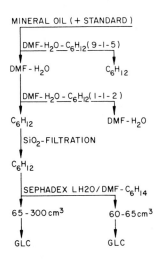

Fig. 24.9. Scheme for concentration of PCA from mineral oil products. (Reproduced from *Chromatographia*, 9 (1976) 30, with permission [45].)

References on p. B542

ficiency glass columns, 10 m × 2 mm ID, packed with 5% (w/w) OV-101 on Gas Chrom Q. However, from the early 1970s onwards [44], there has been an increasing trend towards the use of glass capillary columns, coated with a variety of silicones. A number of papers have been published on the use of liquid-crystal compounds as stationary phases for the separation of PCA mixtures [48, 57–60]. Although they separate some compounds that are not well resolved on silicone phases, the overall improvement is not great, and the separations obtained do not compare favorably with those obtained with silicone-coated high-resolution columns.

24.8. LUBRICATING OILS AND CRUDE OILS

Lubricating oils are complex mixtures of paraffin, naphthene, and aromatic hydrocarbons, and in addition to their chemical complexity they also have high boiling ranges, from about 350°C to 550°C, i.e. molecular weight ranges from about 300 to 600. GC alone cannot separate individual components from mixtures of this range and complexity, but it can still be made to give extremely useful information in the form of a simulated distillation curve. If a nonpolar silicone stationary phase is used in a short, packed column with linear temperature programing, the components of the oil will emerge in the order of their boiling points to give the sort of chromatogram shown in Fig. 24.10. If the gas chromatograph is calibrated with a mixture of compounds of known boiling points, usually a mixture of n-paraffins, then the time axis of the resulting chromatogram can be converted into a temperature scale. When the sample is injected onto the column, an integrator or computer system measures the cumulative areas under equal time slices of the chromatogram,

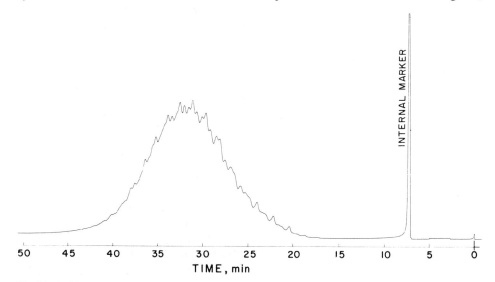

Fig. 24.10. Chromatogram of a lubricating oil on a short, packed column. Column, 1 m × 3.2 mm ID, packed with 5% OV-101 on 60–80-mesh Chromosorb W; temperature, 60°C to 320°C at 5°C/min; carrier gas, He at 25 ml/min; sample size, 1 μl.

usually at about 10-sec intervals. From these results, a graph can then be plotted of cumulative area against boiling point to give a simulated distillation curve. If the sample is very high-boiling, it is necessary to add a known amount of a low-boiling internal marker, such as *n*-dodecane, since quantitative recovery of the whole sample is unlikely. Further details of the technique can be found in ASTM method D2887. Although it has been described here for use with high-boiling mixtures, it can, with suitable modification, be used for lower-boiling mixtures, such as gasolines (ASTM method D3710).

Crude oils, like lubricating oils, are a complex mixture of paraffins, naphthenes, and aromatics, but the compounds in crude oil range in molecular weight from 16 (methane) to 800 and beyond (asphaltenes). The analysis of the components of crude oil with lower molecular weight (up to about C_8) has been described in Chap. 24.4. The analysis of the components with high molecular weight is impossible by GC, but some progress has been made by LC [61]. The routine GC analysis of crude oils is taken up to C_{42}, but even to this point, recovery is only in the range of 80–90%. Since complete recovery is, in general, not possible, the use of an internal marker is necessary, as in the analysis of lubricating oils. However, this technique is difficult to carry out, since there are no gaps in the crude oil chromatogram where a judiciously chosen marker can emerge. The solution is to use a *n*-paraffin as an internal marker and to run the sample twice—once without the marker and a second time with a known percentage of marker. Any convenient peak in the chromatograms can be used to make a correction for variations in sample size; the response per unit weight of marker can then be calculated from the two chromatograms. Worman and Green [62] proposed the use of four paraffins as internal markers to make it easier to use a reasonable total percentage of marker without overloading the FID. McTaggart and Glaysher [63] compared the results obtained by this method with those obtained from a 15-plate distillation and found good agreement. The "four internal marker" method is now widely used throughout the petroleum industry.

24.9. LIQUID CHROMATOGRAPHY

LC has been used for the separation of hydrocarbon mixtures for many years. The work of Criddle and Le Tourneau, published in 1951 [64], led to the establishment of what must be one of the earliest standard methods based on LC [65]. This method uses a glass column, 122 cm × 1.6 mm ID, packed with 100–200-mesh silica gel. A short section near the top of the column contains three fluorescent dyes, which travel with the components of the sample and appear under a UV lamp at the olefin front, the aromatic front, and the end of the aromatic band (hence, the name fluorescent indicator analysis or FIA). The sample is eluted by the continuous addition of 2-propanol to the top of the column until the wetted front, which is the front of the saturates, is within a few centimeters of the column outlet. When this stage is reached, the length of each zone is measured, and from these lengths the percentages of saturates, olefins, and aromatics are calculated. Displacement development LC, of which FIA is an example, is much less common than elution development, but the separation into compound types rather than individual compounds is a common feature of many LC methods. It is the ready ability of LC to effect a type separation that accounts for its frequent use for the preliminary separation of a complex mixture before the final GC, as, for example, in the separation of PCA mixtures described in Chap. 24.7.

ASTM method D2549 for the separation of aromatic and nonaromatic fractions of high-boiling oils may be taken as a typical example of long-established procedures based on elution chromatography. This method uses a column in which the lower part is packed with silica gel and the upper section is packed with alumina. The saturated hydrocarbons are eluted with *n*-pentane, and the aromatics are eluted with ether, followed by chloroform, and finally by ethanol.

Great advances in LC methods have been made in the last few years, following the realization that much higher separating efficiency could be achieved by the use of adsorbents with particle sizes of $5-10$ μm. Such small particles form columns with a large flow impedance, and hence high pressures, of the order of 500–5000 psi, are needed to force the eluents through such columns. This presents no great practical difficulty, since pumps capable of developing these pressures are readily available. Of greater consequence is the fact that these HPLC columns have a sample capacity of only a milligram or less, so that the technique used in methods like D2549 is no longer possible. Since collecting and weighing fractions is laborious and time-consuming, a detector is used to monitor the composition of the column effluent continuously. As in GC, many different principles have been proposed for LC detectors, but again, as in GC, a few detectors have proved to be far better than the rest. The ones that have been used most extensively are UV detectors of fixed and variable wavelength, UV fluorescence detectors, and RI detectors. The RI detector has a universal response and, as such, it is suitable for monitoring the separation of hydrocarbons, but it has a lower sensitivity than UV detectors. In order to obtain the optimum response from the RI detector, Suatoni et al. [66] used Fluorinert FC-78 (a fluorohydrocarbon marketed by the 3M Company) as the eluting solvent in an HPLC equivalent of the FIA analysis. Fluorinert FC-78 has a very low RI of 1.25, allowing even the lower saturated hydrocarbons to be determined. For the separation of mixtures in the boiling range 190–360°C, *n*-hexane can be used as the eluent since there is a significant difference between the RI of the lower and the higher saturates [67].

An interesting analogy between this work and GC is that after the elution of the saturates and olefins the solvent flow is reversed and the aromatics are backflushed off the column. This gives a shorter analysis time and a better and sharper peak shape for the aromatics. The method is currently the object of a study program by an ASTM working group.

Apart from the papers of Suatoni et al., little has been published on the separation of hydrocarbons with the aid of RI detection [68], although recently Colin et al. [69] have described the construction of an RI detector with a considerably better performance than that of previous instruments, which may encourage further work on hydrocarbon separation.

In contrast to the sparse literature on the general separation of hydrocarbons by

HPLC, there are many publications on the HPLC separation of PCA mixtures with UV or UV fluorescence detection. A literature survey of the subject, published in 1977, quoted 43 references [70]; some more recent papers are given in refs. 71–81. As far as the actual separation is concerned, opinion seems to be divided between the use of adsorption columns, packed with silica gel or alumina, columns containing nonpolar materials bonded to microparticulate supports and eluted by a polar solvent (reversed-phase chromatography), and liquid–liquid partition columns where the polar liquid phase is adsorbed on a microparticulate support.

Typical of the first type of separation is the 25 cm × 6-mm OD column, packed with Partisil 5 microparticulate silica gel, which was used by Grant and Meiris [77]. Elution with n-hexane at a flowrate of 2 ml/min gave the chromatogram of a coal tar pitch shown in Fig. 24.11.

An example of reversed-phase separation is the work of Wheals et al. [82], who used a 25 cm × 6-mm OD column, packed with a bonded phase prepared by reacting microparticulate silica gel with octadecachlorosilane. Aqueous methanol was the eluting solvent, and both UV and UV fluorescence detectors were used. The UV fluorescence detector was operated at different excitation wavelengths to give a considerable degree of selectivity.

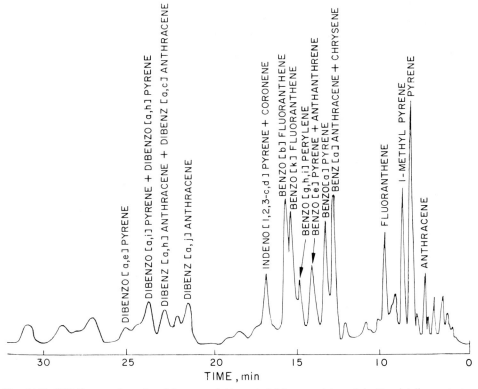

Fig. 24.11. HPLC separation of cyclohexane-soluble material from coal tar pitch. For details, see text. (Reproduced from *J. Chromatogr.*, 142 (1977) 339, with permission [77].)

Durand and Petroff [78] used a 30 cm × 6-mm OD column, packed with a microparticulate silica gel, through which they passed a solution of dimethylsulfoxide (DMSO) in heptane. The eluent was heptane, containing 0.35% DMSO. Again, both UV and UV fluorescence were used for detection.

The authors of the three papers quoted above all obtained good separations and, on the basis of their results, it is difficult to judge which, if any, is the best LC technique to use. It is difficult to prepare liquid–liquid chromatographic columns of high efficiency and to maintain stable liquid–liquid columns without loss of the stationary liquid. Reversed-phase HPLC is now very widely used, partly because great changes in elution can be effected by varying the composition (usually the water content) of the eluent. The choice of LC system depends to some extent on the nature of the sample and on the amount of preliminary clean-up, but beyond this, it seems to depend on individual preference and familiarity with a particular technique. Moreover, the practical consideration must be borne in mind that if a given laboratory is engaged in several different kinds of analysis, all of which are carried out by a particular LC technique, then the natural tendency would be to use this technique for PCA analysis.

Currently LC is a source of considerable frustration for the hydrocarbon chemist. The success of the LC separations achieved on the PCA mixtures described above leaves one to speculate that equally good separations might be possible with other hydrocarbon mixtures, if only a detector as good as the UV detector were available. At the present time, the most promising lines of development for monitoring hydrocarbon separations are Fourier transform infrared (FTIR) spectrometry [83,84] and MS [85,86]. Both of these approaches entail considerable practical difficulties, which have not been completely resolved, and both impose limitations on the LC system for optimum compatibility. Another factor is that both types of equipment are extremely expensive, being more than an order of magnitude dearer than the LC apparatus itself. Under the circumstances, even when all the practical problems have been solved, these techniques are unlikely to be used on a routine basis.

In concluding this section, mention must be made of supercritical-fluid chromatography (SFC). The development of this technique preceded that of HPLC [87–89] and in some respects paved the way for the latter in that the practical difficulties which were encountered and surmounted in SFC were similar in many respects to those encountered in HPLC. SFC differs from other forms of LC in that the mobile phase is a fluid above its critical temperature (T_c), which in the case of n-pentane is 196.6°C. At this temperature, the critical pressure of pentane (P_c) is 33.7 atm. Thus, in addition to the problems of a high pressure (now solved), the technique was faced with a further problem of operation with flammable solvents at elevated temperatures. Because of these difficulties, carbon dioxide has been used frequently as the supercritical fluid. It is obviously much safer, and its T_c is only a little above ambient temperature (31.3°C), although its P_c is higher (73.8 atm). In the supercritical state, compounds like n-pentane and carbon dioxide act as extremely powerful solvents and, in spite of the practical difficulties, the early workers were able to demonstrate elegant separations of high-molecular-weight mixtures which have been confirmed by more recent publications [90,91].

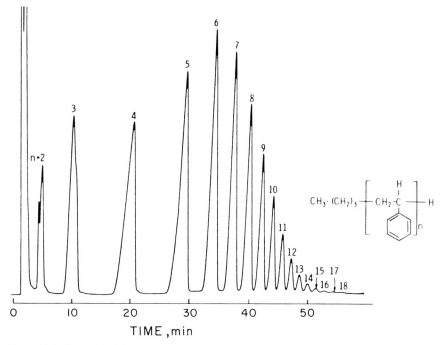

Fig. 24.12. Supercritical-fluid chromatogram of polystyrene oligomers. Column, 4 m × 3.2 mm OD, packed with 120–150-mesh Porasil C (ODCS); temperature, 205°C; pressure, 600 psi for 8 min, then 6 psi/min to 900 psi; mobile phase, 5% methanol in pentane at 3.0–4.6 ml/min; sample size, 40 μl of a 50% solution in benzene. (Reproduced from *J. Polym. Sci., Part B*, 7 (1969) 811, with permission [90].)

Fig. 24.12 shows the separation of a polystyrene of nominal molecular weight 600, in which polymers of molecular weighs up to 1900 can be clearly seen and, although published in 1969, this separation is remarkable even by current standards [82]. Very little work on SFC has been published in recent years. Scientific history is full of examples of discoveries and inventions that have appeared at the "right" time and vice versa. SFC appeared at a time when the thought of operating a column at pressures of a 1000 psi filled most people with trepidation. Now that a decade of experience with HPLC has familiarized chromatographers with these conditions and many of the practical problems have been solved, it would seem to the author that the time is ripe for further effort on SFC.

24.10. THIN-LAYER CHROMATOGRAPHY

TLC has experienced a revival in the last few years with the appearance of HPTLC. This is a term coined to describe TLC in which the plates are coated with the microparticulate materials that are used extensively in HPLC. As a chromatographic method, TLC has the virtues of speed, simplicity, and cheapness, but it

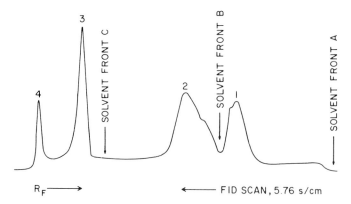

Fig. 24.13. Iatroscan chromatogram of a typical fuel oil. Solvent systems, A = light petroleum, B = toluene–light petroleum (4:1), C = dichloromethane–methanol (19:1). Peaks: 1 = saturated hydrocarbons, 2 = aromatic hydrocarbons, 3 = polar compounds, 4 = highly polar compounds.

cannot compete with GC and HPLC in terms of resolution and quantitative accuracy. However, it does have a further advantage in that the plates may be sprayed with a variety of reagents to effect a degree of selectivity less readily obtained in column chromatography. Hydrocarbons, for example, may be detected by spraying the plates with concentrated sulfuric acid to char the spots of organic material. TLC is frequently used for hydrocarbon separations as a preliminary clean-up before further analysis by HPLC or GC or to supply fractions for analysis by spectrometry [55,93,94]. When applied to PCA analysis, detection of the spots is carried out by viewing the plates under UV light.

Another development that has caused a revival of interest in TLC in recent years is the availability of a commercial instrument—the Iatroscan Analyzer (Iatron Laboratories, Tokyo, Japan) [95,96]. This instrument utilizes rods of quartz, 0.9 mm in diameter and 153 mm long, coated with a layer of silica gel about 75 μm thick. Mixtures are applied to the rod and the chromatogram is developed with a suitable solvent system, as in ordinary TLC. After development and evaporation of the solvent, the rod is passed slowly at a uniform speed through the flame of a specially constructed FID, where each zone burned off produces a signal, which is displayed on a chart recorder or is fed into an integrator for quantitative evaluation. Immediately before use, the rods are cleaned by passing them through the FID flame. The manufacturers claim that each rod can be used for up to 100 determinations, but when high-boiling mixtures are analyzed about 50 would be more realistic. Fig. 24.13 shows the Iatroscan chromatogram of a heavy fuel oil, which was separated into saturates, aromatics, polar compounds, and highly polar compounds by elution with a range of solvents of increasing polarity. Even with a simple type-analysis of this nature it is possible to derive correlations between composition and performance characteristics and some information useful for refinery operations [97]. As with the use of the FID in GC, the best quantitative results are obtained by calibration blends that "bracket" the composition of the unknown sample.

24.11. CONCLUSION

It is always dangerous to discuss the future development of a technique, since a period of steady progress may be followed by a "quantum jump" improvement which, by the nature of things, is impossible to predict. An example of such a jump is the introduction of capillary columns [98].

In GC, the recent developments with fused silica columns [99] should lead to a more fundamental understanding of the physical chemistry of the stationary liquid film, which in turn should lead to more efficient columns. Efficiency has already increased from 2000–3000 to 4000–5000 plates/m, and these figures should improve further in the next few years. The availability of silica tubing of 0.025–0.05 mm ID should encourage further development of the elegant work by Desty et al. [100] in the early 1960s on very rapid GC separation. On the other hand, the availability of silica tubing of 0.7 mm ID may give rise to the gradual replacement of packed columns, since it has already been demonstrated that stainless-steel columns of these dimensions have flow and capacity characteristics similar to those of packed columns, but with superior separating power [101]. There will also be a much greater use of column switching, "heart-cutting", and backflushing [102].

An alternative method to sample introduction by syringe is badly needed, and it is unfortunate that a commercial system which did not use syringe injection [103] is now no longer available. For plant "on-line" analysis, fluid logic switching devices may find favor [104].

If the sensitivity of the FID could be improved substantially it should be possible to work with very low stationary liquid loading at relatively low temperatures to achieve the separation of higher-molecular-weight hydrocarbons beyond the current practical range. Selective detectors are always potentially useful, and on the basis of past work it would seem likely that flame-emission detectors will find wider application [105–108]. The use of simple quadrupole mass spectrometers as GC detectors will increase enormously, if the price of such instruments can be reduced significantly.

As with GC, LC will almost certainly see a gradual increase in column efficiency and a reduction in column diameter, although the merits of microbore columns are the subject of debate at the present time. As indicated in Chap. 24.9, the greatest need in HPLC is for a detector with a high sensitivity for hydrocarbons. As an alternative to very expensive FTIR instruments, videcon tubes are under investigation [109,110]. Although the resolution so far obtained is not very good, there is much potential scope for improvement; the main handicap in this direction lies in the relatively small size of the LC market, which means that advances can only come about as more general developments in electronics appear.

A topic which has not been touched upon is preparative-scale chromatography; this is a subject which is large enough to warrant a chapter of its own. Chromatographic techniques in general are all "infinite dilution" methods and the smaller the sample size the better the separation. There are occasions, however, when it is necessary and desirable to use larger samples to obtain fractions for further analysis. For example, LC is frequently used to obtain "aliphatic" and "aromatic" fractions for GC analysis, and both techniques can be made to yield amounts of up to 100 mg, either by increasing the size of the columns used or by repeated automatic operation of a column little larger than the size normally used for analytical work. However, if preparative-scale work is intended to produce kilogram or even tonnage quantities, there would seem to be little prospect of separations on this scale as far as hydrocarbon mixtures are concerned, since it is difficult to see any commercial incentive for such work.

REFERENCES

1 F.H. Huyten, G.W.A. Rijnders and W. van Beersum, in M. van Swaay (Editor), *Gas Chromatography 1962*, Butterworths, London, 1962, p. 335.
2 O.L. Hollis, *Anal. Chem.*, 38 (1966) 309.
3 J.S. Stufkens and H.J. Bogaard, *Anal. Chem.*, 47 (1975) 383.
4 J.E. Picker and R.E. Sievers, *J. Chromatogr.*, 203 (1981) 29.
5 D.H. Desty, *Chromatographia*, 8 (1975) 452.
6 K. Grob and G. Grob, *J. Chromatogr.*, 90 (1974) 303.
7 K. Grob, K. Grob, Jr. and G. Grob, *J. Chromatogr.*, 106 (1975) 299.
8 K. Grob and G. Grob, *J. Chromatogr.*, 125 (1976) 471.
9 K. Grob, G. Grob and K. Grob, Jr., *Chromatographia*, 10 (1977) 181.
10 K. Grob, J.R. Guenter and A. Portmann, *J. Chromatogr.*, 147 (1978) 111.
11 K. Grob and K. Grob, Jr., *J. Chromatogr.*, 151 (1978) 311.
12 E.R. Adlard, *CRC Crit. Rev. Anal. Chem.*, 5 (1975) 13.
13 R.J. Law, *Marine Poll. Bull.*, 9 (1978) 321.
14 A. Clarke and R.J. Law, *Marine Poll. Bull.*, 12 (1981) 10.
15 W.K. Al-Thamir, R.J. Laub and J.H. Purnell, *J. Chromatogr.*, 142 (1977) 3.
16 W.K. Al-Thamir, R.J. Laub and J.H. Purnell, *J. Chromatogr.*, 188 (1980) 79.
17 F.T. Eggertsen, H.S. Knight and S. Groennings, *Anal. Chem.*, 28 (1956) 303.
18 T. McKenna and J. Idleman, *Anal. Chem.*, 32 (1960) 1299.
19 N.G. McTaggart, C.A. Miller and B. Pearce, *J. Inst. Petrol.*, 54 (1968) 265.
20 E.M. Fredericks and F.R. Brooks, *Anal. Chem.*, 28 (1956) 297.
21 W.N. Sanders and J.B. Maynard, *Anal. Chem.*, 40 (1968) 527.
22 I.M. Whittemore, in K.H. Altgelt and T.H. Gouw (Editors), *Chromatography in Petroleum Analysis*, Dekker, New York, 1979, p. 41.
23 E.R. Adlard, A.W. Bowen and D.G. Salmon, *J. Chromatogr.*, 186 (1979) 207.
24 H. Schulz, B. Gregor, R. Lochmiller and S. San Min, *26th DGMK Meeting, Berlin, 4–6 Oct. 1978, Compend. Deut. Gas Mineralölwiss. Kohlechem.*, 78/79 (1978) 1407.
25 F. Riedo, D. Fritz, G. Tarján and E. sz. Kováts, *J. Chromatogr.*, 126 (1976) 63.
26 F. Vernon and C.O.E. Ogundipe, *J. Chromatogr.*, 132 (1977) 181.
27 J.K. Haken and F. Vernon, *J. Chromatogr.*, 186 (1979) 89.
28 F.I. Onuska, *J. Chromatogr.*, 186 (1979) 259.
29 R.L. Martin, *Anal. Chem.*, 34 (1962) 896.
30 R.E. Robinson, R.H. Coe and J.M. O'Neal, *Anal. Chem.*, 43 (1971) 591.
31 L.L. Stavinola, *J. Chromatogr. Sci.*, 13 (1975) 72.
32 B.W. Bradford, D. Harvey and D.E. Chalkley, *J. Inst. Petrol.*, 41 (1955) 80.
33 S.P. Wasik and W. Tsang, *J. Phys. Chem.*, 74 (1970) 2970.
34 R.J. Cvetanovic, F.J. Duncan, W.E. Falconer and R.S. Irwin, *J. Amer. Chem. Soc.*, 87 (1965) 1827.
35 E. Gil-Av and V. Schurig, *Anal. Chem.*, 43 (1971) 2030.
36 V. Schurig, R.C. Chang, A. Zlatkis, E. Gil-Av and F. Mikes, *Chromatographia*, 6 (1973) 223.
37 O.K. Guha and J. Janák, *J. Chromatogr.*, 68 (1972) 325.

38 L. Soják, J. Hrivňák, I. Ostrovský and J. Janák, *J. Chromatogr.*, 91 (1974) 613.
39 L. Soják, J. Krupčik and J. Janák, *J. Chromatogr.*, 195 (1980) 43.
40 V. Schurig, *Chromatographia*, 13 (1980) 263.
41 E.R. Adlard, in K.H. Altgelt and T.H. Gouw (Editors), *Chromatography in Petroleum Analysis*, Dekker, New York, p. 137.
42 A.G. Butlin, Personal communication.
43 G. Grimmer, A. Hildebrandt and H. Böhnke, *Erdöl Kohle*, 25 (1972) 531.
44 T. Doran and N.G. McTaggart, *J. Chromatogr. Sci.*, 12 (1974) 715.
45 G. Grimmer and H. Böhnke, *Chromatographia*, 9 (1976) 30.
46 D.T. Kaschani, *3rd DGMK Fachgruppentagung, Hannover, GFR, 6–8 Oct. 1975, Compend. Deut. Gas Mineralölwiss. Kohlechem.*, 75/76 (1975) 606.
47 M.L. Lee, K.D. Bartle and M.V. Novotný, *Anal. Chem.*, 47 (1975) 540.
48 G.M. Janini, B. Shaikh and W.L. Zielinski, Jr., *J. Chromatogr.*, 132 (1977) 136.
49 R.F. Severson, M.E. Snook, R.F. Arrendale and O.T. Chortyk, *Anal. Chem.*, 48 (1976) 1866.
50 E.D. John and G. Nickless, *J. Chromatogr.*, 138 (1977) 399.
51 A. Bjørseth, *Anal. Chim. Acta*, 94 (1977) 21.
52 B.P. Dunn, *Environ. Sci. Technol.*, 10 (1976) 1018.
53 W.J. Cretney, P.A. Christensen, B.W. McIntyre and B.R. Fowler, in B.K. Afghan and D. Mackay (Editors) *Hydrocarbons and Halogenated Hydrocarbons in the Marine Environment*, Plenum, New York, 1980, p. 315.
54 G. Grimmer and H. Böhnke, *J. Ass. Offic. Anal. Chem.*, 58 (1975) 725.
55 R. Kadar, K. Nagy and D. Fremstad, *Talanta*, 27 (1980) 227.
56 R.A. Brown and T.D. Searl, in K.H. Altgelt and T.H. Gouw (Editors), *Chromatography in Petroleum Analysis*, Dekker, New York, 1979, p. 367.
57 G.M. Janini, K. Johnston and W.L. Zielinski, Jr., *Anal. Chem.*, 47 (1975) 670.
58 G.M. Janini, G.M. Muschik and W.L. Zielinski, Jr., *Anal. Chem.*, 48 (1976) 809.
59 G.M. Janini, G.M. Muschik, J.A. Schroer and W.L. Zielinski, Jr., *Anal. Chem.*, 48 (1976) 1879.
60 W.L. Zielinski, Jr., *Ind. Res. Dev.*, 19 (1980) 178.
61 S.K. Hajibrahim, P.J.C. Tibbets, C.D. Watts, J.R. Maxwell, G. Eglinton, H. Colin and G. Guiochon, *Anal. Chem.*, 50 (1978) 549.
62 J.C. Worman and L.E. Green, *Anal. Chem.*, 37 (1965) 1620.
63 N.G. McTaggart and P. Glaysher, in D.R. Hodges (Editor), *Recent Analytical Developments in the Petroleum Industry*, Applied Science Publishers, Barking, 1974, p. 79.
64 D.W. Criddle and R.L. le Tourneau, *Anal. Chem.*, 23 (1951) 1620.
65 ASTM Method D1319, Amer. Soc. Testing Mat., Philadelphia, PA; and IP Method 156, Inst. Petroleum, London.
66 J.C. Suatoni, H.R. Garber and B.R. Davis, *J. Chromatogr. Sci.*, 13 (1975) 367.
67 J.C. Suatoni and H.R. Garber, *J. Chromatogr. Sci.*, 14 (1976) 546.
68 S.C.F. Robinson, *Chromatographia*, 12 (1979) 439.
69 H. Colin, A. Jaulmes, G. Guiochon, J. Corno and J. Simon, *J. Chromatogr. Sci.*, 17 (1979) 485.
70 R. Thomas and M. Zander, *Erdöl Kohle*, 30 (1977) 403.
71 D. Kasiske, K.D. Klinkmüller and M. Sonneborn, *J. Chromatogr.*, 149 (1978) 703.
72 G. Goldstein, *J. Chromatogr.*, 129 (1976) 61.
73 A. Radecki, H. Lamparczyk, J. Grzybowski and J. Halkiewicz, *J. Chromatogr.*, 150 (1978) 527.
74 B.S. Das and G.H. Thomas, *Anal. Chem.*, 50 (1978) 967.
75 A. Matsunaga and M. Yagi, *Anal. Chem.*, 50 (1978) 753.
76 R.S. Thomas, R.C. Lao, D.T. Wang, D. Robinson and T. Sakuma, in P.W. Jones and R.I. Freudenthal (Editors), *Carcinogenesis*, Vol. 3, *Polynuclear Aromatic Hydrocarbons*, Raven, New York, 1978, p. 9.
77 D.W. Grant and R.B. Meiris, *J. Chromatogr.*, 142 (1977) 339.
78 J.P. Durand and N. Petroff, *J. Chromatogr.*, 190 (1980) 85.
79 *U.S. Environ. Protect. Agency Fed. Reg.*, 44, No. 233, 3 Dec. 1979, Method 610.
80 H. Boden, *J. Chromatogr. Sci.*, 14 (1976) 391.

81 J.M. Colin and G. Vion, *Analusis,* 8 (1980) 224.

82 B.B. Wheals, C.G. Vaughan and M.J. Whitehouse, *J. Chromatogr.,* 106 (1975) 109.

83 D. Kuehl and P.R. Griffiths, *J. Chromatogr. Sci.,* 17 (1979) 471.

84 D.W. Vidrine, *J. Chromatogr. Sci.,* 17 (1979) 477.

85 P.J. Arpino and G. Guiochon, *Anal. Chem.,* 51 (1979) 682A.

86 W.H. McFadden, *J. Chromatogr. Sci.,* 18 (1980) 97.

87 S.T. Sie, W. van Beersum and G.W.A. Rijnders, *Separ. Sci.,* 1 (1966) 459.

88 S.T. Sie and G.W.A. Rijnders, *Separ. Sci.,* 2 (1967) 729.

89 S.T. Sie and G.W.A. Rijnders, *Separ. Sci.,* 2 (1967) 755.

90 R.E. Jentoft and T.H. Gouw, *J. Polym. Sci., Part B,* 7 (1969) 811.

91 R.E. Jentoft and T.H. Gouw, *Anal. Chem.,* 48 (1976) 2195.

92 T.H. Gouw and R.E. Jentoft, in K.H. Altgelt and T.H. Gouw (Editors), *Chromatography in Petroleum Analysis,* Dekker, New York, 1979, p. 313.

93 R.J. Hurtubise, J.F. Schabron, J.D. Feaster, D.H. Therkildsen and R.E. Poulson, *Anal. Chim. Acta,* 89 (1977) 377.

94 C.A. Gilchrist, A. Lynes, G. Steel and B.T. Whitham, *Analyst (London),* 97 (1972) 880.

95 T. Okumura and T. Kadono, *Bunseki Kagaku,* 22 (1973) 980.

96 J.C. Sipos and R.G. Ackman, *J. Chromatogr. Sci.,* 16 (1978) 443.

97 H.J. Wernicke and E. Lassmann, *26th DGMK Meeting, Berlin, 4–6 Oct. 1978, Compend. Deut. Gas Mineralölwiss. Kohlechem.,* 78/79 (1978) 1484.

98 M.J.E. Golay, in D.H. Desty (Editor), *Gas Chromatography, Amsterdam, 1958,* Butterworths, London, 1958, p. 36.

99 R. Dandeneau, P. Bente, T. Rooney and R. Hiskes, *Amer. Lab.,* 11, No. 9 (1979) 61.

100 D.H. Desty, A. Goldup and W.T. Swanton, in N. Brenner, J.E. Callen and M.D. Weiss (Editors), *3rd ISA Symp. Gas Chromatogr. June 1961,* Academic Press, New York, 1962, p. 105.

101 E.R. Adlard, L.F. Creaser and P.H.D. Matthews, *Anal. Chem.,* 44 (1972) 62.

102 D.R. Deans, *Chromatographia,* 1/2 (1968) 18.

103 E. Otte and D. Jentzsch, in R. Stock, (Editor), *Gas Chromatography 1970,* The Institute of Petroleum, London, 1971, p. 218.

104 R. Annino, M.F. Gonnord and G. Guiochon, *Anal. Chem.,* 51(1979) 379.

105 W.A. Aue and H.H. Hill, *J. Chromatogr.,* 74 (1972) 319.

106 W.A. Aue and H.H. Hill, *J. Chromatogr. Sci.,* 12 (1974) 541.

107 W.A. Aue and H.H. Hill, *J. Chromatogr.,* 122 (1976) 515.

108 W.A. Aue and C.G. Flinn, *J. Chromatogr.,* 186 (1979) 299.

109 R.E. Dessy, W.G. Nunn and C.A. Titus, *J. Chromatogr. Sci.,* 14 (1976) 195.

110 K.L. Ratzlaff, *Anal. Chem.,* 52 (1980) 916.

Subject Index

Page numbers referring to analyses of compounds are printed in italics.

A

Abate, *B446*
Abiatane, *B157*
Abscisic acid, *B156, B157*
Acebutolol, *B318, B319*
Acetaminophen, *B299, B315*
Acetanisidine, *B315*
Acetate, *B471*
Acetylene, *B496*
Acetylneuraminic acid, *see* Sialic acid
ACTH, *B21*
Actinides, A245, A248
Actinium, A246, *B474, B476*
Actinomycins, *B338*
Activity coefficients,
    A54, A55, A58–60, A65, A67, A68
Acylglycerols, neutral, *B111–118*
    argentation chromatography of, *B111–113*
    GLC of, *B113–117*
    HPLC of, *B117, B118*
    isolation of, *B80–87*
Adsorption, A60, A61, A63, A65, A66
Adsorption chromatography, A75–78
Aerogel, A260
Aesculin, *B423*
Affinity chromatography, A8, A320, A321
    of carbohydrates, B245–247
    of proteins, B57, B58
    of RNA, B356, B357
Agarose gels, A260, A263–265
Alanine, *B3, B6, B7, B9, B26, B27, B30, B32*
Aldadiene, *B206*
Aldosterone, *B206, B207*
Aldrin, *B440, B446, B447*
Alkali metals, A273, A275
Alkaline earths, A248
Alkaloids, steroidal, *B198, B199*
Alkyllead, *B501*
Allose, *B261*
Alprenolol, *B299, B306, B307*
Altrose, *B261*
Aluminum, A246, *B466*
Ametryn, *B438, B439*
Amides, A268
Amikacin, *B334*

Amino acid analyzers, B2
Amino acids, *B1–52*
    acyl esters, *B28–31*
    dansyl derivatives, *B11*
    DNP derivatives, *B11, B13*
    enantiomers of, *B13–16*
    fluorescamine derivatives of, *B11, B13*
    GC of, *B21–33*
    HPLC of, *A88, A89, B4–21*
    in biological samples, *B36–38*
    ion-exchange chromatography of, B2–4
    ligand-exchange resolution of, A251, A252
    MS of, *B35–38*
    MTH derivatives of, *B10*
    OPA derivatives of, *B3, B6, B7, B11*
    PTH derivatives of, *B6, B10–13*
    TMS derivatives of, *B24–28*
Aminobutyric acid, *B3, B26, B30*
Aminocarb, *B437*
Aminocyclitols, *B333, B334*
Aminoisobutyric acid, *B3, B26, B30*
Aminonitrazepam, *B302*
Aminopterin, *B309*
Amino sugars, *B227, B237, B238, B244, B251, B254–256, B258, B261–263, B266–268, B271, B273*
Amitriptyline, *B293, B294, B301, B314*
Ammonia, *B492, B498, B501*
Amorphene, *B155*
Amphetamines, *B317, B318*
Ampholine, A364
Amphotericins, *B337, B338*
Ampicillin, *B336*
Amylopectins, *B276*
ANB, *B296*
Androstadienedione, *B203*
Androstane derivatives, *B202–204*
Androstanedione, *B202, B203*
Androstenedione, *B203, B204*
Androstenol, *B203*
Androsterone, *B202–204*
Angiotensin, *B20, B21*
Anions, A243, A274, A276
Anorexicants, *B318*
Antazoline, *B311*